IFIP Advances in Information and Communication Technology 412

Editor-in-Chief

A. Joe Turner, Seneca, SC, USA

Editorial Board

IFIP – The International Federation for Information Processing

IFIP was founded in 1960 under the auspices of UNESCO, following the First World Computer Congress held in Paris the previous year. An umbrella organization for societies working in information processing, IFIP's aim is two-fold: to support information processing within its member countries and to encourage technology transfer to developing nations. As its mission statement clearly states,

> IFIP's mission is to be the leading, truly international, apolitical organization which encourages and assists in the development, exploitation and application of information technology for the benefit of all people.

IFIP is a non-profitmaking organization, run almost solely by 2500 volunteers. It operates through a number of technical committees, which organize events and publications. IFIP's events range from an international congress to local seminars, but the most important are:

- The IFIP World Computer Congress, held every second year;
- Open conferences;
- Working conferences.

The flagship event is the IFIP World Computer Congress, at which both invited and contributed papers are presented. Contributed papers are rigorously refereed and the rejection rate is high.

As with the Congress, participation in the open conferences is open to all and papers may be invited or submitted. Again, submitted papers are stringently refereed.

The working conferences are structured differently. They are usually run by a working group and attendance is small and by invitation only. Their purpose is to create an atmosphere conducive to innovation and development. Refereeing is also rigorous and papers are subjected to extensive group discussion.

Publications arising from IFIP events vary. The papers presented at the IFIP World Computer Congress and at open conferences are published as conference proceedings, while the results of the working conferences are often published as collections of selected and edited papers.

Any national society whose primary activity is about information processing may apply to become a full member of IFIP, although full membership is restricted to one society per country. Full members are entitled to vote at the annual General Assembly, National societies preferring a less committed involvement may apply for associate or corresponding membership. Associate members enjoy the same benefits as full members, but without voting rights. Corresponding members are not represented in IFIP bodies. Affiliated membership is open to non-national societies, and individual and honorary membership schemes are also offered.

Harris Papadopoulos   Andreas S. Andreou
Lazaros Iliadis   Ilias Maglogiannis (Eds.)

Artificial Intelligence Applications and Innovations

9th IFIP WG 12.5 International Conference, AIAI 2013
Paphos, Cyprus, September 30 – October 2, 2013
Proceedings

 Springer

Volume Editors

Harris Papadopoulos
Frederick University
Department of Computer Science and Engineering
1036 Nicosia, Cyprus
E-mail: h.papadopoulos@frederick.ac.cy

Andreas S. Andreou
Cyprus University of Technology
Department of Electrical Engineering/Computer Engineering and Informatics
3603 Limassol, Cyprus
E-mail: andreas.andreou@cut.ac.cy

Lazaros Iliadis
Democritus University of Thrace
Dept. of Forestry and Management of the Environment and Natural Resources
68200 Orestiada, Greece
E-mail: liliadis@fmenr.duth.gr

Ilias Maglogiannis
University of Piraeus
Department of Digital Systems
18534 Piraeus, Greece
E-mail: imaglo@unipi.gr

ISSN 1868-4238   e-ISSN 1868-422X
ISBN 978-3-662-52512-8   e-ISBN 978-3-642-41142-7
DOI 10.1007/978-3-642-41142-7
Springer Heidelberg New York Dordrecht London

CR Subject Classification (1998): I.2.6, I.5.1-4, I.2.1-4, H.2.8, I.4, K.3, F.1.1, F.2.1, H.4.2, J.3

Typesetting: Camera-ready by author, data conversion by Scientific Publishing Services, Chennai, India

Printed on acid-free paper

Springer is part of Springer Science+Business Media (www.springer.com)

Preface

This volume contains the papers accepted for presentation at the 9th IFIP International Conference on Artificial Intelligence Applications and Innovations (AIAI 2013), which was held in Paphos, Cyprus, during September 30 to October 2, 2013. AIAI is the official conference of the IFIP Working Group 12.5 "Artificial Intelligence Applications" of the IFIP Technical Committee on Artificial Intelligence (TC12). IFIP was founded in 1960 under the auspices of UNESCO, following the first World Computer Congress, held in Paris the previous year. The first AIAI conference was held in Toulouse, France in 2004 and since then it has been held annually, offering scientists the chance to present different perspectives on how artificial intelligence (AI) may be applied and offer solutions to real-world problems.

The importance of artificial intelligence is underlined by the fact that it is nowadays being embraced by a vast majority of research fields across different disciplines, from engineering sciences to economics and medicine, as a means to tackle highly complicated and challenging computational as well as cognitive problems. Being one of the main streams of information processing, artificial intelligence may now offer solutions to such problems using advances and innovations from a wide range of sub-areas that induce thinking and reasoning in models and systems.

AIAI is a conference that grows in significance every year attracting researchers from different countries around the globe. It maintains high quality standards and welcomes research papers describing technical advances and engineering and industrial applications of artificial intelligence. AIAI is not confined to introducing how AI may be applied in real-world situations, but also includes innovative methods, techniques, tools and ideas of AI expressed at the algorithmic or systemic level.

In 2013 the AIAI conference was organized and sponsored by IFIP, the Cyprus University of Technology and Frederick University, Cyprus. Additional sponsorship was also provided by Royal Holloway, University of London, UK and by the Cyprus Tourism Organization. The conference was held in the seaside city of Paphos, Cyprus, a city rich in history and culture.

This volume contains a total of 70 papers that were accepted for presentation at the main event (26 papers) and the 8 workshops of the conference after being reviewed by at least two independent academic referees. The authors of these papers come from 24 different countries, namely: Belgium, Brazil, Canada, China, Cyprus, Egypt, France, Greece, Italy, Japan, Luxembourg, Morocco, The Netherlands, Nigeria, Norway, Poland, Portugal, Slovakia, Spain, Sweden, Switzerland, Turkey, The United Kingdom, and The United States.

In addition to the accepted papers, the technical program of the conference featured a symposium titled "Measures of Complexity", which was organized to celebrate the 75th birthday of Professor Alexey Chervonenkis, one of the most important scholars in the field of pattern recognition and computational learning. The symposium included talks by

- Alexey Chervonenkis (Russian Academy of Sciences, Yandex, Russia, and Royal Holloway, University of London, UK)
- Vladimir Vapnik (NEC, USA and Royal Holloway, University of London, UK)
- Richard Dudley (MIT, USA)
- Bernhard Schölkopf (Max Planck Institute for Intelligent Systems, Germany)
- Leon Bottou (Microsoft Research, USA)
- Konstantin Vorontsov (Russian Academy of Sciences)
- Alex Gammerman (Royal Holloway, University of London, UK)
- Vladimir Vovk (Royal Holloway, University of London, UK)

Furthermore, a keynote lecture was given by Tharam Dillon (La Trobe University, Australia) and Elizabeth Chang (Curtin University, Australia) on "Trust, Reputation, and Risk in Cyber Physical Systems".

A total of 8 workshops were included in the technical program of the conference, each related to a specific topic of interest within AI. These were:

- The 3rd Workshop on Artificial Intelligence Applications in Biomedicine (AIAB 2013) organized by Harris Papadopoulos (Frederick University, Cyprus), Efthyvoulos Kyriacou (Frederick University, Cyprus), Ilias Maglogiannis (University of Piraeus, Greece) and George Anastassopoulos (Democritus University of Thrace, Greece).
- The 2nd Workshop on Conformal Prediction and its Applications (CoPA 2013) organized by Harris Papadopoulos (Frederick University, Cyprus), Alex Gammerman (Royal Holloway, University of London, UK) and Vladimir Vovk (Royal Holloway, University of London, UK).
- The 2nd Workshop on Intelligent Video-to-video Communications in Modern Smart Cities (IVC 2013) organized by Ioannis P. Chochliouros (Hellenic Telecommunications Organization – OTE, Greece), Latif Ladid (University of Luxemburg, Luxemburg), Vishanth Weerakkody (Brunel University, UK) and Ioannis M. Stephanakis (Hellenic Telecommunications Organization – OTE, Greece).
- The 2nd Workshop on Applying Computational Intelligence Techniques in Financial Time Series Forecasting and Trading (ACIFF 2013) organized by Spiridon D. Likothanassis (University of Patras, Greece), Efstratios F. Georgopoulos (Technological Educational Institute of Kalamata, Greece), Georgios Sermpinis (University of Glasgow, Scotland), Andreas S. Karathanasopoulos (London Metropolitan University, UK) and Konstantinos Theofilatos (University of Patras, Greece).
- The 1st Workshop on Fuzzy Cognitive Maps Theory and Applications (FCMTA 2013) organized by Elpiniki Papageorgiou (Technological Educational Institute of Lamia, Greece), Petros Groumpos (University of Patras, Greece), Nicos Mateou (Ministry of Defense, Cyprus) and Andreas S. Andreou (Cyprus University of Technology, Cyprus)

- The 1st Workshop on Learning Strategies and Data Processing in Non-stationary Environments (LEAPS 2013) organized by Giacomo Boracchi (Politecnico di Milano, Italy) and Manuel Roveri (Politecnico di Milano, Italy).
- The 1st Workshop on Computational Intelligence for Critical Infrastructure Systems (CICIS 2013) organized by Christos Panayiotou (KIOS/University of Cyprus, Cyprus), Antonis Hadjiantonis (KIOS/University of Cyprus, Cyprus), Demetrios Eliades (KIOS/University of Cyprus, Cyprus) and Andreas Constantinides (University of Cyprus and Frederick University, Cyprus).
- The 1st Workshop on Ethics and Philosophy in Artificial Intelligence (EPAI 2013) organized by Panayiotis Vlamos (Ionian University, Greece) and Athanasios Alexiou (Ionian University, Greece).

We would like to express our gratitude to everyone who has contributed to the success of the AIAI 2013 conference. In particular, we are grateful to Professors Alex Gammerman and Vladimir Vovk for the organization of the "Measures of Complexity" symposium. Special thanks to the symposium and keynote speakers for their inspiring talks. We would like to express our sincere gratitude to the organizers of the eight workshops for enriching this event with their interesting topics. We would also like to thank the members of the Organizing Committee for their great effort in the organization of the conference and the members of the Program Committee who did an excellent job in a timely manner during the review process. Special thanks are also due to Pantelis Yiasemis and Antonis Lambrou for helping us with the formatting of the final proceedings. We are grateful to the Cyprus University of Technology, Frederick University, Royal Holloway, University of London (Computer Science Department) and the Cyprus Tourism Organization for their financial support. We also thank the conference secretariat, Tamasos Tours, for its important support in the organization of the conference. Finally, we would like to thank all authors for trusting our conference and contributing their work to this volume.

August 2013   Harris Papadopoulos
                                                        Andreas S. Andreou
                                                          Lazaros Iliadis
                                                       Ilias Maglogiannis

Organization

Executive Committee

General Chairs

Andreas S. Andreou	Cyprus University of Technology, Cyprus
Harris Papadopoulos	Frederick University, Cyprus

Program Committee Chairs

Harris Papadopoulos	Frederick University, Cyprus
Andreas S. Andreou	Cyprus University of Technology, Cyprus
Lazaros Iliadis	Democritus University of Thrace, Greece
Ilias Maglogiannis	University of Piraeus, Greece

Workshop Chairs

Efthyvoulos Kyriacou	Frederick University, Cyprus
Efi Papatheocharous	University of Cyprus, Cyprus

Tutorials Chair

Sotirios P. Chatzis	Cyprus University of Technology, Cyprus

Publicity Chairs

Christoforos Charalambous	Frederick University, Cyprus
Savvas Pericleous	Frederick University, Cyprus

Sponsorship Chairs

Andreas Konstantinidis	Frederick University, Cyprus
Christos Makarounas	Cyprus University of Technology, Cyprus

Website Administrator

Pantelis Yiasemis	Cyprus University of Technology, Cyprus

Program Committee

Michel Aldanondo
George Anastassopoulos
Ioannis Andreadis
Lefteris Angelis
Plamen Angelov
Costin Badica
Zorana Bankovic
Nick Bassiliades
Nick Bessis
Giacomo Boracchi
Valérie Camps
Georgios Caridakis
Christoforos Charalambous
Aristotelis Chatzioannou
Sotirios Chatzis
Andreas Constantinides
Ruggero Donida Labati
Charalampos Doukas
Demetrios Eliades
Anestis Fachantidis
Javier Fernandez de Canete
Ilias Flaounas
Adina Magda Florea
Charles Fox
Mauro Gaggero
Alex Gammerman
Marinos Georgiades
Christos Georgiadis
Efstratios Georgopoulos
Petros Groumpos
Petr Hajek
Ioannis Hatzilygeroudis
Chrisina Jayne
Jacek Kabzinski
Antonios Kalampakas
Yuri Kalnishkan
Achilleas Kameas
Kostas Karatzas
Vangelis Karkaletsis
Kostas Karpouzis
Ioannis Karydis
Ioannis Katakis
Petros Kefalas

Katia Kermanidis
Kyriakh Kitikidou
Manolis Koubarakis
Kostantinos Koutroumbas
Vera Kurkova
Efthyvoulos Kyriacou
Latif Ladid
Antonis Lambrou
Giorgio Leonardi
Jorge Lopez Lazaro
Nikos Lorentzos
Zhiyuan Luo
Spyridon Lykothanasis
Mario Malcangi
Manolis Maragoudakis
Francesco Marcelloni
Kostantinos Margaritis
Nikos Mateou
Nikolaos Mittas
Stefania Montani
Harris Mouratidis
Nicoletta Nicolaou
Ilia Nouretdinov
Eva Onaindia
Mihaela Oprea
Dominic Palmer Brown
Elpiniki Papageorgiou
Efi Papatheocharous
Ioannis Partalas
Hamid Parvin
Savas Pericleous
Elias Pimenidis
Vassilis Plagianakos
Vijay Rao
Manuel Roveri
Ilias Sakellariou
Christos Schizas
Sabrina Senatore
Kyriakos Sgarbas
Spyros Sioutas
Stephanos Spartalis
Andreas-Georgios Stafylopatis
Ioannis Stamelos

Measures of Complexity

Symposium in Honor of Professor Alexey Chervonenkis
on the Occasion of His 75th Birthday

(Abstracts of Invited Talks)

Professor Alexey Chervonenkis
A Brief Biography

Professor Alexey Chervonenkis has made a long and outstanding contribution to the area of pattern recognition and computational learning. His first book on Pattern Recognition was published in 1974 with Professor Vladimir Vapnik and he has become an established authority in the field. His most important contributions include: The derivation of the necessary and sufficient conditions for the uniform convergence of the frequency of an event to its probability over a class of events. A result that was later developed to the necessary and sufficient conditions for the uniform convergence of means to expectations. The introduction of a new characteristic of a class of sets, later called the VC-dimension. The development of a pattern recognition algorithm called "generalized portrait", which was later further developed to the well-known Support Vector Machine. The development of principles and algorithms for choosing the optimal parameters depending on the available amount of empirical data and the complexity of the decision rule class for the problems of pattern recognition, ridge regression, kernel ridge regression and kriging. Some of these results served as the foundation for many machine learning algorithms.

Professor Chervonenkis obtained his Masters degree in Physics from the Moscow Physical and Technical Institute, Moscow, USSR in 1961 and his PhD in Physical and Mathematical Science from the Computer Centre of the Academy of Sciences of the USSR, Moscow, USSR in 1971. He is currently Head of the Applied Statistical Research Department at the Institute of Control Science, Russian Academy of Sciences. He is also Emeritus Professor at the Computer Learning Research Centre of Royal Holloway, University of London, UK where he has been working as a part-time professor since 2000. Additionally he is a part time Professor at the School of Data Analysis, Moscow, Russia since 2007 and a Scientific Consultant at Yandex, Russia since 2009. Between 1987 and 2005 he served as a Scientific Consultant at the Information Technologies in Geology and Mining (INTEGRA) company (Moscow, Russia). His research interests include the investigation of the properties of set classes and the application of machine learning algorithms to various problems. He has published three monographs and numerous manuscripts in journals and conferences.

Measures of Complexity

Alexey Chervonenkis

Russian Academy of Sciences
chervnks@ipu.ru

Abstract. Even long ago it was understood that the more is the complexity of a model, the larger should be the size of the learning set. It refers to the problem of function reconstruction based on empirical data, learning to pattern recognition or, in general, model construction using experimental measurements. Probably the first theoretical result here was Nikewest criterion (in Russia Kotelnikov theorem). It stated that, if one wants to reconstruct a continuous function on the basis of a set of measurements at discrete points, then the number of measurements should be proportional to the width of the function spectrum. It means that the spectrum width can serve as one of possible metrics of complexity.

In general, for the given amount of learning data one has to limit himself on a certain level of the model complexity depending on the data volume. But for practical implementation of this idea it is necessary to define general notion of complexity and the way to measure it numerically.

In my works with V. Vapnik we reduced the problem of a learning system ability to generalize data to the problem of the uniform convergence of frequencies to probabilities over a class of events (or means to expectations over a class of functions). If such convergence holds, then the system is able to be learned. But not on the contrary. It is possible that uniform convergence does not hold, but the system still has ability to learn.

Conditions of the uniform convergence are formulated in terms of index of evens class over a given sample, growth function and the so called VC-dimension or entropy. VC-dimension allows get estimates of uniform closeness of frequencies to probabilities, which does not depend on probability distribution over input space. Asymptotic entropy per symbol gives necessary and sufficient conditions of the uniform convergence, but they do depend on the probability distribution. In most important cases VC-dimension is equal or close to the number of unknown model parameters. Very important results in this field were gained by M. Talagran, Rademacher and others.

And still there are cases when a decision rule with large number of parameters is searched, but only a few number of examples is sufficient to find. Let us consider an example of two classes in n-dimensional Euclidean space. Each of the classes is formed by a ball having diameter D, and the distance between the centers of the balls is equal to R. If the ratio between the distance R and the diameter D is large enough then it is sufficient to show only two examples to reach 100% of correct answers. And it does not depend on the dimension of the space. A similar situation appears for recognition of two classes under supposition of feature independence (and some other conditions). Boosting algorithms construct very large formulas, and in spite of it they reach good results even for limited amount of learning data. All these facts force us to search new measures of complexity, which are not directly connected to the notion of uniform convergence. It seems that they should depend on the probability distribution. But that is the nature of things.

From Classes of Sets to Classes of Functions

Richard M. Dudley

MIT, USA
rmd@math.mit.edu

Abstract. After some 19th century precursors, the 1968 announcement by A. Chervonenkis and V. N. Vapnik, on a kind of complexity of a class of sets, dramatically expanded the scope of laws of large numbers in probability theory. As they recognized, there were extensions to families of functions. It turned out to be possible to extend also the central limit theorem.

There have been numerous applications to statistics, not only to the original goal of learning theory. Some families of bounded rational functions of bounded degree can be used to give location vector and scatter matrices for observations from general distributions in Euclidean space which may not have finite means or variances.

Causal Inference and Statistical Learning

Bernhard Schölkopf

Max Planck Institute for Intelligent Systems, Germany
bs@tuebingen.mpg.de

Abstract. Causal inference is an intriguing field examining causal structures by testing their statistical footprints. The talk introduces the main ideas of causal inference from the point of view of machine learning, and discusses implications of underlying causal structures for popular machine learning scenarios such as covariate shift and semi-supervised learning. It argues that causal knowledge may facilitate some approaches for a given problem, and rule out others.

Combinatorial Theory of Overfitting: How Connectivity and Splitting Reduces the Local Complexity

Konstantin Vorontsov

Computer Centre, RAN
k.v.vorontsov@gmail.com

Abstract. Overfitting is one of the most challenging problems in Statistical Learning Theory. Classical approaches recommend to restrict complexity of the search space of classifiers. Recent approaches benefit from more refined analysis of a localized part of the search space. Combinatorial theory of overfitting is a new developing approach that gives tight data dependent bounds on the probability of overfitting. It requires detailed representation of the search space in a form of a directed acyclic graph. The size of the graph is usually enormous, however the bound can be effectively estimated by walking through its small localized part that contains best classifiers. We use such estimate as a features selection criterion to learn base classifiers in simple voting ensemble. Unlike boosting, bagging, random forests etc. which learn big ensembles of weak classifiers we learn small ensembles of strong classifiers. Particularly we use two types of base classifiers: low dimensional linear classifiers and conjunction rules. Some experimental results on UCI data sets are also reported.

About the Origins of the Vapnik Chervonenkis Lemma

Leon Bottou

Microsoft, USA
leon@bottou.org

Abstract. Whereas the law of large numbers tells how to estimate the probability of a single event, the uniform law of large numbers explains how to simultaneously estimate the probabilities of an infinite family of events. The passage from the simple law to the uniform law relies on a remarkable combinatorial lemma that seems to have appeared quasi simultaneously in several countries. This short talk presents some material I have collected about the history of this earth shattering result.

Table of Contents

Problem Solving, Planning and Scheduling

Modeling and Decision Support Systems

Robotics

Intelligent Signal and Image Processing

Third Workshop on Artificial Intelligence Applications in Biomedicine (AIAB 2013)

Second Workshop on Conformal Prediction and Its Applications (CoPA 2013)

Second Workshop on Intelligent Video-to-Video Communications in Modern Smart Cities (IVC 2013)

Second Workshop on Applying Computational Intelligence Techniques in Financial Time Series Forecasting and Trading (ACIFF 2013)

First Workshop on Fuzzy Cognitive Maps Theory and Applications (FCMTA 2013)

First Workshop on Learning Strategies and Data Processing in Nonstationary Environments (LEAPS 2013)

First Workshop on Computational Intelligence for Critical Infrastructure Systems (CICIS 2013)

First Workshop on Ethics and Philosophy in Artificial Intelligence (EPAI 2013)

Trust, Reputation, and Risk in Cyber Physical Systems

Elizabeth Chang[1] and Tharam Dillon[2]

[1] Chair IFIP WG2.12/12.4 on Web Semantic
Elizabeth.chang@cbs.curtin.edu.au
[2] Chair IFIP TC12 on Artifical Intelligence Latrobe University,
Digital Ecosystems & Business Intelligence Institute Pty. Ltd, Australia
Tharam.dillon7@gmail.com

Abstract. Cyber Physical Systems (CPS) involve the connections of real world objects into networked information systems including the web. It utilises the framework and architecture for such CPS systems based on the Web of Things previously developed by the authors. This paper discusses the provision of Trust, Reputation and determination of Risk for such CPS systems.

Keywords: Cyber Physical Systems, Trust, Risk, Web of Things, Architecture.

1 Introduction

The National Science Foundation (NSF) CPS Summit held in April 2008 [4] defines CPS as "physical and engineered systems whose operations are monitored, coordinated, controlled and integrated by a computing and communication core". Researchers from multiple disciplines such as embedded systems and sensor networks have been actively involved in this emerging area.

Our vision of CPS is as follows: networked information systems that are tightly coupled with the physical process and environment through a massive number of geographically distributed devices [1]. As networked information systems, CPS involves computation, human activities, and automated decision making enabled by information and communication technology. More importantly, these computation, human activities and intelligent decisions are aimed at monitoring, controlling and integrating physical processes and environment to support operations and management in the physical world. The scale of such information systems range from micro-level, embedded systems to ultra-large systems of systems. Devices provide the basic interface between the cyber world and the physical one.

The discussions in the NSF Summit [4] can be summarized into eleven scientific and technological challenges for CPS solutions. These challenges constitute the top requirements for building cyber-physical systems and are listed below.

- Compositionality
- Distributed Sensing, Computation and Control
- Physical Interfaces and Integration
- Human Interfaces and Integration
- Information: From Data to Knowledge

H. Papadopoulos et al. (Eds.): AIAI 2013, IFIP AICT 412, pp. 1–9, 2013.

- Modeling and Analysis: Heterogeneity, Scales, Views
- Privacy, Trust, Security
- Robustness, Adaptation, Reconfiguration
- Software
- Verification, Testing and Certification
- Societal Impact

Based on the challenges listed above, a new unified cyber-physical systems foundation that goes beyond current computer mediated systems needs to be developed. We explain how this can be achieved, in-line with the challenges to CPS identified by the NSF summit report.

CPS need to stay in constant touch with physical objects. This requires: (1) models that abstract physical objects with varying levels of resolutions, dimensions, and measurement scales, (2) mathematical representation of these models and understanding of algorithmic, asymptotic behavior of these mathematical models, and (3) abstractions that captures the relationships between physical objects and CPS.

Humans have to play an essential role (e.g. influence, perception, monitoring, etc.) in CPS. This requires: (1) seamless integration and adaptation between human scales and physical system scales. (2) support for local contextual actions pertinent to specific users, who are part of the system rather than just being the "users" of the system, (3) new theories on the boundary (e.g. hand-over or switch) between human control and (semi-) automatic control.

Many CPS are aimed at developing useful knowledge from raw data[21]. This requires (1) algorithms for sensor data fusion that also deal with data cleansing, filtering, validation, etc. (2) data stream mining in real-time (3) storage and maintenance of different representations of the same data for efficient and effective (e.g. visualization) information retrieval and knowledge extraction.

CPS needs to deal with massive heterogeneity when integrating components of different natures from different sources. This requires (1) integration of temporal, eventual, and spatial data defined in significantly different models (asynchronous vs. synchronous) and scales (e.g. discrete vs. continuous), (2) new computation models that characterize dimensions of physical objects such as time (e.g. to meet real-time deadline), location, energy, memory footprint, cost, uncertainty from sensor data, etc., (3) new abstractions and models for cyber-physical control that can deal with - through compensation, feedback processing, verification, etc. - uncertainty that is explicitly represented in the model as a "first-class citizen" in CPS, (4) new theories on "design for imperfection" exhibited by both physical and cyber objects in order to ensure stability, reliability, and predictability of CPS, (5) system evolution in which requirements and constraints are constantly changing and need to be integrated into different views of CPS, and (6) new models for dealing with issues in large-scaled systems such as efficiency trade-offs between local and global, emergent behavior of complex systems, etc.

CPS in general reveal a lot of physical information, create a lot of data concerning security (e.g. new types of attacks), privacy (e.g. location), and trust (e.g. heterogeneous resources). This requires: (1) new theories and methods on design principles for resilient CPS, threat/hazard analysis, cyber-physical inter-dependence anatomy, investigation/prediction of gaming plots at different layers of CPS, (2) formal models for privacy specification that allow reasoning about and proof of privacy properties, (3) new mathematical theories on information hiding for real-time streams, (4) lightweight security solutions that work well under extremely limited computational resources (e.g. devices), (5) new theories on confidence and trust maps, context-dependent trust models, and truth/falseness detection capabilities.

Due to the unpredictability in the physical world, CPS will not be operating in a controlled environment, and must be robust to unexpected conditions and adaptable to subsystem failures. This requires: (1) new concepts of robust system design that deals with and lives on unexpected uncertainties (of network topology, data, system, etc.) occurring in both cyber and physical worlds, (2) the ability to adapt to faults through (self-) reconfiguration at both physical and cyber levels, (3) fault recovery techniques using the most appropriate strategies that have been identified, categorized, and selected, (4) system evolvement through learning faults and dealing with uncertainties in the past scenarios, (5) system evolvement through run-time reconfiguration and hot deployment.

One important omission from the above requirements is the need for semantics. In particular semantics that are capable of bridging the real physical world and the virtual world. This is addressed in our earlier paper [IIS Keynote Ref Here].

CPS has recently been listed as the No.1 research priority by the U.S. President's Council of Advisors on Science and Technology [2]. This led the US National Science Foundation to organize a series of workshops on CPS [3]. The CPS framework has the capability to tackle numerous scientific, social and economic issues. The three applications for CPS are in future distributed energy systems, future transportation systems and future health care systems [1,4,14]. We have also investigated their use in collecting information and the control of an offshore oil platform. These applications will require seamless and synergetic integration between sensing, computation, communication and control with physical devices and processes.

In each of the above application areas Trust, Reputation, Security, Privacy and Risk Play a crucial role as they each involve the transfer and utilization of highly sensitive data. This provides the motivation for this paper.

2   Brief Overview of Architectural Framework for CPS Systems

We have previously proposed a Web-of-Things (WoT) framework for CPS systems [1, 20] that augments the Internet-of-Things in order to deal with issues such as information-centric protocol, deterministic QoS, context-awareness, etc. We argue that substantial extra work such as our proposed WoT framework is required before IoT can be utilized to address technical challenges in CPS Systems.

The building block of WoT is Representational State Transfer (REST), which is a specific architectural style [4]. It is, in effect, a refinement and constrained version of the architecture of the Web and the HTTP 1.1 protocol [5], which has become the most successful large-scale distributed application that the world has known to date. Proponents of REST style argue that existing RPC (Remote Procedure Call)-based Web services architecture is indeed not "Web-oriented". Rather, it is merely the "Web" version of RPC, which is more suited to a closed local network, and has serious potential weakness when deployed across the Internet, particularly with regards to scalability, performance, flexibility, and implementability [6]. Structured on the original layered client-server style [4], REST specifically introduces numerous architectural constraints to the existing Web services architecture elements [19, 20] in order to: a) simplify interactions and compositions between service requesters and providers; b) leverage the existing WWW architecture wherever possible.

The WoT framework for CPS is shown in Fig 1, which consists of five layers – WoT Device, WoT Kernel, WoT Overlay, WoT Context and WoT API. Underneath the WoT framework is the cyber-physical interface (e.g. sensors, actuators, cameras) that interacts with the surrounding physical environment. The cyber-physical interface is an integral part of the CPS that produces a large amount of data. The proposed WoT framework allows the cyber world to observe, analyze, understand, and control the physical world using these data to perform mission / time-critical tasks.

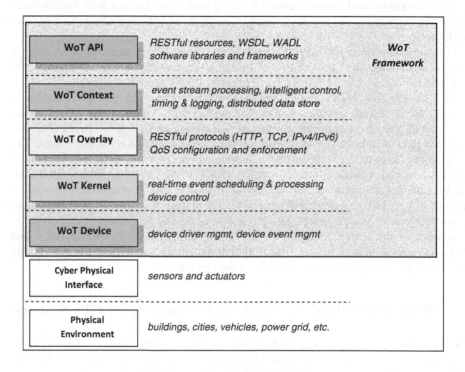

Fig. 1. WoT Framework for CPS

As shown in Fig. 1, the proposed WoT based CPS framework consists of five layers:

1. WoT Device: This layer constitute the cyber-physical interface of the system. It is a resource-oriented abstraction that unifies the management of various devices. It states the device semantics in terms of RESTful protocol.

2. WoT Kernel: This layer provides low level run-time for communication, scheduling, and WoT resources management. It identifies events and allocates the required resources, i.e. network bandwidth, processing power and storage capacity for dealing with a large amount of data from the WoT Device layer.

3. WoT Overlay: This layer is an application-driven, network-aware logical abstraction atop the current Internet infrastructure. It will manage volatile network behavior such as latency, data loss, jitter and bandwidth by allowing nodes to select paths with better and more predictable performance.

4. WoT Context: This layer provides semantics for events captured by the lower layers of WoT framework. This layer is also responsible for decision making and controlling the behaviour of the CPS applications.

5. WoT API: This layer provides abstraction in the form of interfaces that allow application developers to interact with the WoT framework.

Based on the WoT framework in Fig 1, the CPS reference architecture is shown in Fig 2, which aims to capture both domain requirements and infrastructure requirements at a high level of abstraction. It is expected that CPS applications can be built atop the CPS reference architecture.

More details about the CPS Fabric structure and the CPS node structure are given in Dillon et. al. [1].

3   Brief Overview of Our Previous Work on Trust, Reputation and Risk

In this section we give a brief description and definitions of the key ideas of Trust, Reputation and Risk defined in our previous work [15,17].

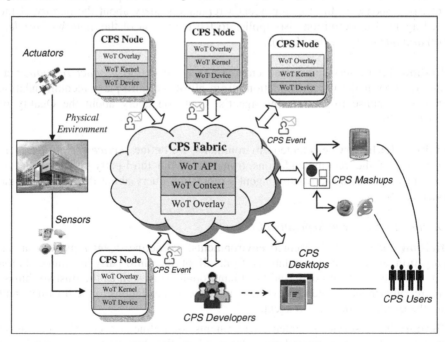

Fig. 2. CPS Reference Architecture

Thus Trust is defined in

Definition: Trust is defined as *the belief* the *trusting agent* has in the *trusted agent's willingness* and *capability to deliver a mutually agreed service* in a given *context* and in a given *timeslot*.

The term *willingness* captures and symbolises the trusted agent's will to act or be in readiness to act honestly, truthfully, reliably and sincerely in delivering on the mutually agreed service.

The term *capability* captures the skills, talent, competence, aptitude, and ability of the trusted agent in delivering on the mutually agreed behaviour. If the trusting agent has low trust in the trusted agent, it may signify that the trusting agent believes that the trusted agent does not have the capability to deliver on the mutually agreed service.

In contrast, if the trusting agent has a high level of trust, then it signifies that the trusting agent believes that the trusted agent has the capability to deliver on the mutually agreed behaviour.

The term *context* defines the nature of the service or service functions, and each *Context* has a name, a type and a functional specification, such as 'rent a car' or 'buy a book' or 'repair a bathroom'. Context can also be defined as *an object* or *an entity* or a *situation* or a *scenario*.

Definition 1 – Basic Reputation Concept

Definition 1a: In service-oriented environments, we define *agent reputation* as an aggregation of the recommendations from all of the third-party recommendation agents, in response to the trusting agent's reputation query about the *quality* of the trusted agent. The definition also applies to the reputation of the *quality of product* (QoP) *and quality of service* (QoS).

Definition 1b: In service-oriented environments, we define *product reputation* as an aggregation of the recommendations from all of the third-party recommendation agents, in response to the trusting agent's reputation query about the Quality of product (QoP).

Definition 1c: In service-oriented environments, we define *service reputation* as an aggregation of the recommendations from all of the third-party recommendation agents, in response to the trusting agent's reputation query about the Quality of the Service (QoS).

Definition 2 – Advanced Reputation Concept

Definition 2a: In service-oriented environments, we define *agent reputation* as an aggregation of the recommendations from all of the third-party recommendation agents and their first-, second- and third-hand opinions as well as the trustworthiness of the recommendation agent in giving correct recommendations to the trusting agent about the quality of the trusted agent.

Definition 2b: In service-oriented environments, we define *service reputation* as an aggregation of the recommendations from all of the third-party recommendation agents and their first-, second-and third-hand opinions as well as the trustworthiness of the recommendation agent in giving correct recommendations to the trusting agent about the Quality of the Product (QoP).

Definition 2c: In service-oriented environments, we define *product reputation* as an aggregation of the recommendations from all of the third-party recommendation agents and their first-, second- and third-hand opinions as well as the trustworthiness of the recommendation agent in giving correct recommendations to the trusting agent about the Quality of the Service (QoS).

Fundamentally, reputation is an aggregated value of the recommendations about the trustworthiness of a *trusted agent* (such as a trusted agent or QoP and services). The reputation value is not assigned but only aggregated by the trusting agent.

There are four primary concepts that should be clearly understood in the reputation definition:

(1) *Reputation*: aggregation of all the recommendations from the third-party recommendation agents about the quality of the trusted agent.

(2) *Recommendation (including opinion and recommendation value)*: submitted by the third-party recommendation agents (recommenders).

(3) *Recommendation agent*: submit the recommendation or opinion to the trusting agent or respond to a reputation query.

(4) *Reputation query*: a query made by the trusting agent about the trusted agent in a given context and timeslot.

The terms 'reputation query', 'third-party recommendation agents', 'first-, second- and third-hand opinions', 'trustworthiness of recommendation agent', 'trusting agent', 'trusted agent' are essential when defining reputation. These new terms can be regarded as the building blocks of reputation, particularly in service-oriented environments. These new terms introduced in the definition of reputation make a fundamental distinction between trust and reputation.

In contrast to Trust and Reputation,when assessing Risk it is important to take into account :[16]

1.   The likelihood of an event taking place
2.   The impact of the event.
        Thus it is not only important to calculate a failure probability but also the the likely financial or other cost of the failure.

We have previously designed a risk measure that takes both of these into account.

In addition, we have developed ontologies for Trust, Reputation and Risk [15,16,18] and Data Mining techniques for analyzing this[21].

4   Extensions of Trust, Reputation and Risk for CPS

In CPS systems we have several issues in relation to Trust and Reputation that have to be addressed.

1.   Previously we determined, forecast, modelled and predicted the Trust and Reputation of a single agent, service or service provider. In CPS systems we frequently are collecting information from a number of sensors, agents, heterogeneous resources and synthesizing or fusing the information to find out

the system state, condition and to provide situation awareness. We need to extend the previous ideas on Trust and Reputation to groups of Agents and Services.

2. The agents in CPS could also be functioning as members of a virtual community which seeks to bargain on their behalf collectively for instance in a smart grid situation. Then the other members of the community need to be able to evaluate and feel confident that any individual member of the community will play their part so as not to jeopardize the community as a whole. This will need not only Trust evaluation but also mechanisms for rewarding good behaviours and penalizing aberrant behaviours.

3. The notion of Trust and Reputation discussed in the previous section essentially dealt belief with the willingness and capability of the agent, service, and service provider to fulfil their commitment. In CPS systems the situation is complicated by the fact that the Data being generated could be noisy and have limitations on its accuracy. This has to be modelled in the Trust models.

4. The Real world environment also has a degree of uncertainty associated with the occurrence of different Events. This uncertainty has to be modelled. Issues 3. and 4. above will provide qualifiers on the evaluation of Trust. Namely The value of Trust will have certain accuracy and confidence level associate with it.

5. The uncertainty in the real world environment also has a major impact on the calculation of Risk.

5   Conclusion

In this keynote we will explain the extensions necessary to apply the concepts of Trust, Reputation and Risk for Cyber Physical Systems in detail. These extensions are of major significance for CPS systems.

References

1. Dillon, T.S., Zhuge, H., Wu, C., Singh, J., Chang, E.: Web-of-things framework for cyber–physical systems. Concurrency and Computation: Practice and Experience 23(9), 905–923 (2011)
2. President's Council of Advisors on Science and Technology (PCAST), Leadership under challenge: Information technology r&d in a competitive world (August 2007), http://www.nitrd.gov/pcast/reports/PCAST-NIT-FINAL.pdf
3. National Science Foundation, Cyber-physical systems (CPS) workshop series, http://varma.ece.cmu.edu/Summit/Workshops.html
4. National Science Foundation, Cyber-physical systems summit report, Missouri, USA (April 24-25, 2008), http://precise.seas.upenn.edu/events/iccps11/_doc/CPS_Summit_Report.pdf
5. Lee, E.: Cyber physical systems: Design challenges. In: IEEE Object Oriented Real-Time Distributed Computing, pp. 363–369 (2008)
6. Lee, E.: Computing needs time. Communications of the ACM 52(5), 70–79 (2009)

7. National Science Foundation. Cyber-physical systems executive summary (2008), http://varma.ece.cmu.edu/Summit/CPS_Summit_Report.pdf
8. Dillon, T.S., Talevski, A., Potdar, V., Chang, E.: Web of things as a framework for ubiquitous intelligence and computing. In: Zhang, D., Portmann, M., Tan, A.-H., Indulska, J. (eds.) UIC 2009. LNCS, vol. 5585, pp. 2–13. Springer, Heidelberg (2009)
9. Dillon, T.: Web-of-things framework for cyber-physical systems. In: The 6th International Conference on Semantics, Knowledge & Grids (SKG), Ningbo, China (2010) (Keynote)
10. Talcott, C.: Cyber-Physical Systems and Events. In: Wirsing, M., Banâtre, J.-P., Hölzl, M., Rauschmayer, A. (eds.) Soft-Ware Intensive Systems. LNCS, vol. 5380, pp. 101–115. Springer, Heidelberg (2008)
11. Tan, Y., Vuran, M.C., Goddard, S.: Spatio-temporal event model for cyber-physical systems. In: 29th IEEE International Conference on Distributed Computing Systems Workshops, pp. 44–50 (2009)
12. Tan, Y., Vuran, M.C., Goddard, S., Yu, Y., Song, M.: Ren, S.: A concept lattice-based event model for Cyber-Physical Systems. Presented at the Proceedings of the 1st ACM/IEEE International Conference on Cyber-Physical Systems, Stockholm, Sweden (2010)
13. Ren, S.: A concept lattice-based event model for Cyber-Physical Systems. Presented at the Proceedings of the 1st ACM/IEEE International Conference on Cyber-Physical Systems, Stockholm, Sweden (2010)
14. Yue, K., Wang, L., Ren, S., Mao, X., Li, X.: An AdaptiveDiscrete Event Model for Cyber-Physical System. In: Analytic Virtual Integration of Cyber-Physical Systems Workshop, USA, pp. 9–15 (2010)
15. Yu, X., Cecati, C., Dillon, T., Godoy Simões, M.: Smart Grids: An Industrial Electronics Perspective. IEEE Industrial Electronics Magazine IEM-02 (2011)
16. Chang, E., Dillon, T.S., Hussain, F.: Trust and Reputation for Service-Oriented Environments. Technologies for Building Business Intelligence and Consumer Confidence. John Wiley & Sons (2006)
17. Hussain, O., Dillon, T.S., Hussain, F.K., Chang, E.: Risk Assessment and Management in the Networked Economy. SCI, vol. 412. Springer, Heidelberg (2012)
18. Chang, E.J., Hussain, F.K., Dillon, T.S.: Fuzzy Nature of Trust and Dynamic Trust Modelling in Service Oriented Environments. In: Proceedings of the 2005 Workshop on Secure Web Services, pp. 75–83 (2005)
19. Wouters, C., Dillon, T., Rahayu, J.W., Chang, E.: A practical Approach to the derivation of a Materialized Ontology View. In: Taniar, D., Rahayu, J. (eds.) Web Information Systems, pp. 191–226. Idea Group Publishing, Hershey (2004)
20. Wu, C., Chang, E.J.: Searching services "on the web: A public web services discovery approach. In: Third International Conference on Signal-Image Technology & Internet-based Systems. IEEE, Shanghai (2007)
21. Dillon, T.S., Wu, C., Chang, E.: Reference architectural styles for service-oriented computing. In: Li, K., Jesshope, C., Jin, H., Gaudiot, J.-L. (eds.) NPC 2007. LNCS, vol. 4672, pp. 543–555. Springer, Heidelberg (2007)
22. Tan, H., Dillon, T.S., Hadzic, F., Feng, L., Chang, E.J.: MB3-Miner: Efficient mining eMBedded subTREEs using tree model guided candidate generation. In: First International Workshop on Mining Complex Data (MCD) in Conjunction with ICDM 2005, November 27. IEEE, Houston (2005)

Designing a Support Tool for Creative Advertising by Mining Collaboratively Tagged Ad Video Content: The Architecture of PromONTotion

Katia Kermanidis[1], Manolis Maragoudakis[2], Spyros Vosinakis[3],
and Nikos Exadaktylos[4]

[1] Department of Informatics, Ionian University, 49100 Corfu, Greece
[2] Department of Information and Communication Systems Engineering,
University of the Aegean, 83200 Karlovasi, Samos, Greece
[3] Department of Product and Systems Design Engineering, University of the Aegean,
84100 Ermoupoli, Syros, Greece
[4] Department of Marketing, Technological Educational Institute of Thessaloniki,
57400 Sindos, Greece
kerman@ionio.gr, {mmarag,spyrosv}@aegean.gr,
nexadakt@mkt.teithe.gr

Abstract. Creative advertising constitutes one of the highest-budget enterprises today. The present work describes the architecture of PromONTotion, an innovative tool for supporting ad designers in setting up a new campaign. The content of existing ad videos is collaboratively tagged, while playing a multi-player web-based game, and the provided annotations populate the support tool-thesaurus, a hierarchical ontological structure. Annotators-players will also provide information regarding the impact the ads had on them. Data mining and machine learning techniques are then employed for detecting interdependencies and correlations between ontology concepts and attributes, and for discovering underlying knowledge regarding the product type, the ad content and its impact on the players. The support tool will make ad videos, terms, statistical information and mined knowledge available to the ad designer, a generic knowledge thesaurus that combines previous ad content with users' sentiment to help the creative process of new ad design.

Keywords: creative advertising, creativity support tool, serious games, collaborative annotation, video annotation, advertisement ontology.

1 Introduction

Creative advertising describes the process of capturing a novel idea/concept for an ad campaign and designing its implementation. It is governed nowadays by significant budget allocations and large investments. Several studies have been published regarding the impact of advertising (Amos, Holmes and Strutton, 2008; Aitken, Gray and Lawson, 2008), as well as creativity in advertising (Hill and Johnson, 2004).

H. Papadopoulos et al. (Eds.): AIAI 2013, IFIP AICT 412, pp. 10–19, 2013.
© IFIP International Federation for Information Processing 2013

A number of creativity support tools have been proposed, to help ad designers come up with novel ideas for setting up a new campaign. Such tools, that could enhance the development of creative ideas, are highly beneficial for the advertising industry. They usually focus on forcing upon the advertiser a certain restricted way of thinking, using creativity templates (Goldenberg et al., 1999). The methodology is based on the hypothesis that total freedom is not the most efficient way for enhancing the creative process, but constraining it with the use of a limited number of idea-forming patterns. Expert decision making systems, like ADCAD (Burke et al., 1990), have been proposed for triggering creative ideas. ADCAD relies on rules and facts that are in reality quite hard to provide as input to the decision making tool. Janusian wording schemata (the use of opposite words in taglines) have been used extensively (Blasko and Mokwa, 1986) in advertising. IdeaFisher (Chen, 1999) is based on a database of predefined concepts and associations between them. GENI (MacGrimmon and Wagner, 1994) guides the user to make connections and transformations between the entities of a brainstorming problem. Idea Expander (Wang et al., 2010) is a tool for supporting group brainstorming by showing pictures to a conversing group, the selection of which is triggered by the conversation context. Opas (2008) presents a detailed overview of several advertising support tools.

Most of the aforementioned creativity support tools make use of static non-expandable databases, term-relation dictionaries, hand-crafted associations and transformations. Static, passive, expert-dependent knowledge models can hurt creativity (Opas, 2008).

The present work describes the architecture and design challenges of PromONTotion, a creative advertising support tool that incorporates

— collaboratively accumulated ad video content annotations
— associations between these annotations derived though reasoning within the ontological structure they form
— knowledge concerning ad type, ad content, ad impact and the relations between them; the knowledge is extracted with mining techniques
— statistical information regarding the impact an ad video has on consumers.

Unlike previous approaches to creativity support tool design, PromONTotion relies on no predefined or hand-crafted elements and associations, apart from an empty ontological backbone structure, that will include ad content concepts, consumer impact data slots, and taxonomic relations between them. The populating of the backbone, as well as the remaining knowledge available by the tool, are data-driven and automatically derived, making PromONTotion generic, dynamic, scalable, expandable, robust and therefore minimally restricting in the creative process and imposing minimal limitations to ideation or brainstorming.

The proposed tool faces a number of significant research challenges. First, crowdsourcing will be employed for the collection of collaborative ad content tags. Following the Games with a Purpose approach (von Ahn, 2006), a multiplayer browser-based action game, 'House of Ads', is implemented for this particular purpose focusing on challenging and entertaining gameplay in order to keep the players' engagement at a high level for a long time. The ontology is an innovation by itself; its content (concepts, features, relations), coverage and representation are very interesting research issues. Mapping the output of the game, i.e. the annotations, to the ontological structure is a very intriguing task, as well as the end-user interface that needs to be friendly and make full use of the resource capabilities. The interdisciplinary nature of the proposed idea is challenging, as the areas of video game design, ontology engineering, human-computer interaction and advertising are linked together to produce an advertising support tool.

In the remainder of this paper, section 2 describes in detail the architecture and the major design challenges of PromONTotion, i.e. the ontological backbone, the serious videogame for content annotation, the data mining phase and the final creativity support tool. The paper concludes in the final section.

2   The Architecture of PromONTotion

The architecture design of the proposed tool relies on several distinct underlying phases, components and techniques. Figure 1 shows the various elements and tasks of PromONTotion.

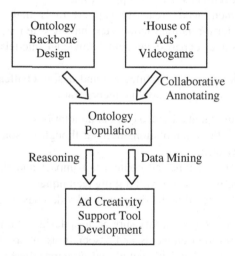

Fig. 1. The Architecture of PromONTotion

2.1   The Ontological Backbone

Marketing and ontology experts have sketched the ontological structure that will include concepts, categories and taxonomic (is-a, part-of etc.) relations between them, that are relevant to an advertising campaign. The parameters that determine the message to the consumer (e.g. tag lines), the ad filming technicalities, the ad content, as well as its artistic features play a significant role in designing the semantic ontology. Also, subjective information regarding the impact of the ad to the consumers is included in the ontology. Table 1 shows a significant part of the ontological backbone.

Table 1. The ontological backbone

Concept	Level 1 Sub-concepts	Level 2 Sub-concepts	Concept Terms/Values
Filming			studio/outdoors
Director			director name
Product Type	Product		houseware/electronic/food/luxury/store/other
	Service		telecom/banking/.../other
Main Character	Human	gender	male/female
		age	age value
		occupation/hobby/activity	employee/businessman/farmer/ housework /.../other
		recognisability	celebrity/everyday person
	Animate	type of animal	pet/wild/...
	Fictitious		cartoon/comics/movie hero
Key participating objects			tool/furniture/vehicle/electronic/technical/other
Location	indoors		home/office/work/other
	outdoors		urban/rural/other
	both		
Genre			non-fiction/tale/cartoon/animated
Art	soundtrack		rock/classical/ethnic/../other
	photography		picturesque/landscape/airphoto/other
	colouring		colour palette
Message Communication Strategy	Linguistic schemata	taglines/word games/paraphrasing/Janusian wording	tagline text / existence of paraphrasing/word games
	Adoption of known elements	book lines	known line text
		movie lines	known line text
		song lines	known line text
	Indirect critique on competition		yes/no
Ad Impact	Convincing power		a lot/some/none
	Consumer sentiment		positive/neutral/negative

As shown in the table, the concepts are organized in a hierarchical taxonomy structure and are represented though a set of features/sub-concepts, thereby forming the desired ontology. For the development of the ontology, as well as the ability to perform reasoning on its content, an ontology editor will be employed, like Protégé[1] or

---

[1] http://protege.stanford.edu

OntoEdit[2]. It is important for the ontology to be scalable, so it can constantly be enriched and updated.

2.2   The 'House of Ads' Game

In order to support the generic, minimal human-expertise-demanding and data-driven nature of the proposed support tool, the ontology backbone is populated through crowdsourcing techniques. It is evident that the success of PromONTotion relies heavily on the plethora of provided annotations; therefore the annotation tool needs to be attractive, engaging, fun and addictive. To this end, 'House of Ads', a browser-based game, is designed and implemented especially for the task at hand.

Several toolkits exist for annotating text (Wang et al., 2010; Chamberlain et al., 2008), images, like Catmaid, Flickr, Riya, ImageLabeler and Imagenotion (Walter and Nagypal, 2007), or video, like VideoAnnex and YouTube Collaborative Annotations have been proposed. Von Ahn (2006) recognized that the high level of game popularity may be taken advantage of, and channeled towards other, more "serious", applications, instead of only pure entertainment. Siorpaes and Hepp (2008) were the first to propose a game for ontology populating. The nature of textual data has not allowed for the design of genuinely entertaining gaming annotation software. The annotation of ad videos, however, inspires the design of software that can keep the player's interest and engagement level active for a very long time. Unlike the Image-notion ontology, the thesaurus aimed at by PromONTotion is comprised of a back-bone of a more elaborate set of concepts and relations, as well as statistical informa-tion regarding the terms inserted by players to populate the backbone, requiring a more elaborate game platform. The design of engaging game scenarios with usable and attractive interfaces has been recognized as one of the key challenges in the de-sign of Games with a Purpose for content annotation (Siorpaes and Hepp, 2008). In the design of 'House of Ads' a first step towards this goal is attempted by adding the fun elements of interaction and competition (Prensky, 2001) in the game and by in-cluding typical action-game challenges rather than simply adopting a quiz-like ga-meplay.

'House of Ads' is an arcade-style, top-down shoot-em-up, and puzzle-like game, accessible to anyone. It supports one to four players and includes two gameplay mod-es: the combat mode and the quiz mode (Figures 2 and 3 show some indicative screenshots of the two gameplay modes). In the former, ontology concepts and sub-concepts are mapped as rooms and each ontology term as a collectable game object within the room representing the respective concept. The atmosphere is retro-like, resembling the arcade-like games of the 80s and 90s. In a TV screen the ad video is reproduced (selected from the over 300,000 available ad videos on *youtube*), and the players, simulated as characters navigating within the house, aim at collecting as many objects that characterize the content of the ad as quickly as possible and exit the house. Players are free to collect another object from the same category if they believe that an object collected by others incorrectly describes the content with respect to this concept. Therefore, contradictive answers and possible cheaters can be easily spotted.

---

[2] www.ontoknowledge.org/tools/ontoedit.shtml

The video reproduction may be controlled by the player (paused, forwarded etc.) through a slider, and every player can block other players from reaching their goal by using a set of available weapons (fence, bombs etc.). The house platform supports different levels of difficulty; as the game evolves the house becomes more complicated, with more rooms and more detailed elements, demanding more fine-grained annotations and more complex attack/defense strategies. At the end of each stage, the player will be asked to comment on the convincing ability and the impact of the ad, providing his personal opinion. There is a large number of available questionnaires for the evaluation of advertising campaigns[3,4], based on which this type of information may be provided.

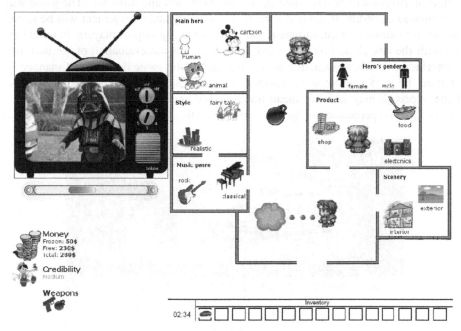

Fig. 2. The 'House of Ads' combat mode

Players earn money at the end of each stage, which can be used to buy new weapons or to improve their abilities. For each object, that can be safely assumed by the system that it is correctly characterizing the ad, the money earned is free to be spent. For objects whose correctness cannot be immediately decided, the money is blocked and will be freed once a positive decision will be made in the future. If a player collects incorrect items, his/her credibility drops and, if it falls below a certain threshold, the game ends and the player has to start over. Annotation correctness is established by taking into account the total number of answers to a question (the greater this

[3] www.surveyshare.com/templates/
   televisionadvertisementevaluation.html
[4] www-sea.questionpro.com/akira/
   showSurveyLibrary.do?surveyID=119&mode=1

number, the safer the drawn conclusions), the majority vote, the popularity of the other votes, the player credibility. A wide popularity (acceptance) of the game will ensure correctness as well as completeness (the complete ontology populating). Incomplete game sessions (sessions that have not provided annotations to all ontology slots) are not as important as attracting as many players as possible.

The quiz mode is dedicated to resolving annotation questions that could not be resolved in the combat mode, e.g. contradictive answers. The questions and their possible answers are posed to all the players simultaneously, and the player who first selects the correct answer receives the money that was blocked so far.

The game will be available online and accessible by anyone. Social media and popular networks will be exploited for its dissemination and diffusion. The game will be evaluated according to several aspects. Its usability and engagement will be tested through interviews and questionnaires handed out to a group of players, in combination with the talk aloud protocol, used extensively for the evaluation of the usability of interfaces. Another evaluation aspect is whether the game does indeed manage to populate the proposed ontology. Further evaluation metrics to determine the success of the approach may include: mean times played and number of successful annotations per player, percentage of incorrect answers, etc.

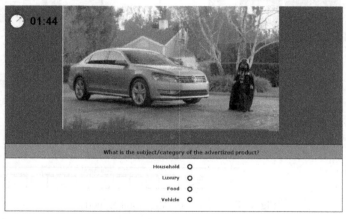

Fig. 3. The 'House of Ads' quiz mode

2.3 Data Mining

All player annotations are used to populate the ontology backbone and are stored in a database for further processing, that includes statistical as well as data mining techniques. The goal of the data processing phase is the detection of co occurrence and correlation information between categories, terms and relations (e.g. in how many ads for cleaning products there is a housewife in the leading role etc.), and its statistical significance, as well as higher-level association information governing them.

Data mining and machine learning techniques will also be employed. The concepts and attributes described earlier will enable the transformation of the annotation set of a given advertisement into a learning feature-value vector. The vectors will, in turn, enable feature selection (through wrapping, filtering or combinations of the two), that will reveal the significance of the selected concepts on advertisement design (and indirectly evaluate the ontology design). Learning will reveal correlations between ad content choices, ad products and attempt prediction of consumer impact. The extracted, mined, knowledge will reveal very interesting and previously unknown information regarding the parameters that directly or indirectly affect ad design and play a role on the consumers' sentiment and how the latter may be influenced.

Due to the anticipated data sparseness and providing that a plethora of attributes will be collected, emphasis will be given to dimensionality reduction algorithms. Singular Value Decomposition (e.g. M. Lee et al., 2010) as well as Support Vector Machines (Vapnik, 1995) will be utilized in order to deal with the aforementioned issues. Visualization of instances and attributes on a new feature space, according to the most significant vectors will reveal clusters of similar ontology mappings together with advertised products. Support Vectors will be used to weight each attribute according to a certain criterion, such as correctness of user responses and could be used to evaluate the most significant ontology nodes.   Finally, another potential use of data mining techniques for the task at hand will be that of association rules discovery, based on the FP-growth algorithm. Association rules are easily interpretable by humans and can be straightforwardly evaluated by them.

2.4   The Creativity Support Tool

The ontological structure with all its content, objective (content annotations) and subjective (ad impact annotations from every player), along with the derived correlations and extracted knowledge mined in the previous phase, will be made usable to professionals in the advertising domain through a user-friendly interface. Advertisers will be able to see the content of old ads for related products, and thereby come up with new ideas, gain insight regarding the impact of previous campaigns from the players' evaluation, look for screenshots of videos using intelligent search, based not only on keywords, but on concepts. More specifically, advertisers will be able

— to have access to a rich library of video ads
— to search the videos by content, based on a query of keywords (e.g. a specific type of product)
— to retrieve statistical data regarding the ads, i.e. see the terms/concepts/attributes his search keyword co-occurs with most frequently
— have access to the consumers' evaluation on the   advertisements' impact.

The innovative generic nature of the tool will allow it to be flexible, scalable and adjustable to the end user's needs. Most importantly, unlike creativity templates, the generic nature does not impose any sort of 'mold' or template to the creative advertiser's way of thinking.

The support tool will be evaluated by a group of advertising experts, who will be handed a product and will be asked to create a hypothetical ad scenario for it. They will evaluate the support tool based on its usability, its completeness, its significance. A combination of evaluation approaches will be employed in order to record the opinion and the impressions of the end users. Questionnaires will be handed out and interviews will be conducted to detect the problems and weaknesses of the support tool. Problems in the tool usability will be identified by evaluating its usage in real time with the think-aloud protocol. A group of ontology experts will evaluate the created ontology based on international ontology evaluation standards for its coverage, classification ability etc.

3   Conclusion

In this work the architecture and the major design issues of the creative advertising support tool PromONTotion have been presented. The tool will facilitate the brainstorming process of advertisers through a semantic thesaurus that includes video ad content annotations collected via crowdsourcing techniques, as well as knowledge relating to ad concepts, product type, annotation terms and the relations governing them. Annotations are objective, describing the content of the video, as well as subjective, denoting the annotator's personal sentiment towards the ad. The annotations will be collected through collaborative game playing. While playing, annotators will tag ad videos crawled from the web, describe the ad content and artistic features, and evaluate the ad impact on themselves. This information (the set of terms, concepts and subjective opinions), as well as statistical co occurrence data regarding concepts, advertised products, and subjective impact, will be structured into a hierarchical ontology. A user-friendly interface will allow ad designers to make full use of the ontology's capabilities, and advertising experts will evaluate the tool's coverage, usability and significance.

Several research questions still remain challenging for future work. The popularity of the game is a major concern, as it will define the number of collected annotations. Future prospects should be open to alterations and improvements on the game to increase its popularity. The use of other crowdsourcing techniques, like the Mechanical Turk, could form another future direction for gathering annotation data. Regarding the machine learning process, various mining techniques and learning schemata could be experimented with, so that the ones more suitable for the specific datasets can be established. The support tool usability is also of research significance, as the underlying knowledge needs to be transparent to the end user, and the tool needs to be as unobtrusive as possible in the creative process. Finally, regarding the impact of a specific ad to consumers, broader sentiment information could be extracted through web sentiment analysis of social media data, and complement the one provided by the annotators.

Acknowledgements. This Project is funded by the National Strategic Reference Framework (NSRF) 2007-2013: ARCHIMEDES III – Enhancement of research groups in the Technological Education Institutes. The authors are thankful for all this support.

References

1. Aitken, R., Gray, B., Lawson, R.: Advertising Effectiveness from a Consumer Perspective. International Journal of Advertising 27(2), 279–297 (2008)
2. Amos, C., Holmes, G., Strutton, D.: Exploring the Relationship between Celebrity Endorser Effects and Advertising Effectiveness. International Journal of Advertising 27(2), 209–234 (2008)
3. Blasko, V., Mokwa, M.: Creativity in Advertising: A Janusian Perspective. Journal of Advertising 15(4), 43–50 (1986)
4. Burke, R., Rangaswamy, A., Wind, J., Eliashberg, J.: A Knowledge-based System for Advertising Design. Marketing Science 9(3), 212–229 (1990)
5. Chamberlain, J., Poesio, M., Kruschwitz, U.: Phrase Detectives: A Web-based Collaborative Annotation Game. Proceedings of I-Semantics (2008)
6. Chen, Z.: Computational Intelligence for Decision Support. CRC Press, Florida (1999)
7. Ericsson, K., Simon, H.: Protocol Analysis: Verbal Reports as Data, 2nd edn. MIT Press, Boston (1993)
8. Goldenberg, J., Mazursky, D., Solomon, S.: The Fundamental Templates of Quality Ads. Marketing Science 18(3), 333–351 (1999)
9. Hill, R., Johnson, L.: Understanding Creative Service: A Qualitative Study of the Advertising Problem Delineation, Communication and Response (apdcr) process. International Journal of Advertising 23(3), 285–308 (2004)
10. Lee, M., Shen, H., Huang, J.Z., Marron, J.S.: Biclustering via sparse singular value decomposition. Biometrics 66, 1087–1095 (2010)
11. MacCrimmon, K., Wagner, C.: Stimulating Ideas Through Creativity Software. Management Science 40(11), 1514–1532 (1994)
12. Opas, T.: An Investigation into the Development of a Creativity Support Tool for Advertising. PhD Thesis. Auckland University of Technology (2008)
13. Prensky, M.: Fun, play and games: What makes games engaging? Digital Game-based Learning, 1–31 (2001)
14. Siorpaes, K., Hepp, M.: Games with a Purpose for the Semantic Web. IEEE Intelligent Systems, 1541–1672 (2008)
15. Vapnik, V.: The Nature of Statistical Learning Theory. Springer (1995) ISBN 0-387-98780-0
16. von Ahn, L.: Games with a Purpose. IEEE Computer 39(6), 92–94 (2006)
17. Walter, A., Nagypal, G.: IMAGENOTION – Collaborative Semantic Annotation of Images and Image Parts and Work Integrated Creation of Ontologies. In: Proceedings of the 1st Conference on Social Semantic Web. LNI, vol. 113, pp. 161–166. Springer (2007)
18. Wang, A., Hoang, C.D.V., Kan, M.Y.: Perspectives on Crowdsourcing Annotations for Natural Language Processing. Technical Report (TRB7/10). The National University of Singapore, School of Computing (2010)
19. Wang, H.-C., Cosley, D., Fussell, S.R.: Idea Expander: Supporting Group Brainstorming with Conversationally Triggered Visual Thinking Stimuli. In: Proceedings of the ACM Conference on Computer Supported Cooperative Work (CSCW), Georgia, USA (2010)

Voting Advice Applications: Missing Value Estimation Using Matrix Factorization and Collaborative Filtering

Marilena Agathokleous and Nicolas Tsapatsoulis

30, Arch. Kyprianos str., CY-3036, Limassol, Cyprus
mi.agathokleous@edu.cut.ac.cy, nicolas.tsapatsoulis@cut.ac.cy
http://nicolast.tiddlyspot.com

Abstract. A Voting Advice Application (VAA) is a web application that recommends to a voter the party or the candidate, who replied like him/her in an online questionnaire. Every question is responding to the political positions of each party. If the voter fails to answer some questions, it is likely the VAA to offer him/her the wrong candidate. Therefore, it is necessary to inspect the missing data (not answered questions) and try to estimate them. In this paper we formulate the VAA missing value problem and investigate several different approaches of collaborative filtering to tackle it. The evaluation of the proposed approaches was done by using the data obtained from the Cypriot presidential elections of February 2013 and the parliamentary elections in Greece in May, 2012. The corresponding datasets are made freely available to other researchers working in the areas of VAA and recommender systems through the Web.

Keywords: Missing values, collaborative filtering, recommender systems, voting advice applications.

1 Introduction

Democracy is the form of government where power emanates from the people, exercised by the people and serves the interests of the people. A central feature of democracy is the decision by voting citizens, in direct democracy, or some representatives, in a representative democracy [16]. However, a fairly large percentage of citizens are accustomed to not exercise their right to vote and many of them are not even informed of the political positions of the candidates. A voting advice application (VAA) is a web application that helps voters to be informed about the political stances of parties and to realize with which of them are closest. VAAs are a new phenomenon in modern election campaigning, which has increased in recent years [5].

Voting Advice Applications pose voters and candidates to answer a set of questions in an online questionnaire. Each question corresponds to the political positions of the parties and their reaction to the developments in the current affairs. Voters and candidates evaluate each issue by giving lower extent to those with which they do not agree at all and higher to those that perfectly expressed

H. Papadopoulos et al. (Eds.): AIAI 2013, IFIP AICT 412, pp. 20–29, 2013.

their position [5][12]. Usually the answering options are 'strongly disagree', 'disagree', 'neither agree nor disagree', 'agree', 'strongly agree' and 'I have no opinion'. In the end, the similarity between voters and candidates is calculated, and, with the aid of a properly designed algorithm every voter is recommended the candidate with whom he/she has the higher similarity.

VAAs are becoming increasingly common in electoral campaigns in Europe. A survey in 2008 showed that VAAs were applied to general elections in 15 European countries, having high impact in most cases [21]. Many argue that VAAs are able to increase and improve democratic participation, since these applications help voters to have easy access in information about the political views of political parties. For better utilization of such applications is essential that advices by VAAs be credible and impartial. Additionally, it is desirable to have access to a broad cross section of society, not only by people of one side of the political and social spectrum.

In VAAs it is highly desirable that both the users and the candidates answer all the questions of the online questionnaire. In this case the recommendation given to the voter is more accurate and the information about candidates' positions are more wide. Unfortunately, this is not the case. Candidates refrain from exposing themselves to controversial issues, choosing answers in the middle of the Likert scale (i.e., 'neither agree nor disagree') while the voters leave unanswered several questions or give answers like 'I have no opinion' or 'neither agree nor disagree' for several reasons, including time constraints, limited information on the corresponding issues, or even due to unclear questions. Sometimes voters forget or avoid answering a question. Also there are questions that are characterized of ambivalence or indecisiveness and those with which the respondent does not agree against the main assumptions of them [17].

Although missing values distort , to some extent, the VAAs. Choosing to ignore them might severely affect the overall sample and it is likely to provide a completely wrong result: it is highly probable the VAA to not be able to estimate the similarity between the voter and the candidates and suggest a wrong candidate to a voter; even if the missing data are not a extended (only few questions left unanswered). As a result the effectiveness and reliability of the corresponding VAA will be deteriorated. In this paper we formulate missing values in VAAs as a recommender system problem and we applied state of the art techniques from this discipline to effectively tackle it. In particular, we investigate both matrix factorization and collaborative filtering techniques as possible ways of missing value estimation. We show that, given the peculiarity of VAA data, even unsuccessful for recommender systems techniques such as SVD perform quite well in estimating missing values in VAAs. We also provide online access, to the research community of VAAs and recommender systems, to two VAA datasets obtained from Cypriot presidential elections of February, 2013 and Greek parliamentary elections of May, 2012. To the best of our knowledge this is the first time the missing value problem in VAAs is approached in this, systematic, way.

2   Background

Missing value estimation is an important problem in several research areas. Whenever the cost of acquiring data or repeating an experiment is high estimating the values of missing data is the method of preference [20]. Expectation maximization (EM) and Maximum Likelihood (ML) are two well-known methods of estimating the parameters of the missing data based on the available ones. ML tries to find the underlying probability distribution of the available data. Once this is done obtaining the values of missing data is straightforward [6]. In EM the logic is similar. However, due to sparseness of available data the method estimates missing values in an iterative way. In the first step the aim is to estimate the parameters of interest from the available data and the probable values for missing data. In continue, the parameters are recalculated using the available data and the estimated values. The new parameters are applied to recalculate the missing values. The process is repeated until the values estimated from one cycle have a high correlation with those of the next [2].

EM and ML are more appropriate for unimodal probability distributions of sample. This is rarely the case in data collected from surveys and especially from VAAs as well as in recommender systems data. In those cases mixture of Gaussians are more appropriate probability distribution methods. In recommender systems this is the basic reason clustering based methods became so popular [19]. Furthermore, both ML and EM are time consuming procedures and they are used in cases where it is critical to make the best estimation irrespectively of the time required to do so. In cases where the response time is more critical than the accuracy of estimation more simple approaches are usually adopted. Single, multiple and hot deck imputation are such methods. The simplest and most common strategy for this method is calculating the value of a missing data, as being the mean for that variable. In multiple imputation approaches, each missing value is replaced by a set of imputed values. The technique which is used to generate these estimates is a simulation-based approach called Markov Chain Monte Carlo (MCMC) estimation [7]. Hot deck imputation replaces a missing value with an observed response from a similar unit [1]. The process includes finding other records in the dataset that are similar in other parts of their responses to the record with the missing values. Usually, there are a lot of data that correlate with those containing missing values. In these cases, the value that will fill the empty cell is selected randomly from the good matches and it is called donor record.

3   Problem Formulation

In this article we formulate the VAA missing value estimation as a recommender system problem. Recommender Systems (RS) are software tools and techniques that make recommendations for items that can be exploited by a user. These systems are particularly useful, since the sheer volume of data, which many modern online applications manage, often hinder users to distinguish what information is related to their interests. So the RSs with the help of special algorithms attempt to predict what products or services a user would find interesting to see or buy.

Formally, the recommendation problem can be formulated as follows: Let C be the set of users (customers) and let S be the set of all possible items that the users can recommend, such as books, movies, restaurants, etc. Let also u be a utility function that measures the usefulness (as may expressed by user ratings) of item s to user c, i.e., $u : C \times S \rightarrow \Re$. The usefulness of all items to all users can be expressed as a matrix U with rows corresponding to users and columns corresponding to items. An entry $u(c, s)$ of this matrix may have either positive value indicating the usefulness (rating) of item s to user c or a zero value indicating that the usefulness $u(c, s)$ has not been evaluated. It should be noted, here, that although there are several cases where the rating scale is different than the one mentioned above and includes negative values (and as a result the non-evaluated items cannot be represented by the zero value) it is always possible to transform the rating scale in an interval $[l_v \ h_v]$ where both l_v and h_v are greater than zero. The recommendation problem can be seen as the estimation of zero values of matrix $U \in \Re^{N_C x N_S}$ from the non-zero ones. The quantities N_C and N_S represent the total number of users and items respectively.

Recommender systems are broadly divided into six main categories: (a) Content-based, (b) Collaborative Filtering, (c) Knowledge-based, (d) Community-based, (e) Demographic and Hybrid Recommender Systems [10]. Collaborative Filtering (CF), which is the most common RS type, assumes that if two users evaluate the same items in a similar way they are likely to have the same 'taste' and, therefore, RSs can make recommendations between them.

Recommendation in the CF approach requires some similarity $sim(c, c')$ between users c and c' to be computed based on the items that both of them evaluated with respect to their usefulness. The most popular approaches for user similarity computation are Pearson correlation and Cosine-based metrics. Both of these methods produce values $sim(c, c') \in [-1 \ 1]$. By computing the similarity of all users in pairs we create the similarity matrix $M \in \Re^{N_C x N_C}$. Zero values of matrix M may correspond to either zero similarity, or, to users with no commonly evaluated items. The influence of a user can be computed by taking the sum across the corresponding row or column of matrix M. The higher this sum is the more influential the user is.

As may one easily understand the zero values of the utility matrix U can be seen as missing values. Furthermore, in VAAs N_C can be seen as the number of users who filled in the online questionnaire while N_S is the number of questions in the questionnaire. In such a setting techniques from the recommender systems can be applied for the estimation of the missing answers.

In commercial recommender systems the utility matrix U is very sparse, that is the non-zero elements are much less than the zero ones [18]. As a result traditional approaches such as nearest neighbor and clustering based recommendation are not so effective. This is because in sparse datasets the 'neighbors' of a user may not actually coincide in preferences with him/her. Therefore, recommendations given by those neighbors are unlikely to be successful. In an effort to tackle this problem alternative collaborative filtering approaches were proposed as well as a new category of methods based on matrix factorization. In the latter methods matrix factorization is applied to the utility matrix U in a effort to identify hidden clusters and get advantage of non-sparse areas in U. Among the matrix factorization techniques the Singular Value Decomposition (SVD) is the most widely known and it has been applied in several fields of study and especially in dimensionality reduction. Unfortunately, in sparse matrices the SVD is not so effective and as a consequence alternative iterative techniques including Alternative Least Squares (ALS) [13] and Stochastic Gradient Descent (SGD) [23] were proposed.

In VAAs the matrix U is not sparse and traditional collaborative filtering techniques as well as SVD are expected to be effective. We investigate the performance of these techniques as well as many others in an effort to estimate missing values in VAAs based on a recommender system perspective as formulated in the previous paragraphs. The results show that the majority of techniques used in recommender systems can be also, successfully, applied for missing value estimation in an elegant and mathematically strict way.

4   Approaches

In this section we briefly report the various approaches we have applied for missing value estimation in VAA data. Emphasis is given to matrix factorization and clustering-based approaches, while comparison against the simple nearest neigbor methods are also provided.

4.1   Clustering Algorithms

Clustering algorithms perform recommendation to the active user by employing a smaller set of highly similar users instead of the entire database [19]. They have been proved as an effective mean to address the well-known scalability problem that recommender system face [18]. A cluster is produced by data objects which are similar to each other. Data belonging to a cluster differ from those which belong to another cluster [8]. In the VAA setting every user is represented by a profile vector c_i which is composed of his/her answers to the online questionnaire. Given that the number of questions in the questionnaire is fixed and the possible answers come from a Likert scale the vector c_i can be encoded as a numerical one. In this way the comparison between profile vectors is straightforward by using and any similarity metric of linear algebra. Clustering of users is obtained by applying such a metric in the profile vectors.

The k-means clustering algorithm is widely used in clustering based recommendation systems mainly due to its simplicity [11]. It aims to partition N_C users, defined by the user profile vectors $\{c_1, c_2, ..., c_{N_C}\}$, into K clusters ($K << N_C$) $\mathbf{S} = \{S_1, S_2, ..., S_K\}$. Each user is classified to a cluster according to the shortest distance between the user profile vector and the cluster's mean vector. The user profile vectors correspond to the rows of the utility matrix U mentioned earlier. The k-means algorithm is similar to the expectation-maximization algorithm for mixtures of Gaussian in that they both attempt to find the centers of natural clusters in the data. The optimization criterion is to find the partition $\mathbf{S}^o$ that minimizes the within-cluster sum of squares (WCSS):

$$\mathbf{S}^o = \underbrace{argmin}_{\mathbf{S}}(\sum_{i=1}^{K} \sum_{c_j \in S_i} \|c_j - \boldsymbol{\mu}_i\|^2) \tag{1}$$

where $\boldsymbol{\mu}_i$ is the mean vector of data points (user profile vectors) in cluster S_i.

By using k-means partitioning method, the missing values of a voter's answers can be estimated by the answers of the cluster members' (where the voter belongs). The facts which negatively affect that method is the sparsity of clusters, that provokes the failure of the prediction on a missing value, and the high cardinality of clusters, which increases the time required for the prediction.

4.2   Nearest Neighbor Methods

In the Nearest Neighbor recommendation the most similar user of an active user is found and provides the recommendation. The main disadvantage of this method is that every time a new voter fills the online questionnaire, the similarity between voters must be recalculated. This is obviously non-scalable and time consuming. Furthermore, if the voter and his/her nearest neighbor have the same unanswered questions then prediction of missing values would be impossible. An alternative is to find the nearest neighbor among those users that have answered the question which the active user left unanswered.

The K-nearest neighbor (K-NN) algorithm is an improved extension to the Nearest Neighbor method. It limits the number of the neighbors we aim to find, by determining the value of K [4]. In case the K is not determined we can use a fixed distance alternative. In this case the K neighbors are those that fall in the hypersphere that (a) is centered on the datapoint of profile vector of the active user, and (b) has a radius equal to the fixed distance. The K nearest neighbors of the c-th voter correspond to the K voters who belong to the c-th row, of the voter similarity matrix M, with the highest values. The prediction of the voter's missing values can be calculated by using either the average of K neighbor voter answers or the median value of these answers. The most common problem with this method is that K is not always obvious since it depends on every voter separately while in the case of fixed distance K can be very small (even equal to zero).

4.3   Matrix Factorization

Matrix factorization techniques proved to be superior to clustering-based methods because they allow integration of additional information about the user. In those methods a matrix is factorized to find out two or more matrices such that when they are multiplied the result is to get back the original matrix [13].

The most common mode of matrix factorization is the calculation of a low-rank approximation to a fully controlled data matrix order to minimize the sum-squared difference of matrix values. A low-rank approximation is given from the viewpoint of the singular value decomposition (SVD) of the data matrix. The SVD expressed on a utility matrix U of size mxn leads to a factorization into three matrices: X is an mxm unitary matrix, S is an mxn rectangular diagonal matrix with nonnegative real numbers on the diagonal, and Y^* (the conjugate transpose of Y) is an nxn unitary matrix [22].

$$U = XSY^* \tag{2}$$

Whenever the utility matrix is large and sparse, we resort to perform Stochastic Gradient Descent (SGD) for matrix factorization. SGD has been successfully applied to large-scale and sparse machine learning problems and can handle problems with more than 10^5 training examples and more than 10^5 features [3]. SGD approximates the true gradient of $E(w, b)$ by considering a single training example at a time. We applied the algorithm for SGD as presented in [23].

Alternating Least Squares (ALS) is another technique used for matrix factorization for sparse matrices [24]. While SGD is easier and faster than ALS, ALS is more appropriate for VAA missing value estimation since the estimated values are guaranteed to real and non-negative.

5   Datasets

In our paper we have used three datasets for experimental evaluation. The first one was collected by a pre-survey for the Cypriot presidential elections 2013 conducted door to door on January 2013, the second one was collected from www.choose4cyprus.com and the third dataset was derived from www.choose4greece.com. The main characteristics of all datasets is the small number of questions and the high number of ratings per question which leads on utility matrices U of low sparsity. In all datasets the ratings are integer values in the range [1 5] corresponding to the answers 'strongly disagree', 'disagree', 'neither agree nor disagree', 'agree', 'strongly agree' respectively. We have set as missing values the instances where voters answered 'I do not have opinion' or not answered, the particular question at all. The main characteristics of these datasets are shown in Table 1. The corresponding datasets can be accessed via the URL: www.preferencematcher.com/datasets

6   Experimental Results and Discussion

Experiments were designed to investigate the performance of the recommendation methods, which have been reported previously, based on the accuracy of prediction of missing values. This measure computes the accuracy of predicting the values of utility matrix U using a variation of Mean Absolute Error (MAE) [9] [15] computed with the aid of Frobenius norm.

Table 1. Dataset Characteristics

	Pre-survey	Choose4Cyprus	Choose4Greece
# voters	815	18,461	75,294
# questions	35	30	30
# ratings	26,419	533,542	2,204,306
# ratings per question (average)	755	17,785	73,477
# sparsity	0.0736	0.0366	0.0247

Table 2. The results of missing value estimation per method. Shown values refer to accuracy A.

Method	Predicted value obtained by:	Pre-survey	Choose4Cyprus	Choose4Greece
SVD		0.0483	0.0492	0.0420
ALS		0.0516	0.0497	0.0438
SGD		0.0615	0.0616	0.0538
Nearest Neighbor		0.0897	0.0638	0.0592
k-Means	Average	0.0580	0.0496	0.0477
	Median	0.0604	0.0504	0.0474
K-NN	Average	0.0640	0.0518	0.0491
	Median	0.0700	0.0592	0.0555
	Weighted Sum	0.0657	0.0503	0.0488
Fixed Distance	Average	0.0710	0.0670	0.0651
	Median	0.0788	0.0707	0.0685
	Weighted Sum	0.0708	0.0661	0.0640

Let $\hat{U}$ be the estimation of utility matrix U, then the accuracy measure A is defined as follows:

$$A = \frac{||U - \hat{U}||}{||U|| + ||\hat{U}||} \tag{3}$$

where $||C||$ denotes the Frobenius norm of matrix C.

Table 2 shows the MAE value of each method which used in datasets of Pre-survey, Choose4Cyprus and Choose4Greece. It can be seen in Table 2 that matrix factorization methods show better prediction accuracy than the clustering based and the nearest neighbor methods. However, the difference between matrix factorization and clustering methods is not so apparent, because of the non-sparse datasets. The best performance achieved by the SVD method, which is something expected as the datasets are not sparse and SVD very effective in non-sparse data. In addition K-NN method achieved similar results with the k-means clustering method. This result is on agreement with the results presented in [18]. We see also that, as expected, the K-NN method achieves better results than the Nearest Neighbor method. Finally, it should be noted the difference in performance of the compared methods to data collected online (Choos4Cyprus and Choose4Greece) and door to door (Pre-survey) with the former to be better. This might happen because the sparsity in the Pre-survey is larger than in the online datasets.

7   Conclusion

In this article we dealt with the problem of missing values in VAAs by using techniques from the recommender systems. We observed that the SVD method, in contrary to commercial recommender systems, is the most effective technique. This is something expected since the datasets we have used are of low sparsity. However, the effectiveness of the other methods are also high indicating that the formulation of VAA missing value problem as recommendation system is successful.

In the near future we plan to implement the SVD approach of missing value estimation in our VAAs so as to improve recommendation effectiveness by filling in unanswered questions. We are also going to investigate the effectiveness of the compared algorithms by artificially increasing the sparsity of our datasets. Finally, comparison with other techniques for missing value estimation such as ML and simple imputation will be also investigated. Comparison with Expectation Maximization is not necessary since the k-Means clustering is based on the same principle and has been already done in this paper.

References

1. Andridge, R.R., Little, R.J.A.: A review of hot deck imputation for survey nonresponse. International Statistical Review 78(1), 40–64 (2010)
2. Bilmes, J.: A gentle tutorial of the EM algorithm and its applications to parameter estimation for Gaussian mixture and hidden Markov models. Technical report, International Computer Science Institute (1998), http://ssli.ee.washington.edu/people/bilmes/mypapers/em.pdf (retrieved) (last access March 2013)
3. Bottou, L.: Large-Scale Machine Learning with Stochastic Gradient Descent. In: Lechevallier, Y., Saporta, G. (eds.) Proceedings of COMPSTAT 2010. Springer, Heidelberg (2010)
4. Bremner, D., Demaine, E., Erickson, J., Iacono, J., Langerman, S., Morin, P., Toussaint, G.: Output- sensitive algorithms for computing nearest-neighbor decision boundaries. Discrete and Computational Geometry 33(4), 593–604 (2005)
5. Cedroni, L., Diego, G. (eds.): Voting Advice Applications in Europe: The State of the Art. ScriptaWeb, Napoli (2010)
6. Enders, C.K.: A primer of maximum likelihood algorithms available for use with missing data. Structural Equation Modeling 8, 128–141 (2001)
7. Gamerman, D., Lopes, H.F.: Markov Chain Monte Carlo: Stochastic Simulation for Bayesian Inference, 2nd edn. Chapman and Hall/CRC (2006)
8. Han, J., Kamber, M., Pei, J.: Data Mining: Concepts and Techniques, 3rd edn. Morgan Kaufmann (2011)

9. Herlocker, J.L., Konstan, J.A., Riedl, J.T.: An empirical Analysis of Design Choices in Neighborhood-Based Collaborative Filtering Algorithms. Information Retrieval 5(4), 287–310 (2002)
10. Jannach, D., Zanker, M., Felfernig, A., Friedrich, G.: Recommender Systems: An Introduction. Cambridge University Press (2010)
11. Kim, K., Ahn, H.: A recommender system using GA K-means clustering in an online shopping market. Expert Systems with Applications: An International Journal 34(2), 1200–1209 (2008)
12. Ladner, A., Pianzola, J.: Do voting advice applications have an effect on electoral participation and voter turnout; Evidence from the 2007 swiss federal elections. In: Tambouris, E., Macintosh, A., Glassey, O. (eds.) ePart 2010. LNCS, vol. 6229, pp. 211–224. Springer, Heidelberg (2010)
13. Salakhutdinov, R., Mnih, A.: Probabilistic Matrix Factorization. In: Advances in Neural Information Processing Systems (NIPS 2007), pp. 1257–1264. ACM Press (2008)
14. Sarwar, B.M., Karypis, G., Konstan, J., Riedl, J.: Application of Dimensionality Reduction in Recommender System - A Case Study. In: Workshop on Web Mining for e-Commerce: Challenges and Opportunities (WebKDD). ACM Press (2000)
15. Shardanand, U., Maes, P.: Social information filtering: Algorithms for automating Word of mouth. In: ACM CHI 1995 Conference on Human Factors in Computing Systems, pp. 210–217. ACM Press (1995)
16. Tansey, S.D., Jackson, N.: Poltics: the basics, 4th edn. Routledge (2008)
17. Triga, V., Serdult, U., Chadjipadelis, T.: Voting Advice Applications and State of the Art: Theory, Practice, and Comparative Insights. International Journal of Electronic Governance 5(3/4) (2012)
18. Tsapatsoulis, N., Georgiou, O.: Investigating the Scalability of Algorithms, the Role of Similarity Metric and the List of Suggested Items Construction Scheme in Recommender Systems. International Journal on Artificial Intelligence Tools 21(4), 19–26 (2012)
19. Ungar, L.H., Foster, D.P.: Clustering Methods for Collaborative Filtering. In: AAAI Workshop on Recommendation Systems, pp. 1–16. AAAI Press (1998)
20. Vogiatzis, D., Tsapatsoulis, N.: Missing Value Estimation for DNA Microarrays with Mutliresolution Schemes. In: Kollias, S., Stafylopatis, A., Duch, W., Oja, E. (eds.) ICANN 2006. LNCS, vol. 4132, pp. 141–150. Springer, Heidelberg (2006)
21. Walgrave, S., Van Aelst, P., Nuytemans, M.: Do the Vote Test: The Electoral Effects of a Popular Vote Advice Application at the 2004 Belgian Elections. Acta Politica 43(1), 50–70 (2008)
22. Wall, M.E., Rechtsteiner, A., Rocha, L.M.: Singular value decomposition and principal component analysis. In: Berrar, D., Dubitzky, W., Granzow, M. (eds.) A Practical Approach to Microarray Data Analysis, pp. 91–109. Kluwer, MA (2003)
23. Zhou, T., Shan, H., Banerjee, A., Sapiro, G.: Kernelized Probabilistic Matrix Factorization: Exploiting Graphs and Side Information. In: SIAM International Conference on Data Mining, pp. 403–414. SIAM / Omnipress (2012)
24. Zhou, Y., Wilkinson, D., Schreiber, R., Pan, R.: Large-Scale Parallel Collaborative Filtering for the Netflix Prize. In: Fleischer, R., Xu, J. (eds.) AAIM 2008. LNCS, vol. 5034, pp. 337–348. Springer, Heidelberg (2008)

A Personalized Location Aware Multi-Criteria Recommender System Based on Context-Aware User Preference Models

Salvador Valencia Rodríguez and Herna Lydia Viktor

School of Electrical Engineering and Computer Science, University of Ottawa
800 King Edward Road, Ottawa, Ontario, Canada
{svale054,hviktor}@uottawa.ca

Abstract. Recommender Systems have been applied in a large number of domains. However, current approaches rarely consider multiple criteria or the level of mobility and location of a user. In this paper, we introduce a novel algorithm to construct personalized multi-criteria Recommender Systems. Our algorithm incorporates the user's current context, and techniques from the Multiple Criteria Decision Analysis field of study to model user preferences. The obtained preference model is used to assess the utility of each item, to then recommend the items with the highest utility. The criteria considered when creating preference models are the user location, mobility level and user profile. The latter is obtained considering the user requirements, and generalizing the user data from a large-scale demographic database. The evaluation of our algorithm shows that our system accurately identifies the demographic groups where a user may belong, and generates highly accurate recommendations that match his/her preference value scale.

Keywords: Recommender Systems, Location Aware, Multi-Criteria, Preference Models, Personalization.

1 Introduction

Consumers often find themselves in situations where they have to choose one item over others. For instance, they may wish to decide which movie to view, which book to read, what items to buy, and so forth. However, due to the advance in technology and the large amount of information available in databases, our options have dramatically increased. We have now reached a point where we have thousands of options at our fingertips, through the use of web-based systems. Consequently, in order to address this information overload, and aid consumers through these daily decision-processes, Recommender Systems have been developed. These systems aim to provide personalized services to each user, showing them only the information that they are most likely to be interested into [1].

A number of techniques to predict the best items have been proposed. While some of these Recommender System implementations have been successful in many domains, a number of challenges still remain. Most implementations consider only

H. Papadopoulos et al. (Eds.): AIAI 2013, IFIP AICT 412, pp. 30–39, 2013.
© IFIP International Federation for Information Processing 2013

single criteria ratings, and consequently are unable to identify *why* a user prefers an item over others. Some systems classify the user into one single group or cluster, an approach that has limited use, since real world users share commonalities in different degrees with diverse types of users. Finally, other systems require a large amount of previously gathered data about users' interactions and preferences, in order to be successfully applied.

This work introduces an algorithm to overcome these previously mentioned disadvantages. Our algorithm, as presented in this paper, builds user preference models considering multiple criteria. That is, we include the users' special needs, and context in the decision making process. This enables our system to clearly identify most important criteria, for a specific user when selecting an item over the others, and correspondingly create accurate recommendations that match his/her preference's value scale. Moreover, by including the user context as part of the recommendation process, the system is able to produce different types of recommendations to the same user, depending on his/her current context. Additionally, our algorithm exploits the information contained in a large-scale demographic database to generalize the information as provided by the user. This is done by clustering the user into one or more demographic groups. This aspect allows our system to leverage commonalities between similar user types, and create richer user profiles without the need of previously stored data about other users' interactions.

This paper is organized as follows. Section 2 details our recommender system algorithm. Section 3 describes our case study and Section 4 concludes.

2   Personalized Location-Aware System

We introduce a Personalized Multi-Criteria Context-Aware Recommender-System, which has been designed to achieve the following goals. Firstly, we aim to identify the user preferences, as based on multiple criteria that lead him/her to select an item over others, and correspondingly potentially generate more accurate recommendations. Further, our objective is to model and consider the user context during the recommendation process, instead of using it only as a filter. That is, we follow a context-aware approach, where the context is used as one of multiple criteria for the creation of preference models [6, 7]. Finally, our ultimate goal is to produce accurate recommendations by generalizing user profiles without gathering data from previous users' interactions. We use a demographic generalization technique to exploit the information contained in large-scale demographic databases that encompass interests and preferences of similar people [8].

Our algorithm is based on the creation of a multiple criteria user preference model considering the following four criteria. These are 1) the probability that a user prefers an item, given the probabilities that the user belongs to each of the demographic clusters identified in a large-scale demographic database; 2) the weight of each item, based on its attributes and the weights the user assigns for each of them; 3) the distance between the user's current location and the item's location; and, 4) the time the user would require to reach each item, based on his/her mobility level, together with

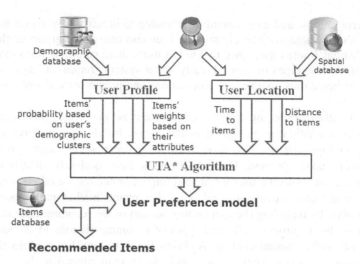

Fig. 1. High-level schematic overview of our algorithm

other geo-spatial constraints (such as a river) and average route times. The user preference model is then used to assess the utility of each item for the user. Finally, the system will recommend to the user the items with the highest estimated utility. The following figure presents the high-level schematic overview of our algorithm.

2.1   Create the User Profile

We create a user profile consisting of two dimensions. The first is the set of all possible demographic clusters where the user potentially belongs to. The second aspect concerns how much importance a specific user assigns to an item. These dimensions may be expressed as the following two vectors. *DemogClusters(u)*, which represents the possible demographic clusters where a user may belong, along with the probability of belonging to each of them. The second is *AttributeWeights(u)*, which contains the possible items' attributes along with the user-assigned weight for each of them.

The set of demographic clusters where a user may belong (*k* clusters), that form the *DemogClusters(u)* vector is obtained by applying a demographic generalization technique. First, the algorithm select the *m* demographic variables that best differentiate each cluster from the others using feature selection techniques. Second, the user is queried to obtain his/her information for the selected demographic variables. Third, our system obtains the probability of a user belonging to each demographic cluster, based on the user given information and the users' distribution (included in the database) for each possible value for each demographic variable. Finally, the set of possible demographic clusters where the user may belong, are those in which the product of the users' distributions for the user given information is higher than a selected threshold. The following equation presents how these clusters are obtained.

$$PossibleClusters(u) = \left\{ C_x : \prod_{j=1}^{m} P_{u,C_x}(DV_j = Val(DV_j, u)) \geq Threshold \right\} \forall 1 \leq x \leq N \tag{1}$$

Here, *DV* stands for *Demographic Variable*; $Val(DV_j, u)$ represents the value of the demographic variable *j* given by user *u*; and finally, $P_{u,c_x}(DV_j = Y) = Z\%$ expresses that there is a *Z*% probability (obtain from the users' distributions) that user *u* belongs to cluster *x*, given that the value of the demographic variable *j* for that user is *Y*.

Once the demographic clusters a user could belong to have been selected, the probabilities already obtained from equation (1) may be considered as the probabilities of belonging to each of them. However, this would imply classifying the user based only on his/her demographic information, and not considering other data that might also be included the database (e.g. information about leisure, shopping, media preferences, hobbies, etc). Therefore, for databases where this type of information is available, we propose the following technique to obtain the probabilities of belonging to each of the selected demographic clusters. First, we obtain the Cartesian product of the non-demographic information included in the database for each of the previously selected clusters (e.g. Leisure X Shopping X Media). Second, we apply the k-means data mining clustering algorithm over the Cartesian product dataset, to learn the demographic clusters and create the same number of groups as previously selected clusters (*k* groups).

The results of the k-means algorithm are *k* groups, each labeled by the most representative non-demographic information it contains. The k-means algorithm also creates a table, with the distribution of instances from each demographic cluster classified into the newly created groups, as shown in Table 1.

Table 1. Instances from demographic clusters classified in the newly created groups

Total instances per demographic clusters

Demographic clusters

	Total instances	Group 1	Group 2	...	Group k	
	$Inst(C_1)$	$Inst(C_1,G_1)$	$Inst(C_1,G_2)$		$Inst(C_1,G_k)$	Dem. Cluster 1
	$Inst(C_2)$	$Inst(C_2,G_1)$	$Inst(C_2,G_2)$		$Inst(C_2,G_k)$	Dem. Cluster 2
						...
	$Inst(C_k)$	$Inst(C_k,G_1)$	$Inst(C_k,G_2)$		$Inst(C_k,G_k)$	Dem. Cluster k
		$Inst(G_1)$	$Inst(G_2)$		$Inst(G_k)$	Total instances

Newly created groups

Total instances per newly created groups

In Table 1, *Inst(Cx)* represents the total number of instances belonging to the demographic cluster *x*, *Inst(Gy)* represents the total number of instances assigned to the newly created group *y*, and *Inst(Cx,Gy)* represents the number of instances from the demographic cluster *x* that were assigned to the newly created group *y*.

Third, we present to the user the obtained newly created groups, so he/she can select the one (or more) group(s) that best define him/her (*p* selected groups), based on the non-demographic information that labels each of them. Fourth, we obtain probabilities for each of the k demographic clusters, using the following equation:

$$P(u,C_x)\forall 1 \le x \le k = \begin{cases} \text{If } \sum_{j=1}^{p} Inst(C_x,G_j) = 0 : 0 \\ \text{Otherwise} : \dfrac{1}{p}\sum_{j=1}^{p}\left[\dfrac{Inst(C_x,G_j)*100}{Inst(G_j)} \right] \end{cases} \qquad (2)$$

Once we have obtained the possible demographic clusters where a user may belong and computed the probability for each of them, the *DemogClusters(u)* vector is complete. Equation (3) shows the structure of this vector, formed by pairs of the form $< C_x, P(u,C_x) >$, where C_x represents the demographic cluster x, and $P(u,C_x)$ the probability that the user u belongs to cluster x.

$$DemogClusters(u) = \left\{ < C_1, P(u,C_1) >, < C_2, P(u,C_2) >, ..., < C_k, P(u,C_k) > \right\} \qquad (3)$$

The second vector included in the user profile (*AttributeWeights(u)*) consists of pairs of the form $< A_x, W(u,A_x) >$, where A_x represents the item's attribute x, and $W(u,A_x)$ is the weight that user u assigns to attribute x. Equation (4) shows the structure of this vector, where r represents the selected number of items' attributes that best differentiate the items between them:

$$AttributeWeights(u) = \left\{ < A_1, W(u,A_1) >, < A_2, W(u,A_2) >, ..., < A_r, W(u,A_r) > \right\} \qquad (4)$$

2.2 Location Awareness and Mobility Level

The second phase of the algorithm consists of obtaining the user's mobility level and current location. Consequently, the user is asked to choose the means of transport (i.e. mobility level) that best describes his/her current situation (e.g. driving a car, riding a bike, wheelchair, or walking). Based on the user selection, the routes to be considered when predicting the distance and time to each item are selected. Subsequently the user's current latitude and longitude spatial coordinates are obtained from Google Maps®, through its application programming interface (API), using the current user address.

2.3 Assessing the Items' Utilities Considering the User Profile

The third phase of our algorithm is divided into two stages. The first stage entails obtaining an utility for each item (*Demographic Utility (DU)*), based on the previously created vector *DemogClusters(u)*. However, in order to do so, all the items in the database need to be mapped to one or more demographic clusters. Consequently, we assign each item to one or more categories, which in turn are linked to one or more demographic clusters.

Once all the items are related to one or more demographic clusters through one or more categories, we compute the probability that a user might prefer each of these categories. As shown in equation (5), these probabilities are calculated based on the probabilities of belonging to each of the k previously selected demographic cluster.

$$P(u, ItemCat_x) = \sum_{i=1}^{k} \begin{cases} \text{If } ItemCat_x \in C_i : P(u, C_i) \\ \text{Otherwise} : 0 \end{cases} \tag{5}$$

Here, $P(u, ItemCat_x)$ represents the probability that user u prefers the item category x. Subsequently, these probabilities are used to obtain the demographic utility (DU) for each item. This DU is computed as the average of the probabilities that the user prefers each category where the item belongs to. The DU of an item is obtained by using equation (6), where q represents the number of identified items' categories.

$$DU(I_x, u) = \frac{1}{Num\,of\,Cat\,of\,I_x} \sum_{i=1}^{q} \begin{cases} \text{If } I_x \in ItemCat_i : P(u, ItemCat_i) \\ \text{Otherwise} : 0 \end{cases} \tag{6}$$

The second stage involves assessing a utility for each item (*Weighted Utility (WU)*), considering now the previously created *AttributeWeights(u)* vector. The *WU* for each item is calculated as the sum of each user given weight for all the attributes of an item, as shown in equation (7).

$$WU(I_x, u) = \sum_{i=1}^{r} \begin{cases} \text{If } A_i \in I_x : W(u, A_i) \\ \text{Otherwise} : 0 \end{cases} \tag{7}$$

2.4 Assessing the Items' Utilities Considering the User Location

The fourth step in the algorithm involves assessing a utility for each item, based on the user's current location and mobility level. In this phase, each item is assessed under two additional criteria: the distance from the user (*Distance Utility (DisU)*), and the time it would take the user to reach the item (*Time Utility (TU)*), considering the user's mobility level. It is important to note that these new utilities, expressed in kilometers and minutes from the user, are inversely proportional to the item utility for the user. That is, the closer an item is the higher its utility for the user. Finally it is noteworthy to mention that while the *TU* is obtained from Google Maps®; the *DisU* is obtained by means of SQL spatial queries, performed over a spatial database using the user coordinates.

2.5 Creating the User Preference Model

The utilities that were obtained during the third and fourth phases represent the four criteria considered to create the user preference model, which in turn represents the importance to the user. To obtain the user preference model we apply the UTA* algorithm. This algorithm is a regression-based technique that infers preference models from given global preferences (e.g. previous user choices) [2, 3]. For more information on this algorithm, we recommend [2-5].

In order to apply this algorithm we first need to obtain the user's weak preference order, which represents the user preferences for some items, after considering the four criterions for each of them. Consequently, the system randomly selects ten items, along with their four utilities, to present to the user, who is then asked to rate them in a non-descending order. The result of the UTA* algorithm is a vector of four values, which represent the weights of each of these four criterions for that user. These weights are calculated using linear programming techniques. The resulting vector from the UTA* algorithm has the following form:

$$PM(u) = \{W_{DU}, W_{WU}, W_{DisU}, W_{TU}\} \qquad (8)$$

2.6   Recommending the Items That Best Match the User Preferences

The last phase of the algorithm involves assessing an integrated, final utility for each item, using the previously obtained user preference model. It follows that, to apply the previously obtained weights from the model, the values of the four different utilities must be normalized so they are in the same range *{0,...,1}*, otherwise the final utility would be biased toward the criterion with the highest scale. Therefore, each utility is divided by its highest value among all the items. Moreover, since the distance and time utilities are inversely proportional to the user utility, after dividing their values by their highest utility the resulting value is subtracted from *1*, and the remainder is the utility to be considered. The final integrated item's utilities are obtained by applying the following equation:

$$U(I_x, u) = W_{DU} \frac{DU(I_x, u)}{\max(DU(u))} + W_{WU} \frac{WU(I_x, u)}{\max(WU(u))} + W_{DisU}\left(1 - \frac{DisU(I_x, u)}{\max(DisU(u))}\right) + W_{TU}\left(1 - \frac{TU(I_x, u)}{\max(TU(u))}\right) \qquad (9)$$

3   Experimental Evaluation

As a case study, we created PeRS, a Personal Recommender System that recommends events to attend to consumers. The source datasets used in this case study are the following. The Ottawa Open Data Events database is our items dataset, which contains the events that will take place within the National Capital Region of Canada. By events we refer to festivals/fairs, film/new media, galleries, music, theater, and so on [9]. Our large-scale demographic database is the PRIZM C2 Database that classifies Canada's neighborhoods into 66 unique lifestyle types, providing insights into the behavior and mindsets of residents [10]. The Ottawa Open Data Geospatial databases contain spatial information to geographically locate different elements of interest (i.e. rivers, buildings, museums, municipalities, wards, roads, cities and country areas) [11].

3.1   Experimental Design

We evaluated our algorithm using an offline experiment. This type of experiment was chosen because it focuses on the recommendation technique that is being used, rather than the system interface, which makes it highly suitable for our environment [1]. Since the design and implementation of a Recommender System depends on the specific requirements, goals, and sources of information, the properties used to evaluate a system should be selected according to each application domain. The selected properties to be evaluated in our offline experiment were the system's prediction accuracy, item and user coverage, and confidence. (For more information on other properties to evaluate and compare Recommender Systems, we recommend [1]).

We performed a pilot study with 30 human subjects to obtain, according to the central limit theorem, a punctual estimation of the standard deviation of the population [11]. Table 2 presents the structure of the questionnaire answered by the human subjects that took part in the experiment. This questionnaire was designed to gather the required data to create the users' preference models, and to evaluate the recommendations.

Table 2. Structure of the questionnaire applied to the users

Section	Gathered Information
Personal Information	The user is asked 9 personal questions, along with his/her mobility level
Preferences	The user is asked to provide weights to the identified events' attributes
Activities	The user is asked to select the groups of activities where he/she identifies the most.
Events	The user is asked to rate 10 randomly selected events, considering the information for all the criteria.

3.2   Experimental Results

Table 3 presents a summary of the results obtained. Our analysis is centered on four aspects. Firstly, our goal was to analyze the distribution of the subjects considered in the experiment. Secondly, we aimed to evaluate the system accuracy and the ability to cluster the subjects into the most adequate demographic groups. Thirdly, we learned the system accuracy from the user profiles. Finally, we evaluated the system coverage, in terms of user and item space coverage.

From the results, as shown in Table 3, we may conclude that, based on their demographic information, 93% of the subjects that partook in experiment have differences in their profiles. This allowed us to test the system for a wide-spread distribution of users, and therefore potentially draw more reliable conclusions for the general population. Additionally, we concluded that the system was able to narrow down the possible matching clusters where a subject may belong, using only the selected demographic variables asked to the subjects. Furthermore, the subjects identified themselves within 86% of the clusters selected by the system, considering the characteristic activities of the people within each group. This indicates that the system accurately classified the subjects into the correct demographic groups.

The system has a predictive accuracy between 75% and 82% with a confidence of 95%. This accuracy represents the effectiveness of the system to identify the user preferences and rank the items in the same way he/she would approach this task. Moreover, for 80% of the subjects considered in our experiment the system was able to rank the items with at least 70% match with the way they would have proceeded (i.e. user space coverage). In addition, the system recommended 81% of the items in the database to at least one user (i.e. item space coverage).

Finally, from the results obtained from the performed classification and statistical analyzes, we concluded that the accuracy of the produced recommendations does not solely depend on the user profile. Therefore, it could potentially be used for a wide range of the population and produce equally accurate recommendations. We will explore the validity of this observation in our future work.

Table 3. Experimental results

Property	Measured Value	Interpretation
Users Distribution	93.33% of the users in the experiment have different profiles (based on their personal information).	The users present an equal distribution in most of the considered personal information. Correspondingly this wide spread distribution of users allows the system to be tested for different user types, and therefore draw reliable conclusions for a wide range of the population.
Narrowing Percentage	The system identified an average of 4 out of 66 possible clusters where a user may belong	The system is able to narrow in a 95.38% the possible demographic groups where a user may belong, during the demographic generalization process.
Accuracy to identify the matching clusters	26 out of 30 users identified themselves, in every one of the selected demographic groups for him/her.	86.66% of the users identified themselves with the activities included in the demographic groups selected by the system.
Prediction Accuracy and Confidence	Accuracy: {75.36 % – 82.96%} Confidence: 95%	The system recommends items that match the user preferences with an average of 79.16%, for 95% of the population.
Learning the System Accuracy	Based on the results obtained from applying classification and statistical regression analyzes, using a decision tree, we conclude that the system accuracy doesn't depend on the user profile. Therefore it follows that the system can produce equally accurate recommendations for a wide range of the population.	
User Space Coverage	For 80% of the users the system created models that can rank the events with at least a 70% match with how the users would have ranked them.	
Item Space Coverage	81.97% of the items in the database were recommended to a user at least once, with at least 50% match of his/her preferences.	

4   Conclusion

This paper presents a Recommender System that utilizes an algorithm to generate user models which includes the user context and preferences in the decision model. Our algorithm, as shown in the experiments, accurately identifies the demographic groups where a user may belong. Our results indicate that we are able to produce highly accurate recommendations for a wide range of the population. Further, we are able to construct accurate user profiles. Our algorithm has the following noteworthy characteristics. It uses more than one decision criteria and is therefore able to identify what it is most important for a user when selecting an item over the others. It creates user preference models, consequently the system is capable to understand the user interests and preferences, and is not biased from previous user actions. It generalizes the user profile, thus the system is capable to reason from the commonalities between demographic groups, and it doesn't require previous gathered information about other users' interactions. Importantly, it is location aware and as a result the system varies the recommendations as the user location or level of mobility changes. Our future work will focus on producing accurate preference models in an imbalanced preference setting. This would imply that our algorithm should distinguish user preferences, even when they may be very similar. Additionally, it would be interesting to evaluate the prediction accuracy of our system considering a larger number of users and different locations. Finally, it would be worthwhile to compare the obtained results from our system, against other linear and non-linear recommendation techniques [12].

References

1. Ricci, F., Rokach, L., Shapira, B., Kantor, P.B.: Recommender Systems Handbook. Springer (2011)
2. Lakiotaki, K., Matsatsinis, N.F., Tsoukias, A.: Multi-Criteria User Modeling in Recommender Systems. Intelligent Systems 26(2), 64–76 (2011)
3. Siskos, Y., Grigoroudis, E., Matsatsinis, N.: UTA Methods. In: Multiple Criteria Decision Analysis: State of the Art Surveys, pp. 297–344. Springer, New York (2005)
4. Siskos, Y., Yannacopoulos, D.: UTASTAR: An ordinal regression method for building additive value functions. Investigação Operacional 5(1), 39–53 (1985)
5. Jacquet-Lagrèze, E., Siskos, J.: Assessing a set of additive utility functions for multicriteria decision making: The UTA method. European Journal of Operational Research 10(2), 151–164 (1982)
6. Lee, J., Mateo, R.M.A., Gerardo, B.D., Go, S.-H.: Location-Aware Agent Using Data Mining for the Distributed Location-Based Services. In: Gavrilova, M.L., Gervasi, O., Kumar, V., Tan, C.J.K., Taniar, D., Laganá, A., Mun, Y., Choo, H. (eds.) ICCSA 2006. LNCS, vol. 3984, pp. 867–876. Springer, Heidelberg (2006)
7. Schilit, B., Theimer, M.: Disseminating active map information to mobile hosts. IEEE Network 8(5), 22–32 (1994)
8. Krulwich, B.: Lifestyle Finder. Intelligent User Profiling Using Large-Scale Demographic Data. AI Magazine 18(2), 37–45 (1997)
9. City of Ottawa, Ottawa Open Data, http://ottawa.ca/en/open-data-ottawa
10. PRIZM C2 Segmentation System, http://www.environicsanalytics.ca/
11. Berenson, M., Levine, D.: Basic Business Statistics: Concepts and Applications. Prentice Hall (1995)
12. Klašnja-Milićević, A., Ivanović, M., Nanopoulos, A.: The use of Nonlinear Manifold Learning in Recommender Systems. In: Proc. of the International Conference on Information Technology (ICIT 2009), Amman, Jordan, pp. 2–6 (2009)

A Dynamic Web Recommender System Using Hard and Fuzzy K-Modes Clustering

Panayiotis Christodoulou[1], Marios Lestas[2], and Andreas S. Andreou[1]

[1] Department of Electrical Engineering / Computer Engineering and Informatics,
Cyprus University of Technology
{panayiotis.christodoulou,andreas.andreou}@cut.ac.cy
[2] Department of Electrical Engineering, Frederick University of Cyprus
eng.lm@frederick.ac.cy

Abstract. This paper describes the design and implementation of a new dynamic Web Recommender System using Hard and Fuzzy K-modes clustering. The system provides recommendations based on user preferences that change in real time taking also into account previous searching and behavior. The recommendation engine is enhanced by the utilization of static preferences which are declared by the user when registering into the system. The proposed system has been validated on a movie dataset and the results indicate successful performance as the system delivers recommended items that are closely related to user interests and preferences.

Keywords: Recommender Systems, Hard and Fuzzy K-Modes Clustering.

1   Introduction

Recommender Systems (RS) have become very common on the World Wide Web, especially in e-Commerce websites. Every day the number of the listed products increases and this vast availability of different options creates difficulties to users of finding what they really like or want. RS can be seen as smart search engines which gather information on users or items in order to provide customized recommendations back to the users by comparing each other [1]. Although RS give a strategic advantage they present some problems that have to be dealt with. Different techniques and algorithms are continuously being developed aiming at tackling these problems.

Existing strategies, despite their wide availability, come with problems or limitations related to the adopted architecture, the implementation of new algorithms and techniques and the considered datasets. For example, some existing RS make recommendations for the interested user utilizing static overall ratings which are calculated based on the ratings or preferences of all other users of the system. Other systems recommend items to the current user based on what other users had bought after they viewed the searched item.

The system proposed in this work makes recommendations based on the preferences of the interested user, which are dynamically changed taking into account previous searches in real time. This approach is enhanced by the utilization of static

H. Papadopoulos et al. (Eds.): AIAI 2013, IFIP AICT 412, pp. 40–51, 2013.

preferences which are declared by the user when registering into the system. The clustering procedure, being the heart of the recommendation engine, is of particular importance and a number of techniques such as Entropy-Based, Hard K-modes and Fuzzy K-modes have been utilized. The proposed system has been tested using the MovieLens1M dataset, which was linked with IMDB.com to retrieve more content information. The final results indicate that the proposed system meets the design objectives as it delivers items which are closely related to what the user would have liked to receive based on how s(he) ranked the different categories depending on what (s)he likes more and her/his previous behavior.

The remainder of the paper is organized as follows: Section 2 provides an overview of the clustering techniques that were used by the proposed system and discusses related work on the topic. Subsequently, section 3 describes the notions behind the technical background of the proposed system. Section 4 focuses on the way the system works, describing the structure of the dataset and discussing briefly the intelligent algorithms that were designed and used. This section also illustrates the experiments carried out followed by a comparison between the two clustering techniques. Finally, section 5 offers the conclusions and some future research steps.

2   Literature Overview

Recommender Systems (RS) are of great importance for the success of Information Technology industry and in e-Commerce websites and gain popularity in various other applications in the World Wide Web [1]. RS in general help users in finding information by suggesting items that s(he) may be interested in thus reducing search and navigation time and effort. The recommendation list is produced through collaborative or content-based filtering. Collaborative filtering tries to build a model based on user past behavior, for example on items previously purchased or selected, or on numerical ratings given to those items; then this model is used to produce recommendations [10]. Content-based filtering tries to recommend items that are similar to those that a user liked in the past or is examining in the present; at the end a list of different items is compared with the items previously rated by the user and the best matching items are recommended [11].

The work described in [5] discusses the combination of collaborative and content-based filtering techniques in a neighbor-based prediction algorithm for web based recommender systems. The dataset used for this system was the MovieLens100K dataset consisting of 100000 ratings, 1682 movies and 943 users, which were also linked with IMDB.com to find more content information. The final results showed that the prediction accuracy was strongly dependent on the number of neighbors taken into account and that the item-oriented implementation produced better prediction results than the user-oriented. The HYB-SVD-KNN algorithm described in this work was four times faster than the traditional collaborative filtering.

Ekstrand and Riedl [6] presented an analysis of the predictions made by several well-known algorithms using the MovieLens10M dataset, which consists of 10 million ratings and 100000 tag applications applied to 10000 movies by 72000 users.

Users were divided into 5 sets and for each user in each partition 20% of their ratings were selected to be the test ratings for the dataset. Therefore, five recommender algorithms were run on the dataset and the predictions of each algorithm for each test rating were captured. The results showed that the item-user mean algorithm predicted the highest percentage of ratings correctly compared to the other algorithms. This showed that the algorithms differed in the predictions they got wrong or right. Item-item got the most predictions right but the other algorithms correctly made predictions of up to 69%.

The work presented in [7] described long-tail users who can play an important role of information sources for improving the performance of recommendations and providing recommendations to the short-head users. First, a case study on MovieLens dataset linked with IMDB.com was conducted and showed that director was the most important attribute. In addition, 17.8% of the users have been regarded as a long tail user group. For the proposed system 20 graduated students were invited to provide two types of recommendations and get their feedbacks. This resulted in 8 users out of 20 to be selected as a long tail user group (LTuG), two times higher than the Movie-Lens case study. Finally, it was concluded that the LTuG+CF outperformed CF by 30.8% higher precision and that user ratings of the LTuG could be used to provide relevant recommendations to the short head users.

The work of McNee et al.[8] suggested that recommender systems do not always generate good recommendations to the users. In order to improve the quality of the recommendations, that paper argued that recommenders need a deeper understanding of users and their information. Human-Recommender Interaction is a methodology of analyzing user tasks and algorithms with the end goal of getting useful recommendation lists and it was developed by examining the recommendation process from an end user's respective. HRI is consisted of three pillars: The Recommendation Dialog, the Recommender Personality and the User Information seeking Tasks and each one contains several aspects.

Karypis et al. [9] presented a class of item-based recommendation algorithms that first determine the similarities between the various items and then use them to identify the set of items to be recommended. In that work two methods were used: The first method modeled the items as vectors in the user space and used the cosine function to measure the similarity between the items. The second method combined these similarities in order to compute the similarity between a basket of items and a recommender item. Five datasets were tested in the experimental part and the results showed that the effect of similarity for the cosine-based scheme improved by 0% to 6.5% and for the conditional probability based by 3% to 12%. The effect of row normalization showed an improvement of 2.6% for the cosine-based and 4.2% for the probability. The model size sensitivity test showed that the overall recommendations accuracy of the item-based algorithms does not improve as we increase the value of k. Finally they concluded that the top-N recommendation algorithm improves the recommendations produced by the user-based algorithms by up to 27% in terms of accuracy and it is 28 times faster.

This paper describes the design and implementation of a new dynamic Web Recommender System using Hard and Fuzzy K-modes clustering. The system provides recommendations based on user preferences that change in real time taking also into account previous searching and behavior and based on static preferences which are declared by the user when registering into the system.

3   Technical Background

3.1   Entropy-Based Algorithm

The Entropy-based algorithm groups similar data objects together into clusters based on data objects entropy values using a similarity measure [2]. The entropy value H_{ij} of two data objects X_i and X_j is defined as follows:

$$H_{ij} = E_{ij} \log_2(E_{ij}) - (1 - E_{ij}) \log_2(1 - E_{ij}) \tag{1}$$

where $i \neq j$.

E_{ij} is a similarity measure between the data objects X_i and X_j and is measured using the following equation:

$$E_{ij} = e^{-aD_{ij}} \tag{2}$$

where D_{ij} is the distance between X_i and X_j and a is calculated by

$$a = \frac{-\ln(0.5)}{\overline{D}} \tag{3}$$

and $\overline{D}$ is the mean distance among all the data objects in the table. From Equation (1) the total entropy value of X_i with respect to all other data objects is computed as:

$$H_i = -\sum_{\substack{j=1 \\ i \neq k}}^{n} [E_{ij} \log_2(E_{ij}) - (1 - E_{ij}) \log_2(1 - E_{ij})] \tag{4}$$

The Entropy-based algorithm passes through the dataset only once, requires a threshold of similarity parameter β and is used to compute the total number of clusters in a dataset, as well as to find the locations of the cluster centers [3]. More specifically, the algorithm consists of the following steps:

1. Select a threshold of similarity β and set the initial number of clusters $c = 0$.
2. Determine the total entropy values H for each data object X as shown by Equation (4).
3. Set $c = c + 1$.

4. Select the data object X_{min} with the least entropy H_{min} and set $Z_c = X_{min}$ as the c_{th} cluster centre.

5. Remove X_{min} and all data objects having similarity β.

6. If X is empty then stop, otherwise go to step 3.

The Entropy-based algorithm was used in this paper to compute the number of clusters in the dataset, as well as to find the locations of the cluster centers.

3.2   Hard K-Modes Clustering

The K-Modes algorithm extends the K-means paradigm to cluster categorical data by removing the numeric data limitation imposed by the K-means algorithm using a simple matching dissimilarity measure or the hamming distance for categorical data objects or replacing means of clusters by their mode [4].

Let X_1 and X_2 be two data objects of X_1 defined by m attributes. The dissimilarity between the two objects is stated as:

$$d(X_1, X_2) = \sum_{j=1}^{m} \delta(x_{1j}, x_{2j}) \qquad (5)$$

where:

$$\delta(x_{1j}, x_{2j}) = \begin{cases} 0, x_{1j} = x_{2j} \\ \\ 1, x_{1j} \neq x_{2j} \end{cases} \qquad (6)$$

In the case of Hard K-Modes clustering, if object X_i in a given iteration has the shortest distance with center Z_l, then this is represented by setting the value of the nearest cluster equal to 1 and the values of the other clusters to 0. For $a = 0$:

$$w_{li} = \begin{cases} 1, \text{if } d(Z_l, X_i) \leq d(Z_{h,} X_i), 1 \leq h \leq k \\ \\ 0, \text{otherwise} \end{cases} \qquad (7)$$

In the above equation w_{li} is the weight degree of an object belonging to a cluster.

3.3   Fuzzy K-Modes Clustering

Fuzzy K-Modes algorithm is an extension of the Hard K-Modes algorithm and it was introduced in order to incorporate the idea of fuzziness and uncertainty in datasets [3].

For $a > 1$:

$$
w_{li} = \begin{cases} 1, & \text{if } X_i = Z_l \\\\ 0, & \text{if } X_i = Z_h, h \neq 1 \\\\ \dfrac{1}{\displaystyle\sum_{h=1}^{k}\left[\dfrac{d(Z_l, X_i)}{d(Zh, X_i)}\right]^{1/a-1}}, & \text{if } X_i \neq Z_l \text{ and } X_i \neq Z_h, 1 \leq h \leq k \end{cases} \tag{8}
$$

If a data object shares the same values of all attributes within a cluster then it will be assigned entirely to that cluster and not to the others. If a data object in not completely identical to a cluster then it will be assigned to each cluster with a membership degree.

4 Methodology and Experimental Results

4.1 Proposed System

Figure 1 shows a schematic representation of the proposed system. First the user registers into the system and ranks the different categories depending on what (s)he likes more using a weight ranking system. The ranking of the categories is called static information because this type of information can only be changed after a certain period of time when the user will be prompted by the system to update her/his rankings as their interest in certain categories may have changed. The system requires a certain number of searches to be conducted first so as to understand the user behavior (dynamic information) and then it starts recommending items. A dynamic bit-string is created after the first searches and is updated with every new search. This string is compared with each movie in the dataset to eliminate those movies that the user is not interested in depending on search profile thus far. If there is at least one 1 in the compared bit string then the movie moves is inserted into a lookup table. After that the system creates the clusters depending on the new dataset size (i.e. the movies in the lookup table) and the entropy threshold similarity value β which is assumed to be constant; however, its value needs to be tuned based on the size of the dataset in order to reach optimal performance. The next step is the update of the clusters to include the static information of the user. This is performed so as to eliminate the problem encountered by the system after having a specific object belonging to two or more clusters. Therefore, the new clusters also include the static information depending on how the user ranks the categories.

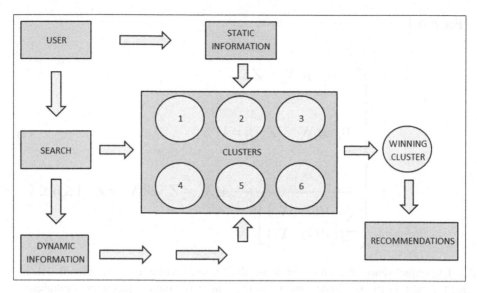

Fig. 1. How the proposed RS works

When the user performs searching queries the keyword used is compared with each cluster center and the proposed system finds the most similar one depending on the searched item (winning cluster); this cluster is then used to provide the recommendations. In the meantime the searched keyword is saved by the system as part of the dynamic information which is thus updated in real time.

4.2   Dataset

The dataset used in the proposed system is an extension of the Movie Lens 1M dataset that can be found at GroupLens.org. The Movie Lens 1M dataset is consisted of 1 million ratings from 6000 users on 4000 movies. The proposed system is not using the user ratings so we deal only with the movies. From the 4000 movies some movies are duplicated so we had to remove them thus concluding with a final dataset numbering a total of 3883 movies. Then we linked the final dataset with IMDB.com, the world's largest movie database, to retrieve more information about the categories of each movie.

Table 1 shows the different movie categories. In total there are 18 different categories. Therefore, the experimental dataset used for the encoding and testing of the proposed system was a matrix of 3883 movies in rows times 18 movie categories in columns.

4.3   Experimental Results

Three users with different characteristics that searched for various items were used to test the proposed recommendation schema on the movie dataset described in section 4.1. As previously mentioned, the system recommended movies based on the Hard and Fuzzy K-Modes clustering. The different characteristics used by the system to predict accurate recommended items were: (i) the total number of the final clusters based on the threshold similarity value β, (ii) the ranking of the movie categories according to the user interests and, (iii) the past history of different searches that each of the users conducted.

Table 1. Movie Categories

Column	Movie Category
1	Animations
2	Children
3	Comedy
4	Adventure
5	Fantasy
6	Romance
7	Drama
8	Action
9	Crime
10	Thriller
11	Horror
12	Sci-Fi
13	Documentary
14	War
15	Musical
16	Mystery
17	Western
18	Film-Noir

The Root Mean Square Error (RMSE) was used as the evaluation metric to assess the accuracy of the results, which is defined as follows:

$$RMSE = \sqrt{\frac{\sum_{i=1}^{n} (x_{obs,i} - x_{model,i})^2}{n}} \qquad (9)$$

where $X_{obs,i}$ and $X_{model,i}$ are the observed and modeled values at the i-th sample respectively.

Table 2 shows how one of our users ranked the different movie categories. This information is inserted into the clusters and changes them to include the weight reflecting the interests of the user.

Table 2. UserA Movie Rankings

1. Drama	10. Crime
2. Adventure	11. Mystery
3. Animation	12. Thriller
4. Action	13. Documentary
5. War	14. Romance
6. Children	15. Film-Noir
7. Musical	16. Sci-Fi
8. Comedy	18. Horror
9. Western	19. Fantasy

Table 3 shows the ten first searches that UserA conducted in order for the system to understand his/her behavior by examining the movie categories searched more. If the value of a specific category exceeds a specific value, in our case this value is equal to three, then the system selects that category as a "frequently searched". In the case of UserA the categories searched more are Adventure and Drama. After the ten first searches the system starts to recommend items to the interested user based on how (s)he ranked the movie categories, what (s)he has searched more and the searched keywords. The analysis of the specific searches follows.

Table 3. UserA – Ten first searches

ID	Movie Title	Movie Categories
23	Assasins	Thriller
31	Dangerous Minds	Drama
86	White Squall	Adventure , Drama
165	The Doom Generation	Comedy , Drama
2	Jumanji	Adventure, Children, Fantasy
167	First Knight	Action, Adventure, Drama, Romance
237	A Goofy Movie	Animation, Children, Comedy, Romance
462	Heaven and Earth	Action, Drama, War
481	Lassie	Adventure, Children
1133	The Wrong Trousers	Animation, Comedy

Close inspection of the results listed in Table 4 reveals the following findings:

   i.   The proposed fuzzy algorithm is more accurate on average than the Hard implementation as it adjusts dynamically the knowledge gained by the RS engine.

  ii.   Both algorithms always suggest movies in ascending order of error magnitude, while there are also cases where movies bear the exact same RMS value; this is quite natural as they belong to the same cluster with the same degree of membership.

 iii.   The error depends on the category of the movie searched for each time; therefore, when this type perfectly matches the clustered ones the error is minimized.

Table 4. UserA recommendation results and evaluations per clustering method and search conducted – RMSE values below each method corresponds to the average of the five searches

Searches and Clustering Methods		Recommendations				
		1	2	3	4	5
#1: Dadetown Documentary	**Hard** 0.3333	Two Family House Drama 0.3333	Tigerland Drama 0.3333	Requime for a Dream Drama 0.3333	Remember the Titans Drama 0.3333	Girlfight Drama 0.3333
	Fuzzy 0.3333	Othello Drama 0.3333	Now and Then Drama 0.3333	Angela Drama 0.3333	Dangerous Minds Drama 0.3333	Restoration Drama 0.3333
#2: A Christmas Tale Comedy and Drama	**Hard** 0.0	Bootmen Comedy and Drama 0.0	Beautiful Comedy and Drama 0.0	Duets Comedy and Drama 0.0	A Knight in New York Comedy and Drama 0.0	Mr.Mom Comedy and Drama 0.0
	Fuzzy 0.0	Waiting to Exhale Comedy and Drama 0.0	To Die For Comedy and Drama 0.0	Kicking and Screaming Comedy and Drama 0.0	Big Bully Comedy and Drama 0.0	Nueba Yol Comedy and Drama 0.0
#3: Yellow Submarine Animation and Musical	**Hard** 0.4588	Digimon Adventure, Animation and Children 0.4082	For the Love of Benji Adventure and Children 0.4714	The Legend of Lobo Adventure and Children 0.4714	Tall Tale Adventure and Children 0.4714	Barneys Adventure and Children 0.4714
	Fuzzy 0.4059	Pete's Dragon Adventure, Animation, Children and Musical 0.3333	Gullivers Travels Adventure, Animation and Children 0.4082	Digimon Adventure, Animation and Children 0.4082	Bedknobs and Broomsticks Adventure, Animation and Children 0.4082	The Lord of the Rings Adventure, Animation, Children and Sci-Fi 0. 4714
#4: Young Guns Action, Comedy and Western	**Hard** 0.2576	Butch Cassidy Action, Comedy and Western 0.2576	Action Jackson Action and Comedy 0.2357	Last Action Hero Action and Comedy 0. 2357	Mars Attack Action, Comedy, Sci-Fi and War 0.4082	Tank Girl Action, Comedy, Musical, Sci-Fi 0.4082
	Fuzzy 0.2357	I Love Trouble Action and Comedy 0.2357	Beverly Hills Cop III Action and Comedy 0.2357	The CowBoy Way Action and Comedy 0.2357	Beverly Hills Ninja Action and Comedy 0.2357	Last Action Hero Action and Comedy 0.2357

Table 5 summarizes the mean RMS error values for four additional users tested on five different searches. Due to size limitations we omitted the details of the searched categories as these resemble the ones presented for UserA. It is once again clear that the algorithm behaves successfully, with average error values below 0.5 with only one exception (first search of UserB) and with consistently better performance being observed for the Fuzzy implementation.

Table 5. Summary of recommendation results and mean evaluations per clustering method for four more users

User	Method	Searches				
		1	2	3	4	5
B	Hard	0.5774	0.3480	0.4572	0.4557	0.2357
	Fuzzy	0.5360	0.3368	0.4082	0.4335	0.2552
C	Hard	0.3135	0.3333	0.3437	0.3714	0.2357
	Fuzzy	0.2552	0.2357	0.3437	0.2747	0.2357
D	Hard	0.0	0.4082	0.3333	0.3999	0.4461
	Fuzzy	0.0	0.4082	0.3333	0.3908	0.4082
E	Hard	0.3782	0.3610	0.3714	0.4885	0.2943
	Fuzzy	0.3782	0.3333	0.3333	0.4673	0.2943

5   Conclusions

Web Recommender Systems are nowadays a powerful tool that promotes e-commerce and advertisement of goods over the Web. High accuracy of recommendations, though, is not an easy task to achieve. This paper examined the utilization of the Hard and Fuzzy K-modes algorithms in the recommendation engine as a means to approximate the interests of users searching for items on the Internet. The proposed approach combines static information entered by the user in the beginning and dynamic information gathered after each individual search to perform clustering of te available dataset. Recommendations are given to the user by comparing only cluster centers with the search string thus saving execution time.

A series of synthetic experiments were executed to validate the RS system using the MovieLens1M dataset, which was linked with IMDB.com to retrieve more content information. More specifically, five different users with various interests on movies performed five different searches and the recommendations yielded by the Hard and Fuzzy K-Modes algorithms were assessed in terms of accuracy using the well-known RMS error. The results revealed small superiority of the Fuzzy over the Hard implementation, while recommendations were quite accurate.

Future work will concentrate on conducting more experiments and comparing with other approaches using collaborative or content-based filtering. In addition, a dedicated website will be developed through which real-world data will be gathered for experimentation purposes. Finally, the system will be enhanced with new functionality which will prompt the user, after a certain period of time, to update her/his static information so as to generate more accurate recommendations as interest in certain categories may have changed.

References

1. de Nooij, G.J.: Recommender Systems: An Overview. MSc Thesis, University of Amsterdam (November 2008)
2. Principia Cybernetica Web, Entropy and Information,
   http://pespmc1.vub.ac.be/ENTRINFO.html
3. Stylianou, C., Andreou, A.S.: A Hybrid Software Component Clustering and Retrieval Scheme, Using an Entropy-based Fuzzy k-Modes Algorithm. In: Proceedings of the 19th IEEE International Conference on Tools with Artificial Intelligence, Washinghton DC, USA, vol. 1, pp. 202–209 (2007) ISBN:0-7695-3015-X
4. Khan, S.S.: Computation of Initial Modes for K-modes Clustering Algorithm. In: Proceedings of the 20th International Joint Conference on Artificial Intelligent, San Francisco, USA, pp. 2784–2789 (2007)
5. Speigel, S., Kunegis, J., Li, F.: Hydra: A Hybrid Recommender System. In: Proceedings of the 1st ACM International Workshop on Complex Networks Meet Information & Knowledge Management, New York, USD, pp. 75–80 (2009)
6. Ekstrand, M., Riedl, J.: When recommender fail: predicting recommending failure for algorithm selection and combination. In: Proceedings of the 6th ACM Conference on Recommender Systems, New York, USA, pp. 233–236 (2012)
7. Jung, J.J., Pham, X.H.: Attribute selection-based recommendation framework for short-head user group: An empirical study by MovieLens and IMDB. In: Proceedings of the Third International Conference on Computational Collective Intelligence: Technologies and Applications, Berlin, Germany, pp. 592–601 (2011)
8. McNee, S.M., Riedl, J., Konstan, J.A.: Making Recommendations Better: An Analytic Model for Human- Recommender Interaction. In: Proceedings of the CHI 2006 Extended Abstracts on Human Factors in Computing Systems (2006)
9. Karypis, G.: Evaluation of Item-Based Top-N Recommendation Algorithms. In: Proceedings of the 10th International Conference on Information and Knowledge Management, New York, USA, pp. 247–254 (2001)
10. Ricci, F., Rokach, L., Shapira, B.: Introduction to Recommender Systems Handbook. Springer (2011) ISBN-13: 978-0387858197
11. Van Meteren, R., van Someren, M.: Using Content-Based Filtering for Recommendation. In: Proceedings of the ECML 2000 Workshop: Machine Learning in New Information Age, Barcelona, Spain, pp. 47–57 (2000)

Fuzzy Equivalence Relation Based Clustering and Its Use to Restructuring Websites' Hyperlinks and Web Pages

Dimitris K. Kardaras[1,*], Xenia J. Mamakou[1], and Bill Karakostas[2]

[1] Business Informatics Laboratory, Dept. of Business Administration,
Athens University of Economics and Business, 76 Patission Street, Athens 10434, Greece
{kardaras,xenia}@aueb.gr
[2] Centre for HCI Design, School of Informatics, City University, Northampton Sq., London
EC1V 0HB, UK
billk@soi.city.ac.uk

Abstract. Quality design of websites implies that among other factors, hyperlinks' structure should allow the users to reach the information they seek with the minimum number of clicks. This paper utilises the fuzzy equivalence relation based clustering in adapting website hyperlinks' structure so that the redesigned website allows users to meet as effectively as possible their informational and navigational requirements. The fuzzy tolerance relation is calculated based on the usage rate of hyperlinks in a website. The equivalence relation identifies clusters of hyperlinks. The clusters are then used to realocate hyperlinks in webpages and to rearrange webpages into the website structure hierarchy.

Keywords: fuzzy equivalence relation, web adaptation, hyperlinks' clustering.

1 Introduction

When designing a website, the way that its content is organised and how efficiently users get access to it, influence the user perceived design quality. The designers' goal is an effective and plain communication of content [2]. Website structure has been identified by many reasearch studies as an important factor that affects web design quality. The users' perception of how different parts of a web site are linked together is a strong indicator of effective design [4], [13]. Thus, the websites' hyperlinks structure should adapt to meet users' changing requirements and priorities depending e.g. on the expertise of users in navigating a website, their familiarity with its content structure, their information needs, etc. Furthermore, hyperlinks structure has been extensivley used in web search engines and web mining [12]. The number of links pointing to a web page is considered as a quality indicator that reflects the authority of the web page the links point at [1]. Many algorithms have been developed to tackle one of the greatest challenges for web design and search engines, i.e. how to specify an appropriate website structure, or how to evaluate webpages quality [8]. This paper suggests the use of fuzzy equivalence relation clustering to restucturing webpages

---

* Corresponding author.

H. Papadopoulos et al. (Eds.): AIAI 2013, IFIP AICT 412, pp. 52–60, 2013.

throught hypelinks popularity. Until recently, the web users' browsing behaviour was often overlooked in approches that attempted to manage websites' structures or determining the quality of webpages [9]. This paper considers the popularity of hypelinks as an indicator of users' browsing behaviour and classifies links into pages and subsequently pages are allocated to different website levels. An illustrative example is provide to expemlify the proposed approach.

2   Hyperlink Analysis

Although hyperlink analysis algorithms that produced significant results have been developed, they still need to tackle challenges such as how to incorporate web user behaviour in hyperlink analysis [8]. A website is considered as a graph of nodes and edges representing webpages and hyperlinks respectively. Based on the hyperlinks analysis, the importance of each node can be estimated thus leading to a website structure that reflects the relative importance of each hyperlink and each web page [7]. Two of the most representative links analysis algorithms are the HITS [5] and the Google's PageRank [1], which assume that a user randomly selects a hyperlink and then they calculate the probabilities of selecting other hyperlinks and webpages. Many other link analysis algorithms stem from these two algorithms. The basic idea behind link analysis algorithms is that if a link points to page (i) from page (j), then it is assumed that there is a semantic relation between the two pages. However, the links may not represent users' browsing behaviour, which is driven by their interests and information needs. Hyperlinks are not clicked by users with equal probabilities and they should not be treated as equally important [9]. Thus, webpages that are visited and hyperlinks that are clicked by users should be regarded as more important than those that are not, even if they belong to the same web page. It is therefore reasonable to use users' preferences to redesign the hyperlink graph of a website [8]. Google Toolbar and Live Toolbar collect user browsing information. User browsing preferences is an important source of feedback on page relevance and importance and is widely adopted in website usability [3], [20], user intent understanding [22] and Web search [9] research studies. By utilising the user browsing behaviour, website structures can be revised by deleting unvisited web pages and hyperlinks and relocating web pages and hyperlinks according to their importance as perceived by the users, i.e. as the number of clicks show. Liu et al. (2008) developed a "user browsing graph" with Web log data [10]. A website representing graph as derived from user browsing information can lead to a website structure closer to users' needs, because links in the website graph are actually chosen and clicked by users. Liu et al. (2008) also proposed an algorithm to estimate page quality, BrowseRank, which is based on continuous-time Markov process model, which according to their study performs better than PageRank and TrustRank.

3   Fuzzy Relations and Fuzzy Classification

3.1   Fuzzy Relations

Fuzzy relations are important for they can describe the strength of interactions between variables [11]. Fuzzy relations are fuzzy subsets of $X \times Y$, that is mapping from $X \to Y$. Let X, Y $\subseteq$ R be universal sets. Then

$$\tilde{R} = \{((x, y), \mu_R(x, y)) \mid (x, y) \in X \times Y\} \tag{1}$$

is called a fuzzy relation on $X \times Y$ [23].

3.2   Fuzzy Classification with Equivalence Relation

Numerous classification methods have been proposed so far including cluster analysis, factor analysis, discriminant analysis [16], k-means analysis [14], c-means clustering [21]. According to Ross (2010) there are two popular methods of classification namely the classification using equivalent relations and the fuzzy c-means. Cluster analysis, factor analysis and discriminant analysis are usually applied in classic statistical problem where large sample or long term data is available. When dealing with small data k-means or c-means methods are preferred [18]. In this paper, we use fuzzy equivalent relations and lambda-cuts (λ-cuts) to classify links, according to their importance in a website. Classification based on λ-cuts of equivalent relations is used in many recent studies [6], [17] and [19]. An important feature of this approach is that for its application it is not required to assume that the number of clusters is known as it is required by other methods such as in the case of k-means and c-means clustering [18].

A fuzzy relation on a single universe X is also a relation from X to X. It is a fuzzy tolerance relation if the two following properties define it:

Reflexivity: $\mu_R(x_i, x_i) = 1$

Symmetry: $\mu_R(x_i, x_j) = \mu_R(x_j, x_i)$

Moreover, it is a fuzzy equivalence relation if it is a tolerance relation and has the following property as well:

Transitivity: $\mu_R(x_i, x_j) = \lambda_1$ and $\mu_R(x_i, x_k) = \lambda_2 \to \mu_R(x_i, x_k) = \lambda$, where $\lambda \geq \min[\lambda_1, \lambda_2]$.

Any fuzzy tolerance relation can be reformed into a fuzzy equivalence relation by at most $(n - 1)$ compositions with itself. That is:

$$\tilde{R}_1^{n-1} = \tilde{R}_1 \circ \tilde{R}_1 \circ ... \circ \tilde{R}_1 = \tilde{R} \tag{2}$$

In fuzzy equivalent relations, their λ-cuts are equivalent ordinary relations.

The numerical values that characterize a fuzzy relation can be developed by a number of ways, one of which is similarity[15]. Min-max method is one of the similarity methods that can be used when attempting to determine some sort of similar pattern or structure in data through various metrics. The equation of this method is given below:

$$r_{i,j} = \frac{\sum\limits_{k=1}^{m} \min(x_{ik}, x_{jk})}{\sum\limits_{k=1}^{m} \max(x_{ik}, x_{jk})} , \tag{3}$$

where i, j = 1,2,...,n

In this paper, we use min-max similarity method due to its simplicity and common use [15].

4   Proposed Methodology

Assume a website (WS) consists of a number (p) of web pages (wp) such as WS={wp}. A website is regarded as a set of partially ordered web pages, according to their popularity in terms of users' preferences, i.e. visits. Thus, assuming a website has three levels of web pages, the web pages are allocated to a website level according to the demand users show for the web page.

The proposed methodology consists of the following steps:
1. Identify the links of each web page in a website.
2. In order to capture the users' browsing behaviour, first measure the clicks made for each link (l_i), then compute and normalise their demand using the following type:

$$Dl_i = \frac{Cl_i}{\sum\limits_{i=1}^{n} Cl_i} , \tag{4}$$

where Dl_i is the demand for link i, i=1,2,...,n and Cl_i are the number of clicks for the link i. $Dl_i \in [0,1]$.
3. Fuzzify the Dl_i using triangular fuzzy numbers (TFNs) and specify their corresponding linguistic variables.
4. Calculate the membership degree to the corresponding linguistic variables for each link, using membership functions. The membership function of a TFN is given by the following formula:

$$\mu_A(x) = \begin{cases} 0, & x \le a \\ (x-a)/(b-a), & a \le x \le b \, and \, a < b \\ (c-x)/(c-b), & b \le x \le c \, and \, b < c \\ 0, & x \ge c \end{cases} \tag{5}$$

5. Calculate the fuzzy tolerance relation ($\tilde{R}_t$), using min-max similarity method, as in Eq. (3).
6. Calculate the fuzzy equivalent relation using Eq. (2).
7. Decide on the λ-cuts to be used.
8. Classify web page links according to λ-cuts. The derived clusters constitute the redesigned web pages.
9. Calculate the new demand for each of the newly formed web pages, according to the type:

$$Dwp_j = \sum_{i=1}^{n} Cl_i \, , \qquad\qquad (6)$$

where Dwp_j is the demand for web page j, with (i) and (n) indicating the hyperlinks and the number of hyperlinks in the webpage respectively. The demand for each web page is calculated according to the total number of the clicks made to all links in this specific page.

10. Normalize the demand for the new web pages and then fuzzify the Dwp_j. Since a website is a partially ordered set of web pages, the newly formed we pages are allocated to a level a website hierarchy level according to their demand. The higher the demand the higher the level they are assigned to.
11. Calculate the validation index as shown in Eq. (7) for different λ-cuts. The λ-cut value that maximises the validation index indicated the optimum number of clusters.

$$\lambda - C(\lambda)/m \qquad\qquad (7)$$

The C(λ) indicates the number of clusters and the (m) shows the number of data sequence that are subject to clustering.

5   Illustrative Example

The following example illustrates the proposed methodology. Let us consider of a website that originally has a total of 10 links (i=10) in all of its 5 web pages as shown in Fig 1.

By the use of cookies we can identify the number of clicks each of these links has had in a specific time period. Then according to Eq. (4) the demand of each link can be calculated. In our example, let's say that the Dl_is are: {0.31, 0.68, 0.45, 0.25, 0.76, 0.59, 0.88, 0.39, 0.77, 0.25}.

To fuzzify the Dl_i we will use three TFNs with their corresponding linguistic variables, which are Low = (0, 0.3, 0.5), Medium=(0.3, 0.5, 0.7) and High=(0.5, 0.7, 1). Then, we calculate the membership degree of each link to the corresponding linguistic variable using Eq. (5), as seen in Table 1.

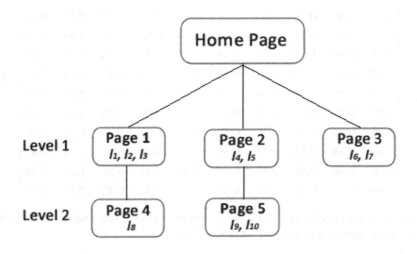

Fig. 1. Original Website with five web pages having links

Table 1. Hyperlinks membership degrees to each of the corresponding linguistic variables

	l_1	l_2	l_3	l_4	l_5	l_6	l_7	l_8	l_9	l_{10}
Low	0.95	0.00	0.25	0.83	0.00	0.00	0.00	0.55	0.00	0.83
Medium	0.05	0.10	0.75	0.00	0.00	0.55	0.00	0.45	0.00	0.00
High	0.00	0.90	0.00	0.00	0.80	0.45	0.40	0.00	0.77	0.00

Using Eq. (3) we form the fuzzy tolerance relation:

$$
\tilde{R}_t = \begin{array}{c} \\ l_1 \\ l_2 \\ l_3 \\ l_4 \\ l_5 \\ l_6 \\ l_7 \\ l_8 \\ l_9 \\ l_{10} \end{array}
\begin{bmatrix}
\begin{array}{cccccccccc}
l_1 & l_2 & l_3 & l_4 & l_5 & l_6 & l_7 & l_8 & l_9 & l_{10} \\
1.00 & 0.03 & 0.18 & 0.83 & 0.00 & 0.03 & 0.00 & 0.43 & 0.00 & 0.83 \\
0.03 & 1.00 & 0.05 & 0.00 & 0.80 & 0.38 & 0.40 & 0.05 & 0.77 & 0.00 \\
0.18 & 0.05 & 1.00 & 0.16 & 0.00 & 0.38 & 0.00 & 0.54 & 0.00 & 0.16 \\
0.83 & 0.00 & 0.16 & 1.00 & 0.00 & 0.00 & 0.00 & 0.43 & 0.00 & 1.00 \\
0.00 & 0.80 & 0.00 & 0.00 & 1.00 & 0.33 & 0.50 & 0.00 & 0.96 & 0.00 \\
0.03 & 0.38 & 0.38 & 0.00 & 0.33 & 1.00 & 0.40 & 0.29 & 0.34 & 0.00 \\
0.00 & 0.40 & 0.00 & 0.00 & 0.50 & 0.40 & 1.00 & 0.00 & 0.52 & 0.00 \\
0.43 & 0.05 & 0.54 & 0.43 & 0.00 & 0.29 & 0.00 & 1.00 & 0.00 & 0.43 \\
0.00 & 0.77 & 0.00 & 0.00 & 0.96 & 0.34 & 0.52 & 0.00 & 1.00 & 0.00 \\
0.83 & 0.00 & 0.16 & 1.00 & 0.00 & 0.00 & 0.00 & 0.43 & 0.00 & 1.00
\end{array}
\end{bmatrix} \quad (8)
$$

To produce the fuzzy equivalent relation ($\tilde{R}_e$) that will be used to classify the links of the website, we use Eq. (2):

$$\tilde{R}_e = \tilde{R}_t^9 = \begin{bmatrix} 1.00 & 0.38 & 0.43 & 0.83 & 0.38 & 0.38 & 0.38 & 0.43 & 0.38 & 0.83 \\ 0.38 & 1.00 & 0.38 & 0.38 & 0.80 & 0.40 & 0.52 & 0.38 & 0.80 & 0.38 \\ 0.43 & 0.38 & 1.00 & 0.43 & 0.38 & 0.38 & 0.38 & 0.54 & 0.38 & 0.43 \\ 0.83 & 0.38 & 0.43 & 1.00 & 0.38 & 0.38 & 0.38 & 0.43 & 0.38 & 1.00 \\ 0.38 & 0.80 & 0.38 & 0.38 & 1.00 & 0.40 & 0.52 & 0.38 & 0.96 & 0.38 \\ 0.38 & 0.40 & 0.38 & 0.38 & 0.40 & 1.00 & 0.40 & 0.38 & 0.40 & 0.38 \\ 0.38 & 0.52 & 0.38 & 0.38 & 0.52 & 0.40 & 1.00 & 0.38 & 0.52 & 0.38 \\ 0.43 & 0.38 & 0.54 & 0.43 & 0.38 & 0.38 & 0.38 & 1.00 & 0.38 & 0.43 \\ 0.38 & 0.80 & 0.38 & 0.38 & 0.96 & 0.40 & 0.52 & 0.38 & 1.00 & 0.38 \\ 0.83 & 0.38 & 0.43 & 1.00 & 0.38 & 0.38 & 0.38 & 0.43 & 0.38 & 1.00 \end{bmatrix} \quad (9)$$

The next step is to decide on the λ-cut. Assuming that the λ-cut that maximises the validation index shown in formula (7) is 0.5 we have:

$$\tilde{R}_{0.5} = \begin{bmatrix} 1 & 0 & 0 & 1 & 0 & 0 & 0 & 0 & 0 & 1 \\ 0 & 1 & 0 & 0 & 1 & 0 & 1 & 0 & 1 & 0 \\ 0 & 0 & 1 & 0 & 0 & 0 & 0 & 1 & 0 & 0 \\ 1 & 0 & 0 & 1 & 0 & 0 & 0 & 0 & 0 & 1 \\ 0 & 1 & 0 & 0 & 1 & 0 & 1 & 0 & 1 & 0 \\ 0 & 0 & 0 & 0 & 0 & 1 & 0 & 0 & 0 & 0 \\ 0 & 1 & 0 & 0 & 1 & 0 & 1 & 0 & 1 & 0 \\ 0 & 0 & 1 & 0 & 0 & 0 & 0 & 1 & 0 & 0 \\ 0 & 1 & 0 & 0 & 1 & 0 & 1 & 0 & 1 & 0 \\ 1 & 0 & 0 & 1 & 0 & 0 & 0 & 0 & 0 & 1 \end{bmatrix} \quad (10)$$

Eq. (10) is then used to classify each of the 10 links in our example. In this case, we find four classes, each of which includes specific links: $class_1=\{l_1,l_4,l_{10}\}$, $class_2=\{l_2,l_5,l_7,l_9\}$, $class_3=\{l_3,l_8\}$ and $class_4=\{l_6\}$. We then re-create the website, having this time only four (out of the original five) web pages. The next step is to calculate the new demand for each of the four newly created web pages, using Eq. (6). In our case we find: $Dwp_1=395$, $Dwp_2=44$, $Dwp_3=442$ and $Dwp_4=62$. Then we normalize each of the Dwp_i and we get $Dwp_1=395/943=0.42$, $Dwp_2=44/943=0.05$, $Dwp_3=442/943=0.47$ and $Dwp_4=62/943=0.07$.

To fuzzify the Dwp_i we use the same three TFNs we used when fuzzifying the Dl_i. That is Low = (0, 0.3, 0.5), Medium = (0.3, 0.5, 0.7) and High = (0.5, 0.7, 1). By calculating the membership functions for each Dwp_i to each TFN, we find that $\mu_{Low}(Dwp_1) = 0.4$, $\mu_{Medium}(Dwp_1) = 0.6$, $\mu_{High}(Dwp_1) = 0$, $\mu_{Low}(Dwp_2) = 0.17$, $\mu_{Medium}(Dwp_2) = 0$, $\mu_{High}(Dwp_2) = 0$, $\mu_{Low}(Dwp_3) = 0.15$, $\mu_{Medium}(Dwp_3) = 0.85$, $\mu_{High}(Dwp_3) = 0$, $\mu_{Low}(Dwp_4) = 0.23$, $\mu_{Medium}(Dwp_4) = 0$ and $\mu_{High}(Dwp_4) = 0$. We can then easily understand that wp_1 and wp_3 belong to the "Medium" category, whereas wp_2 and wp_4 belong to the "Low" category. To decide on the level of each web page to the website, we take all web pages found in the "High" category and put them in Level 1, then the web pages found in the "Medium" category are organized in Level 2 and web pages that belong to the "Low" category are left in Level 3. In case a category does not exist we put in the specific Level, the web pages that are found in the next category. In our example, no web page is found in the "High" category, so Level 1 will include the pages of the "Medium" category, which are wp_1 and wp_3. Wp_2 and wp_4 are then put in Level 2 as shown in Fig. 2.

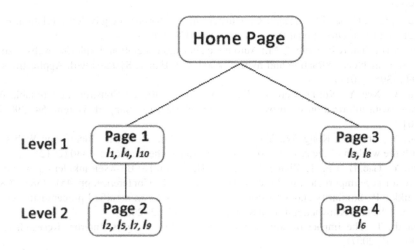

Fig. 2. Restructured website with three web pages having links in two levels

6   Conclusions

This paper suggests an approach to websites structuring. A fuzzy equivalence relation based clustering is been adopted, for it does not assumed a known number of clusters as other clustering techniques do. Further it is a clustering technique that has been recently used in other domains with satisfactory results. In order to derive an appropriate web structure, this paper considers the data that reflect the users' browsing behaviour. Research studies claim that it is important to take into account users' browsing information in determining websites structure and webpages' quality. This paper suggests that clustering of hyperlinks' usage data can be used in order to cluster links of similar popularity into the same web page and then similar demand web page to be grouped into the same web page hierarchy level. An illustrative example has shown the applicability of the proposed approach.

References

1. Brin, S., Page, L.: The anatomy of a large-scale hypertextual web search engine. Computer Networks and ISDN Systems 30(1/7), 107–117 (1998)
2. De Marsico, M., Levialdi, S.: Evaluating web sites: exploiting user's expectations. Int. J. Human Computer Interactions 60(3), 381–416 (2004)

3. Fang, X., Holsapple, C.W.: An empirical study of web site navigation structures' impacts on web site usability. Decision Support Systems 43(2), 476–491 (2007)
4. Henzinger, M., Motwani, R., Silverstein, C.: Challenges in web search engines. Paper Presented at the SIGIR Forum (2002)
5. Kleinberg, J.M.: Authoritative sources in a hyperlinked environment. In: Proceedings of the 9th ACM SIAM Symposium on Discrete Algorithms, Baltimore, MD, USA, pp. 668–677 (1998)
6. Liang, G.S., Chou, T.Y., Han, T.C.: Cluster analysis based on equivalence relation. European Journal of Operational Research 166, 160–171 (2005)
7. Lin, S.H., Chu, K.P., Chiu, C.M.: Automatic sitemaps generation: Exploring website structures using block extraction and hyperlink analysis. Expert Systems with Applications 38, 3944–3958 (2011)
8. Liu, Y., Xue, Y., Xu, D., Cen, R., Zhang, M., Ma, S., Ru, L.: Constructing a reliable Web graph with information on browsing behaviour. Decision Support Systems 54, 390–401 (2012)
9. Liu, Y., Chen, F., Kong, W., Yu, H., Zhang, M., Ma, S., Ru, L.: Identifying Web spam with the wisdom of the crowds. ACM Transaction on the Web 6(1), 1–30 (2012)
10. Liu, Y., Gao, B., Liu, T., Zhang, Y., Ma, Z., He, S., Li, H.: BrowseRank: letting web users vote for page importance. In: Proc. of 31st ACM SIGIR Conference, pp. 451–458 (2008)
11. Luukka, P.: Similarity classifier using similarities based on modified probabilistic equivalence relations. Knowledge-Based Systems 22(1), 57–62 (2009)
12. Mandl, T.: The impact of web site structure on link analysis. Internet Research 17(2), 196–206 (2007)
13. Muylle, S., Moenaertb, R., Despontin, M.: The conceptualization and empirical validation of web site user satisfaction. Information & Management 41(5), 541–560 (2004)
14. Ralambondrainy, H.: A conceptual version of K-means algorithm. Pattern Recognition Letters 16, 1147–1157 (1995)
15. Ross, T.J.: Fuzzy logic with engineering applications, 3rd edn. Wiley-Blackwell (2010)
16. Sharma, S.: Applied multivariate techniques. JohnWiley (1996)
17. Sun, P.G., Gao, L., Han, S.S.: Identification of overlapping and non-overlapping community structure by fuzzy clustering in complex networks. Information Sciences 181, 1060–1071 (2011)
18. Wang, Y.L., Lee, H.S.: A clustering method to identify representative financial ratios. Information Sciences 178, 1087–1097 (2008)
19. Wang, Y.L.: A clustering method based on fuzzy equivalence relation for customer relationship management. Expert Systems with Application 37, 6421–6429 (2010)
20. Wu, H., Gordon, M., DeMaagd, K., Fan, W.: Mining web navigations for intelligence. Decision Support Systems 41(3), 574–591 (2006)
21. Wu, K.L., Yang, M.S.: Alternative c-means clustering algorithms. Pattern Recognition 35, 2267–2278 (2002)
22. Xu, D., Liu, Y., Zhang, M., Ru, L., Ma, S.: Predicting Epidemic Tendency through Search Behavior Analysis. In: Proceedings of the 22nd International Joint Conference on Artificial Intelligence (IJCAI 2011), Barcelona, Spain, pp. 2361–2366 (2011)
23. Zimmermann, H.J.: Fuzzy set theory-and its applications. Springer (2001)

Exploiting Machine Learning for Predicting Nodal Status in Prostate Cancer Patients

Mauro Vallati[1], Berardino De Bari[2], Roberto Gatta[3], Michela Buglione[2],
Stefano M. Magrini[2], Barbara A. Jereczek-Fossa[4], and Filippo Bertoni[5]

[1] School of Computing and Engineering, University of Huddersfield, UK
[2] Department of Radiation Oncology, University of Brescia, Italy
[3] Department of Information Engineering, University of Brescia, Italy
[4] Department of Radiotherapy,
European Institute of Oncology and University of Milan, Italy
[5] Department of Radiation Oncology, Hospital of Modena, Italy

Abstract. Prostate cancer is the second cause of cancer in males. The prophylactic pelvic irradiation is usually needed for treating prostate cancer patients with Subclinical Nodal Metestases. Currently, the physicians decide when to deliver pelvic irradiation in nodal negative patients mainly by using the Roach formula, which gives an approximate estimation of the risk of Subclinical Nodal Metestases.

In this paper we study the exploitation of Machine Learning techniques for training models, based on several pre-treatment parameters, that can be used for predicting the nodal status of prostate cancer patients. An experimental retrospective analysis, conducted on the largest Italian database of prostate cancer patients treated with radical External Beam Radiation Therapy, shows that the proposed approaches can effectively predict the nodal status of patients.

Keywords: Machine Learning, Classification, Medicine Applications.

1 Introduction

Prostate cancer is the second cause of cancer in males [14]. External Beam Radiation Therapy (EBRT) has a well established role in the radical treatment of prostate cancer [1,2], but some issues still remain to be uniformly defined amongst Radiation Oncologists. One of these issues is the role of the pelvic irradiation. The irradiation of pelvic nodes remains controversial in the treatment of intermediate and high risk nodes-negative prostate cancer. The main problem is that microscopic Subclinical Nodal Metestases (SNM) can not be detected by the available technology: the low sensitivity of computed tomography (CT) and magnetic resonance imaging (MRI) limits their usefulness in clinical staging [20]. This results in prostate cancer patients that are considered as nodes-negative even though, in fact, they have SNM.

Even if some retrospective studies [23,24] showed a significant advantage in terms of biochemical control for patients having an extended primary tumor and

H. Papadopoulos et al. (Eds.): AIAI 2013, IFIP AICT 412, pp. 61–70, 2013.

a risk $> 15\%$ (estimated with Roach formula [22]), other retrospective studies, that specifically addressed to establish the role of the pelvic irradiation, [3,16,18] failed in showing any significant advantage in favour of the pelvic irradiation.

In order to help the clinician in the decision to treat or not pelvic nodes, many predictor tools have been proposed in the literature, most of them based on linear regression analyses [22,20,17]. One the most diffused methods to predict the SNM of prostate cancer patients is the Roach formula [22]. The reason of its success is principally its simplicity, as it is based on only 2 parameters: the Gleason Score sum (GSsum) and the Prostate Specific Antigen (PSA) level. GSsum is obtained by the analysis of the specimens of the biopsy: the pathologist assigns a grade to the most common tumor pattern, and a second grade to the next most common tumor pattern. The two grades are added together to get a Gleason Score sum. The PSA is a protein secreted by the epithelial cells of the prostate gland. PSA is present in small quantities in the serum of men with healthy prostates, but it is often elevated in the presence of prostate cancer or other prostate disorders.

The Roach formula is shown in Equation 1. In its last formulation it gives the estimated risk of SNM and when this risk is bigger than 15-25%, Radiation Oncologists often deliver EBRT to the pelvic nodes. Despite that, this method presents some important limits in the more updated and used formulation, the principal one being the absence of information about the clinical stage of the tumor, that has been already showed to be an important factor influencing the risk of nodal involvement [17].

$$\frac{2}{3} \times PSA + ([GSsum - 6] \times 10) \tag{1}$$

Other methods, as the Partins Tables [20], do not share the same simplicity of the Roach formula, but they consider the clinical stage of the tumor. On the other hand their accuracy is still an argument of discussion [21,11]. Moreover, some important limits exists also for the Partins Tables: they consider only a limited part of the population of prostate cancer patients and, furthermore, it has been showed that their algorithm overestimate the presence of lymph node and seminal vesicle involvement in patients with a predicted risk of 40% or more [15]. This could mean that a large part of patients would be uselessly treated on the pelvis and exposed to the potential side effects of the pelvic irradiation (acute and/or chronic colitis, diarrhea, cystitis).

Last but not least, all the proposed algorithms estimates only a rate of risk of SNM (sometimes with large confidence intervals) and their outputs are not based on a clear dichotomic yes/not criteria. This leads to different interpretation of their outcome, that results in different treatments delivered to similar patients.

In this paper we study the exploitation of three Machine Learning (ML) classification algorithms to define patients presenting a SNM status, which can not be detected by MRI or CT, for helping the clinician in deciding the treatment of pelvic nodes. The generated models are based on several pre-treatment parameters of prostate cancer patients.

The possibility to have a reliable predictor to define the nodal status of prostate cancer patients could have a potential interest also in the field of Health

Economy: it has been calculated that to not perform a staging abdominal CT in low risk patients –which would be avoided by exploiting an accurate prediction of the nodal status of the patients– could reduce medical expenditures for prostate cancer management by U.S. $ 20 - 50 million a year [20].

2 Methodology

2.1 Data

The Pattern of Practices II was retrospectively analyzed. It is the largest Italian database including clinical, diagnostic, technical and therapeutic features of prostate cancer patients treated with radical EBRT (+/- hormonal therapies) in 15 Italian Radiation Oncology centers between 1999 and 2003. A total of 1808 patients were included in this analysis.

The pre-treatment features were chosen considering the "real life" situation of the Radiation Oncologist visiting for the first time a prostate cancer patient. Their decision to treat or not the pelvic nodes would be based only on the clinical stage of the tumor, the total Gleason Score (e.g. 3+4) and the Gleason Score sum (e.g. 3+4=7), the initial PSA value, the age of the patient and the presence of an ongoing hormonal therapy. The initial clinical stage of the tumor has been defined with the TNM system, that describes the extent of the primary tumor (T stage), the absence or presence of spread to nearby lymph nodes (N stage) and the absence or presence of distant spread, or metastasis (M stage) [9]. The domains of the considered variables are defined as shown in Table 1. In this study the machine learning techniques were used for predicting the value of N: N+ indicates patients with nodal involvement, N0 indicates that patients that do not have any nodal involvement.

N+ patients were defined as those with a positive contrast enhanced pelvic magnetic resonance imaging (MRI) and/or computed tomography (CT) scan; those showing a nodal only relapse after RT were also classified as N+ (as none of them received pelvic RT). With these criteria, a total of 55 N+ patients were identified out of the 1808 patients considered.

Following the D'Amico [8] classification that takes into account the initial PSA, the GS and the clinical T stage, patients were classified in three prognostically different sets, namely: low, intermediate and high risk. Each category included, 317, 577 and 914 patients, respectively. Finally, 6/317 (1.9%), 10/577 (1.7%), 39/914 (4.2%) N+ patients were found in the 3 D'Amico sets.

2.2 Classification Methods

The selection of the classification techniques to exploit is a fundamental part of this work. Since the objective of the study is to obtain an automated system for supporting the physicians in their work, we looked for something that is reliable, allows accurate predictions and, at the same time, it is easy to explain and to represent. After discussing with Radiation Oncologists and physicians

Table 1. The domains of the considered variables on the selected patients. N represents the nodal status of patients, that is the variable that we want to predict.

Variable name	Domain
GS	1+1, 1+2, 1+3, 2+1, 2+2, 2+3, 2+4, 2+5 3+1, 3+2, 3+3, 3+4, 3+5, 4+2, 4+3 4+4, 4+5, 5+1, 5+2, 5+3, 5+4, 5+5
GSum	2 – 10
Age	46 – 86
Initial PSA	0.1 – 1066
T	1, 1a, 1b, 1c, 2, 2a, 2b, 2c, 3, 3a, 3b, 4
Hormonal therapy	9 possible therapies + no therapy
N	N+, N0

we observed that they will trust and evaluate the suggestions of the automated system only if it is possible to understand how the decisions are taken. Mainly for the latter requirement, we focused our study on decision trees. A model generated by decision tree approaches can be easily represented, and it is intuitive to follow the flow of information and to understand how the suggested outcome is generated.

We considered three machine learning techniques for classification, based on decision trees, included in the well known tool WEKA [12]: J48 [19], Random Tree [10] and Random Forest [5]. Even if we are aware of the objective difficulty for humans to interpret the raw model generated by the Random Forest approach [4], we believe that it will be possible to represent the models generated by this classifier in an understandable format. Moreover they have demonstrated to achieve very good performance in classification [6].

2.3   Experimental Design

The classification methods are used for classifying patients as N0 (not affected by nodal metastasis) and N+ (affected by nodal metastasis). The classes of patients N0 and N+ are very imbalanced through all the D'Amico categories of risk (low, medium and high), and in the whole database. The N+ population represents less then the 4% of the all entire set. For balancing the classes among the considered data sets we applied three different strategies: oversampling, undersampling, and a mix of them, as follows.

- **A** random undersampling of the N0 patients.
- **B** random oversampling of the N+ patients.
- **C** random undersampling of the N0 patients and random oversampling of the N+ patients.

Random oversampling and undersampling techniques were selected due to the improved overall classification performance, compared to a very imbalanced data set, that base classifiers can achieve [13]. Given the fact that it is still not clear which technique could lead to better performance, we decided to experimentally evaluate three combinations of them.

The sampling techniques were applied on the different categories of risk, leading to nine different data sets. The resulting datasets had a distribution of about 40%-60% of, respectively, N+ and N0. Furthermore we generated a further dataset that considers the whole population of the database. The sampling techniques were applied also to this dataset.

Each of the selected classifying techniques was trained on the previously described data sets separately, and the resulting predictive models were then evaluated using a k-fold cross-validation strategy.

For the Roach formula we considered three different cut-offs: 15, 10 and 5 percent. Using a 5% cut-off corresponds to a pessimistic evaluation of the clinical data, and leads to overestimate the risk of a patient to be N+, while the 15% value corresponds to a more optimistic approach. The Roach formula does not need any training on the data sets and is meant to be used as-is, for this reason we applied it directly to the original datasets.

3 Results

In Table 2 the results of the comparison between Roach formula and the selected machine learning algorithms are shown. For the comparison we considered four different metrics: number of patients classified as false negative, specificity, sensitivity and accuracy of the model. The number of false negative, i.e., patients N+ classified as N0, is critical. These are patients that require to be treated with pelvic irradiation but that, according to the classification, will not be treated. Specificity, sensitivity and accuracy are showed as indexes of the overall quality of the generated models. They indicate the proportion of patients classified as N0 which are actually N0 (specificity), the proportion of patients classified as N+ which are actually N+ (sensitivity), and the proportion of the patients correctly classified, regardless to their class (accuracy).

By analyzing the results shown in Table 2, we can derive some interesting observations. Regarding the Roach formula, we can derive a precise behaviour. Its ability in identifying affected patients increases with the category of risk, but at the same time also the number of false positive increases dramatically. On the other hand the accuracy, very good in low risk patients, decreases in medium and high risk patients. The number of false negative rapidly decreases from low risk to high risk categories, because in the latter the Roach formula is classifying as affected a significantly large percentage of the population.

As expected, the performance of the machine learning approaches does not significantly change between the risk categories. On the other hand the sampling technique used on the data set has a great impact on the overall performance. The data set obtained by applying only the undersampling (A) usually leads the algorithms to poor performance, mainly because of the very small populations of the resulting data sets. The only exception is the decision tree generated by the J48 algorithm for the high risk patients, which achieved significant results

Table 2. The results, in terms of number of false negative (2nd column), specificity, sensitivity and accuracy, achieved by the Roach formula with the different cut-offs and the selected classification algorithms on low (upper), medium (middle) and high (lower) risk patients. The numbers in brackets indicate the population of the considered data set.

Roach cut off	False Neg	Specificity (%)	Sensitivity (%)	Accuracy (%)
15%	6	0.0	**100.0**	98.1
10%	6	0.0	**100.0**	98.1
5%	6	0.0	87.0	85.8
ML Approaches	**False Neg**	**Specificity (%)**	**Sensitivity (%)**	**Accuracy (%)**
J48 A (12)	6	0.0	50.0	25.0
J48 B (612)	0	100.0	92.5	96.3
J48 C (354)	0	100.0	94.7	97.2
RandomT A (12)	3	50.0	0.0	25.0
RandomT B (612)	0	100.0	94.8	97.4
RandomT C (354)	0	100.0	95.2	97.5
RandomF A (12)	6	0.0	33.3	16.7
RandomF B (612)	0	100.0	98.7	**99.4**
RandomF C (354)	0	100.0	98.4	99.2

Roach cut off	False Neg	Specificity (%)	Sensitivity (%)	Accuracy (%)
15	6	40.0	64.9	64.6
10	3	70.0	48.6	49.0
5	3	70.0	33.4	34.0
ML Approaches	**False Neg**	**Specificity (%)**	**Sensitivity (%)**	**Accuracy (%)**
J48 A (20)	1	90.0	40.0	65.0
J48 B (966)	0	100.0	93.1	96.0
J48 C (433)	0	100.0	97.7	98.9
RandomT A (20)	4	60.0	**100.0**	80.0
RandomT B (966)	0	100.0	96.5	97.9
RandomT C (433)	0	100.0	97.7	98.9
RandomF A (20)	4	60.0	90.0	75.0
RandomF B (966)	0	100.0	97.7	98.7
RandomF C (433)	0	100.0	99.1	**99.5**

Roach cut off	False Neg	Specificity (%)	Sensitivity (%)	Accuracy (%)
15	3	92.3	20.1	23.2
10	3	92.3	14.5	17.8
5	2	94.9	9.9	13.6
ML Approaches	**False Neg**	**Specificity (%)**	**Sensitivity (%)**	**Accuracy (%)**
J48 A (78)	1	97.4	**100.0**	**98.7**
J48 B (1460)	0	100.0	88.6	93.2
J48 C (701)	0	100.0	92.9	96.4
RandomT A (78)	11	71.8	79.5	75.6
RandomT B (1460)	0	100.0	94.9	96.9
RandomT C (701)	0	100.0	96.9	98.4
RandomF A (78)	2	94.9	97.4	96.2
RandomF B (1460)	0	100.0	96.1	97.7
RandomF C (701)	0	100.0	95.4	97.7

Table 3. The results, in terms of number of false negative (2nd column), specificity, sensitivity and accuracy, achieved by the Roach formula with the different cut-offs and the selected classification algorithms on the complete data set. The numbers in brackets indicate the population of the considered data set.

Roach cut off	False Neg	Specificity (%)	Sensitivity (%)	Accuracy (%)
15	15	75.4	44.2	45.1
10	12	80.3	33.4	34.7
5	11	81.2	32.0	32.7
ML Approaches	**False Neg**	**Specificity (%)**	**Sensitivity (%)**	**Accuracy (%)**
J48 A (110)	10	81.8	81.8	81.8
J48 B (3039)	0	**100.0**	92.1	95.5
J48 C (1489)	0	**100.0**	91.1	95.5
RandomT A (110)	14	74.6	78.2	76.4
RandomT B (3039)	0	**100.0**	96.2	97.8
RandomT C (1489)	0	**100.0**	94.8	97.4
RandomF A (110)	9	83.6	78.2	80.9
RandomF B (3039)	0	**100.0**	**96.9**	**98.2**
RandomF C (1489)	0	**100.0**	96.3	98.1

in both accuracy and sensitivity. This is a surprising result, given the fact that the use of the cross-validation technique should avoid (or, at least, significantly limit) the overfitting issue [13,7].

Usually the best performance are achieved by the classification models generated by the random forest algorithm on either B or C data sets. It is also important to note that *all* the models generated by the selected machine learning approaches lead to 0 false negative while exploiting B or C sampling techniques.

In Table 3 the results of the comparison between Roach formula and the selected machine learning algorithms, on the whole population, are shown. A striking result is that the number of false negative is still 0 for all the selected machine learning approaches on the data sets generated by the B or C sampling techniques. This results is confirmed by the 100.0% specificity value. It should also be noticed that a very few number of patients that are classified as N+ by the proposed techniques, are actually N0; this means only a small number of patients will receive a pelvic irradiation, and it is in contrast with the results obtained with the classical Roach method, even with the more optimistic cut-off (15%). Furthermore the overall accuracy is very high.

Regarding the Roach formula, the high number of high risk patients, that leads the formula to define as N+ a very high percentage of N0 patients (as shown in Table 2), results in poor performance. It is also disappointing that many affected patients are classified as N0 even by the most pessimistic cut-off of the 5%. We experimentally observed that by using smaller cut-offs, almost all the high risk patients will be considered as N+; considering that they are actually the 4.2%, this result is not relevant and the corresponding cut-off cannot be use by physicians for taking every day decisions.

4   Conclusions and Future Work

Prostate cancer is the second cause of cancer in males. Radiotherapy has a well established role in the radical treatment of prostate cancer, but some issues still remain to be uniformly defined amongst Radiation Oncologists. One of these issues is the role of the pelvic irradiation and, in particular, when a prostate cancer patient should or should not receive it. Currently, the decision regarding the pelvic irradiation is still taken mainly by estimating the risk through the Roach formula, which is very inaccurate. The main issue of this formula is related to the small number of pre-treatment parameters that it considers.

In this paper we proposed a different approach, based on three well known classification algorithms: J48, random tree and random forest. The decision trees algorithms have been chosen due to their intuitive representation, which could make them more trustable for the physicians. These techniques have been exploited for generating a predictive model that will support physicians in taking every day decisions. In order to test and experimentally evaluate the proposed approaches we used the largest Italian database of prostate cancer patients treated with radical EBRT. Given the fact that the classes to predict were very imbalanced, we applied random oversampling and undersampling techniques for preprocessing them; this lead to three different data set that were then compared with the Roach formula. The results obtained by the ML algorithms are significantly better; they show that the decision trees classification algorithms can be effectively exploited for supporting the physicians in every day work. The proposed models are usually able to correctly identify all the N+ patients and, furthermore, to limit the number of N0 patients that will receive a pelvic irradiation. Moreover, it seems that the application of both undersampling and oversampling is the best strategy for handling this data.

We see several avenues for future work. Concerning the model generation, we are interested in evaluating the proposed approaches on different, and possibly larger, prostate cancer patients databases. We are also interested in studying the possible differences between different populations of patients (maybe gathered from different geographical regions) and the generated models. It could happen that very different populations lead to the generation of significantly different predictive models; in this case the proposed approach can be explicitly trained for handling different populations.

On the other hand, we are interested in studying a technique for selecting the model to exploit for supporting physicians decisions. A possibility could be the combination of the classifications provided by the models generated by the three exploited classifiers. Additionally, better results could be achieved by clustering the patients and selecting, for each cluster, the best performing approach.

Finally, we are planning to make available a decision support system based on the proposed study. The main features of the DSS that we are designing are: (i) let clinicians to be able to predict the nodal status of new patients; (ii) update the patients database by integrating new patients; (iii) keep the predictive model updated with the new added patients and, (iv) receive feedback from the community. The proposed enhanced DSS will have an intuitive presentation of the models. It will be made available online and via applications for mobile devices; they are currently extensively used by clinicians in every day work.

Acknowledgement. We acknowledge the support provided by the following colleagues: Cynthia Aristei, Enza Barbieri, Giampaolo Biti, Ines Cafaro, Dorian Cosentino, Sandro Fongione, Paola Franzone, Paolo Muto, Pietro Ponticelli, Riccardo Santoni, Alessandro Testolin, Giuseppina Apicella, Debora Beldi, Filippo De Rienzo, Pietro G. Gennari, Gianluca Ingrosso, Fernando Munoz, Andrea Rampini, Enzo Ravo, Ludovica Pegurri, Giovanni Girelli, Alessia Guarnieri, Icro Meattini, Umberto Ricardi, Monica Mangoni, Pietro Gabriele, Rita Bellavita, Marco Bellavita, Alberto Bonetta, Ernestina Cagna, Feisal Bunkheila, Simona Borghesi, Marco Signor, Adriano Di Marco, Marco Stefanacci.

References

1. National collaborating centre for cancer. CG58 prostate cancer: evidence review (2008), http://www.nice.org.uk/nicemedia/pdf/CG58EvidenceReview.pdf
2. National cancer institute. PDQ cancer information summaries (2012), http://www.cancer.gov/cancertopics/pdq/treatment/prostate/HealthProfessional/page4#Reference4.21
3. Asbell, S.O., Martz, K.L., Shin, K.H., Sause, W.T., Doggett, R.L., Perez, C.A., Pilepich, M.V.: Impact of surgical staging in evaluating the radiotherapeutic outcome in RTOG #77-06, a phase III study for T1BN0M0 (A2) and T2N0M0 (B) prostate carcinoma. International Journal of Radiation Oncology, Biology, Physics 40(4), 769–782 (1998)
4. Berthold, M.R., Borgelt, C., Hppner, F., Klawonn, F.: Guide to Intelligent Data Analysis: How to Intelligently Make Sense of Real Data, 1st edn. Springer (2010)
5. Breiman, L.: Random forests. Machine Learning 45, 5–32 (2001)
6. Caruana, R., Karampatziakis, N., Yessenalina, A.: An empirical evaluation of supervised learning in high dimensions. In: Proceedings of the 25th International Conference on Machine Learning, pp. 96–103. ACM (2008)
7. Cawley, G.C., Talbot, N.L.C.: On over-fitting in model selection and subsequent selection bias in performance evaluation. Journal of Machine Learning Research, 2079–2107 (2010)
8. D'Amico, A.V., Whittington, R., Malkowicz, S.B., Schultz, D., Blank, K., Broderick, G.A., Tomaszewski, J.E., Renshaw, A.A., Kaplan, I., Beard, C.J., Wein, A.: Biochemical outcome after radical prostatectomy, external beam radiation therapy, or interstitial radiation therapy for clinically localized prostate cancer. JAMA 280, 969–974 (1998)
9. Edge, S.B., Byrd, D.R., Compton, C.C., Fritz, A.G., Greene, F.L., Trotti, A.: AJCC Cancer Staging Manual, 7th edn. Springer (2010)
10. Fan, W., Wang, H., Yu, P., Ma, S.: Is random model better? on its accuracy and efficiency. In: Proceedings of the 3rd IEEE International Conference on Data Mining, pp. 51–58. IEEE (2003)
11. Graefena, M., Augustina, H., Karakiewiczb, P.I., Hammerera, P.G., Haesea, A., Palisaara, J., Blonskia, J., Fernandeza, S., Erbersdoblerc, A., Hulanda, H.: Can predictive models for prostate cancer patients derived in the united states of america be utilized in european patients? a validation study of the partin tables. European Urology 43(1), 6–11 (2003)

12. Hall, M., Holmes, F.G., Pfahringer, B., Reutemann, P., Witten, I.H.: The WEKA data mining software: An update. SIGKDD Explorations 11(1), 10–18 (2009)
13. He, H., Garcia, E.A.: Learning from imbalanced data. IEEE Transactions on Knowledge and Data Engineering 21(9), 1263–1284 (2009)
14. Jemal, A., Bray, F., Center, M.M., Ferlay, J., Ward, E., Forman, D.: Global cancer statistics. CA: A Cancer Journal for Clinicians 61(2), 69–90 (2011)
15. Kattan, M.W., Stapleton, A.M., Wheeler, T.M., Scardino, P.T.: Evaluation of a nomogram used to predict the pathologic stage of clinically localized prostate carcinoma. Cancer 79, 528–537 (1997)
16. Lawton, C.A., DeSilvio, M., Roach, M., Uhl, V., Kirsch, R., Seider, M., Rotman, M., Jones, C., Asbell, S., Valicenti, R., Hahn, S., Thomas, C.R.: An update of the phase III trial comparing whole pelvic to prostate only radiotherapy and neoadjuvant to adjuvant total androgen suppression: updated analysis of RTOG 94-13, with emphasis on unexpected hormone/radiation interactions. International Journal of Radiation Oncology, Biology, Physics 69, 646–655 (2007)
17. Partin, A.W., Kattan, M.W., Subong, E.N., Walsh, P.C., Wojno, K.J., Oesterling, J.E., Scardino, P.T., Pearson, J.D.: Combination of prostate-specific antigen, clinical stage, and Gleason score to predict pathological stage of localized prostate cancer. A multiinstitutional update. JAMA 277, 1445–1451 (1997)
18. Pommier, P., Chabaud, S., Lagrange, J.L., Richaud, P., Lesaunier, F., Le Prise, E., Wagner, J.P., Hay, M.H., Beckendorf, V., Suchaud, J.P., Pabot du Chatelard, P.M., Bernier, V., Voirin, N., Perol, D., Carrie, C.: Is there a role for pelvic irradiation in localized prostate adenocarcinoma? preliminary results of GETUG-01. International Journal of Radiation Oncology, Biology, Physics 25(34), 5366–5373 (2007)
19. Quinlan, J.R.: C4.5: programs for machine learning. Morgan Kaufmann (1993)
20. Reckwitz, T., Potter, S.R., Partin, A.W.: Prediction of locoregional extension and metastatic disease in prostate cancer: a review. World Journal of Urology 18, 165–172 (2000)
21. Regnier-Couderta, O., McCalla, J., Lothiana, R., Lamb, T., McClintonb, S., N'Dowb, J.: Machine learning for improved pathological staging of prostate cancer: A performance comparison on a range of classifiers. Artificial Intelligence in Medicine 55(1), 25–35 (2012)
22. Roach, M., Marquez, C., Yuo, H.S., Narayan, P., Coleman, L., Nseyo, U.O., Navvab, Z., Carroll, P.R.: Predicting the risk of lymph node involvement using the pre-treatment prostate specific antigen and gleason score in men with clinically localized prostate cancer. International Journal of Radiation Oncology, Biology, Physics 28, 33–37 (1994)
23. Seaward, S.A., Weinberg, V., Lewis, P., Leigh, B., Phillips, T.L., Roach, M.: Identification of a high-risk clinically localized prostate cancer subgroup receiving maximum benefit from whole-pelvic irradiation. The Cancer Journal from Scientific American 4(6), 370–377 (1998)
24. Seaward, S.A., Weinberg, V., Lewis, P., Leigh, B., Phillips, T.L., Roach, M.: Improved freedom from PSA failure with whole pelvic irradiation for high-risk prostate cancer. International Journal of Radiation Oncology, Biology, Physics 42(5), 1055–1062 (1998)

Autoregressive Model Order Estimation Criteria for Monitoring Awareness during Anaesthesia

Nicoletta Nicolaou and Julius Georgiou

KIOS Research Centre and Dept. of Electrical and Computer Engineering,
University of Cyprus, Kallipoleos 75,
1678 Nicosia, Cyprus
{nicolett,julio}@ucy.ac.cy

Abstract. This paper investigates the use of autoregressive (AR) model order estimation criteria for monitoring awareness during anaesthesia. The Bayesian Information Criterion (BIC) and the Akaike Information Criterion (AIC) were applied to electroencephalogram (EEG) data from 29 patients, obtained during surgery, to estimate the optimum multivariate AR model order. Maintenance of anaesthesia was achieved with propofol, desflurane or sevoflurane. The optimum orders estimated from the BIC reliably decreased during anaesthetic-induced unconsciousness, as opposed to AIC estimates, and, thus, successfully tracked the loss of awareness. This likely reflects the decrease in the complexity of the brain activity during anaesthesia. In addition, AR order estimates sharply increased for diathermy-contaminated EEG segments. Thus, the BIC could provide a simple and reliable means of identifying awareness during surgery, as well as automatic exclusion of diathermy-contaminated EEG segments.

Keywords: anaesthesia, AR model order estimation, awareness, EEG.

1 Introduction

Intraoperative awareness is the phenomenon where patients regain consciousness during surgery. It can lead to severe psychological consequences, such as post-traumatic stress disorder, and economic consequences, such as large insurance compensations. The reported incidence of intraoperative awareness ranges from $0.1-0.8\%$ [1]; however, due to the amnesic effect of certain anaesthetics, some patients have no recollection of regaining awareness, therefore it is likely that the real incidence of awareness is much higher. The problem becomes more serious due to the routine administration of muscle relaxants with anaesthetic agents, thus placing patients who regain awareness in a state of 'awake paralysis'. Several indices have been applied to monitor the depth of hypnosis (for a review see [2]) and some devices are commercially available (e.g. Bispectral Index monitor - BIS). These devices convert some combination of various EEG characteristics into a single number from 0-100 representing the level of hypnosis (100 - 'fully awake', 0 - 'isoelectricity'). Commercial devices suffer from a number of issues, e.g. estimation of the level of hypnosis can be affected by the type

H. Papadopoulos et al. (Eds.): AIAI 2013, IFIP AICT 412, pp. 71–80, 2013.

of anaesthetic or the administration of a muscle relaxant [3, 4], and the large inter-subject variability is not captured by the fixed number that represents the level of hypnosis [5].

In this work we propose a method for monitoring awareness based on tracking the anaesthetic-induced changes in EEG signal complexity. The proposed method is based on the hypothesis that anaesthetics cause a decrease in the optimum order necessary for EEG modelling with autoregressive (AR) models, which are widely used in EEG analysis (e.g. [6–8]). Since estimating the optimum AR model order is non-trivial several criteria have been introduced for this purpose, with two of the most widespread being the Bayesian Information Criterion (BIC) [9] and the Akaike Information Criterion (AIC) [10]. The ability of the BIC and the AIC to capture this anaesthetic-induced reduction in EEG complexity is investigated. The choice of these measures for this purpose is motivated by the fact that they can provide a computationally simple method of monitoring awareness, while at the same time addressing some of the issues faced by current monitors.

2   Methods

2.1   Dataset

The data used in the particular study were collected from 29 patients (mean age 42.6 ± 20.5; 1 female patient) undergoing routine general surgery at Nicosia General Hospital, Cyprus. The study was approved by the Cyprus National Bioethics Committee and patients gave written informed consent. A detailed description of the dataset and patient exclusion criteria can be found in previous studies (see, for example, [11]). In summary, anaesthesia was induced with a propofol (1%, 10 mg/ml) bolus. Maintenance of anaesthesia was achieved with (a) intravenous propofol administration (23 patients), or (b) inhalational administration of sevoflurane (4 patients) or desflurane (2 patients). In most patients this was titrated with remifentanil hydrochloride. During induction and/or maintenance some patients received boluses of other drugs, such as neuromuscular blocking agents (cisatracurium, rocuronium, or atracurium) and analgesics, as per surgery requirements.

EEG data were collected using the TruScan32 system (Deymed Diagnostic) at a sampling rate of 256Hz using 19 electrodes according to the 10/20 system (FCz reference). The data were band-pass filtered at 1-100 Hz. Data recording was performed throughout the entire surgical duration (pre-induction, induction, surgery, recovery of consciousness). The point at which the patients stopped responding verbally to commands by the anaesthetist occurred less than a minute after administration of the anaesthetic bolus, depending on patient characteristics. Recovery of consciousness was defined as the point at which the patient responded to verbal commands or tactile stimuli by the anaesthetist. Patient response was expressed either as voluntary muscular movement in response to a command by the anaesthetist or a verbal response. Throughout the recording, time stamps indicating important events, such as anaesthetic induction and recovery of consciousness, were manually inserted in the digital EEG record. These markers were used in subsequent data analysis.

2.2 Multivariate AR Modelling

An AR model can be considered as a linear filter, which describes a current value of a time series with T samples, $\mathbf{X} = \{x(1), x(2), ..., x(T)\}$, using weighted information from p past samples:

$$x(t) = \sum_{i=1}^{p} a_i x(t-i) + e_i(t) \tag{1}$$

where e_i is normal noise. The standard AR model assumes a 1-dimensional time series. However, in many situations it is desirable that N variables are modelled together. This is particularly useful when this N-dimensional information is obtained simultaneously, for example through EEG recordings. For this purpose, the multivariate extension of the AR model (MVAR) can be used [12]. In MVAR models information from all N-dimensional observations are taken into account in a unified model of order p:

$$\mathbf{x}(t) = \sum_{i=1}^{p} \mathbf{A}(i)\mathbf{x}(t-i) + \mathbf{E}(t) \tag{2}$$

where

$$\mathbf{x}(t) = \sum_{i=1}^{p} \begin{bmatrix} a_{11}(i) & \cdots & a_{1N}(i) \\ \vdots & \ddots & \vdots \\ a_{N1}(i) & \cdots & a_{NN}(i) \end{bmatrix} \begin{bmatrix} x_1(t-i) \\ \vdots \\ x_N(t-i) \end{bmatrix} + \begin{bmatrix} e_1(t) \\ \vdots \\ e_N(t) \end{bmatrix} \tag{3}$$

2.3 AR Model Order Estimation

A number of measures have been developed to estimate the optimum model order. The most widespread are the Bayesian Information Criterion (BIC) [9] and the Akaike Information Criterion (AIC) [10]. These criteria measure the relative goodness of fit of a particular AR model. They provide relative measures of the information loss when the model is applied on the observed data. Given a set of models, the preferred one minimizes the information loss (log-likelihood) estimated from the model order criteria.

The AIC is an information-theoretic measure derived by an approximate minimization of the difference between the true data distribution and the model distribution through the Kullback-Leibler divergence. One way of estimating the AIC is via the residual sum of squares after regression on the data with a model of order p,

$$AIC(p) = ln|\sigma_e^2| + 2\frac{k}{T} \tag{4}$$

where $k = N \times N \times p$ is the number of parameters estimated for a multivariate model of order p and N dimensions, σ_e^2 is the variance of the model residuals, and T is the number of samples. From the definition, the first term of the AIC measures the difference between the log-likelihood of the fitted model and the log-likelihood of the model under consideration. The second term is a penalty term, which penalizes the addition of parameters. The combination of these two terms results in large values of AIC when the model has too few or too many parameters. Thus, the best model is one that provides both a good fit, and that has a minimum number of parameters. This avoids the problem of overfitting. The BIC estimation is very similar to the AIC. However, it differs from the AIC in that it is based on Bayesian formalism and is derived from approximating the Bayes Factor, i.e. the evidence ratios of models. It is defined as:

$$BIC(p) = ln|\sigma_e^2| + klnT \qquad (5)$$

Just as the AIC, the BIC also penalizes free parameters, but it does so more strongly. Previous studies have demonstrated that, in general, the BIC is more consistent in contrast to the AIC [13]. The AIC performs well in relatively small samples, but is inconsistent and its performance does not improve in large samples; this is in contrast to the BIC, which appears to perform more poorly in small samples, but remains consistent and its performance is improved with larger sample sizes.

2.4   Methodology

Dimensionality Reduction. The original data space is 19-dimensional (number of electrodes). Fitting a 19-dimensional MVAR model of order p implies estimation of $19 \times 19 \times p$ parameters. The requirement for EEG stationarity does not allow the use of long segments; therefore, such a large number of parameters cannot be estimated reliably. In order to reduce the dimensionality, and subsequently the number of estimated parameters, two approaches were followed:

1. EEG aggregates: five brain areas were defined as the aggregate activity of particular electrodes. These areas were: left frontal (LF: electrodes Fp1, F7, F3, T3, C3), right frontal (RF: Fp2, F8, F4, C4, T4), left posterior (LP: T5, P3, O1), right posterior (RP: T6, P4, O2), and midline (Z: Fz, Cz, Pz). In defining these particular aggregates it was taken into account that fronto-posterior interactions appear to play an important role in (un)consciousness. Electrode impedance is measured automatically; thus, electrodes with high impedance were subsequently excluded from aggregates estimation.
2. Frontal EEG: the majority of work in the literature, and also commercially available depth of anaesthesia monitors, base their analysis on mono- or bipolar frontal EEG activity. Here, we also applied the BIC and AIC measures on activity from electrode locations Fp1 and Fp2 only, in order to investigate the reliability of such features.

Optimum AR Order Estimation. The EEG data of each patient were analysed in non-overlapping windows of 2s duration. The window size should be large enough to ensure adequate number of samples for accurate parameter estimation, but also short enough to ensure data stationarity. The choice of 2-s fulfilled both these requirements. For each window the optimum order from a range of $p = 2, 3, ..., 30$ was estimated using the BIC and AIC measures, for both the EEG aggregates and the frontal EEG. For each patient, the differences in the estimated AR orders at the two conditions were tested for statistical significance (ANOVA F-test, $\alpha = 0.05$).

3   Results

Figure 1 shows the resulting BIC (a-b) and AIC (c-d) boxplots for wakefulness and anaesthesia for all patients, for both EEG aggregates (a, c) and frontal EEG (b, d). Statistical significance was estimated using ANOVA F-test ($\alpha = 0.05$). All differences between wakefulness and anaesthesia were significant. Specifically, for aggregate activity, (1) BIC: $F = 3535.01, p = 0$, and AIC: $F = 5452.03, p = 0$. For frontal activity, (1) BIC: $F = 616.76, p = 0$, and AIC: $F = 253.00, p = 0$. The median AR orders from the aggregate EEG are (a) BIC: 10 (awake) and 7 (anaesthesia); (b) AIC: 30 (awake) and 14 (anaesthesia). For frontal EEG we obtain (a) BIC: 13 (awake) and 10 (anaesthesia); (b) AIC: 30 (awake) and 30 (anaesthesia). Figure 2 shows the median optimum AR orders estimated for each patient using the BIC (a-b) and AIC (c-d) for wakefulness and anaesthesia. All differences were statistically significant (ANOVA F-test, $\alpha = 0.05$), except for patient S22 (indicated with an '*' on figure 2). A general observation is the overestimation of the optimum AR orders using the AIC measure.

4   Discussion

The optimum AR orders estimated from the BIC using the EEG aggregate activity track the transitions between wakefulness and anaesthesia well. Administration of anaesthesia causes a decrease in the estimated optimum AR orders, which remain stable throughout the entire duration of surgery, and increase again after the patient regains consciousness at the end of surgery. This effect was observed only for the BIC, with more robust estimates obtained from the aggregate EEG. Despite a similar effect observed for the AIC, this was not robust and identification between awake and anaesthetized states was difficult. Despite the large inter-subject differences in the raw EEG activity, the optimum AR orders estimated for anaesthesia and wakefulness did not display large inter-subject variability.

The more robust BIC performance could be explained through a representative example for a randomly chosen patient and EEG segment (fig. 4). The estimated BIC values are in general higher than the AIC values, which is expected if we take into consideration the second term of eqs. 4 and 5. From fig. 4 it can be seen that the changes in the AIC values become negligible for AR models

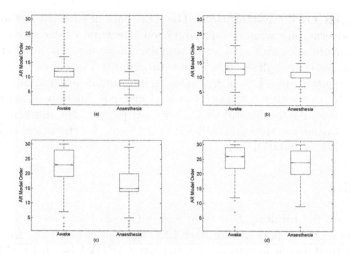

Fig. 1. ANOVA Boxplots over all patients for optimum AR orders estimated during awake and anaesthesia using the BIC (a)-(b), and the AIC (c)-(d). Optimum AR orders in (a) and (c) were estimated from the EEG aggregates, while in (b) and (d) from the frontal EEG. All differences between awake and 'anaesthetized conditions are statistically significant at the 95% significance level.

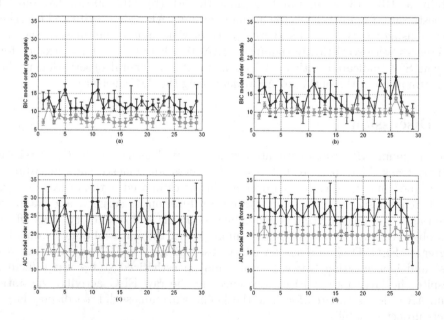

Fig. 2. Median optimum AR orders from BIC (a-b) and AIC (c-d) estimated using EEG aggregate activity (a, c) and frontal activity (b, d), for individual patients. All differences between 'awake' and 'anaesthetized' conditions are statistically significant at the 95% significance level, except from one difference which is not statistically significant; this is indicated by an '*'.

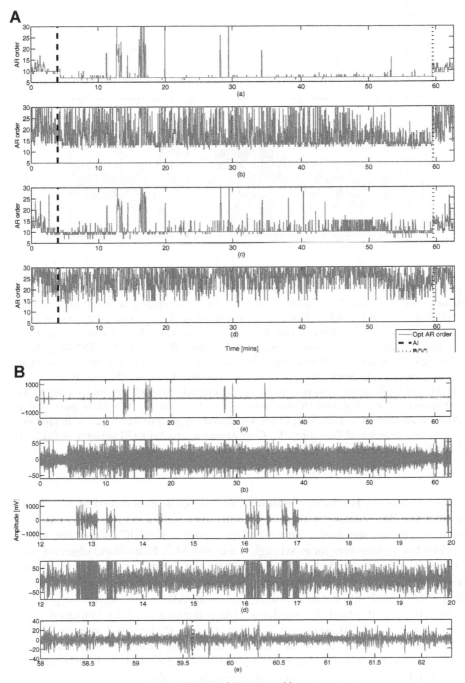

Fig. 2. (*Continued.*)

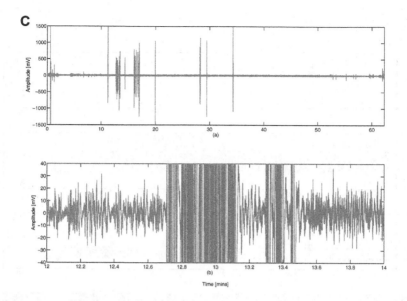

Fig. 3. A: Optimum AR orders for patient S3 throughout entire surgery (maintaince with propofol), estimated by: (a) BIC and (b) AIC from aggregate EEG; and (c) BIC and (d) AIC from frontal EEG. Vertical lines: dashed - induction (AI), dotted - recovery of consciousness (ROC). Sharp peaks in (a) and (c) are caused by diathermy artefacts (see panels B and C for corresponding EEG activity). **B**: Aggregate EEG for patient S3 throughout surgery (62.3 minutes). Average activity at: (a) left frontal location. High frequency peaks with larger amplitude than the EEG are diathermy artefacts. (b) right frontal location, zoomed in to the amplitude level of the underlying EEG. (c) left posterior at 12-20 min (high diathermy contamination). (d) Right posterior at 12-20 min, zoomed in to EEG amplitude level; and (e) midline at 58-62.3 min (patient recovers consciousness at 56.6 min). **C**: Corresponding frontal EEG activity at (a) Fp1, and (b) Fp2 (zoomed in).

with orders $p > 13$. However, the minimum AIC value is obtained at $p = 23$. In contrast, the BIC decreases until a clear minimum value at $p = 10$ is reached, after which it increases again. This is a direct result of the BIC estimation, which penalises free parameters more strongly than the AIC. From these observations: (a) the AIC indicates that the goodness of fit is not substantially improved using models of orders $p > 13$, therefore an appropriate choice of model order without taking into consideration the absolute minimum AIC value would be $p = 13$; (b) the BIC indicates that the goodness of fit is not substantially improved using models of orders $p > 10$, as can be seen by the penalty imposed at such models by the BIC. Therefore, the combination of increasing goodness of fit and larger penalty imposed on the number of free parameters ensures that the BIC attains a clear minimum value at the smallest appropriate order.

   The proposed method has several advantages over existing methods or commercial monitors: (1) the estimated features remain stable over the entire surgical

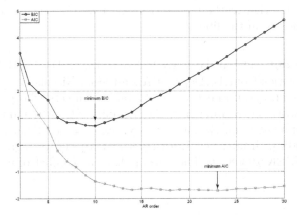

Fig. 4. An example of BIC and AIC values obtained at a randomly chosen EEG segment and a randomly chosen patient. The BIC displays a distinct minimum, as opposed to the AIC, which tends to decrease asymptotically with the model order, resulting into an overestimation of the model order. For this specific example, the optimum AR orders are 10 and 23 as estimated by the BIC and the AIC respectively.

duration; (2) estimations are simple and fast; (3) the features are not affected by the particular anaesthetic protocol; (4) wakefulness/anaesthesia can be identified through simple visual inspection of the estimated features, hence there is no need for conversion to a 'universal' number from 0-100 or the use of a classifier; (5) the estimated AR orders remain stable after anaesthetic administration is switched off, indicating that the AR orders are not simply a reflection of the metabolic decrease of the anaesthetic agent; and (6) diathermy-contaminated EEG segments can be easily identified. In addition, the EEG data were obtained from actual surgery. This allows us to investigate the stability of the estimated AR order patterns and assess the clinical applicability of the proposed method.

5   Conclusion

The feasibility of utilizing autoregressive model order criteria for wakefulness / anaesthesia discrimination from EEG activity during surgery was investigated. The Bayesian Information Criterion (BIC) provided a reliable means of tracking the reduction in EEG complexity during anaesthesia through providing lower optimum AR order estimates. This effect can be tracked more reliably when the BIC is applied on aggregate than frontal EEG activity. The same effect was not clearly identifiable with the Akaike Information Criterion (AIC). However, both measures provided AR orders that were statistically different during wakefulness/anaesthesia. The BIC could also be used for automatic identification of diathermy-contaminated EEG segments. Future work will involve (i) investigation of other preprocessing methods for noise removal, e.g. Independent Component Analysis, and (ii) the comparison of the proposed methodology with other methods, including other methods previously proposed by the authors (e.g. [11]) and other methods from the literature.

Acknowledgement. The authors would like to thank Dr. Pandelitsa Alexandrou and Dr. Saverios Hourris (Depr. of Anaesthesia, Nicosia General Hospital, Cyprus), the hospital staff and the anonymous volunteers who participated in this study. This work was co-funded by the Republic of Cyprus and the European Regional Development Fund through the Cyprus Research Promotion Foundation ('DESMI 2008'). Grants: New Infrastructure Project/Strategic/0308/26 and DIDAKTOR/DISEK /0308/20. The authors have no conflict of interest.

References

1. Bruhn, J., Myles, P., Sneyd, R., Struys, M.: Depth of anaesthesia monitoring: what's available, what's validated and what's next? British Journal of Anaesthesia 97(1), 85–94 (2006)
2. Rampil, I.J.: A primer for EEG signal processing in anesthesia. Anesthesiology 89(4), 980–1002 (1998)
3. Dahaba, A.: Different conditions that could result in the bispectral index indicating an incorrect hypnotic state. Anesth. Analg. 101, 765–773 (2005)
4. Russell, I.F.: The Narcotrend "depth of anaesthesia" monitor cannot reliably detect consciousness during general anaesthesia: an investigation using the isolated forearm technique. British Journal of Anaesthesia 96(3), 346–352 (2006)
5. Voss, L., Sleigh, J.: Monitoring consciousness: the current status of EEG-based depth of anaesthesia monitors. Best. Pract. Res. Clin. Anaesthesiol. 21(3), 313–325 (2007)
6. Anderson, C., Stolz, E., Shamsunder, S.: Multivariate autoregressive models for classification of spontaneous electroencephalographic signals during mental tasks. IEEE T Biomed. Eng. 45(3), 277–286 (1998)
7. McFarland, D., Wolpaw, J.: Sensorimotor rhythm-based brain computer interface (BCI): model order selection for autoregressive spectral analysis. J. Neural Eng. 5(2), 155–162 (2008)
8. Franaszczuk, P., Bergey, G.: An autoregressive method for the measurement of synchronization of interictal and ictal EEG signals. Biological Cybernetics 81(1), 3–9 (1999)
9. Schwarz, G.: Estimating the dimension of a model. Annals of Statistics 6(2), 461–464 (1978)
10. Akaike, H.: A new look at the statistical model identification. IEEE Trans Automatic Control 19(6), 716–723 (1974)
11. Nicolaou, N., Houris, S., Alexandrou, P., Georgiou, J.: EEG-based automatic classification of awake versus anesthetized state in general anesthesia using granger causality. PLoS ONE 7(3), e33869 (2012)
12. Penny, W., Harrison, L.: Multivariate autoregressive models. In: Friston, K., Ashburner, J., Kiebel, S., Nichols, T., Penny, W.D. (eds.) Statistical Parametric Mapping: The Analysis of Functional Brain Images, Elsevier, London (2006)
13. Zhang, P.: On the convergence of model selection criteria. Comm. Stat.-Theory Meth. 22, 2765–2775 (1993)

A Machine-Learning Approach for the Prediction of Enzymatic Activity of Proteins in Metagenomic Samples

Theodoros Koutsandreas, Eleftherios Pilalis, and Aristotelis Chatziioannou

Metabolic Engineering & Bioinformatics Program, Institute of Biology,
Medicinal Chemistry & Biotechnology, National Hellenic Research Foundation,
Athens, Greece
th_koutsandreas@hotmail.com, {epilalis,achatzi}@eie.gr

Abstract. In this work, a machine-learning approach was developed, which performs the prediction of the putative enzymatic function of unknown proteins, based on the PFAM protein domain database and the Enzyme Commission (EC) numbers that describe the enzymatic activities. The classifier was trained with well annotated protein datasets from the Uniprot database, in order to define the characteristic domains of each enzymatic sub-category in the class of Hydrolases. As a conclusion, the machine-learning procedure based on Hmmer3 scores against the PFAM database can accurately predict the enzymatic activity of unknown proteins as a part of metagenomic analysis workflows.

Keywords: Machine-learning, Enzymes, Proteins, Metagenomics.

1 Introduction

The emerging field of Metagenomics comprises the collection and analysis of large amounts of DNA that is contained in an environmental niche [1]. Due to the recent advances in high-throughput sequencing, very large amounts of nucleotide sequences can be generated in short time. Because of the increased volume of data, metagenomics is a promising way to identify novel enzymes and protein functions. However, despite the advances in high-throughput sequencing, the development of appropriate analysis tools remains challenging. Here, we developed a classifier for the prediction of protein enzymatic activity in metagenomic samples. Enzymes are proteins that are used in a wide range of applications and industries, such as Biotechnology and Biomedicine. In order to correlate unknown amino acid/nucleotide sequences with enzyme classes the PFAM database and the Enzyme Nomenclature system were used. PFAM is a database of protein families [2]. Each family is represented by a multiple sequence alignment which is generated by Hidden Markov Models (HMMs) with the Hmmer3 [3] program. Proteins consist of one or more functional regions which are called domains, i.e. the existence of a domain in the tertiary structure of a protein, imply a specific function. Thus, proteins of the same family will include identical or similar domains. The PFAM database contains information about protein families,

H. Papadopoulos et al. (Eds.): AIAI 2013, IFIP AICT 412, pp. 81–87, 2013.

their domains and their architecture. Each entry, represented by a PFAM id, corresponds to a single domain. The similarity of an unknown protein with a protein domain may give great information about its function and its phylogenetic relationships.

The Enzyme Nomenclature (EC) is a numerical classification system for enzymes, based on the chemical reaction that they catalyze. It was developed under the auspices of the International Union of Biochemistry and Molecular Biology during the second half of twenty-first century. Each entry of Enzyme Nomenclature is a four-number code, the enzyme commission number (EC number), which is associated with a specific chemical reaction. Thus each enzyme receives the appropriate EC number according to its chemical activity. The first number of code specifies the major category of catalyzed chemical reaction. There are six major categories of catalysed biochemical reactions: Oxidoreductases: 1.-.-.-, Transferases: 2.-.-.-, Hydrolases: 3.-.-.-, Lyases: 4.-.-.-, Isomerases: 5.-.-.-, Ligases: 6.-.-.-. The next two numbers specify the subclasses of major class and the last one states the substrate of the reaction. For instance, the EC number 3.1.3.- refers to the hydrolysis of phosphoric mono-ester bond and 3.1.3.11 refers to the hydrolysis of fructose-bisphosphatase which contains a phosphoric mono-ester bond. The classifier developed in the current study was able to classify unknown amino acid sequences originating from metagenomic analysis to hydrolases classes pursuant to the results of Hidden Markov Model detection. The classifier consisted of separate classification models, where the classification type was binominal, i.e. is EC number or is not EC number. In order to train the classifier we used well-annotated proteins and analyzed them with Hmmer3. The result was the score of similarity between an examined sequence and a protein domain. As a result, the features of the training data were the PFAM ids and the vector of each training example included its scores to the appropriate fields.

2   Dataset

In order to train the classification models, we used well-annotated proteins from the UniProt database [4]. We specifically selected all the reviewed sequences from the reference proteome set of bacteria (taxonomy: 2). The Uniprot database was used because it is a high quality, manually annotated and non-redundant protein database. A total number of 45612 sequences were collected. During the training of the classification models, we selected an amount of known proteins according to their EC numbers. A separate classification model was trained for each EC number, using binominal training data as positive and negative examples. The positive examples were sequences that belonged to a specific EC number (for example 3.1.3.1). In contrast, sequences that belonged to the same EC upper class, but differ in the last digit (i.e. 3.1.1.-), were the negative examples. In this way, we aimed at the detection of differences in features context, which separated a specific EC number protein family from all the other proteins whose EC number differed in the last digit.

3   Methods

3.1   Training of Enzymatic Classification Models

In order to train separate models for each enzymatic category of the Hydrolases class we implemented a procedure which automatically constructed the corresponding training sets (Fig. 1). In the first step, all sequences that are annotated with the specific EC number were selected from the dataset as positive examples. Each EC number has its upper class number (for EC number 3.-.-.- the upper class is all the no 3.-.-.- classes, i.e. all enzymes which are not hydrolases). In the second step the procedure selected, as negative example set, an equal amount of sequences that belonged to the upper class but not to the specific EC number. Consequently, the difference between positive and negative training data examples was the last EC number digit. Note that there were some conditions to be tested, especially for the EC numbers with three or four digits, during the execution of this step. The procedure stopped if the amount of positive examples was less than five or if negative examples were not found. Thereafter these two sets of examples were analyzed by Hmmer3 against the HMM profiles of the PFAM-A database in order to collect the domain scores as training features. For each enzymatic category an HMM profile library was thus constructed, which contained all the PFAM domains having a score against the sequences in the corresponding training set. For models corresponding to three- or four-digits EC numbers, a custom HMM profile was automatically constructed from the positive example set and was added to the HMM profile library as an additional feature representing the whole sequence length. The training was performed with the k-nearest neighbor (k-NN) algorithm using the Euclidean distance of 4-nearest neighbors, in a10-fold cross-validation process (stratified sampling). The choice of the parameter k=4 was made based on the size of the smaller training sets (12-15 examples). Thus, a stable value 4 was given to k representing approximately the one third of the smallest training data.

3.2   Application of the Trained Models to Unlabeled Sequences

The classification procedure automatically performed Hmmer3 analysis on the unlabeled sequences against the PFAM database and collected their scores as features (Fig. 2). Afterwards, the trained models were applied to the unlabeled feature sets starting from the general classification in Hydrolases class (i.e. EC number 3.-.-.-). Then the procedure continued to the subclasses with two, three and four EC number digits. The classification procedure was not hierarchical top-down as there was not any filtering method during the descending in enzyme subclasses, like the exclusion of sequences which are not annotated as hydrolases in the first classification task. Filtering was avoided because we observed that the classification procedure in main hydrolase class (3.-.-.-) and its subclasses (3.1.-.-, 3.2.-.- etc) had some false negatives that were correctly classified during the next steps in more detailed EC number classes with three or four digits. However, not all sequences passed through the classification procedures. As mentioned above, the training set of each EC number contained a particular set of features (HMM profile library). In the unlabeled sequences classification task, the sequences having no score against this feature set were filtered out as they were considered distant from the upper-level EC number category. The volume of unlabeled sequences to be tested was thus reduced as an execution time optimization of the procedure.

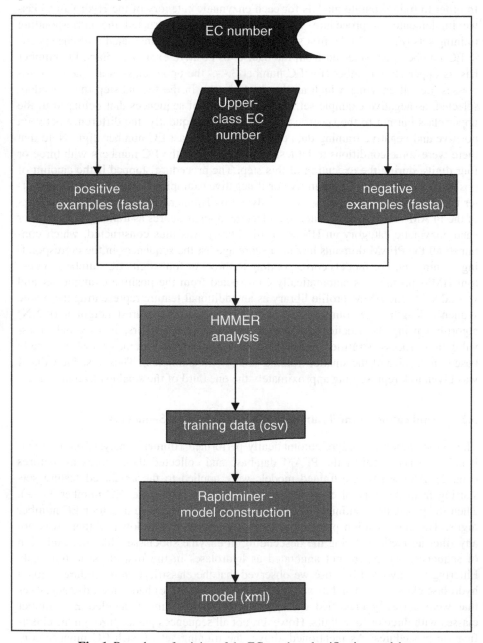

Fig. 1. Procedure of training of the EC number classification models

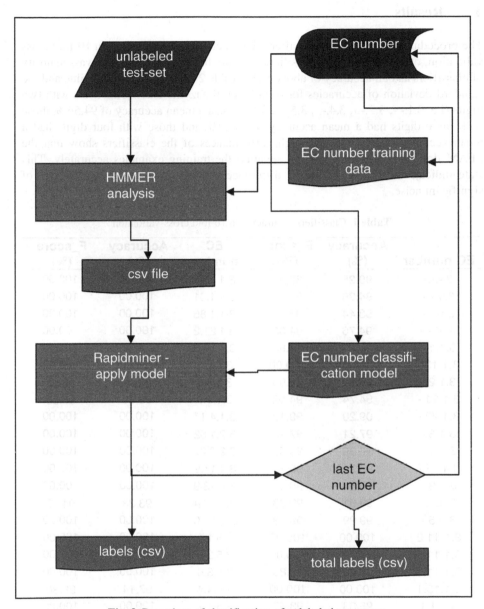

Fig. 2. Procedure of classification of unlabeled sequences

3.3   Software Tools

The aforementioned procedures were implemented in Python scripts. All datasets and results were stored and queried in a MySQL database. The training and the application of the models was performed using RapidMiner (http://rapid-i.com/).

4   Results

The procedure comprised 163 classifiers. The accuracies and F-scores in 10-fold cross validation, of a representative classifiers part, are listed in Table 1. The vast majority of classifiers had an accuracy of above 95%. Table 2 indicates the mean value and the standard deviation of accuracies for each level of enzyme class. Classifiers with two digits (i.e. 3.1.-.-, 3.2.-.-, 3.4.-.-, 3.5.-.-, 3.6.-.-) had a mean accuracy of 94.56%, those with three digits had a mean accuracy of 97.71% and those with four digits had a mean accuracy of 99.39%. The high performances of the classifiers show that the PFAM domain scores were able to separate the training examples accurately. This state indicates a high vector profile difference between two classes and the lack of significant noise.

Table 1. Classifiers accuracies in 10-fold cross-validation

EC number	Accuracy (%)	F_score (%)	EC number	Accuracy (%)	F_score (%)
3.-.-.-	90.25	88.40	3.1.1.45	100.00	100.00
3.1.-.-	94.26	94.42	3.1.1.61	100.00	100.00
3.5.-.-	93.44	93.76	3.1.1.85	100.00	100.00
3.-6.-.-	94.76	94.33	3.1.21.2	100.00	100.00
3.1.1.-	96.03	95.89	3.1.26.3	100.00	100.00
3.1.13.-	100.00	100.00	3.1.2.6	100.00	100.00
3.1.2.-	96.12	96.00	3.1.3.1	95.45	90.91
3.1.21.-	94.74	94.12	3.1.3.5	99.00	98.88
3.1.22.-	99.20	99.13	3.1.4.17	100.00	100.00
3.1.3.-	97.21	97.18	3.2.1.52	100.00	100.00
3.2.1.-	98.39	98.57	3.2.2.27	100.00	100.00
3.4.11.-	95.29	95.12	3.4.11.9	100.00	100.00
3.5.3.-	98.64	98.53	3.4.13.9	100.00	100.00
3.6.1.-	99.31	99.33	3.4.16.4	93.33	94.12
3.6.5.-	99.29	99.23	3.4.21.107	100.00	100.00
3.1.11.2	100.00	100.00	3.5.1.1	100.00	100.00
3.1.11.5	100.00	100.00	3.5.2.5	100.00	100.00
3.1.11.6	98.29	98.86	3.5.3.6	100.00	100.00
3.1.13.1	100.00	100.00	3.5.4.4	97.14	96.30
3.1.1.1	95.83	93.33	3.6.1.7	100.00	100.00
3.1.1.29	99.31	99.25	3.6.3.12	100.00	100.00
3.1.1.3	87.50	82.35	3.6.3.30	100.00	100.00
3.1.1.31	100.00	100.00	3.6.4.12	99.21	99.43

Table 2. Mean and standard deviation in function with the amount of digits in EC numbers

amount of digits	mean value	standard deviation
2	94.86	0.45
3	97.71	0.19
4	99.39	0.09

5   Conclusion

In conclusion, the machine-learning procedure based on Hmmer3 scores against the PFAM database performed well and accurately predicted the enzymatic activity of unknown proteins. Future developments will include the use of other protein motifs databases, like CATH and SCOP and the development of more efficient data mining algorithms. The procedure will also be extended to other enzymatic classes (here we focused on the group of Hydrolases) and will be run-time optimized for its application on very large datasets. Finally, it will be implemented as an independent tool and it will be integrated in more extended metagenomic analysis workflows.

Acknowledgements. The presented work in this paper has been funded by the "Cooperation" program 09SYN-11-675 (DAMP), O.P. Competitiveness & Entrepreneurship (EPAN II).

References

1. Lorenz, P., Eck, J.: Metagenomics and industrial applications. Nat. Rev. Microbiol. 3(6), 510–516 (2005)
2. Finn, R.D., et al.: The PFAM protein families database. Nucleic Acids Res. 36(Database issue), D281–D288 (2008)
3. Finn, R.D., Clements, J., Eddy, S.R.: HMMER web server: interactive sequence similarity searching. Nucleic Acids Res. 39(Web Server issue), W29–W37 (2011)
4. Apweiler, R., et al.: UniProt: The Universal Protein knowledgebase. Nucleic Acids Res. 32(Database issue), D115–D119 (2004)

Modeling Health Diseases Using Competitive Fuzzy Cognitive Maps

Antigoni P. Anninou[1], Peter P. Groumpos[1], and Panagiotis Polychronopoulos[2]

[1] Laboratory for Automation and Robotics, Department of Electrical and
Computer Engineering, University of Patras, Greece
`{anninou,groumpos}@ece.upatras.gr`
[2] Clinic of Neurology, Department of Pathology, General University Hospital of Patras
`ppolychr@yahoo.gr`

Abstract. This paper presents the medical decision support systems (MDSS) and their architecture. The aim of this paper is to present a new approach in modeling knee injuries using Competitive Fuzzy Cognitive Maps (CFCMs). Basic theories of CFCMs are reviewed and presented. Decision Support Systems (DSS) for Medical problems are considered. Finally, it illustrates the development of an MDSS for finding knee injury with the architecture of CFCMs.

Keywords: Fuzzy cognitive maps, medical decision support system, competitive fuzzy cognitive maps, knee injuries.

1 Introduction

A fuzzy cognitive map (FCM) is a soft computing technique, which is capable of dealing with complex systems in situations exactly as a human does using a reasoning process that can include uncertain and ambiguity descriptions [6], [9]. FCM is a promising modeling method for describing particular domains showing the concepts (variables) and the relationships between them (weights) while it encompasses advantageous features [8]. Fuzzy Cognitive Maps (FCM) are fuzzy-graph structures for representing causal reasoning. Their fuzziness allows hazy degrees of causality between causal objects (concepts) [2].

In this paper a decision support system will be implemented. Decision Support System (DSS) is defined as any interactive computer – based support system for making decisions in any complex system, when individuals or a team of people are trying to solve unstructured problems on an uncertain environment. Medical Decision Systems have to consider a high amount of data and information from interdisciplinary sources (patient's records and information, doctors' physical examination and evaluation, laboratory tests, imaging tests) and, in addition to this, medical information may be vague, or missing [1-2]. Furthermore the Medical Diagnosis procedure is complex, taking into consideration a variety of inputs in order to infer the final diagnosis. Medical Decision Systems are complex ones, consisting of

H. Papadopoulos et al. (Eds.): AIAI 2013, IFIP AICT 412, pp. 88–95, 2013.

irrelevant and relevant subsystems and elements, taking into consideration many factors that may be complementary, contradictory and competitive; these factors influence each other and determine the overall diagnosis with a different degree. It is apparent that these systems require a modeling tool that can handle all these challenges and at the same time to be able to infer a decision. Thus, FCMs are suitable to model a Medical Decision Support Systems (MDSS) [2].

A special type of FCM has been introduced for Medical Diagnosis systems, with advanced capabilities, the Competitive Fuzzy Cognitive Map (CFCM) [4-5]. Each decision concept represents a single decision, which means that these concepts must compete against each other so that only one of them dominates and is considered as the correct decision. This is the case of most medical applications, where, the symptoms have to conclude to one diagnosis. The factor-concepts take values from patient data. These concepts represent symptoms, experimental and laboratory tests, and their values are dynamically updated. The decision concepts are considered as outputs in which their calculated values outline the possible diagnosis for the patient [3].

In this paper we present in a simple but illustrative way how useful the FCMs can be in medical problems. In section 2 the basic theories of CFCM, while the decision making support system in knee injuries is described in Section 3. The paper concludes in section 4.

2   Competitive Fuzzy Cognitive Maps

The algorithm that describes the Competitive Fuzzy Cognitive Maps consists of the following steps [3]:

Step 1: Set values A_i of nodes according to values of the factors involved in the decision process. The values Ai are described using linguistic variables.

Step 2: Read the weight matrix W. The interrelationships between nodes may be positive, negative or zero. The existing relationships between the concepts are described firstly, as "negative" or "positive" and secondly, as a degree of influence using a linguistic variable, such as "low", "medium", "high", etc. A positive weight means that the specific factor increases the possibility of diagnosis the interconnected disease. A negative one reduces the possibility of diagnosis the interconnected disease. Lack of interconnection indicates that there is no impact of the factor to the disease. By using a defuzzification method [10], weights between the factor-concepts and the decision-concepts are converted to initial values, which for the current research are between 0 and 1. These are placed in matrix W of size $(n+m)$ x $(n+m)$. The values in the first n columns correspond to the weighted connections from all the concepts towards the n decision-concepts, and the values in the remaining m columns correspond to the weighted connections from all the concepts towards the factor-concepts. This matrix also includes the -1 weight values for competition between output decision-concepts.

Step 3: Update the values: $A_{new} = A_{old} * W$.

Step 4: A=f (A_{new}), where the sigmoid function f belongs to the family of squeezing functions, and the following function is usually used to describe it:

$$f = \frac{1}{1+e^{-\lambda x}}$$

(1)

This is the unipolar sigmoid function, in which $\lambda > 0$ determines the steepness of the continuous function f(x).

Step 5: Repeat steps until equilibrium has been reached and the values of the concepts no longer change.

Step 6: The procedure stops and the final values of the decision-concepts are found, the maximum of which is the chosen decision.

Although the model FCMs-MDSS designed to include all possible symptoms, the causal factors and their relationships, there is a particular situation in which very few of the symptoms are available and taken into account. Thus, in such a diagnostic or prognostic model FCMs-MDSS, the decision will be based using only a small part of the nodes of the entire system, leading to a wrong decision or difficulty convergence since weights of active nodes reflect a small part of the knowledge of specialists.

By using a FCM augmented with a Case-Based-Reasoning (CBR), in such situations, the DSS would draw upon cases that are similar according to distance measures and would use the CBR subsystem to generate a sub-FCM emphasizing the nodes activated by the patient data and thus redistributing the causal weights between the concept-nodes.

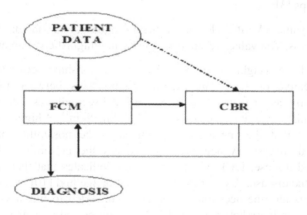

Fig. 1. FCM-CBR

3   Competitive Fuzzy Cognitive Maps for Knee Injuries

Knee pain is a common complaint for many people. There are several factors that can cause knee pain. Awareness and knowledge of the causes of knee pain lead to a more accurate diagnosis. Management of knee pain is in the accurate diagnosis and effective treatment for that diagnosis. Knee pain can be either referred pain or related to the knee joint itself [7].

CFCM distinguish two main types of nodes: decision-nodes (decision-concepts) and factor-nodes [5]. The following figure presents an example model CFC   M that is used in order to determine the medical diagnosis, and includes types of FCMs concepts and causal relationships between them [3].

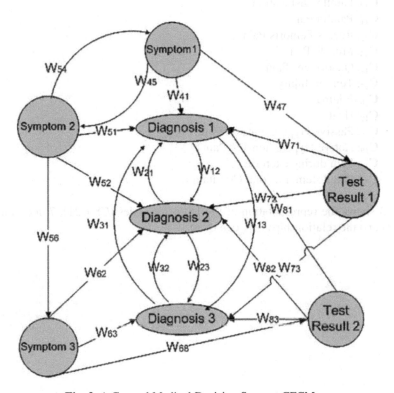

Fig. 2. A General Medical Decision Support CFCM

All concepts may interact with each other and decide the value of diagnosis-nodes that are mutually exclusive in order to show one single diagnosis. In most medical applications according to symptoms, physicians need to complete only one diagnosis and then to determine the appropriate treatment.

In the specific model that describes knee injuries, there are two kinds of nodes:

— n decision-nodes (as outputs):

C_1: Overuse injuries
C_2: Patellofemoral pain syndrome
C_3: Osteochondritis dissecans of the femoral condyles
C_4: Tibial tuberosity apophysitis

C_5: Iliotibial band syndrome
C_6: Biceps femoris tendinitis
C_7: Inferior patellar pole osteochondritis

— m factor-nodes (as inputs):

C_8: Patellar Instability
C_9: Patellar Crepitation
C_{10}: Patella Dislocation
C_{11}: Patella Pain
C_{12}: Rectus Femoris Pain
C_{13}: Muscle Pain
C_{14}: Quadriceps Pain
C_{15}: Tendon Injury
C_{16}: Edema
C_{17}: Heat
C_{18}: Passive Hyperextension Pain
C_{19}: Active Hyperextension Pain
C_{20}: Pain during exercise
C_{21}: Patellofemoral Joint Crepitation

Fig. 3 shows the representation of the 15 factor-nodes (C8-C21), 7 decision-nodes (C1-C7) and the relationships between them.

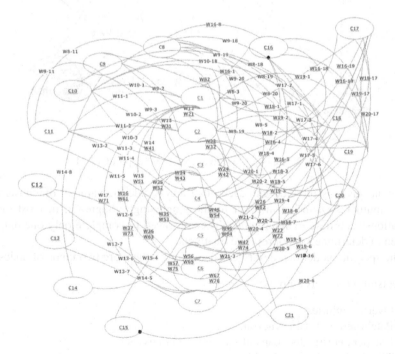

Fig. 3. Competitive Fuzzy Cognitive Map for Knee Injuries

According to the above algorithm and the kind of injuries, the weight matrix W would be as follows using linguistic variables:

Table 1. Linguistic Variables of the Concepts Relationships

	C1	C2	C3	C4	C5	C6	C7	C8	C9	C10	C11	C12	C13	C14	C15	C16	C17	C18	C19	C20	C21
C1		-1	-1	-1	-1	-1	-1														
C2	-1			-1	-1	-1	-1														
C3	-1	-1		-1	-1	-1	-1														
C4	-1	-1	-1		-1	-1	-1														
C5	-1	-1	-1	-1		-1	-1														
C6	-1	-1	-1	-1	-1		-1														
C7	-1	-1	-1	-1	-1	-1															
C8	+M	+VH	+VVH	None	+VVL	None	None				+L							+L	+L	+L	
C9	None	+VL	+VH	None	None	None	None				+L							+L	+L	+L	
C10	+M	+VVL	+VH	None	None	None	None				+L							+L	+L	+L	
C11	+VVH	+VH	+L	+VVH	+M	None	None														
C12	None	None	None	None	None	+VH	+VVH														
C13	None	+M	None	None	None	+VVH	+VH														
C14	None	None	None	None	+VVH	None	None	+L													
C15	None	None	None	None	+VVH	None	None									+L					
C16	+H	None	None	+VH	+VVH	None	None	+L										+L	+L	+L	
C17	+H	+M	+M	+M	+VVH	+VL	None														
C18	+L	+H	+H	+VH	+L	+M	+VH										+L				
C19	+L	+VH	+H	+VVH	+M	+M	+L										+L				
C20	+VH	+VVH	+VH	+VH	+M	+H	+L										+L				
C21	None	+VVL	+M	None	None	None	None														

where:
VVH: Very Very High
VH: Very High
H: High
M: Medium
L: Low
VL: Very Low
VVL: Very Very Low

It's obvious that in order to achieve the competition, the relationships between each of these nodes with the rest should have a very large negative weight (-1). This implies that the higher price of a given node, should lead to a reduction in the values of competing nodes.

After defuzzification method Table 1 is converted to the following table:

Table 2. Numerical values of the Concepts Relationships

	C1	C2	C3	C4	C5	C6	C7	C8	C9	C10	C11	C12	C13	C14	C15	C16	C17	C18	C19	C20	C21
C1		-1	-1	-1	-1	-1	-1														
C2	-1		-1	-1	-1	-1	-1														
C3	-1	-1		-1	-1	-1	-1														
C4	-1	-1	-1		-1	-1	-1														
C5	-1	-1	-1	-1		-1	-1														
C6	-1	-1	-1	-1	-1		-1														
C7	-1	-1	-1	-1	-1	-1															
C8	0,5	0,8	0,9		0,1						0,35							0,35	0,35	0,35	
C9		0,2	0,8								0,35							0,35	0,35	0,35	
C10	0,5	0,1	0,8								0,35							0,35	0,35	0,35	
C11	0,9	0,8	0,35	0,9	0,5																
C12						0,8	0,9														
C13		0,5				0,9	0,8														
C14					0,9			0,35													
C15					0,9											0,35					
C16	0,65	0,5	0,5	0,5	0,9	0,2		0,35										0,35	0,35	0,35	
C17	0,65	0,65	0,65	0,8	0,35	0,5	0,8														
C18	0,35	0,8	0,65	0,9	0,5	0,5	0,35										0,35				
C19	0,8	0,9	0,9	0,8	0,5	0,65	0,35										0,35				
C20		0,1	0,5																		

After using the recursive algorithm each node converges to a final value and the decision node with the maximum value is the most possible diagnosis based on the model. In any case, if the model is correct, even with incomplete data, the result will confirm the experts' diagnosis.

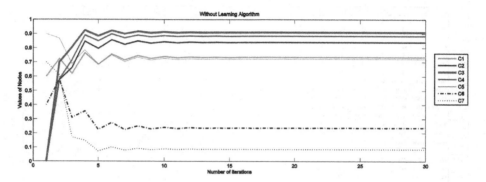

Fig. 4. Subsequent values of concepts till convergence

We observe that in the specific case there is more than one output that has also high convergence values. This means that these types of injuries are very similar because of the adjacent symptoms. So patellofemoral pain syndrome, osteochondritis dissecans of the femoral condyles, and tibial tuberosity apophysitis are the predominant types of injuries that the patient of the example may suffer.

4   Conclusions

There is a need for software tools for Medical Decision Support Systems using theories of FCMs. In addition learning algorithms, such as Hebbian and non-linear, should be considered to further investigate Medical problems. The challenging research is the development of new suitable software tools. An interesting research topic is the development of recursive dynamic state equations for FCMs and the use of them for mathematical modeling of medical problems.

This paper deals with a promising future research area. This approach that exploits the experts' knowledge and experience in a modern medical automation system can be used both for diagnosing diseases and also proposing a treatment. More simulation results and more examples are necessary in order to compute the system's reliability in the diagnosis of a specific injury. The model requires further research. To implement this, the cooperation of both physicians and engineers is necessary. The first ones should determine the right concepts and the rules and the second ones are responsible for the implementation of this problem.

References

1. Groumpos, P.P., Anninou, A.P.: A theoretical mathematical modeling of Parkinson's disease using Fuzzy Cognitive Maps. In: 12th IEEE International Conference on Bioinformatics and Bioengineering, Cyprus, pp. 677–682 (2012)
2. Anninou, A.P., Groumpos, P.P.: Nonlinear Hebbian Learning Techniques and Fuzzy Cognitive Maps in Modeling the Parkinson's Disease. In: 21st Mediterranean Conference on Control and Automation, Chania, pp. 709–715 (2013)
3. Georgopoulos, V.C., Stylios, C.D.: Competitive Fuzzy Cognitive Maps Combined with Case Based Reasoning for Medical Decision Support. In: Magjarevic, R., Nagel, J.H. (eds.) World Congress on Medical Physics and Biomedical Engineering 2006. IFMBE Proceedings, vol. 14, pp. 3673–3676. Springer, Heidelberg (2006)
4. Georgopoulos, V.C., Stylios, C.D.: Augmented fuzzy cognitive maps supplemented with case based reasoning for advanced medical decision support. In: Nikravesh, M., Zadeh, L.A., Kacprzyk, J. (eds.) Soft Computing for Information Processing and Analysis 2005. STUDFUZZ, vol. 164, pp. 391–405. Springer, Heidelberg (2005)
5. Georgopoulos, V.C., Malandraki, G.A., Stylios, C.D.: A fuzzy cognitive map approach to differential diagnosis of specific language impairment. Artif. Intell. Med. 29(3), 261–278 (2003)
6. Stylios, C.D., Groumpos, P.P.: Modeling complex systems using fuzzy cognitive maps. IEEE Transactions on Systems, Man and Cybernetics, Part A: Systems and Humans 34, 155–162 (2004)
7. Majewski, M., Susanne, H., Klaus, S.: Epidemiology of athletic knee injuries: A 10-year study. Knee 13(3), 184–188 (2006)
8. Papageorgiou, E.I., Spyridonos, P.P., Glotsos, D.T., Stylios, C.D., Ravazoula, P., Nikiforidis, G.N., Groumpos, P.P.: Brain tumor characterization using the soft computing technique of fuzzy cognitive maps. Applied Soft Computing 8, 820–828 (2008)
9. Groumpos, P.P., Stylios, C.D.: Modeling supervisory control systems using fuzzy cognitive maps. Chaos Solit. Fract. 11, 329–336 (2000)
10. Saade, J.J.: A unifying approach to defuzzification and comparison of the outputs of fuzzy controller. IEEE Transactions on Fuzzy Systems 4(3), 227–237 (1996)

Automated Scientific Assistant
for Cancer and Chemoprevention

Sotiris Lazarou, Antonis C. Kakas, Christiana Neophytou,
and Andreas Constantinou

Departments of Computer Science and Biological Sciences,
University of Cyprus, P.O. Box 20537 - 1678 Nicosia, Cyprus

Abstract. Logical modeling of cell biological phenomena has the potential to facilitate both the understanding of the mechanisms that underly the phenomena as well as the process of experimentation undertaken by the biologist. Starting from the general hypotheses that Scientific Modeling is inextricably linked to abductive reasoning, we aim to develop a general logical model of cell signalling and to provide an automated scientific assistant for the biologists to help them in thinking about their experimental results and the possible further investigation of the phenomena of interest. We present a first such system, called *ApoCelSys*, that provides an automated analysis of experimental results and support to a biology laboratory that is studying Cancer and Chemoprevention.

Keywords: Abductive Reasoning, Systems Biology, Cancer Modeling.

1 Introduction: Scientific Modeling in Logic

Modeling scientific phenomena requires that we set up a link between our mental understanding of the phenomena and the empirical information that we collect from observing the phenomena. In most cases the scientific effort concentrates on setting up experiments that would provide empirical information suitable for developing further (in the extreme case abandoning) our current model for the phenomena. An abstraction therefore is required both at the object level of understanding the processes that generate the phenomena and at the meta-level of understanding how to set up experiments that would reveal useful information for the further development of our scientific model. Using logic as the basic for scientific modeling facilitates this abstraction at both levels and the connection between them. For Molecular Biology, logic is particularly suited as (at least currently) the theoretical models of cell biology are more at the descriptive level and experiments are developed following a rationale at the qualitative level rather than the quantitative level.

The purpose of the work in this paper is two fold. To develop logical models of biological process through which we can carry out a biologically valid analysis of experimental (in-vivo or in-vitro) results and to provide based on this automated help to the Biologists in thinking about the scientific process of their experiments and their possible further investigation of the phenomena they are studying. The

H. Papadopoulos et al. (Eds.): AIAI 2013, IFIP AICT 412, pp. 96–109, 2013.

aim is to (incrementally) develop an underlying model for signal propagation in a cell that is independent of any particular biological phenomena and process on top of which we can then modularly build a model for specific biological problems. The underlying model needs to capture the propagation of increased as well as that of decreased activation, where the increase and decrease are determined through relative changes with respect to a normal level of activity.

We present a model for cell signaling formulated within the framework of Abductive Logic Programming (ALP), together with a tool to be used as an automated assistant by the biologist. The study is carried out with emphasis on the particular biological process of apoptosis as studied by the group of Cancer Biology and Chemoprevention at the Department of Biological Sciences at the University of Cyprus. Their aim is to investigate the molecular mechanisms through which natural substances or their derivatives participate in the inhibition of carcinogenesis (chemoprevention) or the inhibition of tumor progression.

2   Biological Background

Physiological processes are conducted in cells through sequences of biochemical reactions known as molecular pathways. Each molecular pathway consists of a series of biochemical reactions that are connected by their intermediates: the products of one reaction are the substrates for subsequent reactions, and so on. Apoptosis or programmed cell death (PCD) is a physiological process which is vital for normal development and elimination of damaged cells in living organisms (reviewed in [2]). Apoptosis is a main tumor suppressor mechanism within the body because it gets rid of cells that have extensive DNA damage. Its deregulation can lead to cancer and other diseases.

Chemotherapy primarily refers to the treatment of cancer with an antineoplastic drug. Evidence in the literature suggests that Vitamin E natural and synthetic analogues may target the main survival pathways of the cell (i.e., PI3K and NF-kB), which provides not only an amplification response of the apoptotic pathway, but also kills selectively cancer cells whose survival may depend on activating these pathways [1,6]. The goal of our biological study is to investigate the molecular events by which a Vitamin E synthetic derivative (VitESD) leads to apoptosis in breast cancer. For this purpose, we use two breast cancer cell lines, MCF-7 and MDA-MB-231, which are widely used as tumor models.

In vitro experiments are time-consuming and expensive as we need to screen a large number of possible targets in order to identify proteins that are involved in the signaling cascade affected by the compound under investigation. The use of an automated system will allow for calculated suggestions of molecules likely to be involved in the apoptotic pathway induced by the drug.

3   Scientific Modeling in Abductive Logic Programming

Scientific modeling is inextricably linked to abductive reasoning. Abductive explanations for new observational data, that are based on the current form of our model, generate new information that helps to develop further the model. In effect, abductive reasoning rationalizes the empirical data with respect to the current model with further hypotheses on the incompleteness of the model. Abductive Logic Programming (ALP) [4,3] provides a framework for logical modeling that also directly supports abductive reasoning and hence it is particulary suited for scientific modeling. A model or a theory, T, in ALP is described in terms of a triple (P, A, IC) consisting of a logic program, P, a set of abducible predicates, A, and a set of classical logic formulas IC, called the *integrity constraints* of the theory. The program P contains *definitional knowledge* representing the general laws about our problem domain through a complete definition of a set of *observable predicates* in terms of each other, background predicates (which are again assumed to be completely specified in P) and a set of abducible predicates that are open and which thus carry the incompleteness of the model. The integrity constraints, IC, represent *assertional knowledge* of known properties of the domain that must necessarily be respected and maintained when we complete further our theory. Given such an ALP theory the abductive reasoning that infers an abductive explanation is defined as follows.

Definition 1. *Given an abductive logic theory (P, A, IC), an abductive explanation for an observation O, is a set, Δ, of ground abducible atoms on the predicates A such that:*

- $P \cup \Delta \models_{LP} O$
- $P \cup \Delta \models_{LP} IC$.

where $\models_{LP}$ denotes the logical entailment relation in Logic Programming.

The abductive explanation Δ represents a hypothesis, which together with the model described in P explains how the experimental observations, Q could hold. The role of the integrity constraints IC is to impose additional validity requirements on the hypotheses Δ. They are modularly stated in the theory, separately from the basic model captured in P. As such, they can also be used, as we will see below, to steer the search for specific forms of explanation that our domain experts are looking for.

4   Modeling Apoptosis and Drug Effects

The core of our logical model consists of a fairly general representation of the activity of Signal Propagation in a cell. The task is that given a network of pathways generated through biological experiments, such as the one shown in Figure 1, to logically model the signal propagation over this and the effect of the presence of external drugs on this propagation.

We will consider the propagation of two types of signals: positive signal and negative signal. Normally cell signalling refers to the propagation of a positive signal, namely the iterative increase in activity of molecules along the pathway. In the propagation of a negative signal we refer to the iterative decrease in activity of molecules along the pathway. The increase and decrease of activity are taken relative to some level which is considered to be the normal background level of activity. The propagation of a negative signal is appropriate when we want to follow the inhibition effects of molecules. The core principles of positive and negative signal propagation are respectively the following.

Positive Propagation. The activation of a molecule X will cause the activation of a molecule Y that follows X (i.e. Y is induced by X) in a signalling pathway, provided that Y or this reaction from X to Y is not inhibited by the activation of another molecule.

Negative Propagation. The inactivation of a molecule X will cause the inactivation of a molecule Y that follows X in a signalling pathway, provided there is no other molecule Z that also induces Y and Z is activated.

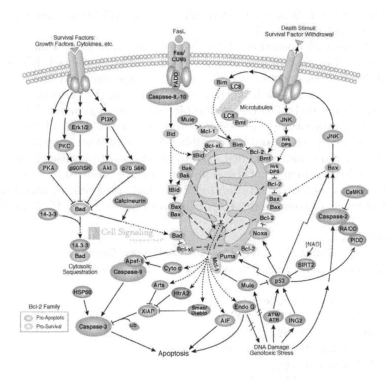

Fig. 1. Apoptosis pathways linked to the Mitochondrial

In these principles we abstract away the details of the actual chemical processes that occur to bring about a single step of propagation and we do not distinguish between the different types of signal propagation that are biologically possible. In general, this high-level abstract view of a single step of propagation is depicted in Figure 2 where the biological pathways are abstracted into a graph with positive or inducement links and negative or inhibitory links. Hence the substance D will be activated when any one of A and B are activated but

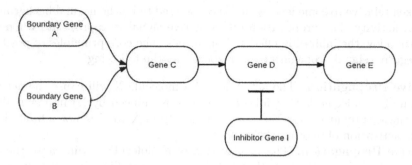

Fig. 2. An abstraction of Cell Signalling

its inhibitory gene, I, is not activated. Similarly, D will be inactive when C is inactive or when I is activated.

Under the normal operation of the cell signals start from some boundary points, such as A or B. When an external drug is added in the cell then we have the possibility that this could activate or inactivate a molecule at any part of the network (or even enable new hitherto unknown links in the network) present in the background network. The task of the biological research in which the logical model and its automation is to assist, is to discover the effect of the drug on these background networks and the possible existence of other pathways relevant to the cell operation that is studied, e.g. apoptosis, as in the case of our study.

The above basic principles of signal propagation are captured by the following recursive rules for $active/2$ and $inactive/2$ in a logic program, P, together with the a set of integrity constraints, IC, of the theory $T =< P, A, IC >$ in ALP.

```
% Program P %
active(Phase, GeneX):- normally_active(Phase,GeneX).
active(Phase, GeneX):- externally_induced(Phase,GeneX).

normally_active(Phase,GeneX) :-
    reaction(rct(GeneY,GeneX)),
    active(Phase,GeneY).
externally_induced( Phase, GeneX ):-
    exists(Phase,Drug),
    possible_drug_effect(Drug,induce,GeneX),
    drug_induced(Phase,Drug,GeneX).

inhibited(Phase,GeneX) :- normally_inhibited(Phase,GeneX).
inhibited(Phase,GeneX) :- externally_inhibited(Phase,GeneX).

normally_inhibited(Phase,GeneX) :-
    inhibitor(GeneX,GeneZ),
    active(Phase,GeneZ).
externally_inhibited(Phase,GeneX) :-
```

```
        exists(Phase,Drug),
        possible_drug_effect(Drug,inhibit,GeneX),
        drug_inhibited(Phase,Drug,GeneX).

inactive(Phase,GeneX) :- inhibited(Phase,GeneX).
inactive(Phase,GeneX) :-
        \+active(Phase,GeneX),
        reaction(rct(GeneY,GeneX)),
        inactive(Phase,GeneY).

% Integrity Constraints IC %
ic :- active(Phase,GeneX),absent(Phase,GeneX).
ic :- inhibited(Phase,GeneX),absent(Phase,GeneX).
ic :- normally_active(Phase,GeneX),inhibited(Phase,GeneX).
ic :- drug_induced(Phase,Drug,Gene),
        drug_inhibited(Phase,Drug,Gene).

% Abducible Predicates A %
abducible(drug_induced(_Phase,_Drug,_GeneX)).
abducible(drug_inhibited(_Phase,_Drug,_GeneX)).
```

The "Phase" variable in the model relations distinguishes experiments where different drugs could be present with different possible effects of these drugs as captured by the lower-level predicate, $possible\_drug\_effect/3$, defined via some biological knowledge that we have a-priori of the drugs or via some hypothetical behavior of the drug to which the biologist wants to restrict their attention of study. The predicate $absent/2$ is also used to describe the variation in the set up of the various experiments where the biologist will arrange for some molecules to be effectively taken out the cell and hence to signal can go through them.

The $reaction/2$ and $inhibitor/2$ relations are given directly from the biological background pathway networks in the literature (see for example figure 1 again) as a database of extensional information of the following form:

```
reaction(rct(ing2,p53)).                inhibitor(bax,bcl2).
reaction(rct(p53,caspase2_raiid_pidd)). inhibitor(bcl2,noxa).
reaction(rct(p53,bax)).                 inhibitor(bcl2,mcl1_puma).
reaction(rct(p53,noxa)).                inhibitor(mcl1_puma,mule).
 ...                                     ...
```

The *observable predicates* in the model are the two predicates of $active/2$ and $inactive/2$ through which the observations in any single experiment are turned into a conjunctive query for an abductive explanation through the model of T. Observations are of two types: activation or inactivation of some molecule from which it is known (or assumed in the model) that some overall operation of the cell, such as apoptosis will ensue or not and the activation or inactivation of intermediate molecules (markers) that the experimenter has measured.

In the above model the only *abducible predicates* refer to assumptions that we can make on the effect of the Drugs on the molecules. Other abducible predicates can be added if we want to extend the model to cover the possibility of some other unknown external effect on the molecules or indeed the possibility of the existence of other unknown background reactions.

5   ApoCelSys: A Scientific Advisor System

ApoCelSys is a system that uses the model described in the previous sections to simulate the essential function of apoptosis, otherwise known as programmed cell death. The main purpose of the system is the creation of explanatory assumptions for validating the phenotype that was observed by the user and presenting a set of hypotheses on the effect of the synthetic substance introduced in the cell.

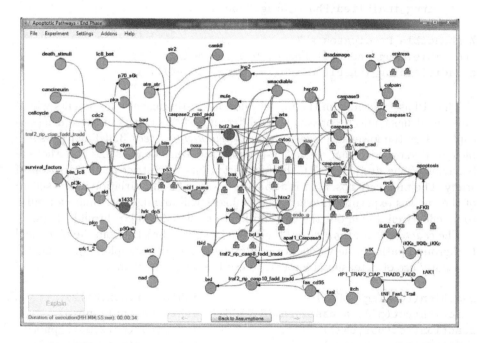

Fig. 3. System Interface of *ApoCelSys* and Propagation of an Explanation

A user interface, shown in figure 3, provides a graphical representation of the molecular pathways leading to apoptosis. This helps to simplify the insertion of the experimental data and the observations to be explained, whether these come from real in-vitro or hypothetical experiments . The user can set the status of any gene to any of the five predefined statuses: "active" either "fixed" by the experimenter or "observed"; "inactive" similarly either "fixed/absent" or "observed" or "neutral" otherwise. Also the possible effect of the drug on the gene can be regulated by the user if and when such information is available. Otherwise, the system assumes that the drug can possibly affect the gene in either way of inducing or inhibiting the gene.

The user can save this experimental data so that s/he can collect and easily revisit a set of experiments that have been analyzed by the system. After passing the experiment's information as input and executing the query, the system displays the first explanation found using distinguishing colors on the corresponding elements (red for the assumption of drug inhibited and green for drug induced).

A tool for enumerating the explanations is also available through which the user can view, filter and present the most interesting, in her opinion, explanation for a more detailed examination. During this single explanation display, the system also allows the propagation of the substances' primary effect on the rest of the cell for an overview of the pathway activation and in particular the way that apoptosis is activated or not. An example of this can be seen in figure 3 where the explanation that the drug induces $traf2\_rip\_ciap\_fadd\_tradd$, aif, $endo\_g$ and inhibits $XIAP$ propagates to activate $jnk, cjun, \ldots$ and this eventually propagates to activate apoptosis. It also shows how this explanation has an inactivation effect on $s1433$ and $bcl2\_bmt$.

Frequency Charts for Multiple Explanations. Due to the multiplicity of the explanations that can be generated, automated analysis tools provide effective views on the likelihood of each assumption on the entire spectrum of explanations. The use of this tool creates graphs that exhibit the probability of the assumption of inducing or inhibiting elements throughout all produced explanations. An example is shown in figure 4 where we can see that aif and $endo\_g$ have the most hits in being induced by the synthetic substance and that as a second choice the most probable effects are the inducements of $bak, xiap, tbid$ and bid. In addition, if requested, the system is in a position to compute, analyze and present charts displaying the likelihood of any gene being activated irrespective of being a direct drug effect or of being a consequence of the latter. This information can be interpreted as displaying all possible pathways along with their appearance frequency, thus enabling the user to better judge the most probable pathways to apoptosis. These visual tools allow the easy comparison by the user between multiple variations of an experiment or even between cross-experiments.

6   Validation and Evaluation of the System

In order to validate the system we have analyzed the experimental results of real in-vitro experiments as described in section 2 for investigating the action of $VitESD$. One such experiment was carried out with the MCF-7 cancel cell-line and another one with the MDA-MB-231 cell-line. The purpose was to judge the appropriateness of the hypotheses produced but also to ascertain the help provided to the biologists in: (1) analyzing the results of their experiments and (2) allowing them to develop further their investigation by carrying out thought experiments that they deemed useful.

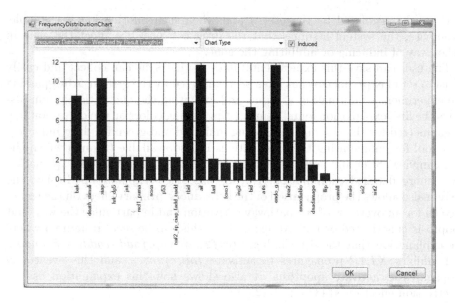

Fig. 4. Frequency Distribution across Explanations

These two experiments were analyzed with the *ApoCelSys* system and then the results were given back to the biologist to evaluate them. For the MCF-7 experiment the results suggest that Bak and tbid are greatly induced and that Smac/Diablo is also moderately activated by the presence of *VitESD*. The ApoCelSys system therefore suggests that most likely *VitESD* helps to activate apoptosis through the intrinsic pathway of apoptosis. The frequency distribution chart, generated by the ApoCelSys system, showing molecules that become inactive in the presence of the drug, suggests that PI3K, the upstream regulator of AKT, is in fact inhibited during apoptosis in our system and that Calcium levels (Ca2)-which control Calpain and subsequently Caspase activation-are also diminished. These findings are in line with the central role that AKT may play and further attenuate the caspase-independent mode of action of the *VitESD*.

Similarly, interesting results were produced for the second experiment with the MDA-MB-231 cells. Furthermore, in both cell lines, Survival factors such as growth proteins are suggested to become inactivated in the presence of the drug. This highlights another unexplored possibility that could explain the induction of apoptosis in our system: the inhibition of Growth Receptors on the cell membrane. The implication of these receptors coincides with the inactivation of PI3K and AKT which as mentioned before seem to play a central role in our system.

6.1   Expanding the Biological Investigation

The development of scientific theories often requires broadening the domain of investigation in several ways. For example, it may be relevant to combine the study of the cell process we are interested in with that of other cell processes that we know or suspect that are related to our primary interest. During the process of developing the *ApoCelSys* system the biologists expanded their investigation to the proliferation process of the cancer cells in addition to that of apoptosis. This required that the underlying pathways network to contain also the main established pathways for cell proliferation, some of which have cross over points with the apoptosis networks. The modular and logical nature of our model allowed this to be carried out in a straightforward and immediate way.

In-silico Thought Experiments. Scientists often use thought experiments for guiding them in their future investigations. These can help in understanding the significance of new ideas and how they can help to expand the horizon of investigation. The *ApoCelSys* system facilitates the examination of experiments that the Biologist may want to hypothetically ask. For example, what would the explanatory effect of the drug be if we also observed some other molecules in the cell? How would the experimental results be explained if we added new information in the model such as new information on the suspected action of the drug in the form of additional integrity constraints, such as that the drug cannot induce together a set of molecules or that it can only induce a molecule if some other molecules are not active.

Also the biologist can examine how would the results be affected if we assume that there are some extra reactions (links) in the background network of pathways? For example, the biologists may suspect that there might exist a new inducement link between two molecules. Adding this extra link and reanalyzing the same observed data can give valuable glues as to whether this is plausible and how it would be experimentally confirmed. This process of hypothesizing in a controlled manner new structures on the underlying network and analyzing their significance can be semi-automated. We are developing tools that would assist the biologist in such thought experiments.

Overall Evaluation of *ApoCelSys* as an Assistant. From a working biologist point of view the overall evaluation of our system is summarized in the following. The ability to input our actual results in an automated assistant that contains all the information needed to describe the cellular microenvironment provides a much simpler alternative than in vitro screening of a large number of proteins. Also the capacity to adjust this microenvironment in order to depict the variability that exists between cell lines greatly enhances the usefulness of this system. The major benefit of this assistant is the frequency distribution charts which provide an indication for which molecules to be investigated next, thus saving time and cutting laboratory costs. Furthermore, the *ApoCelSys* system may serve as a tool for the discovery of unknown reactions between molecules that can reveal novel pathways to be targeted in cancer therapy.

6.2   Evaluation of Computational Behaviour

We have carried out a series of experiments to investigate the computational properties of the current (and first) implementation of the system. This implementation relies on a new implementation of the A-system [5] for computing (minimal) explanations but does not in its current form exploit the constraint solving capabilities of the A-system. The response time of the system is dependent on the users' input and may vary from a few seconds to a couple of hours for a complete search of all minimal explanations.

We have thus tried to investigate to what extend it is necessary to compute all explanations and how we can find a representative sample of these. Under the assumption that the effect of the drug to produce the results on the molecules that we have observed in the experiment cannot be "too far" along the pathways from these observed molecules, we have restricted the depth of search by the A-system and have tried to learn what is a reasonable threshold on the search depth bound. Using the real experiments described above as a basis and running the system with various depth limits we obtain the results shown in figure 5. According to these we can see that no major fluctuation in the nine most probable hypotheses of induced molecules except for a single molecule, *bid*, on the lowest depth limit of 200. The same can also be observed in figure 6 in which all molecules found in figure 5 have been sorted by their inducement probability and labeled with non-zero positive numbers, where one stands for the most probable molecule to be induced by the substance. In this figure, we can observe that most of the molecules retain their positions, which shows that even in low depth searches, the results provided are of satisfactory quality.

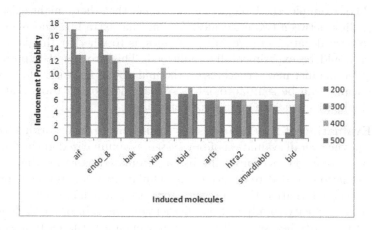

Fig. 5. Search Depth Variation of Explanations

However restricting the search could result in loosing valuable information when our assumption of no "long-distance" effect is not appropriate. The large computational time of an unrestricted search is due to the complex pathways and the fact that the top down search of the A-system contains parts that are unnecessarily duplicated. We plan to develop other implementations that are based on bottom up computational models using tabling or ASP solvers.

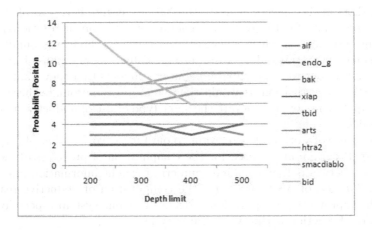

Fig. 6. Probability of drug inducement of molecules

7 Related Work

There are several earlier works of Symbolic Systems Biology which also use a form of logical modeling within Logic Programming (Abductive or Inductive) and rely on abductive and/or inductive reasoning for the analysis of experimental results. Some of the earliest such approaches include [15] for the analysis of genetic networks based on abduction, [7] for generating gene interaction networks from micro-array experiments within ALP and [8,10] for discovering the function of genes from auxotrophic metabolic experiments or for studying lethal synthetic gene mutations using the XHAIL framework and system [9] that supports abductive and inductive reasoning within the representation framework of ALP. In [14] a logical model for capturing the possible toxic effects of drugs in a cell has been developed within ALP and analyzed using the Progol5 system.

Our work can be seen to follow this line of approach, with an emphasis on providing tools with interactive interfaces to help the biologist as an automated scientific assistant. As the underlying network information increases, it is evident that we need to develop a better graphical interface of our system in the form of a map-like interface where the user can focus on the details of the pathways network that s/he is currently investigating. Similarly, new interactive interfaces are needed to facilitate the thought experimentation process by the user.

More recently, [13] have developed an important methodology for qualitative logical modeling in systems biology that rests on underlying quantitative models for the chemical reactions in the signalling pathways. Our approach lacks this link to quantitative computational models and it would benefit from their integration in the logical model as black computational boxes. This work uses a probabilistic method to rank the multiple hypotheses found much in the same spirit as we do with the frequency charts of individual hypotheses over the set of explanations. They then go a step further to rank the full explanations. This is an extension that would be useful to consider in our approach.

The authors in [12] show how using the XHAIL system it is possible to allow abductive reasoning for the revision (deletion or addition of reactions) of the underlying pathways on which a logical model for metabolism is build. In our approach and system this is done semi-automatically by suggestions of the expert biologists in the context of possible thought experiments that they want to carry out for further investigation. It would then be useful to automate these type of thought experiments by incorporate the same type of use of abductive reasoning in our model.

Another piece of work that is relevant to our future work is that of [11] where Answer Set Programming is used for the computation of processes of biochemical reaction networks that combines numerical weight information and logical inference. The use of ASP solvers for the computation of abductive reasoning could help improve the computational behavior of our system especially when the network of reaction pathways contains cycles.

8   Conclusions and Future Work

We have presented a methodology for building automated scientific assistants to help in the research effort of molecular biologists. This is based on logical modelling of the existing background knowledge of the phenomena studied and the automation of abductive reasoning to provide explanations for the observed empirical results of the experiments carried out by the biologists. The resulting tools could also help in assisting with the further development of the research program of the biologists by facilitating the scientific thinking carried out when decisions for the future direction of research are taken. This methodology and resulting automated assistant, called *ApoCelSys*, has been tested in a real life case of a research group that studies the chemoprevention of cancer.

The testing of the *ApoCelSys* system carried out so far provides encouraging results but further systematic experimentation with the system is needed. In particular, it is important to carry out a series of tests to understand better the capabilities of the system for facilitating thought experiments and what extra features and functionality would make it more suitable for this purpose.

References

1. Papas, A., Constantinou, C., Constantinou, A.: Vitamin e and cancer: An insight into the anticancer activities of vitamin e isomers and analogs. Int. J. Cancer 123(4), 739–752 (2008)
2. Elmore, S.: Apoptosis: a review of programmed cell death. Toxicol Pathol. 35(4), 495–516 (2007)
3. Denecker, M., Kakas, A.C.: Abduction in logic programming. In: Kakas, A.C., Sadri, F. (eds.) Computational Logic: Logic Programming and Beyond. LNCS (LNAI), vol. 2407, pp. 402–436. Springer, Heidelberg (2002)
4. Kakas, A.C., Kowalski, R.A., Toni, F.: Abductive Logic Programming. Journal of Logic and Computation 2(6), 719–770 (1993)

5. Ma, J., et al.: A-system v2 (2011),
   http://www-dse.doc.ic.ac.uk/cgi-bin/moin.cgi/abduction
6. Neuzil, J., et al.: Vitamin e analogues as a novel group of mitocans: anti-cancer
   agents that act by targeting mitochondria. Mol. Aspects Med. 28(5-6), 607–645
   (2007)
7. Papatheodorou, I., Kakas, A.C., Sergot, M.J.: Inference of gene relations from
   microarray data by abduction. In: Baral, C., Greco, G., Leone, N., Terracina, G.
   (eds.) LPNMR 2005. LNCS (LNAI), vol. 3662, pp. 389–393. Springer, Heidelberg
   (2005)
8. Ray, O.: Automated abduction in scientific discovery. In: Model-Based Reasoning
   in Science, Technology, and Medicine, pp. 103–116 (2007)
9. Ray, O.: Nonmonotonic abductive inductive learning. J. Applied Logic 7(3),
   329–340 (2009)
10. Ray, O., Bryant, C.H.: Inferring the function of
    genes from synthetic lethal mutations. In: CISIS,
    pp. 667–671 (2008)
11. Ray, O., Soh, T., Inoue, K.: Analyzing pathways using ASP-based approaches. In:
    Horimoto, K., Nakatsui, M., Popov, N. (eds.) ANB 2010. LNCS, vol. 6479, pp.
    167–183. Springer, Heidelberg (2012)
12. Ray, O., Whelan, K.E., King, R.D.: Logic-based steady-state analysis and revision
    of metabolic networks with inhibition. In: CISIS, pp. 661–666 (2010)
13. Synnaeve, G., Inoue, K., Doncescu, A., Nabeshima, H., Kameya, Y., Ishihata,
    M., Sato, T.: Kinetic models and qualitative abstraction for relational learning in
    systems biology. Bioinformatics, 47–54 (2011)
14. Tamaddoni-Nezhad, A., Chaleil, R., Kakas, A.C., Muggleton, S.: Application of ab-
    ductive ilp to learning metabolic network inhibition from temporal data. Machine
    Learning 64(1-3), 209–230 (2006)
15. Zupan, B., Bratko, I., Demšar, J., RobertBeck, J., Kuspa, A., Shaulsky, G.: Ab-
    ductive inference of genetic networks. In: Quaglini, S., Barahona, P., Andreassen,
    S. (eds.) AIME 2001. LNCS (LNAI), vol. 2101, pp. 304–313. Springer, Heidelberg
    (2001)

Task Allocation Strategy Based on Variances in Bids for Large-Scale Multi-Agent Systems

Toshiharu Sugawara

Department of Computer Science and Engineering
Waseda University, Tokyo 1698555, Japan

Abstract. We propose a decentralized task allocation strategy by estimating the states of task loads in market-like negotiations based on an announcement-bid-award mechanism, such as contract net protocol (CNP), for an environment of large-scale multi-agent systems (LSMAS). CNP and their extensions are widely used in actual systems, but their characteristics in busy LSMAS are not well understood and thus we cannot use them lightly in larger application systems. We propose an award strategy in this paper that allows multiple bids by contractors but reduces the chances of simultaneous multiple awards to low-performance agents because this significantly degrades performance. We experimentally found that it could considerably improve overall efficiency.

1 Introduction

Recent technology in the Internet and ambient intelligence has enabled services that are incorporated into daily activities. These services consist of a number of tasks each of which has required capabilities and data, and should be done in agents meeting these requirements. Of course, vast numbers of services are required everywhere in the environment; thus, they are provided in a timely manner by a large number of cooperative agents on computers and sensors through mutual negotiations to appropriately allocate associated tasks among agents. It is, however, unrealistic for agents to know all the capabilities and data other agents have in open environments. Well-known and simple methods of allocating tasks without this information are market-like negotiation protocols, like the contact net protocol (CNP) [7].

However, its characteristics in a large-scale multi-agent system (LSMAS) is poorly understood. Understanding the efficiency of CNP in LSMAS is not easy because if many manager agents simultaneously try to allocate tasks, interference among agents occurs. For example, multiple simultaneous bids are not allowed in a naive CNP. However, this restriction may considerably degrade the efficiency in a busy LSMAS, because bid/allocation processes occur almost sequentially. We can allow multiple bids to enhance concurrency and actually in [7] it is also discussed the possibility of multiple bids. However, allowing multiple bids will result in simultaneous multiple awards begin given to contractors. We can consider two solutions to this situation. First, extra awarded tasks are canceled with certain penalties (e.g. [5]). However, we believe that this makes the task allocation protocol so complicated that it may degrade the overall efficiency of a busy LSMAS. The second approach is to execute all awarded tasks in the

H. Papadopoulos et al. (Eds.): AIAI 2013, IFIP AICT 412, pp. 110–120, 2013.

contracted agents. This is simpler, but awardees cannot complete the tasks within the contracted times, resulting in significant delays in executing tasks in busy environments.

There have been a number of studies on applying it to LSMASs by assuming multiple initiations and bids in CNP [1,6,8]. For example, in [6] the authors addressed the issue of the eager-bidder problem occurring in a LSMAS, where a number of tasks were announced concurrently so that a CNP with certain levels of commitment did not work well. They also proposed a CNP extension based on statistical risk management. However, the types of resources and tasks considered in these papers were quite different, i.e., they focused on exclusive resources such as airplane seats. We focus on divisible resources such as CPUs or network bandwidth, so any number of tasks can be accepted simultaneously but with reduced quality. In [2] it was also discussed the extension of CNP to allow multiple initiations and bids based on colored Petri nets. There have been other extensions that have introduced brokers [4] and personal agents [3]. However, these studies have focused on reducing the communication overhead in LSMAS. In [8] it was proposed an award strategy in which two different strategies were alternatively selected by statistically analysing bid values to improve overall efficiency. However, as their method could only be applied to a set of tasks whose costs were in a certain limited range, its applicability was severely restricted.

We adopt a simpler approach for a busy LSMAS in which all awarded tasks are executed in the awardees. Instead, we reduce the probability of multiple awards given to lower-performance agents, by extending the method in [8], because multiple awards to these types of agents are the major reason for degraded performance. The negative effect of multiple awards on a system's efficiency is strongly affected by the busyness or task loads of LSMAS. If the environment has a low load, hardly any multiple awards occurs; thus, managers can award tasks to the best bidders without having to make careful selections. However, the chance of multiple awards also increases according to the increase in task loads and thus the simple strategy of making awards in which a manager selects the best bidder becomes inappropriate. Lower-performing agents may be identified as the best bidders by multiple managers, particularly in busier situations, and thus be awarded multiple tasks simultaneously; this will considerably reduce performance. We tried to reduce the probability of multiple awards in our approach when the chance of bids going to lower performing agents had increased. Managers have to select the awardees by carefully understanding the states of their local environment to achieve this.

This paper is organized as follows. Section 2 describes the model and raises issues addressed in this paper. We then compare a naive award strategy with a probabilistic award in which non-best agents are selected to allocate tasks. We then propose a method with which managers decide the award strategies based on the statistical analysis only involving light computation. The experimental results revealed that the proposed method could perform well in a busy environment.

2 Problem Description

2.1 Overview and Issues

We define a simple extension of CNP in which simultaneous multiple initiations by different managers and multiple bids to different managers are allowed. First, let $\mathcal{A} = \{1, \ldots, n\}$ be a set of agents and t be a task. $\mathcal{A}$ is the disjoint union of $\mathcal{M} = \{m_1, \ldots, m_{N_m}\}$, which is the set of managers who allocate tasks, and $\mathcal{C} = \{c_1, \ldots, c_{N_c}\}$, which is the set of contractors who execute the allocated tasks. Contracts have different capabilities which affect the time required to complete the assigned tasks. Let us assume that $|\mathcal{A}|$ is large (on the order of thousands) and the agents are widely distributed, like servers on the Internet. This also means that communications between agents incur some delay. For manager $m \in \mathcal{M}$, we define the concept of the *scope*, S_m, which is the set of agents that m knows, and m can announce tasks to all or a part of S_m.

We have to consider a number of issues to avoid inefficiency due to multiple awards. First, task load affects the possibility of multiple awards. When an agent that has outstanding bids receives another task announcement, it will send its bid (multiple bids) and thus may receive multiple award messages. These multiple bids occur at quite a low rate if they are not busy. Even if they occur, multiple tasks are likely to be allocated to relatively efficient agents since efficient contractors are identified as the best bidders by managers, so no significant degradation to performance will occur. However, as the environment becomes busier, the chance of multiple awards being made to low-performing agents increases and overall performance decreases. Therefore, we propose an award strategy that selects awardees to reduce the probability of multiple awards being made to lower-performing agents accomplished by identifying their task load.

Second, longer communication delay (or latency) also increases the chances of multiple awards. However, we will not discuss reduced latency, because this strongly depends on the environmental settings and agents cannot directly handle it. Instead, we focus on the managers' decisions about awardees by identifying situations in which multiple awards to low-performing agents are likely to occur, regardless of high or low latency environments.

Agents have a number of decision-making strategies that may affect overall performance besides the award strategy. First, manager $m \in \mathcal{M}$ having a task announces a call-for-bid message to contractors that are selected from its scope, S_m, on the basis of an *announcement strategy*. Then, agents decide whether they should act on the received call-for-bid messages or not using a *bid strategy*. Here, we assume simple announcement and bid strategies, i.e., an announcement strategy is where m selects all or N agents from S_m randomly, where N is a positive integer, and the bid strategy is where an agent always bids for call-for-bid messages if it can execute associated tasks. These strategies also play important roles in improving overall efficiency, such as restricting the numbers of multiple bids and announcement messages. However, we particularly focused on award strategies since we were concerned about the negotiations for task allocations to prevent multiple awards being made by appropriately selecting the awardees according to the received bid messages.

Because naive CNP involves a timeout basis to receive response messages from others, it is inefficient in large and busy environments. Thus, we made use of *regret messages*, which were sent in the award phase to contractors who were not awarded the contract, to prevent long waits for bids and award messages (as in [9]).

2.2 Model of Agents and Tasks

For task t, let $r_t = \{r_t^1, \ldots, r_t^d\}$ be the set of required resources (or functions) to execute this, where $r_t^k \geq 0$ is integer. Agent i is denoted by tuple, $(\alpha_i, loc_i, S_i, Q_i)$, where $\alpha_i = (a_i^1, \ldots, a_i^d)$ is the agent's capabilities, and a_i^h corresponds to the h-th resource and $a_i^h \geq 0$; $a_i^h = 0$ indicates agent i does not have the h-th resource. Element, loc_i is the location of i, and Q_i is a finite queue where the tasks allocated to i are temporarily stored. The set, $S_i (\subseteq \mathcal{A})$, is i's scope.

We assume a discrete time for any time descriptions. A unit of time is called a *tick*. The execution time of t by i is:

$$\gamma_i(t) = \max_{1 \leq h \leq d} \lceil r_t^h / a_i^h \rceil, \tag{1}$$

where $\lceil x \rceil$ denotes the ceiling function. The *metric* between the agents, $\delta(i, j)$, is based on their locations, loc_i and loc_j. It is only used to calculate communication delay; the time required for message transfer between agents i and j is $\lceil \delta(i, j)/D \rceil$, where D is the *speed factor* for passing messages. Parameter L (≥ 0) is called the *task load*, meaning that L tasks on average are generated according to a Poisson distribution every tick. These are then randomly assigned to different managers.

2.3 Task Assignment

When manager m receives a task, $\tilde{t}$, it immediately initiates the CNP modified for LS-MAS: It first sends announcement messages to the contractors selected from its scope in accordance with the announcement strategy. Each of these contractors sends back a bid message with a certain *bid value*. The bid values might include parameters such as the price for executing the task, the quality of the result, or a combination of these values. We assume that their bid values contain the estimated times for completing $\tilde{t}$, because we are concerned with the efficiency of processing in LSMAS. This time is calculated as follows in contractor c:

$$\gamma_c(\tilde{t}) + \sum_{t \in Q_c} \gamma_c(t) + \beta, \tag{2}$$

where β is the execution time required for the task currently being executed. For multiple bids, c might have a number of outstanding bids. These bids are not considered to calculate the estimated required time because it is uncertain whether they will be accepted[1]. Then, m selects an awardee, on the basis of the *award strategy* and sends it an award message with the announced task, $\tilde{t}$.

When contractor c is awarded a task, it immediately executes it if it has no other tasks. If c is already executing a task, the new task is stored in Q_c, and the tasks in Q_c are executed in turn.

---

[1] Actually, we also examined a case in which agents considered the expected values of execution times for outstanding bids; however no significant differences were observed.

2.4 Performance Measures

We assumed that manager agents could observe the *completion time* for task t, which is the elapsed time from the time the award message is sent to the time the message indicating that t has been completed is received. The completion time thus includes the communication time in both directions, the queue time, and the execution time. We evaluated overall performance using the average completion time observed in all managers; this is denoted by $\wp$. A shorter average completion time is preferable.

Since the queue length of agent i is bounded, some allocated subtasks might be *dropped*. The numbers of dropped tasks are also another measure of performance. However, no dropped tasks occur unless the systems are overloaded so they can only be used in limited situations. Thus, we use them to identify upper limit of the performance for the entire system.

3 Awards to Non-best Bidders and Its Features

It is plausible for a manager to select the best bidder in the award phase. This strategy is called the *best awardee strategy* (BAS). However, multiple awards are likely to occur with the increase in task loads. A simple award strategy for alleviating multiple awards is to allocate some tasks to non-best contractors by introducing randomness to some degree in the award phase [8].

Let $\{c_1, \ldots, c_p\}$ be contractors that bid on the announced task. The estimated completion time in the bid message from c_i is denoted by b_{c_i}. The *probabilistic awardee selection strategy with the fluctuation factor,* f (which is denoted by PAS_f) is the award strategy in which the manager selects the awardee according to the following probability:

$$\Pr(c_i) = \frac{1/(b_{c_i})^f}{\sum_{j=1}^{p} 1/(b_{c_j})^f}, \tag{3}$$

where non-negative integer f is the *fluctuation factor*. A smaller value for f means less randomness when awardees are selected.

We experimentally investigated the characteristics of BAS and PAS_f with various task loads. We set $|\mathcal{C}| = 500$ and $|\mathcal{M}| = 10,000$. Only one type of task, t_{3000}, whose required resource $r_{t_{3000}}$ was $\{3000\}$ (so $d = 1$) was used in this experiment (Exp. 1). For any $c_i \in \mathcal{C}$, a different capability was assigned to c_i so that the values of $3000/a_{c_i}^1$ would be *uniformly distributed* over the range of 25–120; thus, the values of $a_{c_i}^1$ ranged from $25 - 120$. We assumed that the manager agents could not do the tasks themselves forcing them to assign the tasks to agents who could. The agents were randomly placed on a 150×150 grid with a torus topology to which a Euclidean distance was introduced as its metric. Manager m's scope, S_m, was the set of contractor agents whose distance from m was less than 10.0 because it was implausible to allocate tasks to distant agents. We also set $f = 3$ and $D = 2.5$. Thus, tasks maximally took five ticks to send messages to known contractors. We defined these values by assuming that a tick was around 1 ms. The number of announcement messages, N, and the queue length, $|Q_i|$, were set to 20.

Figure 1 (a) plots the average completion times varying according to the values for task load, L. It indicates that average performance with BAS, $\wp(\text{BAS})$, is higher than

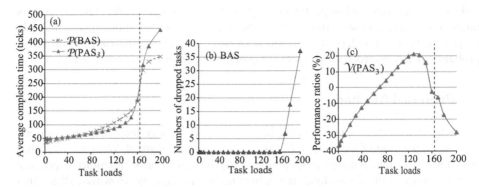

Fig. 1. Results of preliminary experiment (Exp. 1)

that with PAS_f, $\wp(PAS_f)$, when task load was low or quite high but PAS_f outperformed BAS, otherwise. Note that the vertical dotted line when $L = 165$ indicates the limit for the capability of the entire system. Actually, Fig. 1 (b), showing the numbers of dropped tasks per tick with BAS indicates that dropped tasks rapidly increased if $L \geq 165$.[2]

To clearly see the differences between their performance, we define the *performance ratio w.r.t. BAS* as:

$$V(str) = \frac{\wp(BAS) - \wp(str)}{\wp(BAS)} \times 100, \tag{4}$$

where $\wp(str)$ indicates the average completion times with the award strategy specified by the variable, str. Fig. 1 (c), expressing the relationship between $V(PAS_f)$ and the task load, indicates that when L was low ($L \leq 70$), $V(PAS_f)$ was minus and performance with PAS_f was maximally 40% lower than that with BAS. This tendency was also observed when the system was beyond the upper limit of all agents' capabilities ($L \geq$ 160). However, when it was moderately busy and near (but less than) the upper limit for the entire performance ($80 \leq L \leq 160$), performance with PAS_f was maximally 20% higher than that with BAS. We believe that this is an important characteristic: when L is low, any simple strategy can result in acceptable performance and when it is beyond the upper limit for the system, no strategy can do all tasks within acceptable time limits. The system must exert its full capabilities, especially near its upper limit.

The variations in performance in Fig. 1 (c) were caused by the combination of the probabilities of multiple awards and awards to non-best bidders with PAS_f. When L was low, multiple awards rarely occurred. However, as non-best bidders were awarded with PAS_f, $V(PAS_f)$ became minus. All contractors' queues, on the other hand, were almost full if L was beyond the system's limit. Thus, it was better to assign tasks to the best bidders, because (1) lower-performing agents were also so busy that there were no benefits by allowing some tasks to be shifted to these agents, and (2) there were so many tasks that the awards to non-best bidders resulted in other multiple bids in low-performing agents. In contrast, when L was moderate and less than the upper limit, busy and unbusy agents co-existed, i.e., relatively high performing agents had a number of

---

[2] We plotted the numbers of dropped tasks only with BAS in Fig. 1 (b), but they were almost identical to those with PAS_f.

tasks in their queues but low performing agents had few assigned tasks. Thus, managers might assign tasks to low-performing agents, resulting in making simultaneous multiple awards to them. This degraded overall performance. However, managers with PAS$_f$ probabilistically awarded these agents to avoid multiple awards, to some degree.

4   Proposed Method

As PAS$_f$ could attain approximately 20% better performance than BAS in situations near the upper limit of the system, this improvement was not small. We want to emphasize that the system must exert its potential capabilities in busy situations, especially near its upper limit of task load, but achieving this is not easy. However, PAS$_f$ also drastically worsens performance in other situations. Thus, if managers could select an appropriate award strategy by estimating the current degrees of task loads in their local regions, they could take advantage of PAS$_f$.

The variance and standard deviation (SD) of bid values from local contractors can provide the information to estimate task loads in local regions. We introduced *phantom tasks* to achieve this, which are announced by managers even though no contractors are awarded to them. Phantom tasks were proposed in [8], but their use was quite restricted. We used it more actively in our proposed strategy; when a manager announces a task, t, it also announces two phantom tasks. The first one requires much smaller resources, say 1/6, than those of t and this is denoted by p_t^s. The second requires identical to or slightly more resources than those of t and this is denoted by p_t^l. Then, the manager calculates the SDs of bid values for p_t^s and p_t^l received from local agents. These SDs for p_t^s and p_t^l correspond to σ_s and σ_l.

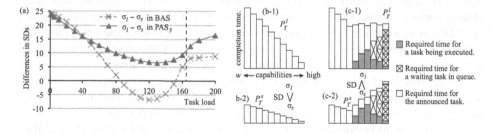

Fig. 2. Bid values in unbusy and moderately busy states

We then investigated changes in the average values of $\sigma_l - \sigma_s$ with variable task loads in an environment identical to that in Exp. 1. Since the task was t_{3000}, the resources required for $p_{t_{3000}}^l$ and $p_{t_{3000}}^s$ were set to $\{3001\}$ for the former and $\{500\}$ for the latter. The results are plotted in Fig. 2 (a).

This graph shows that $\sigma_l > \sigma_s$ in BAS when L is low or beyond the limit, but $\sigma_s > \sigma_l$, otherwise. We can explain this phenomenon as follows. When L is low, the distribution of bid values for p_t^l and p_t^s (see Figs. 2 (b-1) and (b-2), which illustrate the distributions of bid values) directly reflects the capabilities of contractors. Thus, σ_l, the SD of bid values for p_t^l, is larger than σ_s, which is the SD of bid values for p_t^s.

However, in moderately busy situations (L is between 80 and 160) there are chances for lower-performing agents to be awarded, because only higher performing agents have a number of tasks being queued and executed (Fig. 2 (c-1)). Consequently, $\sigma_s > \sigma_l$ held in this situation (cf. Figs. 2 (c-1) and (c-2)). When the system was over the upper limit, all contractors already had a number of tasks waiting in their queues, and their estimated completion times of the tasks at the end of the queues were quite similar. Thus, the estimated completion times for the new announced tasks again reflected the performance of contractors. In contrast, managers with PAS_f allocated tasks with some fluctuations; thus, $\sigma_l > \sigma_s$ was always positive as seen in Fig. 2(a).

From this discussion, we propose a novel award strategy in which BAS or PAS_f is selected using the phantom tasks associated with the requesting task. For task t, we generate phantom tasks, p_t^l and p_t^s, whose required resources are $r_{p_t^l} = \{r_t^1+\alpha, \ldots, r_t^d+\alpha\}$ and $r_{p_t^s} = \{r_t^1/\beta, \ldots, r_t^d/\beta\}$, where α is a small integer equal to or near 0 (in the experiments that follow, we set $(\alpha, \beta) = (1, 6)$). Then, managers announce p_t^l and p_t^s as well as t. Because managers adopt an announcement strategy in which the announcement messages are sent to N agents randomly selected for individual scopes, these announcement messages are not sent to the same set of agents. However, we assumed that the task load states could be estimated by a survey of random sampling.[3] Then, contractors send back the bid messages for the received tasks. Because they cannot distinguish normal from phantom tasks, they adopt the same bid strategy in their decisions. Managers then calculate the SDs, σ_l and σ_s, and select BAS if $\sigma_l - \sigma_s > 0$ and PAS_f if $\sigma_l - \sigma_s \leq 0$, as the award strategy that is used to determine the awardee for t. The proposed method of selecting the award strategy is called *award selection according to bid variances* (ASBV).

5   Experimental Evaluation and Remarks

We experimentally evaluated ASBV. The experimental setting for this experiment (Exp. 2) was identical to that of Exp. 1. The performance ratios w.r.t. the BAS of ASBV, $\wp(ASBV)$, are plotted in Fig. 3 (a).

This graph indicates that ASBV outperformed the selected strategy ratios with BAS when $L \geq 70$ and those with PAS_f near the upper limit for the system (around $L = 150$ to 165). As was previously mentioned, this feature of ASBV is important for all agents in LSMAS to exert all their capabilities.

We also examined the *ratios of selected strategies* (RSS), i.e., the number of managers that selected BAS to that of managers that selected PAS_f. The results are plotted in Fig. 3 (b), which indicates that when task load was low, managers only selected BAS. Then, they increased the chance of selecting PAS_f according to the increase in the task load. However, the selections of BAS and PAS_f in managers balanced at certain ratios to keep the values of $\sigma_l - \sigma_s$ positive. We think that this balanced selection by local autonomous decisions by managers could improve overall performance near the upper limit for the system (around $L = 160$). Fig. 3 (a) indicates that ASBV performed

---

[3] Of course, we allowed multiple bids, where managers could send the same set of agents for these three tasks. The proposed method could slightly reduce the overhead for processing the messages for phantom tasks.

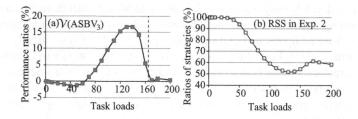

Fig. 3. Performance and selected strategy ratios in Exp. 2

slightly worse when L was low although all managers adopted BAS: The manager selected the awardee in ASBV after all bids for t, p_t^l, and p_t^s had arrived. Therefore, it took slightly longer than that in BAS and PAS$_f$.

We investigated influences with resources required by tasks, the fluctuation factor in PAS$_f$, and communication delay on overall performance with ASBV. We conducted a number of experiments to do this. The generated tasks were t_{5000} (so their required resources were $r_{t_{5000}} = \{5000\}$. The resources required by tasks are denoted by subscripts, such as t_{5000}, after this) in Exp. 3 whose results are plotted in Figs. 4 (a) and (b). Five different tasks, t_{2000}, t_{3000}, t_{4000}, t_{5000}, and t_{6000} were randomly generated in Exp. 4 whose results are plotted in Fig. 4 (c). The generated tasks in Exps. 5 and 6 are identical to those in Exp. 2, but the speed factor for message passing, D, was set to 1.25 and 5 in Exp. 5. Further, fluctuation factor f was set to two and four in Exp. 6. The results for Exp 5 are plotted in Figs. 4 (d) (e) and (f) and those for Exp. 6 are plotted in Figs. 4 (g) (h) and (i).

Figures 4 (a) and (c) indicate that ASBV could also perform better near (but less than) the upper limits of the entire capabilities for tasks that required different resources and for mixtures of different tasks. Note that the upper limits for task loads are approximately 100 and 125 in Exps. 3 and 4. The curves for improvement ratios in these figures are similar to that in Exp. 2. Figure 4 (b) has the values of RSS in Exp. 3 and these are also quite similar to that in Exp. 2. Note that we omitted the figure for RSS in Exp. 4 because it was almost identical to that in Exp. 3.

The length of communication delay affected improvements to the proposed method, because it also affected the chances of simultaneous multiple awards being made. When it was long ($D = 1.25$), the improvement ratios increased more than those in Exp. 2 (cf. Figs. 4 (d) and 3 (a)). Here, as multiple awards occurred more frequently, performance with PAS was degraded more significantly. However, as managers with ASBV selected non-best bidders, improvements with ASBV increased. Of course, when communication delay was short, the improvement ratios decreased as we can see from Fig. 4 (e). Figure 4 (f) also confirms this phenomenon. Because the chance of multiple awards always increased in BAS if communication delay was longer, $\sigma_l - \sigma_s$ lowered. Therefore, the managers with ASBV adopted PAS$_f$ more as award strategy to reduce the chance of multiple awards being made. These experimental results suggest that the proposed strategy is more useful in LSMAS that are widely distributed where their communication delay is significant.

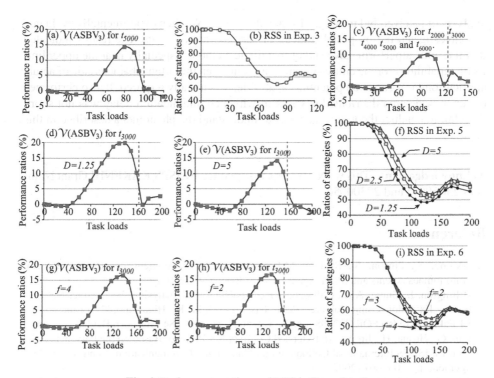

Fig. 4. Performance ratios and RSS in Exps. 3 to 7

The effect of fluctuation factors also yielded interesting features. If we compare the graphs in Figs. 3 (a), 4 (g), and (h), improvements in their performance when L was between 80 and 165 were almost similar. However, when L was beyond the limit of capabilities, performance was slightly lower when fluctuation factor f was two (therefore, the degree of randomness was high). Only a small amount of randomness was appropriate in the latter situations but condition $f = 2$ provided too much randomness to awardee decisions even if RSS was larger. However, when L was between 80 and 165, the proposed strategy could control the values of RSS as shown in Fig. 4 (h); actually, managers with ASBV could adaptively change the values of RSS according to the degrees of randomness provided by PAS_f ($f = 2, 3$, or 4) only when randomness was necessary. Note that the values of RSS always differed according to the values of speed factor in Fig. 4 (f) since communication delay always existed regardless of the task loads. These experiments indicate that ASBV adaptively introduced a degree of randomness, since its selection of award strategy was based on the observed real-time data.

6   Conclusion

We proposed an award strategy called award selection according to bid variances, for CNP-like negotiation protocols, in which two award strategies were alternatively selected by estimating local task loads around individual managers based on the statistical analysis of received bid messages for phantom tasks. We then experimentally evaluated and investigated the proposed strategy in various environments. The results indicated that it could outperform naive and probabilistic methods, especially near the upper limits of capabilities for the entire system. This is quite an important characteristic of actual applications as it can exert potential capabilities when really required. It also performed better in wide-area distributed systems in which communication delay was not short. We plan to conduct theoretical analysis to understand the phenomena described in this paper.

Acknowledgement. This work was in part supported by JSPS KAKENHI Grant Numbers 23650075 and 22300056.

References

1. Aknine, S., Pinson, S., Shakun, M.F.: An extended multi-agent negotiation protocol. Autonomous Agents and Multi-Agent Systems 8(1), 5–45 (2004)
2. Billington, J., Gupta, A.K., Gallasch, G.E.: Modelling and Analysing the Contract Net Protocol - Extension Using Coloured Petri Nets. In: Suzuki, K., Higashino, T., Yasumoto, K., El-Fakih, K. (eds.) FORTE 2008. LNCS, vol. 5048, pp. 169–184. Springer, Heidelberg (2008)
3. Fan, G., Huang, H., Jin, S.: An Extended Contract Net Protocol Based on the Personal Assistant. In: ISECS International Colloquium on Computing, Communication, Control and Management, pp. 603–607 (2008)
4. Kinnebrew, J.S., Biswas, G.: Efficient allocation of hierarchically-decomposable tasks in a sensor web contract net. In: Proc. of IEEE/WIC/ACM Int. Conf. on Web Intelligence and Intelligent Agent Technology, vol. 2, pp. 225–232 (2009)
5. Sandholm, T., Lesser, V.: Issues in automated negotiation and electronic commerce: Extending the contract net framework. In: Lesser, V. (ed.) Proc. of the 1st Int., Conf. on Multi-Agent Systems (ICMAS 1995), pp. 328–335 (1995)
6. Schillo, M., Kray, C., Fischer, K.: The Eager Bidder Problem: A Fundamental Problem of DAI and Selected Solutions. In: Proc. of AAMAS 2002. pp. 599–606 (2002)
7. Smith, R.G.: The Contract Net Protocol: High-Level Communication and Control in a Distributed Problem Solver. IEEE Transactions on Computers C-29(12), 1104–1113 (1980)
8. Sugawara, T., Fukuda, K., Hirotsu, T., Kurihara, S.: Effect of alternative distributed task allocation strategy based on local observations in contract net protocol. In: Desai, N., Liu, A., Winikoff, M. (eds.) PRIMA 2010. LNCS, vol. 7057, pp. 90–104. Springer, Heidelberg (2012)
9. Xu, L., Weigand, H.: The Evolution of the Contract Net Protocol. In: Wang, X.S., Yu, G., Lu, H. (eds.) WAIM 2001. LNCS, vol. 2118, pp. 257–264. Springer, Heidelberg (2001)

Autonomic System Architecture: An Automated Planning Perspective

Falilat Jimoh, Lukáš Chrpa, and Mauro Vallati

School of Computing and Engineering
University of Huddersfield
{Falilat.Jimoh,l.chrpa,m.vallati}@hud.ac.uk

Abstract. Control systems embodying artificial intelligence (AI) techniques tend to be "reactive" rather than "deliberative" in many application areas. There arises a need for systems that can sense, interpret and deliberate with their actions and goals to be achieved, taking into consideration continuous changes in state, required service level and environmental constraints. The requirement of such systems is that they can plan and act effectively after such deliberation, so that behaviourally they appear self-aware. In this paper, we focus on designing a generic architecture for autonomic systems which is inspired by the Human Autonomic Nervous System. Our architecture consists of four main components which are discussed in the context of the Urban Traffic Control Domain. We also highlight the role of AI planning in enabling self-management property of autonomic systems. We believe that creating a generic architecture that enables control systems to automatically reason with knowledge of their environment and their controls, in order to generate plans and schedules to manage themselves, would be a significant step forward in the field of autonomic systems.

Keywords: automated planning, urban traffic control, autonomic systems.

1 Introduction

Autonomic systems (AS) are required to have an ability to learn process patterns from the past and adopt, discard or generate new plans to improve the process control. The ability to identify the task is the most important aspect of any AS element, this enables AS to select the appropriate action when healing, optimising, configuring or protecting itself. Advanced control systems should be able to reason with their surrounding environment and take decision with respect to their current situation and their desired service level. This could be achieved by embedding situational awareness into them and given the ability to generates necessary plans to solve their problem themselves with little or no human intervention - we believe this is the key to embody autonomic properties in systems. Our autonomic system architecture is inspired by the functionality of the Human Autonomic Nervous System (HANS) that handles complexity and

H. Papadopoulos et al. (Eds.): AIAI 2013, IFIP AICT 412, pp. 121–130, 2013.

uncertainty with the aim to realise computing systems and applications capable of managing themselves with minimum human intervention.

The need for planning and execution frameworks has increased interests in designing and developing system architectures which use state-of-the-art plan generation techniques, plan execution, monitoring and recovery in order to address complex tasks in real-world environments [1]. An example of such an architecture is T-Rex (Teleo-Reactive EXecutive): a goal oriented system architecture with embedded automated planning for on-board planning and execution for autonomous underwater vehicles to enhance ocean science [2, 3]. Another recent planning architecture, PELEA (Planning and Execution LEarning Architecture), is a flexible modular architecture that incorporates sensing, planning, executing, monitoring, replanning and even learning from past experiences [4]. The list of system architectures incorporating automated planning is, of course, longer, however, many architectures are designed as domain-specific.

In our previous work, we introduced the problem of self-management of a road traffic network as a temporal planning problem in order to effectively navigate cars throughout a road network. We demonstrated the feasibility of such a concept and discuss our evaluation in order to identify strengths and weaknesses of our approach and point to some promising directions of future research [5, 6]. In this paper, we propose an architecture inspired by Human Autonomic Nervous System (HANS) which embodies AI planning in order to enable autonomic properties such as self-management. This architecture is explained on the example of Urban Traffic Control domain.

2   Urban Traffic Control

The existing urban traffic control (UTC) approaches are still not completely optimal during unforeseen situations such as road incidents when changes in traffic are requested in a short time interval [7, 8]. This increases the need for autonomy in urban traffic control [9–11]. To create such a platform, an AS system needs to be able to consider the factors affecting the situation at hand: the road network, the state of traffic flows, the road capacity limit, accessibility or availability of roads within the network etc. All these factors will be peculiar to the particular set of circumstances causing the problem [6]. Hence, there is a need for a system that can reason with the capabilities of the control assets, and the situation parameters as sensed by road sensors and generate a set of actions or decisions that can be taken to alleviate the situation. Therefore, we need systems that can plan and act effectively in order to restore an unexpected road traffic situation into a normal order. A significant step towards this is exploiting Automated Planning techniques which can reason about unforeseen situations in the road network and come up with plans (sequences of actions) achieving a desired traffic situation.

2.1 Role of AI Planning in Urban Traffic Control

The field of *Artificial Intelligence Planning*, or AI Planning, has evidenced a significant advancement in planning techniques, which has led to development of efficient planning systems [12–15] that can input expressive models of applications. The existence of these general planning tools has motivated engineers in designing and developing complex application models which closely approximate real world problems. Thus, it is now possible to deploy deliberative reasoning to real-time control applications [16–18]. Consequently, AI Planning now has a growing role in realisation of autonomic in control systems and this architecture is a leap towards the realisation of such goal. The main difference between traditional and our autonomic control architecture is depicted in Figure 1. Traditionally, a control loop consists of three steps: sense, interpret and act [19]. In other words, data are gathered from the environment with the use of sensors, the system interprets information from these sensors as the state of the environment. The system acts by taking necessary actions which is feedback into the system in other to keep the environment in desirable state. Introducing deliberation in the control loop allows the system to reason and generate effective plans in order to achieve desirable goals. Enabling deliberative reasoning in UTC systems is important because of its ability to handle unforeseen situations which has not been previously learnt nor hard-coded into a UTC. This helps to reduce traffic congestion and carbon emissions.

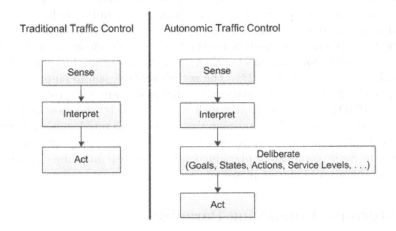

Fig. 1. Illustration of Autonomic System Architecture in Urban Traffic Control

3 Automated Planning

AI Planning deals with the problem of finding a totally or partially ordered sequence of actions whose execution leads from a particular initial state to a state in which a goal condition is satisfied [20]. Actions in plans are ordered in such a way that executability of every action is guaranteed [21]. Hence, an agent is able to transform the environment from an initial state into a desired goal state [22, 23]. A planning problem thus involves deciding "what" actions to do, and "when" to do them [24, 25].

In general, state space of AI planning tasks can be defined as a state-transition system specified by 4-tuple (S, A, E, γ) where:

- S is a set of states
- A is a set of actions
- E is a set of events
- $\gamma : S \times (A \cup E) \to 2^S$ is a transition function

Actions and events can modify the environment (a state is changed after an action or event is triggered). However, the difference between them is that actions are executed by the agent while events are triggered regardless of agent's intentions. Application of an action a (or an event) in some state s results in a state in $\gamma(s, a)$ ($\gamma(s, a)$ contains a set of states). It refers to non-deterministic effects of actions (events).

Classical planning is the simplest form of AI planning in which the set of events E is empty and the transition function γ is deterministic (i.e. $\gamma : S \times A \to S$). Hence, the environment is static and fully observable. Also, in classical planning actions have instantaneous effects.

Temporal planning extends classical planning by considering time. Actions have durative effects which means that executing an action takes some time and effects of the action are not instantaneous. Temporal planning, in fact, combines classical planning and scheduling.

Conformant planning considers partially observable environments and actions with non-deterministic effects. Therefore, the transition function γ is non-deterministic (i.e. $\gamma : S \times A \to 2^S$). The set of events E is also empty.

For describing classical and temporal planning domain and problem models, we can use PDDL [26], a language which is supported by the most of planning engines. For describing conformant planning domain and problem models, we can use PPDDL [27] (an extension of PDDL) or RDDL [28], languages which are supported by many conformant planning engines. Our architecture can generally support all the above kinds of planning.

4   Autonomic Computing Paradigm

We use the term "autonomic system" rather than autonomic computing to emphasise the idea that we are dealing with a heterogeneous system containing hardware and software. Sensors and effectors are the main component of this type of autonomic system architecture [29]. AS needs sensors to sense the environment and executes actions through effectors. In most cases, a control loop is created: the system processes information retrieved from the sensors in order to be aware of its effect and its environment; it takes necessary decisions using its existing knowledge from its domain, generates effective plans and executes those plans using effectors. Autonomic systems typically execute a cycle of monitoring, analysing, planning and execution [30].

System architecture elements are self-managed by monitoring, behaviour analysing and the response is used to plan and execute actions that move or keep the system in desired state. Overall self-management of a system is about doing self-assessment, protection, healing, optimisation, maintenance and other overlapping terms [29]. It is important to stress that these properties are interwoven. The existence of one might require the existence of the others to effectively operate. For instance, a self-optimisation process might not be complete until the system is able to self-configure itself. Likewise, an aspect of a proactive self-healing is an ability of the system to self-protect itself. Any system that is meant to satisfy the above objectives will need to have the following attributes:

− Self-awareness of both internal and external features, processes resources, and constrains
− Self-monitoring it existing state and processes and
− Self-adjustment and control of itself to the desirable/required state
− Heterogeneity across numerous hardware and software architectures

4.1   Case Study of Human Autonomic Nervous System

Human breathing, heartbeat, temperature, immune system, repair mechanisms are all to a great extent controlled by our body without our conscious management. For instance, when we are anxious, frightened, ill or injured, all our bodily functions evolves to react appropriately. The autonomic nervous system has all the organs of the body connected to the body central nervous system which takes decisions in order to optimise the effective functioning of other organs in the body. The sensory cells, senses the state of each organs in relation to the dynamically changing internal and external environment. This information is sent to the brain for interpretation and decision making. The execution of relevant action is sent back to the organs via the linking nerves in real time. This process is repeated trillions of time in seconds within the human body system.

5   System Architecture

Our system architecture is divided into four main blocks. Description of each block is given in the subsequent subsections, however, due to space constraints we cannot go into very details. The entire architecture comprises of a declarative description of the system under study; the individual components that make up such system; the dynamic environment that can influence the performance of such system ; its sensing and controlling capabilities and it ability to reasoning and deliberate.

Our system architecture consists of four main blocks:

− The System Description Block (SD)
− The Sensor and Effector Block (SE)
− The Service Level Checker (SLC)
− The Planning and Action Block (PA)

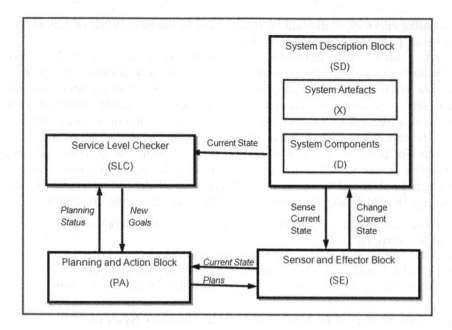

Fig. 2. Diagram of our autonomic system architecture illustrating the relationship between the four main blocks

5.1   The System Description Block (SD)

The system description block stores the information about the environment. Formally, the system description block is a triple $\langle D, X, S \rangle$ where

- D is a finite set of system components
- X is a finite set of artefacts
- S is a set of states, where each state $s_i \in S$ is a set of assignments $f_k(y_l) = v_m$ where $y_l \in D \cup X$ and v_m is a value of y_l

System components are parts of the system which are fully controlled by the system. In human body, components are, for example, legs, kidney, hands, lungs etc. Artefacts refer to external objects which are in the environment where the system operates. In analogy to human body, artefacts are objects the human body can interact with. States of the system capture states of all the components and artefacts which are considered. For example, a stomach can be empty or full, water in a glass can be cold or hot etc.

In Urban Traffic Control (UTC) domain, for instance, SD consists of a declarative description of the road network of an area that need to be controlled while optimising traffic flow pattern. The components are all the objects that can be controlled by the system, for example, traffic control signal heads, variable message signs(VMS) and traffic cameras. The artefacts, objects which cannot be directly controlled by the system such as roads layout and infrastructures.

5.2   The Sensor and Effector Block (SE)

A human body monitors a state of its components (organs), where one can feel pain, hunger, cold etc., and states of artefacts (external objects) with sensors such as skin, eyes, nose etc. SE is responsible for sensing in order to determine a current state of the system (environment) which is stored in SD. Different sensors can provide an information about states of the system components and artefacts. Since the outputs of the sensors might not directly correspond with the representation of states. Formally, we can define a function $\eta : O \to S$ which can transform a vector of outputs of different sensors (O is the set of all such possible vectors) to a state of the system (S is a set of all possible states of the system). For example, in the UTC domain, sensing can be done by road sensors and CCTV cameras.

SE is also responsible for executing plans (sequences of actions) given by the Planning and Action block in Figure 2. Executing actions is done via system components which indirectly affect artefacts as well. In the UTC domain, a plan to change a traffic light from red to green would cause vehicles to start moving through the intersection. Since we have to consider uncertainty, after executing a sequence of actions, the sensors sense the current state, and if the current state is different from the expected one, the information is propagated to the Planning and Action block which provides a new plan.

5.3   The Service Level Checker (SLC)

The service level checker, as the name implies, repeatedly checks the service level of the entire system to see if the system is in a 'good condition'. It gets the current state of the system from SE and compare it with the 'ideal' service level. If the current state is far from being 'ideal', then the system should act in order to recover its state to a 'good condition'. In analogy to a human body, if it is infected by some virus, the body cells detect this anomalies and start acting against the virus. Formally, we can define an error function $\epsilon : S \to \mathbb{R}_0^+$ which determines how far the current state of the system is from being 'ideal'. $\epsilon(s) = 0$ means that s is an 'ideal' state. If a value of the error function in the current state is greater than a given threshold, then SLC generates goals (desired states of a given components and/or artefacts) and passes them to the Planning and Actions block (see below).

An example of this in UTC domain can be seen in an accident scene at a junction. The 'ideal' service level for such a junction is to maximise traffic flow from the junction to neighbouring routes. The present state of the system shows that the traffic at that junction is now static with road sensors indicating static vehicles. This situation is not 'ideal', the error function in such a state is high, hence, new goals are generated by SLC to re-route incoming traffic flowing towards such intersection. It is also possible to alter the error function by an external system controller. For instance, an autonomic UTC system can also accept inputs from the traffic controller (Motoring Agencies). This means that the system behaviour can be altered by the user if needed. This gives the user a superior autonomy over the entire autonomic system. For instance, despite the human body system having the capacity to fight virus (as mentioned before), an expert (doctor) still may have control, so he can, for instance, prescribe anesthetic to reduce the way the body response to pains. This anesthetic act on the brain or peripheral nervous system in other to suppress responses to sensory stimulation. Our architecture is thus designed to have autonomy over it entire component and artefacts and robust enough to accommodate sovereign external user control.

5.4   The Planning and Action Block (PA)

PA can be understood as a core of deliberative reasoning and decision making in the system. This is 'the brain' of our architecture. PA receives the current state of the system from SE, and goals from SLC. Then produces a plan which is then passed to SE for execution. However, it is well known from planning and execution systems [31] that producing a plan given a planning problem does not guarantee its successful execution. As indicated before SE, which is responsible for plan execution, verifies whether executing an action provided a desired outcome. If the outcome is different (from some reason), then this outcome is passed back to PA which re-plans. It is well known that even classical planning is intractable (PSPACE-complete). Therefore, PA might not guarantee generating plans in a reasonable time. Especially, in a dynamic environment a system must be able to react quickly to avoid imminent danger. Therefore, SLC generates goals which are easy to achieve when the time is crucial.

In analogy to human beings, if hungry they, for instance, have to plan how to obtain food (go shopping). This might not be a very time critical task such as if one is standing on the road and there is a car going fast against him he has to act quickly to avoid a (fatal) collision.

In the UTC domain, if some road is congested, then the traffic is navigated through alternative routes. Hence, PA is responsible for producing a plan which may consist of showing information about such alternative routes on variable message signs and optimising signal heads towards such route.

6   Conclusion

In this work, we have describe our vision of self-management at the architectural level, where autonomy is the ability of the system components to configure their interaction in order for the entire system to achieve a set service level. We created an architecture that mimics Human Autonomic Nervous System (HANS). Our architecture is basically made of four main blocks which are discussed in the context of the Urban Traffic Control Domain. We describe the functionality of each block, highlighting their relationship with one another. We also highlight the role of AI planning in enabling self-management property in an autonomic systems architecture. We believe that, creating a generic architecture that enables control systems to automatically reason with knowledge of their environment and their controls, in order to generate plans and schedules to manage themselves, would be a significant step forward in the field of autonomic systems.

In a real-world scenario where data such as road queues are uploaded in real-time from road sensors with traffic signals connected to a planner, we believe that our approach can optimise road traffic with little or no human intervention. In future we plan to embed our architecture into a road traffic simulation platform. We also plan to provide a deeper evaluation of our approach, especially to compare it with traditional traffic control methods, and assess the effort and challenges required to embody such technology within a real-world environment. Other aspects of improvement would include: incorporating learning into our architecture which will store plans for the purpose of re-using valid plans and save the cost of re-planing at every instance.

References

1. Myers, K.L.: Towards a framework for continuous planning and execution. In: Proc. of the AAAI Fall Symposium on Distributed Continual Planning (1998)
2. McGann, C., Py, F., Rajan, K., Thomas, H., Henthorn, R., McEwen, R.: A Deliberative Architecture for AUV Control. In: Intnl. Conf. on Robotics and Automation, ICRA (2008)
3. Pinto, J., Sousa, J., Py, F., Rajan, K.: Experiments with Deliberative Planning on Autonomous Underwater Vehicles. In: IROS Workshop on Robotics for Environmental Modeling (2012)
4. Jimènez, S., Fernández, F., Borrajo, D.: Integrating planning, execution, and learning to improve plan execution. Computational Intelligence 29(1), 1–36 (2013)
5. Shah, M., McCluskey, T., Gregory, P., Jimoh, F.: Modelling road traffic incident management problems for automated planning. Control in Transportation Systems 13(1), 138–143 (2012)
6. Jimoh, F., Chrpa, L., Gregory, P., McCluskey, T.: Enabling autonomic properties in road transport system. In: The 30th Workshop of the UK Planning and Scheduling Special Interest Group, PlanSIG (2012)
7. Roozemond, D.A.: Using intelligent agents for pro-active, real-time urban intersection control. European Journal of Operational Research 131(2), 293–301 (2001)
8. De Oliveira, D., Bazzan, A.L.C.: Multiagent learning on traffic lights control: Effects of using shared information. In: MultiAgent Systems for Traffic and Transportation Engineering, pp. 307–322 (2009)
9. Yang, Z., Chen, X., Tang, Y., Sun, J.: Intelligent cooperation control of urban traffic networks. In: Proceedings of the International Conference on Machine Learning and Cybernetics, pp. 1482–1486 (2005)
10. Bazzan, A.L.: A distributed approach for coordination of traffic signal agents. Autonomous Agents and Multi-Agent Systems 10(1), 131–164 (2005)
11. Dusparic, I., Cahill, V.: Autonomic multi-policy optimization in pervasive systems: Overview and evaluation. ACM Trans. Auton. Adapt. Syst. 7(1), 1–25 (2012)
12. Gerevini, A., Saetti, A., Serina, I.: An approach to temporal planning and scheduling in domains with predictable exogenous events. Journal of Artificial Intelligence Research (JAIR) 25, 187–231 (2006)
13. Coles, A.I., Fox, M., Long, D., Smith, A.J.: Planning with problems requiring temporal coordination. In: Proc. 23rd AAAI Conf. on Artificial Intelligence (2008)

14. Penna, G.D., Intrigila, B., Magazzeni, D., Mercorio, F.: Upmurphi: a tool for universal planning on pddl+ problems. In: Proc. 19th Int. Conf. on Automated Planning and Scheduling (ICAPS), pp. 19–23 (2009)
15. Eyerich, P., Mattmüller, R., Röger, G.: Using the Context-enhanced Additive Heuristic for Temporal and Numeric Planning. In: Proc. 19th Int. Conf. on Automated Planning and Scheduling, ICAPS (2009)
16. Löhr, J., Eyerich, P., Keller, T., Nebel, B.: A planning based framework for controlling hybrid systems. In: Proc. Int. Conf. on Automated Planning and Scheduling, ICAPS (2012)
17. Ferber, D.F.: On modeling the tactical planning of oil pipeline networks. In: Proc. 18th Int. Conf. on Automated Planning and Scheduling, ICAPS (2012)
18. Burns, E., Benton, J., Ruml, W., Yoon, S., Do, M.: Anticipatory on-line planning. In: International Conference on Automated Planning and Scheduling (2012)
19. Gopal, M.: Control systems: principles and design. McGraw-Hill, London (2008)
20. Nau, D., Ghallab, M., Traverso, P.: Automated Planning: Theory & Practice. Morgan Kaufmann Publishers Inc., San Francisco (2004)
21. Fox, M., Long, D.: PDDL2.1: An extension of PDDL for expressing temporal planning domains. Journal of Artificial Intelligence Research (JAIR) 20, 61–124 (2003)
22. Gupta, S.K., Nau, D.S., Regli, W.C.: IMACS: A case study in real-world planning. IEEE Expert and Intelligent Systems 13(3), 49–60 (1998)
23. Garrido, A., Onaindía, E., Barber, F.: A temporal planning system for time-optimal planning. In: Brazdil, P.B., Jorge, A.M. (eds.) EPIA 2001. LNCS (LNAI), vol. 2258, pp. 379–392. Springer, Heidelberg (2001)
24. Gerevini, A., Haslum, P., Long, D., Saetti, A., Dimopoulos, Y.: Deterministic planning in the fifth international planning competition: PDDL3 and experimental evaluation of the planners. Artificial Intelligence 173(5-6), 619–668 (2009)
25. Hoffmann, J., Nebel, B.: The FF planning system: Fast plan generation through heuristic search. Journal of Artificial Intelligence Research (JAIR) 14, 253–302 (2001)
26. McDermott, D., et al.: PDDL–the planning domain definition language. Technical report (1998), http://www.cs.yale.edu/homes/dvm
27. Younes, H.L.S., Littman, M.L., Weissman, D., Asmuth, J.: The First Probabilistic Track of the International Planning Competition. Journal of Artitificial Intelligence Research (JAIR) 24, 851–887 (2005)
28. Sanner, S.: Relational dynamic influence diagram language (RDDL): Language description (2010), http://users.cecs.anu.edu.au/~ssanner/IPPC_2011/RDDL.pdf
29. Ganek, A., Corbi, T.: The dawning of the autonomic computing era. IBM Systems Journal 42(1), 5–5 00188670 Copyright International Business Machines Corporation 2003 (2003)
30. Lightstone, S.: Seven software engineering principles for autonomic computing development. ISSE 3(1), 71–74 (2007)
31. Fox, M., Long, D., Magazzeni, D.: Plan-based policy-learning for autonomous feature tracking. In: Proc. of Int. Conf. on Automated Planning and Scheduling, ICAPS (2012)

MOEA/D for a Tri-objective Vehicle Routing Problem

Andreas Konstantinidis, Savvas Pericleous, and Christoforos Charalambous

Department of Computer Science and Engineering, Frederick University, Nicosia, Cyprus

Abstract. This work examines the Capacitated Vehicle Routing Problem with Balanced Routes and Time Windows (CVRPBRTW). The problem aims at optimizing the total distance cost, the number of vehicles used, and the route balancing, under the existence of time windows and other constraints. The problem is formulated as a Multi-Objective Optimization Problem where all objectives are tackled simultaneously, so as to effect a better solution space coverage. A Multi-Objective Evolutionary Algorithm based on Decomposition (MOEA/D), hybridized with local search elements, is proposed. The application of local search heuristics is not uniform but depends on specific objective preferences and instance requirements of the decomposed subproblems. To test the efficacy of the proposed solutions, extensive experiments were conducted on well known benchmark problem instances and results were compared with other MOEAs.

1 Introduction and Related Work

The Vehicle Routing Problem (VRP) refers to a family of problems in which a set of routes for a fleet of vehicles based at one (or several) depot(s) must be determined for a number of geographically dispersed customers. The goal is to deliver goods to the customers with known demands under several objectives and constraints by originating and terminating at a depot.

The problem has received extensive attention in the literature [1] due to its association with important real-world problems. Several versions and variations of the VRP exist that are mainly classified based on their objectives and constraints [2]. The classic version of the Capacitated VRP (*CVRP*) [1,3] considers a collection of routes, where each vehicle is associated with one route, each customer is visited only once and aims at minimizing the total distance cost of a solution using the minimum number of vehicles while ensuring that the total demand per route does not exceed the vehicle capacity. The extended *CVRP with Balanced Routes (CVRPBR)* [4] introduces the objective of route balancing in order to bring an element of fairness into the solutions. The *CVRP with time windows (CVRPTW)* [5] does not include any additional objective but involves the additional constraint that each customer should be served within specific time windows.

CVRP and its variants are proven *NP-hard* [6]. Optimal solutions for small instances can be obtained using exact methods [2], but the computation time increases exponentially for larger instances. Thus, several heuristic and optimization methods [1] are proposed. More recently, metaheuristic approaches are used to tackle harder CVRP instances including Genetic Algorithms [7] and hybrid approaches [3]. Hybrid approaches, which often include combinations of different heuristic and metaheuristic methods such as the hybridization of Evolutionary Algorithms (EAs) with local search

H. Papadopoulos et al. (Eds.): AIAI 2013, IFIP AICT 412, pp. 131–140, 2013.

(aka *Hybrid* or *Memetic Algorithms*), have been more effective in dealing with hard scheduling and routing problems [3] than conventional approaches in the past.

When real-life cases are considered, it is common to examine the problem under multiple objectives as decision makers rarely take decisions examining objectives in isolation. Therefore, proposed solutions often attack the various objectives in a single run. This can be done by tackling the objectives individually and sequentially [4], or by optimizing one objective while constraining the others [8] or by aggregating all objectives into one single objective function [9] usually via a weighted summation. Such approaches often lose "better" solutions, as objectives often conflict with each other and the trade-off can only be assessed by the decision maker. Therefore, the context of Multi-Objective Optimization (MOO) is much more suited for such problems.

A *Multi-objective Optimization Problem (MOP)* [10] can be mathematically formulated as follows:

$$\text{minimize } F(x) = (f_1(x), \ldots, f_m(x))^T, \quad \text{subject to } x \in \Omega \tag{1}$$

where Ω is the decision space and $x \in \Omega$ is a decision vector. $F(x)$ consists of m objective functions $f_i : \Omega \to \Re, i = 1, \ldots, m$, and $\Re^m$ is the objective space.

The objectives in (1) often conflict with each other and an improvement on one objective may lead to the deterioration of another. In that case, the best trade-off solutions, called the set of Pareto optimal (or non-dominated) solutions, is often required by a decision maker. The Pareto optimality concept is formally defined as,

Definition 1. A vector $u = (u_1, \ldots, u_m)^T$ is said to dominate another vector $v = (v_1, \ldots, v_m)^T$, denoted as $u \prec v$, iff $\forall i \in \{1, \ldots, m\}$, $u_i \leq v_i$ and $u \neq v$.

Definition 2. A feasible solution $x^* \in \Omega$ of problem (1) is called *Pareto optimal solution*, iff $\nexists y \in \Omega$ such that $F(y) \prec F(x^*)$. The set of all Pareto optimal solutions is called the Pareto Set (PS) and the image of the PS in the objective space is called the Pareto Front (PF).

Multi-Objective Evolutionary Algorithms (MOEAs) [11] are proven efficient and effective in dealing with MOPs. This is due to their population-based nature that allows them to obtain a well-diversified approximation of the PF. That is, minimize the distance between the generated solutions and the true PF as well as maximize the diversity (i.e. the coverage of the PF in the objective space). In order to do that, MOEAs are often combined with various niching mechanisms such as crowding distance estimation [12] to improve diversity, and/or local search methods [13] to improve convergence.

In the literature there are several studies that utilized generic or hybrid Pareto-dominance based MOEAs to tackle Multi-Objective CVRPs and variants [14]. For example, Jozefowiez et al. [15] proposed a bi-objective CVRPBR with the goal to optimize both the total route length and routes balancing. In [16], the authors proposed a hybridization of a conventional MOEA with multiple LS approaches that were selected randomly every 50 generations to locally optimize each individual in the population and tackle a bi-objective CVRPTW. In [17], Geiger have tackled several variations of the CVRPTW by optimizing pairs of the different objectives. Over the past decade numerous variants of the investigated problem have been addressed under a MOP setting, involving different combinations of objectives and different search hybridization elements. For the interested reader, indicative examples include ([18], [19]).

Even though the objectives and constraints presented are all important, challenging, and by nature conflicting with each other, to the best of our knowledge no research work has ever dealt with the minimization of the total distance cost, the number of vehicles and the route balancing objectives as a MOP trying to satisfy all side-constraints, simultaneously. Moreover in all the above studies, MOEAs based on Pareto Dominance (such as NSGA-II [12]) are hybridized either with a single local search approach [18,19] or with multiple local search heuristics with one being selected randomly [16] each time a solution was about to be optimized locally.

In this paper, we investigate the **CVRPBRTW**, formulated as a MOP composed of three objectives (minimize the total distance cost, minimize the number of vehicles and balance the routes of the vehicles) and all relevant constraints aiming at increasing its practical impact by making it closer to real-life cases. Solutions are obtained through a hybrid MOEA/D [20] approach that decomposes the proposed MOP into a set of scalar subproblems, which are solved simultaneously using neighborhood information and local search methods each time a new solution is generated. Specifically, the MOEA/D is hybridized with multiple local search heuristics that are adaptively selected and locally applied to a subproblem's solution based on specific objective preferences and instant requirements. We examine our proposition on Solomon's benchmark problem instances [5] against several other MOEA/Ds.

2   Multi-Objective Problem Definition and Formulation

The elementary version of the CVRP [1,3] is often modelled as a complete graph $G(V, E)$, where the set of vertices V is composed of a unique depot $u_0 = o$ and l distinct customers, each based at a prespecified location. The Euclidean distance between any pair of customers is associated with the corresponding edge in E. Each customer must be served a quantity of goods (customer's demand) that requires a predefined service time. To deliver those goods, K identical vehicles are available, which are associated with a maximal capacity of goods that they can transport. Vehicles traverse a unit distance in unit time and time is measured as time elapsed from commencing operations. A solution of the CVRP is a collection of routes, where each route is a sequence of vertices starting and ending at the depot and served by a single vehicle, each customer is visited only once and the total amount of goods transported per route is at most the vehicle's capacity. The CVRP aims at a *minimal total distance cost* of a solution, using *minimum number of vehicles*. In the investigated problem a third objective, that of *route balancing*, is also examined. The balancing objective, which is defined as the difference between the maximum distance traveled by a vehicle and the mean traveled distance of all vehicles [4], brings an element of fairness in solutions. Finally, the well known 'time windows' constraint is imposed. This constraint requires the vehicle serving each customer to arrive within specific time windows.

Note that in the problem variant investigated in this work, if a vehicle arrives at a customer before the earliest arrival time it is allowed to wait until that time is reached, resulting in additional route traveled time. Time windows are treated as a hard constraint in the sense that if the vehicle arrives at a customer after the latest arrival time the solution is considered infeasible.

Therefore, the proposed *CVRP with Balanced Routes and Time Windows (CVRP-BRTW)* can be mathematically formulated as follows:

Given,

V — the set of $l + 1$ vertices (customers) composed of a depot o and for $i = 1, ..., l$ vertices u_i located at coordinates (x_i, y_i).

E — the set of edges (u_i, u_j) for each pair of vertices in $u_i, u_j \in V$ associated with their Euclidean distance $dist(u_i, u_j)$.

$[e_u, e'_u]$ — the time window of customer $u, \forall u \in V$.

q_u — the quantity demand of customer $u, \forall u \in V$; in particular, $q_o = 0$.

t^s_u — the service time of customer $u, \forall u \in V$; in particular, $t^s_o = 0$.

K — the maximum number of vehicles to be used (at most l).

c — the capacity of each vehicle z.

R^m — the *route* followed by the mth vehicle used in the solution. The route is defined as a sequence of customer vertices (excluding the depot vertex).

X — a collection of k routes $X = \{R^1, R^2, ..., R^k\}$ where k is at most K.

$suc(u)$ — given $u \in R^m$, $suc(u)$ is the vertex immediately following u in R^m, if it exists (i.e., u is not the last vertex in R^m), otherwise the depot o.

$pre(u)$ — given $u \in R^m$, $pre(u)$ is the vertex immediately preceding u in R^m, if it exists (i.e., u is not the first vertex in R^m), otherwise the depot o.

$init(R^m)$ — the initial vertex in R^m

t^a_u — the vehicle arrival time at vertex $u \in V \setminus \{o\}$ which can be calculated by the function $\max\{e_u, t^a_{pre(u)} + t^s_{pre(u)} + dist((pre(u), u)\}$, with $t^a_o = 0$.

$D^m(X)$ — the total distance covered by the vehicle serving route R^m in solution X obtained by $dist(o, init(R^m)) + \sum_{\forall u \in R^m} dist(u, suc(u))$

$$\min F(X) = (D(X), B(X), N(X)) \tag{2}$$

$$D(X) = \sum_{m=1}^{k} D^m(X) \tag{3}$$

$$N(X) = k + \left(\min_{1 \leq m \leq k} \left(\frac{|R^m|}{l} \right) \right) \tag{4}$$

$$B(X) = \left(\max_{1 \leq m \leq k} \{D^m(X)\} \right) - \frac{1}{k} D(X) \tag{5}$$

subject to

$$\sum_{\forall u \in R^m} q_u \leq c, \ \forall m = 1, ..., k \tag{6}$$

$$e_u \leq t^a_u \leq e'_u \ \forall u V \setminus \{o\} \tag{7}$$

$$\{u\} \cap \bigcup_{m=1,...,k} R^m = \{u\} \ \forall u \in V \setminus \{o\} \tag{8}$$

$$\sum_{m=1,...,k} |R^m| = l \tag{9}$$

Equation (2) specifies the multi-objective function we wish to minimize, comprising the total distance cost, defined in (3), route balancing, defined in (5), and the number of routes, thus vehicles, used, $k = |X|$. Note that instead of $|X|$, the auxiliary function $N(X)$ defined in (4) is used, as it gives a bias towards solutions with the least customers in the smallest route.

Constraints (6) ensure that the total quantity of goods transported in a route does not exceed the vehicle's capacity, whereas constraints (7) require that the arrival time at all customers is within the corresponding time window. The combination of constraints (8) and (9) guarantee that all customers are served exactly once; constraints (8) ensure that each customer vertex is visited by at least one route, and constraint (9) that the total number of vertices visited is equal to the number of customers.

3   The Proposed Hybrid MOEA/D

3.1   Preliminaries

The problem is tackled by a decomposed MOEA. Before explaining the algorithm, the encoding representation used and the solution evaluation algorithm will be explained.

Encoding Representation: In VRP, solutions are often represented by a variable length vector of size greater than l, which consist of all l customers exactly once and the depot, o, one or more times signifying when each vehicle starts and ends its route. Under such a representation, the solution's phenotype (the suggested routes) can readily be obtained, although several issues of infeasibility arise. In this work however, a candidate solution X is a fixed length vector of size l, composed of all customers only. This solution encoding X is translated to the actual solution using the following algorithm. An empty route R_1 is initially created. The customers are inserted in R_1 one by one in the same order as they appear in solution X. A customer u_j that violates any of the constraints of Section 2 is directly inserted in a newly created route R_2. In the case where more than one route is available, and for the remaining customers, a competitive process starts, in which the next customer u_{j+1} in X is allowed to be inserted in any available route that does not violate a constraint. When more than one such routes exist, the one with the shortest distance to the last customer en route is preferred. If a customer violates a constraint in all available routes, a newly created route is initiated. Note that this process guarantees feasibility irrespective of the actual sequence.

Decomposition: In MOEA/D, the original MOP needs to be decomposed into a number of M scalar subproblems. Any mathematical aggregation approach can serve for this purpose. In this article, the Tchebycheff approach is employed as originally proposed in [20].

Let $F(x) = (f_1, ..., f_m)$ be the objective vector, $\{w_1, ..., w_m\}$ a set of evenly spread weight vectors, which remain fixed for each subproblem for the whole evolution, and z^* the reference point. Then, the objective function of a subproblem i is stated as:

$$g^i(X^i|w^i, z^*) = min\{\sum_{j=1}^{m}(w_j^i \hat{f}_j(X) - z_j^*)\}$$

where $w^i = (w^i_1, ..., w^i_m)$ represents the objective weight vector for the specific decomposed problem i, $\hat{f}$ denotes the min-max normalization of f and $z^* = (z_1, ..., z_m)$ is a vector equal to all best values z_j found so far for each objective f_j. MOEA/D minimizes all these objective functions simultaneously in a single run. As stated in [20], one of the major contributions of MOEA/D is that the optimal solution of subproblem i should be close to that of k if w^i and w^k are close to each other in the weight space. Therefore, any information about these g^ks with weight vectors close to w^i should be helpful for optimizing $g^i(X^i|w^i, z^*)$. This observation will be later utilized for improving the efficiency and the adaptiveness of the newly proposed local search heuristic.

Neighborhoods: In MOEA/D, a neighborhood N^i is maintained for each subproblem i of weight vector w^i. Particularly, N^i is composed of the T subproblems of which the weight vectors are closest to w^i, including i itself. T is a parameter of the algorithm. The Euclidean distance is used to measure the closeness between two weight vectors.

3.2 The Evolutionary Algorithm

The algorithm commences by creating an initial population, named Internal Population (IP) of generation $\gamma = 0$, $IP_0 = \{X^1, ..., X^M\}$. The initial solutions are randomly generated and each individual is evaluated using the process described earlier.

At each step of MOEA/D, for each subproblem i a new solution Y^i is generated through the use of genetic operators. Specifically, using the Neighborhood Tournament Selection (NTS) operator [21], two parent solutions, Pr^1 and Pr^2, are selected from N^i. The two parent solutions are then recombined with a probability rate c_r using the well-known Partially Mapped Crossover (PMX) operator [22] to produce an offspring solution O. Finally, a random mutation operator is utilized to modify each element of solution O with a mutation rate m_r and generate solution Y^i.

After the new solution Y^i is generated for a given subproblem, an attempt is made to improve it through the use of local search. Specifically, one local search heuristic is applied on Y^i, yielding a new solution Z^i. The local search (LS) heuristic used is selected from the following pool [23]:

- *Double Shift (DS):* is a combination of the Backward and Forward Shifts. That is, it initially takes a customer from its current position u_{j_1} and inserts it before a customer u_{k_1}, where $j_1 > k_1$. Then it takes a customer from its current position u_{j_2} and inserts it after a customer u_{k_2}, where $j_2 < k_2$.
- *Lambda Interchange (LI):* First, two routes A and B are chosen. The heuristic starts by scanning through nodes in route A and moves a feasible node into route B. The procedure repeats until a predefined number of nodes are shifted or the scanning ends at the last node of route A.
- *Shortest Path (SP):* attempts to rearrange the order of nodes in a particular route such that the node with the shortest distance from the incumbent is selected.

Central to the proposed approach is the way the local search heuristic (LS) is selected for application each time a new solution is generated. Specifically, in our work the LS is selected based on a weighted probability that is not static among all subproblems but is based on the objective weights each subproblem i holds. Through extensive experimentation on random solution instances we have established an affinity of each LS described above with an objective function and adopt an association between objective and LS. The associations applied are: SP for distance cost, DS for number of vehicles, and LI for route balancing. As a result, we use the weight value w_j^i for subproblem i and objective j as the weighted probability of selecting the associated LS on that subproblem.

Once all solutions Z^i are constructed, the population is updated as follows. Firstly, solution Z^i replaced the incumbent solution X^i for subproblem i iff it achieves a better value for the specific objective function of that subproblem. Subsequently, in an attempt to propagate good characteristics, Z^i is evaluated against the incumbent solutions X^ks of the T closest neighbors of i. For each of these subproblems, if $g^k(Z^i|w^k, z^*) < g^k(X^k|w^k, z^*)$ then Z^i becomes also the incumbent for subproblem k. Finally, a test is made to check whether Z^i is dominated by any solution in the maintained Pareto Front, and if not, it is added to PF. The aforementioned process is repeated for a prespecified number of generations g_m.

4   Experimental Studies

4.1   Experimental Setup and Performance Measures

The experiments were carried out on the well-known Solomon's instances (100-customer problem sets). These instances are categorized into six classes: C1, C2, R1, R2, RC1 and RC2. Category C problems represent clustered data, which means the customers are clustered either geographically or in terms of the time windows. Category R problems represent uniformly randomly distributed data and RC are combinations of the other two classes. Classes C1, R1 and RC1 consider customers with narrower time windows. The algorithmic settings used are as follows: $c_r = 0.9, m_r = 0.01, T = 10, M = 630$ and $g_m = 3000$. Due to the limited space we present results on a subset of instances.

The performance of an MOEA is usually evaluated from two perspectives: the obtained non-dominated set should be (i) as close to the true Pareto Front as possible, and (ii) distributed as diversely and uniformly as possible. No single metric can reflect both of these aspects and often a number of metrics are used [24]. In this study, we use the **Coverage** C [24] and **distance for reference set** I_D [25] metrics:

$$C(A, B) = \frac{|\{x \in B | \exists y \in A : y \prec x\}|}{|B|}; \quad I_D(A) = \frac{\sum_{y \in R}\{min_{x \in A}\{d(x, y)\}\}}{|R|}.$$

Coverage is a commonly used metric for comparing two sets of non-dominated solutions A and B. The $C(A, B)$ metric calculates the ratio of solutions in B dominated by solutions in A, divided by the total number of solutions in B. Therefore, $C(A, B) = 1$ means that all solutions in B are dominated by the solutions in A. Note that $C(A, B) \neq 1 - C(B, A)$.

The distance from reference set is defined by Czyzzak et al. in [25]. This shows the average distance from a solution in the reference set R to the closest solution in A. The smaller the value of I_D, the closer the set A is to R. In the absence of the real reference set (i.e., Pareto Front), we calculate the average distance of each single point to the nadir point since we consider minimization objectives.

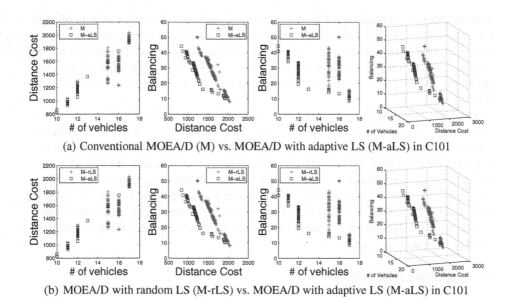

(a) Conventional MOEA/D (M) vs. MOEA/D with adaptive LS (M-aLS) in C101

(b) MOEA/D with random LS (M-rLS) vs. MOEA/D with adaptive LS (M-aLS) in C101

Fig. 1. Evaluation of the proposed MOEA/D with the adaptive Local Search (M-aLS) with respect to the conventional MOEA/D (M) and the MOEA/D with random local search selection (M-rLS)

4.2   Experimental Results

The proposed MOEA/D-aLS (M-aLS) is evaluated with respect to the conventional MOEA/D as proposed by Zhang and Li in [20] and a MOEA/D with a random local search (M-rLS) selection mechanism. To increase the fidelity of our experimental studies we have repeated each experiment of each algorithm for 30 independent runs, having the same number of function evaluations for fairness.

Figure 1 shows that the hybrid M-aLS improves the performance of the conventional MOEA/D and outperforms the M-rLS in test instance C101 in terms of both convergence and diversity. In particular, the M-aLS has obtained a PF that dominates most of the non-dominated solutions obtained by the other MOEA/Ds providing a better approximation towards the nadir point as well. Note that similar results were obtained in most test instances. This is more evident in Table 1 that summarizes the statistical performance of M-aLS and M-rLS in terms of the Coverage (C) and the Distance to the reference set (I_D). The results show that the non-dominated solutions obtained by the M-aLS dominate most (on average 75%) of the non-dominated solutions obtained by M-rLS and performs no worse on average than M-rLS in terms of I_D. Finally, the results in Table 2, which summarize the best objective values obtained by each approach during the evolution, clearly show that the performance of the proposed M-aLS is better than its competitors.

Table 1. MOEA/D with proposed adaptive Local Search (M-aLS) is compared to MOEA/D with random LS selection (M-rLS) based on C and I_D metrics. Best results are denoted in bold.

Test Inst.	C(M-rLS,M-aLS)	C(M-aLS,M-rLS)	I_D(M-rLS)	I_D(M-aLS)
C101:	0.03	**0.74**	43.61	**31.99**
C201:	0	**1**	33.68	**20.14**
R101:	0	**0.86**	**19.16**	22.68
R201:	0.33	**0.36**	**29.04**	36.78
RC101:	0.1	**0.66**	**15.11**	33.47
RC201:	0	**0.88**	42.25	**20.3**

Table 2. M-aLS is compared with conventional MOEA/D and M-rLS in terms of best solutions found for each objective. Best results are denoted in bold.

Test Inst.	M			M-rLS			M-aLS		
	V	D	B	V	D	B	V	D	B
C101:	15.0	1229.150	7.913	**12.0**	1023.474	9.802	**12.0**	**933.462**	**6.824**
C201:	9.0	1153.066	4.605	5.0	706.534	3.459	**3.0**	**625.197**	**1.302**
R101:	23.0	1934.568	12.642	22.0	1906.092	10.979	**21.0**	**1823.122**	**10.835**
R201:	10.0	1699.647	6.133	**8.0**	1351.344	**4.469**	**8.0**	**1350.925**	4.894
RC101:	21.0	2039.398	7.646	**18.0**	1891.283	9.779	**18.0**	**1849.835**	**6.545**
RC201:	10.0	1843.807	9.295	**8.0**	1606.290	**4.767**	**8.0**	**1533.477**	5.311

5   Conclusions and Future Work

The Tri-Objective Capacitated Vehicle Routing Problem with Balanced Routes and Time Windows is proposed and tackled with a Multi-Objective Evolutionary Algorithm based on Decomposition (MOEA/D) hybridized with local search. The MOEAD-aLS decomposes the proposed MOP into a set of scalar subproblems which are solved simultaneously using at each generation multiple LSs adaptively selected based on objective preferences and instant requirements. We evaluate our proposition on a subset of the standard benchmark problem instances. The results show that the MOEA/D-aLS clearly improves the performance of the MOEA/D in all cases and of MOEA/D-rLS in most cases. In the future, we aim at incorporating learning for the selection of a local search approach to further improve the performance of the MOEA/D.

References

1. Goel, A., Gruhn, V.: A general vehicle routing problem. European Journal of Operational Research 191(3), 650–660 (2008)
2. Laporte, G.: Fifty years of vehicle routing. Transportation Science (2009)
3. Prins, C.: A simple and effective evolutionary algorithm for the vehicle routing problem. Computers & Operations Research 31, 1985–2002 (2004)
4. Lee, T.R., Ueng, J.H.: A study of vehicle routing problems with load-balancing. International Journal of Physical Distribution & Logistics Management 29(10), 646–657 (1999)

5. Solomon, M.M.: Algorithms for the vehicle routing problem with time windows. Transportation Science 29(2), 156–166 (1995)
6. Lenstra, J.K., Kan, A.H.G.R.: Complexity of vehicle routing and scheduling problems. Networks 11(2), 221–227 (1981)
7. Reeves, C.R.: A genetic algorithm for flowshop sequencing. Special Issue on Genetic Algorithms in Computers and Operations Research 22(1), 5–13 (1995)
8. Chen, J., Chen, S.: Optimization of vehicle routing problem with load balancing and time windows in distribution. In: 4th International Co. Wireless Communications, Networking and Mobile Computing, WiCOM 2008 (2008)
9. Ombuki, B., Ross, B.J., Hanshar, F.: Multi-objective genetic algorithms for vehicle routing problem with time windows. Applied Intelligence 24, 17–30 (2006)
10. Deb, K.: Multi-Objective Optimization Using Evolutionary Algorithms. Wiley & Sons (2002)
11. Zhou, A., Qu, B.-Y., Li, H., Zhao, S.-Z., Suganthan, P.N., Zhang, Q.: Multiobjective evolutionary algorithms: A survey of the state of the art. Swarm and Evolutionary Computation 1(1), 32–49 (2011)
12. Deb, K., Pratap, A., Agarwal, S., Meyarivan, T.: A fast and elitist multiobjective genetic algorithm: NSGA II. IEEE Transactions on Evolutionary Computation 6(2), 182–197 (2002)
13. Ishibuchi, H., Yoshida, T., Murata, T.: Balance between genetic search and local search in memetic algorithms for multiobjective permutation flowshop scheduling. IEEE Transactions on Evolutionary Computation 7(2), 204–223 (2003)
14. Jozefowiez, N., Semet, F., Talbi, E.G.: Multi-objective vehicle routing problems. European Journal of Operational Research, 293–309 (2008)
15. Jozefowiez, N., Semet, F., Talbi, E.G.: An evolutionary algorithm for the vehicle routing problem with route balancing. European Journal of Operational Research, 761–769 (2009)
16. Tan, K.C., Chew, Y.H., Lee, L.H.: A hybrid multiobjective evolutionary algorithm for solving vehicle routing problem with time windows. Comput. Optim. Appl. 34, 115–151 (2006)
17. Geiger, M.J.: A computational study of genetic crossover operators for multi-objective vehicle routing problem with soft time windows. CoRR (2008)
18. Tan, K., Cheong, C., Goh, C.: Solving multiobjective vehicle routing problem with stochastic demand via evolutionary computation. European Journal of Operational Research 177(2), 813–839 (2007)
19. Ghoseiri, K., Ghannadpour, S.F.: Multi-objective vehicle routing problem with time windows using goal programming and genetic algorithm. Appl. Soft Comput., 1096–1107 (2010)
20. Zhang, Q., Li, H.: MOEA/D: A multi-objective evolutionary algorithm based on decomposition. IEEE Transactions on Evolutionary Computation 11(6), 712–731 (2007)
21. Konstantinidis, A., Yang, K., Zhang, Q., Zeinalipour-Yazti, D.: A multi-objective evolutionary algorithm for the deployment and power assignment problem in wireless sensor networks. New Network Paradigms, Elsevier Computer Networks 54, 960–976 (2010)
22. Goldberg, D.E., Lingle, R.: Alleles, loci, and the traveling salesman problem. In: Grefenstette, J.J. (ed.) Proceedings of the First International Conference on Genetic Algorithms and Their Applications. Lawrence Erlbaum Associates, Publishers (1985)
23. Tan, K.C., Chew, Y.H., Lee, L.H.: A hybrid multi-objective evolutionary algorithm for solving truck and trailer vehicle routing problems. European Journal of Operational Research, 855–885 (2006)
24. Zitzler, E., Thiele, L.: Multiobjective evolutionary algorithms: a comparative case study and the strength pareto approach. IEEE Trans. Evolutionary Computation, 257–271 (1999)
25. Czyzak, P., Jaszkiewicz, A.: Pareto simulated annealing - a metaheuristic technique for multiple-objective combinatorial optimization. Journal of Multi-Criteria Decision Analysis 7(1), 34–47 (1998)

Automatic Exercise Generation in Euclidean Geometry

Andreas Papasalouros

Department of Mathematics, University of the Aegean, 83200 Karlovassi, Greece

Abstract. Automatic assessment has recently drawn the efforts of researchers in a number of fields. While most available approaches deal with the construction of question items that assess factual and conceptual knowledge, this paper presents a method and a tool for generating questions assessing procedural knowledge, in the form of simple proof problems in the domain of the Euclidean Geometry. The method is based on rules defined as Horn clauses. The method enumerates candidate problems and certain techniques are proposed for selecting interesting problems. With certain adaptations, the method is possible to be applied in other knowledge domains as well.

Keywords: exercise generation, knowledge representation.

1 Introduction

Automatic assessment has recently drawn the efforts of researchers in a number of fields including Education Research, Cognitive Psychology and Artificial Intelligence. Automated assessment has a number of potential benefits:

- It facilitates the process of creating question repositories.
- It may support the personalization of questions generated based on student profile and personal learning goals.
- It may become a component of Intelligent Tutoring Systems, where already stored domain knowledge and pedagogic strategies may become the basis for automatic problem construction.
- It can become a crucial component in the process of automation of instructional design.

In this work we envisage a knowledge-based framework where subject domain knowledge will be combined with pedagogic strategies and a knowledge-based description of learning goals, which eventually will drive the whole instructional design process in the sense of generating content knowledge descriptions, examples, and (self-)assessment.

Automatic assessment mainly focuses on generating questions that assess factual or conceptual knowledge. Nevertheless, not much work has been conducted in the direction towards automatic assessment of problem solving skills. The present work aims at generating assessment items for problem solving in the

H. Papadopoulos et al. (Eds.): AIAI 2013, IFIP AICT 412, pp. 141–150, 2013.

form of *proof* exercises in Euclidean Geometry. The domain knowledge that is used is considered to be stored into a knowledge base in the form of *rules* [1] that are used by the learner during the solving process. Problems are stated as a set of *assumptions* (givens) and a *statement* (goal) to be derived from the givens. In the case of Geometry proof problems, which we are dealing with in this paper, assumptions and statement to be proven are related through geometric relationships among geometric elements, i.e. segment lengths and angle sizes, possibly with an associated diagram of the geometric setting under consideration [2].

A student solver is anticipated to successively apply selected rules which are stored in the knowledge base in order to prove the goal statement. These rules correspond to certain kinds of mathematical knowledge: axioms, theorems and definitions [3]. Student knowledge representation with rules is pedagogically sound. For example, Scandura [1] suggests that both lower and higher order knowledge is described in terms of Post-like production rules.

Research in problem solving, either in general or in the field of Mathematics [4, 3], has shown that problem solving skills involve understanding and the application of meta-cognitive and strategic knowledge. This work aims at generating meaningful problems that assess basic understanding and rule application, while the assessment of strategic knowledge is left as a future work.

Furthermore, generated problems are intended to assess *understanding* rather than engage students in a mechanical trial and error procedure for reaching the intended goals. Thus, the presented method has a *cognitive* [5] perspective for the assessment of students. Although exercises generated with the proposed method are simple enough, they are anticipated to *assimilate* exercises that experienced teachers should either generate themselves, or select for application in real educational settings. Although the domain of generated problems is restricted to Euclidean Geometry, the presented method may potentially be applied in other areas of Mathematics, as well in non-mathematical fields.

The structure of this paper is as follows: The relation of the proposed method with other methods in the literature follows in Section 2. Section 3 presents the theoretical background of the proposed method. Section 4 describes the algorithm for problem generation and its implementation, followed by an evaluation in Section 5. The paper ends with some conclusions, focusing on the generalization of the proposed method in other domains.

2   Related Work

Heeren et al. [6] have developed a methodology for describing problem solutions and feedback strategies based on functional programming. The methodology applies in educational problems that engage algorithmic solutions, e.g. matrix manipulation, finding roots of equations, algebraic manipulation of expressions. This methodology can be used in a reverse manner,that is, certain descriptions of problems can be instantiated with particular values, thus, yielding new problems that assess the application of algorithms under consideration. However, this approach is not appropriate for non-routine problems [7], such as the proof exercises in Euclidean Geometry, where no specific algorithm for problem solution is known by the learner.

Holohan et al. [8] describe a method for exercise generation in the domain of database programming. This method is based on an ontology that describes database schemata, which serves as input for generating exercises asking students to form queries based on textual query descriptions. The proposed method can generate exercises from domain specific ontologies, however it is not clear how declarative knowledge described in various ontologies can serve as input for problem generation in a uniform fashion across domains.

Williams [9] proposes the use of domain ontologies for generating mathematical word problems for overcoming the so-called question bottleneck in Intelligent Tutoring Systems (ITS). The semantics of OWL Semantic Web language are utilized. The proposed approach is based on Natural Language Generation (NLG) techniques and aims at exploiting existing Semantic Web knowledge bases in the form of ontologies and linked data. The difficulty of the questions can be specified, based on specific factors such as question length, the existence of distracting information, etc. The approach aims at generating meaningful questions.

The work in [10] describes the generation of multiple choice assessment items for assessing analogical reasoning. Again, domain ontologies are used as input for question generation. Analogies in questions are extracted by identifying certain structural relationships between concepts in knowledge bases in the form of OWL ontologies. Different levels of analogy are defined for extracting correct (key) and false items (distractors).

In [11] a question generation component of an ITS is described. Multiple choice questions are generated by utilizing OWL semantic relationships, subsumption, object /datatype properties and class/individual relationships. The presented method is based on a set of templates in two levels: At the semantic level, implemented as SPARQL queries, and at the syntactic /sentence realization level, implemented as XSL Transformations.

Papasalouros et al. [12] propose a number of strategies for multiple choice question generation, based on OWL semantic relationships. Besides text questions, the authors provide strategies for generating media questions, demonstrating their approach with image hot spot questions. Simple NLG techniques are utilized, so that questions are not always grammatically correct.

The above-mentioned approaches use ontologies for generating questions for mostly assessing declarative knowledge. Declarative (or conceptual) knowledge assessment deals with checking for relationships and meaning,thus ontologies, as formal expressions of semantic networks, are appropriate for expressing knowledge to be assessed. From the above approaches, only the work in [9] deals with problem generation, albeit relatively simple word problems. According to the typology of problem solving proposed in [7], word problems (named story problems) are considered as a different category than the so called 'rule-using problems', which are actually tackled in current work. Current work aims at non trivial problems [13] that assess procedural knowledge in the form of proofs that employ specific execution steps.

3 Theoretical Background

Problem solving has been a field of intensive study in both AI and Cognitive Psychology. Thus, a set of well established principles has been proposed. According to [4]

- problem solving is considered as searching into a problem space;
- a problem space refers to the solver's representation of a task. It consists of
  - set of knowledge states (initial, goal, intermediate),
  - a set of operators for moving from one state to another and
  - local information about the path one is taking through the state

Procedural knowledge, such as problem solving skills, can be modelled by a *production system* with three kinds of productions (knowledge)[14, 1]:

- propositions,e.g. knowledge of rules, etc. that can be applied to a particular situation towards problem solution;
- pattern recognition (matching), in which a particular rule is correctly applied to a given situation yielding a new situation, that is, a new state in the problem space;
- strategic knowledge, which guides the process of rule application through certain heuristics, or engages higher order procedures, such as scripts, that is, proper encodings of already known solutions to intermediate sub-problems.

Thus, given the operators for moving from one state to another, that is, the domain-specific rules, a problem can be defined by identifying the initial and goal states in the above sense. Then, the role of problem solvers is to construct their own path into the problem space, towards the solution of the problem.

Proof exercises in Euclidean Geometry, as well in any other domain, is a kind of procedural knowledge. An exercise can be described through a proof tree, such as the one depicted in Fig. 1 for proving the congruency of two segments. The role of the learner is to correctly apply specific rules, in the form of axioms and theorems, in order to construct the proof. Thus, in the case of proofs such as the above, each state in the problem space is a tentative form of the proof tree, the goal state being the proof tree under consideration, or any tree that proves the statement under consideration under the given sentences.

As an example, we consider a knowledge base on Euclidean Geometry containing the following simple rules, properly encoded:

$$\left.\begin{array}{c} \angle ABC \cong \angle DEF \\ \overline{AB} = \overline{DE} \\ \overline{BC} = \overline{EF} \end{array}\right\} \Rightarrow \triangle ABC \cong \triangle DEF \quad \text{(Side-Angle-Side rule)}$$

$$\triangle ABC \cong \triangle DEF \Rightarrow \overline{AB} = \overline{DE} \quad \text{(Triangle congruency)}$$

$$\angle ABC \cong \angle ACB \Rightarrow \overline{AB} = \overline{AC} \quad \text{(Equilateral triangles)}$$

Then, a proof tree, concerning the above rules is presented in Fig. 1 together with the corresponding diagram. In the following section, a method for generating exercise of this kind will be presented.

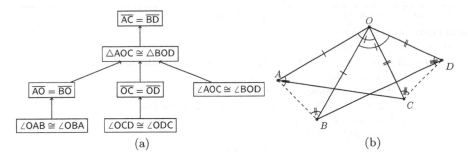

Fig. 1. An example of a generated proof tree (a) and corresponding diagram (b)

4   Problem Generation and Ranking

4.1   Representation of Domain Knowledge

We consider a knowledge base that contains a set C of Horn clauses. Each clause has the form

$$b_1, \ldots, b_n \Rightarrow h,$$

where literal h is the head of the expression,

All literals $b_1, \ldots, b_n, h$ denote relationships between geometry elements, such as congruence, in first order logic.

The following geometry elements are considered: Segments, angles and triangles. A literal example is the following expression:

$$cong(triangle(A, B, C), triangle(D, E, F))$$

In the above, $cong$ is a predicate denoting congruence, $triangle(A, B, C)$ is a term referring to a triangle while $A, B, \ldots, F$ are variables denoting points. The predicate denotes the relationship $\triangle ABC \cong \triangle DEF$.

The root of the proof tree, as depicted in Fig. 1, is the head of some clause $c \in C$. Each node in the tree represents a literal. Every node in the tree represents a head of some clause in the knowledge base. More specifically, for every internal node, t there exists a clause, c in the knowledge base such that the head, h, unifies with t and each child of t unifies with a corresponding clause in the body of c.

4.2   Problem Generation

We present an algorithm that generates trees representing problems. The aim of the algorithm is, given a set of clauses, the enumeration of all *proof* trees of a given maximum depth. The algorithm is presented below.

Algorithm 1 generates all proof trees that can be created by the set C of clauses up to a given depth, namely Max, that satisfy the given constraints. Each literal, L, constitutes a node in the generated proof tree. The above recursive

Algorithm 1. Problem generation algorithm

---

  procedure GENERATE(L,*Depth*,*Path*, *Tree*)
    if *Depth* $\leq$ *Max* **then**
      for all clauses $c \in C$ in the form $b_1, \ldots, b_n \Rightarrow H$ such that
          L' is a unifier of H and L **do**
        Let *Path'* the unification of *Path*
        if $L' \notin Path'$ **and** every node in *Path'* is *valid* **then**
           Add node L' to *Tree*
           Let $Path' = Path' \cup \{L'\}$
           Let $b'_1 \ldots, b'_n$ be the unifications of $b_1 \ldots, b_n$
           for all $b \in \{b'_1 \ldots, b'_n\}$ **do** GENERATE(b, *Depth* + 1, *Path'*, t)
               Add t as a child of L' in the *Tree*
           end for
        end if
      end for
    end if
  end procedure

---

procedure generates the tree in a backward chaining fashion. That is, the root of the tree is a sentence to be proved, while the leaves of the tree are either known literals, such as tautologies and other self-evident predicates, or predicates considered as *assumptions* (givens). The first clause selected by the procedure defines the sentence to be proved. However, the exact form of the predicate to be proved, in terms of the actual points involved, is specified during the execution of the algorithm, since variable substitution may take place during expression unification. During tree construction the algorithm checks whether all nodes in the generated path are *valid*. A sentence is valid if participating terms, angles, triangles, segments, are well formed, e.g. in a segment AB, point A is different from point B.

In Algorithm 1 variables denote points, thus the unification of terms engages only variable substitutions. Unification is executed in the usual sense: if there exists a substitution θ that unifies L and H, i.e. $L\theta = H\theta$, then $b'_1 = b_1\theta, \ldots, b'_n = b_n\theta$ [15].

Term variables are never grounded during problem generation. It is assumed that variables with different names do always refer to different points.

4.3   Problem Ranking

This work aims towards the generation of interesting problems, appropriate for usage in real educational practice. The method described above systematically enumerates proof trees, each defining a corresponding problem. There is a need for characterizing problems on the basis of their pedagogic quality [10], as well as for selecting problems according to specific characteristics.

We consider as interesting exercises those that incorporate specific pedagogic qualities. More specifically, we assume that interesting problems engage learner understanding. Greeno [14] asserts that problems that promote understanding help learners develop internal representations that comprise three characteristics: *coherence* of the internal structure of a problem representation, for example, the coherence of a graph that represents a geometric diagram; *correspondence* between the representation of the problem and a real life representation meaningful to the learner, such as a visual depiction of a geometric diagram; and *connectedness* of general concepts in the learner's knowledge (connections).

Based on the above, a simple measure for identifying interesting exercises is proposed, by identifying certain characteristics of exercise definitions that *may* promote understanding. The degree of connectedness can be measured by the number of different rules involved in the solution of a particular problem. Furthermore, coherence of learner internal structures is promoted by problem definitions which are, by themselves, of high density, that is, engage a large number of connections, i.e., congruency relations, and a small number of related elements. In our problem definition, *points* are the basic elements that are related in the above sense. Thus, we assume that coherence increases with the number of different rules involved in the solution of a particular problem and decreases with the number of points in a particular problem. Regarding correspondence of problem representations, every proposition is a symbolic element that corresponds to an element in a diagram. Although corresponding diagrams are not yet created automatically in our approach, we are currently dealing with issue.

Considering the tree structure of the problem space, it is obvious that the size of the problem space grows exponentially with its depth [16]. Given that deep proof trees correspond to deep problem spaces, we assume that how interesting a problem is increases with the depth of the problem proof tree.

Regarding the estimation of the difficulty of a problem based on internal problem characteristics, we anticipate that difficulty increases with the depth of the problem space, as well as with the number of rules involved.

Given the above, we have identified certain metrics for evaluating generated problems according to their pedagogic *interest*, and *difficulty*, i.e. the maximum depth, M, of the generated problem space, the number, R, of different *rules* that were used for problem generation and the number, P, of points in a particular diagram. We anticipate that generated exercises are ranked for their *interest* according to $I = \frac{M \times R}{P}$, while they are ranked for difficulty according to $D = M \times R$.

4.4   Initial Prototype

A prototype has been implemented in PROLOG. This language seemed a natural choice due to its inherent support for the manipulation of logical expressions, unification and backtracking. The prototype generates the tree (initial, goal, intermediate states) by implementing Algorithm 1. Implementation follows certain conventions for geometric rules representation adopted from [17].

The prototype is able to generate exercises in symbolic form and we are currently working towards the semi-automatic construction of diagrams from exercise descriptions.

5   Evaluation

A pilot evaluation was conducted in order to estimate the feasibility of the whole approach. We entered 7 rules in the system, which correspond to specific geometric theorems: Side-Angle-Side rule of triangle congruency, equal angles in corresponding sides in equilateral triangles and equality of corresponding sides in congruent triangles. The prototype generated 270 problems, some of which were isomorphic, that is, identical from a pedagogical point of view. Note that the problem illustrated in Fig. 1 was among the ones generated. All problems were correct, as checked by the author of this paper. From these, 10 problems were selected for evaluation, covering almost evenly the range of all calculated values in interest and difficulty. The small number of problems was due to evaluators' limited time availability.

Selected problems were given to two experienced High School Mathematics teachers in order to be ranked according to two criteria: interest (*Would you assign the particular problem to your students as an understanding/application exercise?*) and difficulty (*How difficult do you consider the problem in question, related to the other problems?*). For each problem, the assumptions, the sentence to be proved as well as a diagram generated by hand were given to the participants. Teachers were allowed to assign the same rank to specific problems. The concordance of teacher rankings was measured by using Kendall's Tau-b correlation coefficient [18]. Concordance was computed for the rankings of the two participants, as well as between each participant and our proposed measures. Corresponding values are depicted in Table 1.

Table 1. Selected problems evaluation data

Difficulty	
T1 and T2	0.51
T1 and $M \times R$	0.41
T2 and $M \times R$	0.73
Interest	
T1 and T2	0.58
T1 and $\frac{M \times R}{P}$	0.47
T2 and $\frac{M \times R}{P}$	0.62

Although the number of participants is very limited and the report of the above data is anecdotal, we see that for both difficulty and interest, the concordances between each teacher and our metrics is of the same level as the corresponding concordances between teachers. Although these results by no means can be generalized, they are hopeful initial indicators of the potential validity of the proposed measures for exercise selection, proving a useful basis for further justification and /or adjustment.

6 Conclusions

Preliminary evaluation has shown that the proposed method can generate useful exercises in Euclidean Geometry. However, the method can be generalized in other domains as well. The description of rules in the form of clauses is universally accepted in both cognitive psychology and knowledge representation literature as a domain independent formalism. The method generates problems that are evaluated according to search space depth, number of rules and problem element coherence /parsimony. From the above, only the measure of coherence / parsimony is defined in a domain-specific fashion. However, the latter characteristic can be easily adapted to specific problem domains. Nevertheless, the above should be demonstrated by applying this method in other domains as future research.

Solving problems in school by no means is limited to well structured problems such as the ones generated here. Generally, the solution of knowledge rich problems, inside and outside educational environments, involves the identification of patterns of solutions, the development of certain, usually domain specific, heuristics [16], which are not covered by our approach. Furthermore, the difficulty of problems in practice does not depend mainly on internal problem properties, but rather on the knowledge of similar problems by the learner.

Thus, more successful problem generation should involve the implication of heuristics, as well as rules; case based reasoning techniques may also be involved, in the sense that a problem generation agent should be based on a knowledge base containing not only a set of rules but also properly indexed problems together with their solutions.

Acknowledgement. The author would like to thank Mr Christos Tsaggaris for his valuable help in the evaluation presented in this paper.

References

[1] Scandura, J.M.: Knowledge representation in structural learning theory and relationships to adaptive learning and tutoring systems. Tech., Inst., Cognition and Learning 5, 169–271 (2007)

[2] Anderson, J.R., Boyle, C.F., Farrel, R., Reiser, B.J.: Modelling cognition. In: Morris, P. (ed.) Cognitive Principles in the Design of Computer Tutors, pp. 93–133. John Wiley and Sons Ltd. (1987)

[3] Schoenfeld, A.H.: Handbook for research on mathematics teaching and learning. In: Grows, D. (ed.) Learning to Think Mathematically: Problem Solving, Metacognition, and Sense-Making in Mathematics, pp. 334–370. MacMillan, New York (1992)

[4] Novick, L.R., Bassok, M.: Cambridge handbook of thinking and reasoning. In: Holyoak, J., Morrison, R.G. (eds.) Problem Solving, pp. 321–349. Cambridge University Press, New York (2005)

[5] Greeno, J.G., Pearson, P.D., Schoenfeld, A.H.: Implications for NAEP of Research for Learning and Cognition. Institute for Research on Learning, Menlo Park (1996)

[6] Heeren, B., Jeuring, J., Gerdes, A.: Specifying rewrite strategies for interactive exercises. Mathematics in Computer Science 3(3), 349–370 (2010)

[7] Jonassen, D.H.: Toward a design theory of problem solving. ETR&D 48(4), 63–85 (2000)

[8] Holohan, E., Melia, M., McMullen, D., Pahl, C.: The generation of e-learning exercise problems from subject ontologies. In: ICALT, pp. 967–969. IEEE Computer Society (2006)

[9] Williams, S.: Generating mathematical word problems. In: 2011 AAAI Fall Symposium Series (2011)

[10] Alsubait, T., Parsia, B., Sattler, U.: Mining ontologies for analogy questions: A similarity-based approach. In: Klinov, P., Horridge, M. (eds.) OWLED. CEUR Workshop Proceedings, vol. 849. CEUR-WS.org (2012)

[11] Žitko, B., Stankov, S., Rosić, M., Grubišić, A.: Dynamic test generation over ontology-based knowledge representation in authoring shell. Expert Systems with Applications 36, 8185–8196 (2009)

[12] Papasalouros, A., Kotis, K., Kanaris, K.: Automatic generation of tests from domain and multimedia ontologies. Interactive Learning Environments 19(1), 5–23 (2011)

[13] Schoenfield, A.H.: On having and using geometric knowledge. In: Hiebert, J. (ed.) Conceptual and Procedural Knowledge: The Case of Mathematics, pp. 225–264. LEA Publishers, Hillsdale (1986)

[14] Greeno, J.G.: Understanding and procedural knowledge in mathematics instruction. Educational Psychologist 12(3), 262–283 (1987)

[15] Chang, C.-L., Lee, R.C.-T.: Symbolic Logic and Mechanical Theorem Proving, 1st edn. Academic Press, Inc., Orlando (1997)

[16] Chi, M.T., Glaser, R.: Problem solving ability. In: Sternberg, R. (ed.) Human Abilities: An Information-Processing Approach, pp. 227–257. W. H. Freeman & Co. (1985)

[17] Coelho, H., Pereira, L.M.: Automated reasoning in geometry theorem proving with prolog. Journal of Automated Reasoning 2, 329–390 (1986)

[18] Kendall, M.: Rank Correlation Methods. C. Griffin, London (1975)

A Cloud Adoption Decision Support Model Using Influence Diagrams

Andreas Christoforou and Andreas S. Andreou

Department of Electrical Engineering / Computer Engineering and Informatics,
Cyprus University of Technology
ax.christoforou@edu.cut.ac.cy, andreas.andreou@cut.ac.cy

Abstract. Cloud Computing has become nowadays a significant field of Information and Communication Technology (ICT), and this has led many organizations moving their computing operations to the Cloud. Decision makers are facing strong challenges when assessing the feasibility of the adoption of Cloud Computing for their organizations. The decision to adopt Cloud services falls within the category of complex and difficult to model real-world problems. In this paper we propose an approach based on Influence Diagrams modeling, aiming to support the Cloud adoption decision process. The developed ID model combines a number of factors which were identified through litterature review and input received from field experts. The proposed approach is validated against four experimental cases, two realistic and two real-world, and its performance proved to be highly capable of estimating and predicting correctly the right decision.

Keywords: Influence Diagrams, Cloud Adoption, Decision Support.

1 Introduction

Cloud Computing is changing the whole perspective with which we understand computing today. The adoption of Cloud Computing is still a major challenge for organizations daily producing and processing information in the context of their working activities. A constantly increasing number of companies include Cloud Computing in their short or long term planning since sufficient number of services that are available on the Cloud has surpassed infancy and appears to be quite mature and attractive. Many of the major software developers or service providers have already turned their strategy towards Cloud services mostly targeting at increasing their market share. On one hand companies-customers need to consider the benefits, risks and effects of Cloud Computing on their organization in order to proceed with adopting and using such services, and on the other hand Cloud Computing providers need to be fully aware of customers' concerns and understand their needs so that they can adjust and fit their services accordingly.

Although in recent years the research community has increasingly been interested in this field, a review of the literature on Cloud Computing, and especially on Cloud adoption, revealed that there are yet no mature techniques or toolkits to support the decision making process for adopting Cloud services on behalf of customers.

H. Papadopoulos et al. (Eds.): AIAI 2013, IFIP AICT 412, pp. 151–160, 2013.

The ambiguity and uncertainty often surrounding the Cloud adoption, deriving of the multiple, conflicting factors in combination with simplistic assumptions, makes Cloud adoption a highly complex process that cannot be satisfied using classical and linear methods. Furthermore, the extremely fast-moving nature of the Cloud Computing environment changes, both to supply and demand, shows how difficult it may be for any procedure to assist the decision making process timely and correctly. This is the reason why a framework or model which supports Cloud Computing adoption should be quite flexible and dynamically adaptable.

In this paper we propose a methodology based on Influence Diagrams (IDs) which was used to set-up a successful decision model to support Cloud adoption. The model was constructed in a systematic manner: Firstly we performed a study of the most recent and relevant literature on Cloud Computing and particularly on Cloud adoption, through which we identified all possible factors that influence the final Cloud adoption decision. Next, based on the result of this study, we proceeded to categorized those factors, and build and distribute a questionnaire to a group of experts so as to capture their knowledge and expertise as regards approving the list of factors already identified. In addition, the experts were called to define the relation of each factor to Cloud adoption and a corresponding weight on a Likert scale. Finally and using the collected information we developed a novel model based on Influence Diagrams that answers the question "Adopt Cloud Services?" under the current state of the offered service and the associated factors describing each customer's particular situation at the moment of decision.

The rest of the paper is organized as follows: Section 2 presents related work in the area of Cloud Computing adoption based on the existing literature. Section 3 makes a brief description of the theory of Influence Diagrams, while section 4 introduces the Cloud adoption modeling process and discusses and analyses the corresponding experimental results. Finally, section 5 provides our conclusions and suggestions for future research steps.

2   Related Work

Among many definitions of Cloud Computing, a working definition that has been published by the US National Institute of Standards and Technology (NIST) [8], captured the most common agreed aspects. NIST defines Cloud Computing as "a model for enabling ubiquitous, convenient, on-demand network access to a shared pool of configurable computing resources (e.g., networks, servers, storage, applications and services) that can be rapidly provisioned and released with minimal management effort or service provider interaction." This Cloud model promotes availability and is composed of five essential characteristics, three service models and four deployment models as follows:

- Characteristics: on-demand self-service, broad network access, resource pooling, rapid elasticity and measured service.
- Service models: Software as a Service (SaaS), Platform as a Service (PaaS) and Infrastructure as a Service (IaaS).
- Deployment models: private Clouds, community Clouds, public Clouds and hybrid Clouds.

Although it is generally recognized that the adoption of Cloud Services can offer substantial benefits, many organizations are still reluctant to proceed with it. There is a variety of factors that may influence Cloud adoption and it is quite important to properly identify and analyze them aiming to assist customers in taking the right decision. Equally important in this study, from the vendors' point of view, is to define which factors should possibly change so as to revert a current negative adoption decision. Our research is mainly focused on the SaaS model investigating both the Cloud providers' and the customers' point of view.

An investigation of the current literature revealed a relatively small number of papers discussing Cloud adoption from the perspective of decision making and also current feasibility approaches fall short in terms of decision making to determine the right decision. We introduce a summary of these studies, examining the contribution of each work to the decision making problem. In [9] Khajeh-Hosseini et al. presented a Cloud adoption toolkit which provides a framework to support decision makers in identifying their concerns and match them with the appropriate techniques that can be used to address them. Kim et al. [10] examined various issues that impede rapid adoption of Cloud Computing such as cost, compliance and performance. Wu in [6] attempted to contribute to the development of an explorative model that extends the practical applications of combining Technology Acceptance Model (TAM) related theories with additional essential constructs such as marketing effort, security and trust, in order to provide a useful framework for decision makers to assess the issue of SaaS adoption and for SaaS providers to become sensitive to the needs of users. Additionally, Wu [7] explored the significant factors affecting the adoption of SaaS by proposing an analytical framework containing two approaches: the Technology Acceptance Model (TAM) related theories and the Rough Set Theory (RST) data mining. In [11], a solution framework is proposed that employs a modified approach proposed in the 70s named DEMATEL [12] to cluster a number of criteria (perceived benefits and perceived risks) into a cause group and an effect group respectively, presenting also a successful case study. Even though all of the above techniques contribute a significant piece to this new open research field, they may be classified as "traditional", single layer approaches which examine only a specific part of the problem.

Techniques that use IDs in modeling decision process seem to improve the way the problem is approached by offering various strong benefits. IDs offer flexibility representing many dependencies between factors and manage to represent a highly complex problem in a human understandable way. Also, IDs allow interaction of experts through execution of the model with input combinations thus utilizing their expertise in order to calibrate the model and achieve reasonable and helpful answers.

3   Influence Diagrams

An Influence Diagram (ID) [2] is a general, abstract, intuitive modeling tool that is nonetheless mathematically precise [1]. IDs are directed graph networks, with different types of nodes representing uncertain quantities, decision variables, deterministic functions, and value models. IDs were first developed in the mid 1970s as a decision analysis tool to offer an intuitive way to identify and display the essential elements, including decisions, uncertainties, and objectives, and how they influence each other.

In general an ID is a directed acyclic graph with three types of nodes and three types of arcs between nodes. The first is called Decision node, it is drawn as a rectangle and corresponds to some decision to be made. Chance or Uncertainty node is the second type, which is drawn as an oval and represents an uncertainty to be modeled. The third one is the Value node, which is drawn as a hexagon (or octagon) or diamond, and calculates all possible combinations receiving from factors in the modeling environment acting as parent nodes. A Functional arc ends at a value node and represents the contribution of the node at its tail to the calculated value. The second type of arc is the Conditional, which ends at a chance node and indicates that the uncertainty at its head is probabilistically related to the node (oval) at its tail. Finally, an Informational arc ends at a decision node and indicates that the decision at its head is made according to the outcome of the node at its tail, which is known beforehand. A simple example of an ID is presented in Figure 1.

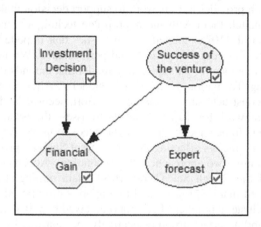

Fig. 1. A simple Influence Diagram

4   Modeling the Cloud Adoption Process

4.1   Model Design

Modeling the Cloud adoption decision-making process was implemented combining two methods: (i) Literature study and (ii) Collection of expert opinion through specially prepared questionnaires followed by interviews. More specifically, a small-scale literature review on the subject was conducted in order to identify a number of factors that potentially influence such a decision which would then be used to form the nodes of our model. The next step involved identifying a group of three experts with strongly related background to the subject (i.e. Cloud Computing related positions). An initial list of factors was then prepared and the experts were asked to evaluate the list and prompted to add or remove factors based on their expertise and working experience. Finally, one more round of discussion with experts was conducted in order to finalize the list of factors. These factors were used to form the nodes of the ID model and are listed in Table 1.

Table 1. Factors Influencing Cloud Adoption

Name	Definition	Reference
Legal Issues	Cloud adoption compliance with all legislative issues. Ability to adjust when legal requirements grow.	[23],[17],[18],[25]
Availability	The amount of time that Cloud Service(s) is operating as the percentage of total time it should be operating.	[14],[9],[15],[10], [23],[18]
Security	Security of service: data transfer, data stores, web servers, web browsers.	[4],[14],[9],[15], [16],[10],[23],[17], [18],[25]
Cost / Pricing	Operational - running costs, migration costs etc. Cost benefits from Cloud adoption.	[4],[13],[14],[9], [15],[16],[24],[10], [23],[25],[26,][27]
ROI	Return on Investment.	[13],[9],[16]
Compliance	Business and Regulatory compliance.	[9],[10],[23],[18]
Performance/Processing	Does Cloud adoption perform the process to the desired quality?	[14],[9],[15],[16], [10],[23]
Scalability	Ability to meet an increasing workload require-ment by incrementally adding a proportional amount of resources capacity.	[15],[17]
Privacy/ Confidentiality	Privacy and confidentiality coverage.	[4],[13],[9],[18], [25],[22]
Elasticity	Ability to commission or decommission resource capacity on the fly.	[14],[9],[25]
Data Access / Import-Export	Access to data in various ways.	[15],[18],[25]
Technology Suitability	Does Cloud technology exhibit the appropriate technological characteristics to support the pro-posed SaaS?	[14],[9],[25]
Hardware Access	Degree of Cloud Service accessibility, on local hardware.	[14],[9]
Audit ability	Ability of Cloud service to provide access and ways for audit.	[15]

Considering the influencing factors that were extracted and processed as described above, we proceeded with the development of the ID shown in Figure 2 using the GeNIe toolbox [3]. The nodes representations and their dependencies are able to model our question and provide a final decision node.

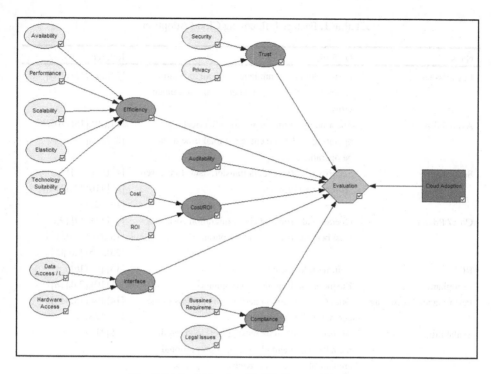

Fig. 2. "Adopt Cloud or Not" Influence Diagram

4.2 Experimental Results

Aiming to test and evaluate the performance of the proposed model, two hypothetical scenarios were first conducted representing the so called "extreme cases", that is, a situation where everything would be in favor of Cloud adoption (positive scenario) and the opposite case (negative scenario). The target was to have a reasonable answer by the evaluation node under known situations and assess the performance of the model demonstrating that the model behaves correctly and as expected to. Next, the model was tested on a number of real-world scenarios, that is, cases collected from real customers of two international Cloud services providers with the aid of the same experts that were utilized to construct our model. The two extreme scenarios and the real-world cases experimentation are described below. Table 2 presents the input values to the model that describe the current situation at the point of time when the decision was about to be made for each of the four scenarios tested expressed in, linguistic terms And transformed to their numerical counterparts.

Table 2. Input values for the four scenarios tested

Factor	Term	Positive	Negative	Real 1	Real 2
Legal	High	0.8	0	0.6	0.2
	Medium	0.2	0.2	0.3	0.6
	Low	0	0.8	0.1	0.2
Availability	High	0.8	0	0.6	0.2
	Medium	0.2	0.2	0.3	0.6
	Low	0	0.8	0.1	0.2
Security	High	0.8	0	0.6	0.2
	Medium	0.2	0.2	0.3	0.6
	Low	0	0.8	0.1	0.2
Cost / Pricing	High	0	0.8	0.3	0.7
	Medium	0.2	0.2	0.6	0.3
	Low	0.8	0	0.1	0
ROI	High	0.8	0	0.3	0
	Medium	0.2	0.2	0.6	0.2
	Low	0	0.8	0.1	0.8
Compliance	High	0.8	0	0.6	0
	Medium	0.2	0.2	0.3	0.2
	Low	0	0.8	0.1	0.8
Performance/Processing	High	0.8	0	0.3	0
	Medium	0.2	0.2	0.6	0.2
	Low	0	0.8	0.1	0.8
Scalability	High	0.8	0	0.8	0
	Medium	0.2	0.2	0.2	0.6
	Low	0	0.8	0	0.4
Privacy/ Confidentiality	High	0.8	0	0.3	0
	Medium	0.2	0.2	0.6	0.2
	Low	0	0.8	0.1	0.8
Elasticity	High	0.8	0	0.8	0
	Medium	0.8	0	0.2	0.6
	Low	0.2	0.2	0	0.4
Data Access / Import-Export	High	0	0.8	0.6	0
	Medium	0.8	0	0.3	0.6
	Low	0.2	0.2	0.1	0.4
Technology Suitability	High	0	0.8	0.6	0
	Medium	0.8	0	0.3	0.6
	Low	0.2	0.2	0.1	0.4
Hardware Access	High	0	0.8	0.3	0.2
	Medium	0	0.8	0.6	0.8
	Low	0.2	0.2	0.1	0
Auditability	High	0.8	0	0.3	0
	Medium	0.8	0	0.6	0.6
	Low	0.2	0.2	0.1	0.4

Scenario 1: Positive Case

This case assumes an ideal environment where the Cloud services offered perfectly match a customer's needs. Thus, the values for each leaf node were chosen so that they reflect this ideal setting and guide the evaluation node to a positive value. In this scenario the diagram executed and the values on the evaluation node were calculated to 0.77 for "Yes" and 0.23 for "No". This means that the model correctly recognized the positive environment and suggested that a decision in favor of Cloud adoption should be taken based on the values "read" in the nodes and the current influences between them.

Scenario 2: Negative Case

Working in the same way as with the positive scenario, appropriate values for each leaf node were chosen this time to guide the evaluation node to a negative value. Executing the model using the values for the negative scenario values the evaluation node yielded 0.75 for "No" and 0.25 for "Yes" which perfectly matched the expected behavior.

By using the above "extreme" scenarios it became evident that the proposed model behaves successfully as it recognized correctly the conditions of the environment and predicted the right decision. Therefore, we may now proceed to test its performance on real-world scenarios.

Real-World Scenarios

As mentioned earlier, we identified two different cases, with the help of Cloud providers. One customer who decided to proceed with Cloud adoption and one customer who rejected it. In order to be able to retrieve the leaf nodes values for each case separately we had a series of interviews both with the Cloud providers and the customers. We managed to record and adjust these values, which essentially reflected the state of the offered service and the associated factors describing each customer's particular situation at the moment of decision.

The first case involved an academic institution with a medium to large size which requested a comprehensive solution for email services. In that case easily someone can discern from the input values that the conditions were in favor of a positive decision. The second case involved a medium insurance broker organization which requested a complete email Cloud package and also a Cloud infrastructure to fit a heavy tailored made owned system. In that case, someone can hardly make an assessment of the final decision, based on the input values.

By executing the proposed model on these two real world scenarios, the results given in Table 3 were produced which were compared with the real decisions. In both cases the model successfully predicted the right decision.

Table 3. Model's decisions compared with real decisions

Real Scenario	ID model's decision (Evaluation Node Values)	Real decision
A	Yes (0.61)	Yes
B	No (0.74)	No

5   Conclusions

Although Cloud Computing has gone from infancy to a new mature state, customers are still facing many challenges with respect to its adoption. The study and understanding of various parameters such as benefits and problems that are involved in this transition is far from an easy and straightforward procedure. This paper proposed a new approach, aiming to assist the Cloud adoption decision process. We demonstrated how a new model based on Influence Diagrams can be constructed and applied to face decision making on the Cloud adoption problem.

The proposed model was experimentally evaluated initially on two extreme scenarios, an ideal setting in favor of Cloud adoption and a completely negative leading to Cloud rejection, showing successful performance. This enabled further experimentation with two real-world scenarios collected form experts/developers in the local software industry. The model succeeded in matching its estimation with the corresponding real decisions.

Although the results may be considered quite encouraging, there are quite a few enhancements that may be performed the model so as to fit the problem more accurately. Further research steps can be separated in two groups. The first involves enriching the knowledge regarding Cloud Computing and its parameters and the second the optimization of the decision tool. The speed with which Cloud Computing and its corresponding technology evolves necessitates the continuous study of the model. In addition, more real-world case scenarios could give a helpful feedback for better calibration of the model and finally, possible expansion and re-identification of the diagram will be investigated so as to include more nodes representing better the real Cloud environment.

References

1. Shachter, R.D.: Evaluating Influence Diagrams. Operations Research 34(6), 871–882 (1986)
2. Howard, R.A.: Readings on the Principles and Applications of Decision Analysis: Professional collection, vol. 2. Strategic Decisions Group (1984)
3. Laboratory, D.S.: Graphical Network Interface (GeNIe). (University of Pittsburgh)
4. GeNIe&SMILE (2013), http://genie.sis.pitt.edu (retrieved January 15, 2013)
5. Khajeh-Hosseini, A., Sommerville, I., Sriram, I.: Research challenges for enterprise Cloud Computing. arXiv preprint arXiv:1001.3257 (2010)
6. Papatheocharous, E., Trikomitou, D., Yiasemis, P., Andreou, S.A.: Cost Modelling and estimation in agile software development environments using influence diagrams. In: 13th International Conference on Enterprise Information Systems (ICEIS(3)), Beijing, pp. 117–127 (2011)
7. Wu, W.-W.: Developing an explorative model for SaaS adoption. Expert Systems with Applications 38, 15057–15064 (2011)
8. Wu, W.-W.: Mining significant factors affecting the adoption of SaaS using the rough set approach. The Journal of Systems and Software 84, 435–441 (2011)
9. Mell, P., Grance, T.: The NIST Definition of Cloud Computing. National Institute of Standards and Technology (2009)
10. Khajeh-Hosseini, A., Greenwood, D., Smith, J.W., Sommerville, I.: The Cloud Adoption Toolkit: supporting Cloud adoption decisions in the enterprise. Software: Practice and Experience 42(4), 447–465 (2012)
11. Kim, W., Kim, S.D., Lee, E., Lee, S.: Adoption issues for Cloud Computing. In: MoMM 2009, pp. 2–5 (2009)
12. Wu, W.W., Lan, L.W., Lee, Y.T.: Exploring decisive factors affecting an organization's SaaS adoption: A case study. International Journal of Information Management 31(6), 556–563 (2011)

13. Gabus, A., Fontela, E.: World problems, an invitation to further thought within the frame-work of DEMATEL. BATTELLE Institute, Geneva Research Centre, Geneva, Switzerland (1972)
14. Yang, H., Tate, M.: Where are we at with Cloud Computing?: a descriptive literature review (2009)
15. Greenwood, D., Khajeh-Hosseini, A., Smith, J.W., Sommerville, I.: The Cloud adoption toolkit: Addressing the challenges of Cloud adoption in enterprise. Technical Report (2010)
16. Armbrust, M., Fox, A., Griffith, R., Joseph, A.D., Katz, R., Konwinski, A., Zaharia, M.: A view of Cloud Computing. Communications of the ACM 53(4), 50–58 (2010)
17. Benlian, A., Hess, T.: Opportunities and risks of software-as-a-service: Findings from a survey of IT executives. Decision Support Systems 52(1), 232–246 (2011)
18. Godse, M., Mulik, S.: An approach for selecting software-as-a-service (SaaS) product. In: IEEE International Conference on Cloud Computing, CLOUD 2009, pp. 155–158. IEEE (2009)
19. Yang, J., Chen, Z.: Cloud Computing Research and security issues. In: 2010 International Conference on Computational Intelligence and Software Engineering (CiSE), pp. 1–3. IEEE (2010)
20. Grandon, E.E., Pearson, J.M.: Electronic commerce adoption: an empirical study of small and medium US businesses. Information & Management 42(1), 197–216 (2004)
21. de Assunção, M.D., di Costanzo, A., Buyya, R.: A cost-benefit analysis of using Cloud Computing to extend the capacity of clusters. Cluster Computing 13(3), 335–347 (2010)
22. Dash, D., Kantere, V.: &Ailamaki, A (2009, An economic model for self-tuned Cloud caching. In . In: IEEE 25th International Conference on Data Engineering, ICDE 2009, pp. 1687–1693. IEEE (2009)
23. Pearson, S.: Taking account of privacy when designing Cloud Computing services. In: ICSE Workshop on Software Engineering Challenges of Cloud Computing, CLOUD 2009, pp. 44–52. IEEE (2009)
24. Dillon, T., Wu, C., Chang, E.: Cloud computing: issues and challenges. In: 24th IEEE International Conference on Advanced Information Networking and Applications (AINA), pp. 27–33. IEEE (2010)
25. Bibi, S., Katsaros, D., Bozanis, P.: Application development: Fly to the clouds or stay in-house? In: 2010 19th IEEE International Workshop on Enabling Technologies: Infrastructures for Collaborative Enterprises (WETICE), pp. 60–65. IEEE (2010)
26. Microsoft Study: Drivers & Inhibitors to Cloud Adoption for Small and Midsize Businesses, http://www.microsoft.com/en-us/news/presskits/ telecom/docs/SMBCloud.pdf
27. de Assunção, M.D., di Costanzo, A., Buyya, R.: A cost-benefit analysis of using cloud computing to extend the capacity of clusters. Cluster Computing 13(3), 335–347 (2010)

Human-Like Agents for a Smartphone First Person Shooter Game Using Crowdsourced Data

Christoforos Kronis, Andreas Konstantinidis, and Harris Papadopoulos

Department of Computer Science and Engineering, Frederick University, Nicosia, Cyprus

Abstract. The evolution of Smartphone devices with their powerful computing capabilities and their ever increasing number of sensors has recently introduced an unprecedented array of applications and games. The Smartphone users who are constantly moving and sensing are able to provide large amounts of *opportunistic/participatory data* that can contribute to complex and novel problem solving, unfolding in this way the full potential of *crowdsourcing*. Crowdsourced data can therefore be utilized for optimally modeling human-like behavior and improving the realizablity of AI gaming. In this study, we have developed an Augmented Reality First Person Shooter game, coined AR Shooter, that allows the crowd to constantly contribute their game play along with various spatio-temporal information. The crowdsourced data are used for modeling the human player's behavior with Artificial Neural Networks. The resulting models are utilized back to the game's environment through AI agents making it more realistic and challenging. Our experimental studies have shown that our AI agents are quite competitive, while being very difficult to distinguish from human players.

1 Introduction

The widespread deployment of Smartphone devices with their powerful computing capabilities and their ever increasing number of sensors has recently introduced an unprecedented array of applications (e.g., Google Play features over 650,000 apps with over 25 billion downloads[1]). A crowd of Smartphone users can be considered as a number of individuals carrying Smartphones (which are mainly used for sharing and collaboration) that are constantly moving and sensing, thus providing large amounts of *opportunistic/participatory data* [1–3]. This real-time collection of data can allow users to transparently contribute to complex and novel problem solving, unfolding in this way the full potential of *crowdsourcing* [4]. There is already a proliferation of innovative applications [5] founded on opportunistic/participatory crowdsourcing that span from assigning tasks to mobile nodes in a given region to provide information about their vicinity using their sensing capabilities (e.g., noise-maps [6]) to estimating road traffic delay [7] using WiFi beams collected by smartphones rather than invoking expensive GPS acquisition and road condition (e.g., PotHole [8].)

The real-time collection of realistic crowdsourced data can also unveil opportunities in the area of Artificial Intelligence, such as modeling human-like behavior [9–11] for social and socio-economic studies as well as for marketing and/or entertainment purposes. The latter is traditionally linked to game AI [12, 13], which as a term is mainly

---

[1] Sept. 26, 2012: Android Official Blog, http://goo.gl/F1zat

H. Papadopoulos et al. (Eds.): AIAI 2013, IFIP AICT 412, pp. 161–171, 2013.

used for describing non player characters (NPCs). The vast majority of AI games mainly relies on old AI technologies, such as A* and finite state machines [13]. More modern AI games, such as Unreal Tournament, Half-life, Warcraft and Supreme Commander that require more sophisticated AI utilize more recent methods such as reinforcement learning, neuroevolution, etc. (see e.g. [10, 11, 14]). However, most of these modern games rely on desktop-like game-playing [11] utilizing data based on the human's decisions only and ignoring spatio-temporal or other sensory information. Here it is important to note that the year 2011 has been extremely important for the Computing field, as the number of smartphones exceeded for the first time in history the number of all types of Personal Computers combined (i.e., Notebooks, Tablets, Netbooks and Desktops), marking the beginning of the post-PC era[2] and therefore the beginning of real-time Smartphone gaming.

In this study, we developed an Augmented Reality First Person Shooter game [11], coined AR Shooter, that allows the crowd to constantly contribute their game play along with various spatio-temporal information. The crowdsourced data are collected by utilizing almost every sensor of the Smartphone device and they are used for training AI agents. Specifically, the AR Shooter is founded on a cloud-based framework that is composed of a back-end that collects the data contributed by the crowd, which are used for training a group of Artificial Neural Networks (ANN) to model the human player's behavior. The resulting models are in turn incorporated back into the game's environment (front-end) through AI agents, making it more realistic and challenging. Our experimental studies have shown that our AI agents are not easy to distinguish from human players and perform much better than non-intelligent agents.

2   Problem Definition

It is straight forward to take a well-defined game, add some control structures for agents and begin teaching them. First person shooter (FPS) games and especially augmented reality first person shooter games are not well defined. There are no defined player tactics in an FPS game. The main purpose of the game is the survival of the fittest, to shoot or to be shot. The player is rewarded higher score the more competitive opponents he/she eliminates. But if the player is "trigger happy" or not aware of his surroundings he might end up getting caught off guard. The penalties for death are severe, letting the player wait for 15 seconds before continuing the competition. Meanwhile the other opponents have the opportunity to pick up geo-located health resources or even finish off a mutual opponent, whom the player has been hunting since the beginning of the game, leaving him/her with no points for the elimination. Even expert players cannot fully explain their tactics, but only give out general tips like not rushing into many enemies or staying in the same position for a large amount of time. Furthermore, there are many exceptions to these rules that make it extremely difficult to hard code a rule based system.

Initially, it is important to figure out the actions to be considered and the data needed for deciding what action to make. Ideally an FPS learning model would consider every

---

[2] Feb. 3, 2012: Canalys Press Release, http://goo.gl/T81iE

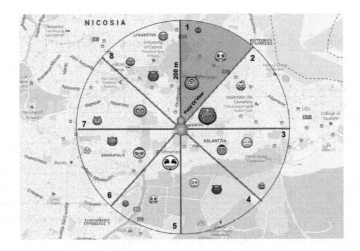

Fig. 1. The surrounding environment of the players is split into eight sectors in order to collect spatio-temporal information. The closest an opponent's avatar is to the user, the larger appears in the augmented reality view.

action made at any given time. In this study, however, we consider the following three basic actions:

- move: north/northeast/east/southeast/south/southwest/west/northwest/don't move
- select weapon: melee/assault/shotgun/sniper
- shoot: shoot/don't shoot

The heart of the problem is to make a choice for any of these decisions. How does an agent know if it's time to move north, change weapon and shoot? In order to achieve human like behavior we must model these decisions after actual players and learn how these decisions are made. The problem now becomes a more traditional machine learning problem. But many of the decisions made by players are reflexive and don't have an explanation behind them. Therefore, a good set of features must be found to adequately represent the environment of the game, but the more the features being considered, the higher the complexity of the model.

2.1 Feature Set and Data Set

The set of actions investigated in this study are movement, weapon selection and whether to shoot or not. These three actions will be the output of the model. The next step is to decide the information needed for making these decisions. One of the most important pieces of information is the enemy locations. Spatial data about enemy location were determined by breaking the surrounding environment of the agent into 8 sectors with a radius of 200 meters as shown in Figure 1. The sectors are relative to the player as actual world position is not as important as information about the player's immediate surroundings. Moreover the distance, health and sector of the closest enemy as well as the distance and sector of the closest resource available were recorded. Furthermore,

Table 1. The data used

Input (current information about the world around the player)	
Player Latitude	Current latitude of the player
Player Longitude	Current longitude of the player
Player Health	Current health of the player
Player weapon	Current weapon of the player
Player heading	Current heading of the player
Player speed	Current speed of the player
Player points	Current points of the player (one point for every kill, reset to zero when player dies)
Player time alive	The current time alive of the player
Closest Resource distance	Distance of the closest resource
Closest Resource sector	Sector to the closest resource
Closest Enemy Distance	Distance to the closest enemy
Closest Enemy Health	Health of the closest enemy
Closest Enemy Sector	Sector of the closest enemy
Enemy Shooting Sector	Sector of the enemy shooting at the player
Enemy Shooting Health	Health of the enemy shooting at the player
Enemy Shooting Distance	Distance of the enemy shooting at the player
Enemy Shooting Weapon	Selected Weapon of the enemy shooting at the player
Enemies per sector	The total number of enemies per sector
Player Shooting	If the player is shooting an enemy
Output (collected at next time step after input is sampled)	
Move direction	0 = do not move, 1 = move north etc.
Weapon	0 = melee, 1 = assault, 2 = shotgun, 3 = sniper
Shoot	0 = do not shoot, 1 = shoot

information about the enemy shooting at the player is useful like distance, sector and health.

In addition to enemy information, the player must be constantly aware of his health, heading, time alive, selected weapon and speed. This will allow us to determine tactics like retreat or advance. Table 1 summarizes the features considered in this study along with a brief description. These data were measured every five seconds. The output was a combination of the three basic actions: (i) move, (ii) select weapon and (iii) shoot.

3   Proposed Framework

In this section, the proposed framework is introduced, beginning with a general overview of the architecture and followed by a description of the ANNs used for modeling the human-like behavior of our AI agents and an introduction to the Graphical User Interface of our Windows Phone prototype system.

3.1   Overview

The proposed framework (see Figure 2) is mainly composed of the cloud-based back-end and the Smartphone front-end. The back-end consist of a Java server with a SQL

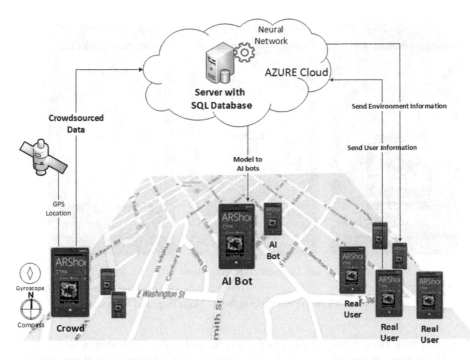

Fig. 2. The proposed architecture. The crowd contributes spatio-temporal information to the server on the clowd (left). The server uses the crowdsourced data to train AI agents that are deployed in the game's environment (center). The real users exchange information with the server regarding their location, the opponents' locations, health etc. (right).

database, which are deployed on the Microsoft's Azure cloud. The server side is responsible for (1) collecting the data contributed by the crowd, which are stored and maintained in the database, as well as (2) coordinating the game-play of the AR Shooter.

1) Collecting the Crowdsourced Data: Each user is able to locally store his/her game-play on his/her Smartphone. The data stored are those summarized in Table 1. In order to collect the users' data almost all sensors and connection modalities of the smarthone are utilized including the compass, gyroscope, accelerometer, GPS, 3G, WiFi and NFC. Note that the Near Field Communication (NFC) technology is only utilized when it is available on the Smartphone for mainly sharing resources with other team-players, such as life, weapons, ammunition etc. Then the user uploads to the server the collected data after finishing playing. Here it is important to note that a permission is granted from the user for sharing his/her data at login.

2) Coordinating the Game-Play: The server is responsible for the communication between the real Smartphone users as well as for coordinating the game play of the NPCs and the "dummy" agents (i.e., non-intelligent agents that mainly rely in random decisions). The communication between the real-users and the server is based on the TCP/IP protocol. The users's utilize their sensors to calculate their heading, location

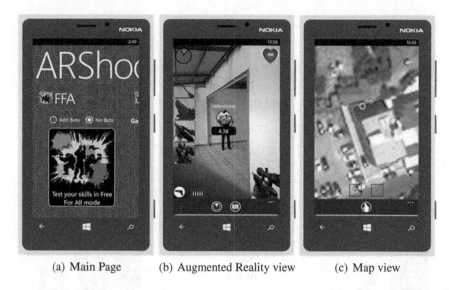

(a) Main Page   (b) Augmented Reality view   (c) Map view

Fig. 3. The GUI of the ARShooter Wars- Main Page and Views

and field of view (this is only needed in the augmented reality view that is introduced next). The server on the other hand, forwards to the users information about nearby users (e.g.,their health and location) and geo-located resources (such as medical packs, ammunition and weapons).

3.2   Graphical User Interface

The Graphical User Interface of our prototype system is interactive as well as easy to learn and use. It includes different view types and game modes and various features. Figure 3 (a) shows its main page. Figure 3 (b) shows the Augmented Reality view of a user, indicating his/her field view, an opponent that is engaged by the user's target (in the center of the smartphone screen) as well as the user's health in the upper right corner, his/her ammunition in the left bottom corner and a selection of weapons in the left side of the screen. Figure 3 (c) shows the map view with the user denoted by a solid circle and the opponents with their selected avatars. Note that the icon in the bottom center of the smartphone screen can be used to switch between the augmented reality and the map views.

Moreover, Figure 4 (a) shows the store where users can choose weapons, ammunition, etc., as well as unlock more advanced features (e.g., grenades) after collecting a specific amount of points (based on their kills). The specifications (i.e., range, damage and angle) of the weapon are summarized just below its image. Figure 4 (b) shows the animation when the user is being shot followed by a vibration of the smartphone and a decrease of his/her health (the decrease is based on the weapon that the opponent used.) Finally, Figure 4 (c) shows the screen that appears when a user is eliminated giving details about the opponent that eliminated him/her along with the user's number of kills. The user automatically re-enters the game after 15 seconds.

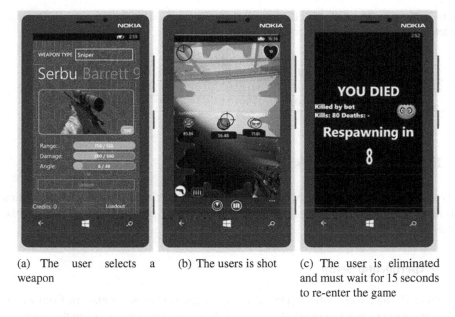

(a) The user selects a weapon

(b) The users is shot

(c) The user is eliminated and must wait for 15 seconds to re-enter the game

Fig. 4. The GUI of the ARShooter Wars - User functionality and actions

The ARShooter supports both a multi-player and a single-player modes. In the multi-player mode, the user plays against other human users, the "Dummy" agents and the AI agents. In the single player mode, the user plays only against the agents. In both modes, the game is in real time and in a real environment.

3.3   Human-Like Behavior Modeling

The developed model comprised of three ANNs, one for each of the three decisions: movement, weapon selection and shooting. The ANN were implemented using the WEKA data mining workbench. They had a two-layer fully-connected structure, with 26 input neurons and eight, four and one output neurons, respectively. Their hidden and output neurons had logistic sigmoid activation functions. They minimized mean squared error with weight decay using BFGS optimization. The network parameters were initialized with small normally distributed random values. All input features were normalized by setting their mean to zero and their variance to one. The combination of outputs produced by the three ANN given the current input formed the set of actions the agent would take next.

For determining the number of hidden neurons to use, a 10-fold cross-validation process was followed on the initially collected data trying out the values 5 to 50. The Root Mean Squared Errors (RMSEs) obtained are summarized in Table 2 with the best values denoted in bold. The number of hidden units that gave the best RMSE for each ANN was then used in the final model.

Table 2. The Root Mean Squared Error for each model (move, select weapon, shoot) with 5-50 hidden units. The best results are denoted in bold.

# of Hidden Units:	5	10	15	20	25	30	35	40	45	50
Movement:	0.119	0.120	0.110	0.110	0.104	0.107	0.107	**0.092**	0.097	0.098
Shoot/Do not Shoot:	0.014	0.012	0.007	0.007	0.013	0.009	0.006	0.005	0.005	**0.004**
Select Weapon:	0.24	0.22	0.213	0.213	0.199	0.19	0.189	**0.182**	0.188	0.187

4 Experimental Studies

This section summarizes the experimental setup used during our experimental studies and introduces two experimental series for evaluating the performance and human-like behavior of our AI agents.

4.1 Experimental Setup

DataSet: The data were collected by sampling the game play of ten players, from which we have collected more than 8000 individual decisions. For each decision the information described in Table 1 were recorded. Data were perpetually collected using crowd-sourcing as explained earlier in Subsection 3.1 and added to the data set.

Evaluation and Settings: The performance of the proposed AI agents was compared against a number of human players and a set of "dummy" agents, which are mainly making random decisions, in terms of time (in seconds) that the agent stayed alive in the game and number of kills (game points). Finally, the last experiment of this section evaluates how well the AI agents mimic human behavior by asking ten real-users to try finding the AI agent through the map view in 60 seconds in an environment composed of one AI agent and several other human users. The results below are averaged over five individual runs.

4.2 Experimental Results

Experimental Series 1 - Performance: In experimental series 1, we have evaluated the average performance of ten AI agents against ten "Dummy" agents and ten human users with respect to the average time they stayed alive in the game (note that there is a 15 seconds penalty each time they are eliminated) and the number of kills they achieved in 120 seconds as well as the time they stayed alive until they are eliminated for the first time. The results of Figure 5 show that the AI agent performs much better than the "Dummy"-random agents and similar to the real users in all cases. Moreover, the AI agents achieve more kills than the human players at the beginning of the game, showing that they adapt faster to the environment. However, the human users manage to stay alive for a longer time than the AI agents on average and achieve more kills after around 120 seconds.

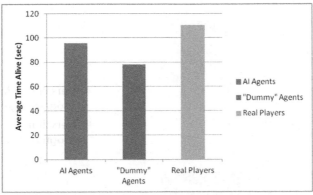

(a) Average Time Alive in 120 seconds (with penalty 15 sec after elimination)

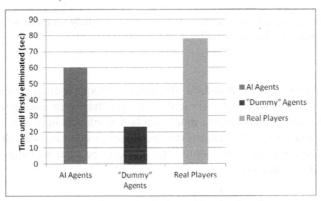

(b) Time Alive until eliminated for the first time

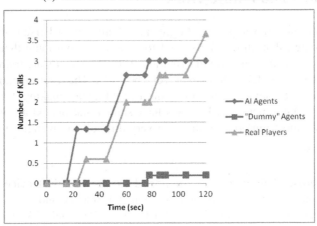

(c) Number of Kills in 120 seconds

Fig. 5. Experimental Series 1 - Performance: AI agents compared against "Dummy" agents and real-players in terms of average time alive, number of kills and average time until they have been eliminated for the first time

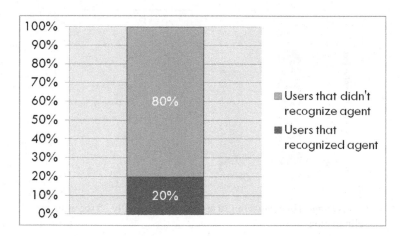

Fig. 6. Experimental Series 2 - Human-Like Behavior: A case study for evaluating how well the AI agent approximates the human behavior in the ARShooter Wars

Experimental Series 2 - Human-Like Behavior: In experimental series 2, we have evaluated how well our AI agents mimic the human behavior by asking ten players to identify the AI agents with respect to nine other human players in 60 seconds using the map view (see Figure 3 (c)). The results that are summarized in Figure 6 show that 80% of the players did not manage to identify the AI agent because its behavior was very close to that of the human users.

5   Conclusions and Future Work

In this study, we have developed an Augmented Reality First Person Shooter game, coined AR Shooter, that allows the crowd to constantly contribute their game play along with various spatio-temporal information. Then three ANN are trained using the crowdsourced data to model the human player's behavior. The resulting model is in turn utilized back into the game's environment through AI agents making it more realistic and challenging. Our experimental studies have shown that our AI agents have good performance, but most importantly they are very difficult to distinguish from human players.

Acknowledgments. This work was supported by the Mobile Devices Laboratory (MDL) grant funded by the Department of Computer Science and Engineering of the Frederick University, Cyprus and Microsoft Cyprus. We would like to thank MTN Cyprus for assisting our experimental studies by providing free 3G Sim Cards and Mr. Valentinos Georgiades, Microsoft Cyprus for his support and valuable comments. We are also grateful to the MDL Developers Constantinos Marouchos, Yiannis Hadjicharalambous and Melina Marangou for helping in the development of the AR Shooter Wars game.

References

1. Campbell, A., Eisenman, S., Lane, N., Miluzzo, E., Peterson, R., Lu, H., Zheng, X., Musolesi, M., Fodor, K., Ahn, G.: The rise of people-centric sensing. IEEE Internet Computing 12(4), 12–21 (2008)
2. Das, T., Mohan, P., Padmanabhan, V.N., Ramjee, R., Sharma, A.: Prism: platform for remote sensing using smartphones. In: MobiSys (2010)
3. Azizyan, M., Constandache, I., Choudhury, R.-R.: Surroundsense: mobile phone localization via ambience fingerprinting. In: MobiCom (2009)
4. Chatzimiloudis, G., Konstantinidis, A., Laoudias, C., Zeinalipour-Yazti, D.: Crowdsourcing with smartphones. IEEE Internet Computing (2012)
5. Konstantinidis, A., Aplitsiotis, C., Zeinalipour-Yazti, D.: Multi-objective query optimization in smartphone social networks. In: 12th International Conference on Mobile Data Management, MDM 2011 (2011)
6. Rana, R.K., Chou, C.T., Kanhere, S.S., Bulusu, N., Hu, W.: Ear-phone: an end-to-end participatory urban noise mapping system. In: IPSN, pp. 105–116 (2010)
7. Thiagarajan, A., Ravindranath, L., LaCurts, K., Madden, S., Balakrishnan, H., Toledo, S., Eriksson, J.: Vtrack: accurate, energy-aware road traffic delay estimation using mobile phones. In: SenSys 2009: Proceedings of the 7th ACM Conference on Embedded Networked Sensor Systems, pp. 85–98. ACM, New York (2009)
8. Eriksson, J., Girod, L., Hull, B., Newton, R., Madden, S., Balakrishnan, H.: The pothole patrol: using a mobile sensor network for road surface monitoring. In: MobiSys, pp. 29–39 (2008)
9. Conroy, D., Wyeth, P., Johnson, D.: Modeling player-like behavior for game ai design. In: Proceedings of the 8th International Conference on Advances in Computer Entertainment Technology, ACE 2011, pp. 9:1–9:8. ACM, New York (2011)
10. Harrison, B., Roberts, D.L.: Using sequential observations to model and predict player behavior. In: Proceedings of the 6th International Conference on Foundations of Digital Games, FDG 2011, pp. 91–98. ACM, New York (2011)
11. Wang, D., Subagdja, B., Tan, A.H., Ng, G.W.: Creating human-like autonomous players in real-time first person shooter computer games. In: 21st Annual Conference on Innovative Applications of Artificial Intelligence (IAAI 2009), Pasadena, California, July 14-16 (2009)
12. Missura, O., Gärtner, T.: Player modeling for intelligent difficulty adjustment. In: Proceedings of the ECML 2009 Workshop From Local Patterns to Global Models, LeGo 2009 (2009)
13. Yannakakis, G.N.: Game ai revisited. In: Proceedings of the 9th Conference on Computing Frontiers, CF 2012, pp. 285–292. ACM, New York (2012)
14. Kienzle, J., Denault, A., Vangheluwe, H.: Model-based design of computer-controlled game character behavior. In: Engels, G., Opdyke, B., Schmidt, D.C., Weil, F. (eds.) MODELS 2007. LNCS, vol. 4735, pp. 650–665. Springer, Heidelberg (2007)

Developing an Electron Density Profiler over Europe Based on Space Radio Occultation Measurements

Haris Haralambous and Harris Papadopoulos

Electrical Engineering and Computer Science and Engineering Departments
Frederick University, 7 Y. Frederickou St., Palouriotisa, Nicosia 1036, Cyprus
{H.Haralambous,H.Papadopoulos}@frederick.ac.cy

Abstract. This paper presents the development of an Artificial Neural Network electron density profiler based on electron density profiles collected from radio occultation (RO) measurements from LEO (Low Earth Orbit) satellites to improve the spatial and temporal modeling of ionospheric electron density over Europe. The significance in the accurate determination of the electron density profile lies on the fact that the electron density at each altitude in the ionosphere determines the refraction index for radiowaves that are reflected by or penetrate the ionosphere and therefore introduces significant effects on signals (navigation and communication). In particular it represents a key driver for total electron content model development necessary for correcting ionospheric range errors in single frequency GNSS applications.

Keywords: Ionosphere, radio occultation, electron density profile.

1 Introduction

In recent times the ionospheric monitoring capability has been significantly enhanced on the basis of a multi-instrument approach in both local and regional scale. Traditionally ionospheric monitoring was carried out by ground-based radars (ionosondes) that provide information on the electron density profile (EDP) within a limited geographical area. There have also been examples of ionospheric services in several parts of the globe where the geographical scope of ionosondes has been extended by combining their measurements therefore improving their spatial validity to facilitate provision of maps of ionospheric parameters [1,2,3]. During the last fifteen years ground-based monitoring capability has been significantly augmented by space-based systems like satellite radio occultation (RO) missions such as CHAMP, and FORMOSAT-3/COSMIC which have increased the spatial scope of these networks.

This paper explores the possibility of utilising this space-based source of ionospheric monitoring with the aim to express the spatial and temporal representation of electron density in the ionosphere over a significant part of Europe. Sections 2 and 3 describe the basic altitude structure of the EDP and its measurement techniques respectively. Section 4 discusses the EDP spatial and temporal characteristics and section 5 outlines the experimental results. Finally, section 6 gives the concluding remarks of the paper.

H. Papadopoulos et al. (Eds.): AIAI 2013, IFIP AICT 412, pp. 172–181, 2013.
© IFIP International Federation for Information Processing 2013

2   Measurement of the Ionospheric Electron Density Profile

The ionosphere is defined as a region of the earth's upper atmosphere where sufficient ionisation exists to affect radio waves in the frequency range 1 to 3 GHz. It ranges in height above the surface of the earth from approximately 50 km to 1000 km. The influence of this region on radio waves is accredited to the presence of free electrons.

The impact of the ionosphere on communication, navigation, positioning and surveillance systems is determined by variations in its EDP and subsequent electron content along the signal propagation path [4]. As a result satellite systems for communication and navigation, surveillance and control that are based on trans-ionospheric propagation may be affected by complex variations in the ionospheric structure in space and time leading to degradation of the accuracy, reliability and availability of their service.

The EDP of the ionosphere (Figure 1) represents an important topic of interest in ionospheric studies since its integral with altitude determines a very important parameter termed as the total electron content which is a direct measure of the delay imposed on trans-ionospheric radiowaves. The bottomside (below the electron density peak) component of the EDP has been routinely monitored by ionosondes compiling an extended dataset of key profile characteristics on a global scale. A significant subset of these measurements (extending over the last three decades) has been provided by the Global Ionospheric Radio Observatory (GIRO) using the ground-based Digisonde network [5].

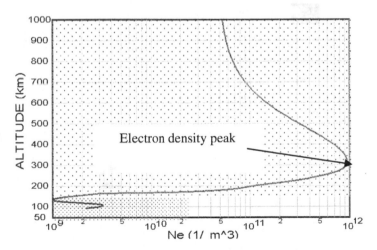

Fig. 1. Typical electron density altitude profile of the ionosphere

The EDP of the topside (above the electron density peak) ionosphere represents an important topic of interest in ionospheric studies since it has been shown that the main contribution to TEC is attributed to an altitude range above the electron density peak. However there is a lack of topside observational data as ground-based ionosondes can probe only up to the electron density peak, and observations from topside sounders are sparse since only a few satellite missions, for example Alouette, ISIS-1 and ISIS-2, have been dedicated to topside EDP measurements in the past [6]. The shape of the profile depends upon the strength of the solar ionising radiation which is a function of time of day, season, geographical location and solar activity [7,8,9]. This paper studies the development of an Artificial Neural Network (ANN) model which describes the temporal and spatial variability of the EDP over a significant part of Europe. The model is developed based on approximately 80000 LEO satellite EDPs from RO measurements recorded from April 2006 to December 2012.

3   Measurement of Electron Density by Ground-Based and Satellite Techniques

Traditionally measurements of electron density were conducted by ionosondes which are special types of radar used for monitoring the electron density at various altitudes in the ionosphere up to the electron density peak. Their operation is based on a transmitter sweeping through the HF frequency range transmitting short pulses. These pulses are reflected at various layers of the ionosphere, and their echoes are received by the receiver giving rise to a corresponding plot of reflection altitude against frequency which is further analysed to infer the ionospheric plasma height-EDP (Figure 2).

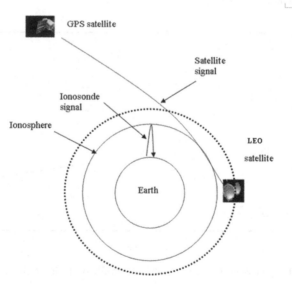

Fig. 2. Schematic illustrating a ground-based (ionosonde) and a space-based technique (satellite RO) for probing the ionosphere

Radio occultation satellite missions of Low Earth Orbit (LEO) satellites are now being widely used for ionospheric studies as they offer an excellent tool for enhancing the spatial aspect of ionospheric monitoring providing information on the vertical electron density distribution on a global scale. An important LEO satellite mission used for RO ionospheric measurements is FORMOSAT-3/COSMIC (a constellation of six satellites, called the Formosa Satellite 3-Constellation Observing System for Meteorology, Ionosphere, and Climate) launched on April 15, 2006 [10,11,12]. The instrument on these satellites that is of interest in this paper is the GPS receiver which is used to obtain atmospheric and ionospheric measurements through phase and Doppler shifts of radio signals. The Doppler shift of the GPS L-band (L_1=1575.42 MHz, L_2=1227.60 MHz) signals received by a LEO satellite is used to compute the amount of signal bending that occurs as the GPS satellite sets or rises through the earth's atmosphere as seen from LEO (Figure 2). The bending angles are related to the vertical gradients of atmospheric and ionospheric refractivity which is directly proportional to ionospheric electron density above 80 km altitude. Through the assumption of spherical symmetry, EDPs can be retrieved from either the bending angles or the total electron content data (computed from the L_1 and L_2 phase difference) obtained from the GPS RO [13]. We also need to emphasise that the RO technique can be applied successfully in retrieving the ionospheric EDP only under the assumption of spherical symmetry in the ionosphere. This assumption is not always satisfied due to significant electron density gradients that give rise to horizontal electron fluxes. This violates the requirement for EDP inversion producing a very unrealistic profile. In order to overcome this limitation and concentrate on good quality EDPs a selection process was applied in order to exclude those measurements where the distortion of the profiles was excessive [14].

Figure 3 demonstrates the uniform distribution of locations where electron density measurements have been recorded during one week of RO. We can verify the uniform geographical sampling they provide therefore complementing the limited spatial, but high temporal sampling rate (as low as 5 min) of the Cyprus ionosonde station (also shown in Figure 3). Measurements from the latter were used to validate the proposed ANN profiler.

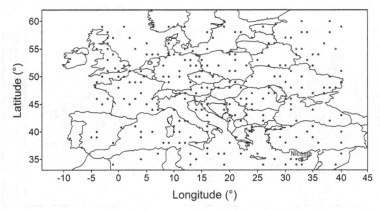

Fig. 3. Map of Europe illustrating the area considered in the model development with positions of one week of RO electron density measurements and location of Cyprus ionosonde station

4   Temporal and Spatial Characteristics of Electron Density and Model Parameters

The temporal variability of the maximum electron density at a single location is well established and has been thoroughly described in previous papers [15,16] primarily based on ionosonde derived electron density datasets. In short, ionospheric dynamics are governed principally by solar activity which in turn influences the electron density of the ionosphere. The EDP exhibits variability on daily, seasonal and long-term time scales in response to the effect of solar radiation. It is also subject to abrupt variations due to enhancements of geomagnetic activity following extreme manifestations of solar activity disturbing the ionosphere from minutes to days on a local or global scale. The most profound solar effect on maximum electron density is reflected on its daily variation as shown in Figures 4 and 5.

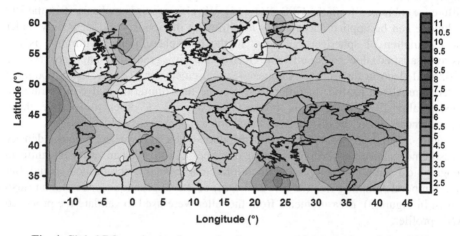

Fig. 4. Global RO maximum electron density map at midnight (universal time-UT)

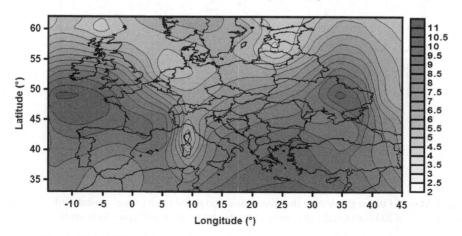

Fig. 5. Global RO maximum electron density map at noon (universal time-UT)

These figures show a map obtained by the superposition of all maximum electron density values obtained around midnight (Figure 4) and noon (Figure 5) with reference to universal time. As it is clearly depicted, there is a strong dependency of maximum electron density which minimises (over Europe) during the night and maximizes around noon emphasising the strong local time dependence of electron density. This is attributed to the rapid increase in the production of electrons due to the photo-ionization process during the day and a gradual decrease due to the recombination of ions and electrons during the night. This is also evident in Figure 6 where a number of EDPs at different times during a day is shown.

The long–term effect of solar activity on the EDP follows an eleven-year cycle and as it is clearly shown by characteristic examples of EDPs obtained over Cyprus around noon (UT) in Figure 7. Clearly electron density levels are lowest during minimum solar activity (indicated by an index of solar activity termed solar flux SF=70) conditions as compared to electron density levels during maximum solar activity conditions (SF=130).

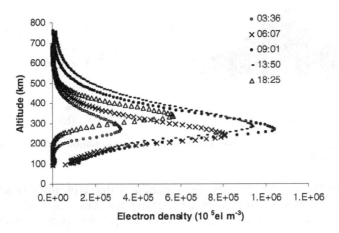

Fig. 6. Examples of EDPs over Cyprus at different hours

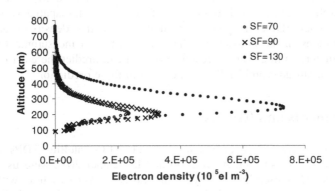

Fig. 7. Examples of EDPs at noon over Cyprus at different solar activity conditions

In addition to the short-term (diurnal) and long-term (solar cycle) effect on EDP we can also identify a clear spatial effect which is registered in the map shown in Figure 5 as decreasing levels with increasing latitude. This is also depicted in Figure 8(a) where all foF2 (maximum signal frequency that can be reflected by the maximum electron density peak - foF2 is proportional to the square root of the maximum electron density) values obtained from RO measurements are plotted as a function of their latitude (positive latitude is along North). It is evident from this figure that not only the average levels but also the variability in maximum electron density is increased as latitude decreases. This spatial characteristic of diminishing maximum electron density with increasing latitude is also observed in Figure 8(b) where the seasonal variation of the median level of Figure 8(a) (for RO foF2 values obtained over Europe at noon) over the three latitude regimes (low, medium and high) is plotted.

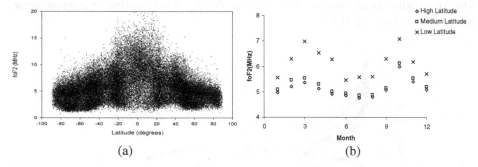

(a)   (b)

Fig. 8. (a) foF2 measured by RO versus latitude (b) seasonal variation of foF2 at noon at low, medium and high latitudes

The plots in Figures 4-8 describe the variabilities that typically characterise the average temporal behaviour of ionospheric electron densities. The model parameters to describe these variabilities have been established in previous papers [15,16] and are annual and daily sinusoidal components as well as a solar activity index (here we use measured daily solar flux). In addition, the year was also used as a temporal parameter as well as latitude and longitude, which express the spatial variability in EDP. Finally, as each EDP corresponds to a number of electron density measurements at different altitudes in the ionosphere, the altitude of each measurement was also used as a parameter; so in effect the resulting model can predict the electron density values at different altitudes and EDP is in fact a set of these predictions.

5   Experiments and Results

As mentioned in Section 2, approximately 80000 LEO satellite EDPs from RO measurements recorded between April 2006 and December 2012 were used for our experiments. The ANN used had a fully connected two-layer structure, with 9 input and 1 output neurons. Both their hidden and output neurons had hyperbolic tangent sigmoid activation functions. The training algorithm used was the Levenberg-Marquardt backpropagation algorithm with early stopping based on a validation set created from 20% of the training examples. In an effort to avoid local minima three ANNs were trained with different random initialisations and the one that performed best on the validation set was selected for being applied to the test examples. The inputs and target outputs of the network were normalized setting their minimum value to -1 and their maximum value to 1. This made the impact of all inputs in the model equal and transformed the target outputs to the output range of the ANN activation functions. The results reported here were obtained by mapping the outputs of the network for the test examples back to their original scale.

First a 2-fold cross-validation process was followed to examine the performance of the proposed approach on the satellite measurements and choose the best number of hidden units to use. Specifically the dataset was randomly divided in two parts consisting of approximately the same number of EDPs and the predictions of each part were obtained from an ANN trained on the other one. Note that the division of the dataset was done in terms of EDPs (groups of values) and not in terms of individual values. The results of this experiment are reported in Table 1 in the form of the Root Mean Squared Error (RMSE) and Correlation Coefficient (CC) between the predicted and true values over the whole dataset.

After the first experiment the proposed approach was further evaluated by training an ANN with 45 hidden units on the whole dataset and assessing its performance on measurements obtained from Cyprus ionosonde station.

Table 1. The Root Mean Squared Error (RMSE) and Correlation Coefficient (CC) between the predicted and true values over the whole dataset

Hidden	RMSE	CC
10	67715	0.9280
15	65385	0.9330
20	63466	0.9370
25	62530	0.9389
30	62189	0.9396
35	61862	0.9403
40	62312	0.9394
45	**61065**	**0.9419**
50	61903	0.9402

The RMSE value as shown in Table 1 lies between 60000 and 68000 10^5el m^{-3}. The table clearly demonstrates superior performance for 45 hidden units. However, we must keep in mind that RMSE is just an average measure of the discrepancy between COSMIC and ANN profiler EDP which varies significantly with altitude. Therefore this RMSE value encapsulates different altitude regimes (bottomside, peak and topside) for the proposed ANN profiler into a single value which could be considered as an over-simplification.

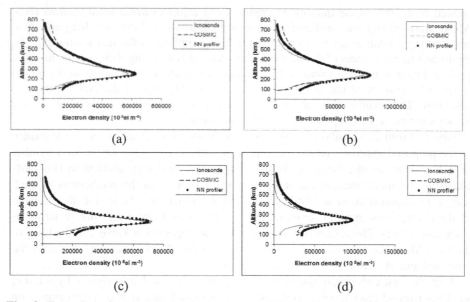

Fig. 9. Examples of measured (by ionosonde and COSMIC) and predicted (by the ANN profiler) EDPs over Cyprus

The examples of measured ionosonde, COSMIC and ANN profiler EDPs shown in Figure 9 demonstrate the good agreement between measurements and modeled values. The ionosonde and COSMIC profiles in these examples were obtained over Cyprus which was used as a validation point over the region considered in the model development (Figure 3). These particular ionosonde and COSMIC profiles where selected so that their peak electron density and corresponding altitude only differed by less than 5 %. In this way we could ensure very good quality profiles in the validation since they were measured by two independent techniques in such a good agreement. We also need to emphasize that in some cases (Figure 9(d)) although the ionosonde and COSMIC EDP profile match very well at the peak they significantly divert at the bottomside. This is due to the presence of very high electron densities in the E-region (100-130 km) due to long-lived metallic ions forming extremely high-ionisation patches. This phenomenon although termed sporadic-E, is quite frequent over Cyprus sometimes causing difficulties in the inversion of occultation measurements into a meaningful EDP around these altitudes. We also need to note that in the case of an ionosonde EDP the topside is actually modeled by a special extrapolation function based on the measured bottomside EDP. This explains the difference with COSMIC EDPs at the topside above 300-400 km.

6   Conclusions

In this paper we have presented the development of an ANN electron density profiler based on satellite electron density profiles to improve the spatial and temporal modeling of ionospheric electron density over Europe. The profiler exhibited promising prospects in profiling electron density over Europe over time and space.

References

1. Belehaki, A., Cander, L., Zolesi, B., Bremer, J., Juren, C., Stanislawska, I., Dialetis, D., Hatzopoulos, M.: DIAS project: The establishment of a European digital upper atmosphere server. Journal of Atmospheric and Solar-Terrestrial Physics 67(12), 1092–1099 (2005)
2. Wilkinson, P., Patterson, G., Cole, D.G., et al.: Australian space weather services-past and present. Adv. Space Res. 26(1), 233–236 (2000)
3. Cander Lj, R.: Towards forecasting and mapping ionosphere space weather under cost actions. Adv. Space Res. 31(4), 4957–4964 (2003)
4. Barclay, L.W.: Ionospheric Effects and Communication Systems performance. In: Keynote paper at the 10th Ionospheric Effects Symposium, Washington DC (2002)
5. Reinisch, B.W., Galkin, I.A., Khmyrov, G.M., Kozlov, A.V., Bibl, K., Lisysyan, I.A., Cheney, G.P., Huang, X., Kitrosser, D.F., Paznukhov, V.V., Luo, Y., Jones, W., Stelmash, S., Hamel, R., Grochmal, J.: The new digisonde for research and monitoring applications. Radio Sci. 44, RS0A24 (2009)
6. Reinisch, B.W., Huang, X.: Deducing topside profiles and total electron content from bottomside ionograms. Adv. Space Res. 27, 23–30 (2001)
7. Goodman, J.: HF Communications, Science and Technology. Nostrand Reinhold (1992)
8. Maslin, N.: The HF Communications, A Systems Approach, San Francisco (1987)
9. McNamara, L.F.: Grid The Ionosphere: Communications, Surveillance, and Direction Finding. Krieger Publishing Company, Malabar (1991)
10. Schreiner, W., Rocken, C., Sokolovsky, S., Syndergaard, S., Hunt, D.: Estimates of the precision of GPS radio occultations from the COSMIC/FORMOSAT-3 mission. Geophys. Res. Lett. 34, L04808 (2007), doi:10.1029/2006GL027557
11. Rocken, C., Kuo, Y.-H., Schreiner, W., Hunt, D., Sokolovsky, S., McCormick, C.: COSMIC system description, Terr. Atmos. Ocean Sci. 11, 21–52 (2000)
12. Wickert, J., Reigber, C., Beyerle, G., König, R., Marquardt, C., Schmidt, T., Grunwaldt, L., Galas, R., Meehan, T.K., Melbourne, W.G., Hocke, K.: Atmosphere sounding by GPS radio occultation: First results from CHAMP, Geophys. Res. Lett. 28, 3263–3266 (2001)
13. Hajj, G.A., Romans, L.J.: Ionospheric electron density profiles obtained with the Global Positioning system: Results from the GPS/MET experiment. Radio Sci. 33, 175–190 (1998)
14. Yang, K.F., Chu, Y.H., Su, C.L., Ko, H.T., Wang, C.Y.: An examination of FORMOSAT-3/COSMIC F peak and topside electron density measurements: data quality criteria and comparisons with the IRI model. Terr. Atmos. Ocean Sci. 20, 193–206 (2009)
15. Haralambous, H., Papadopoulos, H.: A Neural Network Model for the Critical Frequency of the F2 Ionospheric Layer over Cyprus. In: Palmer-Brown, D., Draganova, C., Pimenidis, E., Mouratidis, H. (eds.) EANN 2009. CCIS, vol. 43, pp. 371–377. Springer, Heidelberg (2009)
16. Haralambous, H., Ioannou, A., Papadopoulos, H.: A neural network tool for the interpolation of foF2 data in the presence of sporadic E layer. In: Iliadis, L., Jayne, C. (eds.) EANN/AIAI 2011, Part I. IFIP AICT, vol. 363, pp. 306–314. Springer, Heidelberg (2011)

Fuzzy Classification of Cyprus Urban Centers Based on Particulate Matter Concentrations

Nicky Gkaretsa[1], Lazaros Iliadis[1], Stephanos Spartalis[2], and Antonios Kalampakas[2]

[1] Democritus University of Thrace, Pandazidou 193, Orestias, Greece
garetsa.nicky@gmail.com, liliadis@fmenr.duth.gr
[2] Democritus University of Thrace, Xanthi, Greece
sspart@pme.duth.gr, akalampakas@gmail.com

Abstract. This research aims in the design and implementation of a flexible Computational Intelligence System (CIS) for the assessment of air pollution risk, caused by PM_{10} particles. The area of interest includes four urban centers of Cyprus, where air pollution is a potential threat to the public health. Available data are related to hourly daily measurements for 2006, 2007 and 2008. This Soft Computing (SC) approach makes use of distinct fuzzy membership functions (FMFs) in order to estimate the extent of air pollution. The CIS has been implemented under the MATLAB platform. Some interesting results related to each city are analyzed and useful outcomes concerning the seasonality and spatiotemporal variation of the problem are presented. The effort reveals the severity of air pollution. Risk is estimated in a rather flexible manner that lends itself to the authorities in a linguistic style, enabling the proper design of prevention policies.

Keywords: Z, S, Pi, Gama, Exponential membership functions, Fuzzy Classification, Particulate Matter air pollution.

1 Introduction

The presence of any type of pollutants, noise or radiation in the air, can have a potential harmful effect in the health of all living creatures and might make the environment improper for its desired use. Globally, air pollution is considered responsible for a high number of deaths and it also causes several deceases of the breathing system, mainly in urban centers [2]. PM10 are floating particles that have a diameter higher than 0.0002 μm and smaller than 10μm [8], [13]. Immediate actions have to be taken, as studies in the USA have shown that a slight increase of the PM levels only by 10 μg/m3, can increase mortality by 6%-7% [18].

1.1 Literature Review

Numerous papers describing various Soft Computing approaches have been published in the literature lately. Olej, et al., 2010 [16], have developed a FIS (Mamdani)

H. Papadopoulos et al. (Eds.): AIAI 2013, IFIP AICT 412, pp. 182–194, 2013.

towards air pollution modeling. Garcia, et al., 2010 [7] used a neural system and Artificial Neural Networks (ANNs) for the estimation of air quality related to O3. Iliadis and Papaleonidas, 2009 [12], have developed a distributed multi agent network, employing hybrid fuzzy reasoning for the real-time estimation of Ozone concentration. Aktan and Bayraktar, 2010 [3] used ANNs in order to model the concentration of PM10. Hooyberghs et al., 2005 [9], propose ANN models as tools that forecast average daily PM10 values, in urban centers of Belgium, whereas Iliadis et al., 2007 [11], have done the same forecasting effort for O3 in Athens. A similar research is reported in the literature for Chile by Dı´az-Robles et al., 2009 [5]. Thomas and Jacko, 2007 [20], have conducted a comparison between the application of Soft Computing and typical statistical regression in the case of air pollution. Finally, Mogireddy et al., 2011 [15] and Aceves-Fernandez et al., 2011 [1], have used Support Vector Machines towards air pollution modeling.

1.2   Methodology

As it has already been declared, this paper presents a Soft Computing approach towards the assessment of air pollution levels in Cyprus, by introducing specific fuzzy sets. Soft Computing is an umbrella including neural networks, fuzzy logic, support vector machines and their hybrid approaches [14], [4]. Two types of exponential fuzzy membership functions and also S, Γ (Gama), Pi and Z FMFs were applied to determine the linguistics that characterize the severity of the problem in each case. It should be mentioned that this is the first time that such a wide range of FMFs are employed for the case of air pollution with actual field data obtained from urban centers.

1.2.1.   Fuzzy Membership Functions

Fuzzy Logic (FL) is a universal approximator of real world situations. Several researchers use FL towards systems modeling [18]. The Z, S, Pi spline-based FMFs are named after their shape. They are given by the following functions 1 2 and 3 respectively. Function 4 stands for the Gama, denoted after the Greek letter Γ, where the exponentials $Ex1$ and $Ex2$ are given by functions 5 and 6.

$$f(x;a,b) = \begin{cases} 1, & x \leq a \\ 1 - 2\left(\dfrac{x-a}{b-a}\right)^2, & a \leq x \leq \dfrac{a+b}{2} \\ 2\left(\dfrac{x-b}{b-a}\right)^2, & \dfrac{a+b}{2} \leq x \leq b \\ 0, & x \geq b \end{cases} \tag{1}$$

$$f(x;a,b) = \begin{cases} 0, & x \leq a \\ 2\left(\dfrac{x-a}{b-a}\right)^2, & a \leq x \leq \dfrac{a+b}{2} \\ 1-2\left(\dfrac{x-b}{b-a}\right)^2, & \dfrac{a+b}{2} \leq x \leq b \\ 1, & x \geq b \end{cases} \qquad (2)$$

$$f(x;a,b,c,d) = \begin{cases} 0, & x \leq a \\ 2\left(\dfrac{x-a}{b-a}\right)^2, & a \leq x \leq \dfrac{a+b}{2} \\ 1-2\left(\dfrac{x-b}{b-a}\right)^2, & \dfrac{a+b}{2} \leq x \leq b \\ 1, & b \leq x \leq c \\ 1-2\left(\dfrac{x-c}{d-c}\right)^2, & c \leq x \leq \dfrac{c+d}{2} \\ 2\left(\dfrac{x-d}{d-c}\right)^2, & \dfrac{c+d}{2} \leq x \leq d \\ 0, & x \geq d \end{cases} \qquad (3)$$

$$f(X,a) = \begin{cases} 0 & \text{if } X \leq a \\ \dfrac{k(x-a)^2}{1+k(x-a)^2} & \text{if } X \succ a \end{cases} \qquad (4)$$

$$f(X,a,b) = \begin{cases} e^{-\left(\frac{M-X}{a}\right)^2} & \text{if } X \leq M \\ e^{-\left(\frac{X-M}{b}\right)^2} & \text{if } X \succ M \end{cases} \qquad M = \dfrac{(a+b)}{2} \qquad (5)$$

$$f(Xa,b) = \begin{cases} e^{-\left(\frac{X-C_1}{2Wl}\right)^2} & \text{if } X \prec C_1 \\ e^{-\left(\frac{X-C_r}{2Wr}\right)^2} & \text{if } X \succ C_r \\ 1 \text{ in any other case} \end{cases} \qquad (6)$$

MATLAB has already built in code for the implementation of Z, S and Pi whereas the code for Γ and the two exponential functions have been developed in the form of ".m" executable files under the MATLAB platform. It must be clarified that in the case of the *Z, S FMFs*, parameters *a* and *b* locate the extremes of the sloped portion of the curve, whereas for the function *Pi, a* and *d* locate the "feet" of the curve and *b* and *c* locate its "shoulders" [17]. Finally, in function 6 (Ex2) the parameters *wleft* (*wl*), *cleft* (*Cl*), *cright Cr*), *wright* (*wr*), must be positive numbers and their values must be chosen by the user, following the constraint that *cl<cr*

Actual MATLAB code for the gamamf.m file:

```
function [ y ] = gamamf (x, params)
%    gamaMF(X, PARAMS) returns an array with the degrees
of membership
%    for an input vector X
%    params=[X0 X1] is a vector with 2 elements determining
the break points of the function
if nargin ~= 2,
error ('Two Parameters are required by gamaMF.');
elseif length(params) < 2,
error ('gamaMF requires at least two parameters.');
end
x0 = params(1); x1 = params(2);
y = zeros(size(x)); % Creates table Y
index1 = find(x <= x0);   % If X is less than Xo or equal
if ~isempty(index1),
y(index1) = zeros(size(index1)); %            0
end
index2 = find(x > x0);   % If X is greater than 31
if ~isempty(index2),
y(index2) = (x1*(x(index2)-x0).^2)/(1+x1*(x(index2)-
                  x0).^2);
end
end
```

2   Data and Area of Research

This research effort presents the implementation of a soft computing system (SCS) that is based on fuzzy logic (FL). More specifically *S, Γ, Pi* and *Z FMFs* have been employed in order to determine the proper linguistic that corresponds to the levels of air pollution. The input data include 16,000 data vectors for the four main cities of Cyprus namely: Larnaka, Lemesos, Lefkosia (Nicosia) and Pafos. The field data are related to the period July 2006 to July 2008. Each record contains the following fields: *Date and time, Time, Day, PM10* hourly concentrations (μg/m3), *SEA* (a binary index related to the seasonality of the case), *DoW* (direction of the wind), sin(HoD) and *cos(HoD)* (sine and cosine functions related to the effect of the hour of the day) (Ziomas et al, 1995) [21],   PM10-24 (average *PM10* related to 24 hours), *T* (Surface Temperature) and *RH* (relative humidity). The PM10 concentrations were measured by using the Tapered Element Oscillating Microbalance device and the Filter Dynamics Measurement system. In order to overcome the variations in the magnitude of the data, they were normalized (standardized to zero) by employing the following function 7:

$$Z = \frac{X - \mu}{\sigma} \tag{7}$$

where μ is the average value and σ is the standard deviation [6].

The output of the developed FIS comprises of the fuzzy linguistic values Normal, Alert, Alarm which are related to the level of air pollution caused by PM10 and their assigned fuzzy membership values (FMV).   European Union (EU) considers 50μg/m3 as the limit for the acceptable maximum daily concentration of PM10 whereas the maximum acceptable average annual boundary is 40μg/m3. Also EU has established the average daily values of 90 μg/m3 and 110 μg/m3as the Alert point and the Alarm limit respectively. The employment of specific numbers as the boundaries between Normal, Alert and Alarm are not rational especially when the measured values might not be very accurate. For example how can we accept the fact that 50μg/m3 are satisfactory for the concentration of PM10 whereas the 50.001 μg/m3 are not. On the other hand the main advantage of the fuzzy linguistic model introduced by this research is the fact that it is flexible and innovative, classifying the conditions of each case based on a real value (in the interval [0,1]) that specifies the degree of belonging to the proper linguistic [10].

The minimum and maximum concentrations of the *PM10* for the period 2006-2008 were input to the MATLAB fuzzy toolbox. Based on the range of these values, the parameters *a, b* for the Z and S functions and the actual values of the *parameters a, b, c, d* for the Pi FMF were determined automatically.

3   Results

The following table 1 is a small sample of the hourly classification performed for Lemesos Cyprus, on the 10th of July 2006 based on the *PM10* concentrations. It is interesting that there is a continuous Alert situation after eight o'clock in the morning and at 12 o'clock there is an Alarm signal. The situation goes back to Normal after 18:00 whereas there is a continuous Alert situation in the mean time.   A classification for Lefkosia on the 12th of October 2006 reveals two Alarm situations, one at 7:00 in the morning and one at 18:00 in the afternoon. Pafos has also been assigned Alert Linguistic from 9:00 in the morning till 17:00 and also two Alarm signals from 18:00 till 19:00 for the 13th of August 2006.

Table 1. Sample of hourly classification for the PM_{10} (Lemesos July 06)

City of Lemesos	Pollutant	Degrees of Membership to the three Linguistics			
Date-Time	PM_{10}	Normal Z-FMF	Alert Pi-FMF	Alarm S- FMF	Corres-ponding Linguistic
10/7/2006 1:00	8.56	0.98	0.00	0.00	Normal
10/7/2006 2:00	47.05	0.34	0.01	0.00	Normal
10/7/2006 3:00	9.50	0.97	0.00	0.00	Normal
10/7/2006 4:00	45.22	0.38	0.00	0.00	Normal
10/7/2006 5:00	79.53	0.00	1.00	0.00	Alert
10/7/2006 6:00	98.71	0.00	0.91	0.04	Alert
10/7/2006 7:00	25.84	0.79	0.00	0.00	Normal
10/7/2006 8:00	90.51	0.00	1.00	0.00	Alert
10/7/2006 9:00	82.28	0.00	1.00	0.00	Alert
10/7/2006 10:00	76.54	0.00	0.98	0.00	Alert
10/7/2006 11:00	68.71	0.04	0.79	0.00	Alert
10/7/2006 12:00	**133.82**	**0.00**	**0.00**	**0.85**	**Alarm**

Table 2. Alarm cases for a period of 12 months in Larnaka

Month 2007	S	Γ	Exp1	Exp2	Month 2007	S	Γ	Exp1	Exp2
Jan	4	7	11	12	July	24	4	36	42
Febr	18	13	24	27	August	19	20	17	19
March	15	30	19	21	Sept	30	12	46	50
April	15	49	23	24	Oct	90	64	116	124
May	27	36	85	89	Nov	59	29	81	90
June	29	22	38	38	Dec	11	11	16	20

Table 3. Alarm cases for 6 months of 2006 in Larnaka

Month 2006	S	Γ	Exp1	Exp2
July	7	12	13	14
August	10	10	22	23
Sept	29	11	38	40
Oct	9	34	13	13
Nov	3	58	5	6
Dec	11	30	14	14

Table 4. Aggregation of the hourly Alarm situations in Larnaka for two years

Year	Number of Alarms (S)	Number of Alarms (Γ)	Number of Alarms (Exp₁)	Number of Alarms (Exp₂)
2007	381	337	512	556
2008	388	480	452	476

It is interesting to compare the classifications for two different cities for two different seasons (summer and winter).

An interesting finding is that in a period of 18 months, the Exponential2 function offers the highest number of Alarms in 10 cases (55.5%.) The exponential1 is second in the frequency of Alarms in 55.5% of the classificationς, whereas the Γ FMF is first in five cases (27.75%).

From table 4 it is concluded that the frequencies of the total hourly worst case situations in Larnaka are more or less the same for 2007 and 2008, regardless the FMF.

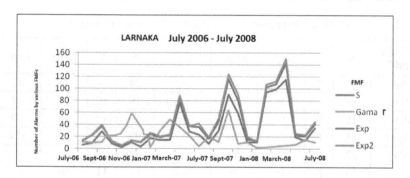

Fig. 1. Evolution of Alarm Signals for Larnaka with various FMFs

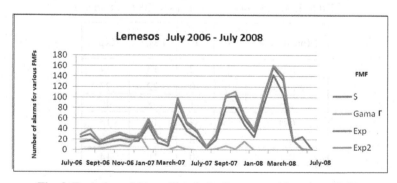

Fig. 2. Evolution of Alarm Signals for Lemesos with various FMFs

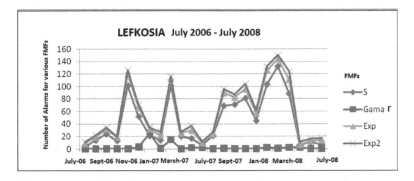

Fig. 3. Evolution of Alarm Signals for Lefkosia with various FMFs

It is obvious from the figures 1,2,3,4, that all of the functions except from the Gama are moving in a parallel mode. The Exp2 FMF always assigns the highest number of Alarms. However (with the exception of the Γ function) the frequency of the Alarms does not have big differences from one membership function to the other. Thus the Exp2 can be employed when the civil protection authorities wish to apply a quite strict policy. Otherwise the S or the Exp1 FMFs can offer a good alternative approach. Another finding is that the Γ function in most of the cases underestimates the risk and it offers rarely only a small number of Alarms. Only for the period October 2006 to January 2007 for Larnaka we have a peak in the number of Alarms introduced by the Γ FMF, which contradicts with the rest FMFs. Thus, the Gama function is not the proper one for this problem.

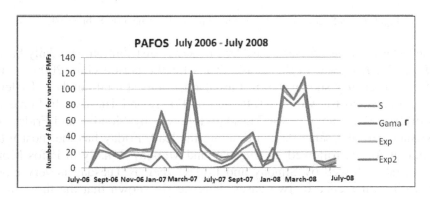

Fig. 4. Evolution of Alarm Signals for Pafos with various FMFs

3.1   Comparative Analysis

The following figures 5, 6 represent a graphical display of the classes Normal, Alert, Alarm for Larnaka and for Lemesos for the same winter period. Number 1 stands for Normal, Number 2 for Alert and number 3 for Alarm.

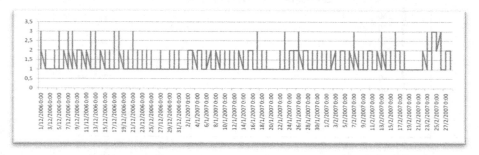

Fig. 5. Classification in 3 classes for Larnaka (December 06-February 07)

It is obvious from figures 5 and 6, that Lemesos has by far the most Alarm and the fewest Normal signals, whereas Larnaka is in an Alert state most of the times and more often than Lemesos.

Fig. 6. Classification in 3 classes for Lemesos (December 06-February 07)

The following table 5 presents the Alerts frequency distribution on a daily basis, for *Larnaka, Lemesos, Lefkosia* and *Pafos* for the period (July 2006 – July 2008). It is clearly shown that in Larnaka the highest total number of Alerts has been obtained for Monday followed by Friday and Tuesday. In Lemesos the corresponding ranking was Tuesday, followed by Monday and Friday. In Lefkosia, again the same days are taking the lead. Friday has the most total number of Alerts, followed by Tuesday and Monday. Finally in Pafos Monday is first, followed by Sunday and Friday whereas Tuesday is very close only with a difference of two cases. It is clearly shown that the three most risky days for all cities (with a small differentiation in the order) are Monday, Friday and Tuesday.

Table 5. Alerts on a daily basis (2006-2008) for four Cyprus cities

	Daily distribution of Alarms for Lemesos						
Year	Mon	Tue	Wed	Thu	Fri	Sat	Sun
2006	21	20	13	18	13	8	2
2007	64	71	55	67	74	57	60
2008	62	80	49	47	56	61	53
TOTAL	147	**171**	117	132	143	126	**115**
	Daily distribution of Alarms for Pafos						
Year	Mon	Tue	Wed	Thu	Fri	Sat	Sun
2006	10	14	16	24	13	6	4
2007	53	37	30	48	65	25	64
2008	54	53	25	24	28	52	46
TOTAL	**117**	104	**71**	96	106	83	114
	Daily distribution of Alarms for Larnaka						
Year	Mon	Tue	Wed	Thu	Fri	Sat	Sun
2006	14	12	12	10	18	2	1
2007	69	45	43	43	78	45	58
2008	82	76	43	35	49	44	59
TOTAL	**165**	133	98	88	145	91	**118**
	Daily distribution of Alarms for Lefkosia						
Year	Mon	Tue	Wed	Thu	Fri	Sat	Sun
2006	4	2	9	7	20	9	4
2007	87	97	60	93	130	59	48
2008	78	110	68	75	74	72	69
TOTAL	169	209	137	175	**224**	140	**121**

The following figure 7 provides clearly, additional arguments to support the conclusions discussed above. Again it is shown that Monday, Friday and Tuesday are the most risky days.

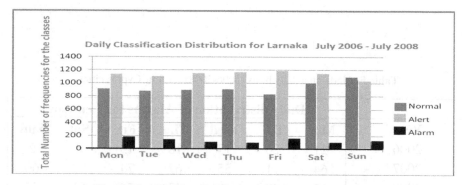

Fig. 7. Classification distribution in 3 classes for Larnaka

Also in figure 7 it is shown that in most of the cases, there is an "Alert situation" regardless the day of the week. Sunday and Saturday have the most Normal days, followed by Thursday.

4   Discussion - Conclusions

A first certain conclusion is that the situation in Cyprus regarding the PM10 concentration is becoming worse every year. A first glance at the graphs 1,2,3,4 reveals that the year 2008 is by far the worst compared to 2006 and 2007. Also it is obvious that there is a continuous and rapid increase of the problem which is detected by the huge differentiations between the peaks. The worst year with the highest number of Alarms is 2008 for Larnaka (150), Lemesos (160), Lefkosia (155). Only for Pafos May 2007 is the most risky with 120 Alarms but the peak of February 2008 is very close with 106. The most dangerous seasons are either October to December or February to April.

Concluding it can be suggested that the problem of the particulate matter concentration in the urban centers of Cyprus is quite serious and the Alert cases are very common, whereas the Alarms are quite frequent. Also it has been shown that the problem is seasonal and the day or the month, play a very serious role. This of course is due to human activities or weather conditions. The S function is a little bit more optimistic, giving the least number of Alarms, whereas the Exp2 FMF is the most strict one.

The Gama function has a very low estimation of the risk (the problem is underestimated significantly) and it is differentiated from all other FMFs that have very similar behavior.

Future research will be the conversion of the CIS to a real time one. Working in a real time mode and offering real time classifications, it can enable the effective handling by the local authorities.

References

1. Aceves-Fernandez, M.A., Sotomayor, O.A., Gorrostieta-Hurtado, E., Pedraza-Ortega, J.C., Ramos-Arreguín, J.M., Canchola-Magdaleno, S., Vargas-Soto, E.: Advances in Airborne Pollution Forecasting Using Soft Computing Techniques. In: Air Quality - Models and Applications, pp. 3–14. INTECH Publications (2001) ISBN 978-953-307-307-1
2. Air Quality in Cyprus, The Department of Labour Inspection (DLI), Ministry of Labour and Social Insurance (2013), http://www.airquality.dli.mlsi.gov.cy
3. Aktan, M., Bayraktar, H.: The Neural Network Modeling of Suspended Particulate Matter with. Autoregressive Structure. Ekoloji 19(74), 32–37 (2010)
4. Chaturvedi, D.K.: Soft Computing Techniques and its Applications in Electrical Engineering. Springer, Berlin (2008)
5. Díaz-Robles, L.A., Fu, J.S., Reed, G.D.: Seasonal Distribution and Modeling of Diesel Particulate Matter in the Southeast US. Environment International 35(6), 956–964 (2009)
6. Dogra, K.: "Autoscaling" QSARword-A Stand Life Sciences Web Resource (2010), http://www.qsarworld.com/qsar-statistics-autoscaling.php
7. European Commission, Instructions 1999/30/EC, related to the limited values of SO2, Particulate Matter in the air (1999),
http://www.modus.gr/site1/gr/000F4240/Data/1999-30-EK.pdf
(April 22, 1999)
8. Garcia, I., Rodriguez, J., Tenorio, Y.: Earth and Planetary Sciences- Oceanography and Atmospheric Sciences. In: Mazzeo, N. (ed.) Earth and Planetary Sciences- Oceanography and Atmospheric Sciences, ch. 3. Air Quality-Models and Applications. INTECH Publications (2010)
9. Gentekakis, I.B.: Air Pollution Consequences, Control and Alternative Technologies, Tziolas edn., Thessaloniki, Greece (1999) (in Greek)
10. Hooyberghs, J., Mensink, C., Dumont, G., Fierens, F., Brasseur, O.: A neural network forecast for daily average PM10 concentrations in Belgium. Atmospheric Environment 39(18), 3279–3289 (2005)
11. Iliadis, L.: Intelligent Information Systems and Applications in Risk estimation Stamoulis Publishers, Thessaloniki, Greece (2007)
12. Iliadis, L., Spartalis, S., Paschalidou, A., Kassomenos, P.: Artificial Neural Network Modelling of the surface Ozone Concentration. International Journal of Computational and Applied Mathematics 2(2), 125–138 (2007) ISSN 1819-4966
13. Iliadis, L.S., Papaleonidas, A.: Intelligent Agents Networks Employing Hybrid Reasoning: Application in Air Quality Monitoring and Improvement. In: Palmer-Brown, D., Draganova, C., Pimenidis, E., Mouratidis, H. (eds.) EANN 2009. CCIS, vol. 43, pp. 1–16. Springer, Heidelberg (2009)
14. Kamataki, A.E.: High scale transfer of Particulate Matter pollution in Thessaloniki. Graduate thesis, Aristotle University of Thessaloniki, Greece (2009) (in Greek)
15. Kecman, V.: Learning and Soft Computing. The MIT Press, Cambridge (2001)
16. Mogireddy, K., Reddy, K.: "Physical Characterization of Particulate Matter Employing Support Vector Machine Aided Image Processing" Master of Science Thesis in Electrical Engineering, University of Toledo, Spain, Electrical Engineering (2011),
http://etd.ohiolink.edu/view.cgi?acc_num=toledo1297358759

17. Olej, V., Obrsalova, I., Krupka, J.: Modelling of Selected Areas of Sustainable Development by Artificial Intelligence and Soft Computing, University of Parduvice, Czech Republic (2009)
18. Sumathi, S., Surekha, P.: Computational Intelligence Paradigms Theory and Applications using Matlab. CRC Press, Taylor and Francis Group, USA (2010)
19. Technical chamber of Greece (2013), http://portal.tee.gr
20. Theocharis, I.: Fuzzy Systems Aristotle University of Thessaloniki, Greece (2006)
21. Thomas, S., Jacko, R.B.: A stochastic model for estimating the impact of highway incidents on air pollution and traffic delay. Journal of Transportation Research Record A Journal of the Transportation Research Board, USA (2007)
22. Ziomas, I., Suppan, P., Rappengluck, B., Balis, D., Tzoumaka, P., Melas, D., Papayiannis, A., Fabian, P., Zerefos, C.: A contribution to the study of photochemical smog in the Greater Athens area Beitr. Phys. Atmosph. 68, 191–203 (1995)

Robots That Stimulate Autonomy

Matthijs A. Pontier[1] and Guy A.M. Widdershoven[2]

[1] VU University Amsterdam,
Center for Advanced Media Research Amsterdam / Network Institute,
De Boelelaan 1081, 1081HV Amsterdam, The Netherlands
m.a.pontier@vu.nl
[2] VU University Amsterdam,
VU University Medical Center, Amsterdam, The Netherlands
g.widdershoven@vumc.nl

Abstract. In healthcare, robots are increasingly being used to provide a high standard of care in the near future. When machines interact with humans, we need to ensure that these machines take into account patient autonomy. Autonomy can be defined as negative autonomy and positive autonomy. We present a moral reasoning system that takes into account this twofold approach of autonomy. In simulation experiments, the system matches the decision of the judge in a number of law cases about medical ethical decisions. This may be useful in applications where robots need to constrain the negative autonomy of a person to stimulate positive autonomy, for example when attempting to pursue a patient to make a healthier choice.

Keywords: Moral Reasoning, Machine Ethics, Cognitive Modeling, Cognitive Robotics, Health Care Applications.

1   Introduction

In view of increasing intelligence and decreasing costs of artificial agents and robots, organizations increasingly use such systems for more complex tasks. In healthcare, the use of robots is necessary to provide a high standard of care in the near future, due to a foreseen lack of resources and healthcare personnel [18]. By providing assistance during care tasks, or fulfilling them, robots can relieve time for the many duties of care workers. Previous research shows that robots can genuinely contribute to treatment and care [4], [15], [17].

As their intelligence increases, the amount of human supervision decreases and robots increasingly operate autonomously. With this development, we increasingly rely on the intelligence of these robots. Because of market pressures to perform faster, better, cheaper and more reliably, this reliance on machine intelligence will continue to increase [2].

When we start to depend on autonomously operating robots, we should be able to rely on a certain level of ethical behavior from machines. As Rosalind Picard [12] nicely puts it: "the greater the freedom of a machine, the more it will need moral standards". Particularly when machines interact with humans, which they

H. Papadopoulos et al. (Eds.): AIAI 2013, IFIP AICT 412, pp. 195–204, 2013.
© IFIP International Federation for Information Processing 2013

increasingly do, we need to ensure that these machines do not harm us or threaten our autonomy. In complex and changing environments, externally defining ethical rules unambiguously becomes difficult. Therefore, autonomously operating care robots require moral reasoning. We need to ensure that their design and introduction do not impede the promotion of values and the dignity of patients at such a vulnerable and sensitive time in their lives [16].

In a recent interview in a multimillion copies free newspaper [10], we presented a humanoid robot for healthcare (a "Caredroid") in which we will implement the moral reasoning system. The caredroid will assist people in finding suitable care, and assist them in making choices concerning healthcare.

Caredroids will encounter moral dilemmas. For example, when supporting a patient in making choices, the caredroid should balance between accepting unhealthy choices or trying to persuade the patient to reconsider them. The caredroid might also have to consider following a previous agreement in which the patients binds himself and asks to be treated in case of a deterioration of the situation due to a psychiatric condition (a Ulysses contract) versus giving up when the patient opposes to the very treatment he has previously agreed upon.

In previous research, Pontier and Hoorn [14] developed a moral reasoning system based on the moral principles developed by Beauchamp & Childress [5]. In simulation experiments, the system was capable of balancing between conflicting principles. In medical ethics, autonomy is the most important moral principle [3].

Often autonomy is equated with self-determination. In this view, people are autonomous when they are not influenced by others. However, autonomy is not just being free from external constraints. Autonomy can also be conceptualized as being able to make a meaningful choice, which fits in with one's life-plan [8]. In this view, a person is autonomous when he acts in line with well-considered preferences. This implies that the patient is able to reflect on fundamental values in life. Core aspects of autonomy as self-determination are mental and physical integrity and privacy. Central in autonomy as ability to make a meaningful choice are having adequate information about the consequences of decision options, the cognitive capability to make deliberate decisions, and the ability to reflect on the values behind one's choices. Autonomy as self-determination can be called negative freedom, or 'being free *of*'. Autonomy as the ability to make a meaningful choice is called positive freedom or 'being free *to*' [6]. In this paper, we will use the notions of 'negative autonomy' and 'positive autonomy' to denote both concepts. To be able to reflect this more complex view of autonomy in the moral reasoning system, we decided to expand the moral principle of autonomy.

The notion of positive autonomy can come close to beneficence. For example, when mental health is facilitated or prevented from worsening, this can be seen as facilitating beneficence, but also as facilitating requirements necessary for making a well-considered choice. Moreover, almost any action stimulating the freedom to choose based on reflection can be seen as facilitating beneficence. Therefore, in the model, when positive autonomy is facilitated, often beneficence also increases. Yet, autonomy as being able to make a meaningful choice is not the same as beneficence. Reflection on and deliberation about values can help people to behave in a more healthy way, but this is not necessarily so. Reflection might result in people taking health risk in favor of other important values. An example is the conscious refusal of blood by Jehovah's witnesses.

In medical practice, a conflict between negative autonomy and positive autonomy can play a role. Sometimes, the self-determination of the patient needs to be constrained on the short-term to achieve positive autonomy on the longer term. For example, when a patient goes into rehab his freedom can be limited for a limited period of time to achieve better cognitive functioning and self-reflection in the future.

2   The Model of Autonomy

In our model we divide autonomy into negative autonomy and positive autonomy. Negative autonomy can be seen as self-determination - or being free *of* others - and consists of the sub-principles physical integrity, mental integrity and privacy. Positive autonomy can be seen as the capability to make a deliberate decision – or being free *to* choose - and consists of having adequate information, being cognitively capable of making a deliberate decision and reflection. The model of autonomy is graphically depicted in Figure 1. As all variables in the model contribute positively to one another, all variables in the model are represented by a value in the domain [0, 1].

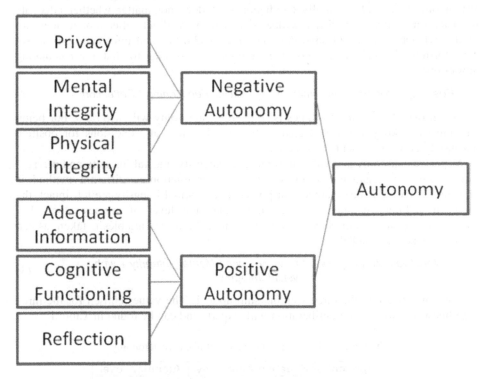

Fig. 1. The Model of Autonomy

To be autonomous, both the conditions for positive autonomy and negative autonomy are relevant [6]. Ideally, both are present to a large extent. When self-determination is compromised, one is not able to make an autonomous decision, because this decision is made by others; the person is not free *of* others to make an own decision. When a person is not able to deliberate, the person is also not autonomous. The person may be free *of* others to make a decision, but not free *to* make an autonomous decision, due to a lack of capabilities to do so. This is reflected in the formula below to calculate the level of autonomy.

$$\text{Autonomy} = \text{Positive\_autonomy} * \text{Negative\_autonomy} \tag{1}$$

When negative autonomy - or self-determination - is 0, autonomy will also be 0. When positive autonomy – or the capability to make a deliberate decision – is 0, autonomy will also be 0. For being autonomous, both negative autonomy and positive autonomy need to have a high value.

Positive autonomy can be divided in having adequate information, cognitive functioning and reflection. For calculating positive autonomy from these three variables, we use the same reasoning as for calculating autonomy. Each should be present to some extent; the higher one of them, the more autonomy. Without any information about the consequences of a decision, it does not matter whether one could have made a well-reflected deliberate decision while having this information. When one is severely mentally handicapped, it does not matter whether adequate information is available. When a decision is made without reflection, it does not matter whether one would have the cognitive capabilities and information to do so. The formula for calculating positive autonomy is similar to that for calculating autonomy.

$$\text{Positive\_Autonomy} = \text{Information} * \text{Cognitive\_Functioning} * \text{Reflection} \tag{2}$$

When one of the three variables is 0, positive autonomy will also be 0. For being capable of making a well-reflected, deliberate decision, all conditions for positive autonomy need to be met to a certain extent.

Negative autonomy is divided into physical integrity, mental integrity and privacy. For calculating negative autonomy, or self-determination, a different method is chosen. If privacy is constrained, but physical and mental integrity are left intact, the level of self-determination can be higher than the level of privacy alone. For calculating negative autonomy, a weighed sum of the three variables is taken, as can be seen in the formula below.

$$\text{Negative autonomy} = wp * \text{Privacy} + wm * \text{Mental\_Integrity} + wph \\ * \text{Physical\_Integrity}; \tag{3}$$

For normalization, the three weights sum up to 1. The values chosen for the three weights were chosen after deliberation with experts and can be found in Table 1.

Table 1. Weights for components of negative autonomy

Component of Negative Autonomy	Ambition level
Privacy	0.20
Mental Integrity	0.30
Physical Integrity	0.50

When making a decision that may influence the autonomy of a patient, the robot will make an estimation of how each of the six variables will change. After doing so, the robot can calculate the resulting autonomy of the patient for every possible decision option. Using the previously developed moral reasoning system [14] the robot can use the outcome to estimate how morally good or bad every decision option is. The calculated level of autonomy simply feeds into 'autonomy' in the moral reasoning system for calculating the morality of each action. The moral reasoning system including the twofold approach of autonomy is graphically depicted in Figure 2.

The agent calculates estimated level of morality of an action by taking the sum of the ambition levels of the three moral principles multiplied with the beliefs that the particular actions facilitate the corresponding moral principles. When moral principles are believed to be better facilitated by an action, the estimated level of Morality will be higher. The following formula, taken from [14], is used to calculate the estimated Morality of an action:

$$\text{Morality}(Action) = \Sigma_{Goal}(\text{Belief}(\text{facilitates}(Action, Goal)) * \text{Ambition}(Goal)) \qquad (4)$$

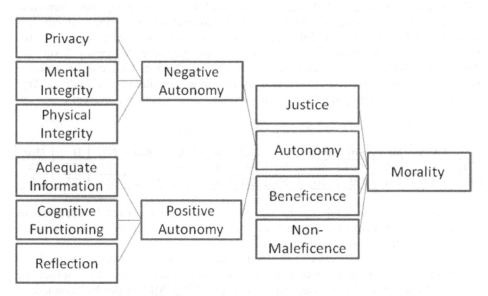

Fig. 2. The Moral Reasoning System

3   Simulation Results

Previously, Pontier and Hoorn [14] developed a moral reasoning system that matched the conclusions of expert ethicists. To control whether the current system, including the twofold approach of autonomy, matches the behavior of the previous system, we repeated one of the previously performed simulation experiments. Additionally, we simulated three law cases of the period 2008-2012 in the Netherlands, to test whether the decisions of the model match those of the judge. The cases did not involve a straightforward application of legal rules, since Dutch law does not regulate situations in which physicians overrule negative autonomy in order to foster positive autonomy. The judgment thus required the use of ethical considerations by the judge, rather than legal rules. Therefore, we used the cases as examples of ethical decision making.

3.1   Experiment: Patient Refuses to Take Medication Due to False Beliefs

A patient with incurable cancer refuses chemotherapy that will let him live a few months longer, relatively pain free. He refuses the treatment because, ignoring the clear evidence to the contrary, he is convinced himself that he is cancer-free and does not need chemotherapy.

According to Buchanan and Brock [7], the ethically preferable answer is to try again. The patient's less than fully autonomous decision will lead to harm (dying sooner) and denies him the chance of a longer life (a violation of the duty of beneficence), which he might later regret. The moral reasoner comes to the same conclusion, as can be seen in Table 2.

In this and the following tables, the fields under the moral principles and subprinciples represent the believed facilitation of the corresponding moral principle by an action. 'Prv' stands for Privacy, 'MI' for Mental Integrity, 'PhI' for Physical Integrity, 'Info' for Adequate Information, 'CF' for Cognitive Functioning, 'Ref' for Reflection, 'A+' for Positive Autonomy, 'A-' for Negative Autonomy, 'Aut' for Autonomy, 'NM' for Non-Maleficence, 'B' for Beneficence. Justice did not play a role in any of the cases and is therefore set to 0 in all simulations. In Table 2, 'Try' stands for the decision option 'Try again', whereas 'Acc' stands for 'Accept'.

Table 2. Simulation results of the example experiment

	Prv	MI	PhI	Info	CF	Ref	A-	A+	Aut	NM	B	Mor
Try	.80	.50	1.0	.50	.50	.50	.81	.50	.64	.75	.75	.70
Acc	1.0	1.0	1.0	.10	.30	.30	1.0	.21	.46	.25	.25	.35

As can be seen in Table 2, the moral reasoner with the twofold approach of autonomy also classifies the action 'Try again' as having a higher level of morality than accepting the decision of the patient.

To test whether the behavior of the model matches actual moral decisions made by judges in cases in which negative autonomy and positive autonomy are in conflict, we simulated a number of law cases from the period 2008-2012 in the Netherlands. In these cases, there was a conflict between respecting the patient's refusal of care or providing care without the patient's consent in cases of serious self-neglect and risk of physical and social deterioration [8].

3.2   Case 1: Assertive Outreach to Prevent Judicial Coercion

In case 1, a man was not taking care of himself and in a state of demise. There was risk of fire and aggression to others. The care-takers decided not to ask for a court order for enforced placement in a psychiatric institution. They saw the man had a quite serious disturbance, but the situation in their eyes did not justify judicial coercion. Although the man was living isolated, the situation was not acute. The care-takers decided to continue offering care, and if necessary make use of assertive outreach. The man complained about the actions of the care-takers, interfering with his freedom. The judge decided that the complaint was not warranted. He considered the assertive outreach justified, given that it aimed to prevent further worsening of the man's situation. Thereby, a future need for judicial coercion was prevented.

This legal argument is reflected in the simulation of this case, as can be seen in Table 3. In Table 3, 'NI' stands for the decision option 'No Intervention', 'AO' stands for 'Assertive Outreach' and 'JC' stand for 'Judicial Coercion'.

With no intervention at all, self-determination is well respected (self-determination = 1.0). However, it does not improve the worrisome situation of the man (beneficence = 0) and there is a risk of worsening of the situation for the man (non-maleficence = 0.30). Still, no intervention at this moment (morality = 0.41) is a slightly ethically better option than judicial coercion (morality = 0.40). Through judicial coercion, self-determination is heavily violated (self-determination = 0.20). The ethically best option in this case was to offer assertive outreach. Hereby, negative autonomy is not violated too heavily (self-determination = 0.58) and positive autonomy may be improved (positive_ autonomy = 0.62), leading to an overall autonomy of 0.60. With this relatively light intervention, the situation of the patient could possibly still be improved (beneficence = 0.40; non-maleficence = 0.40), leading to an overall morality of 0.49.

Table 3. Simulation results of case 1

	Prv	MI	PhI	Info	CF	Ref	A-	A+	Aut	NM	B	Mor
NI	1.0	1.0	1.0	.50	.50	.50	1.0	.50	.71	.30	0.0	.41
AO	.40	.40	1.0	.50	.60	.80	.58	.62	.60	.40	.40	**.49**
JC	.20	.20	.20	.50	.60	.50	.20	.53	.33	.40	.50	.40

3.3   Case 2: Inform Care Deliverers, Not Parents of Adult

In the second case, a psychiatrist contacted the parents of a patient who was in an alarming situation and avoided help. The patient was adult and mentally competent, and did not have regular contact with his parents. Therefore, the judge decided the psychiatrist should have only informed the ambulatory care team, and not the parents.

The legal judgment is in line with the simulation of the case, as can be seen in Table 4. In Table 4, 'NI' stands for the decision option 'No Intervention', 'ICT' stands for 'Inform Care Team' and 'IP' stand for 'Inform Parents'. Informing the parents (privacy = 0.10) is simulated as a heavier violation of privacy than informing the ambulatory care team (privacy = 0.80). In contrast to the ambulatory care team, the parents could spread the information against the patients will and nag the patient. Informing the ambulatory care team could improve the care for the patient and thereby improve his cognitive functioning (cognitive functioning 'do nothing' = 0.50; cognitive functioning 'inform the ambulatory care team' = 0.70). Therefore, because of its advantages for positive autonomy, informing the ambulatory care team is in this situation a better option (autonomy = 0.73) than doing nothing (autonomy = 0.71), even when only taking into account the principle of autonomy.

Table 4. Simulation results of case 2

	Prv	MI	PhI	Info	CF	Ref	A-	A+	Aut	NM	B	Mor
NI	1.0	1.0	1.0	.50	.50	.50	1.0	.50	.71	0.0	0.0	.31
ICT	.80	1.0	1.0	.50	.70	.50	.96	.56	.73	.50	0.0	**.48**
IPar	.10	.80	1.0	.50	.50	.50	.76	.50	.62	0.0	0.0	.27

3.4   Case 3: Negative Autonomy Constrained to Enhance Positive Autonomy

In case 3, a patient signed a self-binding declaration for judicial coercion (a so-called Ulysses contract) when certain circumstances would occur. The patient evaded addiction care several times and had a relapse in alcohol use. Thereby the circumstances of the self-binding declaration were met and, according to the judge, judicial coercion was justified.

As can be seen in Table 5, the decision of the court is in line with the outcome in the simulation of the case. In Table 5, 'NI' stands for the decision option 'No Intervention' and 'JC' stand for 'Judicial Coercion after self-binding'. In the simulation, judicial coercion is a violation of self-determination. However, this violation is smaller (self-determination case 8 = 0.40) than in previous cases, where no self-binding declaration was signed (self-determination previous cases = 0.20). Moreover, without judicial coercion, the patient is likely to diminish in cognitive functioning (0.20) and reflection (0.20) by alcohol misuse, whereas by judicial coercion cognitive functioning and reflection can be recovered (both 1.0) during detoxification. Therefore, even when looking at autonomy alone, judicial coercion is a better option (autonomy = 0.56) than doing nothing (autonomy = 0.52).

Table 5. Simulation results of case 3

	Prv	MI	PhI	Info	CF	Ref	A-	A+	Aut	NM	B	Mor
NI	1.0	1.0	1.0	.50	.20	.20	1.0	.27	.52	0.0	0.0	.23
JC	.40	.40	.40	.50	1.0	1.0	.40	.79	.56	.70	.70	**.63**

4   Discussion

We presented a moral reasoning system including a twofold approach of autonomy. The system extends a previous moral reasoning system [14] in which autonomy consisted of a single variable. The behavior of the current system matches the behavior of the previous system. Moreover, simulation of legal cases for courts in the Netherlands showed a congruency between the verdicts of the judges and the decisions of the presented moral reasoning system including the twofold model of autonomy. Finally, the experiments showed that in some cases long-term positive autonomy was seen as more important than negative autonomy on the short-term.

Case 1 showed that, both according to the judge and to the model, assertive outreach was a morally justifiable option to prevent judicial coercion. By assertive outreach, the mental integrity and privacy of the patient were constrained. However, this prevented worsening of the situation, which would have raised the need for judicial coercion, a measure that would constrain the privacy of the patient more heavily.

In case 2, the psychiatrist should have informed the ambulatory care team instead of the parents of the patient. Informing the ambulatory care team constrained the privacy of the patient less than informing the parents, and had more potential to prevent worsening of the situation and improve the cognitive functioning of the patient. Because of its advantages for positive autonomy, also when only taking into account the principle of autonomy, informing the ambulatory care team is in this situation a better option than doing nothing. Thus, in this case, constraining negative autonomy in benefit of positive autonomy improves the overall level of autonomy.

Finally, case 3 showed that negative autonomy can sometimes be constrained to stimulate positive autonomy. In this case, the patient had agreed to that under certain conditions. Because the conditions of a self-binding declaration were met, judicial coercion was justified. During detoxification, the cognitive function and reflection of the patient could be restored. Because of this stimulation of positive autonomy on the longer term, the constraints of negative autonomy on the short-term in the end positively influence the level of overall autonomy.

The moral reasoning system presented in this paper can be used by robots and software agents to prefer actions that prevent users from being harmed, improve the users' well-being and stimulate the users' autonomy. By adding the twofold approach of autonomy, the system can balance between positive autonomy and negative autonomy.

In future work, we intend to integrate the model of autonomy into Moral Coppélia [13], an integration of the previously developed moral reasoning system [14] and Silicon Coppélia - a computational model of emotional intelligence [9]. Adding the twofold model of autonomy to Moral Coppélia may be useful in many applications, especially where machines interact with humans in a medical context.

After doing so, the level of involvement and distance (cf. [9]) will influence the way the robot tries to improve the autonomy of a patient. In long-term care, nurses tend to be relationally involved with patients, motivating them to accept care [1, 11]. A robot that is more involved with the patient will focus more on improving positive autonomy and especially on reflection. If the robot is relatively little involved with the patient, it will focus more on negative autonomy: physical and mental integrity and privacy.

Acknowledgements. This study is part of the SELEMCA project within CRISP (grant number: NWO 646.000.003).

References

1. Agich, G.J.: Autonomy and long-term care. Cambridge University Press (2003)
2. Anderson, M., Anderson, S., Armen, C.: Toward Machine Ethics: Implementing Two Action-Based Ethical Theories. In: Machine Ethics: Papers from the AAAI Fall Symposium, Association for the Advancement of Artificial Intelligence, Menlo Park (2005)
3. Anderson, M., Anderson, S.: Ethical Healthcare Agents. SCI, vol. 107. Springer (2008)
4. Banks, M.R., Willoughby, L.M., Banks, W.A.: Animal-Assisted Therapy and Loneliness in Nursing Homes: Use of Robotic versus Living Dogs. Journal of the American Medical Directors Association 9, 173–177 (2008)

5. Beauchamp, T.L., Childress, J.F.: Principles of Biomedical Ethics. Oxford University Press, Oxford (2001)
6. Berlin, I.: Two concepts of liberty. Clarendon Press, Oxford (1958)
7. Buchanan, A.E., Brock, D.W.: Deciding for Others: The Ethics of Surrogate Decision Making. Cambridge University Press (1989)
8. Widdershoven, G.A.M., Abma, T.A.: Autonomy, dialogue, and practical rationality. In: Radoilska, L. (ed.) Autonomy and mental disorder, pp. 217–232. Oxford University Press, Oxford (2012)
9. Hoorn, J.F., Pontier, M.A., Siddiqui, G.F.: Coppélius' Concoction: Similarity and Complementarity Among Three Affect-related Agent Models. Cognitive Systems Research Journal 15, 33–49 (2012)
10. Karimi, A. Zorgrobot rukt op. Spits, 5 (October 1, 2012)
11. Moody, H.R.: Ethics in an ageing society. Johns Hopkins UP, Baltimore (1996)
12. Picard, R.: Affective computing. MIT Press, Cambridge (1997)
13. Pontier, M.A., Widdershoven, G., Hoorn, J.F.: Moral Coppélia - Combining Ratio with Affect in Ethical Reasoning. In: Pavón, J., Duque-Méndez, N.D., Fuentes-Fernández, R. (eds.) IBERAMIA 2012. LNCS, vol. 7637, pp. 442–451. Springer, Heidelberg (2012)
14. Pontier, M.A., Hoorn, J.F.: Toward machines that behave ethically better than humans do. In: Miyake, N., Peebles, B., Cooper, R.P. (eds.) Proceedings of of the 34th International Annual Conference of the Cognitive Science Society, CogSci 2012, pp. 2198–2203 (2012)
15. Robins, B., Dautenhahn, K., Boekhorst, R.T., Billard, A.: Robotic Assistants in Therapy and Education of Children with Autism: Can a Small Humanoid Robot Help Encourage Social Interaction Skills? Journal of Universal Access in the Information Society 4, 105–120 (2005)
16. Van Wynsberghe, A.: Designing Robots for Care; Care Centered Value-Sensitive Design. Journal of Science and Engineering Ethics (2012) (in press)
17. Wada, K., Shibata, T.: Social Effects of Robot Therapy in a Care House. JACIII 13, 386–392 (2009)
18. WHO.: Health topics: Ageing (2010),
http://www.who.int/topics/ageing/en/

Simulation of a Motivated Learning Agent

Janusz A. Starzyk[1], James Graham[1], and Leszek Puzio[2]

[1] Ohio University, Athens, OH 45701 USA
{starzykj,jg193404}@ohio.edu
[2] WSIZ, Rzeszow, 35-225 Poland
puzio@wsiz.rzeszow.pl

Abstract. In this paper we discuss how to design a simple motivated learning agent with symbolic I/O using a simulation environment within the NeoAxis game engine. The purpose of this work is to explore autonomous development of motivations and memory of agents in a virtual environment. The approach we took should speed up the development process, bypassing the need to create a physical embodied agent as well as reducing the learning effort. By rendering low-level motor actions such as grasping or walking into symbolic commands we remove the need to learn elementary motions. Instead, we use several basic primitive motor procedures, which can form more complex procedures. Furthermore, by simulating the agent's environment, we both improve and simplify the learning process. There are a few adaptive learning variables associated with both the agent and its environment, and learning takes less time, than it would in a more complex real world environment.

Keywords: Motivated learning, cognitive architectures, simulation, embodied intelligence, virtual agents.

1   Introduction

A significant challenge in robotics is to develop autonomous systems that can reason and perform missions in dynamic, uncertain, and uncontrolled environments [1]. Therefore, recent research efforts are directed towards developing autonomous cognitive systems making a significant progress in this direction [2,3,4,5].

Current cognitive architectures, such as SOAR [6], ACT-R [7], Icarus [8], LIDA [9], Polscheme [10], and CLARION [11], either have to rely on predefined goals (without self-motivated learning) or predefined rules (without autonomous reasoning). Due to their reliance on predefined scripts and heuristic rules, current robotic systems lack autonomy, self-adaptability, and reasoning capabilities either to accomplish complex missions or to handle ever changing missions in uncontrolled environments.

Another important direction in studying development of cognitive systems and robots is based on the idea of embodied intelligence. The principles of designing robots based on the embodied intelligence idea were first described by Brooks [12] and were characterized through several assumptions that would facilitate development of embodied agents.

H. Papadopoulos et al. (Eds.): AIAI 2013, IFIP AICT 412, pp. 205–214, 2013.

Since our aim is to develop intelligent machines we introduce internal motivations, creating abstract goals not previously known to the designer or the robot. Intelligent systems will adapt to unpredictable and dynamic situations in the environment by learning, which will give them a high degree of autonomy, making them a perfect choice for robotics and virtual agents [13]. The recently developed mechanism of motivated learning (ML) has such capacities [14].

With ML, robots can achieve various goals imposed by different challenge scenarios autonomously. They develop higher level abstract goals and increase internal complexity of representations and skills stored in their memory. Our aim in this work is to develop simulation tools of virtual autonomous systems with ML mechanism.

Most current autonomous robot systems concentrate on the cognitive development of individual robots [15,16,17]. They mainly focus on developing simple local behavior control algorithms under heuristic rules, and then seek to emerge global behaviors. Adding intrinsic motivations and advanced reasoning capabilities improves the robots' individual capabilities. In addition, improving robots' learning in complex dynamically changing environments is very important, especially in environments where there may be competition for resources.

Therefore, we work to provide a systematic framework for developing cognitive robots that can autonomously accomplish a wide variety of real-world complex missions in such environments. We have selected NeoAxis to build a virtual 3D environment for embodied motivated agents. That environment is able to simulate a wide scope of robot types, ranging from wheeled, flying or swimming robots, to humanoid robots. The second reason why we use NeoAxis is that it has good support for physics modeling. We can assign static and dynamic friction parameters, mass, bounciness, hardness, etc., to obtain real-world representation of objects and different material types. Objects could be attached to one another to create complex structures, like a car composed of wheels, body, windscreen, and engine. It is also possible to create environment rules, i.e., a tree produces apples in certain time intervals.

The rest of this paper is organized as follows: In section 2, we discuss the motivated learning agent and how it learns to interact with its environment. We discuss how pains are generated and adjusted and how goals are selected, and then we describe action value determination for opportunistic motivated learning (OML) agent. Following this in section 3, we discuss the simulation of a virtual OML agent. Finally, in section 4, we discuss how we integrated the agent into the NeoAxis environment. This includes our current work, and our plans to further advance the simulation tool.

2   Motivated Learning Agent Memory Organization

The motivated learning (ML) agent interacts with the virtual environment changing it by its actions and receiving rewards (external and internal) for its actions. In this implementation we assume that both sensory inputs and motor outputs are symbolic, and provide interface to the virtual environment.

We assume that the ML agent learns in a hostile environment where either there are a limited amount of renewable resources that the agent needs or there are hostile characters that may harm the agent unless it learns to defend itself. A ML agent has a number of predefined basic needs e.g. desired energy level, comfortable temperature zone, or acceptable pressure. The agent can satisfy these needs by taking proper actions. If a need is not satisfied (e.g. energy level is below threshold) the agent feels a pain signal. Thus, a pain signal related to basic needs can be predefined as a deviation from desired levels. If the agent learns how to satisfy its need then an abstract need is introduced in case the agent is unable to perform the desired action. Unsatisfied need manifests itself by a pain signal and the agent is motivated to reduce this pain.

An OML agent uses a neural network where each sensory neuron represents an object and each motor neuron represents an action. The OML system's neural network shown in Fig. 1, in addition to sensory S and motor M neurons, contains pain center neurons P that register the pain signals, and goal neurons G responsible for pain reduction. Selected pain center neurons are connected to the external reward/punishment signals. In RL these neurons receive a reward or punishment signal according to the training algorithm, and in ML they receive primitive pain signals that directly increase or decrease their activation level. In ML, abstract pain centers are created through the goal creation mechanism [14,18] and are activated via an interpretation of the sensory inputs. A goal is an intended action that involves a sensory-motor pair. To implement a goal the agent acts on the observed object. All pain neurons are initially connected to goal neurons with random interconnection weights. All goal neurons and pain neurons are subject to Winner-Take-All (WTA) competition between them. The number of goal neurons is equal to the number of sensory-motor pairs. In the symbolic representation each neuron represents a single symbol, pain, goal or action. Fig. 1 shows symbolically the interconnection structure, between S, P, B, G and M neurons. In Fig. 1 an abstract pain center P_k connections to its sensory, bias, goal and motor neurons are also shown.

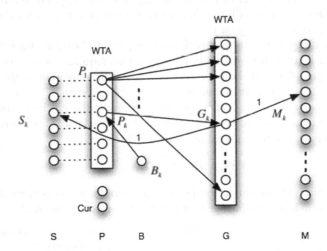

Fig. 1. Connections between sensory, motor, bias, pain and goal neurons

In what follows we explain how various weights of the OML system's neural network are adjusted during the learning process.

2.1   Bias Signals, Weights, and Associated Pains

A bias signal triggers an abstract pain and is defined depending on the type of perceived situation. If the autonomous agent needs to maintain a certain level of resources, the bias reflects how difficult it is to obtain this resource or in a more general case, how difficult it is to perform a desired action. Resources can be either desired if their use can reduce the agent's pain or undesired if they can increase the pain. Thus the agent must first have an experience to determine if the resource is desired or undesired to introduce a resource related bias signal.

The bias signal for desired or undesired resources is calculated from the observed level of resource and its desired/undesired limits as follows:

$$B = \gamma * \left(\frac{\varepsilon + R_d(s_i)}{\varepsilon + R_c(s_i)}\right)^{\delta_r} \tag{1}$$

where R_d is a desired resource value (observed at a sensory input s_i) and R_c is a current resource value. ε is a small positive number to prevent numerical overflow, γ regulates how quickly pain increases, $\gamma > 0$, and $\delta_r = 1$ when the resource is desired, $\delta_r = -1$ when it is not desired, and $\delta_r = 0$ otherwise (when the character of the resource is unknown).

Initially all B-P_k weights w_{bp} are set to 0. Thus, the machine initially responds only to the primitive pain signals P directly stimulated by the environment. Each time a specific pain P is reduced the weight w_{bp} of the B-P_k bias link increases. However, if the goal activated by the pain center P was completed and did not result in reduction of pain P, then the B-P_k weights w_{bp} are reduced. Since the bias weight B-P_k indicates how useful it is to have access to a desired resource S, a bias weight adjustment parameter Δ_b must be properly set to reflect the rate of stimuli applied to a higher order pain center. This rate reflects how often a given abstract pain center P_k was used to reduce the lower order pain signal P.

When a specific goal is not invoked for a long period of time its importance in satisfying a lower level pain is gradually reduced. This requires a reduction of the w_{pg} weight to this goal from all the pain centers. A similar reduction of the B-P_k links indicates a gradual decline in importance of an abstract pain. This mechanism of lowering the weights to an abstract pain center prevents the machine from overestimating its importance to the lower level pain that was responsible for its creation.

Using the bias signal, the pain value is estimated from:

$$P(s_i) = B(s_i) * w_{bp}(s_i) \tag{2}$$

where w_{bp} is a bias to pain weight for a given pain center. w_{bp} is computed incrementally based on signals of pain change that resulted from the action taken as follows:

$$
\begin{aligned}
w_{bp} &= w_{bp} + \Delta_{b+}(\alpha_b - w_{bp}) && \text{if the associated pain was reduced or increased} \\
w_{bp} &= w_{bp}(1 - \Delta_{b+}) && \text{if there was no change in pain} \\
w_{bp} &= w_{bp}(1 - \Delta_{b-}) && \text{when the assoc. sensory input was not involved in this action}
\end{aligned}
\tag{3}
$$

where $\alpha_b = 0.5$, sets the ceiling for w_{bp}; $\Delta_{b-} = 0.0001$, sets the rate of decline for w_{bp} weights; $\Delta_{b+} = 0.08$, sets the rate of increase for w_{bp} weights - increasing Δ_{b+} value increases the rate at which pains increase.

2.2   Changes of the Goal Related and Curiosity Weights

Initial weights between P-G neurons are randomly selected in the 0-α_g interval (a good setting will be between 0.49 and 0.51 of α_g for faster learning).   Assume that the weights are adjusted upwards or downwards by a maximum amount μ_g. In order to keep the interconnection weights within prespecified limits (0< w_{pg} <α_g), the value of the actual weight adjustment applied can be less than μ_g and is computed as

$$\Delta_a = \mu_g \ \min \left(\mid \alpha_g - w_{pg} \mid, w_{pg} \right) \qquad \text{where } \alpha_g \leq 1 \qquad (4)$$

and

$$\mu_g = \mu_0 \left(1 - \frac{2}{\pi} \text{atan} \left(10 * \left(\frac{R_c(s_i)}{R_d(s_i)} \right)^{\delta_r} \right) \right) \qquad \text{where } \mu_0 = 0.3 \qquad (5)$$

Using (4) produces weights that slowly saturate towards 0 or α_g. (For quick learning set ($\mu_g = \alpha_g$ / 2). No other weights from other pain centers to this specific goal are changed, so the sum of weights incoming to the node G is not constant.

If, as a result of the action taken, the pain that triggered this action increased (as determined by pain reduction parameter δ_p), then the w_{pg} weight is decreased by Δ_a, and if the pain decreased, then the w_{pg} weight is increased by Δ_a

$$w_{pg} = w_{pg} + \delta_p * \Delta_a. \qquad (6)$$

2.3   Action Value Determination for OML Agent

In the opportunistic ML agent (OML) the "best action" is determined by the linear heuristic OML model, using action "Value" V_i

$$V_i = \frac{P_i + (\Delta P * (t_{mot} + t_{dist}))}{(t_{mot} + t_{dist})^2} \qquad (7)$$

where ΔP is the estimated change in pain over 1 cycle. P_i is the pain associated with the action under consideration, t_{mot} is the required motor time to complete the action, and t_{dist} is the time required by the agent to travel to perform the action. The action with the highest value of V_i is the one chosen by the OML agent. The selection of actions evaluated in (7) depends on the number of pains above threshold.

3   NeoAxis Implementation

A cognitive architecture organization based on the ML idea was introduced in [19]. This architecture uses a physical body in a physical environment. However, making a physical robot costs money and a design effort, so it is very helpful to use a computer simulation that can imitate real-time physical conditions. Thus we used a simulated environment with a virtual robot.

We implemented the infrastructure of the OML agent in NeoAxis describing the motivated agent functionality in C++ and C#. The virtual environment for OML agents built in NeoAxis is a 3D simulated world governed by realistic physics to present the robots within a complex, challenging world. This simulation environment can be separated into two major components. The first one is the animation controller that handles display tasks and transitions the agent from one action to another. The second component processes the agent's behaviors and defines the potentially sophisticated rules governing the virtual world in which the agent lives. The agent works in the created environment, discovers these rules and learns to use them to its advantage.

In this environment, we created resources that the agent could use, and endowed the agent with the ability to act on these resources (listed in Table 1). The agent's actions are driven by its pains. Only two pains, hunger, which increases over time, and curiosity, are predefined. Curiosity pain makes the agent explore the environment when no other pain is detected. Other pains are learned by the agent using the goal creation methodology [18]. The agent observes which resources it needs and introduces abstract pain(s) if they are unavailable. We also defined world rules, which describe which agent actions make sense and what their results are. Those rules are listed in Table 1 and are unknown to the agent. The agent's actions result in various outcomes like increasing and decreasing resource quantities, as well as in reducing some pains.

Table 1. List of valid Resource-Motor pairs and their outcome

Motor	Resource	Outcome		
		Increase	Decrease	Pain reduce
Eat food from	Bowl		Food in Bowl	Hunger (Primitive)
Take food from	Bucket	Food in Bowl	Food in Bucket	Lack of food in Bowl
Buy food with	Money	Food in Bucket	Money	Lack of food in Stock
Work for money with	Hammer	Money	Hammer	Lack of Money
Study for job with	Book	Hammer	Book	Lack of Job
Play for joy with	Beach ball	Book	Beach ball	Lack of School

3.1   Simulation Algorithm of OML Agent in NeoAxis

To test our agent, we needed to embed it in an environment and provide it with a means to observe the environment and interact with it. To do this, we created a simple, but effective test bed within the NeoAxis engine.

The basic steps of OML agent simulation in NeoAxis contain successive iterations:

Each iteration consists of the Agent Phase, where the agent observes the Environment, updates its internal state, and generates motor output, and an Environment Phase where the environment performs the agent's action and updates itself accordingly, e.g. updates resource quantities, objects' locations, determines motor action outcome, etc.

To visualize resources quantities, current task, pains levels and the agent's memory, we added windows that display the current state of the agent and the environment as shown in Fig. 2.

Fig. 2. Main simulation view with displayed simulation state in windows

When a pain level is above threshold the pains display is red. In this screenshot (Fig. 2), the agent action is driven by the 'Lack of Money' pain. Sometime the agent takes a nonsense action like "Play for joy with hammer." Nevertheless, even actions such as this are used to learn. The memory window, presented in Fig. 2 on the left, displays the memory array state. When the color is gray, then this means that the agent has not learned usefulness of this action yet. When the color is white then this means that the action is useful. Each row in the memory array corresponds to a driving pain, while each column represents a possible action.

Any action performed by the agent is based on the "action tree" that tells the agent what elementary actions must be used to accomplish a desired composed action. For instance, if the agent decided to eat an apple, it must first walk in the direction of the apple, pick it up, and eat. The agent selects the paths to the desired object, while avoiding obstacles along the way. An OML agent is capable of estimating how long it will take for it to approach the object based on the distance information. It uses this information to select the most appropriate action based on the state of the environment and its own internal state. The agent is informed by its sensory inputs when he approached the desired object so it can start the desired action.

We have run multiple simulations where we modified resource quantities and motor action times. When starting resources were sparse, the agent couldn't learn all valid actions because it ran out of resources to test new actions. When resources, like food, were plentiful, the agent did not bother to learn anything new once the hunger pain was under control. Also, when action times were too long, the agent couldn't satisfy all pains. However, with parameters adjusted so that the agent doesn't have access to too many or too few resources the OML agent learns correct behavior outperforming any reinforcement learning agent.

Figs. 3 provides results from the OML agent simulated in Matlab showing dynamic changes in various pain levels.

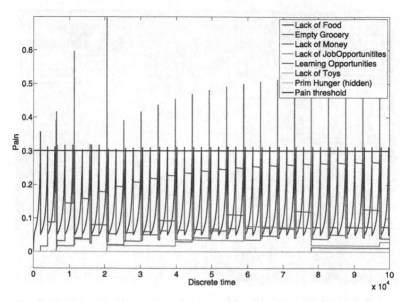

Fig. 3. Matlab simulation results showing dynamic changes in the pain levels

Only the primitive pain (hunger) is predefined and is gradually increasing. As the agent learns how to reduce the pain (by eating food) it introduces an abstract need to have food, and if food is in a short supply the pain of not having it increases. To supply food the agent learns more advanced skills (like buying food) and introduces abstract needs and related abstract pains. These abstract pains were unknown to the agent before it discovered how to use an abstract resource (e.g. money) to its advantage. We can see this in the simulation plot on Fig. 3 by observing that there was no pain of lack of money before 5000 iterations.

The simulation demonstrates that the agent was able to manage its various needs. Once abstract needs were introduced and related pains defined, the agent learned how to keep all pains below threshold, set in our simulation to 0.3.

Most of our code was initially prototyped in Matlab and then ported to NeoAxis, due to the ease to code and familiarity we have with developing Matlab applications. It should be noted, that the Matlab version is not strictly equivalent to the NeoAxis version, since the NeoAxis simulation operate in real time. It adjusts the decision making process to consider time delays required to approach various objects and the time needed to act on them. Thus, initial resource levels and motor times were adjusted to allow the NeoAxis simulation to operate efficiently. As a result of the changes made to the simulation parameters, the NeoAxis agent's pain passes threshold more quickly than in the Matlab simulation, effectively accelerating the rate at which it learns. However, despite the difference in the setup of the two simulations, the underlying model is identical, and as can be observed from the charts, the agent is able to effectively learn the desired actions.

Real time simulation showing the agent's actions and changes in the pains, observed in the display windows (Fig. 2), is in agreement with the Matlab results.

4   Conclusions

In this paper, we have presented our work on motivated learning agents with a focus on simulating the agent within a graphical environment. We included modifications to original ML algorithm [18], with new calculations for bias signals and w_{pg} weights. Additionally, we introduced greater complexity into the environment by introducing undesirable resources. This includes changes in δ_p calculation and the calculation of desired resource levels. By adding these new features we've improved the agent's ability to handle its environment as well as our own ability to implement complex and interesting virtual environments for our agents to interact with.

The simulation results of the embodied OML agent in the virtual 3D environment in NeoAxis prove that our theoretical assumptions for motivated learning agent memory organization, determination of bias signals, weights, goal creation and selection, as well as associated pain calculations, were valid. The OML agent was able to learn all environment rules, and keep the agent's pains under control.

Our further research will focus on the extension of the simulation, specifically, making a more complex environment and to introduce friendly and hostile characters.

Acknowledgements. This work was supported by the grant from the National Science Centre DEC-2011/03/B/ST7/02518.

References

1. Hirukawa, H., et al.: Humanoid robotics platforms developed in HRP. Robotics and Autonomous Systems 48(4), 165–175 (2004)
2. Bakker, B., Schmidhuber, J.: Hierarchical Reinforcement Learning Based on Subgoal Discovery and Subpolicy Specialization. In: Groen, F., Amato, N., Bonarini, A., Yoshida, E., Köse, B. (eds.) Proceedings of the 8th Conference on Intelligent Autonomous Systems, IAS-8, Amsterdam, The Netherlands, pp. 438–445 (2004)
3. Oudeyer, P.-Y., et al.: Intrinsically Motivated Exploration for Developmental and Active Sensorimotor Learning. In: Sigaud, O., Peters, J. (eds.) From Motor Learning to Interaction Learning in Robots. SCI, vol. 264, pp. 107–146. Springer, Heidelberg (2010)
4. Ro, S., et al.: Curiosity-driven acquisition of sensorimotor concepts using memory-based active learning. In: IEEE Intl. Conf. on Robotics and Biometrics, pp. 665–670 (2009)
5. Singh, S., Barto, A.G., Chentanez, N.: Intrinsically motivated learning of hierarchical collections of skills. In: Proc. 3rd Int. Conf. Development Learn., San Diego, CA, pp. 112–119 (2004)
6. Laird, J.E.: Extending the Soar cognitive architecture. In: Artificial General Intelligence 2008, pp. 224–235. IOS Press, Memphis (2008)
7. Anderson, J.R., et al.: An integrated theory of the mind. Psych. Review 111(4), 1036–1060 (2004)
8. Langley, P., Choi, D.: A unified cognitive architecture for physical robots. In: 21st Nat. Conf. Artificial Intelligence, pp. 1469–1474. AAAI Press, Boston (2006)
9. Baars, B.J., Franklin, S.: Consciousness is computational: the LIDA model of global workspace theory. Int. J. Machine Consciousness 1(1), 23–32 (2009)

10. Cassimatis, N., Nicholas, L.: Polyscheme: A Cognitive Architecture for Integrating Multiple Represetnation and Inference Schemes, MIT Ph.D. Diss. (2002)
11. Sun, R.: The importance of cognitive architectures: an analysis based on CLARION. J. Experimental and Theor. Artif. Intell. 19(2), 159–193 (2007)
12. Brooks, R.A.: Intelligence without reason. In: Proc. 12th Int. Conf. on Artificial Intelligence, Sydney, Australia, pp. 569–595 (1991)
13. Pfeifer, R., Bongard, J.C.: How the Body Shapes the Way We Think: A New View of Intelligence. The MIT Press, Bradford Books (2007)
14. Starzyk, J.A.: Motivation in Embodied Intelligence. In: Frontiers in Robotics, Automation and Control, pp. 83–110. I-Tech Education and Publishing, Austria (2008)
15. Clark, A.: Reasons, Robots and the Extended Mind. J. Mind and Language 16(2), 121–145 (2001)
16. Kanda, T., et al.: Development and evaluation of interactive humanoid robots. Proceedings of IEEE 92(11), 1839–1850 (2004)
17. Weng, J., et al.: Autonomous mental development by robots and animals. Science 291(5504), 599–600 (2001)
18. Starzyk, J.A.: Motivated Learning for Computational Intelligence. In: Igelnik, B. (ed.) Computational Modeling and Simulation of Intellect: Current State and Future Perspectives, ch.11, pp. 265–292. IGI Publishing (2011)
19. Starzyk, J.A.: Mental Saccades in Control of Cognitive Process. In: Int. Joint Conf. on Neural Networks, San Jose, CA, July 31-August 5, pp. 495–502 (2011)

Autonomous Navigation Applying Dynamic-Fuzzy Cognitive Maps and Fuzzy Logic

Márcio Mendonça[1], Ivan Rossato Chrun[1], Lúcia Valéria Ramos de Arruda[2], and Elpiniki I. Papageorgiou[3]

[1] Departamento de Engenharia Elétrica, Universidade Tecnológica Federal do Paraná
Av. Alberto Carrazai 1640, CEP 86300-000, Cornelio Procópi, Paraná, Brasil
mendonca@utfpr.edu.br, ivanchrun@gmail.com
[2] Laboratório de Automação e Sistema de Controle Avançado (LASCA),
Universidade Tecnológica Federal do Paraná
Av. Sete de Setembro 3165, CEP 80230-901, Curitiba, Paraná, Brasil
lvarruda@utfpr.edu
[3] Technological Education Institute of Lamia,
Department of Informatics and Computer Tech-nology, Lamia, Greece
epapageorgiou@teilam.gr

Abstract. This work develops a knowledge based system to autonomous navigation using Fuzzy Cognitive Maps (FCM). A new variant of FCM, named Dynamic-Fuzzy Cognitive Maps (D-FCM), is used to model decision tasks and/or make inference in the robot or mobile navigation. Fuzzy Cognitive Maps is a tool that model qualitative structured knowledge through concepts and causal relationships. The proposed model allows representing the dynamic behavior of the mobile robot in presence of environment changes. A brief review of correlated works in the navigation area by use of FCM is presented. Some simulation results are discussed highlighting the ability of the mobile to navigate among obstacle (navigation environment). A comparative with Fuzzy Logic and future works are also addressed.

Keywords: Mobile Robot Navigation, Fuzzy Cognitive Maps, Dynamic-Fuzzy Cognitive Maps, Intelligent decision systems.

1   Introduction

Artificial Intelligence (AI) has applications and development in various areas of knowledge, such as mathematical biology neuroscience, computer science and others. The research area of intelligent computational systems aims to develop methods that try to mimic or approach the capabilities of humans to solve problems. These news methods are looking for emulate human's abilities to cope with very complex processes, based on inaccurate and/or approximated information. However, this information can be obtained from the expert's knowledge and/or operational data or behavior of an industrial system [1].

H. Papadopoulos et al. (Eds.): AIAI 2013, IFIP AICT 412, pp. 215–224, 2013.

Researches in autonomous navigation are in stage of ascent. Autonomous Systems has the ability to perform complex tasks with a high degree of success [2]. In this context, the complexity involved in the task of trajectory generation is admittedly high efficient and, in many cases, requires that the autonomous system is able to learn a navigation strategy through interaction with the environment [3].

There is a growing interest in the development of autonomous robots and vehicles, mainly because of the great diversity of tasks that can be carried out by them, especially those that endanger human health and/or the environment, [4] and [5]. As an example, we can cite Mandow et al. [6], which describes an autonomous mobile robot for use in agriculture, in order to replace the human worker, through inhospitable activities as spraying with insecticides.

The problem of mobile robots control comprises two main sub problems: 1) navigation, determining of robot/vehicle position and orientation at a given time, and 2) guided tours, which refers to the control path to be followed by the robot/vehicle. In some cases have more complexity, example in [7] a Subsumption Architecture to develop dynamic cognitive network-based models it used for autonomous navigation with different goals: avoid obstacles, exploration and reach targets.

This work specifically proposes the development of an autonomous navigation system that uses heuristic knowledge about the behavior of the robot/vehicle in various situations, modeled by fuzzy cognitive maps [8]. In this case, the robot/vehicle determine a planning or generation of sequences of action in order to reach a given goal state from a predefined starting state.

Through cognitive maps, beliefs or statements regarding a limited knowledge domain are expressed through language words or phrases, interconnected by simple relationship of cause and effect (question/non-question). In the proposed model, the FCM relationships are dynamically adapted by rules that are triggered by the occurrence of special events. These events must change mobile behavior. There are various works in the literature that model heuristic knowledge necessary for decision-making in autonomous navigation, by means of Fuzzy systems [9], [10], [11], [12], [13] and [14]. In a similar way, the approach proposed in this paper is to build qualitative models to mobile navigation by means of fuzzy systems. However the knowledge is structured and built as a cognitive map that represents the behavior of the mobile.

Thus, the proposed autonomous navigation system must be able to take dynamic decisions to move through the environment and sometimes it must change the trajectory as a result of an event. For this the proposed FCM model must aggregate discrete and continuous knowledge about navigation. Actions such as the decision to turn left or right when sensors accuse obstacles and accelerate when there is a free path are always valid control actions in all circumstances. In this way, this type of action is modeled as causal relationship in a classical FCM.

However, there are specific situations, such as the need to maintain a trend of motion mainly in curves when the vehicle is turning left and sensors to accuse a new obstacle in the same direction. Due to inertia and physical restrictions, the mobile cannot abruptly change direction; this type of maneuver must be carefully executed. In this context, some specific situations should also be modeled on the map by causal relationships and concepts, but they are valid just as a result of a decision-making task caused by ongoing events. To implement such a strategy, new type of relationships and concepts will be added to the FCM classic model.

This new type of FCM in which the concepts and relationships are valid as a result of decision driven by events modeled by rules and is called Dynamic-FCM. Specifically, the work of Mendonça et al. [15] presents a type of D-FCM, which aggregates the occurrence of events and other facilities that makes appropriate this type of cognitive map, for the development of intelligent control and automation in an industrial environment.

The remainder of the paper is organized as following. Session 2 introduces Fuzzy Cognitive Maps concepts and provides a brief review of its application in autonomous navigation. Session 3 describes the proposed D-FCM and develops the autonomous navigation system. Session 4 presents simulation results obtained with the proposed navigation system and fuzzy logic navigation system and session 5 concludes the paper.

2   Fuzzy Cognitive Maps

Cognitive maps were initially proposed by Axelrod [16] to represent words, thoughts, tasks, or other items linked to a central concept and willing radially around this concept. Axelrod developed also a mathematical treatment for these maps, based in graph theory, and operations with matrices. These maps can thus be considered as a mathematical model of "belief structure" of a person or group, allowing you to infer or predicting the consequences that this organization of ideas represented in the universe.

This mathematical model was adapted for inclusion of Fuzzy logic uncertainty through by Kosko [8] generating widespread fuzzy cognitive maps. Like the original, FCMs are directional graph, in which the numeric values are fuzzy sets or variables. The "graph nodes", associated to linguistic concepts, are represented by fuzzy sets and each "node" is linked with other through connections. Each of these connections has a numerical value (weight), which represents a fuzzy variable related to the strength of the concepts.

The concepts of a cognitive map can be updated through the iteration with other concepts and with its own value. In this context, a FCM uses a structured knowledge representation through causal relationships being calculated mathematically from matrix operations, unlike much of intelligent systems whose knowledge representation is based on rules if-then type. However, due to this "rigid" knowledge representation by means of graph and matrix operation, the FCM based inference models lack robustness in presence of dynamic modifications not a priori modeled [17]. To circumvent this problem, this article develops a new type of FCM in which concepts and causal relationships are dynamically inserted into the graph from the occurrence of events. In this way, the dynamic fuzzy cognitive map model is able to dynamically acquire and use the heuristic knowledge. The proposed D-FCM and its application in autonomous navigation will be developed and validated in the following sections.

2.1   FCM in Intelligent Obstacle Navigation

Some related works which use cognitive maps in the robotics research area can be found in the literature. Among them, we can cite the work in [12] that employs probabilistic FCM in the decision-making of a robot soccer team. These actions are related to the behavior of the team, such as kick the ball in presence of opponents. The probabilistic FCM aggregates a likelihood function to update the concepts of the map. A Fuzzy cognitive map is used by [14] and [13] to guide an autonomous robot. The FCM is designed from a priori knowledge of experts and afterwards it is refined by a genetic algorithm.

Despite of the use of a known trajectory, actions are necessary due to errors and uncertainties inherent in the displacement of the robot, such as slippage, reading errors of the sensors, among others. A review of other related works employing intelligent navigation in robotics can be found in [14]. This paper also presents a Cognitive Map to implement a 3-D representation of the environment where an autonomous robot must navigate. The described architecture use a previously stored neural network based model to implement adjustments and course corrections of the robot in presence of noise and sensor errors. Similar to these works, we also use a fuzzy cognitive map to navigation tasks.

However our navigation system does not use a priori information about the environment. The FCM represent the usual navigation actions as turn right, turn left, accelerate and others. The adaptation ability to environment changes and to take decisions in presence of random events is reached by means of a rule based system. These rules are triggered in accordance of "intensity" of the sensor measurements.

3   The D-FCM Model

The development of a FCM model follows the steps of the work [15]. In this case, we identify 3 inputs related to the description of the environment (presence of obstacles) and 3 outputs describing the mobile's movements: turn left, turn right and move forward. The three inputs take values from the three sensors located at left, right and front side of the mobile.

These concepts are connected by arcs representing the actions of acceleration (positive) and braking (negative). Three decisions are originally modeled, if left sensor accuses an obstacle in this position, the vehicle must turn to the right side and equally if the right sensor accuses an obstacle in the right side, the vehicle turns to the other side. The direction change decision implies smoothly vehicle deceleration. The third decision is related to a free obstacle environment; in this case the mobile follows a straight line accelerating smoothly.

The figure 1 show the D-FCM modeled. The input concepts are SL, SR and SF and the output concepts are OutLeft, OutRigth and OutFront. The values of the concepts are the readings of the corresponding sensors. As a fuzzy number, these values are normalized into the interval [0, 1]. The relationships among these concepts are modeled by weights w1 to w5 which are computed. It is worthwhile to note in figure 1, that the concepts O.L. (-1) and O.R. (-1) are the values of the concepts in the previous state. This representation is equivalent to insert negative values (-1) in the corresponding diagonal positions of matrix W.

We choose to retain this representation to highlight that some concepts has memory. In this case, the mobile can remember the actions taken to turn left or right to prevent a zig-zag motion. As a result the mobile can maintain a movement trend.

In order to model the adaptation ability, we introduce 3 new concepts into the FCM associated to an "intensity" of motion (acceleration or braking) at each direction. There are left factor, right factor and front factor in figure 1. The factor concepts have their values changed according to the current condition of the vehicle motion and the occurrence of events. These events are modeled by the weights ws (ws1, ws2 and ws3) in figure 1 which are obtained by applying the rules of type IF-THEN based on linguistic terms. These rules represent some decisions such as if the mobile is turning right because the left sensor has detected an obstacle and suddenly the right sensor also detects an obstacle then the factor right is small (ws3). The default value of the factor concepts is one. If any rule is triggered the weights ws are null. Finally the outputs of the D-FCM are the product between the factor concepts and the output of classical FCM (OutLeft, OutRigth and OutFront).

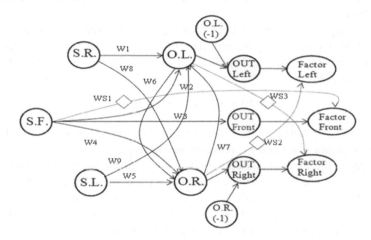

Fig. 1. The proposed D-FCM

In conclusion, the proposed D-FCM navigation system confers to the robot/vehicle the following behavior:

- The mobile is autonomous and it moves into unknown environment from an origin point to an end point.
- If an obstacle is detected by the front, left and/or right sensors the mobile must take a decision about new direction to follow.
- Default navigation position is in a straight line with constant speed, i.e. lateral movements are used only as a result of obstacle detection.
- When the mobile is in motion and the sensors don't identify any obstacle, the mobile accelerates smoothly and then it remains in a constant speed.

- Motion trend corresponds to an average between the current movement values and the values in immediately before instant, which prevents any sharp changes in direction of the mobile navigation.
- When the mobile is turning in left or right direction and the opposite sensor detects an obstacle, the motion trend is maintained but the mobile is softly breaking until to reestablish a straight movement.

For example if the sensors don't detect any obstacle then the mobile will accelerate in a straight line. The output related to turning will have no value related to the decision. If while he is moving forward the right sensor detects an obstacle then the D-FCM will make the decision to turn left and slowly break and turn left. In this case the front factor will be reduced and the output related to turn right will have no effect. And if while he is turning left to deviate from the obstacle the left sensor detects an obstacle the mobile tends to maintain his movement but the mobile will turn less to right since the factor will take new values and the output related to turn right will now take effect in his decision making.

Intelligent control architecture to the navigation system is similar [15]. The input interface read the sensor measurements which are inversely proportional to the distance of obstacles.

In order to dynamic adapt the D-FCM weights we use the hebbian learning algorithm for FCM that is an adaptation of the classic hebbian method. Different proposals and variations of this method applied in tuning or in learning for FCM are known in the literature [18].

$$W_i(k) = W_{ij}(k-1) \pm \gamma \Delta A_i \tag{1}$$

Where: $W_i(k)$ is a new value of the weight, $W_{ij}(k-1)$ old value, γ factor forget and ΔAi is a variation of the same concepts in two steps.

In this paper, the method is used to update the intensity of causal relationships in a deterministic way according to the variation or error in the intensity of the concept or input variable. In [15] one similar proposal for dynamic tuning in FCM uses Reinforcement Learning Algorithm (Q-learning). In this case, the application of Hebbian learning provides control actions as follows: if an obstacle to the right is nearest the causal relationship of exit turn left increases and consequently increases its control action. The others action have same behavior. Forgetting factors were obtained from empirical mode. And finally minimum and maximum limits were placed due to the application of the method is to tune the dynamic D-FCM, thereby varying the intensity of causal relationships should be within a clearly defined range. The range was chosen by observation of the mobile behavior in the environment. This range is defined by observation of the dynamic actions and closed intervals, [0.35, 0.65] for W_F, [0.6, 1] for W_E and [0.6, 1] for W_D.

$$W_F = W_{Fold}(i) - 0.7 * erroF(i) \tag{2}$$

$$W_E = W_{Eold}(i) - 0.7 * erroE(i) \tag{3}$$

$$W_D = W_{Dold}(i) - 0.7 * erroD(i) \tag{4}$$

Where: W_F is the weight related to front proximity sensor, W_E is the weight related to left proximity sensor and W_D is the weight related to the right proximity sensor.

A system of autonomous navigation using fuzzy logic was implemented in order to assess performance, outcomes and differences in acquisition and processing of empirical knowledge used in developing the tool presented fuzzy logic. The Work of [19] is similar and presents a fuzzy control strategy similar. The Fuzzy system is implemented in this work is type Mandani with 3 inputs, 3 outputs and employs 23 rules for abstraction of the same heuristic logic navigation controller inserted in the D-FCM. The inputs are the sensors, right, left and front and outputs are turning right, left and accelerate. These rules were implemented in an intuitive way according to heuristic D-FCM. For example:

- IF the right sensor is strong then turn left strong.
- IF the right sensor is weak then turn left weak.
- IF right sensor and frontal sensor very strong then weak accelerate and turn left very strong
- IF right, left and frontal sensor weak then accelerate Strong.

The strategies of decision making are the same, the difference in this formalism. In D-FCM knowledge is structured in the form of a graph, while the Fuzzy model is based on classical Fuzzy rules.

4   Characterization/Simulation Results

A 2-d animation simulated environment has inspired in real case (figure 2) and designed to test and validate our proposed navigation system.

Fig. 2. Real Environment with Dynamic Obstacle

The kinematic equations simulating the robot dynamic behavior has been inspired by [10] and uses scale in centimeters. In fact, the simulated robot corresponds to a mobile platform with two motors, and three sensors, one frontal, and one in each side. It uses ultrasound sensors and thus the perception of barrier or obstacle exists only within a scope zone of the sensor. Moreover the intensity of the sensor measure is inversely proportional to the distance of the object.

This simulation environment has served initially to knowledge acquisition through observation data input and output, and observation of robot behavior in several situations. Afterwards, two experiments were performed to validate the D-FCM and Fuzzy System navigation systems. In the first and second experiments, two different scenarios with static and dynamics obstacles have been simulated. The first experiment a dynamic obstacle is randomly inserted into the environment, during the robot navigation.

The results are presented in figures 3 and 4, they describe the path take by the fuzzy and D-FCM with hebbian learning algorithm. At Figure 3, the obstacle with coordinates (7.25, 96) is surprise, thus after the mobile executed half of the trajectory, this object enter in scenario. In Figure 4, these figures, the left graphic shows the scenario (x-y plan) with the initial about (15, 0) and end point near (0,160) of the robot trajectory. The graphic shows the dynamic trajectory made by the robot. The apparent flaws in the trajectory represent the speedup, when sensors do not detect an obstacle and the robot accelerates. In all experiments, we consider that the robot successfully attains the target point if its final position is into a horizontal interval [-8, +8] around the desired end point.

In the every scenario, example (figures 3, 4), there is a critical situation with a surprise obstacle around the position y=100. In the figure 3 the robot must to take the decision of to move straight, pass between the two obstacles and immediately to turn left to avoid a frontal barrier and to attain the target point. By analyzing the results in figure 3a and 3b we note that the robot takes the correct decisions.

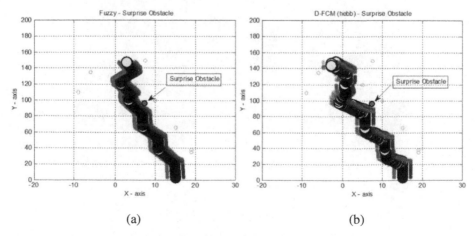

(a)   (b)

Fig. 3. a) Fuzzy Classic Architecture b) D-FCM Architecture with Hebbian Learning Algorithms

In second case, one dynamic obstacle is in the scenario starting about in position (10, 83) and finishing about in position (3, 83), figure 4a and 4b show the correct decisions in all maneuvers. In both cases, the robot motion trend is to move straight to the end point, but an. the robot takes the correct decision to turn in order to avoid a collision but it also maintains the motion trend of follow a straight line.

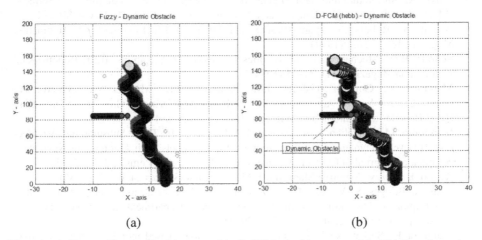

(a)   (b)

Fig. 4. a) Fuzzy Classic Architecture; b) D-FCM Architecture with Hebbian Learning Algorithms

5   Conclusion

This paper developed an autonomous navigation system based on a new type of fuzzy cognitive maps, named dynamic fuzzy cognitive map, D-FCM. The developed D-FCM approach adds new types of relationships and concepts into a classical FCM that allows modeling the human ability of to take decision in presence of random events. The human knowledge is represented by a rule based system that is triggered when critical situation occurs. As a result, the inference engine adds temporally concepts and relationships into the FCM. This approach is a contribution of this paper to the intelligent control area. It is not restricted to navigation systems and can be applied to model intelligent system with online decision making. In accordance with the results presented in this paper, we can conclude that the proposed D-FCM architecture constitutes a flexible and robust tool to navigation system able for navigate in vagueness and uncertainty in environment using dynamic planning. One of the main advantages of the proposed approach is that the knowledge acquisition and representation is simplified by the use of FCM models. Moreover the resulting fuzzy cognitive maps are also easy to implement and run. Thus, it is easily embedded in a hardware robot.

Future works addresses more functions and goals added in the robot, example energy management and reach targets. New scenarios, with more complexity, adding borders and more robots (agents). Finally, due to easy implement real scenarios can be development.

References

1. Passino, M.K., Yourkovich, S.: Fuzzy control. Addison-Wesley, Menlo Park (1997)
2. Russell, S.J., Norvig, P.: Artificial intelligence: a modern approach. Prentice Hall, Englewood Cliffs (1995)
3. Calvo, R.: Arquitetura híbrida inteligente para navegação autônomo de robôs. Dissertação (Mestrado em Ciências de Computação e Matemática Computacional). IMC-USP (2007)
4. Asami, S.: Robots in Japan: Present and Future. IEEE Robotics and Automation Magazine 1(2), 22–26 (1994)
5. Schraff, R.D.: Mechatronics and Robotics for Service Applications. IEEE Robotics and Automation Magazine 1(4), 31–35 (1994)
6. Mandow, A., Gomes-de-Gabriel, J.M., Martinéz, J.L., Muñoz, V.F., Ollero, A., García-Cerezo, A.: The Autonomous Mobile Robot AURORA for Greenhouse Operation. IEEE Robotics and Automation Magazine 3(4), 18–28 (1996)
7. Mendonça, M., Angélico, B.A., Arruda, L.V.R., Neves Jr., F.: A Subsumption Architecture to Develop Dynamic Cognitive Network-Based Models With Autonomous Navigation Application. Journal of Control, Automation and Electrical Systems 1, 3–14 (2013)
8. Kosko, B.: Fuzzy Cognitive Maps. Int. J. Man-Machine Studies 24, 65–75 (1986)
9. Siraj, A., Bridges, S., Vaughn, R.: Fuzzy Cognitive Maps for Decision Support in an Intelligent Intrusion Detection System. Technical Report, Department of Computer Science, Mississippi State University. MS 39762 (2001)
10. Malhotra, R., Sarkar, A.: Development of a fuzzy logic based mobile robot for dynamic obstacle avoidance and goal acquisition in an unstructured environment. In: Proceedings of the 2005 IEEE/ASME International Monterey, California, USA, pp. 24–28 (July 2005)
11. Astudillo, L., Castillo, O., Melin, P., Alanis, A., Soria, J., Aguilar, L.T.: Intelligent Control of an Autonomous Mobile Robot using Type-2 Fuzzy Logic. Engineering Letters 13(2) (2006)
12. Min, H.Q., Hui, J.X., Lu, Y.-S., Jiang, J.: Probability Fuzzy Cognitive Map for Decision-making in Soccer Robotics. In: Proceedings of the IEEE/WIC/ACM International Conference on Intelligent Agent Technology, IAT 2006 (2006), 0-7695-2748-5/06
13. Pipe, A.G.: An Architecture for Building "Potential Field" Cognitive Maps in Mobile Robot Navigation. Adaptive Behavior 8(2), 173–203 (2000)
14. Yeap, W.K., Wong, C.K., Schmidt, J.: Initial experiments with a mobile robot on cognitive mapping. In: Proceedings of the 2006 International Symposium on Practical Cognitive Agents and Robots, Perth, Australia, November 27–28 (2006)
15. Mendonça, M., Arruda, L.V.R., Neves, F.A.: Autonomous Navigation System Using Event Driven-Fuzzy Cognitive Maps. Applied Intelligence (Boston) 37, 175–188 (2011)
16. Axelrod, R.: Structure of Decision: the Cognitive Maps of Political Elites. Princeton University Press, New Jersey (1976)
17. Chun-Mei, L.: Using fuzzy cognitive map for system control. In: WTOS 7, vol. 12, pp. 1504–1515 (December 2008)
18. Papageorgiou, E.: Learning Algorithms for Fuzzy Cognitive Maps. IEEE Transactions on Systems and Cybernetics 42, 150–163 (2012)
19. Harisha, S.K., Ramkanth Kumar, P., Krishna, M., Sharma, S.C.: Fuzzy Logic Reasoning to Control Mobile Robot on Pre-defined Strip Path. In: Proceedings of World Academy of Science, Engineering and Technology 32 (August 2008) ISSN 2070-3740

Induction Motor Control Using Two Intelligent Algorithms Analysis

Moulay Rachid Douiri and Mohamed Cherkaoui

Mohammadia Engineering School, Department of Electrical Engineering,
Avenue Ibn Sina, B.P.765, Agdal-Rabat, Morocco
douirirachid@hotmail.com

Abstract. The main drawbacks of classical direct torque control (C-DTC) are: the high torque and flux ripples and variable switching frequency. To overcome these problems, two intelligent control theories, namely fuzzy logic control (FLC) and neuro-fuzzy control (NFC) are introduced to replace the hysteresis comparators and lookup table of the C-DTC for induction motor drive. The effectiveness and feasibility of the proposed approaches have been demonstrated through computer simulations. A comparison study between the C-DTC, FL-DTC and NF-DTC has been made in order to confirm the validity of the proposed schemes. The superiority of the NF-DTC has been proved through comparative simulation results.

Keywords: direct torque control, fuzzy logic control, induction motor drive, neuro-fuzzy control.

1 Introduction

The direct torque control (DTC) was proposed by M.Depenbrock [1] and I.Takahashi [2]-[3] in 1985. The DTC is an entirely different approach to induction motor control that was developed to overcome field oriented control (FOC) relatively poor transient response and reliance on induction motor parameters [4]-[5]-[6]. Classical DTC is a popular torque control method for induction motors; therefore it is widely used in the area of the EV's motor control. Unfortunately the classical DTC algorithm has some significant limitations. It is difficult to distinguish between small and large variations in reference values. Also the variation of flux and torque over one sector is considerable [7]-[8]. Another problem is that adapting classical DTC to the confines of a DSP's sampling period can significantly deteriorate its performance [9]. To overcome these problems, two intelligent control theories, including fuzzy logic control (FLC) and neuro-fuzzy control (NFC) are introduced to replace the conventional comparators and selection table of direct torque control for induction motor drive.

Fuzzy logic can deal with vague concepts which have relative degrees of truth rather than just the usual true or false, it allows machines to perform jobs that in the past required a human being's ability to think and reason [10]-[11]. Conventional control systems express control contents by using control expressions such as equations or

H. Papadopoulos et al. (Eds.): AIAI 2013, IFIP AICT 412, pp. 225–234, 2013.

logical expressions. This requires a huge amount of information, and some kinds of control are difficult or impossible to model this way. Fuzzy logic control usually requires only one tenth or less of the information required by conventional methods [12]. This is also associated with a high reliability and fast processing speed. Fuzzy logic also deals effectively with a non-linear time varying system.

There is a rapidly growing interest in the fusion of fuzzy systems and neural networks to obtain the advantages of both methods while avoiding their individual drawbacks. The possibility of integration of these two paradigms has given rise to a rapidly emerging field of fuzzy neural networks. There are two distinctive approaches for fuzzy-neural integration. On the one hand, many paradigms that have been proposed simply view a fuzzy-neural system as any ordinary multilayered feed-forward neural network which is designed to approximate a fuzzy control algorithm [13]-[14]. On the other hand, there are those approaches which aim to realize the process of fuzzy reasoning and inference through the structure of a connectionist network [15]. Fuzzy-neural networks are, in general, neural networks whose nodes have 'localized fields' which can be compared with fuzzy rules and whose connection weights can similarly be equated to input or output membership functions. The simplest attempt in merging of fuzzy logic and neural controllers is to make the neural networks (NN) learn the input-output characteristics of a fuzzy controller [16]. The NN in this case imitates the fuzzy controller but the only advantage is that the trained NN output has more smoothing robust actions than that of the fuzzy controller.

This paper is organized as follows: The principle of direct torque control is presented in the second part, the fuzzy logic direct torque control is developed in the third section, section four presents a neuro-fuzzy direct torque control, and the fifth part is devoted to illustrate the simulation performance of this control strategy, a conclusion and reference list at the end.

2   Classical Direct Torque Control

In a C-DTC motor drive, the machine torque and flux linkage are controlled directly without a current control. The principles of C-DTC can be explained by looking at the following torque and current equations of an induction motor:

$$T_e = \frac{3}{2} n_p \mathrm{Im} \left\{ \vec{\lambda}_s^* \vec{i}_s^* \right\} \tag{1}$$

$$\vec{i}_s = \frac{1}{\sigma L_s} \vec{\lambda}_s - \frac{L_m}{\sigma L_s L_r} \vec{\lambda}_r, \text{ with } \sigma = 1 - \frac{L_m^2}{L_s L_r} \tag{2}$$

Substituting Eq. (1) in Eq. (2) we obtain:

$$T_e = \frac{3}{2} n_p \frac{L_m}{\sigma L_s L_r} \left| \vec{\lambda}_s \right| \left| \vec{\lambda}_r \right| \sin \alpha \tag{3}$$

where α is the angle between the stator and rotor flux linkage vectors [8]. The derivative of Eq. (3) can be represented approximately as:

$$\frac{dT_e}{dt} = \frac{3}{2} n_p \frac{L_m}{\sigma L_s L_r} |\vec{\lambda}_s| |\vec{\lambda}_r| \left|\frac{d\alpha}{dt}\right| \cos\alpha \tag{4}$$

The machine voltage equation can be represented and approximated in a short interval of Δt as:

$$\frac{d\vec{\lambda}_s}{dt} = \vec{v}_s - R_s \vec{i}_s \approx \vec{v}_s \text{ implying } \Delta\vec{\lambda}_s = \vec{v}_s \Delta t \tag{5}$$

The rotor flux linkage vector is sluggish in response to a voltage vector during Δt as it is related to the stator flux linkage vector by a first order delay as in

$$\frac{d\vec{\lambda}_s}{dt} + \left(\frac{L_m}{\sigma L_s L_r} - \frac{R_r}{L_r} - j\omega_r\right)\vec{\lambda}_r = \frac{R_r L_m}{\sigma L_s L_r}\vec{\lambda}_s \tag{6}$$

Symbols:

R_s, R_r	stator and rotor resistance [Ω]
i_{sd}, i_{sq}	stator current dq axis [A]
v_{sd}, v_{sq}	stator voltage dq axis [V]
L_s, L_r	stator and rotor self inductance [H]
L_m	mutual inductance [H]
λ_{sd}, λ_{sq}	dq stator flux [Wb]
λ_{rd}, λ_{rq}	dq rotor flux [Wb]
T_e	electromagnetic torque [N.m]
E_{Te}	electromagnetic torque error [N.m]
$E_{\lambda s}$	stator flux error [Wb]
φ_s	stator flux angle [rad]
ω_r	rotor speed [rad/sec]
J	inertia moment [Kg.m^2]
n_p	pole pairs
σ	leakage coefficient

3   Fuzzy Logic Direct Torque Control

The structure of the switching table can be translated in the form of vague rules. Therefore, we can replace the switching table and hysteresis comparators by a fuzzy system whose inputs are the errors on the flux and torque denoted $E_{\lambda s}$ and E_{Te} and the argument φ of the flux. The output being the command signals of the voltage inverter n. The fuzziness character of this system allows flexibility in the choice of fuzzy sets of inputs and the capacity to introduce knowledge of the human expert.

The ith rule R_i can be expressed as:

$$R_i: \text{ if } E_{Te} \text{ is } A_i, E_{\lambda s} \text{ is } B_i, \text{ and } \varphi \text{ is } E_i, \text{ then } n \text{ is } N_i \tag{7}$$

where A_i, B_i and C_i denote the fuzzy subsets and N_i is a fuzzy singleton set.

The synthesized voltage vector n denoted by its three components is the output of the controller.

The inference method used in this paper is Mamdani's [18] procedure based on min-max decision [19]. The firing strength η_i, for i^{th} rule is given by:

$$\eta_i = \min\left(\mu_{A_i}(E_{T_e}), \mu_{B_i}(E_{\lambda_s}), \mu_{C_i}(\varphi)\right) \tag{8}$$

By fuzzy reasoning, Mamdani's minimum procedure gives:

$$\mu'_{N_i}(n) = \min\left(\eta_i, \mu_{N_i}(n)\right) \tag{9}$$

where μ_A, μ_B, μ_C, and μ_N are membership functions of sets A, B, C and N of the variables E_{Te}, $E_{\lambda s}$, φ and n, respectively.

Thus, the membership function μ_N of the output n is given by:

$$\mu_N(n) = \max_{i=1}^{72}\left(\mu'_{N_i}(n)\right) \tag{10}$$

We chose to share the universe of discourse of the stator flux error into two fuzzy sets, that of electromagnetic torque error in five and finally for the flux argument into seven fuzzy sets. However the number of membership functions (fuzzy set) for each variable can be increased and therefore the accuracy is improved. All the membership functions of fuzzy controller are given in Fig. 1.

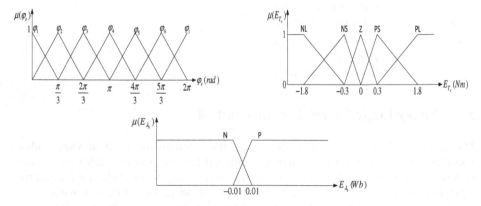

Fig. 1. Membership functions for fuzzy logic controller

Table 1. Fuzzy rules

$E_{\lambda s}$	E_{Te}	φ_1	φ_2	φ_3	φ_4	φ_5	φ_6
	P	V_6	V_2	V_3	V_1	V_5	V_4
PL	Z	V_4	V_6	V_2	V_3	V_1	V_5
	N	V_5	V_4	V_6	V_2	V_3	V_1
	P	V_6	V_2	V_3	V_1	V_5	V_4
PS	Z	V_7	V_0	V_7	V_0	V_0	V_0
	N	V_5	V_4	V_6	V_2	V_3	V_1
	P	V_2	V_3	V_1	V_5	V_4	V_6
NS	Z	V_0	V_7	V_0	V_7	V_0	V_7
	N	V_1	V_5	V_4	V_6	V_2	V_3
	P	V_2	V_3	V_1	V_5	V_4	V_6
NL	Z	V_3	V_1	V_5	V_4	V_6	V_2
	N	V_1	V_5	V_4	V_6	V_2	V_3

4   Neuro-Fuzzy Direct Torque Control

In this section, the Neuro-Fuzzy (NF) model is built using the multilayer fuzzy neural network shown in Fig.1. The controller has a total of five layers as proposed by *Lin and Lee* [17], with two inputs (stator flux error $E_{\psi s}$, electromagnetic torque error E_{Te}) and a single output (voltage space vector) is considered here for convenience. Consequently, there are two nodes in layer 1 and one node in layer 5. Nodes in layer 1 are input nodes that directly transmit input signals to the next layer. The layer 5 is the output layer. The nodes in layers 2 and 4 are "term nodes" and they act as membership functions to express the input/output fuzzy linguistic variables. A bell-shaped function is adopted to represent a membership function, in which the mean value p and the variance χ are adjusted through the learning process. The two fuzzy sets of the first and the second input variables consist of k_1 and k_2 linguistic terms, respectively. The linguistic terms are numbered in descending order in the term nodes; hence, k_1+k_2 nodes and n_3 nodes are included in layers 2 and 4, respectively, to indicate the input/output Linguistic variables.

Layer 1: Each node in this layer performs a MF:

$$y_i^1 = \mu_{Ai}(x_i) = \exp\left\{-\left[\left(\frac{x_i - c_i}{a_i}\right)^2\right]^{b_i}\right\} \tag{11}$$

where x_i is the input of node i, A_i is linguistic label associated with this node and (a_i, b_i, c_i) is the parameter set of the bell-shaped MF. y_i^1 specifies the degree to which the given input belongs to the linguistic label A_i, with maximum equal 1 and minimum equal to 0. As the values of these parameters change, the bell-shaped function varies accordingly, thus exhibiting various forms of membership functions. In fact, any continuous and piecewise differentiable functions, such as trapezoidal or triangular membership functions, are also qualified candidates for node functions in this layer.

Layer 2 - Every node in this layer represents the firing strength of the rule. Hence, the nodes perform the fuzzy AND operation:

$$y_i^2 = w_i = \min\left(\mu_{A\lambda_s}(E_{\lambda_s}), \mu_{BT_e}(E_{T_e}), \mu_{C\varphi_s}(\varphi_s)\right) \tag{12}$$

Layer 3 - The nodes of this layer calculate the normalized firing strength of each rule:

$$y_i^3 = \overline{w}_i = \frac{w_i}{\sum_{i=1}^{n} w_i} \tag{13}$$

Layer 4 - Output of each node in this layer is the weighted consequent part of the rule table:

$$y_i^4 = \overline{w}_i f_i = \overline{w}_i \left(p_i E_{\lambda_s} + q_i E_{T_e} + m_i \varphi_s + n_i\right) \tag{14}$$

where $\overline{w}_i$ is the output of layer 3, and $\{p_i, q_i, m_i, n_i\}$ is the parameter set. Which determine the ith component of vector desired voltage. By multiplying weight y_i by voltage continuous V side of the inverter according to Eq. (15):

$$V^* = y_i V \tag{15}$$

Layer 5 - The single node in this layer computes the overall output as the summation of all incoming signals:

$$y_i^5 = \sum_{i=1}^{n} \overline{w}_i f_i \tag{16}$$

Which determine the vector reference voltage v_s^* (see Fig. 4), from Eq. (17):

$$v_s^* = \sum_{i=1}^{9} y_i V e^{j\xi_i^*} \tag{17}$$

The angle ξ is obtained from the actual angle of stator flux φ_s and angle increment $d\varphi_i$ given by this Eq. (18):

$$\xi_i = \varphi_s + d\varphi_i \tag{18}$$

y_i ($i = 1..9$) are the output signals order i of the third layer respectively.

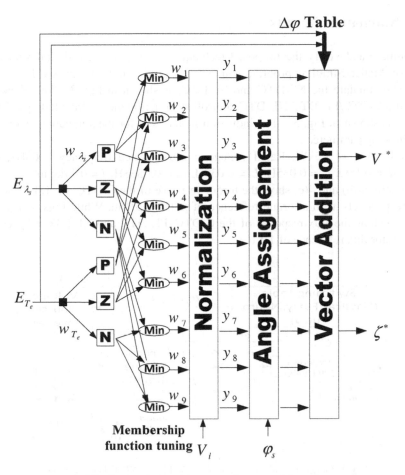

Fig. 2. Topology of the neuro-fuzzy model used

Table 2. Parameters setting for ANFIS model

ANFIS Setting	Details
Input variables	Electromagnetic torque error, and stator flux error
Output response	Space voltage vector
Type of input MFs	Generalized Bell MF
Number of MFs	2,3, 4 and 5
Type of output MFs	Linear and constant
Type inference	Linear Sugeno
Optimization Method	Hybrid of the least-squares and the back propagation gradient descent method.
Number of data	520
Epochs	1000

5   Simulation Results

To compare and verify the proposed techniques in this paper, a digital simulation based on Matlab/Simulink program with a Fuzzy Logic Toolbox and ANFIS Toolbox is used to simulate the NF-DTC and FL-DTC, as shown in Fig. 3. The block diagram of a C-DTC/FL-DTC/NF-DTC controlled induction motor drive fed by a 2-level inverter is shown in Fig. 3. The induction motor used for the simulation studies has the following parameters:

Rated power = 7.5kW, Rated voltage = 220V, Rated frequency = 60Hz, R_r = 0.17Ω, R_s = 0.15Ω, L_r = 0.035H, L_s = 0.035H, L_m = 0.0338H, J = 0.14kg.m^2.

Figs. 4(a), 4(b) and 4(c) show the torque response of the C-DTC, FL-DTC and NF-DTC respectively with a torque reference of [20-10-15]Nm. While Figs. 4(a'), 4(b') and 4(c') show the flux response of the C-DTC, FL-DTC and NF-DTC respectively with a stator flux reference of 1Wb.

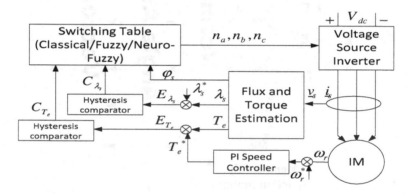

Fig. 3. General configuration of C-DTC/FL-DTC/NF-DTC scheme

Table 3 represents the comparative results in both stator flux and torque ripples percentage for C-DTC, FL-DTC and NF-DTC. The steady state response for the torque in NF-DTC is faster and provided more accuracy compared to other control strategies presented in this paper.

Table 3. Comparative study of C-DTC, FL-DTC and NF-DTC

Control strategies	Torque ripple (%)	Flux ripple (%)	Rise time (sec)	Setting time (sec)
C-DTC	10.6	2.3	0.009	0.01
FL-DTC	3.9	2.1	0.007	0.0085
NF-DTC	2.7	1.4	0.005	0.0071

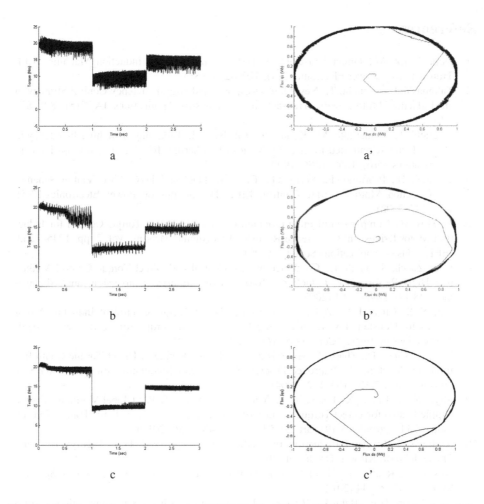

Fig. 4. (a), (b) and (c) torque response of C-DTC, FL-DTC and NF-DTC respectively, (a'), (b') and (c') Stator flux trajectory response of C-DTC, FL-DTC and NF-DTC respectively

6   Conclusions

Two various intelligent torque control schemes worth knowing fuzzy logic direct torque control, and neuro-fuzzy direct torque control have been evaluated for induction motor control and which have been compared with the conventional direct torque control technique. A better precision in the torque and flux responses was achieved with the NF-DTC method with greatly reduces the execution time of the controller; hence the steady-state control error is almost eliminated. The application of neural network techniques simplifies hardware implementation of direct torque control and it is envisaged that NF-DTC induction motor drives will gain wider acceptance in future.

References

1. Depenbrock, M.: Direct Self-Control (DSC) of Inverter-Fed Induction Machine. IEEE Transaction on Power Electronics 3(4), 420–429 (1988)
2. Takahashi, I., Noguchi, T.: New Quick-Response and High-Efficiency Control Strategy of an Induction Motor. IEEE Transactions on Industry Applications IA 22(5), 820–827 (1986)
3. Noguchi, T., Yamamoto, M., Kondo, S., Takahashi, I.: Enlarging Switching Frequency in Direct Torque-Controlled Inverter by Means of Dithering. IEEE Transaction on Industry Applications 35(6), 1358–1366 (1999)
4. Casadei, D., Profumo, F., Serra, G., Tani, A.: FOC and DTC: Two Viable Schemes for Induction Motors Torque Control. IEEE Transactions on Power Electronics 17(5), 779–787 (2002)
5. Le-Huy, H.: Comparison of Field-Oriented Control and Direct Torque Control for Induction Motor Drives. In: Conference Record - IAS Annual Meeting, vol. 2, pp. 1245–1252. IEEE Industry Applications Society (1999)
6. Vaez-Zadeh, S., Jalali, E.: Combined Vector Control and Direct Torque Control Method for High Performance Induction Motor Drives. Energy Conversion and Management 48(12), 3095–3101 (2007)
7. Lai, Y.-S., Chen, J.-H.: A New Approach to Direct Torque Control of Induction Motor Drives for Constant Inverter Switching Frequency and Torque Ripple Reduction. IEEE Transactions on Energy Conversion 16(3), 220–227 (2001)
8. Wei, X., Chen, D., Zhao, C.: Minimization of Torque Ripple of Direct-Torque Controlled Induction Machines by Improved Discrete Space Vector Modulation. Electric Power Systems Research 72(2), 103–112 (2004)
9. Shyu, K.-K., Lin, J.-K., Pham, V.-T., Yang, M.-J., Wang, T.-W.: Global Minimum Torque Ripple Design for Direct Torque Control of Induction Motor Drives. IEEE Transactions on Industrial Electronics 57(9), art. no. 5371908, 3148–3156 (2010)
10. Andrews, P.B.: An Introduction to Mathematical Logic and Type Theory: To Truth Through Proof. Kluwer, Dordrecht (2002)
11. Novák, V.: Reasoning About Mathematical Fuzzy Logic and its Future. Fuzzy Sets and Systems 192, 25–44 (2012)
12. Verbruggen, H.B., Babuška, R.: Fuzzy Logic Control: Advances in Applications. World Scientific Publishing Co Pte Ltd. (1999)
13. Wang, L.-X., Mendel, J.M.: Fuzzy Basis Functions, Universal Approximation, and Orthogonal Least-Squares Learning. IEEE Transactions on Neural Networks 3(5), 807–814 (1992)
14. Hsu, C.-H., Lin, P.-Z., Lee, T.-T., Wang, C.-H.: Adaptive Asymmetric Fuzzy Neural Network Controller Design via Network Structuring adaptation. Fuzzy Sets and Systems 159(20), 2627–2649 (2008)
15. Takagi, H., Suzuki, N., Koda, T., Kojima, Y.: Neural Networks Designed on Approximate Reasoning Architecture and Their Applications. IEEE Transactions on Neural Networks 3(5), 752–760 (1992)
16. Nauck, D., Klawonn, F., Kruse, R.: Foundations of Neuro-Fuzzy Systems, 1st edn. John Wiley (1997)
17. Lin, C.-T., Lee, C.S.G.: Neural-Network-Based Fuzzy Logic Control and Decision System. IEEE Transactions on Computers 40(12), 1320–1336 (1991)

A Long-Range Self-similarity Approach to Segmenting DJ Mixed Music Streams

Tim Scarfe, Wouter M. Koolen, and Yuri Kalnishkan

Computer Learning Research Centre and Department of Computer Science,
Royal Holloway, University of London, Egham, Surrey, TW20 0EX, United Kingdom
{tim,wouter,yura}@cs.rhul.ac.uk

Abstract. In this paper we describe an unsupervised, deterministic algorithm for segmenting DJ-mixed Electronic Dance Music streams (for example; podcasts, radio shows, live events) into their respective tracks. We attempt to reconstruct boundaries as close as possible to what a human domain expert would engender. The goal of DJ-mixing is to render track boundaries effectively invisible from the standpoint of human perception which makes the problem difficult.

We use Dynamic Programming (DP) to optimally segment a cost matrix derived from a similarity matrix. The similarity matrix is based on the cosines of a time series of kernel-transformed Fourier based features designed with this domain in mind. Our method is applied to EDM streams. Its formulation incorporates long-term self similarity as a first class concept combined with DP and it is qualitatively assessed on a large corpus of long streams that have been hand labelled by a domain expert.

Keywords: music, segmentation, DJ mix, dynamic programming.

1 Introduction

Electronic Dance Music (EDM) tracks are usually mixed by a DJ, which sets EDM streams apart from other genres of music. Mixing is the *modus operandi* in electronic music. We first transform the audio file into a time series of features (grouped into tiles) and transform those tiles into a domain where any pair from the same track would be distinguishable by their cosine. Our features are based on a Fourier transform with kernel filtering to accentuate instruments and intended self-similarity. We create a similarity matrix from these cosines and derive from it a cost matrix showing costs of a fitting a track at a given time with a given width. We use Dynamic Programming (DP) to create the cost matrix and again to perform the most economical segmentation of the cost matrix to fit a predetermined number of tracks.

A distinguishing feature of our algorithm is that it focuses on long term self similarity of segments rather than short term transients. Dance music tracks have the property that they are made up of repeating regions, and the ends are almost always similar to the beginning. For this reason we believe that some techniques from structural analysis fail to perform as well for this segmentation task

H. Papadopoulos et al. (Eds.): AIAI 2013, IFIP AICT 412, pp. 235–244, 2013.

because we focus on the concept of self-similarity ranging over a customisable time horizon. Our method does not require any training or tenuous heuristics to perform well.

The purpose of this algorithm is to reconstruct boundaries given a fixed number of tracks known in advance (their names and order are known). This is relevant when one has recorded a show, downloaded a track list and needs to reconstruct the indices given a track list. The order of the indices reconstructed is critical so that we can align the correct track names with the reconstructed indices. If the track list was not known in advance the number of tracks could be estimated in most cases.

To mix tracks DJs always match the speed or *BPM* (beats per minute) of each adjacent track during a transition and align the major percussive elements in the time domain. This is the central concept of removing any dissonance from overlapping tracks. Tracks can overlap by any amount. DJs increase adjacent track compatibility further by selecting adjacent pairs that are harmonically compatible and by applying spectral transformations (EQ).

The main theme of the early literature was attempting to generate a novelty function to find points of change using distance-based metrics or statistical methods. Heuristic methods with hard decision boundaries were used to find the best peaks. A distinguishing feature of our approach is that we evaluate how well we are doing compared to humans for the same task. We compare our reconstructed indices to the ones created by a human domain expert.

J. Foote et al ([1] [2] [3] [4] [5]) have done a significant amount of work in this area and the first to use similarity matrices. Foote evaluated a Gaussian tapered checkerboard kernel along the diagonal of a similarity matrix to create a 1d novelty function. One benefit to our approach is that our DP allows any range of long-term self similarity (which relates to the fixed kernel size in Foote's work).

Goodwin et al. also used DP for segmentation ([6] and [7]). Their intriguing supervised approach was to perform Linear Discriminant Analysis (LDA) on the features to transform them into a domain where segmentation boundaries would be emphasised and the feature weights normalised. They then reformulated the problem into a clustering DP to find an arbitrary number of clusters. We believe the frame of mind for this work was structural analysis, because it focuses on short term transients (mitigated slightly by the LDA) and would find segments between two regions of long term self similarity. Goodwin was the first to discuss the shortcomings of novelty peak finding approaches. Goodwin's approach is not optimized to work for a predetermined number of segments and depends on the parametrization and training of the LDA transform.

Peeters et al ([8] [9]) did some interesting work combining k-means and a transformation of the segmentation problem into Viterbi (a dynamic program).

We compare our error to the relative error of cue sheets created by human domain experts. We focus directly on DJ mixed electronic dance music.

In the coming sections we will describe the Data Set (Section 2), the Evaluation Criteria (Section 3), the Test Set (Section 4), Data Preprocessing (Section 5), Feature Extraction (Section 6), Cost Matrix (Section 7), Computing the Best Segmentation (Section 8), Experiment Methodology and Results (Section 9), and finally Conclusions (Section 10).

2 Data Set

We have been supplied with several broadcasts from three popular radio shows. These are: Magic Island, by Roger Shah (108 shows); A State of Trance with Armin Van Buuren (110 shows); and Trance Around The World with Above and Beyond (99 shows) (Total 317 shows). The show genres are a mix of Progressive Trance, Uplifting Trance and Tech-Trance. We believe this corpus is the largest of its kind used in the literature (see [10]). The music remains uninterrupted after the introduction (no silent gaps). The shows come in 44100 samples per second, 16 bit stereo MP3 files sampled at 192Kbs. We resampled these to 4000Hz 16 bit mono (left+right channel) WAV files to allow us to process them faster. We have used the SoX [11] (Sound eXchange) program to do this. These shows are all 2 hours long. The overall average track length is 5 and a half minutes and normally distributed. The average number of tracks is 23 for ASOT and TATW, 19 for Magic Island. There is a guest mix on the second half of each show. The guest mix DJs show off their skills with some of the most technically convoluted mixing imaginable.

3 Evaluation Criteria

We perform two types of evaluation: average track accuracy (in seconds) given as $\frac{1}{|P|}\sum_{i=1}^{|P|}|P_i - A_i|$ (P is constructed indices, and A is the human indices) and a measure of precision. The precision metric is the percentage of matched tracks within different intervals of time (thresholds) $\{60, 30, 20, 10, 5, 3, 1\}$, in *seconds* as a margin around any of the track indices we have been given. The precisions metric is invariant to alignment of the constructed indexes.

4 Test Set

There is already a large community of people interested in getting track metadata for DJ sets. CueNation ([12]) is an example of this. CueNation is a website allowing people to submit *cue sheets* for popular DJ Mixes and radio shows. A cue sheet is a text file containing time metadata (indices) for a media file.

We had our indices and radio shows provided to us and hand captured by *Dennis Goncharov*; a domain expert and one of the principal contributors to CueNation. As a result of this configuration; we can assume the alignment between the cue sheet and the radio show recording is exact.

Dennis Goncharov provided us with this description of how he captures the indices. To quote from a personal email exchange with Dennis:

The transition length is usually in factors of 8 bars (1 bar is 4 beats. At 135 beats per minute, 8 bars is 14.2 sec). It is a matter of personal preference which point of the transition to call the index. My preference is to consider the index to be the point at which the second track becomes the focus of attention and the first track is sent to the background. Most of the time the index is the point at which the bass line (400Hz and lower) of the previous track is cut and the bass line of the second track is introduced. If the DJ decides to exchange the adjacent tracks gradually over the time instead of mixing them abruptly then it is up to the cuesheet maker to listen further into the second track noting the musical qualities of both tracks and then go back and choose at which point the second track actually becomes the focus of attention.

5   Data Preprocessing

We went through the dataset carefully and removed some of the indices given and the corresponding audio when they did not correspond to actual musical tracks. This was for the show introductions (at the beginning) or for the introductions given to the guest mixes. The algorithm still performs similarly in the case of removing just these indices and leaving the audio intact underneath. When we removed audio from the shows because of extraneous introductions the following indices were nudged accordingly so that they still pointed to the equivalent locations in the audio stream. For those wishing to use this algorithm in practice with pre-recorded shows; the introductions at the start of the shows can be thought of as being fixed length (with a different length for each show type).

6   Feature Extraction

We used SoX [11] to downsample the shows to 4000Hz. We are not particularly interested in frequencies above around 2000Hz because instrument harmonics become less visible in the spectrum as the frequency increases. The Nyquist theorem ([13]) states that the highest representable frequency is half the sampling rate, so this explains our reason to use 4000Hz. We will refer to the sample rate as R. Let L be the length of the show in samples.

Fourier analysis allows one to represent a time domain process as a set of integer oscillations of trigonometric functions. We used the discrete Fourier transform (DFT) to transform the tiles into the frequency domain. The DFT given as $F(x_k) = X_k = \sum_{n=0}^{N-1} x_n \cdot e^{-i2\pi \frac{k}{N} n}$ transforms a sequence of complex numbers $x_0, \ldots, x_N$ into another sequence of complex numbers $X_0, \ldots, X_N$ where $e^{-i2\pi \frac{k}{N} n}$ are points on the complex unit circle. Note that the fftw algorithm that we used to perform this computation (see [14]) operates significantly faster when N is a power of 2 so we zero pad the input to make that the case. Because we are passing real values into the DFT fuction, the second half of the result is a rotational copy of the first half. As we are not always interested in the entire range of the spectrum, we use l to represent a low pass filter (in Hz) and h the high pass filter (in Hz). So we will capture the range from h to l on the first half of the result of F. We always discard the imaginary components of F.

Show samples are collated into a time series $Q_i, i \in \{1, 2, \ldots, \lfloor \frac{L}{M_s} \rfloor\}$ of contiguous, non-overlapping, adjacent *tiles* of equal size. Samples at the end of the show that do not fill a complete tile get discarded. We denote the tile width by M in seconds (an algorithm parameter) and M_s in samples ($M_s = M \times R$). For each tile $t_i \in Q$ we take the DFT $F(t_i)$ and place a segment of it into feature matrix D_i ($|Q|$ feature vectors in D). For each DFT transform we select vector elements $\lceil h \times \frac{M_s}{R} \rceil + 1$ to $\lceil l \times \frac{M_s}{R} \rceil + 1$ to allow effective spectral filtering.

To focus on the instruments and improve performance we perform convolution filtering on the feature vectors in D, using a Gaussian first derivative filter. This works like an edge detection filter but also expands the width of the transients (instrument harmonics) to ensure that feature vectors from the same song appear similar because their harmonics are aligned on any distance measure (we use the cosines). This is an issue because of the extremely high frequency resolution we have from having such large DFT inputs. Typically a STFT approach is used which has smaller DFTs (for example [15]).

The Gaussian first derivative filter is defined as $-\frac{2G}{B^2} e^{-\frac{G^2}{B^2}}$ where $G = \{-\lfloor 2B \rfloor, \lfloor -2B + 1 \rfloor, \ldots, \lfloor 2B \rfloor\}$, $B = b\left(\frac{N}{R}\right)$. b is the bandwidth of the filter in Hz and this is a parameter of the algorithm. After the convolution filter is applied to each feature vector in D, we take the absolute values and normalize each one

$$D_i = |D_i|, \forall i \in D, \qquad D_i = \frac{D_i}{\|D_i\|}, \forall i \in D.$$

Because the application domain is well defined in this setting, we can design features that look specifically for what we are interested in (musical instruments). Typically in the literature; algorithms use an amalgam of general purpose feature extractors. For example; spectral centroid, spectral moments, pitch, harmonicity ([16]). We construct a disimilarity matrix of cosines S from $D \times D^\top$ (dot products).

7   Cost Matrix

We now have a dissimilarity matrix $S(i, j)$ as described in Section 6.

Let w and W denote the minimum and maximum track length in seconds, these will be parameters.

Intuitively, features within the same track are reasonably similar on the whole, while pairs of tiles that do not belong to the same track are significantly more dissimilar. We define $C(f, t)$, the cost of a candidate track from tile f through tile t, to be the sum of the dissimilarities between all pairs of tiles inside it, normalized on track length:

$$C(f, t) = \frac{\sum_{i=f}^{t} \sum_{j=f}^{t} S(i, j)}{\sqrt{t - f + 1}}$$

As a first step, we pre-compute C for each $1 \leq f \leq t \leq T$. Direct calculation using the definition takes $O(TW^3)$ time. However, we can compute the full cost matrix in $O(WT)$ time using the following recursion for the unnormalized quantity $\tilde{C}(f,t) = C(f,t)(t-f)$ (for $f+1 \leq t-1$)

$$\tilde{C}(f,t) = \tilde{C}(f+1,t) + \tilde{C}(f,t-1) - \tilde{C}(f+1,t-1) + S(f,t) + S(t,f).$$

Note that the normalization step can be done independently of the DP procedure. We discovered experimentally that normalizing using the square root of the track length was advantageous. Doing so slightly discourages tracks of a larger length.

8 Computing Best Segmentation

We obtain the cost of a full segmentation by summing the costs of its tracks. The goal is now to efficiently compute the segmentation of least cost.

A sequence $t = (t_1, \ldots, t_{m+1})$ is called an m/T-segmentation if

$$1 = t_1 < \ldots < t_m < t_{m+1} = T+1.$$

m is the number of tracks we are trying to find and is a parameter of the algorithm. We use the interpretation that track $i \in \{1, \ldots, m\}$ comprises times $\{t_i, \ldots, t_{i+1} - 1\}$. Let $\mathbb{S}_m^T$ be the set of all m/T-segmentations. Note that there is a very large number of possible segmentations

$$|\mathbb{S}_m^T| = \binom{T-1}{m-1} = \frac{(T-1)!}{(m-1)!(T-m)!} =$$

$$\frac{(T-1)(T-2)\cdots(T-m+1)}{(m-1)!} \geq \left(\frac{T}{m}\right)^{m-1}.$$

For large values of T, considering all possible segmentations using brute force is infeasible. For example, a two hour long show with 25 tracks would have more than $\left(\frac{60^2 \times 2}{25}\right)^{24} \approx 1.06 \times 10^{59}$ possible segmentations!

We can reduce this number slightly by imposing upper and lower bounds on the song length. Recall that W is the upper bound (in seconds) of the song length, w the lower bound (in seconds) and m the number of tracks. With the track length restriction in place, the number of possible segmentations is still massive. A number now on the order of 10^{56} for a two hour show with 25 tracks, $w = 190$ and $W = 60 \times 15$.

Our solution to this problem is to find a dynamic programming recursion. The loss of an m/T-segmentation t is

$$\ell(t) = \sum_{i=1}^{m} C(t_i, t_{i+1} - 1)$$

We want to compute

$$\mathcal{V}_m^T = \min_{t \in \mathbb{S}_m^T} \ell(t)$$

To this end, we write the recurrence

$$\mathcal{V}_1^t = C(1, t)$$

and for $i \geq 2$

$$\mathcal{V}_i^t = \min_{t \in \mathbb{S}_i^t} \ell(t) = \min_{t_i} \min_{t \in \mathbb{S}_{i-1}^{t_i-1}} \ell(t) + C(t_i, t) =$$

$$\min_{t_i} C(t_i, t) + \min_{t \in \mathbb{S}_{i-1}^{t_i-1}} \ell(t) = \min_{t_i} C(t_i, t) + \mathcal{V}_{i-1}^{t_i-1}$$

In this formula t_i ranges from $t - W$ to $t - w$. We have $T \times m$ values of $\mathcal{V}_m^T$ and calculating each takes at most $O(W)$ steps. The total time complexity is $O(TWm)$.

9   Methodology and Results

We created a validation set of the first 10 episodes from each radio show (30 total) and found the best parameters with a continuous random search optimizing the absolute average accuracy evaluation criterion. We explored 2000 permutations in the search. We searched across the following parameter space:

$$M \in \{1, 2, \ldots, 25\} \qquad w \in \{120, 180, 240\} \qquad b \in \{1, 2, \ldots, 20\}$$

This search did not take long as running time is linear in the parameters. The parameters we found are shown in Table 1. Running the algorithm once on a 2 hour long show takes a couple of seconds on a fast PC and almost all of that time is loading the WAVE file for the show into memory. We had already batch converted the shows into the WAVE files from MP3 and this process took significantly longer, perhaps 30 seconds per show. The parameters we have presented here could be used immediately by an end user so no heavy computation is required. The high and low pass filters h and l were fixed at 0Hz and 2000Hz respectively (effectively were not used but would be useful parameters for specialized implementations). W was fixed at 630 Seconds which we selected by taking the largest track present in the validation set with a 30 second margin added on top.

There are no directly comparable methods in the literature ready to be used for this task. We will construct a simple algorithm to test our algorithm against; the *naive algorithm*. This algorithm constructs indices that are evenly spaced apart across the show.

See Table 2 for the main results. We also provide results for the dataset pruned of any shows with tracks smaller than 180 seconds on Table 3. Ostensibly we would fail to find these tracks as we use 180 as the minimum track length parameter w for the DP algorithm. Having a high value for w allows us to perform robustly most of the time but suffer on the minority of shows that have smaller tracks included. For these pruned results we did not remove the shows used in the validation set.

Table 1. These are the parameter values that were obtained from the parameter search described in Section 9

w	Minimum Track Length (DP)	180	Seconds
W	Maximum Track Length (DP)	617	Seconds
M	Tile Size	9	Seconds
b	Bandwidth Filter	5	Hz
l	Low Pass Filter	2000	Hz
h	High Pass Filter	0	Hz

Table 2. Main results. The accuracy rows show the mean of the absolute differences between the reconstructed tracks and the human indices (our test set). The thresholds indicate the percentage of reconstructed indices that fall within given time horizons centred around the actual indices. This is described in Section 3.

	Dynamic By Show			Overall	
	ASOT	TATW	MAGIC	Dynamic	Naive
Number Shows	101	89	98	288	288
60 Seconds (%)	92.3	96.9	97.9	95.7	42.1
30 Seconds (%)	74.9	90.4	89.8	85.1	22.0
20 Seconds (%)	63.7	82.0	74.4	73.4	14.9
10 Seconds (%)	55.7	70.9	53.2	59.9	7.6
5 Seconds (%)	43.9	51.5	32.9	42.8	4.0
3 Seconds (%)	29.7	35.3	21.2	28.7	2.6
1 Second (%)	11.6	13.9	8.3	11.3	0.9
Accuracy (Seconds)	49.4	40.1	26.5	38.6	112.2

Table 3. Results for the pruned set of shows (that do not contain tracks smaller than 180 seconds). The percentage figure given on the number of shows indicates how many were discarded from the prune. Performance on TATW and Magic Island are robustly improved. Magic Island achieved the improvement with a comparatively small prune of 7.4%.

	Dynamic By Show			Overall	
	ASOT	TATW	MAGIC	Dynamic	Naive
Number Shows	64 (42.3%)	61 (38.4%)	100 (7.4%)	225	288
60 Seconds (%)	93.4	98.2	98.9	96.8	42.1
30 Seconds (%)	75.7	92.3	90.8	86.2	22.0
20 Seconds (%)	63.5	84.0	74.9	74.1	14.9
10 Seconds (%)	56.3	72.8	53.2	60.8	7.6
5 Seconds (%)	44.0	52.8	32.6	43.2	4.0
3 Seconds (%)	30.0	36.5	20.9	29.1	2.6
1 Second (%)	12.1	15.2	8.6	12.0	0.9
Accuracy (Seconds)	32.3	14.9	13.3	20.2	112.2

10   Conclusion and Further Work

We believe our algorithm would be useful for segmenting DJ-mixed audio streams in batch mode. Our overall average is encouraging, taking into account the difficulty of the task at hand. The dissimilarity matrix we use is based solely on instrument features. The most pervasive elements in EDM are the percussion (the beats). We believe on balance that ignoring the percussive information was advantageous, because DJs use percussion primarily to blur boundaries between tracks. We tried to capture percussive based features and found that the transitions between tracks and indeed groups of tracks appeared as stronger self-similar regions in S than the actual tracks.

We would like to improve our cost function with one that has some domain knowledge, perhaps using a machine learning algorithm. Currently our cost function has a weakness that the relative similarity of regions within a song matters slightly, it should be independent. Let us consider the song structure {A,B,A}. The problem is that our cost (summing/normalizing the S square) would somewhat take into consideration the similarity of A and B. Anyone interested in optimizing the algorithm for a one specific radio show could consider modifying the cost function to introduce a parameter $\alpha \in [0,1]$ for fine tuned control over the normalization bias placed on the length of songs; $C(f,t) = \frac{\sum_{i=f}^{t}\sum_{j=f}^{t} S(i,j)}{(t-f+1)^{\alpha}}$.

We would also like to implement some of the methods in the literature (which were mostly designed for scene analysis) to see if we outperform them. It would be tricky to get an exact comparison because we could not find a unsupervised deterministic algorithm which finds a fixed number of strictly contiguous clusters. We could however adapt existing algorithms to get a like for like comparison. We would like to evaluate the performance of J Theiler's contiguous K-means algorithm in particular [17] and also similar algorithms. We have the property of being deterministic but probabilistic methods should be explored. Theiler's algorithm would require some modification to work in this scenario because we require strictly contiguous clusters, not just a contiguity bias.

References

1. Foote, J.: Visualizing music and audio using self-similarity. In: Proceedings of the Seventh ACM International Conference on Multimedia (Part 1), pp. 77–80. ACM (1999)
2. Foote, J.: A similarity measure for automatic audio classification. In: Proc. AAAI 1997 Spring Symposium on Intelligent Integration and Use of Text, Image, Video, and Audio Corpora (1997)
3. Foote, J.: Automatic audio segmentation using a measure of audio novelty. In: 2000 IEEE International Conference on Multimedia and Expo, ICME 2000, vol. 1, pp. 452–455. IEEE (2000)
4. Foote, J.T., Cooper, M.L.: Media segmentation using self-similarity decomposition. In: Electronic Imaging 2003, pp. 167–175. International Society for Optics and Photonics (2003)

5. Foote, J., Cooper, M.: Visualizing musical structure and rhythm via self-similarity. In: Proceedings of the 2001 International Computer Music Conference, pp. 419–422 (2001)

6. Goodwin, M.M., Laroche, J.: Audio segmentation by feature-space clustering using linear discriminant analysis and dynamic programming. In: 2003 IEEE Workshop on Applications of Signal Processing to Audio and Acoustics, pp. 131–134. IEEE (2003)

7. Goodwin, M.M., Laroche, J.: A dynamic programming approach to audio segmentation and speech/music discrimination. In: Proceedings of IEEE International Conference on Acoustics, Speech, and Signal Processing (ICASSP 2004), vol. 4, pp. iv–309. IEEE (2004)

8. Peeters, G., La Burthe, A., Rodet, X.: Toward automatic music audio summary generation from signal analysis. In: Proc. of ISMIR, pp. 94–100 (2002)

9. Peeters, G.: Deriving musical structures from signal analysis for music audio summary generation: "Sequence" and "State" approach. In: Wiil, U.K. (ed.) CMMR 2003. LNCS, vol. 2771, pp. 143–166. Springer, Heidelberg (2004)

10. Peiszer, E., Lidy, T., Rauber, A.: Automatic audio segmentation: Segment boundary and structure detection in popular music. In: Proc. of LSAS (2008)

11. Sox, the swiss army knife of sound processing programs, http://sox.sourceforge.net/

12. Lindgren, M.: Cuenation, website for edm community to share track time metadata, http://cuenation.com/

13. Nyquist, H.: Certain topics in telegraph transmission theory. Transactions of the American Institute of Electrical Engineers 47(2), 617–644 (1928)

14. Frigo, M., Johnson, S.G.: The fftw web page (2004)

15. Tzanetakis, G., Cook, P.: Multifeature audio segmentation for browsing and annotation. In: 1999 IEEE Workshop on Applications of Signal Processing to Audio and Acoustics, pp. 103–106. IEEE (1999)

16. Tzanetakis, G., Cook, F.: A framework for audio analysis based on classification and temporal segmentation. In: Proceedings of 25th EUROMICRO Conference, vol. 2, pp. 61–67 (1999)

17. Theiler, J.P., Gisler, G.: Contiguity-enhanced k-means clustering algorithm for unsupervised multispectral image segmentation. In: Optical Science, Engineering and Instrumentation 1997, pp. 108–118. International Society for Optics and Photonics (1997)

Real Time Indoor Robot Localization
Using a Stationary Fisheye Camera

Konstantinos K. Delibasis[1], Vasilios P. Plagianakos[1], and Ilias Maglogiannis[2]

[1] Univ. of Central Greece, Dept. of Computer Science and Biomedical Informatics,
Lamia, Greece
[2] University of Piraeus, Dept. of Digital Systems, Piraeus, Greece
kdelibasis@yahoo.com, vpp@ucg.gr, imaglo@unipi.gr

Abstract. A core problem in robotics is the localization of a mobile robot (determination of the location or pose) in its environment, since the robot's behavior depends on its position. In this work, we propose the use of a stationary fisheye camera for real time robot localization in indoor environments. We employ an image formation model for the fisheye camera, which is used for accelerating the segmentation of the robot's top surface, as well as for calculating the robot's true position in the real world frame of reference. The proposed robot localization algorithm does not depend on any information from the robot's sensors and does not require visual landmarks in the indoor environment. Initial results are presented from video sequences and are compared to the ground truth position, obtained by the robot's sensors. The dependence of the average positional error with the distance from the camera is also measured.

Keywords: computer vision, indoor robot localization, stationary fisheye camera.

1 Introduction

Robot localization is fundamental for performing any task, such as route planning [1]. Most of the existing mobile robot localization and mapping algorithms are based on laser or sonar sensors, as vision is more processor intensive and good visual features are more difficult to extract and match [2]. Except for input from the sensors, these approaches also require the existence of a map, as well as the relevant software module for navigation. A number of different approaches that use data from images acquired by a camera onboard the robot have been proposed. In [3] visual memory consisting of a set of views of a number of different landmarks is used for robot simultaneous localization and map construction (SLAM). In [4] robot localization is achieved by tracking geometric beacons (visually salient points, visible from a number of locations). Similarly, a vision-based mobile robot localization and mapping algorithm is proposed in [2], which uses scale-invariant image features as natural landmarks in unmodified environments. Specifically for indoor environments, relative localization (dead-reckoning) utilizes information from odometer readings [5]. This class of algorithms, although very fast and simple, present serious drawbacks, since

H. Papadopoulos et al. (Eds.): AIAI 2013, IFIP AICT 412, pp. 245–254, 2013.
© IFIP International Federation for Information Processing 2013

factors like slippage cause incremental error. Absolute localization is often based on laser sensors, or image processing from an onboard camera. For instance, in [6] ceiling lights and door number plates are used.

Very few approaches have been reported that employ the concept of a non fully autonomous robot. The system described in [7] uses a camera system mounted on the ceiling to track persons in the environment and to learn where the people usually walk in their workspace. The robot is tracked by the camera by means of a number of LEDs installed on the robot. In [8], the robot is tracked by a network of cameras, which use the information from a circular shape that has been installed on the robot. Fisheye cameras have been used onboard the robots for visual navigation [9].

In this work we report the use of a single stationary fisheye camera for real time robot localization, with extensions to robot navigation. We propose and utilize a computational model of the fisheye camera, which, in conjunction with the known robot shape can provide reliable location of the robot. More specifically, the top of the robot is segmented in each frame by means of color segmentation. The segmentation is assisted by precalculating all the possible pixels in the video frame that the top of the robot may occupy. The central pixel of the segmented top of the robot is used to obtain the robot's real location on the floor, using the model of the fisheye camera. Our approach is applicable to a very simple robot without any kind of sensors or map of the environment. Furthermore, the proposed approach does not require any visual landmarks in the indoor environment.

2   Methodology

2.1   Block Diagram of the Proposed Algorithm

In this work we present an algorithm for real time robot location in indoor environment, using one stationary fisheye camera. The main component of the proposed robot localization algorithm is the fisheye camera model that relates frame pixels with real world geometry. Thus, the real location of an object with known geometry (such as the robot) can be calculated. This model is also used to accelerate the segmentation of the robot from video frames. The proposed algorithm is very fast, thus the robot is localized in real time, without any additional information from sensors. The main components of the proposed algorithm are shown in Fig. 1, where the online steps are differentiated by the steps performed only once, during the calibration phase.

2.2   Forward and Inverse Fisheye Camera Model

The main characteristic of the fisheye camera is the ability to cover a field of view of 180 degrees. The objective of this subsection is to establish an analytical tool that relates a point in real world coordinates with the image pixel recorded by the fisheye camera and inversely map every frame pixel to the direction of view in the real world.

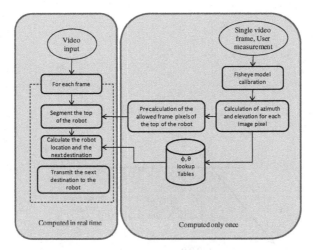

Fig. 1. The overall architecture of the proposed algorithm

The direction of view is defined in spherical coordinates by its azimuth and elevation angles θ, φ. Thus forward fisheye modeling M can be written in the general form by

$$(j,i) = M(x, y, z) \tag{1}$$

where as inverse fisheye modeling is described as:

$$(\theta, \varphi) = M^{-1}(j,i). \tag{2}$$

Forward Fisheye Model

The definition of a model for the fisheye camera is based on the physics of image formation, as described in [9], [10]. We consider a spherical element of arbitrary radius R_0 with its center at $K(0,0,z_{sph})$. For any point P with real world coordinates (x,y,z), we determine the intersection Q of the line KP with the optical element.

The point P (as well as ny point on the KP line of view) is imaged at the central projection (x_{im}, y_{im}) of Q on the image plane with equation $z=z_{plane}$, using the $O(0,0,0)$ as center of projection. The KP line is uniquely defined by its azimuth and elevation angles, θ, φ respectively.

We set z_{plane} to an arbitrary value, less than R_0 and define $z_{sph} = p z_{plane}$, where p is the primary parameter of the fisheye model that defines the formation of the image. To account for possible lens misalignments with respect to the camera sensor that could induce imaging deformations on the imaged frame [11], we introduce two extra model parameters: the X and Y position of the center of spherical lens $K(x_{sph}, y_{sph}, z_{sph})$ with respect to the optical axis of the camera. Now the camera model parameters consist of p, x_{sph}, y_{sph}. Figure 2 shows the geometry of the fisheye camera model for $x_{sph} = 0$ and $y_{sph} = 0$.

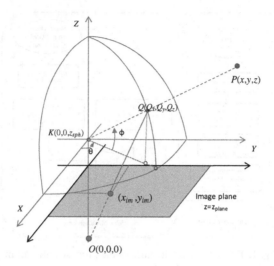

Fig. 2. The geometry of the proposed fisheye camera model

The position of Q is given by

$$\left(Q_x, Q_y, Q_z\right) = \left(\lambda\left(x - x_{sph}\right), \lambda\left(y - y_{sph}\right), \lambda\left(z - z_{sph}\right)\right) \qquad (3)$$

where the parameter λ is obtained by inserting (3) this into the equation of the spherical optical element and requiring $\lambda \in [0,1]$:

$$\left(\lambda\left(x - x_{sph}\right) - x_{sph}\right)^2 + \left(\lambda\left(y - y_{sph}\right) - y_{sph}\right)^2 + \left(\lambda\left(z - z_{sph}\right) - z_{sph}\right)^2 - R_0^2 = 0. \qquad (4)$$

Finally, we calculate the central projection $\left(x_{im}, y_{im}\right)$ of Q on the image plane:

$$\left(x_{im}, y_{im}\right) = \frac{z_{plane}}{z_{sph}}\left(Q_x, Q_y\right) \qquad (5)$$

Thus, any point P with real world coordinate $z > z_{sph}$, will be imaged on the image plane at position $\left(x_{im}, y_{im}\right)$, which is bounded by the radius of the virtual spherical optical element R_0: $-R_0 \le x_{im} - x_{sph} \le R_0$. When $x \to \infty$ then $\left(x_{im} - x_{sph}\right) \to R_0$ (the same holds for the y coordinate as well). In order to calculate the pixel of the video frame, we need to introduce the concept of the center of distortion CoD pixel located as the center of the circular field-of-view, (corresponding to elevation $\varphi = \pi/2$) [12] and the radius R_{FoV} of the field of view. The CoD and R_{FoV} are calculated only once, using user-input and a simple least squares optimization. Now, the image pixel position (i,j) that corresponds to the projection on the image plane (x_{im}, y_{im}) is calculated by a simple linear transform:

$$(j,i) = (x_{im}, y_{im}) \frac{R_{FoV}}{R_0} + (CoD_x, CoD_y) \tag{6}$$

Calibration of Fisheye Camera Model

In order to utilize the fisheye camera model, we need to calibrate the model, i.e. determine the values of the unknown p, x_{sph} and y_{sph} parameters. Initially, the user provides the position of $N_p=18$ landmark points $\{(X_{im}^i, Y_{im}^i)\}, i = 1, 2, ..., N_p$ on any video frame. The real world coordinates of these landmark points were also measured $\{(x_{real}^i, y_{real}^i, z_{real}^i)\}$, with respect to the reference system, (superscripts do not indicate powers). The position of the landmark points (x_{im}^i, y_{im}^i) on the video frame are calculated using (1). The values of the model parameters are obtained by minimizing the error between the expected and the observed frame coordinates of the landmark points. This minimization is performed using exhaustive search. If we allow p to vary from 0.5 to 1.5 with a step of 0.01 and x_{sph}, y_{sph} to vary from in the range of $[-R_0/4, R_0/4]$ with a step of $R_0/32$, the model parameters are obtained in just few minutes using the Matlab programming environment in an average laptop computer. The resulting calibration of the fisheye model is shown in Figure 3, where a virtual grid of points is laid on the floor and on the two walls of the imaged room.

It has to be emphasized that this operation is only performed once after the initial installation of the fisheye camera and it does not need to be repeated in real time.

Fig. 3. Visualization of the resulting fisheye model calibration. The landmark points defined by the user are shown as circles and their rendered position on the frame marked by stars.

Inverse Model of the Fisheye Camera - Azimuth and Elevation Look-Up Tables

To use the model of the fisheye camera to refine the video segmentation, we need to utilize the elevation θ and azimuth φ of the line of view for each segmented pixel. Given the (j,i) coordinates of a pixel of the video frame, the θ and φ angles are calculated as following. Using equation (6), the position of the pixel on the camera sensor is calculated:

$$\left(x_{im}, y_{im}\right) = \frac{R_0}{R_{FOV}}\left(\left(j,i\right) - \left(CoD_x, CoD_y\right)\right) \tag{7}$$

The intersection Q of the spherical optical element with line defined by $O(0,0,0)$ and (x_{im}, y_{im}) is determined, as

$$\left(Q_x, Q_y, Q_z\right) = m\left(x_{im}, y_{im}, z_{plane}\right) \tag{8}$$

where the parameter m is determined by requiring Q to lie on the spherical optical element:

$$m^2\left(x_{im}^2 + y_{im}^2 + z_{plane}^2\right) - 2\left(x_{im}x_{sph} + y_{im}y_{sph} + z_{plane}z_{sph}\right)m + x_{sph}^2 + y_{sph}^2 + z_{sph}^2 - R_0^2 = 0. \tag{9}$$

The required θ and φ are obtained by converting the Cartesian $\left(Q_x, Q_y, Q_z\right)$ to spherical coordinates:

$$\varphi = \cos^{-1}\left(\frac{Q_z - z_{sph}}{R_0}\right), \theta = \sin^{-1}\left(\frac{Q_y - y_{sph}}{\sqrt{\left(Q_x - x_{sph}\right)^2 + \left(Q_x - x_{sph}\right)^2}}\right). \tag{10}$$

The above process is executed only once, after the calibration of the fisheye camera model and the resulting values for the θ and φ parameters for each frame pixel are stored in two look-up tables, of size equal to a single video frame. The look-up tables for the azimuth θ and the elevation φ are shown in Fig. 4(a) and 4(b), respectively. As expected, the azimuth obtains values in $[-\pi,\pi]$, whereas the elevation obtains values in $[0,\pi/2]$, with the maximum value at the CoD pixel of the frame.

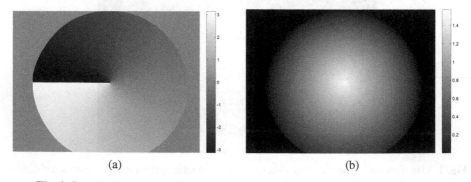

(a)   (b)

Fig. 4. Graphical representation of the azimuth (a) and elevation (b) look-up tables

2.3   Video Acquisition and Robot Localization

Robot localization is achieved utilizing the video stream acquired by the fisheye camera. The first step is segmentation of the robot's top surface. In order to assist segmentation, we have placed a specific color on the top of the robot. Thus, a pixel (i,j) is segmented as top of the robot, according to its color in the *RGB* color system, using the following rule:

$$R(i,j) > 1.8 \cdot G(i,j) > 1.2 \cdot B(i,j) \tag{11}$$

The segmentation achieved by Eq.(11) is accurate and efficient (see Fig. 7), therefore, the use of other color spaces, such as the *HSV* was not considered necessary. Eq.(11) is specific to the color that was used to mark the top of the robot and can be modified if a different color is used. In order to increase the accuracy of the segmentation by excluding possible false pixels, as well as to accelerate its execution, we precalculate and store in a binary mask all the frame pixels in which the top of the robot is possible to be imaged by the specific camera. Since the robot has a constant height h_R, its real position can be calculated unambiguously in the frame, provided that the pixel (i,j) imaging the center of its top surface has been segmented. The azimuth θ and elevation φ of the line of view of this pixel is given by the look-up tables. Then the real position (x_{real}, y_{real}) of the robot on the floor is given by

$$x_{real} = \frac{z_{max} - h_R}{\tan(\varphi_{ij})} \cos(\theta_{ij}),\ y_{real} = \frac{z_{max} - h_R}{\tan(\varphi_{ij})} \sin(\theta_{ij}) \tag{12}.$$

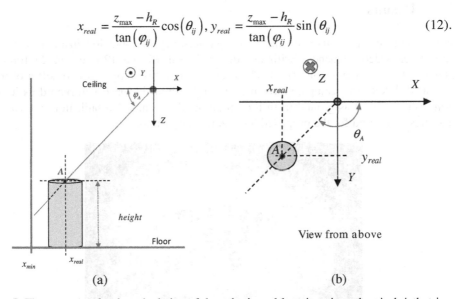

Fig. 5. The geometry for the calculation of the robot's real location given the pixel A that images the center of the robot's flat top surface

The relevant geometric concept is shown in Fig. 5. Given the geometry of the indoor space being imaged by the fisheye camera, we may exclude the non accessible areas. Thus, the binary mask of the allowed pixels is set to 1 only for the pixels for which the corresponding (x_{real}, y_{real}) does not lie on inaccessible areas (Fig. 6). Robot segmentation according to (11) is performed only on the non-zero pixels of the mask. The robot's real location is calculated using (12) and stored in each frame.

Fig. 6. A typical video frame. The pixels where it is possible for the segmented top of the robot to appear have been highlighted.

3   Results

We present results to indicate the proposed accuracy of robot localization. We acquired ten video sequences with duration between 90 and 120 sec, at 25 frames (480x640 pixels) per second (fps). The robot used in this study is a PeopleBot of the Adept MobileRobots company. We used the robot built-in sensors to record its location, to be utilized as ground truth for validating the results. No built-in robot sensor readings were used for the proposed localization algorithm.

Fig. 7. The path of the robot in one of the acquired videos, the segmented top of the robot surface and the determined central point, used for the calculation of real world position

Figure 7 shows the resulting segmentation of the top of the robot from a number of frames of one of the videos, with its center of gravity marked. The path of the robot in the video frame is visible. Figure 8(a) shows the path of the robot, estimated by the proposed algorithm (continuous curve). The ground truth has also been included for comparison (dotted curve). The starting point is marked by a star and the position of the fisheye camera is also shown, since the positional error is expected to vary proportionally to the distance from the camera. As it can be observed, the position estimated by the proposed algorithm is very accurate close to the camera but the error starts to increase as the robot moves away from the camera. This can be attributed to the deterioration of the spatial resolution of the fish-eye image, as well as to inaccuracies of the calibration of the camera model. Notice that the localization error is not accumulated, as in the case of relative measurements (dead reckoning). In Fig. 8(b) the dependence of the error is presented with respect to the distance from the camera.

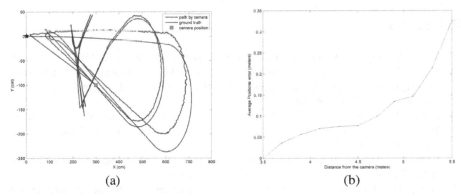

(a)   (b)

Fig. 8. (a) The robot position obtained by the proposed algorithm in a single video, as well as the ground truth positions. (b) The average error in robot localization achieved by the proposed algorithm, as a function of the robot's distance from the fisheye camera.

4   Discussion and Further Work

An algorithm for the localization of a robot in indoor environment has been presented, which is based only on video acquired by a stationary fisheye camera, installed on the ceiling of the imaged room. The proposed algorithm has been implemented using Matlab and executed on an Intel(R) Core i5-2430 CPU @ 2.40 GHz Laptop with 4 GB Ram, under Windows 7 Home Premium. The mean execution time of the proposed localization algorithm was approximately 70 msec per frame of dimension 480x640, with no parallelization or specially optimized source code. In our experiments, the overall procedure including the acquisition of the real time video stream through a WiFi connection resulted in processing 7 frames per second, which can support real time localization because of the low speed of the robot. This rate can be increased using a different development environment.

Regarding the localization accuracy, our experiments showed that the algorithm is able to localize the robot with positional error less than 0.1 meters in distances up to 4.5 meters from the stationary camera. The localization error increases proportionally to the distance of the robot from the camera, due to inaccuracies of the fisheye camera model. For this reason we will explore further the proposed camera model, possibly by including more controlling parameters and evolutionary algorithms for its calibration. Navigation experiments based on the proposed robot localization algorithm will also be performed, to assess its usefulness in assisting environments.

Acknowledgment. The authors would like to thank the European Union (European Social Fund ESF) and Greek national funds for financially supporting this work through the Operational Program "Education and Lifelong Learning" of the National Strategic Reference Framework (NSRF) - Research Funding Program: \Thalis \ Interdisciplinary Research in Affective Computing for Biological Activity Recognition in Assistive Environments.

References

1. Thrun, S., Burgard, W., Fox, D.: Probabilistic Robotics. MIT Press (2005)
2. Se, S., Lowe, D., Little, J.: Mobile Robot Localization and Mapping with Uncertainty using Scale-Invariant Visual Landmarks. The International Journal of Robotics Research 21(8), 735–758 (2002)
3. Royer, E., Lhuillier, M., Dhome, M., Lavest, J.-M.: Monocular Vision for Mobile Robot Localization and Autonomous Navigation, Intern. Journal of Computer Vision 74(3), 237–260 (2007)
4. Leonard, J., Durrant-Whyte, H.: Mobile robot localizations be tracking geometric beacons. IEEE T. on Robotics and Automation 7(3), 376–382 (1991)
5. Aider, O., Hoppenot, P., Colle, E.: A model-based method for indoor mobile robot localization using monocular vision and straight-line correspondences. Robotics and Autonomous Systems 52, 229–246 (2005)
6. Dulimarta, H.S., Jain, A.K.: Mobile robot localization in indoor environment. Pattern Recognition 30(1), 99–111 (1997)
7. Kruse, E., Wahl, F.: Camera-based monitoring system for mobile robot guidance. In: Proc. of the IEEE/RSJ Intern. Conf. on Intelligent Robots and Systems, IROS (1998)
8. Yoo, J.H., Sethi, I.K.: Sethi, Mobile robot localization with multiple stationary cameras. In: Mobile Robots VI. Proc. SPIE, vol. 1613, p. 155 (1992)
9. Courbon, J., Mezouar, Y., Eck, L., Martinet, P.: A Generic Fisheye camera model for robotic applications. In: Proceedings of the 2007 IEEE/RSJ International Conf. on Intelligent Robots and Systems, San Diego, CA, USA, October 29 - November 2 (2007)
10. Computer Generated Angular Fisheye Projections, http://paulbourke.net/dome/fisheye/
11. Shah, S., Aggarwal, J.: Intrinsic parameter calibration procedure for a high distortion fisheye lens camera with distortion model and accuracy estimation. Pattern Recognition 29(11), 1775–1788 (1996)
12. Micusik, B., Pajdla, T.: Structure from Motion with Wide Circular Field of View Cameras. IEEE T. on PAMI 28(7), 1–15 (2006)

Artificial Neural Network Approach for Land Cover Classification of Fused Hyperspectral and Lidar Data

Paris Giampouras[1,2], Eleni Charou[1], and Anastasios Kesidis[3]

[1] Computational Intelligence Laboratory,
Institute of Informatics and Telecommunications,
National Center for Scientific Research "Demokritos",
Athens, Greece
[2] Department of Informatics, University of Athens, Greece
{parisg,exarou}@iit.demokritos.gr
[3] Department of Surveying Engineering,
Technological Educational Institute of Athens, Greece
akesidis@teiath.gr

Abstract. Hyperspectral remote sensing images are consisted of several hundreds of contiguous spectral bands that can provide very rich information and has the potential to differentiate land cover classes with similar spectral characteristics. LIDAR data gives detailed height information and thus can be used complementary with Hyperspectral data. In this work, a hyperspectral image is combined with LIDAR data and used for land cover classification. A Principal Component Analysis (PCA) is applied on the Hyperspectral image to perform feature extraction and dimension reduction. The first 4 PCA components along with the LIDAR image were used as inputs to a supervised feedforward neural network. The neural network was trained in a small part of the dataset (less than 0.4%) and a validation set, using the Bayesian regularization backpropagation algorithm. The experimental results demonstrate efficiency of the method for hyperspectral and LIDAR land cover classification.

Keywords: Hyperspectral images, LIDAR, land cover classification, neural networks, principal component analysis.

1 Introduction

Operational monitoring of land cover/ land use changes using remote sensing is of great importance for environmental Remote sensing due to its repetitive nature and large coverage is a very useful technology to perform such kind of study. This technology, if properly integrated with automatic processing techniques, allows the analysis of large areas in a fast and accurate way. Several studies have been carried out in this field, analyzing the potentialities of different remote sensing sensors such us Passive Multispectral (3-10 spectral bands) or Hyperspectral(several 100s spectral bands). The abundance of spectral information in the hyperspectral image has the potential to differentiate land cover classes with

H. Papadopoulos et al. (Eds.): AIAI 2013, IFIP AICT 412, pp. 255–261, 2013.

similar spectral characteristics that cannot be distinguished with multispectral sensors. Active LIDAR remote sensing sensors are increasingly used in the context of classification and constitute a valuable source of information related to the altitude. Thus hyperspectral and LIDAR data could be used complementary, as they contain very different information: hyperspectral images provide a detailed description of the spectral signatures of classes but no information on the height of ground covers, whereas LIDAR data give detailed information about the height but no information on the spectral signatures[9]. In this study LIDAR and hyperspectral images are integrated for the land cover classification. To show the usefulness and complementarity of LIDAR and hyperspectral image, a land cover classification was performed using hyperspectral image alone. This paper is organized into 4 sections. Section 2 presents the methodology used, the main preprocessing techniques adopted and describes the data set used in our analysis. Section 3 describes and discusses the experimental results obtained. Finally, Section 4 draws the conclusion of this paper.

2   Methodology and Data

Hyperspectral data undoubtedly possess a rich amount of information. Nevertheless, redundancy in information among the bands opens an area for research to explore the optimal selection of bands for analysis. Theoretically, using images with more bands should increase automatic classification accuracy. However, this is not always the case. As the dimensionality of the feature space increases subject to the number of bands, the number of training samples needed for image classification has to increase too. If training samples are insufficient then parameter estimation becomes inaccurate. The classification accuracy first grows and then declines as the number of spectral bands increases, which is often referred to as the Hughes phenomenon [1]. In the proposed method, hyperspectral image data are fused with LIDAR data in order to perform land cover classification. Principal Component Analysis is applied on the Hyperspectral image to perform feature extraction and dimension reduction while preserving information. The first four principal components along with the LIDAR image are used for classification by training a supervised feed-forward neural network. The following sections describe in detail the several steps of the method.

2.1   Principal Component Analysis

Classification performance depends on four factors: class separability, training sample size, dimensionality, and classifier type [2]. Dimensionality reduction can be achieved in two different ways. The first approach is to select a small subset of features which could contribute to class separability or classification criteria. This dimensionality reduction process is referred to as feature selection or band selection. The other approach is to use all the data from original feature space and map[10] the effective features and useful information to a lower-dimensional subspace. This method is referred to as feature extraction. The Principal Component Analysis PCA is a well known feature extraction method in which original data is transformed into a new set of data which may better capture the essential information. When variables are highly correlated the information contained in one variable is largely a duplication of the information contained in another variable. Instead of throwing away the redundant data principal components analysis condenses the information in intercorrelated variables into a few variables, called principal components.PCA analysis have been succesfully used in Hyperspectral data classification[11].

2.2   Neural Networks

The use of Artificial Neural Networks(ANNs) for complex classification tasks of high dimensional data sets, such as hyperspectral images, has been widely spread in the last years. Several studies dealing with pattern recognition/classification problems of remote sensing data, [3,4] have shown that ANNs, in most cases, achieve better results in comparison to conventional classifiers.

Their non-linear properties and the fact that they make no assumption about the distribution of the data, are among the key factors of the NNs classification power. Neurons with nonlinear activation functions are arranged in layers and act like a set of piece-wise nonlinear simulators [5,7]. Neural networks are able to learn from existing examples adaptively, thus, the classification is made objective[8].

2.3   Data Description

The data used in this study is obtained from the data fusion contest 2013, organized by the Data Fusion Technical Committee of the IEEE Geoscience and Remote Sensing Society (GRSS). The current contest involves two sets of data a hyperspectral image (Fig. 1) and a LIDAR derived Digital Surface Model (DSM) (Fig. 2), co-registered and both at the same spatial resolution (2.5 m). The size of the bands of the hyperspectral image as well as the LIDAR is 349x1905 pixels. They were acquired over the University of Houston campus and the neighboring urban area. For the current data set, a total of 144 spectral bands were acquired in the 380 nm to 1050 nm region of the electromagnetic spectrum. The goal is to distinguish among the 15 pre-defined classes. The labels as well as a training set for the classes are also provided (Table 1).

3   Results and Discussion

For the purpose of this study the MATLAB software is used. The Principal Component Analysis is applied on the hyperspectral image. The first four PCA components i.e PCA1,PCA2,PCA3,PCA4 which convey the 99.81% of the information along with the LIDAR opted as our dataset for the processing that follows. A three layer feed-forward ANN is implemented. The input layer of the ANN consists of 5 neurons : 4 neurons for PCA1, PCA2, PCA3, PCA4 and one neuron for the LIDAR image. The hidden layer is composed of 120 neurons and

Fig. 1. RGB combination of hyperspectral image of the study area

Fig. 2. LIDAR image of the study area

Table 1. Distribution of training sample among investigated classes

Class Index	Class Name	Training Samples Number
1	Grass Healthy	198
2	Grass Stressed	190
3	Grass Synthetic	192
4	Tree	188
5	Soil	186
6	Water	182
7	Residential	196
8	Commercial	191
9	Road	193
10	Highway	191
11	Railway	191
12	Parking lot 1	192
13	Parking lot 2	184
14	Tennis Court	181
15	Running Track	187

the tan-sigmoid is selected as the layer's transfer function. The output layer comprised of 15 nodes equal to the total number of the classes. The linear function is opted as the output's layer transfer function. The ANN is trained by the set of 2832 samples of the Table 1. The training set is divided randomly into three sets, namely, the training, validation and test set, respectively. About 70% of the total samples is used as the training set of the ANN. Another 15% is used for validation while the rest 15% is dedicated to testing purposes. The ANN is trained using the Bayesian Regularization back-propagation algorithm [6]. In this case, the weights and bias values are updated according to Levenberg-Marquardt optimization. This minimizes a combination of squared errors and weights, and then determines the correct combination so as to produce a network that generalizes well. After 552 epochs the training procedure is stopped. The test set success is 94.6% , as it can be seen in detail in the confusion matrix of (Fig. 3). The performance achieved, is higher than 95% in the majority of the classes. Specifically, in 7 out of 15 classes (grass healthy, grass stressed, grass synthetic, tree, soil, water, running track), a 100% success is achieved. For 5 classes (commercial, highway, railway, parking lot 1, tennis court) the performance achieved lies between 95% and 99.5%. The classes "residential" and "road" present 85%-90% rate of success. The lowest value (61.5%) is noticed for class parking lot 2 mostly due to the high correlation and the low separability between this class and the parking lot 1 class.

Confusion Matrix

Output Class	1	2	3	4	5	6	7	8	9	10	11	12	13	14	15	
1	30 / 7.1%	0 / 0.0%	0 / 0.0%	0 / 0.0%	0 / 0.0%	0 / 0.0%	0 / 0.0%	0 / 0.0%	0 / 0.0%	0 / 0.0%	0 / 0.0%	0 / 0.0%	0 / 0.0%	0 / 0.0%	0 / 0.0%	100% / 0.0%
2	0 / 0.0%	27 / 6.4%	0 / 0.0%	0 / 0.0%	0 / 0.0%	0 / 0.0%	0 / 0.0%	0 / 0.0%	0 / 0.0%	0 / 0.0%	0 / 0.0%	0 / 0.0%	0 / 0.0%	0 / 0.0%	0 / 0.0%	100% / 0.0%
3	0 / 0.0%	0 / 0.0%	19 / 4.5%	0 / 0.0%	0 / 0.0%	0 / 0.0%	0 / 0.0%	0 / 0.0%	0 / 0.0%	0 / 0.0%	0 / 0.0%	0 / 0.0%	0 / 0.0%	0 / 0.0%	0 / 0.0%	100% / 0.0%
4	0 / 0.0%	0 / 0.0%	0 / 0.0%	33 / 7.8%	0 / 0.0%	0 / 0.0%	0 / 0.0%	0 / 0.0%	0 / 0.0%	0 / 0.0%	0 / 0.0%	0 / 0.0%	0 / 0.0%	0 / 0.0%	0 / 0.0%	100% / 0.0%
5	0 / 0.0%	0 / 0.0%	0 / 0.0%	0 / 0.0%	14 / 3.3%	0 / 0.0%	0 / 0.0%	0 / 0.0%	0 / 0.0%	0 / 0.0%	0 / 0.0%	0 / 0.0%	0 / 0.0%	0 / 0.0%	0 / 0.0%	100% / 0.0%
6	0 / 0.0%	0 / 0.0%	0 / 0.0%	0 / 0.0%	0 / 0.0%	27 / 6.4%	0 / 0.0%	0 / 0.0%	0 / 0.0%	0 / 0.0%	0 / 0.0%	0 / 0.0%	0 / 0.0%	0 / 0.0%	0 / 0.0%	100% / 0.0%
7	0 / 0.0%	0 / 0.0%	0 / 0.0%	0 / 0.0%	0 / 0.0%	0 / 0.0%	32 / 7.5%	0 / 0.0%	0 / 0.0%	0 / 0.0%	0 / 0.0%	0 / 0.0%	0 / 0.0%	1 / 0.2%	0 / 0.0%	97.0% / 3.0%
8	0 / 0.0%	0 / 0.0%	0 / 0.0%	0 / 0.0%	0 / 0.0%	0 / 0.0%	0 / 0.0%	20 / 4.7%	0 / 0.0%	0 / 0.0%	0 / 0.0%	4 / 0.9%	0 / 0.0%	0 / 0.0%	0 / 0.0%	83.3% / 16.7%
9	0 / 0.0%	0 / 0.0%	0 / 0.0%	0 / 0.0%	0 / 0.0%	0 / 0.0%	1 / 0.2%	0 / 0.0%	26 / 6.1%	1 / 0.2%	1 / 0.2%	0 / 0.0%	1 / 0.2%	0 / 0.0%	0 / 0.0%	86.7% / 13.3%
10	0 / 0.0%	0 / 0.0%	0 / 0.0%	0 / 0.0%	0 / 0.0%	0 / 0.0%	1 / 0.2%	0 / 0.0%	0 / 0.0%	33 / 7.8%	0 / 0.0%	0 / 0.0%	2 / 0.5%	0 / 0.0%	0 / 0.0%	91.7% / 8.3%
11	0 / 0.0%	0 / 0.0%	0 / 0.0%	0 / 0.0%	0 / 0.0%	0 / 0.0%	0 / 0.0%	0 / 0.0%	0 / 0.0%	0 / 0.0%	25 / 5.9%	0 / 0.0%	0 / 0.0%	0 / 0.0%	0 / 0.0%	100% / 0.0%
12	0 / 0.0%	0 / 0.0%	0 / 0.0%	0 / 0.0%	0 / 0.0%	0 / 0.0%	2 / 0.5%	0 / 0.0%	1 / 0.2%	0 / 0.0%	0 / 0.0%	34 / 8.0%	3 / 0.7%	0 / 0.0%	0 / 0.0%	85.0% / 15.0%
13	0 / 0.0%	0 / 0.0%	0 / 0.0%	0 / 0.0%	0 / 0.0%	0 / 0.0%	0 / 0.0%	1 / 0.2%	1 / 0.2%	0 / 0.0%	0 / 0.0%	1 / 0.2%	16 / 3.8%	0 / 0.0%	0 / 0.0%	84.2% / 15.8%
14	0 / 0.0%	0 / 0.0%	0 / 0.0%	0 / 0.0%	0 / 0.0%	0 / 0.0%	1 / 0.2%	0 / 0.0%	1 / 0.2%	0 / 0.0%	0 / 0.0%	0 / 0.0%	0 / 0.0%	31 / 7.3%	0 / 0.0%	93.9% / 6.1%
15	0 / 0.0%	0 / 0.0%	0 / 0.0%	0 / 0.0%	0 / 0.0%	0 / 0.0%	0 / 0.0%	0 / 0.0%	0 / 0.0%	0 / 0.0%	0 / 0.0%	0 / 0.0%	0 / 0.0%	0 / 0.0%	34 / 8.0%	100% / 0.0%
	100% / 0.0%	100% / 0.0%	100% / 0.0%	100% / 0.0%	100% / 0.0%	100% / 0.0%	86.5% / 13.5%	95.2% / 4.8%	89.7% / 10.3%	97.1% / 2.9%	96.2% / 3.8%	97.1% / 2.9%	61.5% / 38.5%	96.9% / 3.1%	100% / 0.0%	94.6% / 5.4%

Target Class

Fig. 3. Test Set Confusion Matrix

By applying the ANN on the whole dataset, we get the resulting image that can be seen in Fig. 4, where each class is depicted with a different color. A quantitative assessment on the whole image, is not possible since no ground truth data are provided. However, a qualitative assessment on the whole classified image reveals promising results. It can be noticed that most pixels which belong to classes where high success rates are achieved, are correctly classified (compared to the RGB image of Fig. 2). In contrast, poorer results are achieved for pixels in classes characterized by lower classification rates, such as "parking lot 2".

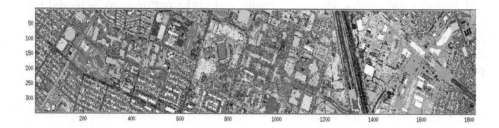

Fig. 4. Resulting Image

4   Conclusion

A fully automated process based on hyperspectral and LIDAR data for efficient land cover classification is presented. The fusion of hyperspectral image and LIDAR data is shown that can be used complementary, for effective land cover classification based on ANNs. Even in highly correlated classes such as grass healthy, grass stressed (unhealthy) and grass synthetic (ANN's) performance is very satisfactory. The methodology presented can also be used in other classification tasks.

Acknowledgements. This research has been co-funded by the European Union (European Social Fund) and Greek national resources under the framework of the Archimedes III: Funding of Research Groups in T.E.I. of Athens project of the Education & Lifelong Learning Operational Programme. The authors would also like to acknowledge the Data FusionContest 2013 Technical Committee of the IEEE Geoscience and Remote Sensing Society (GRSS) for providing the dataset.

References

1. Hughes, G.F.: On the mean accuracy of statistical pattern recognizers. IEEE Trans. Inform. Theory 14, 55–63 (2003)
2. Hsien, P.F., Landgrebe, D.: Classification of High Dimensional Data. Ph. D. Dissertation, School of Electrical and Computer Engineering, Purdue University, West Lafayette, Indiana (1998)

3. Mernyi, Erzsbet, et al: Classification of hyperspectral imagery with neural networks: Comparison to conventional tools. Photogrammetric Eng. Remote Sens. (2007)
4. Mernyi, E., et al.: Quantitative comparison of neural network and conventional classifiers for hyperspectral imagery. In: Summaries of the Sixth Annual JPL Airborne Earth Science Workshop, Pasadena, CA, vol. 1 (1996)
5. Hornik, K., Stinggchombe, M., Whitee, H.: Multilayer feedforward networks are universal approximators. Neural Networks 2, 359–366 (1989)
6. http://www.mathworks.com/help/nnet/ref/trainbr.html (last access: June 22 , 2013, GMT 9:58:12 a.m)
7. Haykin, S.: Neural Networks: A comprehensive foundation. Macmillan, New York (1994)
8. Beale, M., Hagan, M.T., Demuth, H.B.: Neural network toolbox. Neural Network Toolbox, The Math Works, 5–25 (1992)
9. Dalponte, B.L., Gianelle, D.: Fusion of Hyperspectral and LIDAR Remote Sensing Data for Classification of Complex Forest Areas. IEEE Transactions on Geoscience and Remote Sensing 46(5), 1416–1437 (2008)
10. Petridis, S., Charou, E., Perantonis, S.J.: Non redundant feature selection of multiband remotely sensed images for land cover classification. In: Tyrrhenian International Workshop on remote sensing Elba, Italy, pp. 657–666 (2003)
11. Rodarmel, C., Shan, J.: Principal Component Analysis for Hyperspectral Image Classification. Surveying and Land Information Systems 62(2), 115–122 (2002)

A Linear Multi-Layer Perceptron for Identifying Harmonic Contents of Biomedical Signals

Thien Minh Nguyen and Patrice Wira

Université de Haute Alsace, Laboratoire MIPS, Mulhouse, France
{thien-minh.nguyen,patrice.wira}@uha.fr
http://www.mips.uha.fr/

Abstract. A linear Multi Layer Perceptron (MLP) is proposed as a new approach to identify the harmonic content of biomedical signals and to characterize them. This layered neural network uses only linear neurons. Some synthetic sinusoidal terms are used as inputs and represent a priori knowledge. A measured signal serves as a reference, then a supervised learning allows to adapt the weights and to fit its Fourier series. The amplitudes of the fundamental and high-order harmonics can be directly deduced from the combination of the weights. The effectiveness of the approach is evaluated and compared. Results show clearly that the linear MLP is able to identify in real-time the amplitudes of harmonic terms from measured signals such as electrocardiogram records under noisy conditions.

Keywords: frequency analysis, harmonics, MLP, linear learning, ECG.

1 Introduction

Generally, decomposing a complex signal measured through time into simpler parts in the frequency domain (spectrum) facilitate analysis. So, different signal processing techniques have been widely used for estimating harmonic amplitudes, among them Fourier-based transforms like the Fast Fourier Transform (FFT), wavelet transforms or even time-frequency distributions. The analysis of the signal can be viewed from two different standpoints: Time domain or frequency domain. However, they are susceptible to the presence of noise in the distorted signals. Harmonic detection based on Fourier transformations also requires input data for one cycle of the current waveform and requires time for the analysis in next coming cycle.

Artificial Neural Networks (ANNs) offer an alternative way to tackle complex and ill-defined problems [1]. They can learn from examples, are fault tolerant and able to deal with nonlinearities and, once trained, can perform generalization and prediction [2]. However, the design of the neural approach must necessarily be relevant, i.e., must take into account a priori knowledge [3].

This paper presents a new neural approach for harmonics identification. It is based on a linear Multi Layer Perceptron (MLP) whose architecture is able to fit any weighted sums of time-varying signals. The linear MLP is perfectly able

H. Papadopoulos et al. (Eds.): AIAI 2013, IFIP AICT 412, pp. 262–271, 2013.

to estimate Fourier series by expressing any periodic signal as a sum of harmonic terms. The prior knowledge, i.e., the supposed harmonics present in the signal, allows to design the inputs. Learning consists in finding out optimal weights according to the difference between the output and the considered signal. The estimated amplitudes of the harmonic terms are obtained from the weights. This allows to individually estimate the amplitude of the fundamental and high-order harmonics in real-time. With its learning capabilities, the neural harmonic estimator is able to handle every type of periodic signal and is suitable under noise and time-varying conditions. Thus, it can be used to analyze biomedical signals, even non-stationary signals. This will be illustrated by identifying harmonics of electrocardiogram (ECG) recordings.

2   Context of This Study

An ECG is a recording of the electrical activity of the heart and is used in the investigation of heart diseases. For this, the conventional approach generally consists in detecting the P, Q, R, S and T deflections [4] which can be achieved by digital analyses of slopes, amplitudes, and widths [5]. Other well-known approaches use independent components analysis (for example for fetal electrocardiogram extraction) or time-frequency methods like the S-transform [6].

Our objective is to develop an approach that is general and therefore able to process various types biomedical and non-stationary signals. Its principle is illustrated by Fig. 1. Generic and relevant features are first extracted. They are the harmonic terms and statistical moments and will be used to categorize the signals in order to help the diagnosis of abnormal phenomenons and diseases.

The following study focuses on the harmonic terms extraction from ECG. A harmonic term is a sinusoidal component of a periodic wave or quantity having a frequency that is an integer multiple of the fundamental frequency. It is therefore a frequency component of the signal. We want to estimate the main frequency components of biomedical signals, and specially non-stationary signals. Neural approaches are therefore used. They have been applied successfully for estimating the harmonic currents of power system [7, 8].

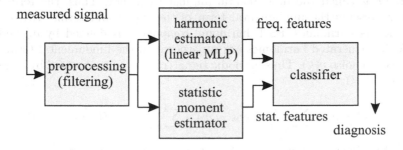

Fig. 1. General principle for characterizing ECG records

Estimating harmonics can be achieved with an Adaline [9] whose mathematical model directly assumes the signal to be a sum of harmonic components. As a result, the weights of the Adaline represent the coefficients of the terms in the Fourier series [8, 10, 11]. MPL approaches have also been proposed for estimating harmonics. In [12], a MLP is trained off-line with testing patterns generated with different random magnitude and phase angle properties that should represent possible power line distortions as inputs. The outputs are the corresponding magnitude and phase coefficient of the harmonics. This principle has also been applied with Radial Basis Functions (RBF) [13] and feed forward and recurrent networks [14].

In these studies, the neural approaches are not on-line self-adapting. The approach introduced thereafter is simple and compliant with real-time implementations.

3   A Linear MLP for Estimating Harmonic Terms

A linear MLP is proposed to fit Fourier series. This neural network takes synthetic sinusoidal signals as its inputs and uses the measured signal as a target output. The harmonics, as Fourier series parameters, are obtained from the weights and the biases at the end of the training process.

3.1   Fourier Analysis

According to Fourier, a periodic signal can be estimated by

$$f(k) = a_0 + \sum_{n=1}^{\infty} a_n \cos(n\omega k) + \sum_{n=1}^{\infty} b_n \sin(n\omega k) \tag{1}$$

where a_0 is the DC part and n is called the n-th harmonic. Without loss of generalization, we only consider sampled signals. The time interval between two successive samples is $T_s = 1/f_s$ with a sampling frequency of f_s, k is the time. The sum of the terms $a_n \cos(n\omega k)$ is the even part and the sum of the terms $b_n \sin(n\omega k)$ is the odd part of the signal. If T (scalar) is the period of the signal, $\omega = 2\pi/T$ is called the fundamental angular frequency. Thus, the term with $n = 1$ represents the fundamental term of the signal and terms with $n > 1$ represents its harmonics. Each harmonic component is defined by a_n and b_n. Practically, generated harmonics are superposed to the fundamental term with an additional noise $\eta(k)$. Thus, periodic signals can be approximated by a limited sum (to $n = N$):

$$\hat{f}(k) = a_0 + \sum_{n=1}^{N} a_n \cos(n\omega k) + \sum_{n=1}^{N} b_n \sin(n\omega k) + \eta(k) \tag{2}$$

The objective is to estimate coefficients a_0, a_n and b_n and for this we propose a linear MLP.

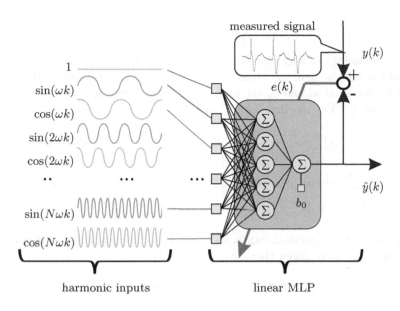

Fig. 2. The linear MLP with 5 neurons in one hidden layer for harmonic estimation

3.2   The Linear MLP

A linear MLP consists of a feedforward MLP with three layers of neurons. Its inputs are the values of the sine and cosine terms of all harmonic terms to be identified. There is only one output neuron in the output layer. A desired output is used for a supervised learning. This reference is the measured signal whose harmonic content must be estimated. All neurons of the network are with a linear activation function, i.e., identity function. The MLP is therefore linear and nonlinearities are introduced by the input vector. This architecture is shown by Fig. 2.

$\hat{f}(k)$ is a weighted sum of sinusoidal terms and is therefore a linear relationship that can be fitted by a linear MLP taking sine and cosine terms with unit amplitude as its inputs. Thus,

$$\hat{f}(k) = \begin{bmatrix} a_0 \ b_1 \ a_1 \ b_2 \ a_2 \ ... \ b_N \ a_N \end{bmatrix}^T \mathbf{x}(k) \tag{3}$$

with

$$\mathbf{x}(k) = \begin{bmatrix} 1 \ \sin(\omega k) \ \cos(\omega k) \ ... \ \sin(N\omega k) \ \cos(N\omega k) \end{bmatrix}^T \tag{4}$$

can be estimated by a linear MLP with only one hidden layer with M neurons and with one output neuron. The linear MLP takes R inputs, $R = 2N + 1$, N is the number of harmonics.

The output of the i-th hidden neuron $\hat{y}_i(k)$ $(i = 1, ...M)$ and the output of the network are respectively

$$\hat{y}_i(k) = w_{i,1} \sin(\omega k) + w_{i,2} \cos(\omega k) + ...$$
$$+ w_{i,R-1} \sin(N\omega k) + w_{i,R} \cos(N\omega k) + b_i, \tag{5}$$

$$\hat{y}(k) = \sum_{i=1}^{M} w_{o,i} \hat{y}_i(k) + b_o, \tag{6}$$

with $w_{i,j}$ the weight of i-th hidden neuron connected to the j-th input, $w_{o,i}$ the weight of the output neuron connected to the i-th hidden neuron, b_i the bias of the i-th hidden neuron and b_o the bias of the output neuron.

The output $\hat{y}(k)$ of the linear MLP therefore writes:

$$\begin{aligned}
\hat{y}(k) &= \left(\sum_{i=1}^{M} w_{o,i} w_{i,1}\right) \sin(\omega k) + \left(\sum_{i=1}^{M} w_{o,i} w_{i,2}\right) \cos(\omega k) \\
&+ \ldots \\
&+ \left(\sum_{i=1}^{M} w_{o,i} w_{i,R-1}\right) \sin(N\omega k) + \left(\sum_{i=1}^{M} w_{o,i} w_{i,R}\right) \cos(N\omega k) \\
&+ \left(\sum_{i=1}^{M} w_{o,i} b_i\right) + b_o.
\end{aligned} \tag{7}$$

The output of the network can be expressed by (8) with $\mathbf{x}$ from (4) and with $\mathbf{c}_{weight}$ and c_{bias} introduced thereafter:

$$\hat{y}(k) = \mathbf{c}_{weight} \mathbf{x}(k) + c_{bias}. \tag{8}$$

Definition 1 (The weight combination). *The weight combination of the linear MLP is a row-vector (with R elements) that is a linear combination of the hidden weights with the output weights which writes:*

$$\mathbf{c}_{weight} = \left[c_{weight(1)} \cdots c_{weight(R)} \right] = \mathbf{w}_o^T . \mathbf{W}_{hidden} \tag{9}$$

where $\mathbf{w}_o$ is the weight vector of the output neuron (with M elements) and $\mathbf{W}_{hidden}$ is a $M \times R$ weight matrix of all neurons of the hidden layer.

Definition 2 (The bias combination). *The bias combination of the linear MLP, c_{bias}, is a linear combination of all biases of hidden neurons with the weights of output neuron which writes:*

$$c_{bias} = \mathbf{w}_o^T . \mathbf{b}_{hidden} + b_o \tag{10}$$

where $\mathbf{b}_{hidden} = \left[b_1 \ldots b_M \right]^T$ is the bias vector of the hidden layer.

In order to update the weights, the output $\hat{y}(k)$ of the linear MLP is compared to the measured signal $y(k)$. After learning [1, 2], the weights $\mathbf{c}_{weight}$ and bias c_{bias} converge to their optimal values, respectively $\mathbf{c}_{weight}^*$ and c_{bias}^*. Due to the linear characteristic of the expression, $\mathbf{c}_{weight}^*$ converges to:

$$\mathbf{c}_{weight}^* \to \left[a_0 \ b_1 \ a_1 \ b_2 \ a_2 \ \ldots \ b_N \ a_N \right]^T. \tag{11}$$

The signal $y(k) = s(k)$ is thus estimated by the linear MLP with optimal values of $\mathbf{c}_{weight}^*$ and c_{bias}^*. Furthermore, the amplitudes of the harmonic terms are obtained from the weight combination (11). After convergence, the coefficients come from the appropriate element of $\mathbf{c}^*_{weight}$, i.e., $a_0 = c^*_{weight(1)} + c^*_{bias}$ and the a_n and b_n from $c^*_{weight(j)}$ for $1 < j < R$:

$$c^*_{weight(j)} = \sum_{i=1}^{M} (w^*_{o,i} . w^*_{i,j}). \tag{12}$$

Linear activation functions have been used for the neurons of the MLP so that the mathematical expression of the network's output looks like a sum of harmonic terms if sinusoidal terms have been provided as the inputs at the same time. Indeed, the output of the linear MLP has therefore the same expression than a Fourier series. As a consequence, the neural weights of the MLP have a physical representation: Combined according to the two previous definitions, they correspond to the amplitudes of the harmonic components.

4   Results in Estimating Harmonics of Biomedical Signals

The effectiveness of the linear MLP is illustrated in estimating the frequency content of ECG signals from the MIT-BIH Arrhythmia database [15]. A linear MLP with initial random weights is chosen. The fundamental frequency of the signal is on-line extracted from the ECG signal with a zero-crossing technique based on the derivative of the signal. Results are presented by Fig. 3 a). In this study, tracking the frequency is also used detect abnormal heart activities. If the estimated frequency is within in a specific and adaptive range, it means that the heard activity is normal. This range is represented on Fig. 3 b) by a red area. It is centered on the mean value of the estimated fundamental frequency. If the estimated frequency is not included in the range (corresponding to the orange squares on Fig. 3 b)), than the fundamental frequency is not updated and data will not be used for the learning of the linear MLP.

Based on the estimated main frequency, sinusoidal signals are generated to synthesize the input vector $\mathbf{x}_{1-20}$ to take into account harmonics of ranks 1 to 20 at each sampled time k. The desired output of the network is the digital ECG with a sampling period $T_s = 2.8$ ms. The Levenberg-Marquardt algorithm [2] with a learning rate of 0.7 is used to train the network and allows to compute the values of the coefficients a_0, a_n and b_n of (11). The amplitudes of the harmonic terms are obtained from the weights after convergence.

Results over three periods of time for the record 104 are shown on Fig. 4 with 3 hidden neurons and $\mathbf{x}_{1-20}$ for the input. The estimated signal is represented in Fig. 4 a) and its frequency content on Fig. 4 b). This figure provides comparisons to an Adaline (with the same input) and FFT calculated over the range 0-50 Hz. Harmonics obtained by the neural approaches are multiples of the fundamental frequency $f_o = 1.2107$ Hz while FFT calculates all frequencies directly. It can be seen that the estimation of the linear MLP is very close to the one obtained by the FFT.

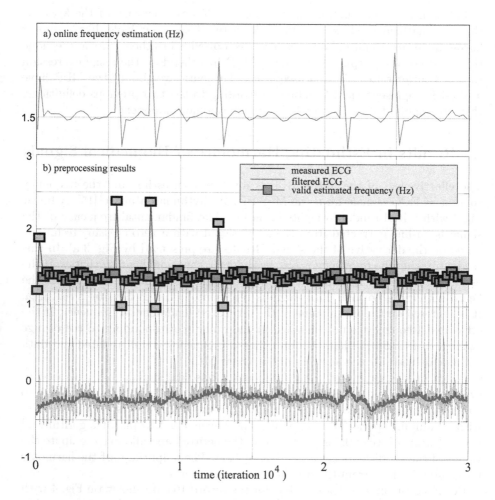

Fig. 3. On-line fundamental frequency tracking of an ECG

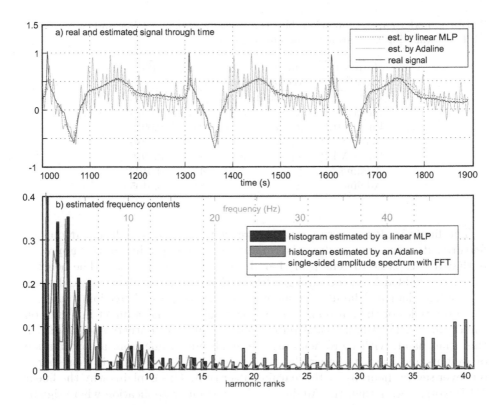

Fig. 4. Performances of a linear MLP with 3 hidden neurons, an Adaline and the FFT in identifying harmonics of an ECG

The MSE (Mean Square Error) of the estimation is used as a measure of overall performance. The resulting MSE is less than $1.6 \ 10^{-3}$ with the linear MLP with 3 hidden neurons. The MSE represents $1.4 \ 10^{-3}$ with the FFT and $10.2 \ 10^{-3}$ with the Adaline. The estimated coefficients obtained with the linear MLP therefore perfectly represent the harmonic content of the ECG. Results are similar for other signals from the database. Additional results with an input vector $\mathbf{x}_{1-40}$ that takes into account harmonics of ranks 1 to 40 and with more hidden neurons are presented in Tab. 1. The linear MLP approach is the best compromise in terms of performance and computational costs evaluated by the number of weights. The computing time required by a linear MLP with 3 hidden neurons is less than for the FFT.

The robustness against noise has been evaluated by adding noise to the signal. Even with a signal-to-noise ratio up to 10 dB, the harmonic content of the ECG is estimated by a linear MLP with 3 hidden neurons with a MSE less than $2 \ 10^{-3}$ compared to $4 \ 10^{-3}$ for the FFT and to $8 \ 10^{-3}$ for the Adaline. The linear MLP has been applied to the other records of the MIT-BIH database for training and validation. The MSE calculated after the initial phase of learning is in all cases less than $2.5 \ 10^{-3}$ with 3 hidden neurons.

Table 1. Performance comparison between the linear MLP, Adaline and conventional FFT in estimating the harmonic content of an ECG

Harmonic estimator	Input vector	Nb of neurons	Nb of weights	MSE
FFT	0 to 50 Hz	-	-	0.0014
linear MLP	x_{1-20}	3+1	127	0.0016
linear MLP	x_{1-40}	3+1	247	0.0016
linear MLP	x_{1-20}	6+1	253	0.0016
linear MLP	x_{1-40}	6+1	493	0.0016
Adaline	x_{1-20}	1	41	0.0102
Adaline	x_{1-40}	1	81	0.0105

The linear MLP is a very generic approach that performs efficient frequency feature extraction even under noisy conditions. One byproduct of this approach is that it is capable to generically handle various types of signals. The benefits of using a hidden layer, i.e., using a linear MLP, is that it allows more degrees of freedom than a Adaline. For an Adaline, the degrees of freedom represent the amplitudes of the harmonics. The weight adaption has a direct influence on their values. The Adaline is therefore more sensitive to outliers and noise. On the other hand, with more neurons, the amplitudes come from a combination of weights and are not the weights values. The estimation error is thus shared out over several neurons by the learning algorithm. This explains why the linear MLP works better than the Adaline in this particular application where signals are noisy and non-stationary.

5   Conclusion

This paper presents a linear Multi Layer Perceptron for estimating the frequency content of signals. Generated sinusoidal signals are taken for the inputs and a measured signal is used as a reference that is compared to its own output. The linear MLP uses only neurons with linear activation functions. This allows the neural structure to express the signal as a sum of harmonic terms, i.e., as a Fourier series. The learning algorithm determines the optimal values of the weights. Due to the architecture of the MLP, the amplitudes of the harmonics can be written as a combination of the weights after learning. The estimation of the frequency content is illustrated on ECG signals. Results show that the linear MLP is both efficient and accurate in characterizing sensory signals at a given time by frequency features. Furthermore, the linear MLP is able to adapt itself and to compensate for noisy conditions. With its simplicity and facility of implementation, it consists of a first step in order to handle various biomedical signals subject to diseases and abnormal rhythms.

References

1. Haykin, S.: Neural Networks: A Comprehensive Foundation, 2nd edn. Prentice Hall (1999)
2. Bishop, C.M.: Neural Networks for Pattern Recognition. Oxford (1995)
3. Haykin, S., Widrow, B.: Least-Mean-Square Adaptive Filters. Wiley Interscience (2003)
4. Rangayyan, R.: Biomedical Signal Analysis: A Case-Study Approach. Wiley-IEEE Press (2002)
5. Pan, J., Tompkins, W.: A real-time qrs detection algorithm. IEEE Transactions on Biomedical Engineering BME-32(3), 230–236 (1985)
6. Moukadem, A., Dieterlen, A., Hueber, N., Brandt, C.: A robust heart sounds segmentation module based on s-transform. Biomedical Signal Processing and Control (2013)
7. Ould Abdeslam, D., Wira, P., Mercklé, J., Flieller, D., Chapuis, Y.A.: A unified artificial neural network architecture for active power filters. IEEE Trans. on Industrial Electronics 54(1), 61–76 (2007)
8. Wira, P., Ould Abdeslam, D., Mercklé, J.: Learning and adaptive techniques for harmonics compensation in power supply networks. In: 14th IEEE Mediterranean Electrotechnical Conference, Ajaccio, France, pp. 719–725 (2008)
9. Dash, P., Swain, D., Liew, A., Rahman, S.: An adaptive linear combiner for on-line tracking of power system harmonics. IEEE Trans. on Power Systems 11(4), 1730–1735 (1996)
10. Vázquez, J.R., Salmerón, P., Alcantara, F.: Neural networks application to control an active power filter. In: 9th European Conference on Power Electronics and Applications, Graz, Austria (2001)
11. Wira, P., Nguyen, T.M.: Adaptive linear learning for on-line harmonic identification: An overview with study cases. In: International Joint Conference on Neural Networks, IJCNN 2013 (2013)
12. Lin, H.C.: Intelligent neural network based fast power system harmonic detection. IEEE Trans. on Industrial Electronics 54(1), 43–52 (2007)
13. Chang, G., Chen, C.I., Teng, Y.F.: Radial-basis-function neural network for harmonic detection. IEEE Trans. on Industrial Electronics 57(6), 2171–2179 (2010)
14. Temurtas, F., Gunturkun, R., Yumusak, N., Temurtas, H.: Harmonic detection using feed forward and recurrent neural networks for active filters. Electric Power Systems Research 72(1), 33–40 (2004)
15. Moody, G.B., Mark, R.G.: A database to support development and evaluation of intelligent intensive care monitoring. Computers in Cardiology, 657–660 (1996)

Ultrasound Intima-Media Thickness and Diameter Measurements of the Common Carotid Artery in Patients with Renal Failure Disease

Christos P. Loizou[1,*], Eleni Anastasiou[2], Takis Kasparis[2], Theodoros Lazarou[3], Marios Pantziaris[4], and Constandinos S. Pattichis[5]

[1] Department of Computer Science, Intercollege, P.O. Box 51604, CY-3507, Limassol, Cyprus
panloicy@logosnet.cy.net
[2] Cyprus University of Technology, Departement of Electical,
Computer Engineering and Informatics, Limassol, Cyprus
takis.kasparis@cut.ac.cy
[3] Nephrology Clinic, Limassol General Hospital, Limassol, Cyprus
t.lazarou@cytanet.com.cy
[4] Cyprus Institute of Neurology and Genetics, Nicosia, Cyprus
pantzari@cing.ac.cy
[5] Departement of Computer Science, University of Cyprus, Nicosia, Cyprus
pattichi@ucy.ac.cy

Abstract. Although the intima-media thickness (IMT) of the common carotid artery (CCA) is an established indicator of cardiovascular disease (CVD), its relationship with renal failure disease (RFD) is not yet established. In this study, we use an automated integrated segmentation system based on snakes, for segmenting the CCA, perform measurements of the IMT, and measure the CCA diameter (D). The study was performed on 20 longitudinal-section ultrasound images of healthy individuals and on 22 ultrasound images acquired from 11 RFD patients. A neurovascular expert manually delineated the IMT and the D in all RFD subjects. All images were intensity normalized and despeckled, the IMC and the D, were automatically segmented and measured We found increased IMT and D measurements for the RFD patients when compared to the normal subjects, but we found no statistical significant differences for the mean IMT and mean D measurements between the normal and the RFD patients.

Keywords: Intima-media thickness, carotid diameter, renal failure, ultrasound image, carotid artery.

1 Introduction

Cardiovascular disease (CVD), is the consequence of increased atherosclerosis, and is accepted as the leading cause of morbidity and mortality in patients with end-stage

---

* Corresponding author.

H. Papadopoulos et al. (Eds.): AIAI 2013, IFIP AICT 412, pp. 272–281, 2013.

renal failure disease who undergo hemodialysis [1], [2]. Atherosclerosis, which is a buildup on the artery walls is the main reason leading to CVD and can result to heart attack, and stroke [1], [2]. Carotid intima-media-thickness (IMT) is a measurement of the thickness of the innermost two layers of the arterial wall and provides the distance between the lumen-intima and the media-adventitia. The IMT can be observed and measured as the double line pattern on both walls of the longitudinal images of the common carotid artery (CCA) [2] (see also Fig. 1) and it is well accepted as a validated surrogate marker for atherosclerosis disease and endothelial dependent function [2].

Carotid IMT measurements have also been widely performed in hemodialysis patients with renal failure disease (RFD), and have shown that those patients have increased carotid IMT [3], impaired endothelium dependent, but unimpaired endothelium independent dilatation [4], compared to the age- and gender-matched normal controls. Furthermore, the traditional risk factors for increased atherosclerosis in the general population, such as dyslipidemia, diabetes mellitus, and hypertension, are also frequently found in RFD patients, and it was shown that those factors affect their vascular walls [5], [6]. Noninvasive B-mode ultrasound imaging is used to estimate the IMT of the human CCA [5]. IMT can be measured through segmentation of the intima media complex (IMC), which corresponds to the intima and media layers (see Fig. 1) of the arterial CCA wall. There are a number of techniques that have been proposed for the segmentation of the IMC in ultrasound images of the CCA, which are discussed in [7]. In two recent studies performed by our group [8], [9], we presented a semi-automatic method for IMC segmentation, based on despeckle filtering and snakes segmentation [10], [11]. In [9], we presented an extension of the system proposed in [8], where also the intima- and media-layers of the CCA could be segmented.

There have been a relatively small number of studies performed, for investigating the effect of increased IMT in the CCA and its relation to the progression of RFD. More specifically, in [12], carotid and brachial IMT were evaluated in diabetic and non-diabetic RFD patients undergoing hemodialysis. In [13], a correlation between the IMT and age in RFD patients was found, while in [14], kidney dysfunction was associated with carotid atherosclerosis in patients with mild or moderate chronic RFD. Increased IMT values in chronic RFD patients were also reported in [15], while in [16] it was shown that hemodialysis pediatric patients had reduced endothelial function with a development of carotid arteriopathy.

In this study, we use an automated integrated segmentation system based on snakes, for segmenting the CCA, perform measurements of the IMT and measure the CCA diameter (D). Our objective was to estimate differences for the IMT as well as for the D between RFD patients and normal subjects. Our study showed that the IMT and D are larger in ERD patients when compared with normal individuals. The study also showed that, there were no statistical significant differences for the IMT and the D measurements between the normal and the RFD patients. The findings of this study may be helpful in understanding the development of RFD and its correlation with atherosclerosis in hemodialysis patients.

The following section presents the materials and methods used in this study, whereas in section 3, we present our results. Sections 4 and 5 give the discussion, and the concluding remarks respectively.

2   Materials and Methods

2.1   Recording of Ultrasound Images

A total of 42 B-mode longitudinal ultrasound images of the CCA which display the vascular wall as a regular pattern (see Fig. 1a) that correlates with anatomical layers were recorded (task carried out by co-authors T. Lazarou and M. Pantziaris). The 20 images were from healthy individuals at a mean±SD age of 48±10.47 years and the 22 images were acquired from 11 RFD patients (1 woman and 10 men) at a mean±SD age of 55±12.25 years (5 patients had CVD symptoms). The images were acquired by the ATL HDI-5000 ultrasound scanner (Advanced Technology Laboratories, Seattle, USA) [17] with a resolution of 576X768 pixels with 256 gray levels. We use bicubic spline interpolation to resize all images to a standard pixel density of 16.66 pixels/mm (with a resulting pixel width of 0.06 mm). For the recordings, a linear probe (L74) at a recording frequency of 7MHz was used. Assuming a sound velocity propagation of 1550 m/s and 1 cycle per pulse, we thus have an effective spatial pulse width of 0.22 mm with an axial system resolution of 0.11 mm [17]. A written informed consent from each subject was obtained according to the instructions of the local ethics committee.

2.2   Ultrasound Image Normalization

Brightness adjustments of ultrasound images were carried out in this study based on the method introduced in [18], which improves image compatibility by reducing the variability introduced by different gain settings, different operators, different equipment, and facilitates ultrasound tissue comparability. Algebraic (linear) scaling of the images were manually performed by linearly adjusting the image so that the median gray level value of the blood was 0-5, and the median gray level of the adventitia (artery wall) was 180-190 [18]. The scale of the gray level of the images ranged from 0-255. Thus the brightness of all pixels in the image was readjusted according to the linear scale defined by selecting the two reference regions. Further details of the proposed normalization method can be found in [8]-[11].

2.3   Manual Measurements

A neurovascular expert (task carried out by co-author M. Pantziaris) delineated manually (in a blinded manner, both with respect to identifying the subject and delineating the image by using the mouse) the IMC and the D [8], on all longitudinal ultrasound images of the CCA after image normalization (see subsection 2.2) and speckle reduction filtering (see subsection 2.4). The IMC was measured by selecting

20 to 40 consecutive points for the intima and the adventitia layers, while by selecting 30 to 60 consecutive points for the near and far wall of the CCA respectively. The manual delineations were performed using a system implemented in Matlab (Math Works, Natick, MA) from our group. The measurements were performed between 1 and 2 cm proximal to the bifurcation of the CCA on the far wall [18] over a distance of 1.5 cm starting at a point 0.5 cm and ending at a point 2.0 cm proximal to the carotid bifurcation. The bifurcation of the CCA was used as a guide, and all measurements were made from that region. The IMT and the D were then calculated as the average of all the measurements. The measuring points and delineations were saved for comparison with the snake's segmentation method.

2.4   Speckle Reduction Filtering (DsFlsmv)

In this study the linear scaling filter (despeckle filter linear scaling mean variance-DsFlsmv) [19], utilizing the mean and the variance of a pixel neighborhood was used to filter the CCA ultrasound images from multiplicative noise prior the IMC and D segmentation. The filter may be described by a weighted average calculation using sub region statistics to estimate statistical measurements over 5x5 pixel windows applied for two iterations [10], [11]. Further implementation details of the DsFlsmv despeckle filter may be found in [10].

2.5   Automatic IMC and D Snakes Segmentation

The IMC and the D were automatically segmented after normalization and despeckle filtering (see section 2.2 and section 2.4), using an automated segmentation system proposed and evaluated in [8] and [9] based on a Matlab® software developed by our group. We present in Fig. 1b the automated results of the final snake contours after snakes deformation for the IMT at the far wall and the D, which were automatically segmented by the proposed segmentation system. An IMT and D initialization procedure was carried out for positioning the initial snake contour as close as possible to the area of interest, which is described in [8]-[10].

The Williams & Shah snake segmentation method [20] was used to deform the snake and segment the IMC and D borders. The method was proposed and evaluated in [8] and [9] for the IMC segmentation, in 100 ultrasound images of the CCA and more details about the model can be found there. For the Williams & Shah snake, the strength, tension and stiffness parameters were equal to $\alpha = 0.6$, $\beta_s = 0.4$, and $\gamma_s = 2$ respectively. The extracted final snake contours (see Fig. 1b), corresponds to the adventitia and intima borders of the IMC and the D at the near wall. The distance is computed between the two boundaries (at the far wall), at all points along the arterial wall segment of interest moving perpendicularly between pixel pairs, and then averaged to obtain the mean IMT (IMT_{mean}). The near wall of the lumen was also segmented for calculating the mean lumen of D (D_{mean}). Also the maximum (IMT_{max}, D_{max}), minimum (IMT_{min}, D_{min}), and median (IMT_{median}, D_{median}) IMT and D values, were calculated. Figure 1b shows the detected IMT_{mean}, and D_{mean} on an ultrasound image of the CCA.

276   C.P. Loizou et al.

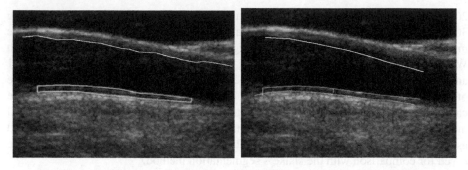

(a) Manual CCA IMT and D segmentations   (b) Automated CCA IMT and D segmentations

Fig. 1. a) Manual IMC and D segmentation measurements (IMT_{mean}=0.73 mm, IMT_{max}=0.85 mm, IMT_{min}=0.55 mm, IMT_{median}=0.66 mm, D_{mean}=5.81 mm, D_{max}=5.95 mm, D_{min}=4.61 mm, D_{median}=5.73 mm), and b) automated IMC and D segmentation measurements (IMT_{mean}=0.72 mm, IMT_{max}=0.83 mm, IMT_{min}=0.51 mm, IMT_{median}=0.67 mm, D_{mean}=5.71 mm, D_{max}=5.75 mm, D_{min}=4.69 mm, D_{median}=5.75 mm) from an RFD patient at the age of 54.

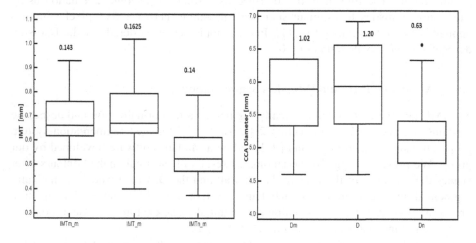

Fig. 2. Box plots for the mean values of the CCA for: (a) IMT, and (b) D. From left to right, mean manual IMT and Diameter (IMTn_m, Dm), mean automated IMT and Diameter (IMT_m, D) on RFD patients and mean manual IMT and diameter on normal subjects (IMTn_m, Dn) respectively. Inter-quartile rage values are shown above the box plots. Straight lines connect the nearest observations with 1.5 of the inter-quartile range (IQR) of the lower and upper quartiles. Unfilled rectangles indicate possible outliers with values beyond the ends of the 1.5xIQR.

2.6   Statistical Analysis

The Wilcoxon rank sum test was used in order to identify if for each set of normal and RFD patients measurements as well as for the manual and automated segmentation measurements, a significant difference (S) or not (NS) exists between the extracted IMC and D measurements, with a confidence level of 95%. For significant differences, we require p<0.05. Furthermore, box plots for the different measurements, were plotted. Bland–Altman plots [21], with 95% confidence intervals, were also used to further evaluate the IMT measurement agreement between the normal subjects and RFD patients. Also, the correlation coefficient, ρ, between the normal and RFD IMT and D measurements, which reflects the extent of a linear relationship between two data sets.

3   Results

We have evaluated 20 ultrasound images of the CCA from healthy subjects at a mean±SD age of 48±10.47 years and 22 ultrasound images acquired from 11 RFD patients (1 woman and 10 men) at a mean±SD age of 55±12.25 years, out of which 5 had CVD symptoms.

Figure 1 illustrates the manual (see Fig.1a) and the automated (see Fig. 1b) IMC and D segmentation measurements performed on a normalized despeckled ultrasound image of the CCA of an RFD patient at the age of 54.

Table 1 presents the IMC and D, segmentation measurements for the mean, maximum, minimum, and median IMT and D values on all normalized despeckled images of the CCA investigated in this study. These were performed on all images acquired from normal subjects (IMTn, Dn), as well as on all images acquired from the RFD patients, manually by the neurovascular expert (IMTm: 0.69±0.11 mm, Dm: 5.81±0.65 mm) and automated by the segmentation system (IMT: 0.71±0.13 mm, D: 5.96±0.72 mm).

After performing the non-parametric Wilcoxon rank sum test, we found no statistical significant differences between the aforementioned IMT and D measurement groups. More specifically we found $p = 0.44, p = 0.87$, and $p = 0.66$, for the IMT between the IMTm vs IMT, IMTm vs IMTn and IMT vs IMTn) and $p = 0.78, p = 0.61$, and $p = 0.89$, for the D between the Dm vs D, Dm vs Dn and D vs Dn). We also found no statistical significant difference between the manual and the automated IMT measurements ($p = 0.11$) of the CCA.

The correlation coefficient ρ, between the RFD manual and the automated IMT and D measurements (- / -) was ρ=0.30 / ρ=0.22 ($p = 0.15$ / $p = 0.91$), between the RFD manual and normal subjects IMT measurements was ρ=0.67 / ρ=0.56 ($p = 0.058$ / $p = 0.67$), and between the automated RFD and normal subjects IMT and D measurements was ρ=0.25 / ρ=0.46 ($p = 0.41$ / $p = 0.22$) respectively.

Figure 2a) presents box plots for the manual (IMTm_m) and the automated (IMT_m) mean IMT segmentation measurements for the RFD patients as well as for the automated IMT measurements (IMTn_m) from normal subjects. In Fig. 2b) we present box plots for the the CCA diameter measurements for the manual RFD patients (Dm), for the automated RFD patients (D), and for the manual segmentations (Dn) on the normal subjects. The IMT difference between the RFD automated and the normal subjects was (0.16±0.032) mm (see also Fig. 3). The difference between the manual and the automated IMT measurements for the RFD patients was (0.02±0.01) mm. Figure 3 illustrates a Bland–Altman plot between the automated mean IMT for

Table 1. Manual and Automated Mean, Maximum, Minimum And Median±(Standard Deviation) Measurements For The IMT And the D For 20 Normal and 22 RFD Patients. Values Are in mm. The Standard Deviations Are Given In Parentheses.

	Mean (mm)	Maximum (mm)	Minimum (mm)	Median (mm)
CCA IMT				
IMT_n	0.55(0.10)	0.67(0.13)	0.41(0.09)	0.55(0.12)
IMT_m	0.69(0.11)	0.93(0.16)	0.57(0.17)	0.68(0.11)
IMT	0.71(0.13)	0.90(0.23)	0.53(0.23)	0.69(0.12)
CCA Diameter				
D_n	5.10(0.58)	5.23(0.61)	4.73(0.64)	5.12(0.56)
D_m	5.81(0.65)	5.99(0.62)	5.58(0.69)	5.66(0.62)
D	5.96(0.72)	6.08(0.68)	5.59(0.74)	5.78(0.71)

IMT_n, D_n: Automated IMT and D measurements from normal subjects. IMT_m, D_m, IMT, D: Manual and automated IMT and D measurements from RFD subjects.

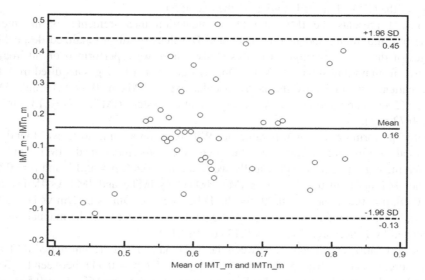

Fig. 3. Regression lines (Bland–Altman plots) between the automated mean IMT for the RFD patients (IMT_m) versus the automated mean normal IMT (IMTn_m) of the CCA for all subjects. The middle line represents the mean difference, and the upper and lower two outside lines representthe limits of agreement between the two measurents, which are the mean of the data±2sd for the estimated difference between the two measurements.

all RFD patients (IMT_m) and the normal mean IMT (IMTn_m) segmentation measurements of the CCA for all subjects. As it is shown from Fig. 3, the difference of the two measurement methods was (0.16+0.45) mm and (0.16-0.13) mm.

4   Discussion

The results of this study showed that the IMT and D in the CCA of ERD patients are larger when compared with the values found for the normal subjects. The study also showed that, there are no statistical significant differences for the IMT and D measurements between the normal subjects and the RFD patients. We also found no statistical significant differences between the manual and the automated segmentation measurements for the IMC and the D. A difference was found between the automated segmented IMT mean (IMT_m) measurements of the RFD patients and the automated segmented mean IMT measurements of the normal subjects (IMTn_m) (see also Fig. 2a and Table 1). Although the aforementioned difference was estimated, we found no statistical significant differences for both the IMT and the carotid diameter, D, between those two measurements, as it was shown with the Wilcoxon rank sum test performed in this study. From Table 1, it is also obvious that the mean IMT and D for the normal subjects, exhibit both lower values when compared to the values of the RFD patients group. This is also the case for all other measurements of the IMT (mean, maximum, minimum, and median). Our findings show that kidney dysfunction may probably thus be associated with increased carotid IMT and carotid diameter, D. The subjects investigated in this study included a high proportion of patients with some atherosclerotic risk factors, indicating that kidney dysfunction is an atherosclerotic risk factor, especially in patients at high risk for CVD disease.

Atherosclerosis is the most common cause of RFD disease, resulting in hypertension and ischemic nephropathy [2], [5]. A regular follow up was suggested in [5], for selected patients with atherosclerotic renal artery stenosis before they develop severe hypertension or renal insufficiency. Randomized clinical trials will also be necessary to define the role of early renal revascularization in the management of these patients.

In a large study [22], where 1351 male individuals were investigated in order to estimate whether chronic RFD was associated with carotid IMT thickening, it was shown that chronic RFD may be associated with early carotid atherosclerosis in low-risk individuals, such as those undergoing general health screening, who have hypertension and/or impaired glucose metabolism.

In [12] the authors evaluated CCA IMT and brachial arteries, flow-mediated dilatation (FMD), and nitroglycerin-mediated dilatation (NMD) in diabetic and non-diabetic hemodialysis patients. In all patients a positive correlation was found with carotid and brachial IMT, and a negative one with FMD and NMD. With respect to hypertension as well as diabetes, a negative correlation was found with FMD and NMD. It was shown that age is the most important factor that significantly affected all studied markers of atherosclerosis in hemodialysis patients. According to their results, intensive antihypertensive treatment is recommended in hypertensive chronic hemodialysis patients. The authors in [13] found a correlation between IMT and age in RFD patients and that IMT values were correlated with total cholesterol. It was furthermore reported in [14], that kidney dysfunction was associated with increased carotid IMT independent of the classical atherosclerotic risk factors but they found no association between kidney dysfunction and calcified plaque. Carotid IMT may be a predictive marker of CVD disease that reflects early kidney dysfunction, especially in high-risk patients.

In another study [23], the relation between carotid IMT and hemodialysis patients in chronic RFD was investigated, where 78 RFD patients were examined. A significant positive correlation was found between IMT and hemodialysis ($p=0.045$) independent of traditional risk factors. It was also shown that hemodialysis patients had significantly higher mean IMT (1.136 ± 0.021 mm) compared to that of normal (0.959 ± 0.023 mm) patients. Similar findings were also reported in this study.

There are also some limitations for the present study which arises mostly from the small number of subjects included and that the duration of diabetes and RFD were not included. The material used in this study was acquired from RFD patients undergoing hemodialysis, where the 5 patients with CVD disorders were not studied separately. Additionally, the study should be further applied on a larger sample, a task which is currently undertaken by our group. Additional variables, such as age, sex, weight, blood pressure and others should be taken into account for better evaluating the RFD in hemodialysis patients.

5   Concluding Remarks

This paper presents a study on the carotid IMT and D measurements of the CCA in RFD patients undergoing haemodialysis. It was shown that the degree of atherosclerosis is greater in RFD patients when compared to that of normal subjects. We anticipate that the above comparison may have some clinical value in retrospectively screening of the dysfunction of the CCA through time. It would be furthermore interesting in the future to study the atherosclerotic carotid plaque changes in the aforementioned group of subjects and estimate whether and how kidney dysfunction is associated with the build-up of carotid plaques. It may also be possible to identify and differentiate those individuals into high and low risk groups according to their CVD risk before the development of plaques. The use of texture features will also be utilised in order to provide new feature sets, which can be used successfully for the classification of the IMC structures in normal and abnormal subjects. Future work will incorporate a new texture image retrieval system that uses texture features extracted from the IMC, to retrieve images that could be associated with the same level of the risk for developing RFD. Further research is required for estimating IMT and diameter differences between normal and RFD individuals in the CCA as well as their correlation with texture features extracted from these areas, and the progression of RFD.

References

1. Mendis, S., Puska, P., Norrving, B.: Global Atlas on cardiovascular disease prevention and control. WHO (2012) ISBN: 978-92-4-156437-3
2. Foley, R.N., Parfrey, P.S., Sarnak, M.J.: Clinical epidemiology of cardiovascular disease in chronic renal disease. Am. J. Kidney Dis. 32, 112–119 (1998)
3. Hojs, R.: Carotid intima-media thickness and plaques in hemodialysis patients. Artif. Organs 24, 691–695 (2000)
4. van Guldener, C., Lambert, J., Janssen, M.J., Donker, A.J., Stehouwer, C.D.: Endothelium-dependent vasodilatation and distensibility of large arteries in chronic hemodialysis patients. Nephrol. Dial. Transplant 12(Suppl. 2), 14–18 (1997)

5. Zielner, E.R.: Atherosclerotic renovascular disease: Natural history and diagnosis. Pers. Vasc. Surg. Endovasc. Ther. 16(4), 299–310 (2004)
6. Maeda, N., Sawayama, Y., Tatsukawa, M., et al.: Carotid artery lesions and atherosclerotic risk factors in Japanese hemodialysis patients. Atherosclerosis 169, 183–192 (2003)
7. Molinari, F., Zeng, G., Suri, J.S.: A state of the art review on intima-media thickness measurement and wall segmentation techniques for carotid ultrasound. Comp. Meth. and Progr. in Biomed. 100, 201–221 (2010)
8. Loizou, C.P., Pattichis, C.S., Pantziaris, M., Tyllis, T., Nicolaides, A.N.: Snakes based segmentation of the common carotid artery intima media. Med. Biolog. Eng. Comput. 45, 35–49 (2007)
9. Loizou, C.P., Pattichis, C.S., Nicolaides, A.N., Pantziaris, M.: Manual and automated media and intima thickness measurements of the common carotid artery. IEEE Trans. Ultras. Ferroel. Freq. Contr. 56(5), 983–994 (2009)
10. Loizou, C.P., Pattichis, C.S.: Despeckle filtering algorithms and Software for Ultrasound Imaging. Synthesis Lectures on Algorithms and Software for Engineering. Ed. Morgan & Claypool Publishers, 1537 Fourth Street, Suite 228, San Rafael, CA 94901 USA (2008)
11. Loizou, C.P., Pattichis, C.S., Christodoulou, C.I., Istepanian, R.S.H., Pantziaris, M., Nicolaides, A.: Comparative evaluation of despeckle filtering in ultrasound imaging of the carotid artery. IEEE Trans. Ultras. Ferroel. Freq. Contr. 52(10), 1653–1669 (2005)
12. Rus, R., Butotovic-Ponikvar, J.: Intima media thickness and endothelial function in chronic hemodialysis patients. Therapeutic Apheresis and Dialysis 13(4), 249–299 (2009)
13. Bevc, S., Hojs, R., Ekart, R., Hojs-Fabjan, T.: Atherosclerosis in hemodylaisis patienst: traditional and non-traditional risk factors. Acta Dermatoven, APA 15(4), 151–157 (2006)
14. Makiko, T., Abe, Y., Furukado, S., Miwa, K., et al.: Chronic kidney disease and carotid atherosclerosis. J. Stroke Cerebrov. Diseas 21(1), 47–51 (2012)
15. Somnath, D., Jayantha, P., Kanti, G.M., Kumar, A.D., et al.: Carotid artery intima-media thickness in chronic renal failure patients with diabetes and without diabetes. Int. J. Med. Scienc. 1(3), 65–70 (2012)
16. Muscheites, J., Meyer, A.A., Drueckler, E., et al.: Assessment of the cardiovascular system in pediatric chronic kidney disease: a pilot study. Pediatr. Nephrol. 23, 2233–2239 (2008)
17. A Philips Medical System Company, Comparison of image clarity, SonoCT real-time compound imaging versus conventional 2D ultrasound imaging. ATL Ultrasound, Report (2001)
18. Elatrozy, T., Nicolaides, A.N., Tegos, T., Zarka, A., Griffin, M., Sabetai, M.: The effect of B-mode ultrasonic image standardization of the echodensity of symptomatic and asymptomatic carotid bifurcation plaque. Int. Angiol. 17(3), 179–186 (1998)
19. Lee, J.S.: Refined filtering of image noise using local statistics. Comp. Graph. and Image Process 15, 380–389 (1981)
20. Williams, D.J., Shah, M.: A fast algorithm for active contours and curvature estimation. Int. J. on Graph, Vision and Imag. Proc. Image Underst. 55, 14–26 (1992)
21. Bland, J.M., Altman, D.G.: Statistical methods for assessing agreement between two methods of clinical measurement. Lancet 1(8476), 307–310 (1986)
22. Ishizaka, N., Ishizaka, Y., Toda, E., Koike, K., et al.: Assosiation between chronic kidney disease and carotid intim-media thickness in individuals with hybertension and impaired glucose metabolism. Hypertens. Res. 30(11), 1035–1041 (2007)
23. Paul, J., Dasgupta, S., Ghosh, K.M.: Carotid artery IMT as a surrogate marker of atherosclerosis in patients with chronic renal failure on hemodialysis. North Amer. J. Med. Sciens. 4(2), 77–80 (2012)

Texture Analysis in Ultrasound Images of Carotid Plaque Components of Asymptomatic and Symptomatic Subjects

Christos P. Loizou[1,*], Marios Pantziaris[2], Marilena Theofilou[3],
Takis Kasparis[3], and Efthivoulos Kyriakou[4]

[1] Department of Computer Science, Intercollege, P.O. Box 51604, CY-3507,
Limassol, Cyprus
panloicy@logosnet.cy.net
[2] Cyprus Institute of Neurology and Genetics, Nicosia, Cyprus
pantzari@cing.ac.cy
[3] Cyprus University of Technology, Departement of Electical,
Computer Engineering and Informatics, Limassol, Cyprus
takis.kasparis@cut.ac.cy
[4] Dept. of Computer Science and Engineering, Frederick University Cyprus,
Limassol, Cyprus
e.kyriacou@frederick.ac.cy

Abstract. There are indications that the texture of certain components of atherosclerotic carotid plaques in the common carotid artery (CCA), obtained by high resolution ultrasound imaging, may have additional prognostic implication for the risk of stroke. The objective of this study was to perform texture analysis of the middle component of atherosclerotic carotid plaques in 230 CCA plaque ultrasound images (115 asymptomatic and 115 symptomatic). These were manually delineated by a neurovascular expert after normalization and despeckle filtering using the linear despeckle filter (DsFlsmv). Texture features were extracted from the middle plaque component. We found statistical significant differences for some of the texture features extracted, between asymptomatic and symptomatic subjects. The results showed that it may be possible to identify a group of patients at risk of stroke (asymptomatic versus symptomatic) based on texture features extracted from the middle component of the atherosclerotic carotid plaque in ultrasound images of the CCA.

Keywords: Atherosclerotic carotid plaque components, texture analysis, ultrasound image, carotid artery.

1 Introduction

Cardiovascular disease (CVD) is the third leading cause of death and major disability in the world and is the consequence of increased atherosclerosis in the walls of the common-carotid artery (CCA) [1]. Atherosclerosis is a buildup on the artery walls and

---

* Corresponding author.

H. Papadopoulos et al. (Eds.): AIAI 2013, IFIP AICT 412, pp. 282–291, 2013.

is the main reason leading to CVD and can result to heart attack, and stroke [1], [2]. It is a disease of the large and medium size arteries, and it is characterized by plaque formation (see also Fig. 1a) due to progressive intimal accumulation of lipid, protein, and cholesterol esters in the blood vessel wall [1]-[3], which reduces blood flow significantly. The risk of stroke increases with the severity of carotid stenosis and is reduced after carotid endarterectomy [2], [3]. There have been clinical indications that different components of the atherosclerotic carotid plaque of the CCA may contribute differently to the development of the risk of stroke. The objective of this study was to investigate the middle component of the atherosclerotic carotid plaque between asymptomatic and symptomatic subjects at risk of atherosclerosis, based on texture features analysis extracted from this area, and provide information that might indicate the importance of this structure in the development of stroke or of future neurological events.

There are only few studies reported in the literature, for the segmentation of the CCA plaque from ultrasound images [4], where the manual delineation of the plaques was used to be the golden standard [2], [3]. Recently a full automated system for the full segmentation of the CCA in ultrasound images based on snakes was proposed [5], as well as a system for the video segmentation of the CCA plaque [6]. There is only one study found in the literature [7], where the association between the presence of juxtaluminal hypoechoic areas and clinical symptoms was investigated. Furthermore, a relatively small number of plaque and classification studies based on B-mode ultrasonic imaging have been proposed and used in several cross-sectional and longitudinal natural history studies [8]-[10], attempting to correlate ultrasonic plaque features with the development of neurological events. More specifically, in [8] an automated segmentation method was proposed by our group based on image normalization, despeckle filtering and initial plaque contour estimation for segmenting the atherosclerotic carotid plaque from ultrasound images of the CCA. In [9], texture features were extracted from asymptomatic and symptomatic subjects at risk of atherosclerosis and features were found which may classify these groups of subjects. Finally, in [10] morphological features extracted from plaques areas were used for classifying plaques in asymptomatic and symptomatic.

In this study, we use the manual segmentations from a neurovascular expert, for investigating the middle component of the atherosclerotic carotid plaque in ultrasound images of the CCA (see also Fig. 1) in 115 asymptomatic and 115 symptomatic subjects and, perform texture analysis of this structure. Our objective in this study was to estimate significant differences between the aforementioned groups of patients, which may help in the correct evaluation of the risk of stroke. The study showed that there are texture features that may be used to classify the two different groups of subjects as well as to follow up the progression of the atherosclerosis disease. The findings of this study may additionally be helpful in understanding the development of stroke in asymptomatic and/or asymptomatic subjects at risk of atherosclerosis.

2   Materials and Methods

2.1   Recording of Ultrasound CCA Plaque Images

A total of 230 (115 asymptomatic and 115 symptomatic) B-mode longitudinal ultrasound images of the CCA bifurcation were acquired by the ATL HDI-3000 ultrasound scanner (Advanced Technology Laboratories, Seattle, USA) [11], equipped with a high linear array transducer (4-7 MHz), and were recorded digitally on a magneto optical drive, with a resolution of 768x576 pixels with 256 gray levels. Asymptomatic images were recorded from patients at risk of atherosclerosis in the absence of clinical symptoms. Symptomatic images were recorded from patients at risk of atherosclerosis, who already have developed clinical symptoms, such as a stroke or a transient ischemic attack (TIA). Digital images were resolution normalized

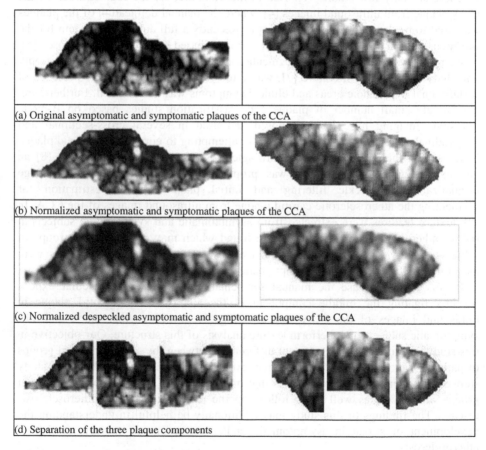

(a) Original asymptomatic and symptomatic plaques of the CCA

(b) Normalized asymptomatic and symptomatic plaques of the CCA

(c) Normalized despeckled asymptomatic and symptomatic plaques of the CCA

(d) Separation of the three plaque components

Fig. 1. a) Manually segmenetd ultrasound CCA plaque images of an asymptomatic (left colum) and a symptomatic subject (right column), b) Intensity normalised plaque images from a), c) Normalised and despeckled carotid plaque images with the filter DsFlsmv, d) Automatted segmentions of the three different plaque components.

at 16.66 pixels/mm (see ultrasound image normalization section). This was carried out due to the small variations in the number of pixels per mm of image depth (i.e. for deeply situated carotid arteries, image depth was increased and therefore digital image spatial resolution would have decreased) and in order to maintain uniformity in the digital image spatial resolution [8], [10]. The images were recorded at the Institute of Neurology and Genetics, Nicosia, Cyprus.

2.2   Ultrasound Image Normalization

Brightness adjustments of ultrasound images were carried out in this study based on the method introduced in [12], which improves image compatibility by reducing the variability introduced by different gain settings, different operators, different equipment, and facilitates ultrasound tissue comparability. Algebraic (linear) scaling of the images were manually performed by linearly adjusting the image so that the median gray level value of the blood was 0-5, and the median gray level of the adventitia (artery wall) was 180-190 [12]. The scale of the gray level of the images ranged from 0-255. Thus the brightness of all pixels in the image was readjusted according to the linear scale defined by selecting the two reference regions. Further details of the proposed normalization method can be found in [4]-[10].

2.3   Manual Segmentation and Separation of Plaques

A neurovascular expert (coauthor Marios Pantziaris) manually delineated (using the mouse) the 230 CCA plaque ultrasound images (115 symptomatic + 115 asymptomatic). The plaques were used for normalization, speckle reduction filtering, and texture features extraction. The manual delineations were performed using a system implemented in Matlab (Math Works, Natick, MA) from our group. The delineations were performed between 1 and 2 cm proximal to the bifurcation of the CCA at the far wall [12] over a distance of 1.5 cm starting at a point 0.5 cm and ending at a point 2.0 cm proximal to the carotid bifurcation. The bifurcation of the CCA was used as a guide, and all measurements were made from that region. The plaque contours were saved in order to be used for the texture analysis (see subsection 2.5). All sets of manual segmentation measurements were performed by the expert in a blinded manner, both with respect to identifying the subject and delineating the image. The 230 CCA plaques were then automatically separated in three different equidistant components. The horizontal major axis of the segmented plaque was estimated and perpendicular lines (starting from the left component of the plaque) were selected, at the 33%, and 66% of the plaque horizontal major axis (see also Fig. 1d), and thus separating the CCA plaques into three different components of equal major axis length sizes. The middle component of the plaque was then used for texture features analysis.

2.4   Speckle Reduction Filtering (DsFlsmv)

In this study the linear scaling filter (despeckle filter linear scaling mean variance-DsFlsmv) [13], utilizing the mean and the variance of a pixel neighborhood was used in order to filter the CCA ultrasound plaque images from multiplicative noise prior the texture features extraction. The filter may be described by a weighted average calculation using sub region statistics to estimate statistical measurements over 5x5 pixel windows applied for two iterations on each image. The DsFlsmv filter was applied on 440 CCA ultrasound images in [14], compared with 10 other different despeckle filters, and showed the best performance.

Table 1. Texture Features (mean±IQR) Extracted from the Asymptomatic (N=115) and the Symptomatic (N=115) CCA Plaques for the Original and the Despeckled Images (- / -). Inter-quartile range values are given in parentheses (±IQR)

Texture Features	Asymptomatic (N=115)	Symptomatic (N=115)	Wilcoxon test[1]
Statistical Features			
Median	58.7(39.1) / 49.5(50.1)	39.7(34.9) / 28(40.7)	0.007 / 0.005
Spatial Gray Level Dependence matrix (SGLDM)			
Contrast	23(13) / 29(3)	51(61) / 78(40)	0.02 / 0.02
IDM	0.33(0.22) / 0.28(0.18)	0.48(0.31) / 0.41(0.29)	0.01 / 0.001
SA	117(77.8) / 132(81)	79.5(70.3) / 96.5(71.7)	0.006 / 0.001
Gray level Difference Statistics (GLDS)			
Entropy	5.7(2.1) / 7.2(1.7)	4.4(2.54) / 5.9(2.68)	0.002 / 0.006
ASM	0.005(0.001) / 0.003(0.002)	0.01(0.01) / 0.005(0.003)	0.005 / 0.007
Statistical Feature Matrix (SFM)			
Coarseness	16.4 (5.2) / 36.8(23.1)	7.2(5.2) / 21.4(16.4)	0.001 / 0.03
Complexity	29180(16832) / 99876(47230)	25463(23123) / 07838(67293)	0.001 / 0.004
Laws Texture Energy Measures			
Energy LL	177502(55567)/ 185517(59639)	148720(94936) / 157578(97482)	0.013 / 0.005
Fractal Dimension (FD)			
Radial Sum	3723(1752) / 5140(1723)	2831(1656) / 4367(1579)	0.002 / 0.0001
ASM	3120(1402) / 3930(1645)	2435(1533) / 3288(1394)	0.001 / 0.002
Grey Level Run Length (GLRL)			
SRE	0.17(0.01) / 0.15(0.01)	0.16(0.01) / 0.14(0.009)	0.0001 / 0.002
LRE	38.2(2.03) / 38.6(1.9)	38.3(1.82) / 38.8(1.67)	0.003 / 0.006

IDM: Inverse difference moment, SA: Sum average, ASM: Angular second moment, SRE: Short run emphasis, LRE: Long-run emphasis. [1]Test Carried out at $p<0.05$ between the asymptomatic and the symptomatic despeckled CCA plaques.

2.5   Texture Analysis

Texture provides useful information for the characterization of plaque images in the CCA [9]. In order to estimate textural characteristics extracted from the different plaque components, a total of 71 different texture features were extracted, where only the most significant are shown in this work. The following texture feature sets were used: (i) First Order Statistics (FOS) [9], [15]: a) mean, b) variance, c) median, d)

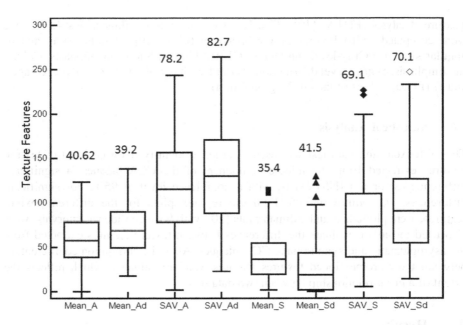

Fig. 2. Box plots for selected plaque texture features extracted from the middle component of the asymptomatic and symptomatic CCA plaques from all patients for the normalised despeckled images. We show the texture features: a) mean and sum average (SAV) for asymptomatic (_A) and symptomatic (_S) subjects as well as for original and despeclked (_d) images. In each plot we display the median, lower, and upper quartiles and confidence interval around the median. Straight lines connect the nearest observations within 1.5 of the IQR of the lower and upper quartiles. Unfilled triangles indicate possible outliers with values beyond the ends of the 1.5 x IQR. IQR values are shown above each boxplot.

skewness, e) kurtosis, f) energy, and g) entropy. (ii) Spatial Gray Level Dependence Matrices (SGLDM) as proposed by Haralick et al. [15]: a) angular second moment, b) contrast, c) correlation, d) sum of squares variance (SOSV), e) inverse difference moment (IDM), f) sum average, g) sum variance, h) sum entropy, i) entropy, j) difference variance, k) difference entropy, and l) information measures of correlation. For a chosen distance d (in this work d=1 was used) and for angles $\theta = 0^0, 45^0, 90^0$, and 135^0, we computed four values for each of the above texture measures. (iii) Gray Level Difference Statistics (GLDS) [16]: a) homogeneity, b) contrast, c) energy, d) entropy, and e) mean. The above features were calculated for displacements δ=(0, 1), (1, 1), (1, 0), (1, -1), where $\delta \equiv (\Delta x, \Delta y)$, and their mean values were taken. (iv) Neighborhood Gray Tone Difference Matrix (NGTDM) [17]: a) coarseness, b) contrast, c) busyness, d) complexity, and e) strength. (v) Statistical Feature Matrix (SFM) [18]: a) coarseness, b) contrast, c) periodicity, and d) roughness. (vi) Laws Texture Energy Measures (LTEM) [18]: LL-texture energy from LL kernel, EE-texture energy from EE-kernel, SS-texture energy from SS-kernel, LE-average texture energy from LE and EL kernels, ES-average texture energy from ES and SE kernels, and LS-average texture energy from LS and SL kernels. (vii) Fractal Dimension

Texture Analysis (FDTA) [19]: The Hurst coefficients for dimensions 4, 3 and 2 were computed. (viii) Fourier Power Spectrum (FPS) [19]: a) radial sum, and b) angular sum. (ix) Gray-level run length (GLRL) [10]: a) Short run emphasis, b) long run emphasis, c) gray level distribution, d) run length distribution), e) run percentage), and f)-j) mean values of the GLRL group for a)-e).

2.6   Statistical Analysis

The Wilcoxon rank sum test was used in order to identify if for each set of texture features extracted from the middle component of the CCA plaques, a significant difference (S) or not (NS) exists, with a confidence level of 95%. For significant differences, we require $p<0.05$. Furthermore, box plots for the different texture features, were plotted. Bland–Altman plots [20], with 95% confidence intervals, were also used to further evaluate the differences between texture features extracted from the asymptomatic and symptomatic CCA plaques. Also, the correlation coefficient, ρ, between the aforementioned textures features were investigated, which reflects the extent of a linear relationship between two data sets.

3   Results

Figure 1 illustrates a manually segmented asymptomatic (see left column of Fig. 1) and a symptomatic (see right column of Fig. 1) plaque from an ultrasound image of the CCA. We also present the normalized image (see Fig. 1b), normalized despeckled images, (see Fig. 1c) and the three different plaque components (see Fig. 1d), respectively.

Table 1 presents selected texture features (mean±IQR) extracted from the middle component of the plaques, that showed significant differences between the asymptomatic and symptomatic CCA plaques for the original and despeckled images (- / -). The IQR values for each feature are given in parentheses (±IQR). The non-parametric Wilcoxon rank-sum test was performed between the asymptomatic and symptomatic despeckled CCA plaques and the p-values are given in the last column of Table 1, showing statistical significant differences between the two different groups.

In Fig. 2 we present box plots for selected plaque texture features that showed significant differences between asymptomatic and symptomatic subjects, extracted from the middle component of the CCA plaques from all subjects for the original and the despeckled images with filter DsFlsmv. We show the texture features: a) mean and sum average (SAV) for asymptomatic (_A) and symptomatic (_S) subjects as well as for the original and the despeclked (_d) images.

The correlation coefficient, ρ, between the texture feature Mean_A and Mean_S for the original images was $\rho=0.034$ (p=0.74), and between the Mean_Ad and Mean_Sd for the despeckled images was $\rho=0.0037$ (p=0.94). The correlation coefficient, ρ, between the texture feature SAV_A and SAV_S for the original images was $\rho=0.04$ (p=0.65), and between the SAV_Ad and SAV_Sd for the despeckled images was $\rho=0.007$ (p=0.94).

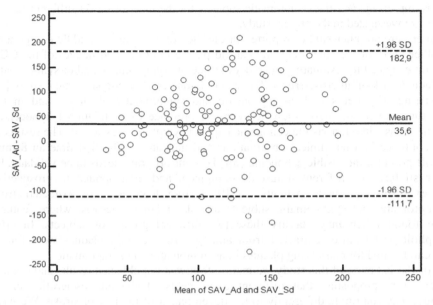

Fig. 3. Regression lines (Bland–Altman plot) between the texture feature sum average for the asymptmatic and the symptomatic normalised despeckled images. The middle line represents the mean difference, and the upper and lower two outside lines representthe limits of agreement between the two measurents, which are the mean of the data±2sd for the estimated difference between the two measurements.

Figure 3 presents a Bland-Altman plot for the texture feature sum average (SAV) extracted from the middle component of the carotid plaque for the asymptomatic and the symptomatic normalised despeckled images for all subjects investigated in this study. The difference of the two measurements between the asymptomatic and symptomatic images was 35.6+182.9 and 35.6-111.7 (ρ=0.007 at p=0.94).

4   Discussion and Conclusions

In this study, we use the manual segmentations from a neurovascular expert to perform texture analysis of the middle component of the atherosclerotic carotid plaque in ultrasound images of the CCA, in 115 asymptomatic and 115 symptomatic subjects. Our objective in this study was to estimate significant differences between the aforementioned groups of patients which may lead to a more accurate evaluation of the risk of stroke. The study showed that there are some texture features that might be used to classify the two different groups of subjects as well as to follow up the progression of the atherosclerosis disease. The findings of the present study may be helpful in understanding the development of stroke in asymptomatic and/or symptomatic subjects at risk of atherosclerosis. We showed in the present study, that the middle component of the atherosclerotic carotid plaque in ultrasound images may be helpful in classifying, possibly more reliably, asymptomatic and symptomatic subjects and might contribute to the research for the evaluation of the risk of stroke and its association with the vulnerable atherosclerotic carotid plaque. We also found strong statistical significant differences for the asymptomatic and the symptomatic subjects for the most of the 71 different texture features investigated in the present study.

There are no other studies reported in the literature where the middle component, or other parts, of the atherosclerotic carotid plaque in ultrasound images of the CCA was investigated for evaluating the risk of stroke in asymptomatic and/or symptomatic subjects at risk of atherosclerosis. There is only one cross-sectional clinical study [7], consisting of patients with asymptomatic and symptomatic plaques found in the literature, where the association between the presence of juxtaluminal hypoechoic plaque areas without a visible echogenic cap and symptoms was found. The results in [7], confirmed earlier clinical observations and indicate for the first time an 8 mm^2 cut-off point for the visible echohenic cap. This cut-off point needs to be validated by other studies and different ultrasonic equipment and then applied to prospective studies of asymptomatic patients. In [9] texture features were extracted from asymptomatic and symptomatic subjects at risk of atherosclerosis, where features were found which may classify these two different groups of subjects. In [10], morphological features extracted from atherosclerotic carotid plaques were found, that can be used for classifying plaques in asymptomatic and symptomatic.

For the initiation of this study we mainly followed unpublished results and clinical observations proposing, that different components of the atherosclerotic carotid plaque, may contribute differently to the development of the risk of stroke. We have concentrated our current investigation in the middle component of the plaque, which according to unpublished results may contribute stronger to the development of a neurological episode in asymptomatic or symptomatic subjects at risk of stroke. To overcome much of the subjectivity for plaque characterization, image normalization has been proposed using linear scaling and two reference points: blood and adventitia [12]. Various studies have validated this method for image normalization with measurement of reproducibility of the overall plaque echodensity using the gray-scale median [4]-[12] and two subsequent studies have demonstrated that the risk of stroke increases with decreased plaque gray scale median.

A limitation of our study is the lack of precise information on the duration between the onset of symptoms and the time of the ultrasound examination in individual symptomatic patients. It is therefore possible that plaque image appearance may have changed during this interval as a result of medical intervention. Additionally, the study should be further applied on a larger sample of subjects, a task which is currently undertaken by our group.

In a future study, additional variables, such as age, sex, weight, blood pressure and others should be taken into account for better evaluating the risk of stroke in the aforementioned group of patients. Furthermore, a classifier may be employed in order to provide a classification score that could provide a better insight of the best features to select for asymptomatic and symptomatic subjects.

References

1. Mendis, S., Puska, P., Norrving, B.: Global Atlas on cardiovascular disease prevention and control. WHO (2012) ISBN 978-92-4-156437-3
2. Executive committee for the asymptomatic carotid atherosclerosis study: Endarterectomy for asymptomatic carotid stenosis. J. Am. Med. Assoc. 273, 142–1428 (2002)

3. Nicolaides, A.N., Sabetai, M., Kakkos, S.K., Dhanjil, S., Tegos, T., Stevens, J.M.: The asymptomatic carotid stenosis and risk of stroke study. Int. Angiol. 22(3), 263–272 (2003)
4. Loizou, C.P., Pattichis, C.S., Pantziaris, M., Nicolaides, A.N.: An integrated system for the segmentation of atherosclerotic carotid plaque. IEEE Trans. Inform. 11(6), 661–667 (2007)
5. Loizou, C.P., Kasparis, T., Spyrou, C., Pantziaris, M.: Integrated system for the complete segmentation of the carotid artery bifurcation in ultrasound images. In: 9th Int Conf. Artif. Intell. Applic. & Innov. (AIAI 2013), Pafos, Cyprus, September 26-28, pp. 1–10 (2013)
6. Loizou, C.P., Pattichis, C.S., Petroudi, S., Kasparis, T., Pantziaris, M., Nicolaides, A.N.: Segmentation of atherosclerotic carotid plaque in ultrasound video. In: 34th Ann. Int. Conf. IEEE Eng. Med. Biol., San Diego, USA, August 28-September1, pp. 53–56 (2012)
7. Griffin, M., Kyriakou, E., Pattichis, C.S., Bond, D., et al.: Juxtaluminal hypoechoic area in ultrasonic images of carotid plaques and hymispheric symptoms. J. Vasc. Surg. 52(1), 69–76 (2010)
8. Geroulakos, G., Ramaswami, G., Nicolaides, A.N., James, K., Labropoulos, N., Belcaro, G., et al.: Characterization of symptomatic and asymptomatic carotid plaques using high-resolution real-time ultrasonography. Br. J. Surg. 80, 1274–1277 (1993)
9. Christodoulou, C., Pattichis, C.S., Pantziaris, M., Nicolaides, A.N.: Texture-based classification of atherosclerotic carotid plaques. IEEE Trans. Med. Imag. 22, 902–912 (2003)
10. Kyriakou, E., Pattichis, M.S., Christodoulou, C.I., Pattichis, C.S., et al.: Ultrasound imaging in the analysis of carotid plaque morphology for the assessment of stroke. In: Suri, J.S., Yuan, C., Wilson, D.L., Laxminarayan, S. (eds.) Plaque Imaging: Pixel to Molecular level, pp. 241–275. IOS Press (2005)
11. A Philips Medical System Company. Comparison of image clarity, SonoCT real-time compound imaging versus conventional 2D ultrasound imaging. ATL Ultrasound, Report (2001)
12. Elatrozy, T., Nicolaides, A.N., Tegos, T., Zarka, A., Griffin, M., Sabetai, M.: The effect of B-mode ultrasonic image standardization of the echodensity of symptomatic and asymptomatic carotid bifurcation plaque. Int. Angiol. 17(3), 179–186 (1998)
13. Lee, J.S.: Digital image enhancement and noise filtering by using local statistics. IEEE Trans. Pattern Anal. Mach. Intellig. PAMI-2(2), 165–168 (1980)
14. Loizou, C.P., Pattichis, C.S., Christodoulou, C.I., Istepanian, R.S.H., Pantziaris, M., Nicolaides, A.: Comparative evaluation of despeckle filtering in ultrasound imaging of the carotid artery. IEEE Trans. Ultras. Ferroel. Freq. Contr. 52(10), 1653–1669 (2005)
15. Haralick, R.M., Shanmugam, K., Dinstein, I.: Texture features for image classification. IEEE Trans. Systems, Man., Cyber. SMC-3, 610–621 (1973)
16. Weszka, J.S., Dyer, C.R., Rosenfield, A.: A comparative study of texture measures for terrain classification. IEEE Trans. Syst. Man. Cyber. SMC-6, 269–285 (1976)
17. Amadasun, M., King, R.: Textural features corresponding to textural properties. IEEE Trans. Syst. Man. Cyber. 19(5), 1264–1274 (1989)
18. Wu, C.M., Chen, Y.C., Hsieh, K.-S.: Texture features for classification of ultrasonic images. IEEE Trans. Med. Imag. 11, 141–152 (1992)
19. Chen, T.-J., et al.: A novel image quality index using Moran I statistics. Ph. in Medic. Biol. 48, 131–137 (2003)
20. Bland, J.M., Altman, D.G.: Statistical methods for assessing agreement between two methods of clinical measurement. Lancet 1(8476), 307–310 (1986)

Integrated System for the Complete Segmentation of the Common Carotid Artery Bifurcation in Ultrasound Images

Christos P. Loizou[1,*], Takis Kasparis[2], Christina Spyrou[2], and Marios Pantziaris[3]

[1] Department of Computer Science, Intercollege, P.O. Box 51604, CY-3507, Limassol, Cyprus
panloicy@logosnet.cy.net
[2] Cyprus University of Technology,
Dept. of Electical, Computer Engineering and Informatics, Limassol, Cyprus
takis.kasparis@cut.ac.cy
[3] Cyprus Institute of Neurology and Genetics, Nicosia, Cyprus
pantzari@cing.ac.cy

Abstract. The complete segmentation of the common carotid artery (CCA) bifurcation in ultrasound images is important for the evaluation of atherosclerosis disease and the quantification of the risk of stroke. This requires the extraction of the intima-media complex (IMC), the delineation of the lumen the atherosclerotic carotid plaque and measurement of the artery stenosis. The current research proposes an automated segmentation system for the complete segmentation of the CCA bifurcation in ultrasound images, which is based on snakes. The algorithm was evaluated on 20 longitudinal ultrasound images of the CCA bifurcation with manual segmentations available from a neurovascular expert. The manual mean±SD measurements were for the IMT: (0.96±0.22) mm, lumen diameter: (5.59±0.84) mm and ICA origin stenosis (48.1±11.52) %, while the automated measurements were for the IMT: (0.93±0.22) mm, lumen diameter: (5.77±0.99) mm and ICA stenosis (51.05±14.51) % respectively. We found no significant differences between all manual and the automated segmentation measurements.

Keywords: Intima-media thickness, lumen diameter, atherosclerotic plaque, carotid segmentation, ultrasound image, carotid artery.

1 Introduction

Atherosclerosis of the carotid artery is a pathological process mainly affecting the common carotid artery (CCA) bifurcation and is one of the major clinical manifestations leading to cardiovascular disease (CVD). It causes thickening of the artery walls, which affects blood flow, and may develop atherosclerotic carotid plaques, which causes stenosis in the artery lumen, thus affecting the normal blood flow [1]. Atherosclerosis can result to heart attack, and stroke [1].

---

* Corresponding author.

H. Papadopoulos et al. (Eds.): AIAI 2013, IFIP AICT 412, pp. 292–301, 2013.
© IFIP International Federation for Information Processing 2013

Carotid intima-media-thickness (IMT) is a measurement of the thickness of the innermost two layers of the arterial walls and provides the distance between the lumen-intima and the media-adventitia. The IMT can be observed and measured as the double line pattern on both walls of the longitudinal images of the CCA [2], [3] (see also Fig. 1) and it is well accepted as a validated surrogate marker for atherosclerotic disease [1]. Furthermore, the arterial stenosis is a significant marker of atherosclerosis and can be evaluated through the segmentation of atherosclerotic carotid plaques [4].

There are a number of segmentation methods proposed in the literature for the segmentation of the IMT, [2], [3], carotid diameter [4], [5], and the atherosclerotic carotid plaque [6], from ultrasound images of the CCA bifurcation, but there are no other studies reported where a complete segmentation of the CCA artery bifurcation has been attempted. More specifically, in [2] a review of different automated or semi-automated techniques for the segmentation of the IMT was reported, while in [3] a snakes based system for the segmentation of IMT in ultrasound images of the CCA was proposed. In [7], the system proposed in [3] was further extended for segmenting the intima layer and the media layer in ultrasound images of the CCA were additionally speckle reduction [8], [9] was applied in the image as a preprocessing step. Different carotid artery diameter indices were proposed in [4] and [5] where the lumen diameter was segmented and measured. Finally, in [6] a snakes based segmentation system was proposed for the segmentation of the atherosclerotic carotid plaque from ultrasound images of the CCA. The system requires the blood flow image, from which an initial contour estimation for the carotid plaque borders is estimated, which is then used as an input to the snake's segmentation algorithm.

All above integrated systems were proposed either for the segmentation of the IMC, lumen diameter or atherosclerotic plaque and there is no other system proposed earlier in the literature for integrating all above segmentation techniques in one software application. It is therefore desirable for the clinical expert to be able to use an integrated segmentation system, in which all the above aforementioned techniques could be integrated together in order to evaluate in a more precise and objective way the risk of stroke in asymptomatic and symptomatic subjects at risk of stroke .

In this study, our objective was to develop and evaluate an automated integrated segmentation system based on snakes, for segmenting the IMC, the atherosclerotic carotid plaque and the lumen diameter in longitudinal ultrasound images of the CCA bifurcation.

The segmentation system proposed in this study was evaluated on 20 ultrasound images acquired from symptomatic subjects and were compared with the manual delineations made by a neurovascular expert. We found no statistical significant differences between the manual and the automated segmentation measurements. The findings of this study indicate that the proposed system may also be effectively applied in the clinical praxis.

The following section presents the materials and methods used in this study, whereas in section 3, we present our results. Sections 4 and 5 give the discussion, and the concluding remarks respectively.

2   Materials and Methods

2.1   Recording of Ultrasound Images

A total of 20 B-mode longitudinal ultrasound images of the CCA bifurcation, with atherosclerotic plaques, were recorded (task carried out by co-author M. Pantziaris). The images display the vascular wall as a regular pattern (see Fig. 1a) that correlates with anatomical layers. The images were collected from symptomatic subjects, which have already developed clinical symptoms, such as a stroke or a transient ischemic attack at a mean±SD age of 48±10.47 years. The images were acquired by the ATL HDI-5000 ultrasound scanner (Advanced Technology Laboratories, Seattle, USA) [10] with a resolution of 768x576 pixels with 256 gray levels. We use bicubic spline interpolation to resize all images to a standard pixel density of 16.66 pixels/mm (with a resulting pixel width of 0.06 mm). For the recordings, a linear probe (L 7-4) at a recording frequency of 4-7 MHz was used. Assuming a sound velocity propagation of 1550 m/s and 1 cycle per pulse, we thus have an effective spatial pulse width of 0.22 mm with an axial system resolution of 0.11 mm [10]. Consent from each subject was obtained according to the instructions of the local ethics committee.

2.2   Ultrasound Image Normalization

Brightness adjustments of ultrasound images were carried out in this study based on the method introduced in [11], which improves image compatibility by reducing the variability introduced by different gain settings, different operators, different equipment, and facilitates ultrasound tissue comparability. Algebraic (linear) scaling of the images were manually performed by linearly adjusting the image so that the median gray level value of the blood was 0-5, and the median gray level of the adventitia (artery wall) was 180-190 [11]. The scale of the gray level of the images ranged from 0-255. Thus the brightness of all pixels in the image was readjusted according to the linear scale defined by selecting the two reference regions. Further details of the proposed normalization method can be found in [6]-[9].

2.3   Manual Measurements

A neurovascular expert (task carried out by co-author M. Pantziaris) manually delineated (in a blinded manner, both with respect to identifying the subject and delineating the image by using the mouse) the IMC, the atherosclerotic plaque and the lumen diameter [3], on all longitudinal ultrasound images of the CCA bifurcation after image normalization (see subsection 2.2) and speckle reduction filtering (see subsection 2.4). The IMT was measured by selecting 10 to 20 consecutive points for the intima and the adventitia layers, while by selecting 20 to 30 consecutive points for the near and far wall of the CCA respectively. The atherosclerotic plaque was measured by selecting 20-30 consecutive points forming a closed contour, while the lumen was measured by selecting the adventitia at the near wall of the CCA and the

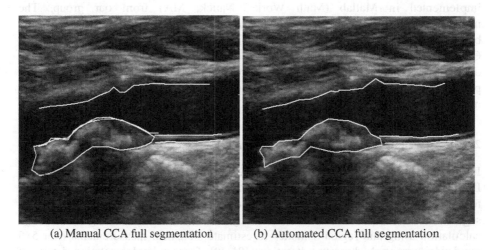

(a) Manual CCA full segmentation   (b) Automated CCA full segmentation

Fig. 1. a) Manual full CCA segmentation (IMT_{mean}=0.73 mm, IMT_{max}=0.85 mm, IMT_{min}=0.55 mm, IMT_{median}=0.66 mm, D_{mean}=5.51 mm, D_{max}=6.80 mm, D_{min}=3.31 mm, D_{median}=5.78 mm), and b) automated full CCA segmentation (IMT_{mean}=0.72 mm, IMT_{max}=0.83 mm, IMT_{min}=0.51 mm, IMT_{median}=0.67 mm, D_{mean}=5.65 mm, D_{max}=6.92 mm, D_{min}=3.45 mm, D_{median}=5.75 mm) from a symptomatic subject at risk of stroke the age of 53.

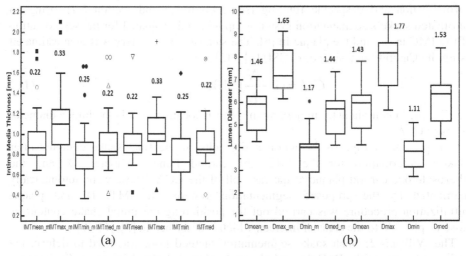

(a)   (b)

Fig. 2. Box plots for the mean, maximum, minimum and median segmentation measurements of the CCA for the: (a) IMT, and (b) Lumen diameter (D). From left to right, we present the mean manual and automated IMT (IMT_m, IMT), and D (D_m, D) for all 20 patients investigated. Inter-quartile rage values are shown above the box plots. Straight lines connect the nearest observations with 1.5 of the inter-quartile range (IQR) of the lower and upper quartiles. Unfilled rectangles indicate possible outliers with values beyond the ends of the 1.5xIQR.

origin of the ICA. The manual delineations were performed using a system implemented in Matlab (Math Works, Natick, MA) from our group. The measurements were performed on the far wall between 2 cm proximal to the bifurcation of the CCA [11] ending at a point 2.0 cm distal to the bifurcation of the CCA [12]. The IMT, the atherosclerotic plaque geometrical measures, and the lumen diameter, were then calculated as the average of all the measurements. The measuring points and delineations were saved for comparison with the snake's segmentation method.

2.4   Speckle Reduction Filtering (DsFlsmv)

In this study the linear scaling filter (despeckle filter linear scaling mean variance-DsFlsmv) [13], utilizing the mean and the variance of a pixel neighborhood was used to filter the CCA ultrasound images from multiplicative noise prior the full segmentation of the CCA.   The filter may be described by a weighted average calculation using sub region statistics to estimate statistical measurements over 5x5 pixel windows applied for two iterations [8], [9]. Further implementation details of the DsFlsmv despeckle filter may be found in [8].

2.5   Automatic IMC, Lumen and Plaque Snakes Segmentation

The IMC, lumen and plaque of the CCA were automatically segmented after image normalization and despeckle filtering (see section 2.2 and section 2.4), using an automated snake's segmentation system proposed and evaluated for the segmentation of the IMC in [3] and the plaque in [6]. The stenosis of the artery was then estimated using the Carotid Stenosis Index (CSI) [4], [12] as follows:

$$CSI = (1 - \frac{N}{D}) * 100 \tag{1}$$

with N, the lumen diameter in maximum stenosis location, and D the lumen diameter of distal CCA.

We present in Fig. 1b the automated results of the final snake contours after snake's deformation for the IMT, the carotid lumen diameter, D, and the atherosclerotic carotid plaque at the far wall of the CCA. These were automatically segmented by the proposed segmentation system. An IMT, D, and plaque initialization procedure was carried out for positioning the initial snake contour as close as possible to the areas of interest, which is described in [3], [6], [7].

The Williams & Shah snake segmentation method [14], was used to deform the snake and segment the IMC, D, and plaque borders. For the Williams & Shah snake, the strength, tension and stiffness parameters were equal to $\alpha = 0.6$, $\beta_s = 0.4$, and $\gamma_s = 2$ respectively. The extracted final snake contours (see Fig. 1b), corresponds to the adventitia and intima borders of the IMC, the carotid diameter, D, as well as the borders of the atherosclerotic carotid plaque at the far wall of the CCA. The distance for the IMT is computed between the two boundaries (at the far wall), at all points along the arterial wall segment of interest moving perpendicularly between pixel pairs, and then averaged to obtain the mean IMT (IMT_{mean}). The plaque is segmented at the far wall of the CCA. The near wall of the lumen was also segmented for calculating the mean lumen of D (D_{mean}). Also the maximum (IMT_{max}, D_{max}), minimum (IMT_{min}, D_{min}), and median (IMT_{median}, D_{median}) IMT and D values, were calculated. Figure 1b illustrates the segmented IMT_{mean}, and D_{mean} on an ultrasound image of the CCA.

Table 1. Manual and Automated Mean, Maximum, Minimum And Median±(Standard Deviation) Measurements For The IMT, D, and the ICA Stenosis Performed on 20 Subjects. Values Are in [mm]. The Standard Deviations Are Given In Parentheses.

	Mean [mm]	Maximum [mm]	Minimum [mm]	Median [mm]
Manual Measurements				
IMT	0.96(0.22)	1.14(0.37)	0.87(0.32)	0.95(0.33)
D	5.59(0.84)	7.43(0.89)	3.81(0.99)	5.45(0.91)
Stenosis	48.1(11.52)%			
Automated Measurements				
IMT	0.93(0.25)	1.05(0.21)	0.81(0.25)	0.92(0.26)
D	5.77(0.99)	7.86(1.21)	3.75(0.66)	6.15(1.06)
Stenosis	51.05(14.51)			

IMT D: Automated IMT and D measurements from all subjects.

Table 2. ROC Analysis on TPF, TNF, FPF, FNF, Overlap Index, Sp, P and F=1-E For The Atherosclerotic Carotid Plaque Snakes Segmentation Method on 20 Ultrasound Images of the CCA Bifurcation

System Detects	Expert Detects no plaque	Expert Detects plaque	KI	Overlap Index	Sp	P	F
No Plaque	TNF=98.1% FPF=3.2%	FNF=3.6% TPF=95.2%	86%	79.3%	0.98	0.945	0.917

TPF, TNF, FPF, FNF: True-positive fraction, true-negative fraction, false-positive fraction, false-negative fraction, KI: Similarity kappa index, Sp: Specificity, P: Precision, F: Effectiveness measure

Table 3. Geometric Measurements for the Manual and the Automated Segmentation Mean±(Standard Deviation) Measurements for the CCA Bifurcation Plaque for All the 20 Ultrasound Images of the CCA. The Standard Deviations Are Given In Parentheses.

Geometric Measures	Manual [mm]	Automated [mm]
Perimeter	50(20.66)	48.88(20.57)
Area	527.51(302)	520.79(30.44)
Diameter x-axis	22.5(9.80)	21.66(9.62)
Diameter y-axis	4.31(1.58)	4.19(1.57)

2.6   Statistical Analysis

The Wilcoxon rank sum test was used in order to identify if for each set of manual and automated segmentation measurements, a significant difference (S) or not (NS) exists between the extracted IMT, D, and plaque segmentation measurements, with a confidence level of 95%. For significant differences, we require p<0.05. Furthermore, box plots for the different measurements, were plotted. Also, the correlation coefficient, ρ, between the manual and the automated measurements, which reflects the extent of a linear relationship between two data sets, was calculated. Furthermore, we estimated the true-positive fraction (TPF), true-negative fraction (TNF), false-positive fraction (FPF) and false-negative fraction (FNF) between the manual and the automated plaque segmentation measurements. We also calculated the similarity kappa index, KI, the overlap index between the manual and the automated segmentations, as well as the specificity, Sp, precision, P, and the effectiveness measure, F [15].

3   Results

We have evaluated in this study 20 ultrasound images of the CCA bifurcation from symptomatic subjects at risk of stroke at a mean±SD age of 48±10.47 years. Figure 1 illustrates the manual (see Fig.1a) and the automated (see Fig. 1b) IMC, plaque and D segmentation measurements performed on a normalized despeckled ultrasound image of the CCA from a symptomatic subject at the age of 53.

Figure 2 presents box plots for the mean, maximum, minimum and medina segmentation measurements of the CCA for the IMT (see Fig. 2a), and the lumen diameter, D (see Fig. 2b). From left to right, we present the mean manual and automated IMT (IMT_m, IMT), and D (D_m, D), segmentation measurements for all 20 patients investigated in this study. We observe that manual and automated segmentation measurements are very close. After performing the non-parametric Wilcoxon rank sum test, we found no statistical significant differences between the manual and the automated segmentation measurements for the IMT and the D. More specifically, we've found $p = 0.41, p = 0.86, p = 0.21$ and $p = 0.52$ for the mean, maximum, minimum and median IMT manual and automated segmentation measurements measurements, respectively. For the CCA diameter, D, we've found $p = 0.50, p = 0.52, p = 0.96$ and $p = 0.50$ for the mean, maximum, minimum and median D measurements, respectively. We finally estimated maximum ICA lumen stenosis according to (1), with (48.1±11.52)% and (51.05±14.51)% of stenosis for the manual and the automated segmentations, respectively.

Table 1 presents the manual and automated mean, maximum, minimum and mean±(Standard deviation) measurements for the IMT, D and ICA lumen stenosis performed on the 20 subjects investigated in this study.

In Table 2, we show the TPF, TNF, FPF, FNF, KI, overlap index, Sp, P, and F for the proposed snakes segmentation method, performed on all the 20 ultrasound images of the CCA bifurcation. We've found a TPF of 95.2% and an overlap index of 79.3% between the manual and the automated segmentation methods.

Finally in Table 3, we illustrate the geometric measurements for the CCA bifurcation plaque for both the manual and the automated segmentation mean±(Standard Deviation) for all the 20 images investigated.

The correlation coefficient ρ, between the mean manual and the mean automated IMT and D measurements were $\rho=0.13$ ($p = 0.56$) for the IMT and $\rho=0.63$ ($p = 0.001$) for the D, respectively.

4   Discussion

We proposed in this study a new snake's based segmentation system that is able to segment all three different structures of interest in an ultrasound image of the CCA, bifurcation, namely the IMC, the lumen diameter, and the atherosclerotic carotid plaque. All three different structures are important for the clinician in order to correctly evaluate the degree of the severity of the atherosclerosis disease in an asymptomatic or symptomatic individual at risk of stroke. It should be noted that this is the first automated segmentation system proposed in the literature for complete segmentation of ultrasound images of the CCA bifurcation.

The results of this study showed that there are no significant differences between the manual and the automated segmentation measurements for the IMT, lumen diameter, and the atherosclerotic carotid plaque features. The differences found between the manual and the automated segmentation measurements (see also Table 1) are in general small, as also reported in other studies for the IMT [2], [3], [16], the lumen diameter [4], [5], and the atherosclerotic carotid plaque [6]. Furthermore, the automated segmentation measurements for the IMT and the diameter, D, in this study are very close to the manual measures (see Table 1) as also reported in other studies [2]-[5]. In Table 1 we found relatively high IMT values ($IMT_{mean}=0.96mm$) and this may be attributed to the fact that the patient group was from symptomatic subjects at risk of atherosclerosis with an age of 48±10.47 years [2], [16]. It has been reported in the literature [2], [3], [11], [16] that normal IMT values lies between 0.6 mm and 0.8 mm and that symptomatic subjects at risk of atherosclerosis exhibit higher IMT values [16]. The lumen diameter, D, as well as the CSI found in this study is consistent with the normal values of the lumen diameter and stenosis found in the literature [4], [5] and it is consider to lie within the normal range of values.

From Table 2, we may also observe that the agreement between the manual and the automated plaque segmentation is very high and that the system agrees with the expert in 95% of the cases (TPF=95%) for non-detecting a plaque and 98.1% of the cases (TNF=98.1%) for detecting a plague. We can thus conclude that the automated measurements performed in this study are as much reliable and accurate as the manual measurements performed by the neurovascular expert and therefore the system may be used in the clinical praxis with confidence.

Atherosclerosis is the most common cause of stroke [5], [16]. A regular follow up is suggested for selected patients with atherosclerotic artery stenosis as well as randomized clinical trials will also be necessary to define the role and the progression of atherosclerosis in the management of these patients [5], [11], [16], [17].

The vulnerable arterial plaque may cause atherothrombotic events, myocardial infarction and stroke, which are responsible for approximately 35% of the total mortality in the western population, and are the leading causes of morbidity world-wide [17]. The first indication of CVD disease is a thickening of the intimal and medial layers of the CCA wall. It involves lipid accumulation and the migration and proliferation of many cells in the sub-intimal and medial layers, which results in the formation of plaques [16]. It is the rupture of such plaques that causes myocardial infarcts, cerebrovascular events, peripheral vascular disease and kidney infarcts. The impact of the IMT and the atherosclerotic plaques, on the incidence of CVD events in the Rotterdam study [17] using B-mode ultrasound, indicates that the risk of myocardial infarction increases by 43% per standard deviation increase (0.163 mm) in the common carotid IMT. The main conclusions resulting from this study were supported by other independent investigations which reveal that an IMT higher than 0.9-1.0 mm [4]-[7] indicates potential atherosclerotic disease, which translates into an increased risk of a CVD event. Hence, the robust segmentation and measurement of the IMT as well as the atherosclerotic carotid plaque by B-mode ultrasound has a considerable impact in the early diagnosis of atherosclerosis, prognosis evaluation and prediction, and in the monitoring of response to lifestyle changes and to prescribed pharmacological treatments.

However, there are also some limitations for the present study, which arises mostly from the small number of subjects included and that the duration of the follow up of these patients. The study should be therefore, further applied on a larger sample of subjects, a task which is currently undertaken by our group. Additional variables, such as age, sex, weight, blood pressure and other atherosclerotic risk factors should be taken into account for better evaluating the atherosclerotic process and disease and the risk of stroke in this group of patients.

5   Concluding Remarks

In this paper we present a new snakes based segmentation system for the complete segmentation of the CCA bifurcation in ultrasound images. It should be noted that this is the first semi-automated integrated segmentation system proposed in the literature for the complete segmentation of the CCA. We anticipate in the future to apply the proposed system in a larger group of subjects and to extract texture characteristics from the segmented areas (IMC and carotid plague). These texture characteristics can then be used to classify and or separate subjects with increased IMC structures, as it was shown in [18], in different risk groups. Additionally, features extracted from the atherosclerotic carotid plaque, may be used for the separation between asymptomatic and symptomatic subjects. However, a larger scale study is required for evaluating the system before its application in the real praxis.

References

1. Mendis, S., Puska, P., Norrving, B.: Global Atlas on cardiovascular disease prevention and control. WHO (2012) ISBN: 978-92-4-156437-3
2. Molinari, F., Zeng, G., Suri, J.S.: A state of the art review on intima-media thickness measurement and wall segmentation techniques for carotid ultrasound. Comp. Meth. and Progr. Comp. Meth. and Progr. in Biomed. 100, 201–221 (2010)

3. Loizou, C.P., Pattichis, C.S., Pantziaris, M., Tyllis, T., Nicolaides, A.N.: Snakes based segmentation of the common carotid artery intima media. Med. Biolog. Eng. Comput. 45, 35–49 (2007)
4. Bladin, C.F., Alexandrov, A.V., Murphy, J., Maggisano, R., Norris, J.W.: Carotid Stenosis Index. A New Method of Measuring Internal Carotid Artery Stenosis. Stroke 26, 230–234 (1995)
5. Eigenbrodt, M.L., Sukhija, R., Rose, K.M., Tracy, R.E., et al.: Common carotid artery wall thickness and external diameter as predictors of prevalent and incident cardiac events in a large population study. Cardiovascular Ultrasound 5(11), 1–11 (2007)
6. Loizou, C.P., Pattichis, C.S., Pantziaris, M., Nicolaides, A.N.: An integrated system for the segmentation of atherosclerotic carotid plaque. IEEE Trans. Inform. Techn. Biomed. 11(6), 661–667 (2007)
7. Loizou, C.P., Pattichis, C.S., Nicolaides, A.N., Pantziaris, M.: Manual and automated media and intima thickness measurements of the common carotid artery. IEEE Trans. Ultras. Ferroel. Freq. Contr. 56(5), 983–994 (2009)
8. Loizou, C.P., Pattichis, C.S.: Despeckle filtering algorithms and Software for Ultrasound Imaging. Synthesis Lectures on Algorithms and Software for Engineering. Ed. Morgan & Claypool Publishers, 1537 Fourth Street, Suite 228, San Rafael, CA 94901 USA (2008)
9. Loizou, C.P., Pattichis, C.S., Christodoulou, C.I., Istepanian, R.S.H., Pantziaris, M., Nicolaides, A.: Comparative evaluation of despeckle filtering in ultrasound imaging of the carotid artery. IEEE Trans. Ultras. Ferroel. Freq. Contr. 52(10), 1653–1669 (2005)
10. A Philips Medical System Company. Comparison of image clarity, SonoCT real-time compound imaging versus conventional 2D ultrasound imaging. ATL Ultrasound, Report (2001)
11. Elatrozy, T., Nicolaides, A.N., Tegos, T., Zarka, A., Griffin, M., Sabetai, M.: The effect of B-mode ultrasonic image standardization of the echodensity of symptomatic and asymptomatic carotid bifurcation plaque. Int. Angiol. 17(3), 179–186 (1998)
12. North American Symptomatic Carotid Endarterectomy Trial Collaborators: Beneficial effect of carotid endarterectomy in symptomatic patients with high-grade carotid stenosis. N. Engl. J. Med. 325, 445–453 (1991)
13. Lee, J.S.: Refined filtering of image noise using local statistics. Comp. Graph. and Image Process 15, 380–389 (1981)
14. Williams, D.J., Shah, M.: A fast algorithm for active contours and curvature estimation. Int. J. on Graph., Vision and Imag. Proc. Image Underst. 55, 14–26 (1992)
15. Metz, C.: Basic principles of ROC analysis. Semin. Nucl. Medic. 8, 283–298 (1978)
16. Bots, M.L., Hofma, A., Diederick, E.G.: Increased common carotid Intima-media thickness. Adaptive response or a reflection of atherosclerosis? Findings from the Rotterdam Study. Stroke 28, 2442–2447 (1997)
17. van der Meer, I.M., Bots, M.L., Hofman, A., del Sol, A.I., van der Kuip, D.A., Witteman, J.C.: Predictive value of noninvasive measures of atherosclerosis for incident myocardial infarction: The Rotterdam Study. Circulation 109(9), 1089–1094 (2004)
18. Loizou, C.P., Pantziaris, M., Pattichis, M.S., Kyriakou, E., Pattichis, C.S.: Ultrasound image texture analysis of the intima and media layers of the common carotid artery and its correlation with age and gender. Comput. Med. Imag. Graph. 33(4), 317–324 (2009)

Diagnostic Feature Extraction on Osteoporosis Clinical Data Using Genetic Algorithms

George C. Anastassopoulos[1], Adam Adamopoulos[2], Georgios Drosos[3],
Konstantinos Kazakos[3], and Harris Papadopoulos[4,5]

[1] Medical Informatics Laboratory, Medical School,
Democritus University of Thrace, Greece
anasta@med.duth.gr
[2] Medical Physics Laboratory, Medical School,
Democritus University of Thrace, Greece
adam@med.duth.gr
[3] Department of Orthopedics, University Hospital of Alexandroupolis, Medical School,
Democritus University of Thrace, GR-68100,
drosos@otenet.gr, kazakosk@yahoo.gr
[4] Frederick Research Center, Filokyprou 7-9, Palouriotisa, Nicosia 1036, Cyprus
[5] Computer Science and Engineering Department, Frederick University,
7 Y. Frederickou St., Palouriotisa, Nicosia 1036, Cyprus
h.papadopoulos@frederick.ac.cy

Abstract. A medical database of *589* women thought to have osteoporosis has been analyzed. A hybrid algorithm consisting of Artificial Neural Networks and Genetic Algorithms was used for the assessment of osteoporosis. Osteoporosis is a common disease, especially in women, and a timely and accurate diagnosis is important for avoiding fractures. In this paper, the 33 initial osteoporosis risk factors are reduced to only 2 risk factors by the proposed hybrid algorithm. That leads to faster data analysis procedures and more accurate diagnostic results. The proposed method may be used as a screening tool that assists surgeons in making an osteoporosis diagnosis.

Keywords: osteoporosis, risk factor, artificial neural networks, genetic algorithm.

1 Introduction

Physicians use medical protocols in order to diagnose diseases. Medical diagnosis is the accurate decision of upon the nature of a patient's disease, the prediction of its likely evolution and the chances of recovery, based on a set of clinical and laboratorial data. When dealing with clinical data provided by a medical protocol that takes account of a large number of clinical features, one has to consider methods for feature selection. These methods could tentatively decrease the number of clinical features that are used for clinical assessment of the patients [1, 2]. Four are the main benefits of a successful feature selection: detection of clinical data redundancy, smaller clinical data sets, faster data analysis procedures, more accurate diagnostic

H. Papadopoulos et al. (Eds.): AIAI 2013, IFIP AICT 412, pp. 302–310, 2013.

results. As far as the first benefit, clinical features that do not play any significant role in patients' assessment could be detected as redundant and therefore could be ignored, or omitted thereafter. As a result, we step to the second benefit: the elimination of redundant clinical features that are not considered for the clinical assessment of the patients could lead to smaller clinical protocols that could incorporate fewer clinical parameters. Consequently, since the number of the clinical parameters that are considered is decreased, the time for data analysis and assessment is decreased in an analogous fashion. That is the third benefit we obtain. Finally, a robust and efficient feature extraction procedure, not only minimizes the size of the data set that is considered for analysis, but at the same time leads to more accurate diagnostic results, since it helps practitioners to take account of the diagnostic parameters that are essential for patients' assessment and prevent any confusion that could be caused from redundant data. At its best, a feature extraction method could lead to an optimum, by providing an optimal combination of the smallest set of clinical data that, when analyzed, could give the highest diagnostic results.

In the present paper, the method that was developed and applied for the detection of the optimal set of clinical osteoporosis data is a hybrid algorithm that incorporated two main algorithmic tools provided by the field of Computational Intelligence: Artificial Neural Networks (ANN) and Genetic Algorithms (GA) [3]. The main idea behind the proposed hybrid algorithm is based on two steps: First, to apply ANN for the osteoporosis data classification. This could provide an estimation of the classification error of the ANN when the whole clinical data set, with all of the clinical parameters that it incorporates, is considered as input for the ANN. Second, to apply a GA that could perform a heuristic search, in order to investigate for the optimally minimized clinical data set, that could be used as an input to the ANN, leading to even higher performance and even smaller classification error. Specifically, the GA is designed to accomplish a three-fold task: to find the optimally minimum set of clinical parameters that could be used as input to the ANN, to find the optimal ANN architecture by investigation for the minimum number of neurons in the input and the hidden layers of the ANN, and finally, to investigate for the combination of the input data set and ANN architecture that provides the optimal classification error, that is the minimum classification error possible.

When the GA investigation procedure comes to a successful end, the input parameters that are not included in the optimal (smallest) subset of the essential input parameters of the ANN, are considered as redundant and therefore can be omitted and eliminated during ANN training and testing.

The paper is structured as follows: In the next section the osteoporosis disease, as well as the risk factors that affect osteoporotic fractures are presented. In section 3 the used data and method are described. Section 4 details the results, while in section 5 a comparison with the other screening tools is made. Finally, section 6 gives our conclusions.

2   Osteoporosis

Osteoporosis is defined as a systemic skeletal disease characterized by low bone mass and microarchitectural deterioration of bone tissue, with a consequent increase in bone fragility and susceptibility to fracture [4]. Fractures related to osteoporosis usually affect the hip, spine, distal forearm and proximal humerus and account for a relatively high percentage of the total number of fractures [5, 6]. These fractures are associated with a high societal and personal cost and some of these fractures –hip and vertebra fractures- are also associated with increased mortality, morbidity or permanent disability [5 – 7].

2.1   Diagnosis of Osteoporosis

Osteoporosis is characterized by low Bone Mineral Density (BMD); this is the amount of bone mass per unit volume (volumetric density), or per unit area (areal density). BMD is measured by different techniques like Dual-Energy X-ray Absorptiometry (DEXA), Quantitative UltraSound (QUS), Quantitative Computed Tomography (QCT) [8 – 10]. DEXA of the proximal femur and lumbar spine (central DEXA) is the most widely used bone densitometric technique [8, 11] since there are many prospective studies that have documented a strong gradient of risk for fracture prediction [7].

According to the European Society for Clinical and Economic Aspects of Osteoporosis and Osteoarthritis, the objectives of bone mineral measurements are to provide diagnostic criteria, prognostic information on the probability of future fractures, and a baseline on which to monitor the natural history of the treated or untreated patient [7].

2.2   Risk Factors for Low BMD

The identification of the high-risk subset of patients (women) is of great importance for the prevention of osteoporotic fractures, but routine BMD measurement of all women is not feasible for most populations [7].

Several osteoporosis risk assessment instruments have been proposed to identify women who are likely to have osteoporosis and should therefore undergo bone densitometry [12 – 20]. Nevertheless at present there is no universally accepted policy for population screening in Europe to identify patients with osteoporosis or those at high risk of fracture.

3   Patients and Methods

This is a prospective study including 589 women that underwent a measurement of bone mineral density (BMD) with Dual-Energy X-ray Absorptiometry (DEXA) of the lumber spine. Patients receiving treatment for osteoporosis were excluded. Also, race is not included as a parameter, since all the women were Caucasian. The data set consisted of 33 parameters, which are presented in Table 1.

One more parameter, the estimated T-score was used for the classification of each subject. The T-score data were preprocessed and expressed to integer values ranging from 0 (absence of osteoporosis) to 3 (severe osteoporosis).

For the specific problem described above, two steps were considered: At the first step, the data set was divided in two parts: 80% of the data were used for ANN training and the rest 20% was used for ANN testing. A large number of computer experiments were performed in this step in order to investigate the effectiveness of some ANN parameters on the obtained performance in terms of the classification error. The main aim of this step was to conclude the internal architecture of the ANN in terms of the number of hidden layers, the number of neural nodes per hidden layer and the type of the transfer function of neurons. The obtained classification error that was provided by this procedure was kept for the purpose of comparison with the corresponding results that were obtained by the GA pruned ANN in the second step of the method.

Table 1. Parameters of the data set

A/A	Osteoporosis risk factor
1	Occupation
2	Allergies
3	Age
4	Body weight
5	Height
6	Menarche
7	Menopause
8	Number of pregnancies
9	Smoking
10	Alcohol consumption
11	Coffee intake
12	History of fracture
13	Spinal fracture
14	Carpal fracture
15	Sports
16	Parents with osteoporosis
17	Loss of height more than 3 cm
18	Kyphosis
19	Amenorrhea (pause of menstruation) for more than 12 months
20	Rheumatoid arthritis
21	Dairy consumption
22	History of diarrhea
23	Cortisone intake
24	Thyroxin intake
25	Estrogen therapy
26	Anorexia nervosa
27	Hyper parathyroidism
28	Insulin depended diabetes
29	Ovariectomy
30	Paget disease
31	Steroids intake
32	Diuretics intake
33	Chemotherapy

In the second step a GA was invoked for the selection of input parameters and investigation of the internal architecture of the ANN. The GA was utilized to search for the optimal subset of the input parameters that should be used for training and testing the pruned ANN. The individuals of the GA population consisted of binary strings of length equal to the number of parameters of the original data set plus six more bits that represented the binary expression of the number of neurons in the hidden layer. Since we considered a total of $N = 33$ clinical parameters for the evaluation of the patients, the GA individuals' chromosome consisted of $33 + 6 = 39$ genes, with each gene to be represented by a binary digit (bit). For the first 33 genes of the chromosome, the allele 0 denotes that the corresponding clinical parameter is not included in the subset of the parameters that will be used for ANN training and testing and therefore is omitted. On the opposite, the allele 1 for the first 33 genes of the chromosome denotes that the corresponding parameter will be considered for ANN training and testing. The total number of the considered parameters was denoted by I. The last 6 genes of the chromosome were considered as the binary representation of the number H of neurons in the hidden layer. Starting counting from 1, the 6-bit binary strings could represent up to a number of 64 neurons in the hidden layer. Since binary representation is adopted for the chromosome of the individuals of the GA, all the well known genetic operators for selection, crossover and binary mutation could be applied on the GA population.

By the 33 input parameters, there were performed 10 individual computer experiments. For each experiment, the training and the testing set were constructed randomly (80% and 20% of the cases respectively). The mean value of these 10 experiments denoted by $<MSE_t>$ was kept for comparison with the obtained results of the GA pruned ANN.. This mean value is denoted as $<MSE_t>$. On the other hand, the application of the GA concluded to pruned ANN with performance that was related to the MSE that in this case is denoted as MSEp. For the evaluation of the individuals of the GA the fitness function that was used consisted of three terms, referring to the performance, the size of the input subset and the size of the hidden layer, that is, MSE_p over $<MSE_t>$, I over N and H over 64, respectively. Thus, the fitness function can be written as:

$$f = \frac{MSE_p}{<MSE_t>} + \frac{I}{N} + \frac{H}{64} \qquad (1)$$

As it is obvious in Eq. (1), the GA searches for the optimal combination of performance, input parameter data set size and number of hidden neurons. This is done by trying to minimise the fitness function of Eq. (1) by using the Matlab GA tool For each individual of the GA, an ANN is constructed, with the number I of input nodes that is indicated by the individual's chromosome first 33 genes and the number of neurons H that is denoted by the decimal representation of the last 6 genes of the chromosome. The input data subset is also constructed by considering the alleles 1 of the first 33 genes of the chromosome. Subsequently, the constructed ANNs are trained using the 80% of the cases and tested using the rest 20% of the cases, and the generated MSE is recorded.

4   Results

The proposed methodology was applied on the osteoporosis clinical data of *589* patients that were described in Section 3. In the first step of our methodology, ten individual computer experiments were performed with ANN that were trained and tested by the full clinical data set, that is, ANN with *33* input nodes in the input layer that correspond to the 33 clinical parameters of the medical protocol used to collect the data and *12* nodes in the hidden layer. This step resulted to a mean value of the MSE of these ten individual experiments which is $<MSE_t> = 3.75 \cdot 10^{-4}$. The results that were obtained by the computer experiments of the application of the GA are presented in Table 2, whereas a typical evolution of the fitness function in those experiments is shown in Fig. 1. Each row on Table 2 presents the obtained results of a specific computer experiment. The first column of Table 2 refers to the experiment number, the second column refers to the generation number, the third column refers to the MSE_p of the fittest individual (pruned ANN) of that generation, the fourth column refers to the number of input nodes *I* of the ANN of the fittest individual, the fifth column refers to the features used by that ANN (as numbered in Table 1), and finally, the sixth column refers to the number *H* of neurons in the hidden layer of the ANN of the fittest individual.

Table 2. Results of GA search and ANN pruning

#Exp.	#Gen.	MSEp	I	Features	H
1	1	$1.63 \cdot 10^{-4}$	15	1 3 7 8 10 11 13 15 17 20 23 24 28 30 32	2
1	17	$1.43 \cdot 10^{-4}$	5	2 6 7 10 25	2
1	29	$1.37 \cdot 10^{-4}$	3	3 7 10	1
1	49	$1.37 \cdot 10^{-6}$	2	7 10	1
2	1	$1.85 \cdot 10^{-6}$	12	3 7 8 10 14 15 16 18 23 25 28 32	6
2	19	$1.65 \cdot 10^{-6}$	2	7 10	1

The first four rows in Table 2 refer to experiment Nr. 1, presenting four instances of the GA evolution at four different generations. For each generation it is shown the performance of the best individual. Thus, the GA starts with *15* inputs in generation Nr. 1, to settle to just 2 inputs in generation Nr. 49. It is noteworthy that the number of hidden nodes (*H*) is relatively small, ranging from 2 for the 1st generation, to *1* for the 29th and 49th generation. The last two rows of Table 2 refer to the results of a second computer experiment. In that experiment the GA starts with *12* inputs at generation Nr. 1, to settle down to 2 inputs at generation Nr. 19. The results presented in the third column of Table 2 indicate that assuming the performance of the pruned ANN is also improved, in terms of the MSE, since MSE_p is decreased down to the *36.5%* of the original $<MSE_t>$ of the non-pruned ANN. According to the results presented in fourth column, the smallest number of inputs is just 2, which is only the *6%* of the total number of diagnostic features, (which is $N = 33$). Even by using these small data subsets as input, and only one neuron in the hidden layer (as it is shown in the last column of Table 2), the pruned ANN maintain improved performance in terms of the MSE that they achieve, which is roughly the *1/3* of the $<MSE_t>$.

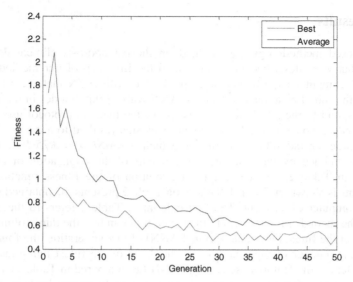

Fig. 1. GA evolution of mean and best individual fitness for 50 generations

5   Discussion

A common feature that the vast majority of the obtained results share is that they reveal a high order of redundancy in the original osteoporosis data set. The results presented in Table 2 indicated that just *2* to *3* input parameters, out of a total of *33*, that is a portion that varies from *6%* down to *9%* of the original data set, are essential in order of high performance to be achieved. In other words, the ANNs that were used for the classification of the osteoporosis data were pruned to less than *90%*. At the same time, the performance of these ANNs was highly improved, in terms of the MSE that is achieved, which for the pruned ANN is decreased almost to the *1/3* of the corresponding one that was achieved by the non-pruned ANN.

The two parameters that were common in all experiments is menopause (*7*) and alcohol intake (*10*), followed by a third one, the age (*3*). The osteoporosis risk assessment instruments mentioned earlier have been used to identify women that should undergo bone densitometry [12 - 17]. Apart from one instrument that uses only one parameter (body weight) [12] all other instruments include age and body weight, and some also include hormone replacement therapy [13 – 15]. History of fracture is included in two [13, 15] while one of them includes also history of rheumatoid arthritis and race [15].

In all but one osteoporosis risk assessment instruments, age is one of the most significant parameter. In our study, menopause, alcohol intake and age are the most common parameters. Menopause is related to age; actually it is related to a certain age group. Alcohol intake is not an important parameter in any of these instruments. It is surprising that alcohol intake is an important factor in our study since in Mediterranean countries alcohol intake is not common in women; unless alcohol intake is related to other dietary habits. One could expect that other parameters (like weight, previous fracture) to be more significant.

6 Conclusion

Simpler ANN architecture with decreased number of neural nodes and synaptic connections may result in less complicated and less time-consuming training and testing procedures and at the same time to performance improvement. Nowadays increased computer power and the contemporary development of ANN training algorithms, provide the essential means for fast and accurate implementation of ANN techniques to solve problems of various types, in different scientific fields, (classification, prediction, system identification, to name a few). Despite the fact that ANN training and testing even for complicated problems and large data sets is accomplished with low computational cost, it is legitimate, if not desirable, to investigate for even faster, more reliable and more accurate ANN training and testing methods. The present work was focused on the detection of any kind of redundancy in the data sets that are used as inputs during ANN training and testing. Redundancy and overlapping in data sets that are used for ANN training, when exists, definitely increase computational cost for ANN training, while at the same time, may mislead, or suppress the training procedure. This could generate problems in terms of ANN ability to accomplish successfully and with the desired accuracy the task that it was designed for.

From the clinical point of view, the results of our study are novel. Apart from menopause, age (well known parameters that are related to bone mineral density) and alcohol intake; we would expect some other parameters to be more common in our experiments. We will continue the study with more cases, in an effort to extract the most significant risk factors that affect the osteoporosis.

Acknowledgements. This work was supported by the European Regional Development Fund and the Cyprus Government through the Cyprus Research Promotion Foundation "DESMI 2009-2010", research contract TPE/ORIZO/0609(BIE)/24 ("Development of New Venn Prediction Methods for Osteoporosis Risk Assessment").

References

1. Papatheocharous, E., Papadopoulos, H., Andreou, A.S.: Feature Selection Techniques for Software Cost Modelling and Estimation: A Comparative Approach. Engineering Intelligent Systems 18(3-4), 233–246 (2010)
2. Papatheocharous, E., Papadopoulos, H., Andreou, A.S.: Software Effort Estimation with Ridge Regression and Evolutionary Attribute Selection. In: Proceedings of the 3rd Workshop on Artificial Intelligence Techniques in Software Engineering, AISEW 2010, CoRR abs/1012.5754 (2010)
3. Michalewicz, Z.: Genetic Algorithms + Data Structures = Evolution Programs. Springer (1996)
4. Consensus Development Conference: Diagnosis, prophylaxis and treatment of osteoporosis. Am. J. Med. 94, 646–650 (1993)

5. Cooper, C., Atkinson, E.J., Jacobsen, S.J., O'Fallon, W.M., Melton, L.J.: A population based study of survival after osteoporotic fractures. Am. J. Epidemiol. 137, 1001–1005 (1993)
6. Johnell, O., Kanis, J.A.: An estimate of the worldwide prevalence and disability associated with osteoporotic fractures. Osteoporos Int. 17, 1726–1733 (2006)
7. Kanis, J.A., Burlet, N., Cooper, C., Delmas, P.D., Reginster, J.Y., Borgstrom, F., Rizzoli, R.: European Society for Clinical and Economic Aspects of Osteoporosis and Osteoarthritis (ESCEO). European Guidance for the Diagnosis and Management of Osteoporosis in Postmenopausal Women. Osteoporos Int. 19(4), 399–428 (2008)
8. World Health Organization: Assessment of fracture risk and its application to screening for postmenopausal osteoporosis. Technical Report Series 843. WHO, Geneva (1994)
9. Blake, G.M., Fogelman, I.: Role of dual-energy X-ray absorptiometry in the diagnosis and treatment of osteoporosis. J. Clin. Densitom. 10, 102–110 (2007)
10. Engelke, K., Gluer, C.C.: Quality and performance measures in bone densitometry. I. Errors and diagnosis. Osteoporos Int. 17, 1283–1292 (2006)
11. Marshall, D., Johnell, O., Wedel, H.: Meta-analysis of how well measures of bone mineral density predict occurrence of osteoporotic fractures. Br. Med. J. 312, 1254–1259 (1996)
12. Michaëlsson, K., Bergström, R., Mallmin, H., Holmberg, L., Wolk, A., Ljunghall, S.: Screening for osteopenia and osteoporosis: selection by body composition. Osteoporos Int. 6(2), 120–126 (1996)
13. Lydick, E., Cook, K., Turpin, J., Melton, M., Stine, R., Byrnes, C.: Development and validation of a simple questionnaire to facilitate identification of women likely to have low bone density. Am. J. Manag. Care. 4(1), 37–48 (1998)
14. Cadarette, S.M., Jaglal, S.B., Kreiger, N., McIsaac, W.J., Darlington, G.A., Tu, J.V.: Development and validation of the Osteoporosis Risk Assessment Instrument to facilitate selection of women for bone densitometry. CMAJ 162(9), 1289–1294 (2000)
15. Sedrine, W.B., Chevallier, T., Zegels, B., Kvasz, A., Micheletti, M.C., Gelas, B., Reginster, J.Y.: Development and assessment of the Osteoporosis Index of Risk (OSIRIS) to facilitate selection of women for bone densitometry. Gynecol. Endocrinol. 16(3), 245–250 (2002)
16. Koh, L.K., Sedrine, W.B., Torralba, T.P., Kung, A., Fujiwara, S., Chan, S.P., Huang, Q.R., Rajatanavin, R., Tsai, K.S., Park, H.M., Reginster, J.Y.: Osteoporosis Self-Assessment Tool for Asians (OSTA) Research Group. A Simple Tool to Identify Asian Women at Increased Risk of Osteoporosis. Osteoporos Int. 12(8), 699–705 (2001)
17. Geusens, P., Hochberg, M.C., van der Voort, D.J., Pols, H., van der Klift, M., Siris, E., Melton, M.E., Turpin, J., Byrnes, C., Ross, P.: Performance of risk indices for identifying low bone density in postmenopausal women. Mayo. Clin. Proc. 77(7), 629–637 (2002)
18. Weinstein, L., Ullery, B.: Identification of at-risk women for osteoporosis screening. Am. J. Obstet. Gynecol. 183(3), 547–549 (2000)
19. McLeod, K.M., Johnson, C.S.: Identifying Women with Low Bone Mass: A Systematic Review of Screening Tools. Geriatric Nursing 30(3), 164–173 (2009)
20. Anastassopoulos, G., Mantzaris, D., Iliadis, L., Kazakos, K., Papadopoulos, H.: Osteoporosis Risk Factor Estimation Using Artificial Neural Networks. Engineering Intelligent Systems 18(3/4), 205–211 (2010)

Gene Prioritization for Inference of Robust Composite Diagnostic Signatures in the Case of Melanoma

Ioannis Valavanis[1], Kostantinos Moutselos[2], Ilias Maglogiannis[3], and Aristotelis Chatziioannou[1,*]

[1] Institute of Biology, Medicinal Chemistry & Biotechnology,
National Hellenic Research Foundation, Athens, Greece
[2] Department of Biomedical Informatics, University of Central Greece, Lamia, Greece
[3] University of Piraeus, Dept. of Digital Systems, Piraeus, Greece
{ivalavan,achatzi}@eie.gr,
kmouts@ucg.gr, imaglo@unipi.gr

Abstract. An integrated dataset originating from multi-modal datasets can be used to target underlying causal biological actions that through a systems level process trigger the development of a disease. In this study, we use an integrated dataset related to cutaneous melanoma that comes from two separate sets (microarray and imaging) and the application of data imputation methods. Our goal is to associate low-level biological information, i.e. gene expression, to imaging features, that characterize disease at a macroscopic level. Using an average Spearman correlation measurement of a gene to a total of 31 imaging features, a set of 1701 genes were sorted based on their impact to imaging features. Top correlated genes, comprising a candidate set of gene biomarkers, were used to train an artificial feed forward neural network. Classification performance metrics reported here showed the proof of concept for our gene selection methodology which is to be further validated.

Keywords: multi-modal, microarray, gene, imaging feature, data imputation, correlation, artificial neural network.

1 Introduction

The use of biomedical data from different sources, so called multi-modal datasets, is of known importance in the context of personalized medicine and future electronic health record management. Different data linked together can help towards a holistic approach. Especially, in cancer research data from clinical studies (age, sex, size or grade of tumor size) can be integrated with gene expression data from microarray experiments [1].

Integration can take place at different levels, e.g. across sub-systems (musculoskeletal, cardiovascular, etc.), or across temporal and dimensional scales (body, organ, tissue, cell) [2]. In the context of Virtual Physiological Human (VPH), an integrated framework should promote the interconnection of predictive models

---

* Corresponding author.

H. Papadopoulos et al. (Eds.): AIAI 2013, IFIP AICT 412, pp. 311–317, 2013.

pervading different scales, with different methods, characterized by different granularity. An integrated framework could produce system level information and enable formulation and testing of hypotheses, facilitating a holistic approach [3-4]. The framework should make it possible to interconnect predictive models defined at different scales, with different methods, and with different levels of detail, into systemic networks that provide a concretization of those systemic hypotheses [2,5].

An integrated framework studying multi-modal datasets can target underlying causal biological actions that through a systems level disease manifestation are translated to macroscopic disease related phenotypes. Motivated by this, in this study we aim to associate low-level biological information, i.e. gene expression, to imaging features using two different datasets related to cutaneous melanoma. The datasets used here come from two different sets of subjects that are described either by molecular features (gene expression) or imaging features. These sets have been previously used by authors in [3] to produce an integrated data set by applying data imputation methods to handle missing values in each of the sets. We actually re-use the produced dataset and our aim here is to find a robust gene signature that in whole influences the set of imaging features in the derived integrated dataset. We thus use spearman correlation measurements to derive the gene subsets that mostly affect the imaging features. The selected molecular features are then used to construct and evaluate artificial neural network classifiers that are trained to distinguish cutaneous melanoma cases from controls. Results show that the statistical selection of gene features using the multi-modal features' correlation can provide a robust signature that generalizes well when inputted to the classifiers.

2   Dataset

Two different datasets, one corresponding to microarray data and one to imaging data, were used. Since both sets are related to cutaneous melanoma, a brief introduction is firstly done in this section to the disease and then the two sets are described. Finally, the integrated dataset and how it has been produced is described.

2.1   Cutaneous Melanoma

Cutaneous Melanoma (CM) is considered a complex multigenic and multifactorial disease that involves both environmental and genetic factors. CM tumorigenesis is often explained as a progressive transformation of normal melanocytes to nevi that subsequently develop into primary cutaneous melanomas. The molecular pathways involved have been although little studied [6] and despite that genomic markers or gene signatures have been defined for various cancers (such as breast cancer), there has been no similar progress for malignant melanoma. Genomic studies that have been performed on CM exploit different microarray technological platforms applied in highly heterogeneous patient sets. These differences hurdle significantly comparisons, yielding cohorts of reduced total size and diversity.

Regarding the clinical diagnosis of melanoma, several approaches for analysis and diagnosis of lesions exist that use images for the analysis and diagnosis of lesions. The Menzies scale, the Seven-point scale, the Total Dermoscopy Score based on the

ABCD rule, and the ABCDE rule (Asymmetry, Border, Color, Diameter, Evolution) are some examples of these. As human interpretation of image content is fraught with contextual ambiguities, advanced computerized techniques can assist doctors in the diagnostic process [7].

2.2   Microarray Data

The microarray dataset was taken from the Gene Expression Omnibus (GEO) [8], GDS1375. RNA isolated from 45 primary melanoma, 18 benign skin nevi, and 7 normal skin tissue specimens was used for gene expression analysis, using the Affymetrix Hu133A microarray chip containing 22,000 probe sets. Signal intensities were globally scaled so that the average intensity equals 600. The gene expression values across all categories were log transformed, and the mean values of all genes in the normal skin were calculated. Subsequently, the mean gene vector concerning the normal skin categories was subtracted from all replicate vectors of the other two categories (due to log-transformation the by normal skin category was replaced with a subtraction). The initial signal intensities provided thus ratios of differential expression, calculated by dividing the signal intensities of each category by the respective gene value of the normal category. The differentially expressed gene values of the melanoma versus skin, and nevi versus skin, were then analyzed. An FDR for multiple testing adjustment, p-value 0.001 and a 2-fold change thresholds were applied and thus 1701 genes were statistically preselected.

2.3   Imaging Data

The dataset derived from skin lesion images contained 972 instances of nevus skin lesions and 69 melanoma cases. The following three types of features were analyzed: Border Features which cover the A and B parts of the ABCD-rule of dermatology, Color Features which correspond to the C rules, and Textural Features which are based on D rules [9]. A total of 31 features were produced (one feature was removed due to having zero variation across the samples). The relevant pre-processing for all features is described in [9].

2.4   Integrated Set

Microarray and imaging data sets were unified into one dataset using missing value imputation, as already described in [3]. The dataset prior to missing value imputation corresponded to a sparse matrix containing 1104 samples (benign or malignant samples, either from microarray data or imaging data) and a total of 1732 features (differential gene expression or imaging features). Prior to missing value imputation, examples originating from microarray dataset included missing values for imaging features, and examples originating from imaging dataset included missing values for gene expression measurements. A uniform missing value imputation methodology was used and a final integrated dataset (including no missing values) was produced. The procedure was followed twice, and two integrated datasets (Set 1 and Set 2) were created. One dataset (Set 1) was used for a triple scope: i) the selection of genes based on their correlation to imaging features ii) training an artificial neural network classifier to distinguish disease status (benign vs. malignant) when inputted by selected genes or alternatively by all imaging features (see Section 4) and iii) testing the classifier using 3-fold cross validation. Set 2 was used as an independent testing set for testing the classifier.

3   Methods

Using the pool of 1701 statistically pre-selected genes, we identified here the genes that are mostly correlated with the imaging features in whole. For correlation measurements, Spearman correlation was used (-1 implies negative correlation, 1 implies positive correlation). For a gene i, its correlation to an imaging feature j was calculated and marked as $Corr_{i,j}$ ($1 \leq i \leq 1701$, $1 \leq j \leq 31$, $-1 \leq Corr_{i,j} \leq 1$). The average values of absolute correlation measurements of gene i to all $N=31$ imaging features was used as a total correlation measurement ($Total\_Corr_i$) of gene i to imaging features (eq. 1). All genes were sorted in descending order according to total correlation and the most correlated genes were used as input to a feed-forward artificial neural network (ANN) that was trained and evaluated in distinguishing malignant from benign samples. Serially, the most correlated gene was used as input, than the two most correlated ones and the three most correlated ones. Top 5 and top 10 genes were also used as input, while the total set of imaging features was also used as input to the ANN for comparison reasons.

$$Total\_Corr_i = \frac{\sum_{j=1..N=31} abs(Corr_{i,j})}{N} \tag{1}$$

The ANN used here was trained using the back-propagation algorithm for 1000 epochs with a learning rate equal to 0.3 and momentum equal to 0.2. The hidden layer used sigmoid activation function and contained ((num of features+num of classes)/2 + 1) nodes. ANN was trained using Set 1 and classifier's performance in terms of total accuracy (number of samples correctly classified), and class sensitivity (number of true positives in a class that were correctly classified in this class) was measured using this set and 3-cross validation. The ANN was also trained using the whole Set 1 and was then used to classify samples in Set 2 and performance metrics were calculated as well. The classification and testing protocol was implemented within the stand-alone Rapidminer platform [10-11].

4   Results and Discussion

Table 1 presents the performance metrics (total accuracy, benign class sensitivity, malignant class sensitivity) measured when differential gene expression of various subsets of genes from the pool of 1701 were fed to ANN. Specifically, ANN was fed by the top 1, top 2, top 3, top 5 and top 10 genes according to total correlation

Table 1. Performance metrics obtained by ANN when fed by top gene(s) based on correlation to imaging features or the set of imaging features

ANN input features (gene(s) based on correlation to imaging features or the set of imaging features) (vector dimension)	Set 1 - Total Accuracy (3-cross validation)	Set 1 - Benign Class Sensitivity (3-cross validation)	Set 1 - Malignant Class Sensitivity (3-cross validation)	Set 2 – Total Accuracy	Set 2 – Benign Class Sensitivity	Set 2 – Malignant Class Sensitivity
Top 1 gene (n=1)	96.47	97.88	83.33	95.74	99.6	62.82
Top 2 genes (n=2)	98.73	99.49	92.11	98.37	99.49	88.6
Top 3 genes (n=3)	99.46	99.9	95.61	98.82	99.6	92.11
Top 5 genes (n=5)	99.95	100	95.61	99.55	99.7	98.25
Top 10 gens (n=10)	100	100	100	99.64	99.7	99.12
Gene with the median Total_Corr value (n=1)	89.58	93.13	58.77	93.48	100	36.84
Gene Least correlated (n=1)	83.79	87.98	47.37	89.67	100	0
Imaging Features (n=31)	59.69	61.62	42.98	89.67	100	0

measurements to imaging features as described above (Table 1, Rows 2-6). Performance of ANN when inputted by the worst gene according to total correlation measurement, the gene featured the median total correlation measurement (sorted as top 50% in the sorted gene list) and the total of imaging features are reported as well for comparison reasons (Table 1, Rows 7-9) .

Results show that top genes can provide very good performance metrics and when serially adding top genes performance gets better. Eventually, almost all samples can be classified correctly when top 10 genes are used and this happens also for Set 2 that was not used in training process (see further discussion below). In general, a little worse performance is obtained when ANN is evaluated in Set 2, while sensitivity measurements for malignant class are worse than the corresponding ones for benign class. This has to do with the much greater abundance of benign samples in the integrated dataset. The performance obtained when genes less correlated to imaging features are fed to ANN are much lower, showing the proof of concept for selecting gene features by taking into account their impact to imaging features. Results in Table 1 show also that the performance metrics obtained here by the top genes in terms of their correlation to imaging features are much higher than the ones obtained when imaging features are fed to the ANN. This shows that selected genes, actually being involved in the biological actions beneath melanoma phenotype, could comprise a molecular signature and a potential set of molecular biomarkers/predictors for the disease. This feature set describing low-level biomedical information seems to perform better than the set of macroscopic imaging features, but of course this is to be cross-validated by further tests.

It is to be noted that the high ANN performances observed here are possibly a result of the data imputation scheme in conjunction to the formation of the specific dataset. This has to do with the fact that similar patterns of features may exist within Set 1 and Set 2 or across these two sets, since missing data imputation has taken place to a great extent as regards the signal population of the integrated dataset (features and disease phenotype). This could not be avoided since the integrated dataset has originated from two separate datasets (microarray and imaging), while a multi-modal dataset based on a single set of subjects forming an epidemiological cohort yet remains elusive to the best of our knowledge. However, further cross-validation tests and the application of more missing value imputation methods represent tangible goals for future work.

References

1. Martin, C., grosse Deters, H., Nattkemper, T.W.: Fusing Biomedical Multi-modal Data for Exploratory Data Analysis. In: Kollias, S.D., Stafylopatis, A., Duch, W., Oja, E. (eds.) ICANN 2006. LNCS, vol. 4132, pp. 798–807. Springer, Heidelberg (2006)
2. Viceconti, M., Clapworthy, G., Testi, D., Taddei, F., McFarlane, N.: Multimodal fusion of biomedical data at different temporal and dimensional scales. Comp. Mtds. and Progs. Biomed. 102(3), 227–237 (2010)
3. Moutselos, K., Chatziioannou, A., Maglogiannis, I.: Feature Selection Study on Separate Multi-modal Datasets: Application on Cutaneous Melanoma. AIAI (2), 36–45 (2012)

4. Fenner, J.W., Brook, B., Clapworthy, G., Coveney, P.V., Feipel, V., Gregersen, H., Hose, D.R., Kohl, P., Lawford, P., McCormack, K.M., Pinney, D., Thomas, S.R., Van Sint Jan, S., Waters, S., Viceconti, M.: The EuroPhysiome, STEP and a roadmap for the virtual physiological human. Philos. Transact. A Math. Phys. Eng. Sci. 366, 2979–2999 (2008)
5. STEP Consortium. Seeding the EuroPhysiome: A Roadmap to the Virtual Physiological Human (July 5, 2007), http://www.europhysiome.org/roadmap
6. Balázs, M., Ecsedi, S., Vízkeleti, L., et al.: Genomics of Human Malignant Melanoma. In: Tanaka, Y. (ed.) Breakthroughs in Melanoma Research. InTech (2011)
7. Ogorzałek, M., Nowak, L., Surowka, G., et al.: Modern Techniques for Computer-Aided Melanoma Diagnosis. In: Murph, M. (ed.) Melanoma in the Clinic - Diagnosis, Management and Complications of Malignancy. InTech (2011)
8. Barrett, T., Troup, D.B, Wilhite, S.E., et al.: NCBI GEO: archive for functional genomics data sets - 10 years on. Nucleic Acids Res. 39(Database issue), D1005–D1010 (2011); Chevalier, R.L.: Obstructive nephropathy: towards biomarker discovery and gene therapy. Nat. Clin. Pract. Nephrol. 2(3), 157–168 (2006)
9. Maragoudakis, M., Maglogiannis, I.: Skin lesion diagnosis from images using novel ensemble classification techniques. In: 10th IEEE EMBS International Conference on Information Technology Applications in Biomedicine, Corfu, Greece (2010)
10. Mierswa, I., Wurst, M., Klinkenberg, R., Scholz, M., Euler, T.: YALE: Rapid Prototyping for Complex Data Mining Tasks. In: Proceedings of the 12th ACM SIGKDD International Conference on Knowledge Discovery and Data Mining, KDD-2006 (2006)
11. http://rapid-i.com/

On Centered and Compact Signal and Image Derivatives for Feature Extraction

Konstantinos K. Delibasis[1], Aristides Kechriniotis[2], and Ilias Maglogiannis[3]

[1] Univ. of Central Greece, Dept. of Computer Science and Biomedical Informatics,
Lamia, Greece
[2] TEI of Lamia, Dept. of Electronics, Lamia, Greece
[3]University of Piraeus, Dept. of Digital Systems, Piraeus, Greece
kdelibasis@yahoo.com; kechrin@teilam.gr; imaglo@unipi.gr

Abstract. A great number of Artificial Intelligence applications are based on features extracted from signals or images. Feature extraction often requires differentiation of discrete signals and/or images in one or more dimensions. In this work we provide two Theorems for the construction of finite length (finite impulse response -FIR) masks for signal and image differentiation of any order, using central differences of any required length. Moreover, we present a very efficient algorithm for implementing the compact (implicit) differentiation of discrete signals and images, as infinite impulse response (IIR) filters. The differentiator operators are assessed in terms of their spectral properties, as well as in terms of the performance of corner detection in gray scale images, achieving higher sensitivity than standard operators. These features are considered very important for computer vision systems. The computational complexity for the centered and the explicit derivatives is also provided.

Keywords: Signal analysis, computer vision, image feature extraction, signal discrete derivatives, centered image derivatives, compact (implicit) image derivatives.

1   Introduction

Analysis and feature extraction of discrete signals and images is often an integral part of any signal or image based artificial intelligence system. Local features in signals and images are very frequently extracted using information about the derivatives of the single or multidimensional signal. For instance, image edge enhancement is based on derivatives and allows better segmentation and pattern recognition results for computer vision systems. The Harris corner detection [1] is a simple and reliable image feature point detector that uses the image gradient information. A number of quite popular feature extraction methods [2-4] utilize the image Hessian matrix, which is computed for each image pixel using first and second order partial image derivatives. The Scale Invariant Feature Transform (SIFT) of an image [5] approximates the required image Laplacian operator with difference of Gaussians (DoG).

The derivatives of any given sequence of numbers may be approximated by convolution of one-dimensional masks using centered finite differences [6]. Most image

H. Papadopoulos et al. (Eds.): AIAI 2013, IFIP AICT 412, pp. 318–327, 2013.
© IFIP International Federation for Information Processing 2013

processing textbooks provide a set of small size two-dimensional masks, such as the Robert's or Sobel masks which are convolved with the image to provide its two partial derivatives (with respect to rows or columns). The use of the Laplacian, as well as the Laplacian of Gaussian operator (LoG) operator is very common for second order derivatives. However, as the size of the convolving mask increases, the more accurate the numerical approximation becomes.

The contribution of this work concerns the provision of an algorithm for calculating the masks to be used for approximating the derivatives of any order applied on a discrete signal/image. The size of the convolving mask may be arbitrarily chosen. We also consider the use of compact or implicit derivatives, as described in the work of Lele [7]. The advantage of implicit derivatives is that they achieve better spectral properties the centered differences, while they require less support points. Although compact derivatives have been applied to areas of numerical analysis, their application in image processing has been only recently investigated [8]. In this work we propose an implementation scheme based on an IIR prefiltering step as described in Unser et al [9, section IIA]. The proposed scheme is computationally more efficient than the standard matrix formulation, as it will be discussed in detail.

2   Proposed Methodology

In this section, we present the theorems for generating the finite impulse response (FIR) filters for signal derivatives of odd and even order. These filters are essentially symmetric finite differences; therefore we will use the terms, central, centered of FIR interchangeably. We will also present the Equations for generating the 1st and 2nd order compact derivatives. A very fast numerical implementation of central derivatives is also presented.

2.1   Central Derivatives

Consider the $n \times n$ matrices O_n, E_n defined by

$$
O_n = \begin{bmatrix} 1 & 2 & \cdots & n \\ 1 & 2^3 & \cdots & n^3 \\ \vdots & & & \\ 1 & 2^{2n-1} & & n^{2n-1} \end{bmatrix} \quad E_n = \begin{bmatrix} 1 & 2^2 & \cdots & n^2 \\ 1 & 2^4 & \cdots & n^4 \\ \vdots & \vdots & & \\ 1 & 2^{2n} & & n^{2n} \end{bmatrix} \tag{1}
$$

For $i = 1,...,n$; $k = 1,..., n-1$ we denote by $DO_{i,k}$ and $DE_{i,k}$ the determinant of order $n-1$, which follows by deleting the ith row, by replacing the kth column with the nth column and then deleting the nth column of O_n and E_n respectively. Further, we denote by $DO_{i,n}$ and $DE_{i,n}$ the negative of the determinant of order $n-1$, obtained by deleting the ith row and nth column of O_n and E_n respectively.

Example. For $n=5$, we present the 5x5 matrix O_5 according to (1) and the exemplar matrices $DO_{2,4}$ and $DO_{2,5}$:

$$O_5 = \begin{bmatrix} 1 & 2 & 3 & 4 & 5 \\ 1 & 2^3 & 3^3 & 4^3 & 5^3 \\ 1 & 2^5 & 3^5 & 4^5 & 5^5 \\ 1 & 2^7 & 3^7 & 4^7 & 5^7 \\ 1 & 2^9 & 3^9 & 4^9 & 5^9 \end{bmatrix}, \quad DO_{2,4} = \begin{vmatrix} 1 & 2 & 3 & 5 \\ 1 & 2^5 & 3^5 & 5^5 \\ 1 & 2^7 & 3^7 & 5^7 \\ 1 & 2^9 & 3^9 & 5^9 \end{vmatrix}, \quad DO_{2,5} = -\begin{vmatrix} 1 & 2 & 3 & 4 \\ 1 & 2^5 & 3^5 & 4^9 \\ 1 & 2^7 & 3^7 & 4^9 \\ 1 & 2^9 & 3^9 & 4^9 \end{vmatrix}$$

We can now present the theorems for approximating the odd and even-order derivative of any given sequence.

Theorem 1. Given a real sequence $f(x)$, its n-point derivative $f^{(2i-1)}(x)$ of any odd order $(2i-1)$ with $i=1,....,n-1$, is given by

$$f^{(2i-1)}(x) = \frac{(2i-1)!}{2h^{2i-1}\sum_{k=1}^{n} k^{2i-1} DO_{i,k}} \sum_{k=1}^{n} DO_{i,k} \left(f(x+kh) - f(x-kh) \right) + O\left(h^{2n+1}\right) \quad (2)$$

Theorem 2. Given a real sequence $f(x)$, its n-point derivative $f^{(2i)}(x)$ of any even order $(2i)$ with $i=1,....,n-1$, is given by:

$$f^{(2i)}(x) = \frac{(2i)!}{2h^{2i}\sum_{k=1}^{n} k^{2i} DE_{i,k}} \left[\frac{1}{2} \sum_{k=1}^{n} DE_{i,k} \left(f(x+kh) - f(x-kh) \right) + f(x) \sum_{k=1}^{n} DE_{i,k} \right] + O\left(h^{2n+1}\right) \quad (3)$$

Assuming that $f(x)$ is sampled with a constant sampling frequency, we may set $h=1$. Now, equations (2) and (5) can be written as linear convolution:

$$f^{(2i-1)}(x) = (f * M_{odd})(x) \quad (4)$$

$$f^{(2i)}(x) = (f * M_{even})(x) \quad (5)$$

where $M_{even}(k)$ and $M_{odd}(k)$ for $k=-n,...,n$ are defined as

$$M_{even}(k) = \frac{(2i)! DE_{i,k}}{2\sum_{k=1}^{n} k^{2i} DE_{i,k}} \left[-1, -1, ..., -1, \frac{1}{2}, 1, ..., 1, 1 \right] \quad (6)$$

$$M_{odd}(k) = \frac{(2i-1)! DO_{i,k}}{2\sum_{k=1}^{n} k^{2i-1} DO_{i,k}} \left[-1, -1, ..., -1, 0, 1, ..., 1, 1 \right] \quad (7)$$

The one-dimensional masks for the first and second order derivatives, created by Equations (6) and (7) using up to 11 points ($n=5$), are provided in Table 1, in rational form. The 3x3 Sobel operator can be generated by the following linear convolution, by exploiting its separability property, using the first mask of Table 1, as $[1/2 \ \ 0 \ \ -1/2]*[1 \ \ 2 \ \ 1]^T$. The well-known Laplacian operator can be generated by the following linear convolution: $[1 \ \ -2 \ \ 1]*[1 \ \ -2 \ \ 1]^T$, using the 3-point 2nd order derivative mask from Table 1.

Table 1. The resulting masks M_{even} and M_{odd} for first and second order differentiation

order of deriv.	n	Num. of points in Mask	Mask
1	1	3	$[1/2, 0, -1/2]$
1	2	5	$[-1, 8, 0, -8, 1]/12$
1	3	7	$[1, -9, 45, 0, -45, 9, -1]/60$
1	4	9	$[-1/280, 4/105, -1/4, 0, 1/4, -4/105, 1/280]$
1	5	11	$[4, -5, +30, -120, +430, 0, -430, +120, -30, 5, -4]/504$
2	1	3	$[1, -2, 1]$
2	2	5	$[-1/12, 4/3, -2.5, 4/3, -1/12]$
2	3	7	$[1/90, -3/20, 1.5, -49/18, 1.5, -3/20, 1/90]$
2	4	9	$[-1/560, 8/315, -1/5, 8/5, 205/144, 8/5, -1/5, 8/315, -1/560]$

2.2   Compact (Implicit) Derivatives

Let us denote by $f_n = f(n)$ the values of a discrete signal with n integer and by f' the values of the signal's derivative of the first order. According to Eq. (2.1) in the work of Lele [7], the following holds:

$$q_2 f'_{i-2} + q_1 f'_{i-1} + f'_i + q_1 f'_{i+1} + q_2 f'_{i+2} = \left(\frac{c_3}{6}, \frac{c_2}{4}, \frac{c_1}{2}, 0, -\frac{c_1}{2}, -\frac{c_2}{4}, -\frac{c_3}{6} \right) * f \quad (8)$$

where q_1, q_2 and c_1, c_2, c_3 are coefficients, whose value defines the order of accuracy of the 1st derivative approximation and the symbol $*$ stands for linear convolution. The order of accuracy for derivative approximation is not to be confused with the order of the derivative itself. The second order derivative f''_i can also be described in manner similar to (8), according to (2.2) in [7], for the one-dimensional case:

$$q_2 f''_{i-2} + q_1 f''_{i-1} + f''_i + q_1 f''_{i+1} + q_2 f''_{i+2} = \left(\frac{c_3}{9}, \frac{c_2}{4}, \frac{c_1}{2}, -2\left(\frac{c_3}{9} + \frac{c_2}{4} + \frac{c_1}{2} \right), \frac{c_1}{2}, \frac{c_2}{4}, \frac{c_3}{9} \right) * f \quad (9)$$

According to Equations (8), (9), the values of f' and f''_i are not obtained explicitly with a simple convolution (as in the case of centered difference (4), (5)), thus the term *implicit*. More specifically, the right part of (8) and (9) is a linear convolution with a symmetric kernel of up to 7 points, whereas the left hand side involves inverse convolution with a symmetric kernel $Mask_1$ of up to 5 points, where $Mask_1[k]=(q_2, q_1, 1, q_1, q_2)$, $k=-2,-1,\ldots,2$.

The 6th, 8th and 10th order of accuracy of the 1st derivative can be obtained by setting the q_1, q_2 and c_1, c_2, c_3 parameters in (8) according to (2.1.12) and (2.1.14) in Lele's work [7]. Similarly, the 4th, 6th, 8th and 10th order of accuracy of the 2nd derivative can be obtained by setting the q_1, q_2 and c_1, c_2, c_3 parameters in (9) according to (2.2.7) - (2.1.11) in [7]. All the necessary parameters in (8) and (9) are provided in Table 2.

Table 2. The values of the parameters of Eq.(8) and Eq.(9) for the 1st and 2nd order derivative

Parameters	1st order Derivative (8) with order of accuracy			2nd order Derivative (9) with order of accuracy			
	6	8	10	4	6	8	10
q_1	1/3	4/9	1/2	0	2/11	344/1179	334/899
q_2	0	1/36	1/20	0	0	23/2358	43/1798
c_1	14/9	40/27	17/12	4/3	12/11	320/393	1065/1798
c_2	1/9	25/54	101/105	-1/3	3/11	310/393	1038/899
c_3	0	0	1/100	0	0	1/213	79/1798

In matrix form, the problem obtaining the required values f' and f_i'' from (8) and (9) respectively, is reformulated as solving a linear system of simultaneous equations, whose matrix is symmetric pentadiagonal. Thus, it can be solved efficiently using Gauss elimination with complexity $O(10N)$ [10]. The contribution of this work concerning implicit derivatives, is the implementation based on the Infinite impulse response (IIR) *prefiltering* algorithm described in the work of Unser et al [9, section IIA]. This algorithm was proposed as a very efficient implementation of the b-spline interpolation in signals and images and as it will be shown, it achieves complexity of $O(8N)$. According to this approach, the left hand side of (8) and (9) may be written as filtering with the following transfer function,

$$H(z) = \frac{1}{q_2 z^2 + q_1 z + 1 + q_1 z^{-1} + q_2 z^{-2}} \tag{10}$$

The roots of the denominator in (11) are real pairs of reciprocals: $\rho_1, \rho_1^{-1}, \rho_2, \rho_2^{-1}$ with $|\rho_1|, |\rho_2| < 1$. In this case, the algorithm (2.5) in [9] may be applied directly, according to which, the calculation of the unknown f' and f_i'' is performed as a cascade of a first order causal and an anti-causal linear filter with infinite impulse response (IIR). Let us denote by $x(k)$ the signal/image values (which is considered as input) and by $y^+(k)$ and $y(k)$ the output of the causal and the anti-causal part of (10), respectively (see Fig. 1). The linear difference equations that implements the $H(z)$ is shown in two steps (forward and inverse):

$$y^+(k) = x(k) + \rho_i y^+(k-1), k = 2,3,...,K$$
$$y(k) = \rho_i \left(y(k+1) - y^+(k) \right), k = K-1,...,1 \tag{11}$$

The necessary initial conditions for (11) are:

$$y^+(1) = \sum_{k=1}^{K_0} z_i^{|k-1|} x(k), y(K) = -\frac{z_i}{1-z_i^2}(2y^+(K)-x(K))$$ (12)

where K_0 a small positive integer [9]. Equations (11) and (12) are repeated for $i=1,2$, which correspond to the two roots $|\rho_1|,|\rho_2|$ of the denominator of (10).

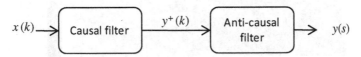

Fig. 1. The implementation of the implicit derivative as an IIR filter, decomposed into a cascade of a causal (forward) and an anticausal (inverse) filters

The application of image partial derivatives takes place in a straightforward manner, applying the one-dimensional derivative approximation along image lines and subsequently along columns. For instance, in order to calculate the quantity $\dfrac{\partial^3 I}{\partial x \partial y^2}$ (image 2nd partial derivative along rows and 1st partial derivative along columns), we apply the one-dimensional calculation of first order derivative along each image line and subsequently the result is used as input column-wise into the 1-D approximation of the second order derivative.

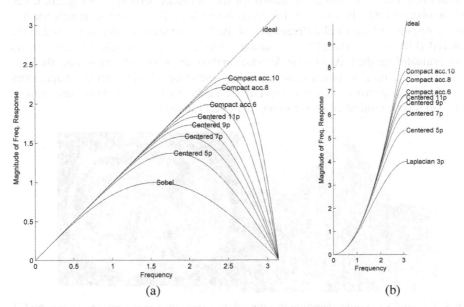

Fig. 2. Magnitude of Frequency response of the 1st and 2nd order central derivatives and compact derivatives (the order of accuracy for the later is also provided)

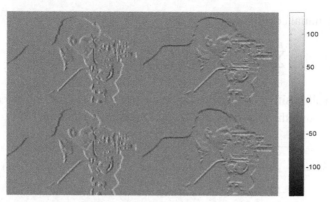

Fig. 3. The gradient along columns and lines of the well-known cameraman image, using the implicit (order of accuracy=8) and the centered (7-point) filters

3   Experimental Results

The frequency response of the central derivatives with 3 (Sobel), 5, 7, 9 and 11 points (equivalently $n=2, 3, 4$ and 5 in (4)) is show in Fig. 2(a). In the same Figure, the frequency response of the compact derivative with order of accuracy equal to 6, 8 and 10 (equivalently 5, 5 and 7-point stencil) is also shown. It is clear that the central difference (FIR) derivatives follow closer the ideal differentiator as the number of points in the finite length mask increases. The compact derivatives exhibit better spectral properties than the central ones. The same results are shown for the 2nd order derivative in Fig. 2b. It can also be observed that the compact (IIR) derivative filter follows more accurately the frequency response of the ideal differentiator of the first and second order, compared to the centered (FIR) filters. The FIR convolution masks as well as the IIR derivative filters were evaluated for the tasks of edge detection and corner points extraction using the Harris detector [1]. Initial results, assessed visually, were obtained using various images from [http://www.imageprocessingplace.com/DIP-3E/dip3e_book_images_downloads.htm] and from routine clinical computer tomography (CT) studies.

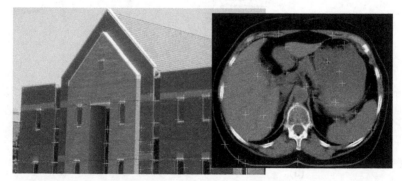

Fig. 4. The results of corner detection in a typical gray scale photograph and in a transverse CT of the abdomen, using the implicit derivative mask with order of accuracy=10 (1st raw), the 7-point FIR derivative mask generated by the proposed algorithm (2nd raw) and the 7-point derivative proposed in [12] (3rd raw)

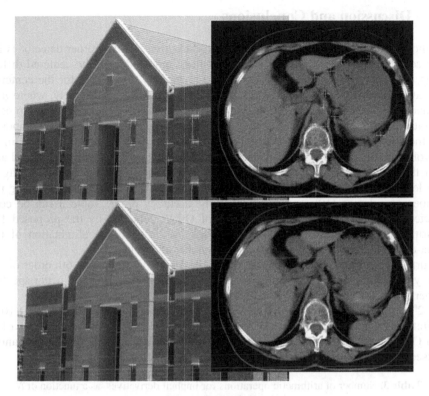

Fig. 4. (*Continued.*)

In Fig. 3 the first order differentiator filters (first row: implicit with 8th order of accuracy, second row: 7-point centered) are applied for gradient calculation for the well-known cameraman image. Figure 4 shows the result of the application of the Harris corner detection [1], on two different gray scale images: a photograph of a building (routinely used for algorithm evaluation) and a typical clinical transverse slice of a CT abdomen study. The Harris corner detection is a simple and popular method for discovering salient points in images. Salient points are very useful for performing image-based robot localization, image spatial registration, scene recognition etc. The Harris corner detection is based on the image Hessian matrix, whose 1st and 2nd order derivatives are calculated using the implicit derivative mask with order of accuracy=8 (1st raw), the 7-point FIR derivative mask generated by the proposed algorithm (2nd raw) and the 7-point derivative proposed in [12] (3rd raw). The source code provided in [11] is used for the Harris detector. As it can be observed, the detector using the IIR-based implicit derivative is significantly more sensitive to detecting corners in images than the 7-point FIR differentiator generated by (7), whereas the detector based on the 7-point differentiator [12] is the least sensitive.

4   Discussion and Conclusions

As shown in the previous section (Fig. 4), the sensitivity of the corner detector is increased when using the IIR differentiation filter, compared with the centered differences (FIR) filter. The number of arithmetic operations per sample for the centered derivative filters of even and odd order is $2(n+1)$ and $2n+1$ respectively, where n is defined as in Theorem 1 and 2 (the symmetry of the masks has been taken into consideration). The computational complexity of the implicit derivatives is decomposed into the cost of the left hand side (implicit part) and the right hand side (explicit linear convolution) of (8) and (9). The former is 2 arithmetic operations for the forward and 2 for the inverse filtering for each root of the denominator (pole) in (10) with magnitude less than 1, for each image pixel (for each image dimension). By observing the number of poles of (10) as a function of the accuracy order, Table 3 can be constructed. Thus, the arithmetic complexity of $O(8N)$ achieved by the proposed IIR algorithm is superior to the $O(10N)$ achieved by the Gaussian elimination of the penta-diagonal matrix, with N denoting the number of pixels.

Furthermore, the 1st order compact (implicit) derivative filter with 6th order of accuracy with computational complexity of $O(9N)$, exhibits significantly better spectral properties than the 11-point FIR differentiator with $O(11N)$, as depicted in Fig. 2(a). The 2nd order implicit differentiator with order of accuracy equal to 6 with $O(10N)$ exhibits marginally better spectral properties, compared to 11-point centered filter with $O(12N)$. Finally, it has to be reminded that the IIR differentiation filters cannot be used in real-time signal processing, due to their non-causal component.

Table 3. Number of arithmetic operations for implicit derivatives, as a function of N

	1st order Derivative with order of accuracy			2nd order Derivative with order of accuracy			
	6	8	10	4	6	8	10
Left side of (8) and (9)	$4N$	$8N$	$8N$	0	$4N$	$8N$	$8N$
Right side of (8)	$5N$	$5N$	$7N$	N/A			
Right side of (9)	N/A			$6N$	$6N$	$8N$	$8N$
Total	$9N$	$13N$	$15N$	$6N$	$10N$	$16N$	$16N$

Appendix

We will present the proof for the odd-order derivatives. The proof for the even-order is in total analogy.

Proof of Theorem 1

Using the Taylor's Theorem with an integral form of remainder we get

$$f(x+t) = \sum_{j=1}^{2n} \frac{t^j f^{(j)}(x)}{j!} + R_{2n}(t)$$

where $R_{2n}(t) = \frac{1}{(2n+1)!} \int_{x}^{x+t} f^{(2n+1)}(u)(x+t-u)^{2n} du$

We apply the previous relation $2n$ times for $t=kh$, $k=\pm1$, ±2, ...,$\pm n$. Then for any real constants c_1, ..., c_n we get

$$\frac{1}{2}\sum_{k=1}^{n}c_k\left(f\left(x+kh\right)-f\left(x-kh\right)\right)=\sum_{j=1}^{n}\frac{h^{2j-1}f^{(2j-1)}\left(x\right)}{\left(2j-1\right)!}\sum_{k=1}^{n}c_k k^{2j-1}+\frac{1}{2}\sum_{k=1}^{n}c_k\left(R_{2n}\left(kh\right)-R_{2n}\left(-kh\right)\right)$$

If we choose $c_n=-1$ and find c_k for $k=1,2,...,n-1$ by solving the following system of linear equations, $\sum_{k=1}^{n}c_k k^{2i-1}=n^{2j-1}$, $j=1,2,...,n, i\neq j$, we obtain the required result.

References

1. Harris, C., Stephens, M.: A combined corner and edge detector. In: ALVEV Vision Conference, pp. 147–151 (1988)
2. Lindeberg, T., Garding, J.: Shape-adapted smoothing in estimation of 3-D shape cues from affine deformations of local 2-D brightness structure. Image and Vision Computing 15(6), 415–434 (1997)
3. Mikolajczyk, K., Schmid, C.: Scale and affine invariant interest point detectors. International Journal of Computer Vision 1(60), 63–86 (2004)
4. Bay, H., Ess, A., Tuytelaars, T., Gool van, L.: Speeded-up robust features (SURF). International Journal on Computer Vision and Image Understanding 110(3), 346–359 (2008)
5. Lowe, D.: Distinctive Image Features from Scale-Invariant Keypoints. International Journal of Computer Vision 60(2), 91–110 (2004)
6. Keller, H.B., Pereyra, V.: Symbolic Generation of Finite Difference Formulas. Mathematics of Computation 32(144), 955–971 (1978)
7. Lele, S.K.: Compact difference Schemes with Spectral-like Resolution. Journal of Computational Physics 103, 16–42 (1992)
8. Belyaev, A.: On implicit image derivatives and their applications. In: Hoey, J., McKenna, S., Trucco, E. (eds.) BMVC 2011, Dundee, Scotland, UK (2011)
9. Unser, M., Aldroubi, A., Eden, M.: B-spline signal processing: Part II - Efficient Design and Applications. IEEE Trans. Sighal Process 41(2), 834–848 (1993)
10. Benkert, K., Fischer, R.: An Efficient Implementation of the Thomas-Algorithm for Block Penta-diagonal Systems on Vector Computers. In: Shi, Y., van Albada, G.D., Dongarra, J., Sloot, P.M.A. (eds.) ICCS 2007, Part I. LNCS, vol. 4487, pp. 144–151. Springer, Heidelberg (2007)
11. MATLAB and Octave Functions for Computer Vision and Image Processing, http://www.csse.uwa.edu.au/~pk/research/matlabfns/
12. Farid, H., Simoncelli, E.: Differentiation of Discrete Multi-Dimensional Signals. IEEE Trans. Image Processing 13(4), 496–508 (2004)

Detecting Architectural Distortion in Mammograms Using a Gabor Filtered Probability Map Algorithm

O'tega Ejofodomi, Michael Olawuyi, Don Uche Onyishi, and Godswill Ofualagba

Department of Electrical Engineering, Federal University of Petroleum Resources, P.M.B. 1221, Effurun, Nigeria
tegae@yahoo.com

Abstract. Breast Cancer is a disease that is prevalent in many countries. Computer-Aided detection (CAD) systems have been developed to assist radiologists in detecting breast cancer. This paper discusses an algorithm for architectural distortion (AD) detection with a better sensitivity than the current CAD systems.

19 images containing ADs were preprocessed with a median filter and Gabor filters to extract texture information. AD probability maps were generated using a maximum amplitude map and histogram analysis on the orientation map of the Gabor filter response. AD maps were analyzed to select ROIs as potential AD sites.

AD map analysis yielded a sensitivity of 79% (15 out of 19 cases of AD were detected) with a false positive per image (FPI) of 18. Future work involves the development of a second stage in the algorithm to reduce the FPI value and application of the algorithm to a different set of database images.

Keywords: Breast Cancer, Architectural Distortion, Gabor Filters.

1 Introduction

Breast cancer is a disease that is prevalent in many countries. In the United States of America, breast cancer is the most common type of cancer, side from skin cancer. The American Cancer Society estimates that there will be 234, 580 new cases of breast cancer (232,340 female, 2,240 male) and that 40,030 people (39,620 female, 410 male) will die from breast cancer in 2013 [1]. In the United Kingdom, it is estimated that there were 49,961 new cases (49,564 female, 397 male) of breast cancer, and 11, 633 deaths (11,556 female, 77 male) related to breast cancer in 2010 [2]. In some countries, the statistics for breast cancer is unreliable.

The current gold standard for detecting breast cancer is screening mammography, which is X-ray imaging of the breast. Mammography can often detects breast cancer at an early stage, when treatment is more effective and a cure is more likely, especially for women over 50 years. This reduces the mortality rate of the disease. Numerous studies have shown that early detection with mammography saves lives and increases treatment options [1, 3-4]. In mammography, two types of X-ray images are usually obtained: the Mediolateral Oblique (MLO) view, which is a sideways view of the breast, and the Craniocaudal view (CC), which is a top to bottom view of the breast.

H. Papadopoulos et al. (Eds.): AIAI 2013, IFIP AICT 412, pp. 328–335, 2013.

Radiologists search each image for signs of lesions and abnormalities that may represent breast cancer. Typically, they search for masses, calcifications and architectural distortions. Breast lesions are described and reported according to the Breast Imaging Reporting and Data System (BIRADS ™), which is a terminology developed by the American College of Radiology (ACR), for the description of mammographic lesions [5]. A mass is defined as a space-occupying lesion seen in at least two different projections. Calcifications are calcium deposits in the breasts that show up as white, bright spots in mammograms. Architectural distortions (AD) are abnormal breast lesions in which the breast parenchyma is distorted with no visible mass. The distortion includes spiculations radiating from a point, and focal retraction or distortion of the edge of the parenchyma [5].

Breast cancer detection using mammograms is far from perfect. It is estimated that radiologists fail to detect about 10-30% of breast cancers [6-8]. Computer-Aided Detection (CAD) systems has been developed to enable radiologists detect lesions that may be indicative of cancer. These systems serve as support to the radiologist and are not used by themselves as a breast cancer detection tool. Commercial CAD systems accurately identified 90% cases of masses and microcalcifications [9, 10]. However, the sensitivity for AD is low. The R2 Image Checker system successfully identified 49% cases containing ADs with 0.7 false positives per image. The CADx Second Look system [9] successfully detected 33% cases of ADs with 1.27 false positives per image [11].

There is need to improve the detection of ADs in CAD systems. The goal of this work is to develop an algorithm that can detect ADs with higher sensitivity than the currently existing CAD systems. Details of the algorithm are presented in the methods section. The results of the algorithm are shown in the results section. Analysis of the performance of the algorithm is provided in the discussion section. Future improvements to the algorithm are highlighted in the conclusion section.

2 Materials and Methods

2.1 Materials

All of the images used in this study were obtained from the Mammographic Image Analysis Society (MIAS) Digital Mammogram Database [12]. All the images were digitized at 50 micron pixel edge. The sizes of all images were 1024 by 1024 pixels. There were a total of 19 images containing ADs. The pixel coordinates of the AD sites, the pathology of the AD, and the nature of the background breast tissue are also provided. Image views were both MLO and CC views of the breast. These 19 images were used in this pilot study. Figure 1 shows a MIAS image with a circle marking the site of the AD. The ground truth provided by the database was used in determining the effectiveness of the algorithm.

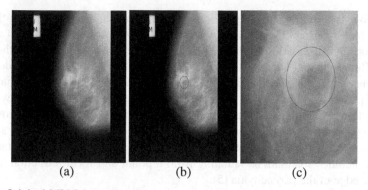

(a)   (b)   (c)

Fig. 1. (a) Original MIAS image (b) Ground truth: Original image with a boundary containing AD overlaid. (b) Enlarged ground truth image.

2.2   Methods

Most of the reported automated detection of AD in literature consists of several steps. There may be some preprocessing to remove noise from the mammograms and to select the breast region within the mammogram. The next stage involves the extraction of the texture orientation (line structures oriented at different angles within the image). The texture orientation is then analyzed to produce regions of interest (ROI) which are potential AD sites. Estimation of the characteristic features of ADs are then used to narrow down the number of potential AD sites, thereby reducing the number of false positives in automated detection. The methodology employed in this work consists of four major steps: preprocessing, Gabor filtering to extract texture information, generation of AD maps, and finally, analysis of AD maps to select ROIs. The flow chart for the algorithm is given in Figure 2.

2.2.1   Image Preprocessing

The images were filtered with a median filter to reduce the noise in the image. No region growing methods were employed to segment the breast region from the background.

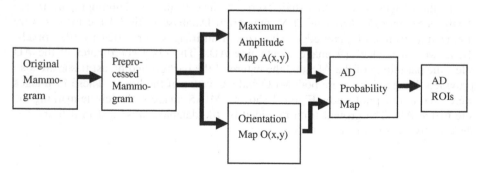

Fig. 2. Flow Chart for AD probability map algorithm

2.2.2   Gabor Filtering

There are many different techniques employed to extract the texture orientation from an image. The method used in this study is the Gabor Filters. The Gabor filter is a sinusoidally modulated Gaussian function. Mathematically, a 2D Gabor function g, is the product of a 2D Gaussian and a complex exponential function. The general expression is given by:

$$g_{\theta, \lambda, \sigma1, \sigma2}(x, y) = \exp -1/2\{x\ y\}M\ \{x\ y\}^T \exp\{\underline{j\pi} (x\cos\theta + y\sin\theta) \qquad (1)$$
$$\lambda$$

where $M = \text{diag}(,\sigma_1^{-2}, \sigma_2^{-2})$. The parameter θ represents the orientation, λ is the wavelength, and σ_1 and $\sigma2$ represent scale at orthogonal directions. When the Gaussian part is symmetric, we obtain the isotropic Gabor function:

$$g_{\theta, \lambda, \sigma}(x, y) = \exp\{-x2 + y2/2\sigma^2)\exp\ \{\ \underline{j\pi} (x\cos\theta + y\sin\theta) \qquad (2)$$
$$\lambda$$

A bank of Gabor filters was applied to the images. The angle of orientation in the bank of filters was varied from $\theta \in [-\pi/2; \pi/2]$, the frequency of sinusoidal function was set to $f = 0.022$, and the variance S was set to 0.05. These parameters were chosen experimentally. A total of 19 Gabor filters were applied to the images.

2.2.3   AD Map Generation

Two maps were created from the output of the Gabor filtration: a maximum amplitude map A(x, y) and a dominant orientation map O(x,y). The maximum amplitude for each point (x,y) in the responses to the Gabor filters was extracted to form the amplitude map A(x,y). The angle of orientation of the maximum amplitude was considered to be the dominant angle and was used in creating the orientation map O(x,y). Both maps are used to create the final Architectural Distortion map AD(x,y).

To create the AD map, histogram analysis was performed on O(x,y). Both the amplitude and the orientation map were divided into blocks of 16 by 16 pixels. The maximum amplitude for each block was obtained from A(x,y). Histogram analysis was performed on each block of O(x,y). It is expected that blocks containing ADs should show several dominant peaks (corresponding to dominant orientation angles), while blocks of normal tissue should only have one dominant peak. Therefore, in the histogram analysis an experimentally determined threshold T_1 was set to determine if a peak is dominant or not. If a peak was within a certain percentage of the maximum orientation angle, it was classified as a dominant orientation angle.

The AD map was created by multiplying the maximum grayscale amplitude in block with the number of dominant angles and then dividing by a characteristic number N. The characteristic number was established experimentally. Mathematically,

$$AD_i = \underline{Dn_i * A_{max}} \qquad (3)$$
$$N$$

Where i is the index number of the block, Dn is the number of dominant angles in block i, A_{max} is the maximum grayscale amplitude in block i, and N is the characteristic number. AD is a value that quantifies the probability of the presence of AD in block i [13].

2.2.4   AD Map Analysis

In the final stage, the AD maps were analyzed. The blocks with the highest AD values were selected as regions of interest (ROI). The image containing ROIs was then compared with the ground truth mammogram. A true positive (TP) was recorded when a ROI was within the region identified as AD by the ground truth. A false positive (FP) was recorded when a ROI was not within the region identified as AD by the ground truth.

3   Result

Our data consisted of the 19 images identified with AD in the MIAS database. Table 1 shows the number of true positives (TP) and false positives (FP) identified in each image. Images in which ADs were unidentified have 0 TPs. In one case (Image 18) the AD has been marked twice (2 TPs). We achieved a sensitivity of 78.9% (15 cases of AD were detected) with a false positive per image (FPI) of 18.4 (see Table 2). 10 out of 19 ADs were classified as malignant, while 9 was classified as benign AD in the ground truth data. Of the 10 malignant ADs, the AD Map analysis correctly identified 80% (8 out of 10) malignant ADs, and 78% (7 out of 9) of benign ADs. The total number of ROIs obtained was 291, of which 16 ROIs positively identified AD. The results of the AD Map analysis algorithm are compared with two commercial CAD systems in Table 2. The sensitivity for the AD Map analysis algorithm is much higher than that of the commercial CAD systems. However, the FPI is also relatively high. Most AD detection algorithms employ two stages: Identification of ROIs in the first stage and reduction of the number of FPs in the second stage [13, 14]. The AD map analysis algorithm is currently a one stage detection algorithm. A second stage would need to be included to reduce the FPI to a more reasonable number.

Table 1. True Positives (TP) and False Positives (FP) identified in Images

Image No.	No. of True Positives (TP)	No. of False positives (FP)
1	1	41
2	1	5
3	1	5
4	1	10
5	1	17
6	0	>> 41
7	1	10
8	1	17
9	1	40
10	1	40
11	1	33
12	1	13
13	1	19
14	0	>> 41
15	0	>> 41
16	1	11
17	0	>>41
18	2	7
19	1	7

Table 2. Comparison of AD Map Analysis Algorithm to Commercial CAD systems

	AD Map Analysis	Image Checker R₂	CADx Second Look
Sensitivity (%)	79	38	21
No of Images	19	45	45
FPI	18	0.7	1.27
% of malignant ADs detected	80%	-	-
% of benign ADs detected	70%	-	-
Total No of ROIs	291	-	-
Total No of ROIs with AD	16	-	-

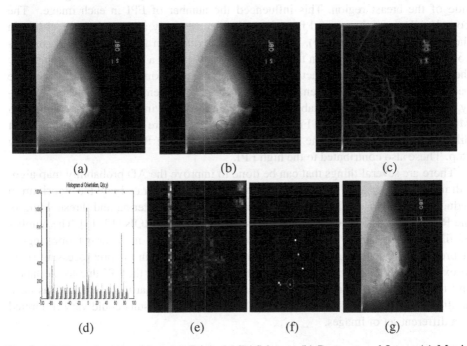

(a)   (b)   (c)

(d)   (e)   (f)   (g)

Fig. 3. AD Detection Algorithm. (a) Original MIAS image (b) Preprocessed Image (c) Maximum Amplitude Map A(x,y) (d) Orientation Map O(x,y) (e) AD map (f) ROIs with highest AD values (g) AD detection results overlaid over original image.

Figure 3 shows the outputs of the algorithm for one of the 19 cases. The original image from the MIAS database is shown in Figure 3a. The median filtered image is shown in Figure 3b. The maximum amplitude map A(x,y) obtained after Gabor filtering is shown in Figure 3c. The histogram of the orientation map O(x,y) is shown in

Figure 3d. The generated AD map is shown in Figure 3e. The selected ROI from the AD map analysis is shown in Figure 3f. Finally, the selected ROIs overlaid over the original image is shown in Figure 3f. The ground truth is also labeled in Figure 3f. In this particular case, one of the ROIs is located within the region identified as AD by the ground truth.

4   Discussion

The sensitivity of the Gabor filtered probability map algorithm is far greater than that of the two commercial CAD systems shown in Table 2. The high number of FPI can be attributed to several factors. First as was duly noted, no segmentation to separate the breast region was performed on the images in the preprocessing step. Non-identification of the breast boundary (skin-air) caused the algorithm to flag ROIs outside of the breast region. This influenced the number of FPI in each image. The boundary between breast and film showed up consistently as a bright vertical line in the AD maps (see Figure 3e). This had to be excluded in the AD map analysis. Secondly, the pectoral muscle in the MLO view shows up as a bright white region. Since the algorithm relies on the fact that AD sites possess maximum grayscale values, the pectoral muscle region is often flagged as a ROI even when AD is not present. Thirdly, images in the MIAS database contain labels identifying the image as either an MLO view or CC view (see Figure 1a). These labels were imprinted on the films at different locations and so the algorithm did not remove them in the preprocessing step. These also contributed to the high FPI.

There are several things that can be done to improve the AD probability map algorithm. Most importantly, a second stage needs to be incorporated into the algorithm to reduce the number of FPI. This can be accomplished by filtering and thresholding in the Radon Domain, feature extraction and classification of ROIs [12, 13]. The number of filters in the Gabor banks can be increased to include more orientation angles. A breast segmentation algorithm can also be implemented in the preprocessing stage to extract the breast region from the surrounding to reduce the FPI due to the labels and the breast-film boundary. Finally, it should be noted that the algorithm has been applied to a single dataset from the MIAS database. The results could vary if applied to a different set of images.

5   Conclusion

We have developed an algorithm to detect ADs in mammograms. The Gabor filtered probability map algorithm has a sensitivity of 79% and an FPI of 18, more than two times that of commercial CAD systems. Future work involves the development of a second stage in the algorithm to reduce the FPI value and application of the algorithm to a different set of database images.

References

1. American Cancer Society: Cancer Facts and Figures (2013). American Cancer Society, Inc., http://www.cancer.org (accessed February 28, 2013)
2. Cancer Research, UK., http://www.cancerresearchuk.org (accessed February 28, 2013)
3. Magnus, M.C., Ping, M., Shen, M.M., Bourgeois, J., Magnus, J.H.: Effectiveness of mammography screening in reducing breast cancer mortality in women aged 39-49 years: a meta-analysis. Journal of Women's Health 20(6), 845–852 (2011)
4. Mandelblatt, J.S., Cronin, K.A., Bailey, S., et al.: Effects of mammography screening under different screening schedules: model estimates of potential benefits and harms. Annals of Internal Medicine 151(10), 738–747 (2009)
5. American College of Radiology ACR BI-RADS - Mammography, Ultrasound & Magnetic Resonance Imaging, 4th edn. Reston, VA (2003)
6. Kerlikowske, K., Carney, P.A., Geller, B., Mandelson, M.T., Taplin, S.H., Malvin, K., Ernster, V., Urban, N., Cutter, G., Rosenberg, R., Ballard-Barbash, R.: Performance of screening mammography among women with and without a first-degree relative with breast cancer. Annals of Internal Medicine 133, 855–863 (2000)
7. Kolb, T.M., Lichy, J., Newhouse, J.H.: Comparison of the performance of screening mammography, physical examination, and breast US and evaluation of factors that influence them: an analysis of 27,825 patient evaluations. Radiology 225, 165–175 (2002)
8. Bird, R.E., Wallace, T.W., Yankaskas, B.C.: Analysis of cancers missed at screening mammography. Radiology 184, 613–617 (1992)
9. The CADx Second Look system, http://www.icadmed.com (accessed February 20, 2013)
10. The R2 Technology's Image Checker, http://www.r2tech.com (accessed February 20, 2013)
11. Sampat, M.P., Markey, M.K., Bovik, A.C.: Computer-aided detection and diagnosis in mammography. IEEE Publication (2004)
12. Suckling, J., et al.: The Mammographic Image Analysis Society Digital Mammogram Database. In: Exerpta Medica. International Congress Series, vol. 1069, pp. 375–378 (1994)
13. Jasionowska, M., Przelaskowski, A., Rutczynska, A., Wroblewska, A.: A two-step method for detection of architectural distortions in mammograms. In: Piętka, E., Kawa, J. (eds.) Information Technologies in Biomedicine. AISC, vol. 69, pp. 73–84. Springer, Heidelberg (2010)

Breast Cancer Detection in Mammogram Medical Images with Data Mining Techniques

Konstantinos Kontos and Manolis Maragoudakis

University of the Aegean,
Department of Information and Communication Systems Engineering,
Samos, Greece
{icsdm11027,mmarag}@aegean.gr

Abstract. A domain of interest for data mining applications is the study of biomedical data which, in combination with the field of image processing, provide thorough analysis in order to discover hidden patterns or behavior. Towards this direction, the present paper deals with the detection of breast cancer within digital mammography images. Identification of breast cancer poses several challenges to traditional data mining applications, particularly due to the high dimensionality and class imbalance of training data. In the current approach, genetic algorithms are utilized in an attempt to reduce the feature set to the informative ones and class imbalance issues were also dealt by incorporating a hybrid boosting and genetic sub-sampling approach. As regards to the feature extraction approach, the idea of trainable segmentation is borrowed, using Decision Trees as the base learner. Results show that the best precision and recall rates are achieved by using a combination of Adaboost and k-Nearest Neighbor.

Keywords: Image Processing, Trainable Segmentation, Data Mining, Genetic Algorithms.

1 Introduction

The term breast cancer refers to the development of malignant tumor in the breast. It is one of the most common cancers worldwide and accounts for 22.9% of all cancers (excluding non-melanoma skin cancers) in women. Caused by uncontrolled proliferation of pathological cells, breast cancer results in the formation of malignant tumor [1]. Pathological cells have the potential of spreading to adjacent tissues in hostile consequences for the entire organization. Conversely, the incidence of the disease in males is real but very small. Unfortunately, little are known about the causes of breast cancer, despite the fact that researchers have identified several risk factors, such as age, heredity, disorders of menstruation, alcohol consumption, obesity, smoking, contraceptive pills, history of cancer, exposure to radiation and sedentary life. A very important method of detecting breast cancer is mammography. The goal of mammography is the early detection of breast cancer, typically through detection of characteristic masses and/or microcalcifications. It can be done either by the classical method of radiography or digital mammography. In any case, the doctor's opinion is necessary, based on the findings, so if the results of mammography indicate a tumor further tests must be done.

H. Papadopoulos et al. (Eds.): AIAI 2013, IFIP AICT 412, pp. 336–347, 2013.

Based on the outcome of this method, this paper proposes a classification system for identifying segments of cancer cells from mammography images in order to assist professional radiologists and gynecologists in their diagnosis. Emphasis has been given to the feature extraction process, in which we borrow the idea of [2] and apply a trainable image segmentation process. Since breast cancer cases are much rarer than healthy ones, it is evident that the classification problem will face a large imbalance between the two classes. In the discussed domain, the percentage of the positive class (i.e. tumor) is only 2.6%, therefore, traditional classification approaches are very likely to fail towards producing a robust model, able to generalize well for future, unseen cases.

Apart from the class imbalance problem, there is also a large number of input features, obtained from the feature extraction phase. Since not all of them are actually informative for the class separation process, a feature selection approach is followed, using Genetic Algorithms and a performance metric as fitness function. Boosting is also used together with a deterministic filter in order to eliminate noisy and redundant instances from the majority class. Experimental results support our claim that feature selection and sub-sampling using Genetic Algorithms and boosting are able to result in a robust and accurate breast cancer identification system, providing classification outcomes in terms of precision and recall that are within the current State-of-the-Art.

The structure of the article is as follows: in Section 2, previous work is described in order to obtain a general perspective of the task at hand. In Section 3, the preprocessing phase is analyzed, with emphasis given to the segmentation and feature extraction steps. In Section 4, a theoretical background of Genetic Algorithms, Adaboost and k-NN is provided. In Section 5, the detailed description of the proposed approach is contained, along with the experimental results. Finally, in section 6, the conclusions are presented, followed by the acknowledgements.

2   Previous Work

Throughout recent years, there have been several studies the field of data mining in oncology:

- The work of [3] presented a model that was designed to automatically detect breast cancer through a Bayesian network. In this article, the number of dataset instances (taken from Wisconsin breast cancer database)[1] was decreased, in order to improve results. For the reduction of the dataset, the technique of feature ranking was applied, thus, features that were not important were eliminated. The correct classification rate that achieved by using Bayesian networks was 98.15%. Although the performance is quite good, the system has some evident weaknesses. At first, the large amount of preprocessing steps along with the inherent slow learning rate of Bayesian networks result in a fairly slow process. Moreover, the dataset is not automatically extracted from mammogram photos but used the Wisconsin breast cancer database. This fact can cause variations in system performance, if the attributes of mammogram photos that has to be tested, are quite different from the Wisconsin dataset. In other words, this technique cannot be used in photos with different attributes from the database of Wisconsin.

---

[1] http://orange.biolab.si/doc/datasets/breast-cancer-wisconsin-cont.htm

- The article presented by Rani [4] dealt with the creation of a system that automatically detects breast cancer with neural networks, executed in a parallel manner. The experimental results showed that the best accuracy of the model is 92% in a multilayer neural network, with 300 training samples and 50 test samples. As we shall see in later paragraphs, this outcome is far lower than that of our proposed work.
- The paper of Pin Wei and Ming Der [5], presented three different algorithms for detecting breast cancer and essentially highlights the fact that the genetic algorithm can provide better results than other techniques. The accuracy of the proposed algorithms was achieved by the method of 10-fold cross validation. The first algorithm is a variation of decision trees. The average correct classification rate is 94.35%. Regarding the neural network algorithm, its performance is slightly higher than that of decision trees, in the range of 95.02%. In contrast to these two algorithms, the genetic algorithm that applied in the present paper provides not only better results but also a thorough analysis of class-imbalance and feature selection process. Additionally, another drawback of this approach is that dataset is not automatically extracted from mammogram photos but used the Wisconsin breast cancer database, as in the paper of [3]. As denoted above, this fact can cause variations in system performance, if the attributes of mammogram photos that has to be tested, are quite different from the Wisconsin dataset.

A general deduction from the above studies is that performance should not be measured as the total number of correct predictions. While in many classification tasks this may seem reasonable, for the task at hand, where examples of the healthy class overwhelm the examples of the cancer class, this performance metric is not informative since one can obtain good performance simply by favoring the majority class.

3   Data Pre-processing

In order to perform identification of breast cancer, a set of 45 mammography images was kindly provided by a private diagnostic center, upon having manually annotated malignant tumors by a specialist. Images were cleaned in order to remove handwritten codes that were found within (e.g. type of the chest, image code, etc.). Since these types of annotations varied in size and position, an automated filtering approach was adopted, which operated as follows: initially, a Gaussian blur filter was applied to ensure that spurious high-frequency information does not appear again and smooth the image. The Statistical Region Merging (SMR) algorithm [6] was then used to extract segments within the image. After several experiment with various bin sizes, a small set of the resulted segments was labeled as those that contained the handwritten annotations and a decision tree learner (using Information Gain as splitting criterion) was used in order to classify the rest of the segments for all of the remaining training images. Segments that were labeled as containing the handwritten annotations were eliminated (masked as black to be more precise) and the remaining segments were merged together so that a clean image was produced. Fig. 1 depicts a sample of an original and a cleaned image. It is noteworthy to mention that the SMR algorithm together with the Decision Tree classifier achieved 100% accuracy, meaning that in all of the input images, the area with the annotations was successfully identified and eliminated without any human interference. In the next sub-sections, a detailed analysis of the image segmentation and feature extraction methodologies is provided.

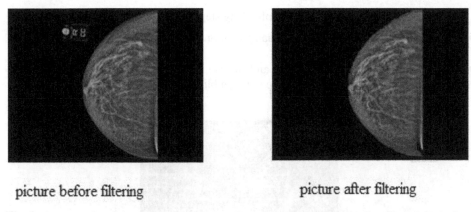

picture before filtering   picture after filtering

Fig. 1. An example of the cleaning process, using statistical region merging for segmentation and decision trees for identifying segments that contained manual annotations

3.1   Image Segmentation

As described in the introductory section, upon cleaning of the images, a segmentation step was carried out, by using trainable segmentation with low-level features, an approach initially inspired by [2].

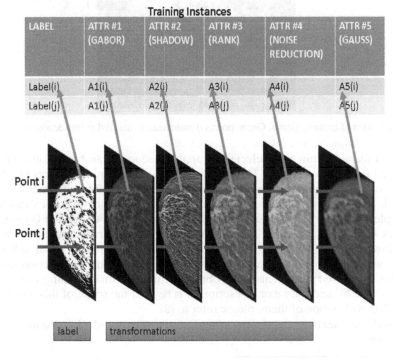

Fig. 2. The main context of trainable segmentation

The exact procedure contains the following steps:

- Histograms of all images were equalized in order to diminish variation in intensity between layers.
- Points from both classes (*healthy* and *tumor*) were selected, giving importance to doubtful areas where healthy tissue is resembling problematic cases and vice-versa (Fig. 3).

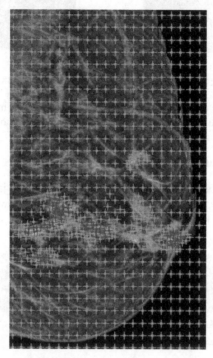

Fig. 3. Selection of training points. Green points denote healthy and red points denote tumor tissues.

- Extract features from each selected by applying several transformations in a multilayered fashion (see Fig. 2). The majority of the aforementioned filters use the value of each local pixel and also the values of surrounding pixels in order to be computed. Thus, by using a variety of transformations, it is possible to extract a more complex perspective about each image area. In our case, we applied Gaussian blur filters with σ varying from 2 to 16, image convolution transform using the north shadow kernel, a Gabor filter using a radius of 4 and wavelength equal to 8, a Rank filter with type set to mean and radius to 4 and finally a noise reduction filter. All of the considered transforms were obtained from the ImageMining extension[2] of RapidMiner®[3] and their exact description is beyond the scope of this article. For a detailed explanation of them, please refer to [2].
- Train the segmentation algorithm using Decision Trees with Gain Ratio as splitting criterion.

---

[2] http://splab.cz/en/research/data-mining/articles
[3] http://www.rapid-i.com

3.2   Feature Extraction

The completion of the former phase resulted in creating segments from each training image. Segments were manually evaluated using a hand-labeled set of ground-truth segmentations. The measure of comparison was the Rand index, which is commonly used in statistics Segment-level features were considered based on color and shape. The following table tabulates the parameters obtained from each Region of Interest (ROI). Furthermore, an additional set of 64 features was also extracted by calculating the histogram of each segment. Therefore, the total number of extracted features was 64+18=82. The final datasets, upon segmentation and feature extraction contained about 35.000 examples, 920 of them (an analogy of 2.6%) belonging to the tumor class and the rest belonging to the healthy class, with 82 attributes. It is clear that the presented dataset poses significant challenges to traditional classification algorithms, due to the class imbalance issues and also due to the large (however not huge) number of input attributes. The proposed methodology uses Genetic algorithms to deal with feature selection.

Table 1. Segment-Based Features

Name	Comments
Mean	The average gray value in the image.
Kurtosis	The fourth order moment about the mean.
Skewness	The third order moment about the mean.
Standard deviation	Standard deviation of the gray values used to generate the mean gray value.
Min/max gray value	
Area	Area of ROI in pixels.
Area fraction	The percentage of non-zero pixels.
Center of mass	the brightness-weighted average of the x and y coordinates of all pixels in the image.
Centroid of ROI	
Major/Minor Axis	Length of major/minor axis of fitted ellipse.
Angle	Angle in degrees of fitted ellipse.
Perimeter	The length of the outside boundary of the selection.
Feret /Projected Area diameter	
Circularity	4pi(area/perimeter^2).
Eccentricity	

4   Theoretical Background

4.1   Genetic Algorithms

The basic idea behind Genetic Algorithms (GA) is the imitation of the mechanisms of nature. The basic idea behind Genetic Algorithms (GA) is the imitation of the mechanisms of nature. Take, for example, hares and how they reproduce and evolve from generation to generation. Suppose we begin to observe a certain population of hares. Naturally, some of them will be faster and glibber than others. These faster and smarter hares are less likely to be the meal of a fox and so by the time they manage to survive, will deal with reproduction of their kind. Of course, there will be a small number of slow and less glib hares, who will manage to survive just because they were lucky. All these hares that have managed to survive, will begin production of the next generation, a generation that combines all the features of its members, combined in various ways to each other. So, slow hares will mixed with some fast, some fast with other fast, and some glib hares with less glib and so on. The hare of the next generation will be, on average, faster and smarter than their ancestors since from the previous generation survived more quick and clever hares. Fortunately, for the preservation of the natural balance, foxes replaced in the same replication process, otherwise the hares will become too fast and smart to be caught. By analogy with living creatures, referred to persons or genotype in a population. Very often these persons also called chromosomes. This can lead to some wrong conclusions if a parallel with the natural organisms, where each cell of any specific type containing a specified number of chromosomes (human cells, for example containing 46 chromosomes). In GA almost always refer to individuals with a single chromosome. The chromosomes that are composed of different elements called genes and are arranged in linear sequence. Each gene affects the inheritance of one or more characteristics. Genes affecting specific features of the individual person and in specific positions of a chromosome called places (loci). Each feature of the individual (such as the hair color) has the ability to display in various forms, depending on the situation in which the corresponding gene that affects. These different situations that can get the gene, called alleles (attribute values). Each genotype (which in most cases is only one chromosome) represents a possible solution to a problem. The translated content of a given chromosome is called phenotype and specified by the user, depending on their needs and requirements. A development process applied on a population of chromosomes represents an extensive search in a space of possible solutions. A prerequisite for the successful outcome of such balancing is groping two processes are obviously contradictory, exploitation and conservation of the best solutions and the best possible exploration of the whole space.

A GA searches in many directions by maintaining a population of potential solutions and by supporting recording and exchanging information between these orientations. The population undergoes a simulated genetic evolution. In each generation, relatively "good" solutions reproduce, while relatively "bad" removed. The separation and evaluation of different solutions achieved by means of a fitness function), which plays the role of the environment in which the population evolves [7]. A GA for a given problem should be composed of the following steps:

1. Generate an initial population consisting of population_size individuals. Each attribute is switched on with probability p_initialize.
2. For all individuals in the population

   — Perform mutation, i.e. set used attributes to unused with probability p_mutation and vice versa.
   — Choose two individuals from the population and perform crossover with probability p_crossover. The type of crossover can be selected by crossover_type.

3. Perform selection, map all individuals to sections on a roulette wheel whose size is proportional to the individual's fitness and draw population size individuals at random according to their probability.
4. As long as the fitness improves, go to step 2.

4.2   Adaboost

Boosting [8], is a Machine Learning iterative process that can adaptively change the distribution of selected training examples, focusing on those that are particularly difficult to be classified. As mentioned before, for the task at hand, examples of the positive class are much less than the negative and thus, traditional learners fail to classify them correctly. That is exactly where boosting contributes, i.e. to change the weights (increase) of instances that are difficult to be correctly labeled and then favor the selection of them on the next iteration using these weights. The most representative implementation of this process is the Adaboost [9] algorithm. Adaboost creates training sets by sampling with replacement, according to a weighting factor W. Initially, all instances are equally weighted. In each iteration, a base classifier (also called a weak learner) is used for training and testing, and the error rate is calculated for each testing instance. Those that are correctly classified lower their weights and those that are not increase their weight, so that in the next round, they are more probable to be selected. The final outcome of Adaboost is a majority voting over all weak learners trained before. The exact process is mentioned below:

1. Initialize all weight equally (w=1/N, where N is the number of instances)
2. Let T be the number of iterations.
3. For i=1 to T do:
   (a) Create a training set Di by sampling (with replacement) from all available examples, according to w.
   (b) Train a base classifier Ci on Di.
   (c) Apply Ci to all examples in the original dataset D.
   (d) Calculate the error, $\varepsilon_i = \frac{1}{N}\left[\sum_j w_j \delta(C_i(x_j) \neq y_j)\right]$ (where $\delta(C_i(x_j) \neq y_j$ equals to 1 if the prediction is correct and 0 otherwise).
   (e) If εi>0.5 then
       (i) Reset all weights for all Examples.
       (ii) Go back to step 3a.

(f) End if

(g) $a_i = \dfrac{1}{2}\ln\dfrac{1-\varepsilon_i}{\varepsilon_i}$

(h) Update   weights   according   to   equation:

$$\dfrac{w_i^{(j+1)}}{W} = \dfrac{w_i^{(j)}}{z_j} \times \begin{cases} exp^{-\alpha j} \, if \; classified \; correctly \\ exp^{\alpha j} \, if \; classidied \; incorrectly \end{cases}$$

(i) End for

- Output the prediction as: $C(x) = argmax(\sum_{j=1}^{T} a_j \delta(C_i(x) = y))$.

4.3   k-Nearest Neighbor

The k-Nearest Neighbor algorithm (k-NN) is a method for classifying objects based on the closest training examples in the feature space. k-NN is a type of instance-based learning, or lazy learning where the function is only approximated locally and all computation is deferred until classification [10]. The k-nearest neighbor algorithm is amongst the simplest of all machine learning algorithms: an object is classified by a majority vote of its neighbors, with the object being assigned to the class most common amongst its k nearest neighbors (k is a positive integer, typically small). If k = 1, then the object is simply assigned to the class of its nearest neighbor. Figure 4 depicts the basic operation of the algorithm.

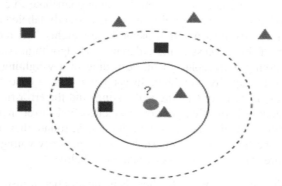

Fig. 4. An Example of k-NN classification. The test instance (green circle) should be labelled either as red triangle or blue squares. If K equals 3 (the solid circle) it is assigned to the first class, because there are 2 triangles and only 1 square inside the inner circle. If K equals 5 (the dashed line circle) it is assigned to the second class, since 3 squares outperform the 2 triangles inside this circle.

The neighbors are taken from a set of objects for which the correct classification is known. This can be thought of as the training set for the algorithm, though no explicit training step is required. The k-nearest neighbor algorithm is sensitive to the local structure of the data. Usually Euclidean distance is used as the distance metric; however this is only applicable to continuous variables. In cases such as text classification, another metric such as the overlap metric (or Hamming distance [11]) can be used. Often, the classification accuracy of "k"-NN can be improved significantly if the distance metric is learned with specialized algorithms such as Large Margin Nearest Neighbor [12] or Neighborhood Component Analysis [13].

5   Methodology and Experimental Results

As mentioned above, the class imbalance problem was dealt using Adaboost and k-NN with k=3 as a weak learner. Since Adaboost is sensitive to noisy data, a filtering approach was applied though a 2D visual inspection of the training points that resulted in the elimination of more than 15% of the majority class instances. A set of 10 iterations for Adaboost proved to balance between both complexity and accuracy. In order to deal with the feature selection problem, a genetic approach was used to find the best subset of input features. The representation of genes was straightforward, in the sense that a bit of 0 in the j-th position symbolized the absence of the j-th input feature and vice-versa. The f-measure [14] metric was chosen as fitness function, since accuracy may always favor the majority class and is not a clear depicter of the classification significance. A population size of 15 was used and selection was performed by Boltzman tournament [15]. The maximum number of generations was set to 50. Probability of mutation was set to 1/100 and probability of crossover (uniform) was set to 0.5. Experiments were carried out using the 10-fold cross validation method. In the following tables, the performance was measured in terms of precision P and recall R (or Sensitivity) for both classes. Precision is given by: $P - \frac{tp}{tp+fp}$, while recall equals to: $R = \frac{tp}{tp+fn}$. The first experiment aims at denoting the usefulness of the boosting process, which confronts imbalance data effectively. Therefore, in the following table, results obtained from a set of basic classifiers, namely, Decision Trees, k-NN using k=3 and Naïve Bayes are tabulated.

Table 2. Precision and Recall metrics for the base classifiers

	Decision Trees		k-NN		Naïve Bayes	
	Precision	Recall	Precision	Recall	Precision	Recall
Healthy Class	97.68%	98.05%	98.75%	98.4%	99.3%	91.67%
Tumor Class	71.79%	69.23%	74.87%	78.6%	45.61%	91.97%

As seen by the results, recall and precision for the tumor class are quite low, which is not adequate for a robust tumor identification/decision support system. By using the proposed boosting strategy, even without the inclusion of the feature selection module, results are far better. In order to depict the contribution of the feature selection genetic approach, results with and without this module are tabulated in Tables 3 and 4 respectively:

Table 3. Performance of the system using Adaboost and k-NN, without feature selection

Total Accuracy: 97,32%	Precision	Recall
Healthy Class	98.73%	98.38%
Tumor Class	79.55%	83.28%

Table 4. Performance of the system using Adaboost and k-NN, with Genetic feature selection

Total Accuracy: 99,24%	Precision	Recall
Healthy Class	99.69%	97.28%
Tumor Class	99.53%	88.29%

As seen from the above tables, performance is significantly improved in both precision and recall for the positive class (tumor). The genetic algorithm feature selection subsystem results in an increase of 20% in precision and 5% in recall for the minority class. The total accuracy reaches 99.24% which is, according to our knowledge, is within the current state of the art in the field and slightly better. Nevertheless, results need to be compared to a common set of training and testing mammograms in order to fully support the previous claim.

6   Conclusion and Future Work

One of the best methods of detecting breast cancer is mammography but in some cases, radiologists cannot clearly detect tumors despite their experience. Based on the utility of this method, this paper proposes a system for detecting cancer that could assist medical staff and improve the accuracy of identification of tumor tissues. In the current approach, two main problems were faced. The former deals with class imbalance within the training set, which in our case reached a tiny 2.6% of the rare class distribution compared to the majority class. The latter deals with a plethora of extracted input features and the need for a selection scheme over them. Towards the first direction, a boosting approach was utilized, along with manual removal of noisy examples with k-NN learner. For the second issue, a Genetic feature selection method was incorporated which led to a significant improvement in precision and recall metrics for the minority class, while maintaining high performance scores for the majority class as well. An accuracy of 99.24% was reached, in a dataset of 45 mammograms that generated approximately 11.500 instances. Results were obtained using a 10-fold cross validation methodology. In the future, we intend to use a larger mammographic database (with more than 100 mammograms) and to extract more features from the images. More specifically, features that can be extracted from the images such as dilate (increases the size of bright objects) etc., could be of particular interest and relevant for classification. The influence and the performance of these new attributes will be thoroughly studied. An additional future research direction is the study of parameter selection for the K-NN algorithm as well as the choice of the distance metric.

Acknowledgement. The authors would like to thank the Diagnostic Center, Magnitiki of Patras for the ongoing support and free access to digital mammography photos and the radiologist Mrs. Kokoli Maria for her valuable collaboration and her continuous assistance she provided throughout the annotation and evaluation phase.

References

1. Altman, M.B., Flynn, M.J., Nishikawa, R.M., Chetty, I.J., Barton, K.N., Movsas, B., Kim, J.H., Brown, S.L.: The potential of iodine for improving breast cancer diagnosis and treatment. Medical Hypotheses 80(1) (2013)
2. Burget, R., Uher, V., Masek, J.: Trainable Segmentation Based on Local-level and Segment-level Feature Extraction. In: ISBI Conference (2012)
3. Fallahi, A., Jafari, S.: An Expert System for Detection of Breast Cancer Using Data Preprocessing and Bayesian Network. International Journal of Advanced Science and Technology 34 (2011)
4. Rani, U.K.: Parallel Approach for Diagnosis of Breast Cancer using Neural Network Technique. International Journal of Computer Applications 10(3) (2010)
5. Chang, W.P., Liou, D.L.: Comparison of Three Data Mining Techniques with Genetic Algorithm in the Analysis of Breast Cancer Data.
6. Nock, R., Nielsen, F.: Statistical Region Merging. IEEE Transactions of Pattern Analysis and Machine Intelligence 26(11) (2004)
7. Lykothanasis, S.: Genetic Algorithms and Applications (Greek: Γενετικοί αλγόριθμοι και εφαρμογές) (2001)
8. Schapire, R.E., Freund, Y.: Boosting: foundations and algorithms (2012)
9. Freund, Y., Schapire, R.E.: A Short Introduction to Boosting. Journal of Japanese Society for Artificial Intelligence (1999) (In Japanese, translation by Naoki Abe.)
10. Bremner, D., Demaine, E., Erickson, J., Iacono, J., Langerman, S., Morin, P., Toussaint, G.: Output-sensitive algorithms for computing nearest-neighbor decision boundaries. Discrete and Computational Geometry (2005)
11. Hamming, R.W.: Error detecting and error correcting codes. Bell System Technical Journal 29 (1950)
12. Weinberger, B.K.Q., Saul, L.K.: Distance Metric Learning for Large Margin Nearest Neighbor Classification. In: Advances in Neural Information Processing Systems, NIPS, vol. 18 (2006)
13. Goldberger, J., Roweis, S., Hinton, G., Salakhutdinov, R.: Neighbourhood Component Analysis. In: Advances in Neural Information Processing Systems, vol. 17, pp. 513–520 (2005)
14. Hripcsak, G., Rothschild, A.S.: Agreement, the F-Measure, and Reliability in Information Retrieval. American Medical Informatics Association (2005)
15. Goldberg, D.E.: A Note on Boltzmann Tournament Selection for Genetic Algorithms and Population-Oriented Simulated Annealing. Complex Systems 4, 445–460 (1990)

Transductive Conformal Predictors

Vladimir Vovk

Department of Computer Science,
Royal Holloway, University of London,
Egham, Surrey, UK
v.vovk@rhul.ac.uk
http://vovk.net

Abstract. This paper discusses a transductive version of conformal predictors. This version is computationally inefficient for big test sets, but it turns out that apparently crude "Bonferroni predictors" are about as good in their information efficiency and vastly superior in computational efficiency.

Keywords: Conformal prediction, transduction, Bonferroni adjustment.

1 Introduction

The most standard learning problems are inductive: given a training set of labelled objects, the task is to come up with a predictor with a reasonable performance on unknown test objects. In typical transductive problems ([7], Chapter VI, Sections 10–13, [6], Chapter 8) we are given both a training set of labelled objects and a test set of unlabelled objects; the task is to come up with a predictor, which may depend on both sets, with a reasonable performance on the test set. Conformal predictors ([9], Chapter 2) are not transductive in this sense, although they do have a transductive flavour (see, e.g., [9], pp. 6–7).

The goal of this paper is to introduce a fully transductive version of conformal predictors. The basic definitions are given in Section 2. Section 3 introduces Bonferroni predictors, a simple and computationally efficient modification of conformal predictors adapted to the transductive framework. Sections 4 and 5 contain simple theoretical results about transductive conformal predictors and Bonferroni predictors. Section 6 reports on experimental results. Finally, Section 7 concludes.

The expression "transductive conformal predictors" has been used before (see, e.g., [3]) to refer to what is called "conformal predictors" in [9] and this paper. This agrees with our terminology, since conformal predictors are a special case of our transductive conformal predictors corresponding to a test set of size 1.

2 Transductive Conformal Predictors

Let $z_1 = (x_1, y_1), \ldots, z_l = (x_l, y_l)$ be a training set and $x_{l+1}, \ldots, x_{l+k}$ be a test set. The test set consists of *objects* $x_j \in \mathbf{X}$ and the training set consists of

H. Papadopoulos et al. (Eds.): AIAI 2013, IFIP AICT 412, pp. 348–360, 2013.
© IFIP International Federation for Information Processing 2013

labelled objects, or *examples*, $z_i = (x_i, y_i) \in \mathbf{Z} := \mathbf{X} \times \mathbf{Y}$. The *object space* $\mathbf{X}$, *label space* $\mathbf{Y}$, and *example space* $\mathbf{Z} := \mathbf{X} \times \mathbf{Y}$ are fixed throughout the paper; they are assumed to be measurable spaces.

Transductive conformal predictors are determined by their transductive nonconformity measures, which are defined as follows. A *transductive nonconformity measure* is a measurable function $A : \mathbf{Z}^* \times \mathbf{Z}^* \to \mathbb{R}$ such that $A(\zeta_1, \zeta_2)$ does not depend on the ordering of ζ_1. (For the specific transductive nonconformity measures used in this paper $A(\zeta_1, \zeta_2)$ will not depend on the ordering of ζ_2 either.) The intuition is that $A(\zeta_1, \zeta_2)$ (the *transductive nonconformity score*) measures the lack of conformity of the "test set" ζ_2 to the "training set" ζ_1.

The *transductive conformal predictor* (*TCP*) corresponding to A finds the prediction region for the test set $x_{l+1}, \dots, x_{l+k}$ at a *significance level* $\epsilon \in (0, 1)$ as follows:

- For each possible set of labels $(v_1, \dots, v_k) \in \mathbf{Y}^k$:
  - set $y_j := v_{j-l}$ and $z_j := (x_j, y_j)$ for $j = l+1, \dots, l+k$;
  - compute the transductive nonconformity scores

$$\alpha_S := A(z_{\{1,\dots,l+k\}\setminus S}, z_S),$$

  where S ranges over all $(l+k)!/l!$ ordered subsets $(s_1, \dots, s_k)$ of $\{1, \dots, l+k\}$ of size k, z_S stands for the sequence $(z_{s_1}, \dots, z_{s_k})$ (when $S = (s_1, \dots, s_k)$), and $z_{\{1,\dots,l+k\}\setminus S}$ stands for z_B, B being any ordering of $\{1, \dots, l+k\} \setminus S'$ and S' being the set of all elements of S (It does not matter which ordering is chosen, by the definition of a transductive nonconformity measure);
  - compute the p-value

$$p(v_1, \dots, v_k) := \frac{\left|\{S \mid \alpha_S \geq \alpha_{(l+1,\dots,l+k)}\}\right|}{(l+k)!/k!}, \qquad (1)$$

  where S ranges, as before, over all $(l+k)!/l!$ ordered subsets of $\{1, \dots, l+k\}$ of size k, and $|\dots|$ stands for the size of a set.
- Output the prediction region

$$\Gamma^\epsilon(z_1, \dots, z_l, x_{l+1}, \dots, x_{l+k}) := \left\{(v_1, \dots, v_k) \in \mathbf{Y}^k \mid p(v_1, \dots, v_k) > \epsilon\right\}. \qquad (2)$$

Smoothed TCPs are defined in the same way except that (1) is replaced by

$$p(v_1, \dots, v_k) := \frac{\left|\{S \mid \alpha_S > \alpha_{(l+1,\dots,l+k)}\}\right| + \theta\left|\{S \mid \alpha_S = \alpha_{(l+1,\dots,l+k)}\}\right|}{(l+k)!/k!},$$

where θ are random variables distributed uniformly on $[0, 1]$ (no independence between different sets of postulated labels $v_1, \dots, v_k$ is required, but later on when we consider the online prediction protocol we will assume that θ are independent between different trials).

A *nonconformity measure* can now be defined as the restriction of a transductive nonconformity measure to the domain $\mathbf{Z}^* \times \mathbf{Z}$ (we identify a 1-element sequence with its only element). Nonconformity measures are well studied and there are many useful examples of them (see, e.g., [9]).

An interesting class of transductive nonconformity measures can be obtained from nonconformity measures. Let $\mathbb{R}$ be the set of real numbers. A *simple nonconformity aggregator* is a function $M : \mathbb{R}^* \to \mathbb{R}$ that is symmetric and increasing in each argument. (The requirement that M be symmetric, i.e., $M(\zeta)$ not depend on the ordering of ζ, is not necessary but convenient for the following discussion. The requirement that M be increasing in each argument is not necessary either but very natural.) With each nonconformity measure A and simple nonconformity aggregator M we can associate the transductive nonconformity measure

$$A_M((z_1, \ldots, z_l), (z_{l+1}, \ldots, z_{l+k})) := M(\alpha_{l+1}, \ldots, \alpha_{l+k}),$$

where

$$\alpha_j := A((z_1, \ldots, z_l, z_{l+1}, \ldots, z_{j-1}, z_{j+1}, \ldots, z_{l+k}), z_j),$$
$$j = l+1, \ldots, l+k. \quad (3)$$

Our experiments in Section 6 use the *Nearest Neighbour nonconformity measure*

$$A(((x_1, y_1), \ldots, (x_l, y_l)), (x, y)) := \frac{\min_{i=1,\ldots,l:y_i=y} d(x, x_i)}{\min_{i=1,\ldots,l:y_i \neq y} d(x, x_i)}, \quad (4)$$

where d is a distance, and the *max nonconformity aggregator*

$$M(\alpha_1, \ldots, \alpha_k) := \max(\alpha_1, \ldots, \alpha_k). \quad (5)$$

Remark 1. Alternatively, we could set $\alpha_j := A((z_1, \ldots, z_l), z_j)$ in (3) (cf. (12) below), but this would adversely affect the already low computational efficiency of TCPs in our experiments in Section 6.

Rank-Based Transductive Conformal Predictors

The notion of a simple nonconformity aggregator can be generalized as follows. A *nonconformity aggregator* is a function $M : \mathbb{R}^* \times \mathbb{R}^* \to \mathbb{R}$ such that $M(\zeta_1, \zeta_2)$ depends neither on the ordering of ζ_1 nor on the ordering on ζ_2. (The most natural class of nonconformity aggregators is where $M(\zeta_1, \zeta_2)$ is decreasing in every element of ζ_1 and increasing in every element of ζ_2, but it is too narrow for our purposes.) With each nonconformity measure A and nonconformity aggregator M we associate the transductive nonconformity measure

$$A_M((z_1, \ldots, z_l), (z_{l+1}, \ldots, z_{l+k})) := M((\alpha_1, \ldots, \alpha_l), (\alpha_{l+1}, \ldots, \alpha_{l+k}))$$

where

$$\alpha_i := A((z_1, \ldots, z_{i-1}, z_{i+1}, \ldots, z_l, z_{l+1}, \ldots, z_{l+k}), z_i), \quad i = 1, \ldots, l, \quad (6)$$

and $\alpha_{l+1}, \ldots, \alpha_{l+k}$ are defined by (3). We identify each simple nonconformity aggregator M with the nonconformity aggregator

$$M^{\dagger}((\alpha_1, \ldots, \alpha_l), (\alpha_{l+1}, \ldots, \alpha_{l+k})) := M(\alpha_{l+1}, \ldots, \alpha_{l+k}).$$

For transductive nonconformity measures obtained from nonconformity measures and nonconformity aggregators, the p-value (1) as function of the nonconformity scores α_i of individual examples reduces to the well-known notion of a one-sided permutation test (see, e.g., [1], Section 1.7.E). In classical nonparametric statistics, the most popular permutation tests are rank tests, and we will give corresponding definitions in our current context. Let $\mathbb{N} := \{1, 2, \ldots\}$. A (simple) *rank aggregator* is a function $M : \mathbb{N}^* \to \mathbb{N}$ that is symmetric and increasing in each argument. The corresponding nonconformity aggregator is

$$M'((\alpha_1, \ldots, \alpha_l), (\alpha_{l+1}, \ldots, \alpha_{l+k})) := M(R_{l+1}, \ldots, R_{l+k}), \qquad (7)$$

where $R_1, \ldots, R_{l+k}$ are the ranks of $\alpha_1, \ldots, \alpha_{l+k}$, respectively, in the multiset $\wr \alpha_1, \ldots, \alpha_{l+k} \wr$. Formally, R_i is defined as

$$R_i := |\{j = 1, \ldots, l + k \,|\, \alpha_j < \alpha_i\}| + 1.$$

If there are no ties (i.e., equal elements in $\wr \alpha_1, \ldots, \alpha_{l+k} \wr$), this is the usual notion of a rank; in the presence of ties, our definition is somewhat non-standard giving each tie the smallest of the ranks that it spans. (And this definition causes a counterintuitive behaviour of the definition (7), where M' is not necessarily increasing in α_j, $j \in \{l + 1, \ldots, l + k\}$, even in the case where M is the max nonconformity aggregator (5).)

The most popular rank aggregator in classical nonparametric statistics is the *ranksum aggregator*

$$M(R_1, \ldots, R_k) := R_1 + \cdots + R_k, \qquad (8)$$

which is used in the Wilcoxon ranksum test (see [11] or [1], Section 1.2). Using the ranksum aggregator, however, produces very poor results (see Appendix B of [10] and the right panel of Figure 4) when the efficiency of TCPs is measured by the number of multiple predictions that they produce, as in this paper (see Section 6 below).

Notice that the nonconformity aggregator (5) is equivalent (in the sense of leading to the same TCP) to the rank aggregator $M'(R_1, \ldots, R_k) := \max(R_1, \ldots, R_k)$. The corresponding TCP will be called the *rankmax TCP* (and the TCP corresponding to (8) will be called the *ranksum TCP*).

It is easy to give an explicit representation of the rankmax TCP. Remember that the size of the training set is l and the size of the test set is k and suppose that the value of the *rankmax test statistic* $\max(R_{l+1}, \ldots, R_{l+k})$ is t. The probability that a random subset $\{s_1, \ldots, s_k\}$ of $\{1, \ldots, l + k\}$ of size k will lead to a value of the test statistic $\max(R_{s_1}, \ldots, R_{s_k})$ of at least t can be found as 1 minus the probability that a random subset of $\{1, \ldots, l + k\}$ of size k is covered by a fixed subset of $\{1, \ldots, l+k\}$ of size $t-1$ (namely, by the set of indices i with $R_i < t$). In other words, the p-value is

$$1 - \frac{\binom{t-1}{k}}{\binom{l+k}{k}} = 1 - \frac{(t-1)!\, l!}{(t-1-k)!\, (l+k)!} \qquad (9)$$

(which is understood to be 1 when $t \leq k$).

3   Bonferroni Predictors

Unfortunately, transductive conformal predictors are computationally inefficient, especially if we want to predict many test objects at once: we have to go over all $|\mathbf{Y}|^k$ combinations of labels for the test set. (Even if $A(\zeta_1, \zeta_2)$ does not depend on the ordering of ζ_2, there are no computational savings unless the test set contains many identical objects.) We next introduce a family of region predictors based on the idea of the Bonferroni adjustment of p-values. In brief, a Bonferroni predictor computes a p-value for each test object separately and then combines the k p-values into one p-value using the Bonferroni formula

$$p := \min(kp_1, \ldots, kp_k, 1). \qquad (10)$$

The full description of the *Bonferroni predictor* (*BP*) corresponding to a non-conformity measure A is as follows:

- For each object x_j, $j \in \{l+1, \ldots, l+k\}$, in the test set and each possible label $v \in \mathbf{Y}$:
  - set $y_j := v$ and $z_j := (x_j, y_j)$;
  - compute the nonconformity scores

$$\alpha_i := A((z_1, \ldots, z_{i-1}, z_{i+1}, \ldots, z_l, z_j), z_i), \qquad i = 1, \ldots, l, \qquad (11)$$
$$\alpha_j := A((z_1, \ldots, z_l), z_j); \qquad (12)$$

  - compute the p-value

$$p_{j-l}(v) := \frac{|\{i = 1, \ldots, l \mid \alpha_i \geq \alpha_j\}| + 1}{l+1}. \qquad (13)$$

- Output the prediction region

$$\Gamma^\epsilon(z_1, \ldots, z_l, x_{l+1}, \ldots, x_{l+k}) := \prod_{j=l+1}^{l+k} \{v \mid p_{j-l}(v) > \epsilon/k\}, \qquad (14)$$

  where $\epsilon \in (0, 1)$ is the significance level.

Notice that the prediction region (14) output by the BP can be rewritten in the form (2) if we define

$$p(v_1, \ldots, v_k) := \min(kp_1(v_1), \ldots, kp_k(v_k), 1) \qquad (15)$$

(cf. (10)).

It is difficult to compare the rankmax TCP and the corresponding BP theoretically, but the following intermediate notion facilitates a comparison. The *semi-Bonferroni predictor (SBP)* is defined as follows:

- For each possible set of labels $(v_1, \ldots, v_k) \in \mathbf{Y}^k$ for the test set:
  - set $y_j := v_{j-l}$ and $z_j := (x_j, y_j)$ for $j = l+1, \ldots, l+k$;
  - compute nonconformity scores α_i, $i = 1, \ldots, l+k$, by

$$\alpha_i := A((z_1, \ldots, z_{i-1}, z_{i+1}, \ldots, z_{l+k}), z_i), \quad i = 1, \ldots, l+k \qquad (16)$$

  (cf. (6) and (3); the main difference of (16) from (6) and (3) is that (16) involves the true training examples and test objects whereas (6) and (3) involve arbitrary subsets of size l and k of the union of the training set and the test set with postulated labels);
  - compute the p-value (13) for each $j = l+1, \ldots, l+k$ and merge these p-values using (15).
- Output the prediction region (2).

Notice that the SBP becomes identical to the BP when (16) is replaced by (11) and (12) for $j = l+1, \ldots, l+k$.

The following lemma shows that SBPs are usually weaker than the corresponding rankmax TCPs. (However, in Remark 2 and Section 6 we will see that the difference can be surprisingly small.)

Lemma 1. *Suppose all nonconformity scores (16) are different. The p-value (9) produced by a rankmax TCP does not exceed the p-value (15) produced by the corresponding SBP.*

Proof. Let t be the value of the rankmax test statistic, as defined at the end of Section 2. We are required to prove

$$1 - \frac{\binom{t-1}{k}}{\binom{l+k}{k}} \leq k \frac{l+k-t+1}{l+1} . \qquad (17)$$

Indeed, the left-hand side of (17) is identical to (9), and the ratio on the right-hand side of (17) is the smallest of the p-values (13) over j (cf. (15)). The statement that the ratio on the right-hand side of (17) is the smallest of the p-values (13) depends on (16) being all different (in fact, it is sufficient to assume that the maximum in the definition of the rankmax test statistic t is attained on only one test object). Notice, however, that the right-hand side of (17) is always an upper bound on the SBP p-value; this fact will be used in our discussions below.

We will prove a slightly stronger inequality that (17) replacing the denominator $l+1$ by $l+k$. In principle, t can take any value in $\{1, \ldots, l+k\}$, but we can assume, without loss of generality, that $t \in \{k+1, \ldots, l+k\}$: if $t \leq k$, the left-hand side of (17) is 1 by definition and the right-hand side is at least 1 (even when $l+1$ is replaced by $l+k$). Rewriting (17) (with $l+k$ in place of $l+1$) as

$$1 - \frac{(t-1)(t-2)\cdots(t-k)}{k!\binom{l+k}{k}} \le k\frac{l+k-t+1}{l+k}, \tag{18}$$

we can assume that $t \in [k+1, l+k]$. Since the fraction on the left-hand side of (18) is a convex function of t (the second derivative is obviously nonnegative) and for $t := k+l$ (18) holds (it becomes $k/(1+k) \le k/(l+k)$), it suffices to prove that the derivative in t of the left-hand side of (18) at the point $l+k$ is equal to or exceeds the derivative of the right-hand side:

$$-\frac{(\Gamma(t)/\Gamma(t-k))'_{t=l+k}}{k!\binom{l+k}{k}} \ge k\frac{-1}{l+k},$$

where Γ is the gamma function, $\Gamma(n) = (n-1)!$ for $n \in \mathbb{N}$. By the definition of the digamma function ψ, the last inequality can be rewritten as

$$\frac{\Gamma(l+k)}{\Gamma(l)}\left(\psi(l+k) - \psi(l)\right) \le \frac{k}{l+k}k!\binom{l+k}{k},$$

which simplifies to

$$\psi(l+k) - \psi(l) \le \frac{k}{l}.$$

The well-known expression for ψ at the integer values of its argument (see, e.g., http://dlmf.nist.gov/5.4.14) allows us to rewrite the last inequality as

$$\frac{1}{l} + \frac{1}{l+1} + \cdots + \frac{1}{l+k-1} \le \frac{k}{l},$$

which is obviously true.   □

Remark 2. The proof of Lemma 1 shows that the inequality (17) is strict whenever $k > 1$ (for $k = 1$ the two p-values coincide). Three factors contribute to its being strict: the SBP p-value is larger than the rankmax TCP p-value at $t = l+k$; as function of t, the SBP p-value has a steeper (negative) slope at $t = l+k$; besides, to the left of $t = l+k$ the SBP p-value goes in a straight line whereas the rankmax TCP p-value veers down. This is illustrated in Figure 1 for $l = 1000$ and $k = 2$ (typical values for our experiments reported in Section 6); the first two factors, however, are not noticeable.

It is plausible that a BP usually produces somewhat smaller p-values (and, therefore, somewhat smaller prediction regions) than the corresponding SBP: the only difference is that, when computing p-values, the SBP uses more test objects with arbitrarily assigned labels, and this may lead to a greater distortion of the nonconformity scores.

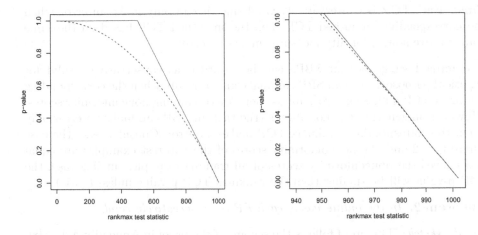

Fig. 1. Left panel: SBP p-values (the solid red line) and rankmax TCP p-values (the dashed blue line) for $l = 1000$ and $k = 2$ as functions of t. Right panel: the lower right corner of the left panel.

4   Validity

The strongest notion of validity for conformal and related predictors can be stated in the online mode. Suppose we are given a sequence of positive integer numbers $k_1, k_2, \ldots$ and the incoming sequence of examples is $z_1 = (x_1, y_1), z_2 = (x_2, y_2), \ldots$; set $l_n := \sum_{i=1}^{n-1} k_i$ (in particular, $l_1 := 0$). At trial $n = 1, 2, \ldots$ of the online prediction protocol, Predictor predicts the k_n labels $y_{l_n+1}, \ldots, y_{l_n+k_n}$ given the l_n examples $z_1, \ldots, z_{l_n}$ and k_n objects $x_{l_n+1}, \ldots, x_{l_n+k_n}$. The prediction is a subset Γ_n of $\mathbf{Y}^{k_n}$. It can be *multiple* ($|\Gamma_n| > 1$), *singleton* ($|\Gamma_n| = 1$), or *empty* ($|\Gamma_n| = 0$). Predictor *makes an error* if $(y_{l_n+1}, \ldots, y_{l_n+k_n}) \notin \Gamma_n$.

In this section we assume either that the sequence of examples $z_1, z_2, \ldots$ is infinite and the examples are produced independently from the same probability distribution on $\mathbf{Z}$, or that the sequence of examples is finite, $z_1, \ldots, z_N$, and produced from an exchangeable probability distribution on $\mathbf{Z}^N$.

The following simple result states the validity of TCPs in the online mode; its proof is standard (see, e.g., [8] or [9], Section 8.7) and is omitted.

Theorem 1. *In the online mode, a smoothed TCP makes errors with probability ϵ (the significance level) independently at different trials.*

A suitable version of validity in the absence of smoothing is *conservative validity*, i.e., being dominated by a sequence of independent Bernoulli trials with probability of success equal to the significance level; for details, see [9], p. 21. By Theorem 1, TCPs are conservatively valid:

Corollary 1. *In the online mode, each TCP is conservatively valid.*

Proof. Each TCP is conservatively valid since it can only make an error when the corresponding smoothed TCP (i.e., the smoothed TCP based on the same transductive nonconformity measure) makes an error.   □

Lemma 1 suggests that SBPs can be regarded as conservatively valid for practical purposes, since an SBP can make an error only when the corresponding rankmax TCP makes an error, unless there are ties among nonconformity scores. (However, in general, it is not always true that an SBP can make an error only when the corresponding rankmax TCP makes an error. Consider, e.g., the case where $k = 2$ and the nonconformity scores of the two test examples are equal and exceed the nonconformity scores of all training examples; in this case, the SBP p-value will be smaller than the rankmax TCP p-value unless $l = 1$.)

Theorem 2. *In the online mode, each BP is conservatively valid.*

Proof (sketch). The proof follows the scheme of the proof in Appendix A.1 of [8]. Given the bag $\lfloor z_1, \ldots, z_{l_n+k_n} \rceil$ and under the assumption of exchangeability, the probability that the BP will make an error at trial n for a given test example (e.g., for the second example in the test set) is at most ϵ/k_n. Therefore, the probability that it will make an error for some of the k_n test examples is at most ϵ. We can increase the indicator function of making an error to obtain a Bernoulli random variable with probability of success equal to ϵ (this might involve extending the probability space). It remains to notice that whether an error is made at trial n is determined by the bag $\lfloor z_1, \ldots, z_{l_n} \rceil$ and examples $z_{l_n+1}, \ldots, z_{l_n+k_n}$ (cf. Lemma 2 in [8]).   □

5   Universality

A *transductive confidence predictor* is a measurable strategy for Predictor in the online prediction protocol (as described in the previous section) depending on a parameter $\epsilon \in (0, 1)$ (the significance level) in such a way that for each training set and each test set the prediction at a larger significance level is a subset of the prediction at a smaller significance level. We say that the transductive confidence predictor is conservatively valid if the sequence of errors that it makes at any significance level ϵ is dominated by a sequence of independent Bernoulli trials with probability of success ϵ. We say that it is *invariant* if, when fed with examples $z_1, \ldots, z_{l_n}$ and objects $x_{l_n+1}, \ldots, x_{l_n+k_n}$ at any trial n, it issues the same prediction regardless of the ordering of $z_1, \ldots, z_{l_n}$. And we say that a transductive confidence predictor Γ' is *at least as good as* another transductive confidence predictor Γ'' if at any significance level ϵ the prediction region issued by Γ' is completely covered by the prediction region issued by Γ''. The following result, whose proof is omitted in this version of the paper, can be proved similarly to Theorem 2.6 in [9].

Theorem 3. *Suppose* **Z** *is a Borel space. For any conservatively valid transductive confidence predictor Γ there exists a transductive conformal predictor Γ' that is at least as good as Γ.*

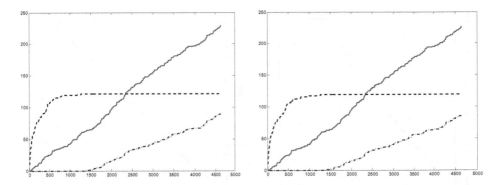

Fig. 2. Left panel: the performance of the rankmax TCP based on Nearest Neighbour for tangent distance on the USPS data set (randomly permuted) for the size $k = 2$ of test sets and significance level 5%. The cumulative errors are shown with a solid red line, multiple predictions with a dashed black line, and empty predictions with a dash-dot blue line. Right panel: the analogous picture for the BP.

Theorem 3 says that TCPs are universal in the sense of dominating all conservatively valid transductive confidence predictors. In particular, for any BP there is a TCP that is at least as good as that BP. However, in the next section we will see that the rankmax TCP corresponding to the same nonconformity measure does not always satisfy this property.

6 Experiments

In our experiments we will use the standard USPS data set of hand-written digits. The training set (7291 examples) is merged with the test set (2007 examples) and the resulting data set of 9298 examples is randomly permuted, to make sure the assumption of exchangeability is satisfied. The prediction protocol is online. In a typical scenario the digits might arrive in batches of $k = 5$ digits and represent American zip codes (in this case, however, the exchangeability assumption is only a crude approximation, since the digits within the same zip code are likely to be written by the same person). However, the TCP and SBP are too computationally inefficient to be applied in this case, and for comparing them with the BP we first consider online prediction of batches of $k = 2$ digits; intuitively, our task is to recognize a two-digit number.

We always use the Nearest Neighbour nonconformity measure (4), where d is tangent distance [5], and study empirically the corresponding rankmax TCP, SBP, and BP. As the significance level we always take 5%. The left panel of Figure 2 shows the performance of the rankmax TCP using three functions: the cumulative number of errors made over the trials $1, \ldots, n$ as function of n, the cumulative number of multiple predictions made over the trials $1, \ldots, n$ as function of n, and the cumulative number of empty predictions over the trials $1, \ldots, n$ as function of n. The performance of the SBP and BP as measured

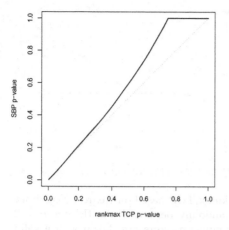 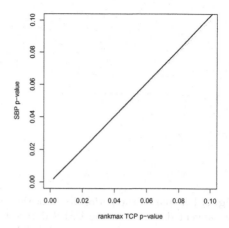

Fig. 3. Left panel: the p-values produced by an SBP vs the p-values produced by the corresponding rankmax TCP (the solid blue line) for the size $l = 1000$ of the training set and $k = 2$ of the test set. Right panel: the lower left corner of the left panel.

by these functions is very similar; only the latter is shown in the right panel of Figure 2, but all three graphs are visually indistinguishable. The BP even makes 2 fewer multiple predictions than the rankmax TCP, which confirms the claim made in Section 5 that the rankmax TCP corresponding to the same nonconformity measure as a given BP is not always at least as good as that BP. (It is not true in general that the BP always makes fewer multiple predictions than the corresponding rankmax TCP. It just happens to be true for tangent distance and seed 0 for the MATLAB pseudorandom number generator; e.g., the BP makes slightly more multiple predictions for Euclidean distance and seed 0.) The SBP makes one more multiple prediction than the rankmax TCP, which agrees with Lemma 1.

The cause of the similarity between the plots of the type shown in Figure 2 for the rankmax TCP and corresponding SBP is illustrated by Figure 3 (essentially a version of Figure 1), which shows the p-values produced by the SBP plotted against the respective p-values produced by the corresponding rankmax TCP, assuming there are no ties among the nonconformity scores. When the p-values are small, they are remarkably close to each other. And even without making any assumptions, we can still see that the SBP p-values are never significantly worse than the respective rankmax TCP p-values, assuming the latter are not too large.

The main advantage of BPs is that they are much more computationally efficient than both TCPs and SBPs. Because of their computational efficiency, it is very easy to produce the analogue of the right panel of Figure 2 for $k = 5$ (as in American zip codes): see the left panel of Figure 4; but it is not clear at all how to make the computations for rankmax TCPs and SBPs feasible, even for moderately large k.

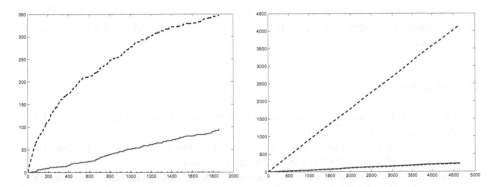

Fig. 4. Left panel: the performance of the BP for the size $k = 5$ of test sets. Right panel: the performance of the ranksum TCP for $k = 2$ (very poor). The setting is as in Figure 2: the prediction algorithms are based on Nearest Neighbour and tangent distance; the cumulative errors are shown with a solid red line, multiple predictions with a dashed black line, and empty predictions with a dash-dot blue line; the significance level is 5%.

7 Conclusion

Based on our theoretical and empirical results, the preliminary recommendation is to use Bonferroni predictors in transductive problems: as compared to rankmax TCPs and SBPs, they enjoy the same theoretical validity guarantees, have comparable predictive performance empirically, but are much more computationally efficient.

The conclusion is preliminary since our empirical comparison in Section 6 only covers TCPs for a small size k of the test set, namely $k = 2$. The computational inefficiency of TCPs greatly complicates their empirical comparison with the BPs and SBPs for large values of k.

The comparison is much more straightforward in the case of transductive and Bonferroni extensions of inductive conformal predictors ([4]; [9], Section 4.1), and it can be shown that the two extensions produce similar p-values in practically important cases: see [10], Appendix A, for details.

Acknowledgments. I am grateful to Harris Papadopoulos for a discussion at COPA 2012 that rekindled my interest in transduction. Thanks to Wouter Koolen for illuminating discussions and for writing a MATLAB program for the Wilcoxon ranksum test (the standard programs in MATLAB and R produce unsatisfactory results in the default mode and are prohibitively slow in the exact mode). My thanks also go to the COPA 2013 reviewers, whose comments have helped me in improving the presentation and suggested new directions of research. In my experiments I used the C program for computing tangent distance written by Daniel Keysers and adapted to MATLAB by Aditi Krishn.

This work was partially supported by the Cyprus Research Promotion Foundation (research contract TPE/ORIZO/0609(BIE)/24) and by EPSRC (grant EP/K033344/1).

References

1. Lehmann, E.L.: Nonparametrics: Statistical Methods Based on Ranks, revised 1st edn. Springer, New York (2006)
2. National Institute of Standards and Technology: Digital library of mathematical functions (October 1, 2012), http://dlmf.nist.gov/
3. Nouretdinov, I., Costafreda, S.G., Gammerman, A., Chervonenkis, A., Vovk, V., Vapnik, V., Fu, C.H.Y.: Machine learning classification with confidence: Application of transductive conformal predictors to MRI-based diagnostic and prognostic markers in depression. Neuroimage 56, 809–813 (2011)
4. Papadopoulos, H., Vovk, V., Gammerman, A.: Qualified predictions for large data sets in the case of pattern recognition. In: Proceedings of the First International Conference on Machine Learning and Applications, pp. 159–163. CSREA Press, Las Vegas (2002)
5. Simard, P., LeCun, Y., Denker, J.: Efficient pattern recognition using a new transformation distance. In: Hanson, S., Cowan, J., Giles, C. (eds.) Advances in Neural Information Processing Systems, vol. 5, pp. 50–58. Morgan Kaufmann, San Mateo (1993)
6. Vapnik, V.N.: Statistical Learning Theory. Wiley, New York (1998)
7. Vapnik, V.N., Chervonenkis, A.Y.: Theory of Pattern Recognition. Nauka, Moscow (1974) (in Russian); German translation:Theorie der Zeichenerkennung, Akademie, Berlin (1979)
8. Vovk, V.: On-line Confidence Machines are well-calibrated. In: Proceedings of the Forty Third Annual Symposium on Foundations of Computer Science, pp. 187–196. IEEE Computer Society, Los Alamitos (2002)
9. Vovk, V., Gammerman, A., Shafer, G.: Algorithmic Learning in a Random World. Springer, New York (2005)
10. Vovk, V.: Transductive conformal predictors. On-line Compression Modelling project (New Series), Working Paper 8, an extended version of this paper (2013), http://alrw.net
11. Wilcoxon, F.: Individual comparisons by ranking methods. Biometrics Bulletin 1, 80–83 (1945)

PyCP: An Open-Source Conformal Predictions Toolkit

Vineeth N. Balasubramanian, Aaron Baker, Matthew Yanez,
Shayok Chakraborty, and Sethuraman Panchanathan

Center for Cognitive Ubiquitous Computing
School of Computing, Informatics and Decision Systems Engineering
Arizona State University, Tempe AZ 85282 USA
{vineeth.nb,aaron.j.baker,mmyanez,schakr10,panch}@asu.edu

Abstract. The Conformal Predictions framework is a new game-
theoretic approach to reliable machine learning, which provides a
methodology to obtain error calibration under classification and
regression settings. The framework combines principles of transductive
inference, algorithmic randomness and hypothesis testing to provide
guaranteed error calibration in online settings (and calibration in offline settings supported by empirical studies). As the framework is being
increasingly used in a variety of machine learning settings such as active learning, anomaly detection, feature selection, and change detection,
there is a need to develop algorithmic implementations of the framework
that can be used and further improved by researchers and practitioners. In this paper, we introduce PyCP, an open-source implementation
of the Conformal Predictions framework that currently provides support
for classification problems within transductive and Mondrian settings.
PyCP is modular, extensible and intended for community sharing and
development.

Keywords: Conformal predictions, Open-source software.

1 Introduction

The Conformal Predictions (CP) framework is a game-theoretic approach to reliable machine learning, which provides a methodology to obtain error calibration
under classification and regression settings [1]. The framework combines principles of transductive inference, algorithmic randomness and hypothesis testing to
provide guaranteed error calibration in online settings [2] (and calibration in offline settings supported by empirical studies [3]). The framework can be applied
to various classification and regression algorithms, making it very generalizable.
In recent years, the framework has also been applied to other machine learning
settings including active learning [4], anomaly detection [5], feature selection [6],
change detection [7] and quality estimation [8]. The framework has found application in a variety of domains including medical diagnosis [9], bioinformatics
[10], biometrics [11], and network analysis [12].

H. Papadopoulos et al. (Eds.): AIAI 2013, IFIP AICT 412, pp. 361–370, 2013.

The growing relevance of the CP framework has generated a need to develop algorithmic implementations of the framework that can be shared among researchers and practitioners in the community. Machine learning researchers have recently begun to promote development of usable, inter-operable, flexible, and scalable machine learning software. As the discipline of machine learning has matured, the community has developed an extensive body of learning algorithms that are useful in diverse applications. Sharing source code improves usability and interoperability because it leads to quicker detection and correction of errors, better experimental reproducibility, incremental advancements of previous work, combinations of distinct theoretical advances, and more percolation of technical methods into other disciplines and industry [13].

In this paper, we present PyCP, the first open-source implementation of the CP framework, in the Python scripting language. Python has several advantages for use in an open-source implementation for scientific application. As a scripting language with white space syntax, it is relatively simple and has many widely-used libraries for scientific computing (e.g., NumPy, SciPy, Matplotlib). These libraries have contributed to a strong increase in the popularity of Python as a programming tool in machine learning research [14]. In order to avoid duplication of code of machine learning methods, PyCP has been developed on top of the popular SciKitLearn toolkit[1] [15], an open-source machine learning software. PyCP is currently available for use of the CP framework in classification settings with options for classifiers (k-Nearest Neighbor, Support Vector Machines, Decision Trees, Logistic Regression) as well as the framework setting (Transductive, Mondrian). It has been designed in a modular fashion to allow for easy extensibility and further development. The remainder of this paper is organized as follows: a brief background of the CP framework is presented in Section 2, the design of PyCP is presented in Section 3, examples of its use and availability are provided in Section 4, and we conclude with a discussion of future work in Section 5.

2   Background

2.1   The Conformal Predictions Framework

The theory of conformal predictions was developed by Vovk, Shafer and Gammerman [1] based on the principles of algorithmic randomness, transductive inference and hypothesis testing. This theory is based on the relationship derived between transductive inference and the Kolmogorov complexity of an i.i.d. (identically independently distributed) sequence of data instances. Hypothesis testing is subsequently used to construct conformal prediction regions, and obtain reliable measures of confidence. The CP framework brings together principles of hypothesis testing and traditional machine learning algorithms through the definition of a non-conformity score, which is a measure that quantifies the conformity of a data point to a particular class label, and is defined suitably for

---

[1] http://scikit-learn.org/stable/

each classifier. As an example, the non-conformity measure of a data point x_i for a k-Nearest Neighbor classifier is defined as:

$$\alpha_i^y = \frac{\sum_{j=1}^k D_{ij}^y}{\sum_{j=1}^k D_{ij}^{-y}} \tag{1}$$

where D_i^y denotes the list of sorted distances between a particular data point x_i and other data points with the same class label, and D_i^{-y} denotes the list of sorted distances between x_i and data points with any other class label. D_{ij}^y is the jth shortest distance in the list of sorted distances, D_i^y. Figure 1 illustrates the idea.

The methodology for applying the CP framework in a classification setting is as follows. Given a new test data point, say x_{n+1}, a null hypothesis is assumed that x_{n+1} belongs to the class label, say, $y^{(p)}$. The non-conformity measures of all the data points in the system so far are recomputed assuming the null hypothesis is true. A p-value function is defined as:

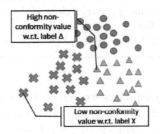

Fig. 1. An illustration of the non-conformity measure defined for k-NN

$$p(\alpha_{n+1}^{y^{(p)}}) = \frac{count\left\{i \in \{1,\dots,n+1\} : \alpha_i^{y^{(p)}} \geq \alpha_{n+1}^{y^{(p)}}\right\}}{n+1} \tag{2}$$

where $\alpha_{n+1}^{y^{(p)}}$ is the non-conformity measure of x_{n+1}, assuming it is assigned the class label $y^{(p)}$. It is evident that the p-value is highest when all non-conformity measures of training data belonging to class $y^{(p)}$ are higher than that of the new test point, x_{n+1}, which points out that x_{n+1} is *most conformal* to the class $y^{(p)}$. This process is repeated with the null hypothesis supporting each of the class labels, and the highest of the p-values is used to decide the actual class label assigned to x_{n+1}, thus providing a transductive inferential procedure for classification. If p_j is the highest p-value and p_k is the second highest p-value, then p_j is called the *credibility* of the decision, and $1 - p_k$ is the *confidence* of the classifier in the decision. Given a user-specified confidence level, ϵ, the output conformal prediction regions, Γ_ϵ, contain all the class labels with a p-value greater than $1 - \epsilon$. These regions are *conformal* i.e. the confidence threshold, $1 - \epsilon$ directly translates to the frequency of errors, ϵ in the online setting [2]. The methodology is summarized in Algorithm 1. The CP framework can be used in association with any classifier, with the suitable definition of a non-conformity measure. Sample non-conformity measures for various classification algorithms are presented in Table 1 [1].

2.2 Mondrian Conformal Predictors

The prediction regions output by the CP framework are *valid*, i.e. the frequency of errors is calibrated by the user-defined confidence level. However, in cases of

---

Algorithm 1. Conformal Predictors for Classification

Require: Training set $T = \{(x_1, y_1), ..., (x_n, y_n)\}$, $x_i \in X$, Number of classes M, $y^{(i)} \in Y = \left\{y^{(1)}, y^{(2)}, \ldots, y^{(M)}\right\}$, Classifier Ξ, Confidence level ϵ

1: Get new unlabeled example x_{n+1}.
2: **for** all class labels, $y^{(j)}$, where $j = 1, \ldots, M$ **do**
3:    Assign label $y^{(j)}$ to x_{n+1}.
4:    Update the classifier Ξ, with $T \cup \left\{x_{n+1}, y^{(j)}\right\}$.
5:    Compute non-conformity measure value, $\alpha_i^{y^{(j)}} \forall i = 1, \ldots, n+1$ to compute the p-value, p_j, w.r.t. class $y^{(j)}$ (Equation 2).
6: **end for**
7: Output the conformal prediction regions $\Gamma_\epsilon = \left\{y^{(j)} : p_j > 1 - \epsilon, y_j \in Y\right\}$.

---

imbalanced datasets, where the number of data instances in one class is significantly greater than the numbers in the others, it is possible that all errors may manifest within a single minority class (e.g., cancerous class in cancer prediction problems), limiting the practical applicability of the obtained predictions. In order to overcome this issue, Vovk et al. [1] proposed the Mondrian conformal predictors, where the p-value is computed by comparing the non-conformity score of the new test data instance against only training instances from the same label class as the current hypothesis for the new instance (3) (we address the earlier method as captured in Algorithm 1 as the *Transductive* setting.)

$$p(\alpha_{n+1}^{y^{(p)}}) = \frac{count\left\{i \in \{1, \ldots, n+1\} : y_i = y^{(p)} \text{ and } \alpha_i^{y^{(p)}} \geq \alpha_{n+1}^{y^{(p)}}\right\}}{count\left\{i \in \{1, \ldots, n+1\} : y_i = y^{(p)}\right\}} \quad (3)$$

3   PyCP: An Open-Source Conformal Predictions Toolkit

We now introduce *PyCP*, the first open-source implementation of the CP framework. In the currently available implementation (please see Section 4 for availability), we chose to use the Python 2.7 programming language. (Although adoption of the more recent Python 3 is high, certain critical Python 2 libraries are not yet supported in Python 3.) As mentioned earlier, in order to avoid code duplication of standard machine learning algorithms, we developed *PyCP* on top of the popular *scikit-learn* toolkit [15], an open-source machine learning library for the Python programming language. *scikit-learn* features various classification, regression and clustering algorithms including support vector machines, logistic regression, naive Bayes, k-means and DBSCAN; and is designed to interoperate with the Python numerical and scientific libraries, *NumPy* and *SciPy*. *scikit-learn* is under active development, and is being used by machine learning researchers around the world.

The overall design of *PyCP* is presented in Figure 2. *PyCP* was implemented with a modular design to allow for easy extensibility. The `prepareData` module

Table 1. Non-conformity measures for various classifiers

Classifier	Non-conformity measure	Description
k-NN	$$\frac{\sum_{j=1}^{k} D_{ij}^{y}}{\sum_{j=1}^{k} D_{ij}^{-y}}$$	Ratio of the sum of the distances to the k nearest neighbors belonging to the same class as the hypothesis y, and the sum of the distances to the k nearest neighbors belonging to all other classes (See [1] Chapter 4).
Support Vector Machines	Lagrange multipliers, or a suitable function of the distance of a data point from the hyperplane	Data points on the correct side of the hyperplane have low non-conformity scores, while data points on the margin or the incorrect side have high non-conformity scores [16].
Decision Trees	$$\frac{1}{N} \sum_{i \in J} d(x_j, x_i)$$	Average distance of a data point to all other points sharing the same label. The distance between two points is defined as the length of the shortest path between the terminal nodes containing them. If two points share the same terminal node the distance is 0 (adapted from [1]).
Logistic Regression	$$\begin{cases} -w.x & ,y=1 \\ w.x & ,y=0 \end{cases}$$	A monotonic transformation of the reciprocal of the estimated probability of the observed y given the observed x for a given data instance. w is the weight vector typically computed using Maximum Likelihood Estimation (See [1] Chapter 4).

randomly samples the data by class and returns a partition of the original dataset into two new datasets: a training set and a test set, based on a user-specified percentage split. Each dataset contains the same proportion of class labels as the original data. The computeNcs module computes the non-conformity scores for each of the classifiers as listed in Table 1, by invoking the respective modules for each of the classifiers (sklearn.neighbors.classification, sklearn.svm.classes,sklearn.tree.tree, and sklearn.linear_model .logistic) from *scikit-learn*. In case of Support Vector Machines, we use the one-versus-one model for multi-class classification, i.e., if $\alpha_i^{j,k}$ denotes the non-conformity score of point x_i in the $y^{(j)}$ vs $y^{(k)}$ model, the non-conformity score $\alpha_i^{y^{(j)}}$ is computed as the average of the non-conformity scores from all the models involving $y^{(j)}$, i.e. $\frac{1}{n-1} \sum_{k \neq j} \alpha_i^{j,k}$. The ncsToPVal module computes the p-value of a new data instance with respect to each class label, based on a user-specified choice between the *Transductive* and *Mondrian* settings of the framework. The computePVal module iterates over the test instances, invoking the computeNcs and ncsToPVal modules with appropriate input

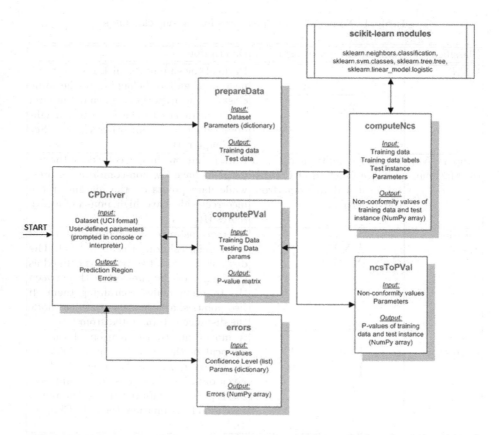

Fig. 2. Overall design of *PyCP*

parameters, and consolidates the p-values for each class label and each of these test instances. The `CPdriver` module serves as a primary reference point, and initiates the implementation of the complete framework by calling all the modules as required. The `errors` module is used to present the final output including errors, empty predictions, multiple predictions and graph plots. We note that the current version of *PyCP* implements CPs in the online setting. While the *PyCP* toolkit supports only the classifiers listed above, the underlying *scikit-learn* toolkit provides a robust set of options for these and other classifiers. To incorporate a classifier not listed here, the modular design of *PyCP* only requires a user to compute an $n \times 1$ non-conformity vector (where n is the number of data points) with a desired non-conformity measure for the new classifier, and store it in a *Numpy* array object within the `computeNcs` module. The classifier can then be invoked from the options in the `CPdriver` module.

4   Availability, Use and Examples

PyCP is currently available for download at: http://www.public.asu.edu/~v-nallure/conformalpredictions/pycp.html, and is distributed as open-source under the BSD-2 license. It can be downloaded as a .zip file containing the aforementioned .py modules, which can be compiled and run through IDLE (the Integrated Development Environment developed for Python) or a similar shell program. On IDLE, *PyCP* is initiated by specifying the options and parameters in quickRun.py file. Figure 3 illustrates a sample quickRun.py file. As evident, the user can specify the framework setting in cpType (*Mondrian* = 0 or *Transductive* = 1), the list of confidence levels at which results are sought in confList, the portion of the dataset to be used for training in trainPortion, the classifier option in classifier (k-NN = 1; SVM = 2; Decision Trees = 3; Logistic Regression = 4), and other classifier-specific paramters such as k for the k-NN classifier. Once the parameters are specified, quickRun.py can be run on IDLE to obtain the predictions for the test data points. Instructions for using *PyCP* are also available in the README.txt included in the downloaded .zip file. Figure 4 shows the IDLE shell when quickRun.py is executed to obtain conformal predictions for the test data points. *PyCP* currently operates with

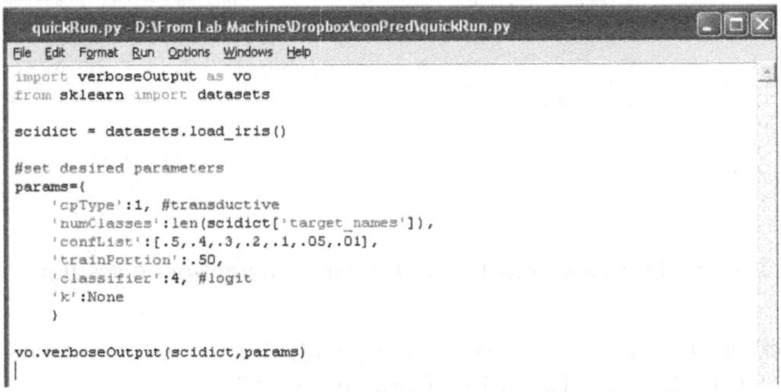

Fig. 3. Specifying options and parameters on quickRun.py within an IDLE shell

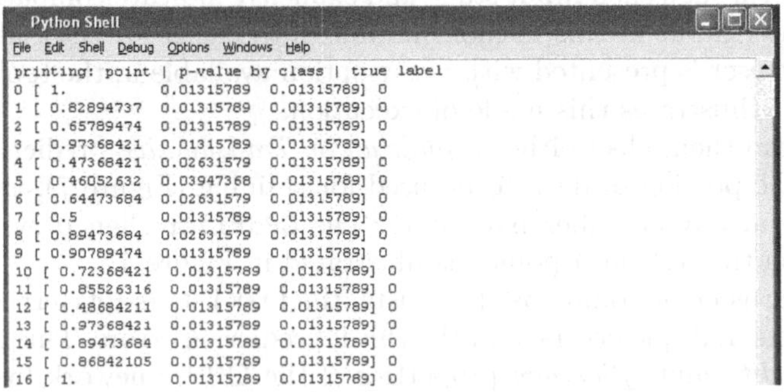

Fig. 4. Running quickRun.py within an IDLE shell

```
                              *Python Shell*
 File  Edit  Debug  Options  Windows  Help
 Python 2.7.3 (default, Sep 26 2012, 21:53:58)
 [GCC 4.7.2] on linux2
 Type "copyright", "credits" or "license()" for more information.
 ==== No Subprocess ====
 >>>
 Welcome to PyCP, the Python Conformal Predictor.
```

Fig. 5. Running the *CPdriver* module from within an IDLE shell

```
The program supports Mondrian and transductive conformal frameworks.
Which flavor would you like?
Enter '0' for Mondrian prediction.
Enter '1' for transductive prediction.
0
```

Fig. 6. The user selects the Mondrian conformal prediction framework

```
Testing with UCI Iris dataset.
What portion of the data is to be used for training?
0.8

Preparing data using a training portion of 80.0%.
Data is ready.
```

Fig. 7. The user selects a portion of data to be used for training

```
The program supports k-nearest neighbors, support vector machines,
and decision trees.
Which classifer would you like to use?
Enter '1' for k-nearest neighbor.
Enter '2' for support vector machine.
Enter '3' for decision tree.
1
```

Fig. 8. The user selects the *k*-nearest neighbors classifier

datasets specified in the .csv (comma-separated value) format, as used commonly in the UCI Machine Learning Repository [17].

PyCP can also be executed using the IDLE shell console, where the program is driven by a series of prompts requesting user input. Organized console prints walk the user through each input stage and explicitly indicate compatible inputs. Prompting user input in this fashion minimizes the need for tutorial documentation, as the user is presented with every option available in the toolkit at each step. Figure 5 illustrates this mode of execution.

The user can then select either *Mondrian* or *Transductive* prediction settings (Figure 6), the portion of data to be used for training (Figure 7), the classifier (Figure 8), and any classifier paramters. The CPdriver then provides output predictions on the test data points as illustrated in Figure 4.

PyCP returns error counts as three lists; the first lists empty errors by class, the second lists multiple errors, and the third, prediction errors. This helps study both the *validity* and *efficiency* properties of the CP framework (as described in [1]). Finally, *PyCP* also provides plotting capabilities using the *Matplotlib* library. Figure 9 shows a sample plot obtained using *PyCP*.

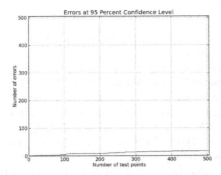

Fig. 9. Sample plot obtained in *PyCP* on the Breast Cancer dataset from the UCI Machine Learning repository [17] using Decision Trees classifier

5   Conclusions and Future Work

In this paper, we have introduced *PyCP*, the first open-source implementation of the Conformal Predictions (CP) framework. *PyCP* is currently available under the BSD-2 license for machine learning researchers and practitioners to apply the CP framework to datasets in their respective domains, as well as for further development by the community. *PyCP* currently provides support for applying the CP framework under classification settings, with four classifiers: *k*-Nearest Neighbors, Support Vector Machines, Decision Trees and Logistic Regression. *PyCP* also supports the Mondrian setting (for class-conditional validity), in addition to the standard transductive setting of the framework. *PyCP* has been developed using Python 2.7 on top of the popular *scikit-learn* toolkit, an open-source machine learning software developed in Python and used by machine learning researchers across the world. It has been designed in a modular fashion to allow for easy extensibility, and will lead to better experimental reproducibility, incremental advancements of previous work, and more percolation of technical methods into other disciplines and industry.

Our future efforts will include: (i) Extend emphPyCP to include inductive conformal predictors; (ii) Extend *PyCP* to other classifiers and regression settings; (iii) Allow different output options for users to choose from (text output of prediction regions, error graphs at different confidence levels, regions with a single prediction, etc.); (iv) Comprehensive documentation including an informative tutorial and in-line code-level comments; (v) Include a setup script (`setup.py`) and create a source distribution using Python Distutils (distribution utilities), which will allow *PyCP* to be hosted on the Python Package Index (PyPI) online repository; (vi) Develop a graphical interface or visual programming framework for the toolkit; (vii) Support other popular dataset formats such as the `.arff` format from *Weka*; (viii) Port the code to Python 3 when the necessary libraries are available on that platform; and (ix) Make *PyCP* available on `mloss.org`, a web portal exclusively maintained for machine learning open-source software.

References

1. Vovk, V., Gammerman, A., Shafer, G.: Algorithmic Learning in a Random World. Springer, New York (2005)
2. Vovk, V.: On-line confidence machines are well-calibrated. In: 43rd Symposium on Foundations of Computer Science, Washington, DC, USA, pp. 187–196 (2002)
3. Vanderlooy, S., van der Maaten, L., Sprinkhuizen-Kuyper, I.: Off-line learning with transductive confidence machines: An empirical evaluation. In: Perner, P. (ed.) MLDM 2007. LNCS (LNAI), vol. 4571, pp. 310–323. Springer, Heidelberg (2007)
4. Ho, S.S., Wechsler, H.: Query by transduction. IEEE Transactions on Pattern Analysis and Machine Intelligence 30(9), 1557–1571 (2008)
5. Laxhammar, R., Falkman, G.: Conformal prediction for distribution-independent anomaly detection in streaming vessel data. In: Proceedings of the First International Workshop on Novel Data Stream Pattern Mining Techniques, StreamKDD 2010, pp. 47–55. ACM, New York (2010)
6. Bellotti, T., Luo, Z., Gammerman, A.: Strangeness minimisation feature selection with confidence machines. In: Corchado, E., Yin, H., Botti, V., Fyfe, C. (eds.) IDEAL 2006. LNCS, vol. 4224, pp. 978–985. Springer, Heidelberg (2006)
7. Ho, S.S., Wechsler, H.: A martingale framework for detecting changes in data streams by testing exchangeability. IEEE Transactions on Pattern Analysis and Machine Intelligence 32(12), 2113–2127 (2010)
8. Kukar, M.: Quality assessment of individual classifications in machine learning and data mining. Knowledge Information Systems 9(3), 364–384 (2006)
9. Lambrou, A., Papadopoulos, H., Gammerman, A.: Reliable confidence measures for medical diagnosis with evolutionary algorithms. IEEE Transactions on Information Technology in Biomedicine 15(1), 93–99 (2011)
10. Shao, H., Yu, B., Nadeau, J.H.: Strangeness-based feature weighting and classification of gene expression profiles. In: Wainwright, R.L., Haddad, H. (eds.) SAC, pp. 1292–1296. ACM (2008)
11. Li, F., Wechsler, H.: Open set face recognition using transduction. IEEE Trans. Pattern Anal. Mach. Intell. 27(11), 1686–1697 (2005)
12. Dashevskiy, M., Luo, Z.: Network traffic demand prediction with confidence. In: GLOBECOM, pp. 1453–1457 (2008)
13. Sonnenburg, S., Braun, M.L., Ong, C.S., Bengio, S., Bottou, L., Holmes, G., LeCun, Y., Müller, K.R., Pereira, F., Rasmussen, C.E., Rätsch, G., Schölkopf, B., Smola, A., Vincent, P., Weston, J., Williamson, R.: The need for open source software in machine learning. Journal of Machine Learning Research 8, 2443–2466 (2007)
14. Schaul, T., Bayer, J., Wierstra, D., Sun, Y., Felder, M., Sehnke, F., Rückstieß, T., Schmidhuber, J.: PyBrain. J. Mach. Learn. Res. 11, 743–746 (2010)
15. Pedregosa, F., Varoquaux, G., Gramfort, A., Michel, V., Thirion, B., Grisel, O., Blondel, M., Prettenhofer, P., Weiss, R., Dubourg, V., Vanderplas, J., Passos, A., Cournapeau, D., Brucher, M., Perrot, M., Duchesnay, D.: Scikit-learn: Machine learning in python. Journal of Machine Learning Research 12, 2825–2830 (2011)
16. Balasubramanian, V., Gouripeddi, R., Panchanathan, S., Vermillion, J., Bhaskaran, A., Siegel, R.: Support vector machine based conformal predictors for risk of complications following a coronary drug eluting stent procedure. Computers in Cardiology, 5–8 (2009)
17. Bache, K., Lichman, M.: UCI machine learning repository (2013)

Conformal Prediction under Hypergraphical Models

Valentina Fedorova, Alex Gammerman, Ilia Nouretdinov, and Vladimir Vovk

Computer Learning Research Centre, Royal Holloway, University of London,
Egham, Surrey, TW20 0EX, United Kingdom

Abstract. Conformal predictors are usually defined and studied under the exchangeability assumption. However, their definition can be extended to a wide class of statistical models, called online compression models, while retaining their property of automatic validity. This paper is devoted to conformal prediction under hypergraphical models that are more specific than the exchangeability model. Namely, we define two natural classes of conformity measures for such hypergraphical models and study the corresponding conformal predictors empirically on benchmark LED data sets. Our experiments show that they are more efficient than conformal predictors that use only the exchangeability assumption.

1 Introduction

The method of conformal prediction was introduced and is usually used for producing valid prediction sets under the exchangeability assumption; the validity of the method means that the probability of making a mistake is equal to (or at least does not exceed) a prespecified significance level ([6], Chapter 2). However, the definition of conformal predictors can be easily extended to a wide class of statistical models, called online compression models (OCMs; [6], Chapter 8). OCMs compress data into a more or less compact summary, which is interpreted as the useful information in the data. With each "conformity measure", which, intuitively, estimates how well a new piece of data fits the summary, one can associate a conformal predictor, which still enjoys the property of automatic validity. Numerous machine learning algorithms have been used for designing efficient conformity measures: see, e.g., [6] and [2].

This paper studies conformal prediction under the OCMs known as hypergraphical models ([6], Section 9.2). Such models describe relationships between data features. In the case where every feature is allowed to depend in any way on the rest of the features, the hypergraphical model becomes the exchangeability model. More specific hypergraphical models restrict the dependence in some way. Such restrictions are typical of many real-world problems: for example, different symptoms can be conditionally independent given the disease. A popular approach to such problems is to use Bayesian networks (see, e.g., [3]). The definition of Bayesian networks requires a specification of both the pattern of dependence between features and the distribution of the features.

H. Papadopoulos et al. (Eds.): AIAI 2013, IFIP AICT 412, pp. 371–383, 2013.

Usual methods guarantee a valid probabilistic outcome if the used distributions of features are correct. Several algorithms (see, e.g., [3], Chapter 9) are known for estimating the distribution of features; however, the accuracy of such approximations is a major concern in applying Bayesian networks. The conformal predictors constructed from hypergraphical OCMs use only the pattern of dependence between the features but do not involve their distribution. This makes conformal prediction based on hypergraphical models more robust and realistic than Bayesian networks. (The notion of a hypergraphical model can be regarded as more general than that of a Bayesian network: the standard algorithms in this area transform Bayesian networks into hypergraphical models by "marrying parents", forgetting the direction of the arrows, triangulation, and regarding the cliques of the resulting graph as the hyperedges; see, e.g., [3], Section 3.2.)

As far as we know, conformal prediction has been studied, apart from the exchangeability model and its variations, only for the Gauss linear model and Markov model (see [6], Chapter 8, and [4]). Hypergraphical OCMs have been used only in the context of Venn rather than conformal prediction (see [6], Chapter 9).

The rest of the paper is organised as follows. Section 2 formally defines hypergraphical OCMs and briefly reviews their basic properties. Section 3 describes the method of conformal prediction in the context of hypergraphical models and introduces two conformity measures for hypergraphical OCMs. Section 4 reports the performance of the corresponding conformal predictors on benchmark LED data sets. Section 5 concludes.

2 Background

Consider two measurable spaces $\mathbf{X}$ and $\mathbf{Y}$; elements of $\mathbf{X}$ are called *objects* and elements of $\mathbf{Y}$ are called *labels*. Elements of the Cartesian product $\mathbf{X} \times \mathbf{Y}$ are called *examples*. A *training set* is a sequence of examples $(z_1, \ldots, z_l)$, where each example $z_i = (x_i, y_i)$ consists of an object x_i and its label y_i. The general prediction problem considered in this paper is to predict the label for a new object given a training set. We focus on the case where $\mathbf{X}$ and $\mathbf{Y}$ are finite.

2.1 Hypergraphical Structures

In this paper we assume that examples are structured, consisting of variables. Hypergraphical structures describe relationships between the variables. Formally a *hypergraphical structure*[1] consists of three elements $(V, \mathcal{E}, \Xi)$:

1. V is a finite set; its elements are called *variables*.
2. $\mathcal{E}$ is a finite collection of subsets of V whose union covers all variables: $\bigcup_{E \in \mathcal{E}} E = V$. Elements of $\mathcal{E}$ are called *clusters*.

---

[1] The name reflects the fact that the components $(V, \mathcal{E})$ form a hypergraph, where a hyperedge $E \in \mathcal{E}$ can connect more than two vertices.

3. Ξ is a function that maps each variable $v \in V$ into a finite set (of the values that v can take).

A *configuration* on a set $E \subseteq V$ (we are usually interested in the case where E is a cluster) is an assignment of values to the variables from E; let $\Xi(E)$ be the set of all configurations on E. A *table*[2] on a set E is an assignment of natural numbers to the configurations on E. The *size* of the table is the sum of values that it assigns to different configurations. A *table set* is a collection of tables on the clusters $\mathcal{E}$, one for each cluster $E \in \mathcal{E}$. The number assigned by a table set σ to a configuration on E is called its σ-*count*.

2.2   Hypergraphical Online Compression Models

The example space $\mathbf{Z}$ associated with the hypergraphical structure is the set of all configurations on V. One of the variables in V is singled out as the *label variable*, and the configurations on the label variable are denoted $\mathbf{Y}$. All other variables are *object variables*, and the configurations on the object variables are denoted $\mathbf{X}$. Since $\mathbf{Z} = \mathbf{X} \times \mathbf{Y}$, this is a special case of the prediction setting described at the beginning of this section.

An example $z \in \mathbf{Z}$ *agrees* with a configuration on a set $E \subseteq V$ (or the configuration agrees with the example) if the restriction $z|_E$ of z to the variables in E coincides with the configuration. A table set σ *generated* by a sequence of examples $(z_1, \ldots, z_n)$ assigns to each configuration on each cluster the number of examples in the sequence that agree with the configuration; the size of each table in σ will be equal to the number of examples in the sequence, and this number is called the *size* of the table set. Different sequences of examples can generate the same table set σ, and we denote $\#\sigma$ the number of different sequences generating σ.

The *hypergraphical online compression model* (HOCM) associated with the hypergraphical structure $(V, \mathcal{E}, \Xi)$ consists of five elements $(\Sigma, \square, \mathbf{Z}, F, B)$, where:

1. The *empty table set* $\square$ is the table set assigning 0 to each configuration.
2. The set Σ is defined by the conditions that $\square \in \Sigma$ and $\Sigma \setminus \{\square\}$ is the set of all table sets σ with $\#\sigma > 0$. The elements $\sigma \in \Sigma$ are called *summaries*.
3. The *forward function* $F(\sigma, z)$, where σ ranges over Σ and z over $\mathbf{Z}$, updates σ by adding 1 to the σ-count of each configuration which agrees with z.
4. The *backward kernel* B maps each $\sigma \in \Sigma \setminus \{\square\}$ to a probability distribution $B(\sigma)$ on $\Sigma \times \mathbf{Z}$ assigning the weight $\#(\sigma \downarrow z)/\#\sigma$ to each pair $(\sigma \downarrow z, z)$, where z is an example such that, for all configurations which agree with z, the corresponding σ-counts are positive, and $\sigma \downarrow z$ is the table set obtained by subtracting 1 from the σ-counts of the configurations that agree with z. Notice that $B(\sigma)$ is indeed a probability distribution, and it is concentrated on the pairs $(\sigma \downarrow z, z)$ such that $F(\sigma \downarrow z, z) = \sigma$.

---

[2] Generally, a table assigns real numbers to configurations. In this paper we only consider *natural tables*, which assign natural numbers to configurations, and omit "natural" for brevity.

We will use "hypergraphical models" as a general term for hypergraphical structures and HOCMs when no precision is required. When discussing hypergraphical models we will always assume that the examples $z_1, z_2, \ldots$ are produced independently from a probability distribution Q on $\mathbf{Z}$ that has a decomposition

$$Q(\{z\}) = \prod_{E \in \mathcal{E}} f_E(z|_E) \tag{1}$$

for some functions $f_E : \Xi(E) \to [0,1]$, $E \in \mathcal{E}$, where z is an example and $z|_E$ its restriction to the variables in E.

2.3 Junction Tree Structures

An important type of hypergraphical structures is where clusters can be arranged into a "junction tree". For the corresponding HOCMs we will be able to describe efficient calculations of the backward kernels. If one wants to use the calculations for a structure that cannot be arranged into a junction tree it can be replaced by a more general junction tree structure before defining the HOCM.

Let (U, S) denote an undirected tree with U the set of vertices and S the set of edges. Then (U, S) is a *junction tree* for a hypergraphical structure $(V, \mathcal{E}, \Xi)$ if there exists a bijective mapping C from the set of vertices U of the tree to the set $\mathcal{E}$ of clusters of the hypergraphical structure that has the following property: $C_u \cap C_w \subseteq C_v$ whenever a vertex v lies on the path from a vertex u to a vertex w in the tree (we let C_x stand for $C(x)$). Not every hypergraphical structure has a junction tree, of course: an example is a hypergraphical structure with three clusters whose intersection is empty but whose pairwise intersections are not. See, e.g., [3], Section 4.3, for further information on junction trees; intuitive examples of junction trees will be given in Section 4.

If $s = \{u, v\} \in S$ is an edge of the junction tree connecting vertices u and v then C_s stands for $C_u \cap C_v$. It is convenient to identify vertices u and edges s of the junction tree with the corresponding clusters C_u and sets C_s, respectively.

If $E_1 \subseteq E_2 \subseteq V$ and f is a table on E_2, the *marginalisation* of f to E_1 is the table f^* on E_1 assigning to each $a \in \Xi(E_1)$ the number $f^*(a) = \sum_b f(b)$, where b ranges over the configurations on E_2 such that $b|_{E_1} = a$. If σ is a summary then for $u \in U$ denote σ_u the table that σ assigns to C_u, and for $s = \{u, v\} \in S$ denote σ_s the marginalisation of σ_u (or σ_v) to C_s. We will use the shorthand $\sigma_u(z)$ for the number assigned to the restriction $z|_{C_u}$ by the table for the vertex u and $\sigma_s(z)$ for the number assigned to $z|_{C_s}$ by the marginal table for the edge s. Consider the HOCM corresponding to the junction tree (U, S). We use the notation $P_\sigma(z)$ for the weight assigned by $B(\sigma)$ to $(\sigma \downarrow z, z)$. It has been proved ([6], Lemma 9.5) that

$$P_\sigma(z) = \frac{\prod_{u \in U} \sigma_u(z)}{n \prod_{s \in S} \sigma_s(z)}, \tag{2}$$

where n is the size of σ. If any of the factors in (2) is zero then the whole ratio is set to zero.

3   Conformal Prediction for HOCM

Consider a training set $(z_1, \ldots, z_l)$ and an HOCM $(\Sigma, \square, \mathbf{Z}, F, B)$. The goal is to predict the label for a new object x.

A *conformity measure* for the HOCM is a measurable function $A : \Sigma \times \mathbf{Z} \to \mathbb{R}$. The function assigns a *conformity score* $A(\sigma, z)$ to an example z w.r. to a summary σ. Intuitively, the score reflects how typical it is to observe z having the summary σ.

For each $y \in \mathbf{Y}$ denote $\sigma^* \in \Sigma$ the table set generated by the sequence $(z_1, \ldots, z_l, (x, y))$ (the dependence of σ^* on y is important although not reflected in our notation). For $z \in \mathbf{Z}$ such that $\sigma^* \downarrow z$ is defined denote the conformity scores as $\alpha_z := A(\sigma^* \downarrow z, z)$ (notice that $\alpha_{(x,y)}$ is always defined). The *p-value* for y, denoted $p^{(y)}$, is defined by

$$p^{(y)} := \sum_{z : \alpha_z < \alpha_{(x,y)}} P_{\sigma^*}(z) + \theta \cdot \sum_{z : \alpha_z = \alpha_{(x,y)}} P_{\sigma^*}(z) \tag{3}$$

(cf. (8.4) in [6]), where $\theta \sim \mathbf{U}[0, 1]$ is a random number from the uniform distribution on $[0, 1]$, $P_{\sigma^*}(z)$ is the backward kernel, as defined above, and the sums involve only those $z \in \mathbf{Z}$ for which α_z is defined. Then for a significance level ϵ the *conformal predictor* Γ based on A outputs the prediction set

$$\Gamma^\epsilon(z_1, \ldots, z_l, x) := \{y \in \mathbf{Y} : p^{(y)} > \epsilon\}.$$

(Such randomized conformal predictors were referred to as "smoothed" in [6].)

We will describe two conformity measures for HOCMs in Subsection 3.1. These conformity measures optimise different criteria for the quality of conformal predictors. The following subsection describes the criteria used in this paper.

3.1   Conformity Measures for HOCM

Consider a summary σ and an example (x, y). The *conditional probability conformity measure* is defined by

$$A(\sigma, (x, y)) := P_{\sigma^*}(y \mid x) := \frac{P_{\sigma^*}((x, y))}{\sum_{y' \in \mathbf{Y}} P_{\sigma^*}((x, y'))}, \tag{4}$$

where $\sigma^* := F(\sigma, (x, y))$ and $P_{\sigma^*}((x, y))$ is the backward kernel. In other words, $A(\sigma, (x, y))$ is the conditional probability $P_{\sigma^*}(y \mid x)$ of y given x under P_{σ^*}. The conditional probability $P_{\sigma^*}(y \mid x)$ can be easily computed using (2).

Define the *predictability* of an object $x \in \mathbf{X}$ as

$$f(x) := \max_{y \in \mathbf{Y}} P_{\sigma^*}(y \mid x), \tag{5}$$

the maximum of conditional probabilities. If the predictability of an object is close to 1 then the object is "easily predictable". Fix a *choice function* $\hat{y} : \mathbf{X} \to \mathbf{Y}$ such that

$$\forall x \in \mathbf{X} : f(x) = P_{\sigma^*}(\hat{y}(x) \mid x).$$

The function maps each object x to one of the labels at which the maximum in (5) is attained. The *signed predictability conformity measure* is defined by

$$A(\sigma, (x, y)) := \begin{cases} f(x) & \text{if } y = \hat{y}(x) \\ -f(x) & \text{otherwise.} \end{cases} \tag{6}$$

3.2 Criteria for the Quality of Conformal Prediction

In this paper we study the performance of conformal predictors in the online prediction protocol (Protocol 1). Reality generates examples (x_n, y_n) from a probability distribution Q satisfying (1) for some hypergraphical structure. Predictor uses a conformal predictor Γ to output the prediction set $\Gamma_n^\epsilon :=$ $\Gamma^\epsilon(x_1, y_1, \ldots, x_{n-1}, y_{n-1}, x_n)$ at each significance level ϵ.

Protocol 1. Online prediction protocol

---
  for $n = 1, 2, \ldots$ **do**
    Reality outputs $x_n \in \mathbf{X}$
    Predictor outputs $\Gamma_n^\epsilon \subseteq \mathbf{Y}$ for all $\epsilon \in (0, 1)$
    Reality outputs $y_n \in \mathbf{Y}$
  end for

Two important properties of conformal predictors are their validity and efficiency; the first is achieved automatically and the second is enjoyed by different conformal predictors to a different degree. Predictor *makes an error* at step n if y_n is not in Γ_n^ϵ. The validity of conformal predictors means that, for any significance level ϵ, the probability of error $y_n \notin \Gamma_n^\epsilon$ is equal to ϵ. It has been proved that conformal predictors are automatically valid under their models ([6], Theorem 8.1). In this paper we study problems where the hypergraphical model used for computing the p-values is known to be correct; therefore, the predictions will always be valid, and there is no need to test validity experimentally.

The efficiency of valid predictions can be measured in different ways. The standard way is to count the *number of multiple predictions* Mult_n^ϵ over the first n steps defined by

$$\text{mult}_n^\epsilon := \begin{cases} 1 & \text{if } |\Gamma_n^\epsilon| > 1 \\ 0 & \text{otherwise} \end{cases} \quad \text{and} \quad \text{Mult}_n^\epsilon := \sum_{i=1}^{n} \text{mult}_i^\epsilon$$

at each significance level $\epsilon \in (0, 1)$ (cf. [6], Chapter 3). Another way is to report the *cumulative size of the prediction sets*

$$\text{Size}_n^\epsilon := \sum_{i=1}^{n} |\Gamma_i^\epsilon|$$

at each significance level $\epsilon \in (0, 1)$. We will also consider two ways to measure the efficiency of conformal predictors that do not depend on the significance level. Let $p_n^{(y)}$, $y \in \mathbf{Y}$, be the p-values (3) used by the conformal predictor for computing the prediction set Γ_n^ϵ at the nth step of the online prediction protocol. The *cumulative unconfidence* Unconf_n over the first n steps is defined by

$$\mathrm{unconf}_n := \inf \left\{ \epsilon : |\Gamma_n^\epsilon| \leq 1 \right\} \quad \text{and} \quad \mathrm{Unconf}_n := \sum_{i=1}^{n} \mathrm{unconf}_i;$$

the *unconfidence* unconf_n at step n can be equivalently defined as the second largest p-value among $p_n^{(y)}$, $y \in \mathbf{Y}$. (Unconfidence is a trivial modification of the standard notion of confidence: see [6], (3.66).) Finally, the efficiency can be measured by the *cumulative sum of p-values*

$$\mathrm{pSum}_n := \sum_{i=1}^{n} \sum_{y \in \mathbf{Y}} p_i^{(y)}.$$

All four criteria work in the same direction: the smaller the better. As already mentioned, the number of multiple predictions is a standard criterion; the three other criteria are first used in this paper and [5].

In our experiments we will use the following more intuitive versions of the first two criteria: the *percentage of multiple predictions* $\mathrm{Mult}_n^\epsilon/n$ and the *average size of predictions* $\mathrm{Size}_n^\epsilon/n$; we would like the former to be close to 0 and the latter to be close to 1 for small significance levels.

It can be shown that, in a wide range of situations:

- the signed predictability conformity measure is optimal in the sense of Mult_n^ϵ and in the sense of Unconf_n;
- the conditional probability conformity measure is optimal in the sense of Size_n^ϵ and in the sense of pSum_n.

See [5] for precise statements and proofs.

4   Experimental Results

4.1   LED Data Set

For our experiments we use benchmark LED data sets generated by a program from the UCI repository [1]. The problem is to predict a digit from an image in the seven-segment display.

Figure 1 shows several objects in the data set (these are "ideal images" of digits; there are also digits corrupted by noise). The seven leds (light emitting diodes) can be lit in different combinations to represent a digit from 0 to 9. The program generates examples with noise. There is an ideal image for each digit. An example has seven binary attributes $s_0, \ldots, s_6$ (s_i is 1 if the ith led is lit) and a label c, which is a decimal digit. The program randomly chooses a label (0 to 9 with equal probabilities), inverts each of the attributes of its ideal image with probability $p_{\mathrm{noise}} = 1\%$ independently, and adds the noisy image and the label to the data set.

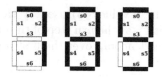

Fig. 1. LED images for digits 7, 8, and 9 in the seven-segment display

Let $(S_0, \ldots, S_6, C)$ be the vector of random variables corresponding to the attributes and the label, and let $(s_0, \ldots, s_6, c)$ be an example. According to the generating mechanism the probability of the example decomposes as

$$Q\left(\{(s_0, \ldots, s_6, c)\}\right) = Q_7\left(C = c\right) \cdot \prod_{i=0}^{6} Q_i\left(S_i = s_i \mid C = c\right), \qquad (7)$$

where Q_7 is the uniform distribution on the decimal digits and

$$Q_i\left(S_i = s_i \mid C = c\right) := \begin{cases} 1 - p_{\text{noise}} & \text{if } s_i = s_i^c \\ p_{\text{noise}} & \text{otherwise,} \end{cases} \qquad i = 0, \ldots, 6, \qquad (8)$$

$(s_0^c, \ldots, s_6^c, c)$ being the attributes of the ideal image for the label c. As usual, examples are generated independently.

4.2 Hypergraphical Assumptions for LED Data Sets

We consider two hypergraphical models that agree with the decomposition (7). These models make different assumptions about the pattern of dependence between the attributes and the label; they do not depend on a particular probability of noise p_{noise} or the fact that the same value of p_{noise} is used for all leds. For both hypergraphical structures the set of variables is $V := \{s_0, \ldots, s_6, c\}$.

Nontrivial Hypergraphical Model. Consider the hypergraphical structure with the clusters $\mathcal{E} := \{\{s_i, c\} : i = 0, \ldots, 6\}$. A junction tree for this hypergraphical structure can be defined as a chain with vertices $U := \{u_i : i = 0, \ldots, 6\}$ and the bijection $C_{u_i} := \{s_i, c\}$. By saying that U is a chain we mean that there are edges connecting vertices 0 and 1, 1 and 2, 2 and 3, 3 and 4, 4 and 5, and 5 and 6 (and these are the only edges). It is clear that this is a junction tree and that $C_s = \{c\}$ for each edge s. It is also clear from (7) that the assumption (1) is satisfied; e.g., we can set

$$f_{\{s_0, c\}}\left(s_0, c\right) := Q_7\left(C = c\right) \cdot Q_0\left(S_0 = s_0 \mid C = c\right);$$
$$f_{\{s_i, c\}}\left(s_i, c\right) := Q_i\left(S_i = s_i \mid C = c\right), \quad i = 1, \ldots, 6.$$

Exchangeability Model. The hypergraphical model with no information about the pattern of dependence between the attributes and the label is the exchangeability model. The corresponding hypergraphical structure has one cluster, $\mathcal{E} := \{V\}$. The junction tree is the one vertex associated with V and no edges.

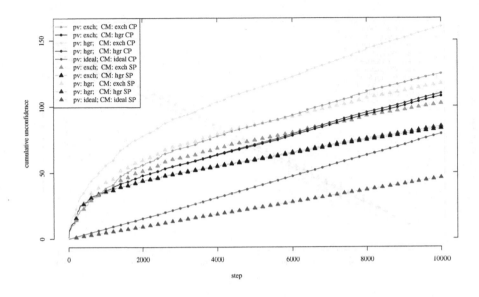

Fig. 2. Cumulative unconfidence for online predictions. The results are for the LED data set with 1% of noise and 10000 examples.

4.3   Experiments

For our experiments we create a LED data set with 10000 examples. The data are generated according to the model (7) with the probability of noise $p_{\text{noise}} = 1\%$. The data generation programs are written in C, and our data processing programs are written in R; in both cases we set the seed of the pseudorandom number generator to 0. The text below assumes that the reader can see Figures 2–5 in colour; the colours become different shades of grey in black-and-white. We hope our descriptions will be detailed enough for the reader to identify the most important graphs unambiguously.

Each of the figures corresponds to an efficiency criterion for conformal predictors; namely, Figure 2 plots Unconf_n versus $n = 1, \ldots, 10000$ in the online prediction protocol, Figure 3 plots pSum_n versus $n = 1, \ldots, 10000$, Figure 4 plots $\text{Mult}_{10000}^\epsilon / 10000$ (the percentage of multiple predictions) versus $\epsilon \in [0, 0.05]$, and Figure 5 plots $\text{Size}_{10000}^\epsilon / 10000$ (the average size of predictions) versus $\epsilon \in [0, 0.05]$. We consider two conformity measures: the conditional probability (CP) conformity measure (4) and the signed predictability (SP) conformity measure (6). The graphs corresponding to the former are represented in our plots as lines with dots, and the graphs corresponding to the latter are represented as lines with triangles.

Two of the plots in each figure correspond to idealized predictors and are drawn only for comparison, representing an unachievable ideal goal. In the

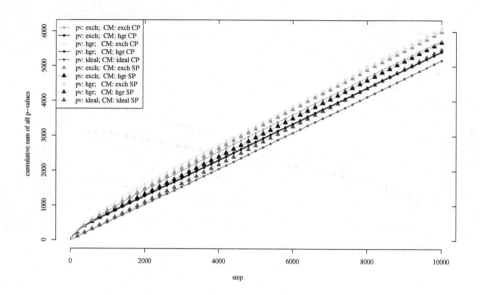

Fig. 3. Cumulative sum of p-values for online predictions. The results are for the LED data set with 1% of noise and 10000 examples.

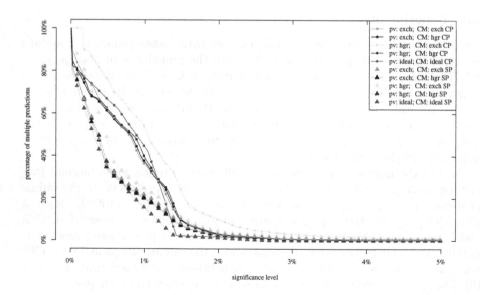

Fig. 4. The final percentage of multiple predictions for significance levels between 0% and 5%. The results are for the LED data set with 1% of noise and 10000 examples.

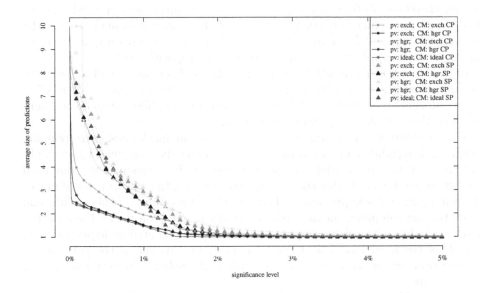

Fig. 5. The final average size of predictions for significance levels between 0% and 5%. The results are for the LED data set with 1% of noise and 10000 examples.

idealized case we know the true distribution for the data (given by (7), (8), and $p_{\text{noise}} = 1\%$). The true distribution is used instead of the backward kernel P_{σ^*} in both (3) and (4) for the CP conformity measure and in both (3) and (6) for the SP conformity measure. It gives us the ideal results (the two red lines in our plots) for the two conformity measures, CP and SP. At least one of them gives the best results in each of the figures (remember that for all our criteria the lower the better).

For each of the two conformity measures we also consider four realistic predictors (which are conformal predictors, unlike the idealized ones). The *pure hypergraphical conformal predictor* (represented by blue lines in our plots) is obtained using the nontrivial hypergraphical model both when computing p-values (see (3)) and when computing the conformity measure ((4) in the case of CP and (6) in the case of SP). Analogously we use the exchangeability model to obtain the *pure exchangeability conformal predictor* (green lines in our plots). The two *mixed conformal predictors* (black and yellow lines) are obtained when we use different models to compute the p-values and the conformity scores.

The intuition behind the pure and mixed confomal predictors can be explained using the distinction between hard and soft models made in [7]. The model used when computing the p-values (see (3)) is the hard model; the validity of the conformal predictor depends on it. The model used when computing conformity scores (see (4) and (6)) is the soft model; when it is violated, validity is not affected, although efficiency can suffer. The true probability distribution (7) conforms to both the exchangeability model and the nontrivial hypergraphical model; therefore, all four conformal predictors are automatically valid, and we study only their efficiency. (In the context of this paper, it is obvious that the exchangeability model is more general than the nontrivial hypergraphical model, but we can also apply the criterion given in [6], Proposition 9.2.)

In the legends of Figures 2–5, the hard model used is indicated after "pv" (the way of computing the p-values), and the soft model used is indicated after "CM" (the conformity measure); "exch" refers to the exchangeability model, and "hgr" refers to the nontrivial hypergraphical model.

The most interesting graphs in Figures 2–5 are the black ones, corresponding to the exchangeability model as the hard model and the nontrivial hypergraphical model as the soft model. The performance of the corresponding conformal predictors is typically better than, or at least close to, the performance of any of the remaining realistic predictors. The fact that the validity of these conformal predictors only depends on the exchangeability assumption makes them particularly valuable. The yellow graphs correspond to the nontrivial hypergraphical model as the hard model and the exchangeability model as the soft model; the performance of the corresponding conformal predictors is very poor in our experiments.

Now we will comment on each of the figures individually. We only discuss the results for the seed 0 of the pseudorandom numbers generators, but we have observed similar results for other seeds as well.

Figure 2 shows the cumulative unconfidence $Unconf_n$, and so the right conformity measure to use is SP, as discussed at the end of Section 3; and indeed, all SP graphs lie below their CP counterparts. The two bottom graphs are the ones corresponding to idealized predictors; the graph corresponding to the CP idealized predictor, however, has a suboptimal slope. Of the realistic predictors, the lowest graph is the black SP one (but the blue SP graph, corresponding to the pure hypergraphical conformal predictor, is very close).

Figure 3 shows the cumulative sum of p-values $pSum_n$. For this criterion the predictors based on the CP conformity measure outperform the predictors based on the SP conformity measure (the lines with dots are below the lines of the same colour with triangles), as expected. The bottom graph corresponds to the idealized CP predictor; the idealized SP predictor is the second best most of the time, but at the end it is overtaken by the black and blue graphs corresponding to the conformal predictors based on the CP conformity measure using the nontrivial hypergraphical model. The black and blue graphs are very close; the blue one is slightly lower but the conformal predictor corresponding to the black one still appears preferable as its validity only depends on the weaker exchangeability assumption. Notice that even for the best predictors the cumulative sum of p-values exceeds 5000: this is to be expected, as summing only the p-values for the true labels would already give 5000, up to statistical fluctuations.

Figure 4 shows the percentage of multiple predictions after observing 10000 examples as function of the significance level. For small significance levels the percentage of the multiple predictions is smaller for the predictors based on the SP conformity measure, again as expected. The performance of the conformal predictor corresponding to the black SP graph is again remarkably good, better than that of any other realistic predictor.

Figure 5 shows the average size of predictions after observing 10000 examples as function of the significance level. For small significance levels the predictors based on the CP conformity measure perform better, again confirming the theoretical results mentioned earlier. The black CP graph is very close (or even better) than the blue CP graph, corresponding to the pure hypergraphical predictor, except for very low significance levels when the average size exceeds 2.

5   Conclusion

The main finding of this paper is that nontrivial hypergraphical models can be useful for conformal prediction when they are true. More surprisingly, in our experiments they only need to be used as soft models; the performance does not suffer much if the exchangeability model continues to be used as the hard model. This interesting phenomenon deserves a further empirical study.

Acknowledgements. We thank the COPA 2013 reviewers for comments that improved the results of the paper (in particular, for suggesting the CP conformity measure). We are indebted to Royal Holloway, University of London, for continued support and funding. This work has also been supported by: the EraSysBio+ grant SHIPREC from the European Union, BBSRC and BMBF; a VLA grant on machine learning algorithms; a grant from the National Natural Science Foundation of China (No. 61128003); a grant from the Cyprus Research Promotion Foundation (research contract TPE/ORIZO/0609(BIE)/24); grant EP/K033344/1 from EPSRC.

References

1. Bache, K., Lichman, M.: UCI machine learning repository. School of Information and Computer Sciences. University of California, Irvine (2013), http://archive.ics.uci.edu/ml
2. Balasubramanian, V.N., Ho, S.S., Vovk, V. (eds.): Conformal Prediction for Reliable Machine Learning: Theory, Adaptations, and Applications. Elsevier, Waltham (to appear, 2013)
3. Cowell, R., Dawid, P., Lauritzen, S., Spiegelhalter, D.: Probabilistic Networks and Expert Systems. Springer, New York (1999) (reprinted in 2007)
4. Fedorova, V., Nouretdinov, I., Gammerman, A.: Testing the Gauss linear assumption for on-line predictions. Progress in Artificial Intelligence 1, 205–213 (2012)
5. Vovk, V., Fedorova, V., Nouretdinov, I., Gammerman, A.: Optimality criteria for conformal prediction. Manuscript (2013)
6. Vovk, V., Gammerman, A., Shafer, G.: Algorithmic Learning in a Random World. Springer, New York (2005)
7. Vovk, V., Nouretdinov, I., Gammerman, A.: On-line predictive linear regression. Annals of Statistics 37, 1566–1590 (2009)

Defensive Forecast for Conformal Bounded Regression

Ilia Nouretdinov[1] and Alexander Lebedev[2]

[1] Computer Learning Research Centre, Royal Holloway University of London
[2] Stavanger University Hospital, Centre for Age-Related medicine

Abstract. The paper considers a conformal prediction method for bounded regression task. A predictor was based on the Defensive Forecast algorithm and has been applied for a medical prognostic problem. These empirical results are compared and discussed.

1 Introduction

The conformal prediction has been applied to regression estimation in [1,2,3,4], under assumption that label y is approximately linearly dependent on feature vector x. This was also extended for non-linear dependency using a non-linear transformation (kernel mapping) Φ of x into a higher dimensional space – see, for example, [9].

In this paper we consider the non-linear problem of bounded regression. A typical problem that requires a bounded regression is a prediction of examination marks, bounded from 0% to 100% or some problems in medical prognosis that have a range from healthy individuals to patients with a completely developed disease after a time delay. We apply an inductive conformal regression method to this type of problem to make valid regression estimations.

In Section 2 we recall key notions of machine learning and conformal prediction. In particular, what functions can be used as non-conformity measures and what changes if we apply conformal prediction in the inductive form of data processing that will be needed further. In Section 3 we describe a non-conformity measure for inductive conformal predictor based on K29 algorithm from [8], that was initially developed as game-theoretical approach (Defensive Forecast) to regression in bounded intervals. In Section 4 we give an example of application.

2 Machine Learning Background

2.1 Conformal Classification and Regression

The core element of a conformal predictor is a Non-Conformity Measure (NCM) that is a function A satisfying the equation

$$(\alpha_1,\ldots,\alpha_n) = A(z_1,\ldots,z_n) \implies (\alpha_{\pi(1)},\ldots,\alpha_{\pi(n)}) = A(z_{\pi(1)},\ldots,z_{\pi(n)}).$$

NCM can be also undestood as a distance between a set $\{z_1,\ldots,z_n\}$ and one of its elements z_i, reflecting a relative strangeness α_i of the element with respect to the others. The NCM values (non-conformity scores) are converted to p-values by the formula

$$p(z_1,\ldots,z_n) = \frac{\#\{i=1,\ldots,n \mid \alpha_i^z \geq \alpha_n^z\}}{n}$$

H. Papadopoulos et al. (Eds.): AIAI 2013, IFIP AICT 412, pp. 384–393, 2013.

A conformal predictor checks each of a set of hypotheses (possible labels) when presented with a new example and assigns it a p-value (Algorithm 1). Here $z_1, \ldots, z_{n-1}$ are examples with known classification, each z_i consists of a feature vector $x_i \in X$ and the label $y_i \in Y$, and y is a hypothetical label for a new example with the feature vector x_n.

Algorithm 1. A step of conformal prediction

Input Non-conformity measure A
Input $z_1 = (x_1, y_1), \ldots, z_{n-1} = (x_{n-1}, y_{n-1}), x_{new}$
for $y \in Y$ **do**
$\quad z_n = (x_{new}, y)$
$\quad (\alpha_1, \ldots, \alpha_n) = A(z_1, \ldots, z_n)$
$\quad p(y) = \frac{\#\{i=1,\ldots,n | \alpha_i^z \geq \alpha_n^z\}}{n}$
end for

One of the ways to interpret p-values output by the conformal predictor is to find the prediction set R^γ is a list of labels that are not discarded at a given significance level γ:

$$R^\gamma = \{y : p(z_1, \ldots, z_{n-1}, (x_n, y)) > \gamma\}.$$

Conformal predictors are region predictors: their output is a prediction set R - a list of possible lanbales that are not discarded at a given significance level γ.

The prediction set should cover the true label y_n with probability at least $1 - \gamma$, if i.i.d. assumption is true.

Alternatively the prediction can be done by comparing the different p-values and selecting more likely hypothesis. This makes results of conformal prediction comparable to standard ones if needed.

2.2 Inductive form of Conformal Prediction

The approach discussed above is a transductive version of conformal prediction. Inductive conformal predictor was proposed in order to make calculations more computationally efficient. Some previous applications of it can be found in [5,6] and other works. Usually they use same non-conformity measures as standard (transductive) conformal predictors, but for this work we will need some extension of this.

The idea is to use a fixed additional set $u_1, \ldots, u_h$ and to define the NCM $A(z_1, \ldots, z_n) = (\alpha_1, \ldots, \alpha_n)$ in such way that

$$\alpha_i = A_0(u_1, \ldots, u_h, z_i).$$

Usually $u_1, \ldots, u_h$ (called *proper training set*) and $z_1, \ldots, z_{n-1}$ (called *calibration set*) are taken from the same data set with a random split. This interpretation of the inductive conformal framework is analogous to one given in [7] for inductive probabilistic (Venn) predictor.

A general scheme of Inductive Conformal Prediction (ICP) can be found in Algorithm 3.

3 Approach for Bounded Regression

Aim of this section is to present a conformal predictor based on K29 algorithm from the work [8] that develops a game-theoretic approach to machine learning. In principle, details related to this theory are not necessary to understand how the algorithms works and how non-conformity scores are calculated. However, we remind some of them in order to have some intuitive justification for the choice of non-conformity measure.

3.1 Prediction as a Game

The Protocol 1 describes a simple form of prediction game with 3 players: Nature, Predictor and Sceptic. Nature generates examples x: say x_n on n-th round, Predictor gives a forecast $\hat{y}_n$ of Nature's move, and once he has done the prediction, Nature announces the real label y_n. Sceptic has an initial capital C_0 and bets s_n at round n.

Protocol 1
for $n = 1, 2, \ldots$ **do**
   NATURE: x_n
   PREDICTOR: $\hat{y}_n \in [-1, 1]$
   SCEPTIC: s_n
   NATURE: y_n
   Sceptic's capital: $C_n = C_{n-1} + s_n(y_n - \hat{y}_n)$
end for

Usually in machine learning the Predictor tries to predict some value given by the nature (such a the new example's label) and his preformance is assessed by a loss function. However Nature does not have any interest to fail Predictor. Therefore game-theoretic approach to prediction usually assumes that Predictor has an antagonist, called Sceptic, whose win is Predictor's loss.

3.2 Defensive Forecast with Non-conformity Measure

In [8], Sceptic has to show in advance his potential reaction as a betting function S_n of Predictor's move, so that $s_n = S_n(\hat{y}_n)$.

Predictor develops the following strategy. After seeing the object x_n on round n Predictor has to solve the equation

$$\sum_{j=1}^{n-1} K((x_j, \hat{y}_j), (x_n, y))(y_j - \hat{y}_j) = 0$$

where K is a Mercer kernel.

Algorithm 2 follows K29 game protocol for the Defensive Forecast [8] and shows how the non-conformity scores $\alpha_n = |S_n(y_n)|$ could be extracted.

If Predictor's move (the prediction) is different from Nature's move (the label), then the discrepancy is measured not directly by their difference (as is it usually done in regression), but by the difference of Sceptic's reaction to them. This follows a general idea of game-theoretic probability: an event is rare or strange if someone with reasonable

Algorithm 2. K29 algorithm with players' strategies and NCM

Input: Kernel function $K((x,y),(x',y')) = \Phi(x,y) \cdot \Phi(x',y')$ where Φ is a continouous mapping to a Hilbert space.

for $n = 1, 2, \ldots$ **do**

    NATURE: x_n

    SCEPTIC: $S_n(y) = \sum_{j=1}^{n-1} K((x_j, \hat{y}_j), (x_n, y))(y_j - \hat{y}_j)$

    PREDICTOR: $\hat{y}_n \in [-1, 1]$ is either y such that $S_n(y) = 0$ or the sign of S_n if it never reaches zero on $[-1, 1]$.

    NATURE: y_n

    Sceptic's capital: $C_n = C_{n-1} + S_n(\hat{y}_n)(y_n - \hat{y}_n)$

    NCM: $A_0((x_1, y_1), \ldots, (x_{n-1}, y_{n-1}), (x_n, y_n)) = |S_n(y_n)|$

end for

strategy may make a profit from betting for it. Therefore Sceptic's move showing his betting intention is used to measure the strangeness. If for example $S_n(y_n) = S_n(\hat{y}_n) = 0$, Sceptic prefers not to play in both cases, then the difference between y_n and $\hat{y}_n$ is not considered as an essential one.

3.3   Using Defensive Forecast in Inductive Mode

NCM defined above can be used only in the *inductive* conformal prediction because otherwise, for transductive conformal predictors, the assumption of exchangeability does not hold: the order of the examples follows the protocol.

In the inductive mode of conformal prediction, the data are split into three parts of sizes h (proper training set $u_1, \ldots, u_h$), m (calibration set $z_1, \ldots, z_m$) and $N - h - m$ (testing set $z_{m+1}, \ldots, z_{N-h}$). For an individual testing example the prediction is done as in Algorithm 3.

The only examples we deal with are the ones in the calibration or testing set, while proper training set can be considered as a parameter. That way the exchangeability property is satisfied.

If the non-conformity measure defined in Section 3.2 is applied in inductive mode, this means that Protocol 1 is run on examples $u_1, u_2, \ldots, u_h$ as usally, but the step $n =$

Algorithm 3. A step of inductive conformal prediction

Input Non-conformity measure A_0

Input $u_1 = (x_{u.1}, y_{u.1}), \ldots, u_h = (x_{u.h}, y_{u.h})$

Input $z_1 = (x_1, y_1), \ldots, z_m = (x_m, y_m), x_{new}$

for $i = 1, \ldots, m$ **do**

    $\alpha_i = A_0(u_1, \ldots, u_h, z_i)$

end for

for $y \in Y$ **do**

    $z_{m+1} = (x_{new}, y)$

    $\alpha_{m+1} = A_0(u_1, \ldots, u_h, z_{m+1})$

    $p(y) = \frac{|\{i=1,\ldots,m+1 | \alpha_i^z \geq \alpha_{m+1}^z\}|}{m+1}$

end for

$h+1$ is repeated many times starting from the same point. Each of calibration and testing examples in turn plays the role of x_{h+1} in Protocol 1 in order to get its non-conformity score. As for the testing examples, each of them is used also with different hypotheses about y_n, in this context a Nature's move on the step n may mean a hypothesis about this move.

3.4   Kernels

In Algorithm 2 there is a parameter K: a kernel function (or a scalar product) after a *feature mapping* to a Hilbert space. This is analogous to the well-known kernels [9] $K(x,x') = \Phi(x) \cdot \Phi(x')$ but in K29 the kernels are dependent on y as well as on x.

This is useful for bounded regression problem because it allows to consider a non-linearity in a wider sense: non-linearity in y (labels) as well as labels rather than in x (feature vectors). An example is the polynomial kernel:

$$K_{Poly(d,e)}((x,y),(x',y')) = (x \cdot x' + 1)^d + (y \cdot y' + 1)^e.$$

where d and e are degrees of non-linearity in x and in y.

$S_n(y)$ can be represented as $\Phi(x_n,y) \cdot w_{n-1}$ where

$$w_{n-1} = \sum_{j=1}^{n-1} (y_j - \hat{y}_j) \Phi(x_j, \hat{y}_j)$$

plays a role similar to the slope w of separating hyperplane in Support Vector Machines [10]. But in SVM one can find w by solving a quadratic optimization problem, while in K29 calculation of w is separated into $n-1$ easy steps of on-line update.

Kernels depending on y were also used in a generalized form of SVM for structured output space [11] but in this algorithms optimization problem is even harder than in a standard SVM.

4   Application

4.1   Data

In our application, we use the data obtained from the Alzheimer's Disease Neuroimaging Initiative (ADNI) database, ADNI-1 cohort [17]. The database includes more than 800 subjects with up to 5 years annual follow-up with comprehensive clinical, neuropsychological, imaging and laboratory evaluations, performed at the specialized research centers. For the present study, we used 1.5 Tesla 3D T1 magnetic resonance imaging (MRI) brain scans from patients with Alzheimer's Disease (AD), with Mild Cognitive Impairment (MCI) and Healthy Controls (HC), who had long term follow-up information and met the inclusion criteria (see *Diagnosis* below).

In earlier applications of conformal method to other MRI data (see [19]) the diagnostic was considered as a classification problem, while now we observe the data ordered by these labels reflecting the following disease stages.

- 164 healthy examples;
- 17 examples known to be healthy at the time of earliest measument who became Mild cognitive impairment (MCI) patients in less than 5 years time;
- 119 with Mild cognitive impairment (MCI);
- 156 known to be MCI at the measurement time and to convert to Dementia (AD) in less than 5 years time; this includes 62 examples will convert in at most 1 year;
- 169 with Dementia (AD).

4.2 Diagnosis

All AD patients met NINCDS/ADRDA criteria for probable AD, had mild level of dementia, defined as Mini-Mental State Examination (MMSE) score between 20 and 26, Clinical Dementia Rating Scale score of 1.0. Inclusion criteria for MCI were: 1) MMSE score between 24 and 30, 2) memory complaints and objective memory impairment measured by Logical Memory II subscale of the Wechsler Memory Scale (education adjusted), 3) CDR of 0.5, 4) absence of significant levels of impairment in other cognitive domains, 5) preserved activities of daily living, and 6) absence of dementia. MCI converters had to meet the criteria for Alzheimer's disease during at least two sequential evaluations (e.g., at 24 and 36 month follow ups). Controls (general inclusion/exclusion criteria): 1) MMSE scores between 28 and 30, 2) CDR of 0, 3) they did not meet criteria for clinical depression at baseline, MCI or dementia within 3 years of follow-up.

4.3 Image Post-processing

Raw 3D T1 MRI data underwent `Freesurfer v5.1` (`http://surfer.nmr.mgh.harvard.edu`) steps for surface-based cortex reconstruction and volumetric segmentation. As a result, 68 measures of brain cortical thickness (32 for each hemisphere) averaged by parcellation as described in [15] and 41 volumetric measurements of subcortical structures (corrected for intracranial volume) acquired for every subject were combined with apoE-allele carrying information, basic clinical evaluations (MMSE and Word-recall) and demographics (age, gender, education). Each example therefore contained 109 brain morphometric measurements combined with 6 non-imaging features. Originally they were serial: same patient can have several measurements at different follow-up timepoints. For each patient, we will use its first (earliest) measurement. The label is based on the current diagnosis at that moment together with information about later dynamics of the disease.

4.4 Prediction Intervals

According to the data structure, we consider the following 21 hypotheses related to ADNI.

- Healthy ($y = -1$);
- 4.5,4,...,1,0.5 years before Mild Cognitive Impairment (MCI) ($y = -0.9, \ldots, -0.1$);
- MCI non-converter ($y = 0$);
- MCI converter 4.5,4,...,1,0.5 years before conversion to Dementia ($y = 0.1, \ldots, 0.9$);
- Dementia ($y = +1$).

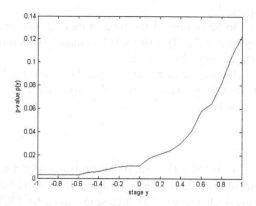

Fig. 1. Prediction for one of the examples: p-value as a function showing likelihood of a stage

In order to apply K29 algorithm we use 109+6 features as vectors x and the stage numbers as their labels y. They are ranging from -1 to 1 with step 0.1 as shown in the list above.

Conformal predictor assigns p-value to each hypothesis about the diagnosis. A standard interpretation of conformal prediction is done in terms of intervals. Suppose that for one of examples, each possible y is assigned a p-value by the conformal predictor.

Fig. 1 presents a typical individual prediction made for an example. Its true label 0.9 meaning: MCI in 6 months before its conversion to AD.

Examples of corresponding prediction sets (intervals) are:

- for the significance level $\gamma = 10\%$, $R = \{y : p(y) > 0.1\} = [0.9; 1]$ that covers the true value with probability at least 90%;
- for the significance level $\gamma = 5\%$, $R = \{y : p(y) > 0.05\} = [0.6; 1]$ that covers the true value with probability at least 95%.
- for the significance level $\gamma = 1\%$, $R = \{y : p(y) > 0.1\} = [0.2; 1]$ that covers the true value with probability at least 99%;

4.5 Accuracy of Two-Class Problems

In addition to prediction intervals, we can use p-values obtained form a conformal predictor for some two-class problems. The following ones were selected because of their popularity in the literature [13,14,16]:

- (A) Healthy vs Dementia;
- (B) Healthy vs MCI;
- (C) MCI non-converters vs (0.5–4.5 year) MCI converters
- (C_1) MCI non-converters vs (0.5-year and 1-year) MCI converters.

These problems can be solved by comparing highest p-values reached on corresponding intervals. For example, if we restrict our interest to the problem (C_1) then the interpretation of p-values is following:

- a prediction is correct in one of the following cases:

$$y = 0, p(0) > \max_{0.8 \leq y < 1} p(y);$$

$$y \in \{0.8, 0.9\}, p(0) \leq \max_{0.8 \leq y < 1} p(y);$$

- wrong predictions:

$$y = 0, p(0) \leq \max_{0.8 \leq y < 1} p(y);$$

$$y \in \{0.8, 0.9\}, p(0) > \max_{0.8 \leq y < 1} p(y);$$

- examples with true labels $y < 0; 0 < y < 0.8; y = 1$ are irrelevant for the accuracy although they are still used for training.

The best results are presented in Table 1. The accuracy is averaged over 50 random splits with ICP parameters $h = 500$ and $m = 100$ (see Sec.3.3). We also compare K29 with a simpler approach based on linear regression extended with a T-test feature selection step used in our previous work [19] applied in leave-one-out mode.

Table 1. Results with two-class accuracy

Underlying algorithm	Parameter	Task			
		(A)	(B)	(C)	(C_1)
Linear regression with feature selection		0.94 (best)	0.72 (best)	0.70	0.76
K29	trivial kernel	0.91	0.65	0.69	0.75
K29	polynomial kernel $K_{Poly(3,1)}$	0.50	0.42	0.57	0.32
K29	polynomial kernel $K_{Poly(1,3)}$	0.92	0.63	0.72 (best)	0.78 (best)

5   Discussion and Conclusions

This bounded conformal regression method has been applied to a problem of medical prognosis. A development of Alzheimer's disease has several stages before the actual dementia onset. Neurodegeneration usually starts from the entorhinal cortex and hippocampal formation and subsequently spreads throughout the brain. This pattern is consistent with our results. Thus, the most important features for prediction were volumes of the Left and Right Hippocampi, Left Amygdala, thickness of the Left Entorhinal cortex, apoE-genotype (known genetic biomarker associated with different risks for Alzheimer's disease [18], and the result of Mini-Mental State Examination (screening tool to assess cognitive functions).

We have proposed a conformal predictor based on a new kind of non-conformity measure, based on the ideas of game-theoretic defensive forecasting method, originally developed for a bounded regression. This techniques has some advantages that were discussed in the theoretical part of the paper. The experimental results are especially interesting as an illustration of a generalized kernel technique in the context of bounded regression.

Acknowledgements. This work was supported by EPSRC grant EP/K033344/1 ("Mining the Network Behaviour of Bots"); by Thales grant ("Development of automated methods for detection of anomalous behaviour"); by EraSysBio+ grant funds from the European Union/BBSRC Shiprec project: "Living with uninvited guests"; by Veterinary Laboratories Agency (VLA) of Department for Environment, Food and Rural Affairs (Defra) through the project: "Machine learning algorithms for analysis of large veterinary data sets"; by the National Natural Science Foundation of China (No.61128003) grant; and by grant 'Development of New Venn Prediction Methods for Osteoporosis Risk Assessment' from the Cyprus Research Promotion Foundation. We are indebted to Alex Gammerman for setting up the problem and the idea of application. We would like to express our sincere thanks to Vladimir Vovk and Alexey Chervonenkis for useful discussions and help. Alexander Lebedev was supported by the Helse Vest Strategic Funding 2013. We would also like to thank Andrew Simmons (Institute of Psychiatry, King's College London) and Eric Westman (NVS, Karolinska Institute, Stockholm, Sweden) for their contributions to image collection and post-processing and Alzheimer's Disease Neuroimaging Initiative (ADNI) for making available neuroimaging data.

References

1. Vovk, V., Gammerman, A., Shafer, G.: Algorithmic Learning in a Random World. Springer (2005)
2. Melluish, T., Saunders, C., Nouretdinov, I., Vovk, V.: Comparing the Bayes and typicalness frameworks. In: Flach, P.A., De Raedt, L. (eds.) ECML 2001. LNCS (LNAI), vol. 2167, pp. 360–371. Springer, Heidelberg (2001)
3. Vovk, V., Nouretdinov, I., Gammerman, A.: On-line predictive linear regression. Annals of Statistics 37(3), 1566–1590 (2009)
4. Gammerman, A., Vovk, V.: Hedging Predictions in Machine Learning. Computer Journal 50(2), 151–172 (2007)
5. Papadopoulos, H., Haralambousm, H.: Reliable Prediction Intervals with Regression Neural Networks. Neural Networks 24(8), 842–851 (2011)
6. Papadopoulos, H., Vovk, V., Gammerman, A.: Regression Conformal Prediction with Nearest Neighbours. J. Artif. Intell. Res (JAIR) 40, 815–840 (2011)
7. Lambrou, A., Papadopoulos, H., Nouretdinov, I., Gammerman, A.: Reliable Probability Estimates Based on Support Vector Machines for Large Multiclass Datasets. In: Iliadis, L., Maglogiannis, I., Papadopoulos, H., Karatzas, K., Sioutas, S. (eds.) Artificial Intelligence Applications and Innovations, Part II. IFIP AICT, vol. 382, pp. 182–191. Springer, Heidelberg (2012)
8. Vovk, V.: On-line regression competitive with reproducing kernel Hilbert spaces, arXiv:cs/0511058v2
9. Kernel methods. Wiki for On-Line Prediction,
http://onlineprediction.net/?n=Main.KernelMethods
10. Cristianini, N., Shawe-Taylor, J.: An Introduction to Support Vector Machines and other kernel-based learning methods. Cambridge University Press (2000)
11. Tsochantaridis, I., Hofmann, T., Joachims, T., Altun, Y.: Support Vector Learning for Interdependent and Structured Output Spaces. In: ICML (2004)
12. Alzheimer's Disease Neuroimaging Initiative. Sharing Alzheimer's Research Data with the World, http://adni.loni.ucla.edu/

13. Liu, M., Zhang, D., Shen, D.: Alzheimer's Disease Neuroimaging Initiative. Ensemble sparce classification of Alzheimer's disease. Neuroimage 60, 1106–1116 (2012)
14. Zhang, D., Wang, Y., Zhou, L., Yuan, H., Shen, D.: Alzheimer's Disease Neuroimaging Initiative. Multimodal Classification of Alzheimer's Disease and Mild Congitive Impairment. Neuroimage 55(3), 856–867 (2011)
15. Destriux, C., Fischl, B., Dale, A., Halgren, E.: Automatic parcellation of human cortical gyri and sulci using standard anatomical nomenclature. Neuroimage 53, 1–15 (2010)
16. Ye, J., Farnum, M., Yang, E., Verbeeck, R., Lobanov, V., Raghavan, N., Novak, G., DiBernard, A., Narayan, V.A.: Sparse learning and stability selection for predicting MCI to AD conversion using baseline ADNI data. BMC Neurology 12, 46 (2012), doi:10.1186/1471-2377-12-46.
17. Aisen, P.S., Petersen, R.C., Donohue, M.C., Gamst, A., Raman, R., Thomas, R.G., Walter, S., Trojanowski, J.Q., Shaw, L.M., Beckett, L.A., Jack, C.R., Jagust, W., Toga, A.W., Saykin, A.J., Morris, J.C., Green, R.C., Weiner, M.W.: Alzheimer's Disease Neuroimaging Initiative. Clinical Core of the Alzheimer's Disease Neuroimaging Initiative: Progress and Plans. Alzheimers Dement, 239–246 (2010)
18. Alonso Vilatela, M.E., Lopez-Lopez, M., Yescas-Gomez, P.: Genetics of Alzheimer's disease. Arch. Med. Res. 43, 622–631 (2012)
19. Nouretdinov, I., Costafreda, S.G., Gammerman, A., Chervonenkis, A., Vovk, V., Vapnik, V., Fu, C.H.Y.: Machine learning classification with confidence: Application of transductive conformal predictors to MRI-based diagnostic and prognostic markers in depression. Neuroimage 56(2), 809–813 (2011)

Learning by Conformal Predictors with Additional Information

Meng Yang, Ilia Nouretdinov, and Zhiyuan Luo

Computer Learning Research Centre, Royal Holloway, University of London
Egham Hill, Egham, Surrey TW20 0EX, UK
{m.yang,ilia,zhiyuan}@cs.rhul.ac.uk

Abstract. In many supervised learning applications, the existence of additional information in training data is very common. Recently, Vapnik introduced a new method called LUPI which provides a learning paradigm under privileged (or additional) information. It describes the SVM+ technique to process this information in batch mode. Following this method, we apply the approach to deal with additional information by conformal predictors. An application to a medical diagnostic problem is considered and the results are reported.

Keywords: LUPI, additional information, conformal predictor.

1 Introduction

In machine learning classification problems, in batch setting, we usually work with a set of training and testing examples. In a data-rich world, there often exist some "pieces" of information about the data that we can add and use it. But, this information may be available at a training stage and not for the new examples at the testing stage. For example, usually doctors try to make diagnosis using all available information, but if at the end of an investigation the diagnosis is still unclear, they may send the patient for some additional tests such as pathological reports, blood test, MRI scans, etc. This is additional or privileged information and can be used to improve the quality of training set and hence, the decision rules. However, the same additional information may not be available for new patients. The question is: can this additional information at the training stage improve the accuracy of diagnosis for the new patients? Traditional learning methods cannot use the additional information directly when it is not available in test set – it is summarised Table 1. Recently, Vapnik proposed a

Table 1. Data set with additional information

Data Set	Content		
	'Usual' information	Additional information	Label
Training examples	Known	Known	Known
Test examples	Known	Unknown	To be predicted

H. Papadopoulos et al. (Eds.): AIAI 2013, IFIP AICT 412, pp. 394–400, 2013.

general approach to deal with this problem, known as Learning Using Privileged Information (LUPI) [10]. However, LUPI approach does not allow us to estimate confidence in the prediction. This paper extends the Conformal Predictors method [2] to include some additional information available in the training set in order to make prediction, estimate confidence of the prediction and apply it in batch and on-line mode.

2   Learning Using Privileged Information

Learning using privileged information (LUPI) is a recently proposed learning paradigm and the aim is to incorporate that type of information into learning [10]. An example of privileged information, according to Vapnik, is when teachers provide students with extra knowledge which exists in explanation, comments, comparisons and so on. There is no formal definition of "privileged" information, but we shall interpret it as information that exists only in the training set.

Let's consider a sequence of examples x with their labels y:

$$(x_1, x_1^*, y_1), (x_2, x_2^*, y_2), ..., (x_{n-1}, x_{n-1}^*, y_{n-1}), x_i \in X, x_i^* \in X^*, y_i \in Y.$$

Here $x_i \in X$ is an example i that is a vector of attributes of "usual" or "available" information and $x_i^* \in X^*$ is a vector of additional (or "privileged") attributes; y_i is a corresponding label.

In the classical SVM a prediction for the new example x_n can be calculated by the following equation:

$$\hat{y}_n = \sum_{i=1}^{n-1} \alpha_i y_i (x_i \cdot x_n)$$

where weighting coefficients α_i are calculated on the basis of the examples $x_1, \ldots, x_{n-1}$ and x_n is a new example from the test set. A new method, SVM+ is an extension of SVM and Lagrange multipliers α_i are replaced with α_i^* calculated from $x_1^*, \ldots, x_{n-1}^*$, while the dot product $(x_i \cdot x_n)$ is not changed to $(x_i^* \cdot x_n^*)$ because x_n^* is unavailable.

3   Conformal Approach

3.1   Conformal Predictors

Conformal predictor is a general learning framework to make well-calibrated predictions, and provides predictions with reliable measures of confidence. The prediction is based on the statistical p-value, which is derived from the strangeness (or non-conformity) measure α_i, that indicates how "strange" a particular example is. Any strangeness measure can be used, as long as it holds the exchangeability property. Strangeness measures may be constructed from almost any existing learning algorithms, such as Neural Networks [3], Random Forests [6] and SVMs [11]. In this paper, we consider the Nearest Centroid method [8] to derive the strangeness measure. In general, given a strangeness measure A, the corresponding values are computed for each hypothetic label $y \in |Y|$ as

$$\alpha_i = A(\langle (x_1, y_1), ..., (x_{i-1}, y_{i-1}), (x_{i+1}, y_{i+1}), ...(x_n, y) \rangle, z_i), i = 1, ..., n-1$$

Given a strangeness measure we can compute p-values:

$$p(y) = \frac{\#\{i = 1, ..., n : \alpha_i \geq \alpha_n\}}{n}$$

Obviously, $0 < p_y \leq 1$. The lower p-value is, the more "strange" the example is in relation to the entire training set.

3.2   Learning with Additional Information

Let's consider a sequence of independent and identical examples x with additional information x^* and their labels y. For the prediction of new object x_n, we firstly assign it an hypothetic label (y) and hypothetic values (x^*) of additional attributes and then measure how "strange" the new example is by calculating $p(y, x^*)$. The more likely the hypothetic label is, the higher extended p-value $p(y, x^*)$ is. However, the number of possible combinations will affect the speed of the processing.

The advantage of Conformal Predictors is its validity, which means:

$$Prob\{p(y) \leq \varepsilon\} \leq \varepsilon$$

for any $0 < \varepsilon < 1$. Therefore, our next task is how to combine a number of extended p-value $p(y, x^*)$ into $p(y)$ and to maintain the validity property. Since only one of the hypotheses is true, selecting the maximum extended p-values is the only way to hold the validity:

$$\max_{x^*} p(y, x^*) \geq p(y, x^*_{true}), y \in Y, x^* \in X^*$$

Thus:

$$Prob\{\max_{x^*} p(y, x^*) \leq \varepsilon\} \leq Prob\{p(y, x^*_{true}) \leq \varepsilon\}$$

So:

$$Prob\{\max_{x^*} p(y, x^*) \leq \varepsilon\} \leq \varepsilon$$

Excluding x^* from it we would get a standard conformal predictor that ignores additional information. Algorithm 1 summarises the procedure:

This method could be applied both in the on-line mode and the off-line mode. In the on-line mode, the examples are presented one by one. Each time, we observe the object and predict its label. We could assume that after the prediction is done, both the label y_i and the attribute value x^*_i will be revealed,

Algorithm 1. Learning With Additional Information

Require: training example sequence $z_1 = (x_1, x_1^*, y_1), z_2 = (x_2, x_2^*, y_2), ..., z_{n-1} = (x_{n-1}, x_{n-1}^*, y_{n-1})$

Require: new example x_n

Require: strangeness measure A

    for $y \in Y$ **do**

        for $x^* \in X^*$ **do**

        $z_n = (x_n, x^*, y)$

        for i in $1, 2, ..., n$ **do**

            $\alpha_i = A(\wr z_1, z_2 ..., z_{i-1}, z_{i+1}, ... z_n \wr, z_i)$

        end for

        $p(y, x^*) = \frac{\#\{i=1,...,n : \alpha_i \geq \alpha_n\}}{n}$

        end for

        $p(y) = \max_{x^*} p(y, x^*)$

    end for

---

see in the following description of the on-line prediction with additional information protocol. At the n-th step, we have observed the previous examples $(x_1, x_1^*, y_1), ..., (x_{n-1}, x_{n-1}^*, y_{n-1})$ and new object x_n and our task is to predict y_n without x_n^*. The new example will be added to the training examples and used to generate a new rule for next prediction. On-line mode is a simple form of the slow learning from [11] where the feedback is given with a delay. In this protocol we assume that some symptoms may also come with a delay. For example, if a prediction algorithm is designed to classify whether a patient has a disease or not by some symptoms and blood test in on-line mode, but the blood test result is not available (will be given, maybe, one day later).

    On-line prediction with additional information Protocol:

FOR $n = 1, 2....$;

Reality outputs $x_n \in X$;

Predictor outputs $\Gamma_n^\varepsilon \subseteq Y$ for all $\varepsilon \in (0, 1)$;

Reality outputs $x_n^* \in X^*$, $y_n \in Y$;

END FOR

4   Applications and Experiments

The conformal prediction method with additional information has been applied to Abdominal Pain dataset [1]. The data set consists of 6387 patient records with 9 categories of diseases and 135 symptoms [1,4,5]. The 9 diseases for diagnosis are: Appendicitis (APP, 844 examples), Diverticulitis (DIV, 143 examples), Perforated Peptic Ulcer (PPU, 130 examples), Non-Specific Abdominal Pain (NAP, 2835 examples), Cholecystitis (CHO, 572 examples), Intestinal Obstruction(INO, 417 examples), Pancreatitis(PAN, 96 examples), Renal Colic(RCO, 473 examples) and Dyspepsia(DYS, 877 examples).

    Each symptom has two values, 1 and 0: either the patient has the symptom or not. For each disease group, experts suggest a sequence of symptoms which are more relevant for its diagnosis. Suppose that some of these symptoms are known for the collected training data but are unknown for a testing example, then they play the role of privileged information in this paper.

If we now choose, for example, the Nearest Centroid algorithm as an underlying algorithm to derive the corresponding strangeness measure, by using the ratio of distances as a strangeness measure:

$$\alpha_i = \frac{\{D(x_i, \mu_y)|y_i = y\}}{\min\{D(x_i, \mu_i)|y_i \neq y\}}$$

where D is the Euclidean distance measure and μ_i is the centroid (the averaged example) of the class i. Then, we can label a new example the same way as the examples of the nearest class.

Table 2. Single prediction by Conformal Predictor on Abdominal Pain dataset

Diagnostic Group	With additional information Average accuracy	No additional information Average accuracy	Size of additional attributes
APP	0.89±0.014	0.85±0.042	3
DIV	0.97±0.004	0.93±0.052	3
PPU	0.98±0.014	0.96±0.045	8
CHO	0.97±0.038	0.93±0.014	4
INO	0.95±0.016	0.91±0.009	3
RCO	0.94±0.022	0.93±0.016	6
DYS	0.89±0.080	0.86±0.029	2

Table 3. Predictions by SVM+ and SVM on Abdominal Pain dataset

Diagnostic Group	SVM+ Average accuracy	SVM Average accuracy
APP	0.88±0.009	0.86±0.021
DIV	0.63±0.015	0.60±0.019
PPU	0.54±0.010	0.53±0.013
CHO	0.82±0.005	0.79±0.028
INO	0.69±0.022	0.62±0.027
RCO	0.68±0.033	0.68±0.041
DYS	0.78±0.005	0.75±0.016

Experimental results are given in Table 2 where the binary classification is performed in one against all other classes. In batch learning mode, we only care about accuracies of predictions. To avoid the influence of redundant attributes, we use some selected symptoms here. For each disease group, we use 5 most relevant symptoms selected in [7] as "usual" attributes because these 5 selected symptoms could provide the similar confidence level as whole set of symptoms. The features provided by experts in [1] are used as privileged attributes. The dataset is randomly divided into training set (4387 examples) and test set (2000 examples). The average accuracy and the corresponding standard deviation are shown for 7 diagnostic groups as the experts do not give any relevant information for the other two diagnostic groups (NAP and PAN). We then apply SVM and SVM+ on the same data, results are shown in Table 3. The kernel used here is Radial Basis Function(RBF), $K(x_i, x_j) = exp(-\gamma||x_i - x_j||^2)$, $\gamma \geq 0$. Cross-validation is applied on the training examples to find the optimal parameters. We can see that both SVM+ and our approach utilize additional information to improve classification accuracy. Due to the unbalance size of classes for prediction, accuracies of SVM and SVM+ are not as good as that of the conformal prediction approach.

5 Conclusion and Discussion

In this paper, we extend Conformal Predictors to deal with additional information. Experiments show that our approach successfully utilize additional information to improve the performance of classification as we expected. However, some more work need to be completed in the future.

We only used the Abdominal Pain dataset in this paper. Further experiments need be carried out on various databases. It would be interesting to consider and apply on-line predictions and slow learning where the feedback is given with an n-step delay. We would like to find out what kind of information could be defined as privileged.

Acknowledgements. This work was supported by EPSRC grant EP/K033344/1 ("Mining the Network Behaviour of Bots"); by EraSys-Bio+ grant funds from the European Union/BBSRC Shiprec project: "Living with uninvited guests"; by the National Natural Science Foundation of China (No.6112803) grant; and by grant "Development of New Venn Prediction Methods for Osteoporosis Risk Assessment" from the Cyprus Research Promotion Foundation. We would like to express our sincere thanks to Alex Gammerman, Vladimir Vovk and Vladimir Vapnik (Royal Holloway, University of London) for setting the problem, useful discussions and help.

References

1. Gammerman, A., Thatcher, A.R.: Bayesian Diagnostic Probabilities without Assuming Independence of Symptoms. Methods Inf. Med. 30(1), 15–22 (1991)
2. Gammerman, A., Vovk, V.: Hedging Predictions in Machine Learning. The Computer Journal 50(2), 151–163 (2007)
3. Papadopoulos, H.: Inductive conformal prediction: Theory and application to neural networks. Tools in Artificial Intelligence, 315–329 (2008)
4. Papadopoulos, H., Gammerman, A., Vovk, V.: Reliable Diagnosis of Acute Abdominal Pain with Conformal Prediction. Engineering Intelligent Systems 17(2-3), 127–137 (2009)

5. Papadopoulos, H., Gammerman, A., Vovk, V.: Confidence Predictions for the Diagnosis of Acute Abdominal Pain. In: Iliadis, L., Vlahavas, I., Bramer, M. (eds.) Artificial Intelligence Applications and Innovations III. IFIP, vol. 296, pp. 175–184. Springer, Boston (2009)
6. Yang, F., Wang, H.-Z., Mi, H., Lin, C.-D., Cai, W.-W.: Using random forest for reliable classification and cost-sensitive learning for medical diagnosis. BMC Bioinformatics 10(1), S22 (2009)
7. Yang, M., Nouretdinov, I., Luo, Z., Gammerman, A.: Feature selection by conformal predictor. In: Iliadis, L., Maglogiannis, I., Papadopoulos, H. (eds.) EANN/AIAI 2011, Part II. IFIP AICT, vol. 364, pp. 439–448. Springer, Heidelberg (2011)
8. Yu, L., Liu, H.: Feature Selection for High-Dimensional Data: A Fast Correlation-Based Filter Solution. In: Proceedings of the Twentieth International Conference on Machine Learning, Washington DC (2003)
9. Vapnik, V., Vashist, A.: A New Learning Paradigm: Learning Using Privileged Information. Neural Networks 22, 544–557 (2009)
10. Vapnik, V., Vashist, A., Pavlovitch, N.: Learning using hidden information: Master class learning. In: Proceedings of NATO Workshop on Mining Massive Data Sets of Security, vol. 19, pp. 3–14. IOS Press (2008)
11. Vovk, V., Gammerman, A., Shafer, G.: Algorithmic Learning in a Random World. Springer (2005)

Conformity-Based Transfer AdaBoost Algorithm

Shuang Zhou, Evgueni N. Smirnov, Haitham Bou Ammar, and Ralf Peeters

Department of Knowledge Engineering, Maastricht University,
P.O. BOX 616, 6200 MD Maastricht, The Netherlands
{shuang.zhou,smirnov,haitham.bouammar,
ralf.peeters}@maastrichtuniversity.nl

Abstract. This paper proposes to consider the region classification task in the context of instance-transfer learning. The proposed solution consists of the conformal algorithm that employs a nonconformity function learned by the Transfer AdaBoost algorithm. The experiments showed that our approach results in valid class regions. In addition the conditions when instance transfer can improve learning are empirically derived.

Keywords: Region classification, Conformal framework, Transfer learning.

1 Introduction

Most of the research in machine learning is concentrated on the task of point classification: estimating the correct class of an instance given a data sample drawn from some unknown *target* probability distribution. However, in applications with high misclassification costs, region classification is needed [1,10]. The task of region classification is to find a region (set) of classes that contains the correct class of the instance to be classified with a given probability of error $\varepsilon \in [0, 1]$. Thus by employing region classification we can control the error in a long run which however has practical sense if the class regions are efficient; i.e., small.

This paper proposes to consider the region classification task in a new context, in the context of instance-transfer learning [4,7]. This means that in addition to the *target* data sample we have a second data sample generated by some unknown *source* probability distribution. The main assumption is that the target and source distributions are different but somehow similar. Thus, the region-classification task in this case is to find class regions according to the target probability distribution, given the target data sample, by transferring relevant instances from the source data sample.

To solve the region-classification task in the instance-transfer learning setting we note that (1) the conformal framework [1,10] is a general framework for the region classification task, (2) the Transfer AdaBoost algorithm is a base algorithm for instance-transfer learning [7], and (3) the AdaBoost algorithm was used under the conformal framework [5,10]. Thus, our solution for the task is a combination of these techniques.

To compare our research with relevant work we note that instance-transfer learning has been applied so far only for the task of point classification [4,7]. Thus our region classification task considered in instance-transfer learning and the approach that we propose for the task are novel and they form the main contributions of this paper.

H. Papadopoulos et al. (Eds.): AIAI 2013, IFIP AICT 412, pp. 401–410, 2013.

The remaining of the paper is organized as follows. Section 2 formalizes the tasks of point classification and region classification in traditional learning and instance-transfer learning. The conformal framework and Transfer AdaBoost algorithm are given in Sections 3 and 4, respectively. Section 5 proposes the weights-based nonconformity function for implementing the conformity framework using the Transfer AdaBoost algorithm. The experiments are given in Section 6. Finally, Section 7 concludes the paper.

2 Point and Region Classification

This section formalizes the tasks of point classification and region classification. The formalizations are given separately for the traditional-learning setting and instance-transfer learning setting in the next two subsections.

2.1 Traditional Learning Setting

Let $\mathbf{X}$ be an instance space and $\mathbf{Y}$ a class set. We assume an unknown probability distribution over the labeled space $\mathbf{X} \times \mathbf{Y}$, namely the *target* distribution $p^t(x, y)$. We consider training sample $D_n^t \subseteq \mathbf{X} \times \mathbf{Y}$ defined as a bag $\wr (x_1, y_1)^t, (x_2, y_2)^t, ..., (x_{n^t}, y_{n^t})^t \wr$ of n instances $(x_i, y_i)^t \in \mathbf{X} \times \mathbf{Y}$ drawn from the probability distribution $p^t(x, y)$.

Given training sample D_n^t and an instance $x_{n+1}^t \in \mathbf{X}$ drawn according to $p^t(x)$,

- the point-classification task is to find an estimate $\hat{y} \in \mathbf{Y}$ of the class of the instance x_{n+1} according to $p^t(x, y)$;
- the region-classification is to find a class region $\Gamma^\varepsilon(D_n^t, x_{n+1}) \subseteq \mathbf{Y}$ that contains the class of x_{n+1} according to $p^t(x, y)$ with probability at least $1 - \varepsilon$, where ε is a significance level.

In point classification estimating the class of any instance $x \in \mathbf{X}$ assumes that we learn a point classifier $h(D_n^t, x)$ in a hypothesis space H of point classifiers h ($h : (\mathbf{X} \times \mathbf{Y})^{(*)} \times \mathbf{X} \rightarrow 2^{\mathbb{R}})^1$ using the target sample D_n^t. The classifier $h(D_n^t, x)$ outputs for any instance x a posterior distribution of scores $\{s_y\}_{y \in \mathbf{Y}}$ over all the classes in $\mathbf{Y}$. The class y with the highest posterior score s_y is the estimated class $\hat{y}$ for the instance x. In this context we note that the point classifier $h(D_n^t, x)$ has to be learned such that it performs best on new unseen instances $(x, y)^t \in \mathbf{X} \times \mathbf{Y}$ drawn according to the target probability distribution $p^t(x, y)$.

In region classification (according to the conformity framework [1,10]) computing class region $\Gamma^\varepsilon(D_n^t, x_{n+1}) \subseteq \mathbf{Y}$ for any instance $x_{n+1} \in \mathbf{X}$ requires two steps. First we derive a nonconformity function A that given a class $y \in \mathbf{Y}$ maps the sample D_n^t and the instance (x_{n+1}, y) to a nonconformity value $\alpha \in [0, R \cup \{\infty\}]$. Then we compute the p-value p_y of class y for the instance x_{n+1} as the proportion of the instances in $D_n^t \cup \wr (x_{n+1}, y) \wr$ of which the nonconformity scores are greater than or equal to that of the instance (x_{n+1}, y). The class y is added to the final class region for the instance x_{n+1} if $p_y \geq \varepsilon$. In this context we note that the nonconformity function A has to be learned such that it performs best on new unseen instances $(x, y)^t \in \mathbf{X} \times \mathbf{Y}$ drawn according to the target probability distribution $p^t(x, y)$.

---

[1] $(\mathbf{X} \times \mathbf{Y})^{(*)}$ denotes the set of all bags defined over $\mathbf{X} \times \mathbf{Y}$.

2.2   Instance-Transfer Learning Setting

In instance-transfer learning in addition to the instance space $\mathbf{X}$, the class set $\mathbf{Y}$, the *target* distribution $p^t(x, y)$, and the training sample D_n^t, we have a second unknown probability distribution over $\mathbf{X} \times \mathbf{Y}$, namely the *source* distribution $p^s(x, y)$, and training sample D_m^s defined as a bag of m instances $(x_i, y_i)^s \in \mathbf{X} \times \mathbf{Y}$ drawn from $p^s(x, y)$. Assuming that the target distribution $p^t(x, y)$ and source distribution $p^s(x, y)$ are different but somehow similar we define:

- the instance-transfer point-classification task as a point classification task for which we learn the point classifier $h(D_n^t, x)$ by transferring relevant instances from the source sample D_m^s in addition to the target sample D_n^t;
- the instance-transfer region-classification as a region-classification task for which we learn the nonconformity function A by transferring relevant instances from the source sample D_m^s in addition to the target sample D_n^t.

3   The Conformal Framework

This section introduces the conformal framework [1,10]. It formalizes the framework, provides possible options, and introduces metrics for evaluating region classifiers.

3.1   Formal Description

The conformal framework has been proposed in [1,10] for developing region classifiers. The framework is proven to be valid when the target sample D_n^t and each instance $\mathbf{x}_{n+1} \in \mathbf{X}$ to be classified are drawn from the same unknown target distribution $p^t(x, y)$ under the exchangeability assumption. The exchangeability assumption holds when different orderings of instances in a bag are equally likely.

Applying the conformal framework is a two-stage process. Given a point classifier $h(D_n^t, x)$, a nonconformity function is constructed for $h(D_n^t, x)$ capable of measuring how unusual an instance is for other instances in the data. Then, the conformal algorithm employing this nonconformity function is applied to compute the class regions.

Formally, a nonconformity function is of type $A : (\mathbf{X} \times \mathbf{Y})^{(*)} \times (\mathbf{X} \times \mathbf{Y}) \to \mathbb{R} \cup \{\infty\}$. Given a bag $D_n^t \in (\mathbf{X} \times \mathbf{Y})^{(*)}$ and instance $(x, y) \in (\mathbf{X} \times \mathbf{Y})$ it returns a value α in the range $[0, R \cup \{\infty\}]$ indicating how unusual the instance (x, y) is with respect to the instances in D_n^t. In general, the function A returns different scores for instance (x, y) depending on whether (x, y) is in the bag D_n^t (added prediction) or not (deleted prediction): if $(x, y) \in D_n^t$, then the score is lower; otherwise it is higher.

The general nonconformity function was defined in [10] for any point classifier $h(D_n^t, x)$. Given training bag $D_n^t \in (\mathbf{X} \times \mathbf{Y})^{(*)}$ and instance (x, y_r), it outputs the sum $\sum_{y \in \mathbf{Y}, y \neq y_r} s_y$ where s_y is the score for class $y \in \mathbf{Y}$ produced by $h(D_n^t, x)$.

The conformal algorithm is presented in Algorithm 1. Given significance level $\varepsilon \in [0, 1]$, target sample D_n^t, instance $x_{n+1} \in \mathbf{X}$ to be classified, and the nonconformity function A for a point classifier $h(D_n^t, x)$, the algorithm constructs a class region $\Gamma^\varepsilon(D_n^t, x_{n+1}) \subseteq \mathbf{Y}$ for the instance x_{n+1}. The class-region construction is realized separately for each class $y \in \mathbf{Y}$. To decide whether to include the class y in the class region $\Gamma^\varepsilon(D_n^t, x_{n+1})$ the instance x_{n+1} and class y are first combined into labelled instance (x_{n+1}, y). Then, the algorithm computes the nonconformity score α_i for each instance (x_i, y_i) in the bag D_{n+1}^t, using the nonconformity function A for the point classifier $h(D_n^t, x)$. The nonconformity scores are used for computing the p-value p_y of the class y for the instance x_{n+1}. More precisely, p_y is computed as the proportion of the instances in the bag D_{n+1}^t of which the nonconformity scores α_i are greater or equal to that of the instance (x_{n+1}, y). Once p_y is set, the algorithm includes the class y in the class region $\Gamma^\varepsilon(D_n^t, x_{n+1})$ if $p_y > \varepsilon$. The conformal algorithm was originally designed for the online learning setting. This setting assumes initially an empty data set D_n^t. Then for each integer n from 0 to $+\infty$ we first construct class region $\Gamma^\varepsilon(D_n^t, x_{n+1})$ for the new instance x_{n+1} being classified, and then add the instance (x_{n+1}, y_r) to the bag where y_r is the correct class of x_{n+1}. In this context we note that the conformal algorithm is proven to be valid [1,10], i.e., it constructs for any object x_{n+1} class region $\Gamma^\varepsilon(D_n^t, x_{n+1}) \subseteq \mathbf{Y}$ containing the correct class $y \in \mathbf{Y}$ for x_{n+1} with probability at least $1 - \varepsilon$, if (a) the data are drawn from the same target distribution $p^t(x, y)$ under the exchangeability assumption; and (b) the learning setting is online.

Algorithm 1. Conformal algorithm

Input: Significance level ϵ, Target sample D_n^t, Instance x_{n+1} to be classified,
              Non-conformity function A for a point classifier $h(D_n^t, x)$.
Output: Class region $\Gamma^\epsilon(D_n^t, x_{n+1})$

1: $\Gamma^\epsilon(D_n^t, x_{n+1}) = \emptyset$.
2: **for each** class $y \in Y$ **do**
3:      $D_{n+1}^t = D_n^t \cup \{(x_{n+1}, y)\}$.
4:      **for** $i := 1$ to $n + 1$ **do**
5:          **if** using deleted prediction **then**
6:              Set nonconformity score $\alpha_i := A(D_{n+1}^t \setminus \{(x_i, y_i)\}, (x_i, y_i))$.
7:          **else if** using added prediction **then**
8:              Set nonconformity score $\alpha_i := A(D_{n+1}^t, (x_i, y_i))$.
9:          **end if**
10:     **end for**
11:     Calculate $p_y := \frac{\#\{i=1,...,n | \alpha_i \geq \alpha_{n+1}\}}{n+1}$.
12:     Include y in $\Gamma^\epsilon(D_n^t, x_{n+1})$ if and only if $p_y > \epsilon$.
13: **end for**
14: Output $\Gamma^\epsilon(D_n^t, x_{n+1})$.

3.2   Possible Options

Applying the conformal framework is not a trivial task. One has to make a set of choices concerning the nonconformity function used and the learning setting.

The conformal algorithm outputs valid class regions for any real-valued function used as nonconformity function [1]. The class regions will be efficient if the function estimates well the difference of any instance with respect to the training data. In this context we note that the general nonconformity function is not always the most efficient. Therefore, one of the main issues when applying the conformal framework is how to design a specific nonconformity function for the point classifier used.

The conformal algorithm is proven to be valid when the learning setting is online [1]. However, there are reported experiments in the offline (batch) setting [6]. In contrast to the online setting, the offline setting assumes a target training sample that is non-empty initially. Furthermore, the target sample remains the same throughout the classification process. These experiments show that the conformal algorithm produces valid class regions in the offline setting.

3.3   Evaluation Metrics

Any class region $\Gamma^\varepsilon(D_n^t, x_{n+1})$ is valid if it contains the correct class $y \in \mathbf{Y}$ of the instance $x_{n+1} \in \mathbf{X}$ being classified with probability of at least $1 - \varepsilon$. To evaluate experimentally the validity of the class regions provided by the conformal algorithm we introduce the error metric. The error E is defined as the proportion of the class regions that do not contain the correct class. Thus, in order to prove experimentally that the conformal algorithm is valid we have to show that for all significance levels $\varepsilon \in [0, 1]$ the error E is less than or equal to ε.

Any class region $\Gamma^\varepsilon(D_n^t, x_{n+1})$ is efficient if it is non-empty and small. Thus, to evaluate experimentally the efficiency of the class regions provided by the conformal algorithm we introduce three metrics: the percentage P_e of empty-class regions, the percentage P_s of single-class regions, and the percentage P_m of multiple-class regions. The empty-class regions, single-class regions, and multiple-class regions can be characterized by their own errors. The percentage P_e of empty-class regions is essentially an error, since the correct classes are not in the class regions. The error E_s on single-class regions is defined as the proportion of the invalid single-class regions among all the class regions. The error E_m on multiple-class regions is defined as the proportion of the invalid multiple-class regions among all the class regions.

The errors P_e, E_s, and E_m are components of the error E. More precisely, it is easy to prove that $E = P_e + E_s + E_m$. The error E has its own upper bound E^u representing the worst case when we are not able to pick up correct classes from valid multi-class regions. In this case we will err on all the multi-class regions and, thus, E^u is defined equal to $P_e + E_s + P_m$. We note that for any significance level $\varepsilon \in [0, 1]$ there is no guarantee that E^u is less than or equal to ε unless $P_m = 0$.

4   Transfer AdaBoost Algorithm

The Transfer AdaBoost algorithm [7] is a learning method for the instance-transfer point-classification task (see Algorithm 2). The algorithm itself is an extension of the well-known AdaBoost algorithm [8]. It treats the target sample D_n^t and source sample D_m^s differently. For the target sample D_n^t, the Transfer AdaBoost algorithm uses the same re-weighting scheme as AdaBoost. It decreases the weights of *correctly classified instances* and through normalization increases the weights of *incorrectly classified instances.* On the other hand, for the source sample D_m^s, the Transfer AdaBoost algorithm uses opposite re-weighting scheme. It decreases the weights of *incorrectly classified instances* and through normalization increases the weights of *correctly classified instances.* This means that source instances that are less likely to be generated by the target distribution receive lower weights and source instances that are more likely to be generated by the target distribution receive higher weights. Thus, the Transfer AdaBoost algorithm focuses on more difficult (high-weight) target instances and on more similar (high-weight) source instances in the next iterations.

Algorithm 2. Transfer AdaBoost Algorithn, adapted from Dai et al.[7]

Input: Two labeled data sets D_n^t and D_m^s
      Weak point classifier $h(B, x)$,
      Number of iterations T.

1: for any instance $(x_i, y_i) \in D_m^s \bigcup D_n^t$ initialize weight $w_1(x_i) = 1$. Let p be vector of the normalized weights of instances in $D_m^s \bigcup D_n^t$ and p^t be vector of the normalized weights of instances in D_n^t.
2: **for** $k = 1$ to T **do**
3:    Train weak classifier $h_k : X \to Y$ on $D_m^s \bigcup D_n^t$ using normalized weights from p_k.
4:    Calculate the weighted error ϵ_k of h_k on D_n^t using normalized weights from p_k^t
5:    **if** $\epsilon_k = 0$ or $\epsilon_k \geq \frac{1}{2}$ **then**
6:      Set $T = k - 1$.
7:      **Abort loop.**
8:    **end if**
9:    Set $\beta = \frac{1}{1 + \sqrt{2 \ln m / T}}$ and $\beta_k = \frac{\epsilon_k}{1 - \epsilon_k}$.
10:    Update the weight for any instance $(x_i, y_i) \in D_m^s \bigcup D_n^t$:

$$\begin{cases} w_{k+1}^s(x_i) = w_k^s(x_i) \beta^{[h_k(x_i) \neq y_i]} & \text{if } (x_i, y_i) \in D_m^s; \\ w_{k+1}^t(x_i) = w_k^t(x_i) \beta_k^{-[h_k(x_i) \neq y_i]} & \text{if } (x_i, y_i) \in D_n^t. \end{cases}$$

11: **end for**
12: **Output** the strong classifier: $h_f(x) = sign(\sum_{k=T/2}^T \ln(\frac{1}{\beta_k}) h_k(x))$

---

5   Weights-Based Non-conformity Function

To solve the instance-transfer region-classification task we propose to apply the conformal framework that employs a nonconformity function based on the Transfer AdaBoost algorithm. An obvious option in this context is use the general nonconformity function (see subsection 3.1) given in [1]. However in this paper we go further and propose a new nonconformity function called the weight-based nonconformity function. As the name suggests it is based on the weights of the training instances from the target sample calculated by the Transfer AdaBoost algorithm. Since they indicate the classification difficulty of the instances, we interpret them as nonconformity values.

To theoretically justify the weight-based nonconformity function we note that for any target instance $(x_i, y_i) \in D_n^t$, given the initial weight equals 1, we have that:

$$w_{T+1}^t(x_i) = w_1^t(x_i) \prod_{k=1}^{T} (\frac{1}{\beta_k})^{[h_k(x_i) \neq y_i]} = \prod_{k=1}^{T} (\frac{1}{\beta_k})^{[h_k(x_i) \neq y_i]} \qquad (1)$$

In addition we note that $\epsilon_k \in (0, 0.5)$. Thus, $\frac{1}{\beta_k} > 1$ and $\prod_{k=1}^{T} (\frac{1}{\beta_k})^{[h_t(x_i) \neq y_i]} \geq 1$. Since $[h_f(x_i) \neq y_i] \leq 1$, it follows that:

$$[h_f(x_i) \neq y_i] \leq \prod_{k=1}^{T} (\frac{1}{\beta_k})^{[h_k(x_i) \neq y_i]} \qquad (2)$$

Combining equations (1) and (2) results in:

$$[h_f(x_i) \neq y_i] < w_{T+1}^t(i)$$

Thus, the weight of any target instance is found to be the upper bound on the training error of that instance [2]. This explains why we propose the use of weights for nonconformity values.

Formally the weights-based nonconformity function is defined as follows: given a target sample set D_n^t and an instance (x, y_r), the function returns the weight $w_{T+1}(x)$ for the instance (x, y_r) calculated by the Transfer AdaBoost algorithm after T iterations. We note that, since the Transfer AdaBoost algorithm computes weights only for target instances, the instance (x, y_r) has to belong to the data D_n^t. This implies that the conformal algorithm is used with the option "added prediction" only. We note that in this case computing p-value p_y for one class $y \in \mathbf{Y}$ requires only one run of the Transfer AdaBoost algorithm. Thus, the time complexity for constructing one class region is $O(|\mathbf{Y}|C)$ (where C is the time complexity of the Transfer AdaBoost algorithm).

6   Experiments and Discussion

This section presents our experiments with the conformal algorithm presented in subsection 3.1. Given a sample D_n^t and a sample D_m^s, the algorithm was instantiated using three types of nonconformity functions:

- the weight-based nonconformity function based on the AdaBoost algorithm trained on the sample D_n^t,
- the weight-based nonconformity function based on the Transfer AdaBoost algorithm trained on the sample D_n^t as a target sample and sample D_m^s as a source sample,
- the weight-based nonconformity function based on the AdaBoost algorithm trained on the sample $D_n^t \cup D_m^s$.

The conformal algorithm based on the first function is denoted as CAdaBoostT. The conformal algorithm based on the second function is denoted as CTrAdaBoostTS. The conformal algorithm based on the third function is denoted as CAdaBoostTS.

The generalization performance of these three algorithms is given in terms of the validity and efficiency of the final class regions. We note that CAdaBoostT is used as a base-line algorithm. The algorithms CTrAdaBoostTS and CAdaBoostTS are used in order to decide whether we need to transfer or add source instances.

6.1   Data Sets

The datasets for our experiments were taken from the UCI Machine Learning Repository [9]. In order to fit transfer-learning scenario each data set was split into target sample and source sample with different probability distributions. For example, the breast-cancer data set was split using the attribute *irradiat*. The target sample in this case consisted of all the instances having the attribute value *irradiate=no*, while the source sample consisted of all the instances having the attribute value *irradiate=yes*. Table 1 given below shows the description of each data-set split. It also provides the KL-divergence estimate [3] on the difference of class distributions for each pair of target sample and source sample. We note that two datasets, namely breast cancer and auto were split twice using different binary attributes.

Table 1. The descriptions of the data sets

Data Set	Number of Classes	KL-divergence	Size	
			D^t	D^s
hepatitis	2	0.041	79	76
colic	2	0.078	272	92
heart-c	2	0.148	74	77
heart-c2	2	0.204	96	207
dermatology	6	0.224	79	294
breast-cancer	2	0.285	56	115
auto	4	0.302	59	112
auto2	4	0.391	83	97
lymph	4	0.621	73	75
vote	2	1.135	264	171

6.2   Validation Setup

Experiments were performed on ten data sets from Table 1 using all the three conformal algorithms CAdaBoostT, CTrAdaBoostTS, and CAdaBoostTS. The weak point classifier for these three algorithms was the decision-stump classifier for the first 8 datasets, while for the last two datasets was the Naive-Bayes classifier [2]. The class regions of the algorithms were evaluated in terms of validity and efficiency using five measures (defined in subsection 3.3): the error E, the percentage P_e of empty-class regions, the percentage P_s of single-class regions, the percentage P_m of multiple-class regions, and the upper-bound error E^u. The method of evaluation was repeated stratified 10-fold cross validation. Results for each data set were generated for 10 to 100 boosting iterations (with step 10). The best results for each algorithm over the ten different iteration numbers are reported in Table 2 on two significance levels $\varepsilon = 0.05$ and $\varepsilon = 0.1$.

---

[2] The reason for employing NaiveBayes is that for the last two datasets the decision-stump classifier resulted in error greater than 0.5 on the first iteration.

Table 2. Performance of the CAdaBoostT, CTrAdaBoostTS, and CAdaBoostTS algorithms

		$\epsilon = 0.05$					$\epsilon = 0.1$				
		E	P_e	P_s	P_m	E^u	E	P_e	P_s	P_m	E^u
hepatitis	CAdaBoostT	0.061	0.006	0.775	0.219	0.280	0.119	0.054	0.854	0.092	0.210
	CAdaBoostTS	0.054	0.001	0.776	0.223	0.277	0.115	0.023	0.911	0.066	0.181
	CTrAdaBoostTS	0.051	0.008	0.771	0.221	0.272	0.116	0.039	0.889	0.072	0.189
colic	CAdaBoostT	0.052	0.000	0.517	0.483	0.534	0.098	0.000	0.789	0.211	0.309
	CAdaBoostTS	0.050	0.000	0.572	0.428	0.478	0.098	0.000	0.823	0.177	0.275
	CTrAdaBoostTS	0.049	0.000	0.565	0.435	0.483	0.098	0.001	0.820	0.178	0.276
heart-c	CAdaBoostT	0.049	0.004	0.689	0.307	0.356	0.093	0.032	0.872	0.096	0.189
	CAdaBoostTS	0.053	0.000	0.802	0.198	0.251	0.098	0.017	0.951	0.032	0.130
	CTrAdaBoostTS	0.048	0.007	0.787	0.206	0.254	0.097	0.029	0.889	0.072	0.169
heart-c2	CAdaBoostT	0.058	0.006	0.674	0.320	0.378	0.102	0.023	0.838	0.139	0.241
	CAdaBoostTS	0.058	0.000	0.683	0.317	0.375	0.097	0.005	0.944	0.051	0.148
	CTrAdaBoostTS	0.056	0.001	0.758	0.241	0.297	0.099	0.025	0.877	0.098	0.197
dermotology	CAdaBoostT	0.080	0.020	0.301	0.679	0.758	0.092	0.021	0.307	0.672	0.733
	CAdaBoostTS	0.104	0.067	0.674	0.259	0.364	0.101	0.053	0.554	0.393	0.481
	CTrAdaBoostTS	0.014	0.001	0.374	0.625	0.639	0.014	0.003	0.429	0.568	0.575
breast-cancer	CAdaBoostT	0.059	0.000	0.184	0.816	0.875	0.113	0.000	0.355	0.645	0.757
	CAdaBoostTS	0.048	0.000	0.105	0.895	0.943	0.108	0.000	0.229	0.771	0.873
	CTrAdaBoostTS	0.044	0.000	0.173	0.827	0.871	0.089	0.000	0.338	0.662	0.752
auto	CAdaBoostT	0.054	0.000	0.137	0.863	0.880	0.098	0.002	0.266	0.732	0.769
	CAdaBoostTS	0.010	0.000	0.007	0.993	0.997	0.092	0.000	0.341	0.659	0.729
	CTrAdaBoostTS	0.044	0.000	0.166	0.834	0.849	0.094	0.002	0.312	0.686	0.729
auto2	CAdaBoostT	0.049	0.000	0.080	0.920	0.933	0.102	0.002	0.161	0.836	0.881
	CAdaBoostTS	0.017	0.000	0.007	0.993	0.995	0.108	0.000	0.157	0.834	0.901
	CTrAdaBoostTS	0.034	0.000	0.093	0.907	0.912	0.084	0.001	0.165	0.833	0.857
lymph	CAdaBoostT	0.026	0.000	0.242	0.758	0.774	0.070	0.001	0.407	0.592	0.660
	CAdaBoostTS	0.034	0.000	0.660	0.340	0.374	0.085	0.000	0.723	0.277	0.362
	CTrAdaBoostTS	0.034	0.000	0.667	0.333	0.367	0.057	0.000	0.745	0.255	0.312
vote	CAdaBoostT	0.050	0.011	0.972	0.017	0.067	0.099	0.074	0.926	0.000	0.099
	CAdaBoostTS	0.050	0.015	0.976	0.009	0.059	0.098	0.082	0.918	0.000	0.098
	CTrAdaBoostTS	0.048	0.014	0.957	0.029	0.077	0.096	0.070	0.930	0.000	0.096

6.3 Results and Discussion

The performance results of the three conformal algorithms are given in Table 2. The table shows that the class regions computed by the algorithms are valid. This is due to the fact that the error E is close to the significance level ε up to some neglectable statistical fluctuation. Thus we can derive one of the main results of this paper, namely that the conformal algorithm is capable of obtaining valid class regions for the instance-transfer region-classification task. This is possible (at least for now) when we employ our weight-based nonconformity function learned by the Transfer AdaBoost algorithm.

In addition, Table 2 shows how instance transfer can help learning. When the target and source distributions are very close (small KL-divergence number) and the size of the target sample is relatively big, the performance statistics of the conformal algorithms CAdaBoostT, CTrAdaBoostTS, and CAdaBoostTS are close. This observation is illustrated by the performance of the algorithms for the datasets hepatitis and colic. Thus in this case any of these algorithms can be applied; i.e., instance transfer does not improve significantly the results. When the distance between the target and source distributions increases (KL-divergence number in the range $[0.1, 0.9]$) and the size of the target sample is smaller, the percentage P_m of multiple-class regions and the upper-bound error E^u of the conformal algorithm CTrAdaBoostTS are smaller than those of CAdaBoostTS on most of the datasets. This observation is illustrated best by the performance of the algorithms for the datasets auto, auto2, and lymph. Thus in this case the conformal algorithm CTrAdaBoostTS has to be applied; i.e., instance transfer does improve the final results. When the distance between the target and source distributions becomes high (KL-divergence number in the range $[0.9, +\infty]$), again the performance statistics of the conformal algorithms CAdaBoostT, CTrAdaBoostTS, and CAdaBoostTS are close (see dataset vote). Thus in this case any of these algorithms can be applied; i.e., instance transfer does not improve the results.

7   Conclusion

This paper showed that the region classification task can be solved in the context of instance-transfer learning [4,7]. The proposed solution consists of the conformal algorithm that employs a nonconformity function based on the instance weights learned by the Transfer AdaBoost algorithm. The experiments showed that the approach results in valid class regions. In addition the conditions when instance transfer can improve learning are empirically derived.

References

1. Shafer, G., Vovk, V.: A Tutorial on Conformal Prediction. Journal of Machine Learning Research 9, 371–421 (2008)
2. Schapire, R.E., Singer, Y.: Improved boosting algorithms using confidence-rated predictions. Machine learning 37(3), 297–336 (1999)
3. Kullback, S., Leible, R.A.: On information and sufficiency. The Annals of Mathematical Statistics 22(1), 79–86 (1951)
4. Pan, S.J., Yang, Q.: A Survey on Transfer Learning. IEEE Trans. Knowl. Data Eng. 22(10), 1345–1359 (2010)
5. Moed, M., Smirnov, E.N.: Efficient AdaBoost Region Classification. In: Perner, P. (ed.) MLDM 2009. LNCS, vol. 5632, pp. 123–136. Springer, Heidelberg (2009)
6. Vanderlooy, S., van der Maaten, L., Sprinkhuizen-Kuyper, I.G.: Off-Line Learning with Transductive Confidence Machines: An Empirical Evaluation. In: Perner, P. (ed.) MLDM 2007. LNCS (LNAI), vol. 4571, pp. 310–323. Springer, Heidelberg (2007)
7. Dai, W., Yang, Q., Xue, G., Yu, Y.: Boosting for transfer learning. In: 24th International Conference on Machine Learning, pp. 193–200. ACM (2007)
8. Freund, Y., Schapire, R.E.: Experiments with a New Boosting Algorithm. In: International Conference on Machine Learning, pp. 148–156 (1996)
9. UCI Machine Learning Repository, http://archive.ics.uci.edu/ml
10. Vovk, V., Gammerman, A., Shafer, G.: Algorithmic Learning in a Random World. Springer-Verlag New York, Inc., Secaucus (2005)

Enhanced Conformal Predictors for Indoor Localisation Based on Fingerprinting Method

Khuong An Nguyen and Zhiyuan Luo

Dept. of Computer Science, Royal Holloway, University of London
Egham, Surrey TW20 0EX, United Kingdom
khuong@cantab.net, zhiyuan@cs.rhul.ac.uk

Abstract. We proposed the first Conformal Prediction (CP) algorithm for indoor localisation with a classification approach. The algorithm can provide a region of predicted locations, and a reliability measurement for each prediction. However, one of the shortcomings of the former approach was the individual treatment of each dimension. In reality, the training database usually contains multiple signal readings at each location, which can be used to improve the prediction accuracy. In this paper, we enhance our former CP with the Kullback-Leibler divergence, and propose two new classification CPs. The empirical studies show that our new CPs performed slightly better than the previous CP when the resolution and density of the training database are high. However, the new CPs performs much better than the old CP when the resolution and density are low.

Keywords: indoor localisation, fingerprinting, conformal prediction.

1 Introduction

The purpose of indoor localisation is to identify and observe a user inside a building. Global Positioning System (GPS) has long been an optimal solution for outdoor localisation, yet the indoor counterpart remains an open research problem, because of the harsh and complex indoor building structure. Current indoor localisation systems remain either too expensive or not accurate enough [4]. In our previous work [2], we proposed the first Conformal Predictor (CP) for indoor localisation based on classification with the weighted K-nearest neighbours algorithm, which performed well in our test sets. However, in reality, the prediction accuracy depends on the resolution, and the density of the training database. In this paper, we enhance our former CP with the Kullback-Leibler divergence, which is a better way to compare two signal strength distributions. We propose two new conformal predictors for classification. The empirical studies show that our new CPs perform slightly better than the previous CP when the resolution and density of the training database are high. However, the new CPs performs much better than the previous CPs when the the resolution and density are low.

The paper begins with a brief introduction of the indoor localisation problem, and the concept of location fingerprinting. The next section describes our

H. Papadopoulos et al. (Eds.): AIAI 2013, IFIP AICT 412, pp. 411–420, 2013.

implementation of the conformal prediction with two new nonconformity measures. The performance of our implementations is evaluated in Section 4. We summarise our contributions and future work in Section 5.

2   The Indoor Localisation Problem

An indoor localisation system is a network of devices used to wirelessly locate objects or people inside a building. A user can be coarsely identified at room-level or precisely localised at sub-room level. Different approaches have been developed in recent years, however, most precise sub-room level tracking systems are expensive, while most affordable tracking systems are not accurate enough [1,7]. Many attempts to improve the location accuracy might work in one environment, but did not work in others because of the signal attenuation. We consider a popular indoor localisation method known as Location Fingerprinting and discuss why CP is a suitable approach for such problem.

The idea of Location Fingerprinting is to use a pre-surveyed database containing a mapping of wireless signal properties to 3-dimensional physical location. The signal properties can be the signal strength (RSSI) or the link quality (LQ). Based on the fact that the wireless signal attenuates and gets weaker as it travels in the air, with many wireless transmitter sources, each location in the tracking zone can have a unique combination of signal properties. To predict the physical location of a known signal properties, the properties are compared with each entry in the database to find a closest match, which will predict possible physical location. The wireless signal properties are regarded as the object set, while the physical location is regarded as the label set.

The main challenge is that two distinct locations might not have a linear relationship in terms of RSSI/LQ and the distance between them. This phenomenon is caused by the human movements, humidity, furniture re-arrangement, as well as the multi-path fading of the indoor environment [3]. In this paper, we show how to deal with such problem using the Conformal Prediction algorithm.

The Location Fingerprinting method can be mathematically formulated as follows. A single location L is modelled in a 3-dimensional space $L = (d^x, d^y, d^z)$. The signal strength RSSI between the user at a location L and all N stations is modelled as an N-tuple $RSSI_L = (s_1, \ldots, s_N)$, where s_i is the signal strength received from the station i ($1 \leq i \leq N$). The task is that, given an RSSI tuple $RSSI_{unknown} = (s_1, \ldots, s_N)$ of an unknown location inside the tracking zone, the system uses the database B to predict the co-ordinate (d^x, d^y, d^z) of the unknown location. This is a multi-label problem.

3   Conformal Prediction for Indoor Localisation

In our previous work, we proposed the first Conformal Predictor for the indoor localisation based on the weighted K-nearest neighbours and the location fingerprinting method [2]. The algorithm provided better location accuracy than existing methods with a similar approach. Further, our method added a confidence parameter for each prediction to solve the uncertainty problem of the indoor localisation.

3.1 Conformal Prediction

Conformal Prediction (CP) is a relatively new machine learning framework, which uses experiences in the past to confidently and precisely predict the outcome of a new sample [5,6]. It has been mathematically proved that the prediction region generated by CP is valid in the on-line setting [6]. However, CP demands a weak assumption that the training database and the new sample to be classified are generated from the same distribution independently.

The most important component of CP is the 'nonconformity measure', which is a real-valued function $A(B, z)$ measuring how different a sample z is to the training database B. Ideally, for a wrong test label, we would prefer this sample to be completely different from all training samples in B. Therefore, the sequence's randomness can be maintained. This is also the core idea of CP, which can be seen as a test of randomness. Whenever a new sample needs to be classified, we exhaustedly test every label recorded in the training data. A p-value function compares the non-conformity score α_{l+1} of the new sample to all remaining α_i of each example in the training data. If the new sample's label is wrong, the returned p-value is small because, α_{l+1} will be bigger than most α_i. The assumed label of the new sample is accepted if p-value is greater than the significance level ε we choose. Regardless of the chosen nonconformity measure, the set of locations predicted by CP is always valid in the on-line setting.

3.2 Enhancement of Classification Indoor Localisation

In this paper, we enhance our work with two key improvements. First, the Kullback-Leibler divergence (KL) approach will emphasize the similarity between the 2 distributions, rather than calculating the difference in distance in the Euclidean space, as we used in the previous approach. Second, we propose 2 new CPs to include more information with the x, y and z co-ordinates for our nonconformity measure. The new CP algorithm is summarised as follows.

Giving a training database B mapping the signal strength to the physical location co-ordinate (d^x, d^y, d^z), and a signal strength fingerprint at an unknown location, we will predict a set of possible locations in the database, which likely matches this new signal fingerprint. The task can be formulated as a classification problem, because we divide the locations into grid points, and the label set is finite. The measured signal strengths at these grid points are regarded as the object set $\mathbf{X}$, and the physical locations are regarded as the label set $\mathbf{Y}$. We will apply CP using both the old examples - the training database $B = (z_1, z_2, \ldots, z_l)$, and the signal fingerprint of the unknown location (as a new object of z_{l+1}). Each example z_i is a combination of the signal strength $RSSI_i = (s_1^i, s_2^i, \ldots, s_N^i)$ and the co-ordinate $L_i = (d_i^x, d_i^y, d_i^z)$. A prediction region of K examples is $R^\varepsilon(L_1, L_2, \ldots, L_K) \subset \mathbf{Y}$.

To calculate the similarity between 2 signal strength distributions P_X and P_Y, we use the symmetrised KL formula, with M is the number of bins in the histogram, and N is the number of stations.

$$Sym\_D_{KL}(P_X, P_Y) = D_{KL}(P_X||P_Y) + D_{KL}(P_Y||P_X) \tag{1}$$

where

$$D_{KL}(P_X \| P_Y) = \sum_{j=1}^{N} \sum_{i=1}^{M} P_X^j[i] \, log_2 \frac{P_X^j[i]}{P_Y^j[i]} \tag{2}$$

According to the CP algorithm, we will assign each of the labels in the training database to the new sample, and calculate the nonconformity measure for such label. We propose 2 new nonconformity measures based on the above KL divergence, one version is simple and easy to implement, while the other includes more information of the dimensions, but requires more computation.

Using the nonconformity measure, we calculate the nonconformity score α_i, with $i = 1, \ldots, l$, for every example in the database B. We then count the number of α_i which is larger than or equal to the nonconformity score α_{l+1} of the new sample, and divide the total number of examples in the database B to have the p-value for a possible label $\hat{y}$.

$$p(\hat{y}) = \frac{\#\{i = 1, \ldots, l+1 : \alpha_i \geq \alpha_{l+1}\}}{l+1} . \tag{3}$$

Given a significance level ε beforehand (such as $\varepsilon = 0.05$), the current assumed co-ordinate label is accepted as the label for the new sample, if and only if the p-value $> \varepsilon$. All accepted locations form a prediction region, which guarantees to contain the correct position 95% of the time (when $\varepsilon = 0.05$) in the on-line setting. We explain our 2 new CPs below.

3.3   The First Nonconformity Measure

For the first nonconformity measure, we first find K nearest examples in the training database with **different location labels** (d_i^x, d_i^y, d_i^z) to the label of the new sample $(d_{l+1}^x, d_{l+1}^y, d_{l+1}^z)$. We applied the KL divergence $D_{KL}(P_{l+1}, P_i)$, with $i = 1, \ldots, l$, to compare two signal strength distributions, as presented in Equation (2).

Once we obtain a set of K examples, we combine them into one weighted average example with the label $L_{diff} = (d_{diff}^x, d_{diff}^y, d_{diff}^z)$, with ϵ is a small constant to prevent division by zero. Our assumption is that the majority of the similar signal strengths are close to each others. By considering the KL weights, we penalise the isolated neighbours, which appear because of the indoor attenuation problem.

$$d_{diff}^{co\_ord} = \frac{\sum_{i=1}^{K} \frac{1}{D_{KL}(P_{l+1}, P_i) + \epsilon} d_{diff}^{co\_ord}}{\sum_{i=1}^{K} \frac{1}{D_{KL}(P_{l+1}, P_i) + \epsilon}}, co\_ord = x, y, z. \tag{4}$$

We then find another set of K nearest examples, this time with the **same location labels** $(d_{l+1}^x, d_{l+1}^y, d_{l+1}^z)$ with the new sample. Another weighted average example $L_{same} = (d_{same}^x, d_{same}^y, d_{same}^z)$ is calculated as follows.

$$d_{same}^{co\_ord} = \frac{\sum_{i=1}^{K} \frac{1}{D_{KL}(P_{l+1}, P_i) + \epsilon} d_{same}^{co\_ord}}{\sum_{i=1}^{K} \frac{1}{D_{KL}(P_{l+1}, P_i) + \epsilon}}, co\_ord = x, y, z. \tag{5}$$

The nonconformity measure is defined as the distance between these two locations L_{same} and L_{diff}. We use an Euclidean approach to calculate such difference.

$$A_1 = \sqrt{(d_{same}^x - d_{diff}^x)^2 + (d_{same}^y - d_{diff}^y)^2 + (d_{same}^z - d_{diff}^z)^2} . \tag{6}$$

3.4   The Second Nonconformity Measure

For the second nonconformity measure, we implement a multi-label approach. Since we have 3 different dimensions (d^x, d^y, d^z) for each location, there are 8 possibilities for any 2 locations (d_1^x, d_1^y, d_1^z) and (d_2^x, d_2^y, d_2^z).

1. $(d_1^x = d_2^x), (d_1^y = d_2^y), (d_1^z = d_2^z)$
2. $(d_1^x \neq d_2^x), (d_1^y = d_2^y), (d_1^z = d_2^z)$
3. $(d_1^x = d_2^x), (d_1^y \neq d_2^y), (d_1^z = d_2^z)$
4. $(d_1^x = d_2^x), (d_1^y = d_2^y), (d_1^z \neq d_2^z)$
5. $(d_1^x \neq d_2^x), (d_1^y \neq d_2^y), (d_1^z = d_2^z)$
6. $(d_1^x \neq d_2^x), (d_1^y = d_2^y), (d_1^z \neq d_2^z)$
7. $(d_1^x = d_2^x), (d_1^y \neq d_2^y), (d_1^z \neq d_2^z)$
8. $(d_1^x \neq d_2^x), (d_1^y \neq d_2^y), (d_1^z \neq d_2^z)$

We find a set of K nearest examples for each of the 8 possibilities, using the Equation (2). For each of the 8 sets, we combine all K examples into one weighted average location, $L_i = (d_i^x, d_i^y, d_i^z)$, with $i = 1, \ldots, 8$, using the Equation (5), as similar to how we did in the previous CP above.

Our nonconformity measure is the difference between a combination of the first 7 cases (where at least one co-ordinate is similar), and the 8th case (where all co-ordinates of the label are different). We combine the first 7 cases into one average location $L_{one\_same} = (d_{one\_same}^x, d_{one\_same}^y, d_{one\_same}^z)$.

$$L_{one\_same} = (\frac{\sum_{i=1}^{7} d_i^x}{7}, \frac{\sum_{i=1}^{7} d_i^y}{7}, \frac{\sum_{i=1}^{7} d_i^z}{7}) . \tag{7}$$

The nonconformity measure is defined as follows

$$A_2 = \sqrt{(d_{one\_same}^x - d_8^x)^2 + (d_{one\_same}^y - d_8^y)^2 + (d_{one\_same}^z - d_8^z)^2} . \tag{8}$$

The algorithm outline is presented in Algorithm 1.

Algorithm 1. Classification Conformal Predictor for Indoor Localisation

---

Input: Training database $B = \{z_1, \ldots, z_l\}$, significance level ε, new example $z_{l+1} = (P_{l+1})$.

Output: Prediction region R.

Function D_{KL} is defined in Section 3.2

function WEIGHTED(KNN_{set}, z)
    for $coordinate = \{x, y, z\}$ **do**
        for $i = 1 \to K$ **do**
            $weight1 = 1/(D_{KL}(P_{l+1}, P_i) + \epsilon)* \text{KNN}_{set}(d_i^{coordinate})$
            $weight2 = 1/(D_{KL}(P_{l+1}, P_i) + \epsilon)$
            $d_{weighted}^{coordinate} = weight1/weight2$
        end for
    end for
    return $d_{weighted}$
end function

function NONCONFORMITY_1(B, z)
    for $i = 1 \to L$ **do**
        if $(d_i^{\{x,y,z\}} \neq d_z^{\{x,y,z\}})\&D_{KL}(B_i, z)$ is the smallest **then**
            $\text{KNN}_{diff} = \text{KNN}_{diff} + \{B_i\}$
        end if
        if $(d_i^{\{x,y,z\}} = d_z^{\{x,y,z\}})\&D_{KL}(B_i, z)$ is the smallest **then**
            $\text{KNN}_{same} = \text{KNN}_{same} + \{B_i\}$
        end if
    end for
    $L_{diff} = weighted(\text{KNN}_{diff}, z)$; $L_{same} = weighted(\text{KNN}_{same}, z)$
    $\alpha_1 = sqrt(L_{diff} - L_{same})$
    return α_1
end function

function NONCONFORMITY_2(B, z)
    for $i = 1 \to L$ **do**
        if $(d_i^x = d_z^x\&d_i^y = d_z^y\&d_i^z = d_z^z)or(d_i^x \neq d_z^x\&d_i^y = d_z^y\&d_i^z = d_z^z)or(d_i^x = d_z^x\&d_i^y \neq d_z^y\&d_i^z = d_z^z)or(d_i^x = d_z^x\&d_i^y = d_z^y\&d_i^z \neq d_z^z)or(d_i^x \neq d_z^x\&d_i^y \neq d_z^y\&d_i^z = d_z^z)or(d_i^x \neq d_z^x\&d_i^y = d_z^y\&d_i^z \neq d_z^z)or(d_i^x = d_z^x\&d_i^y \neq d_z^y\&d_i^z \neq d_z^z)andD_{KL}(B_i, z)$ is the smallest **then**
            $\text{KNN}_{one\_same} = \text{KNN}_{one\_same} + \{B_i\}$
        end if
        if $(d_i^{\{x,y,z\}} \neq d_z^{\{x,y,z\}})\&D_{KL}(B_i, z)$ is the smallest **then**
            $\text{KNN}_8 = \text{KNN}_8 + \{B_i\}$
        end if
    end for
    $L_{one\_same} = WEIGHTED(\text{KNN}_{one\_same}, z)$; $L_8 = WEIGHTED(\text{KNN}_8, z)$
    $\alpha_2 = sqrt(L_{one\_same} - L_8)$
    return α_2
end function

Algorithm 1. (*Continued.*)

for each possible label $y \in Y$ do
   $z_{l+1} = (P_{l+1}, y)$
   $\alpha_{l+1} = NONCONFORMITY\_1/\_2(B, z_{l+1})$
   for $i = 1 \rightarrow L$ do
      $\alpha_i = NONCONFORMITY\_1/\_2(B - \{z_i\} + \{z_{l+1}\}, z_i)$
      if $(\alpha_i \geq \alpha_{l+1})$ then $count = count + 1$
      end if
   end for
   $p = count/(l + 1)$
   Predictive set $R = \{y : p(y) > \varepsilon\}$
end for

4 Performance Evaluation

Having explained our enhancements, we apply the new algorithms for our two Bluetooth fingerprinting testbeds presented in [4]. At each location, we collected the signals in 8 orientations (North, Northest, East, Southest, South, Southwest, West and Northwest). The signal readings are measured multiple times for each orientation.

Testbed 1 is just a single room of 15 square metres (5m x 3m). The resolution and the density of the training dataset is high, with 30-40 signal readings for each orientation at every location. There are 64,000 samples for the training set, and 1,000 samples for the test set.

Testbed 2 has lower resolution and signal density across a 170 square metre (17m x 10m) corridor, with over 48,000 samples for the training set, and 500 samples for the test set. There are just 10-15 signal readings for each of the 8 orientations.

4.1 Performance of the First Nonconformity Measure

Tables 1 and 2 show the performance of the first CP on our 2 data sets, compared to our previous CP. The CP error rate is the percentage in which CP does not produce a prediction region containing the exact location. At the significance level $\varepsilon = 0.15$ and $K = 3$, the new CP has fewer than 1.65m location error for testbed 1, and fewer than 2.3m error for testbed 2.

Table 1. Average system accuracy with 1st nonconformity measure for Testbed 1

Confidence level	Significance level ε	Pred. error	Pred. size	CP error rate	Old Pred. error	Old Pred. size
90%	0.10	$\geq$2m	61	8.3%	$\geq$2m	62
85%	0.15	1.65m	27	13.7%	1.7m	29
70%	0.30	$\geq$1.8m	8	28.2%	$\geq$1.8m	8

Table 2. Average system accuracy with 1st nonconformity measure for Testbed 2

Confidence level	Significance level ε	Pred. error	Pred. size	CP error rate	Old Pred. error	Old Pred. size
90%	0.10	≥2.9m	39	9.7%	≥3.1m	44
85%	0.15	2.3m	17	13.8%	2.5m	19
70%	0.30	≥3m	13	29%	≥3m	13

We observed that the nonconformity measures did not improve the accuracy much on Testbed 1, compared to our previous CP. Our assumption is because of the high resolution and signal density in this test bed, two close locations may have a very different signal readings because of the signal fluctuation and attenuation, which introduced many errors in the training data. We did observe a slight improvement on Testbed 2, compared to our old CP. Testbed 2 has lower resolution than Testbed 1 with few signal readings at each orientation.

4.2   Performance of the Second Nonconformity Measure

Tables 3 and 4 show the performance of the second CP on our 2 data sets, compared to our previous CP. The CP error rate is the percentage in which CP does not produce a prediction region containing the exact location. At a significance level $\varepsilon = 0.15$ and $K = 3$, the new CP has fewer than 1.6m location error for testbed 1, and fewer than 1.9m error for testbed 2.

We observed that the nonconformity measure did not improve the accuracy much on Testbed 1, compared to our old CP. However, we did observe a better performance accuracy on Testbed 2, compared to our old CP. Not only does the new CP produce a more accurate prediction, the prediction region is also smaller than that in our old CP for Testbed 2. This finding implies that when the tracking zone is large, and the observed locations are spread out as in our

Table 3. Average system accuracy with 2nd nonconformity measure for Testbed 1

Confidence level	Significance level ε	Pred. error	Pred. size	CP error rate	Old Pred. error	Old Pred. size
90%	0.10	≥1.9m	58	8.3%	≥2m	62
85%	0.15	1.6m	24	13.7%	1.7m	29
70%	0.30	≥1.75m	8	28.2%	≥1.8m	8

Table 4. Average system accuracy with 2nd nonconformity measure for Testbed 2

Confidence level	Significance level ε	Pred. error	Pred. size	CP error rate	Old Pred. error	Old Pred. size
90%	0.10	≥2.6m	37	9.7%	≥3.1m	44
85%	0.15	1.9m	13	13.8%	2.5m	19
70%	0.30	≥2.4m	12	29%	≥3m	13

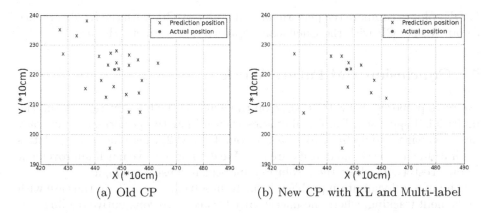

(a) Old CP   (b) New CP with KL and Multi-label

Fig. 1. Prediction Region of new CP with Multi-label Approach on Testbed 2

Testbed 2, the correlation among the co-ordinates has an impact on the prediction accuracy (Figure 1).

4.3 Overall Performance Discussion

Compared to our first classification CP with weighted K-nearest neighbours, our second nonconformity measure with the Kullback-Leibler divergence and the multi-label approach has reduced the prediction error by roughly 25% for Testbed 2 - from 2.5m to 1.9m at 85% confidence level. We still observed at least 16% improvement in different confidence levels.

Unfortunately, we did not see much improvement on Testbed 1 with both nonconformity measures. Our assumption is because of the high resolution and signal density in Testbed 1, two close locations may have a very different signal readings because of the signal fluctuation and attenuation, which introduced many errors in the training data. We use a Cumulative Distribution Function

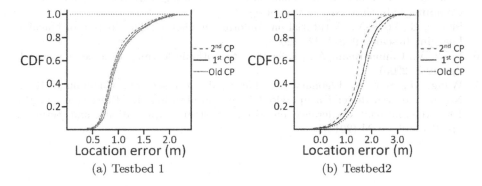

(a) Testbed 1   (b) Testbed2

Fig. 2. Performance Gain of 2 new CPs with KL and Multi-label

(CDF) plot to compare the performance of both new CPs with the KL divergence to our former CP with the Euclidean approach in 2 test sets. (Figure 2).

5 Conclusion and Future Work

We have proposed 2 new CPs based on the Kullback-Leibler divergence, and the multi-label approach. Empirical studies showed that the performance accuracy has been enhanced up from 16%-25% for our Testbed 2, where the training data's resolution and the signal density are low. We did not observe much improvement in our Testbed 1, where the resolution and signal density are high.

The indoor localisation problem is far from solved, especially in the case with movement tracking, where the uncertainty between two consecutive readings are high. We are researching how to apply Conformal Prediction for such problem.

Acknowledgements. We are grateful to the Computer Science Department, Royal Holloway, University of London for continued support and funding. This work has also been supported by a grant from the National Natural Science Foundation of China (No. 61128003) and EPSRC grant EP/K033344/1 ("Mining the Network Behaviour of Bots").

References

1. Chen, Y., Lymberopoulos, D., Liu, J., Priyantha, B.: Fm-based indoor localization. In: Proceedings of the 10th International Conference on Mobile Systems, Applications, and Services, pp. 169–182. ACM (2012)
2. Nguyen, K., Luo, Z.: Conformal prediction for indoor localisation with fingerprinting method. In: Iliadis, L., Maglogiannis, I., Papadopoulos, H., Karatzas, K., Sioutas, S. (eds.) AIAI 2012 Workshops, Part II. IFIP AICT, vol. 382, pp. 214–223. Springer, Heidelberg (2012)
3. Nguyen, K., Luo, Z.: Evaluation of bluetooth properties for indoor localisation. In: Progress in Location-Based Services, pp. 127–149. Springer (2013)
4. Nguyen, K.A.: Robot-based evaluation of bluetooth fingerprinting. Master's thesis, Computer Lab, University of Cambridge (2011)
5. Shafer, G., Vovk, V.: A tutorial on conformal prediction. The Journal of Machine Learning Research 9, 371–421 (2008)
6. Vovk, V., Gammerman, A., Shafer, G.: Algorithmic learning in a random world. Springer (2005)
7. Wang, H., Sen, S., Elgohary, A., Farid, M., Youssef, M., Choudhury, R.R.: No need to war-drive: Unsupervised indoor localization. In: Proceedings of the 10th International Conference on Mobile Systems, Applications, and Services, pp. 197–210. ACM (2012)

Local Clustering Conformal Predictor
for Imbalanced Data Classification

Huazhen Wang[1], Yewang Chen[1], Zhigang Chen[2], and Fan Yang[2]

[1] School of Computer Science and Technology, Huaqiao University, Xiamen, 361021, China
[2] Department of Automation, Xiamen University, Xiamen, 361005, China

Abstract. The recently developed Conformal Predictor (CP) can provide calibrated confidence for prediction which is out of the traditional predictors' capacity. However, CP works for balanced data and fails in the case of imbalanced data. To handle this problem, Local Clustering Conformal Predictor (LCCP) which plugs a two-level partition into the framework of CP is proposed. In the first-level partition, the whole imbalanced training dataset is partitioned into some *class-taxonomy* data subsets. Secondly, the majority class examples proceed to be partitioned into some *cluster-taxonomy* data subsets by clustering method. To predict a new instance, LCCP selects the nearest cluster, incorporated with the minority class examples, to build a re-balanced training data. The designed LCCP model aims to not only provide valid confidence for prediction, but significantly improve the prediction efficiency as well. The experimental results show that LCCP model presents superiority than CP model for imbalanced data classification.

Keywords: conformal predictor, imbalanced data, local clustering.

1 Introduction

Traditional pattern classification methods focus on the improvement of the accuracy on the test set while neglects confidence analysis of the results [1,2]. Suffered from this weakness, the machine learning methods are constantly criticized by traditional statisticians and shows inapplicable in many realistic practice. Moreover, traditional pattern recognition algorithms are generally designed based on the balanced distributed data, and thus always deteriorate terribly on imbalanced datasets. In other words, traditional pattern recognition algorithms tend to prefer the majority class examples to the minority class examples, regardless of the fact that the minority class examples might be important for the users [3, 4]. Thus, the cross-study in these two areas seems changeable and significant.

The recently developed Conformal Predictor (CP) can provide confidence analysis of results and output reliable prediction [5]. However, the CP model works for the evenly distributed dataset and cannot effectively solve the problem of imbalanced data learning. It is worth noting, in order to address the cost sensitive learning, a modified model named MCP (Mondrian Conformal Predictor) can provide label-conditional valid confidence [6]. It shows that CP can be incorporated with multi-partition technology to fit a variety of particular learning settings. This encourages our

H. Papadopoulos et al. (Eds.): AIAI 2013, IFIP AICT 412, pp. 421–431, 2013.

exploring on the possibility of multi-partition technology in imbalanced data learning to improve the feasibility of CP.

In this paper, we introduce a two-level partition method into the framework of the CP model, and then build a modified model named Local Clustering Conformal Predictor (LCCP) for classification of imbalanced data. LCCP model adopts cluster technology to explore the local construction and selects the nearest cluster to be the representative of majority class examples without potential loss. And then constitutes a re-balanced train dataset for confidence prediction. The designed LCCP model aims to not only provide valid confidence for prediction, but significantly improve the prediction efficiency as well.

2   Related Work

2.1   Conformal Predictor(CP)

To address the problem of reliable prediction, Professor Vovk proposed the Conformal Predictor (CP) which can output prediction tailed by valid confidence[7]. According to the CP, the i.i.d assumption is equivalent to the Kolmogorov algorithmic randomness statistic test[8]. When carried on a new test instance, CP applies transductive inference learning to incorporate the train data with the test instance and thus establish a *test data sequence*. Then the algorithmic randomness test is carried out in the test data sequence and subsequently the p value of it is applied to response the prediction. CP model has aroused increased interest in the literature of machine learning and has been applied successfully in the classification of medical data[9], image data[10], and so on. Besides classification, CP has been extended to regression[11-12],feature selection[13], and so on.

At the aspect of the framework of CP, some modified models have been proposed to improve the flexibility of CP. In order to improve the computation efficiency of CP model, Papadopoulos proposed ICP (Inductive CP) for large data sets [14]. In our previous work, the HCCP (Hybrid-Compression CP) not only improves the computational efficiency but preserves the prediction efficiency as well [15]. In order to address the cost sensitive learning, Vovk proposed MCM (Mondrian Confidence Machine) model, which is renamed MCP (Mondrian Conformal Predictor) nowadays [6] and has applied interesting implementations on the gene expression data [16] and breast cancer data[17]. According to all the modifications, Vovk proposed OCM (On-line Compression Model), which is a universal framework that can regulate all the existing CP-related models[6].

2.2   Classification of Imbalanced Data

To address the particular problem of imbalanced data, the solutions can be divided into two categories: data-level methods and algorithms-level methods. The former applies over-sampling or under-sampling to build a re-balanced training data. The SMOTE model is one of the typical approaches [18]. On the other hand, some particular algorithms have been designed based on the assumptions of imbalanced data distribution, such as cost-sensitive learning, active learning and so on[19].

Here we give a brief review of cluster-based under-sampling methods for imbalanced data, because it shows more related to our work[20-23]. These algorithms differ on whether the clustering is done on the whole training data or inside each category. The first one clusters the whole imbalanced dataset into several groups, and then selects some representatives from each group to rebuild the majority class examples[20]. The latter performs clustering inside each class and then tapes pseudo-class labels for those sub-groups. After that, they expanded the two-class learning problem in the multi-class classification setting [21-23].

3   Local Clustering Conformal Predictor for Imbalanced Data

3.1   The Framework of CP

It is necessary to present the framework of CP model because our LCCP model is derived from it. The reality outputs the training data sequence $Z^{(n-1)} = (Z_1, Z_2, ..., Z_{n-1})$, and now a new instance x_n is given to be recognized. CP exhausts all the labels $y \in Y = \{1, 2, ..., C\}$ (C is the number of classes) to be the candidate label for x_n, and thus forms the corresponding *test example* $Z_n^y = (x_n, y)$. Next, CP incorporates each Z_n^y with $Z^{(n-1)}$ to construct the *test data sequence*. Consequently, there are C test data sequences, such as:

$$z^{(n)y} = \{(z_1, z_2, ..., z_{n-1}, z_n^y), y = 1, 2, ..., C\} \tag{1}$$

Subsequently, CP designs a function $\Lambda: Z^{(n)y} \to \alpha^{(n)y}$, which maps each example Z_i to a single nonconformity point α_i, and thus conforms a one-dimension *nonconformity measurement sequence*:

$$\alpha^{(n)y} = \{(\alpha_1, \alpha_2, ..., \alpha_{n-1}, \alpha_n^y), y = 1, 2, ..., C\} \tag{2}$$

where α_i measures the degree of the nonconformity between Z_i an $Z^{(n)y}$. Based on $\alpha^{(n)y}$, the p value which serves as the probability of y being the true label y_n is computed as follows:

$$p_n^y = \frac{\left|\{i = 1, 2, ..., n-1 : \alpha_i \geq \alpha_n^y\}\right| + 1}{n} \tag{3}$$

Given a significance level ε, CP outputs the prediction for x_n as follows:

$$\tau_n^\varepsilon = \{y : p_n^y > \varepsilon, y = 1, 2, ..., C\} \tag{4}$$

where τ_n^ε is a region prediction rather than a point prediction. An error occurs when the prediction set τ_n^ε does not contain the true label y_n. Thus, CP has been proven, in the online setting, the error rate is not greater than the significance level ε, i.e.,

$$P\{p_n^y(z_1, z_2, ..., z_{n-1}, z_n^y) \leq \varepsilon\} \leq \varepsilon \tag{5}$$

The above equation (5) is known as the *validity theorem* which has been theoretically proved [6] and tested in practice.

3.2   Local Clustering Conformal Predictor (LCCP)

a)   Binary Classification Setting

Consider a typical binary classification setting, i.e. $y \in \{1,2\}$. Moreover, $y = 1$ refers to the minority class, $y = 2$ corresponds the majority class. Thus the process of LCCP can be designed as follows :

1) *first-level partition*: dividing the whole training data sequence $z^{(n-1)}$ into two *class-taxonomy* data subsets, and then produces the *minority-class examples sequence* $Z^{(n_1)} = (Z_1, Z_2, ..., Z_{n_1})$ and the *majority-class examples sequence* $Z^{(n_2)} = (Z_1, Z_2, ..., Z_{n_2})$.

2) *second-level partition*: clustering $Z^{(n_2)}$ into J *cluster-taxonomy* data subsets, i.e., $Z^{(n_{2j})} = \{(Z_1, Z_2, ..., Z_{n_{2j}}), j = 1,2,...,J\}$. J is set to be same as the ratio of the majority examples to the minority examples; Subsequently, the J class centroids $Cen_j \, j = 1,2,...,J$ can be carried out.

3) *re-imbalanced training data building*:   computing   the   distance $D_j \, j = 1,2,...,J$ between x_n and $Cen_j \, j = 1,2,...,J$; extracting $Z^{(n_{2k})}$ corresponding to the minimum distance D_k ; combining the $Z^{(n_{2k})}$ with $Z^{(n_1)}$ to build the re-imbalanced training data $Z^{(n_1 + n_{2k})} = (Z^{(n_1)} \bigcup Z^{(n_{2k})})$

4) *reliable prediction*: executes reliable prediction for x_n based on $Z^{(n_1 + n_{2k})}$.

b)   Multi-class Classification Setting

Next, we further discuss the three-class classification setting. If the learning setting is made of two minority classes and one majority class, only the majority examples should be imposed to the clustering algorithm. Subsequently, one of *cluster-taxonomy* data subsets is selected to build the re-imbalanced training data.

If the learning setting is made of one minority class and two majority classes, both of the majority examples should be clustered. Afterwards, one of cluster-taxonomy data subsets is selected to build the re-imbalanced training data. According to the above scheme, we can extend the LCCP algorithm to the multi-class classification setting in general.

4   Experimental Setup

4.1   The Selection of Clustering Algorithm

In order to contribute the merits of LCCP to the compact clusters, we set up the clustering process in reverse form, i.e., we assemble some compact clusters to get the majority examples. At the second stage that provides a prediction based on the re-balanced training dataset, we apply Random Forest to design the nonconformity measurement, the detail process can be seen in our previous work [24].

4.2   Datasets

a)   Synthetic Imbalanced Dataset: Simplex

The base Simplex dataset is depicted in   Fig. 1, which shows the good separability distribution. The number of classes is 4, and the number of examples per class is 2500. One of the class examples are used to be minority examples while the rest to be the majority examples, i.e., two imbalanced ratios of 1:2 and 1:3 are available .

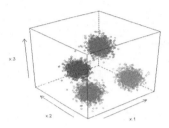

Fig. 1. The scatter distribution of Simplex

b)   Real Imbalanced Dataset: TEP

Tennessee Eastman Process (TEP) was a government sponsored program for the evaluation of fault detection in the large scale industrial process[25]. There are 52 monitors (i.e. , features) to diagnose 21 faults in the process. In our experiments, total 17600 points with 800 points per fault and an additional 800 points as the normal class are sampled. In the experiment, we select the normal form as the minority class, and then compile the remainder of the examples to be the majority class, which creates a series of   different imbalanced ratios, such as 1:2,1:3,..., 1:13.

5   Experimental Results

5.1   The Validity of Confidence of LCCP

Compared with the classical CP , the experimental results for Simplex is illustrated in Fig. 2, and the experimental results for TEP is demonstrated in Fig. 3.

In the upper zone of Fig.2 and Fig.3, with the significance level 5% (the corresponding confidence level 0.95), the x-axis represents the size of test data and the y-axis represents the number of errors. It can be seen that with the expansion of the size of test data, the errors increase accordingly. But the slope of the curve (i.e., *error calibration line*) is constant and close to the significance level 5%, which reveals the validity theorem seen in formula (5). Notable is, the slope of LCCP is apparently smaller than that of CP on TEP dataset, i.e., about 3.8% for LCCP while 5% of CP. It shows that the error rate of LCCP is less than the significance level, which is also applicable in formula (5).

In the lower zone of Fig.2 and Fig.3, the x-axis represents the confidence and the y-axis represents accuracy rate. The diagonal line with legend *'base calibration'* exhibits the optimal relationship between the accuracy rate and the confidence. As clearly shown in Fig.2, for the Simplex data, the *accuracy calibration line* of LCCP and CP both closely attached to the "*base calibration line*", which reveals that the accuracy of LCCP can be calibrated by the confidence. In addition, as shown in Fig.4, the *accuracy calibration line* of LCCP is slightly higher than the *base calibration line*, which implies that the accuracy rates of LCCP are always greater than the corresponding confidence, which also corresponds with the formula (5) to demonstrate the validity of LCCP.

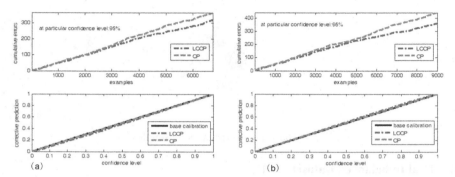

Fig. 2. The comparison of calibration on Simplex dataset at different imbalanced ratio (a) 1:2 (b) 1: 3

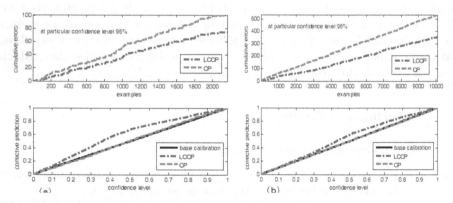

Fig. 3. The comparison of calibration on TEP dataset at different imbalanced ratio (a) 1:2 (b) 1: 3

5.2   The Prediction Efficiency of LCCP

The *favorite prediction* which contains not only one label but also being the true label has been recognized as a key index to exhibits the prediction efficiency. The performance of LCCP is illustrated in Fig.4 and Fig.5.

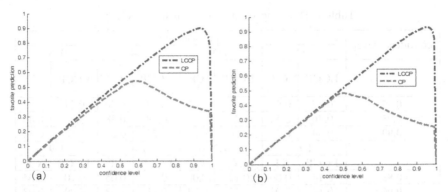

Fig. 4. The comparison of favorite prediction on Simplex dataset at different imbalanced ratio (a)1:2 (b)1: 3

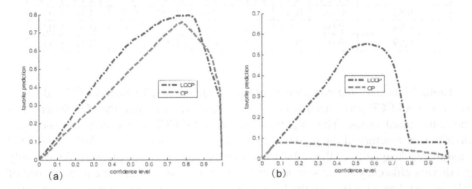

Fig. 5. The comparison of favorite prediction on TEP dataset at different imbalanced ratios (a)1:2 (b)1:13

As can be seen from Fig.4, the favorite prediction of LCCP performs significantly higher than the CP with the confidence level within [0.5 1] which is preferred in practice. On the other hand, we can find that , on TEP dataset, the gap of favorite prediction ratio becomes higher and higher along with the increasing of the imbalanced ratio. This means a large gap between the randomness levels of the C test data sequences. Furthermore, the corresponding label for the higher one must always be the true label. The superiority of LCCP comes from the mechanism that LCCP selects the nearest cluster for the test instance.

Next, Given the confidence level 0.85、 0.95、 0.99 which are more interested in real practice, all the favorite prediction rates at the different imbalanced ratios on TEP dataset are shown in table 1.

Table 1. The comparison of the predictive efficiency

confidence level	imbalanced ratio							
	1:2		1:3		1:4		1:5	
	LCCP	CP	LCCP	CP	LCCP	CP	LCCP	CP
0.85	0.77	0.68	0.67	0.57	0.79	0.50	0.24	0.77
0.95	0.46	0.54	0.47	0.45	0.49	0.38	0.16	0.46
0.99	0.34	0.38	0.26	0.29	0.22	0.27	0.16	0.34
	1:6		1:7		1:8		1:9	
	LCCP	CP	LCCP	CP	LCCP	CP	LCCP	CP
0.85	0.14	0.06	0.12	0.06	0.11	0.05	0.10	0.04
0.95	0.14	0.04	0.12	0.04	0.11	0.03	0.09	0.03
0.99	0.14	0.02	0.12	0.02	0.11	0.01	0.09	0.01
	1:10		1:11		1:12		1:13	
	LCCP	CP	LCCP	CP	LCCP	CP	LCCP	CP
0.85	0.16	0.03	0.09	0.03	0.07	0.03	0.07	0.03
0.95	0.08	0.02	0.08	0.02	0.07	0.02	0.07	0.02
0.99	0.08	0.01	0.08	0.01	0.07	0.01	0.07	0.01

Table 1 illustrates the comparison of *favorite prediction* between LCCP and CP. It is clear that LCCP performs distinctly higher favorite prediction than CP under all of the imbalanced ratios. This highlights again that LCCP can deeply dig the local distribution structure of the majority examples through *second-level clustering*. LCCP guarantees the quality of the re-balanced training data, and thus promote significantly prediction efficiency in the subsequent learning. On the contrary, the performance of CP decline dramatically with the high imbalanced ratio, because it has to execute the prediction in the whole imbalanced dataset.

5.3   The Performance under the Domain-Related Indices

Considering the particular imbalanced learning problem, It is essential to evaluate LCCP by some domain-related indices. That is, given the TP (the number of correct predictions among the minority examples), TN (the number of correct predictions among the majority examples), FN (the number of error predictions among the minority examples) and FP (the number of error predictions among the majority examples), some specific indices, such as *Recall, Precision, F, G-man's* and *AUC* value, are designed to demonstrate the power of classification for minority class or the integrated ability of classification [19].

Nonetheless, the indices above can only be set based on the point prediction setting, which seems incompatible with the region prediction with the LCCP model. Thus, the prediction method of LCCP has to be changed to be the point prediction mode, which selects a single label corresponding the maximum *p value* based on formula (3). In such setting, the performance of LCCP is illustrated in table 2.

Table 2. The performance of LCCP under domain-specified indices

dataset	Imbalanced ratio	Recall	Precision	F	G-means	AUC
Simplex	1:2	1.00	1.00	1.00	1.00	1.00
	1:3	1.00	0.97	0.99	0.99	0.99
TEP	1:2	1.00	0.99	1.00	1.00	1.00
	1:3	1.00	0.97	0.98	0.99	0.99
	1:4	1.00	0.92	0.96	0.99	0.99
	1:5	1.00	0.52	0.69	0.903	0.91
	1:6	1.00	0.38	0.55	0.85	0.86
	1:7	1.00	0.30	0.46	0.814	0.83
	1:8	1.00	0.31	0.47	0.851	0.86
	1:9	1.00	0.29	0.45	0.86	0.87
	1:10	1.00	0.31	0.47	0.88	0.89
	1:11	1.00	0.26	0.42	0.864	0.87
	1:12	1.00	0.21	0.34	0.826	0.84
	1:13	1.00	0.22	0.36	0.85	0.86

As shown in table 2, on the Simplex dataset, all of the indices are approximately to be 1, which indicates that LCCP can successfully address the problem of the imbalanced data. On the contrary, the performance of LCCP on the TEP data set fluctuates across these indices. The values of *Recall* index show very high to be around 1, but the performance of the *Precision* index gradually decreases verse the increasing of imbalanced ratio. However, the three indices, *F, G-means, AUC value,* demonstrate high values all over 0.8.

It is clearly that LCCP can recognize the minority examples well and performs quite well on the whole dataset. But LCCP seems poor in the classification of majority examples, especially in the case of the higher imbalanced ratio. The underlying cause of the situation lies in the high overlap of the based TEP dataset[26]. It indicates that the clustering algorithm plays a significant influence on the *second-level partition*. Proper selection, the clustering algorithm can widen the divergence among the clusters and thus boost the ability of classification for the majority examples.

6   Conclusions

In this paper we propose Local Clustering Conformal Predictor (LCCP) to provide valid prediction for the imbalanced data. The experimental results show that LCCP not only provide valid confidence for prediction, but significantly improve the prediction efficiency as well. Furthermore, the LCCP model seems virtually a general framework to deeply dig the local distribution structure of the dataset and thus can promote the prediction efficiency in other application.

Acknowledgements. The work is supported by the Natural Science Foundation of Fujian Province, under Grant No 2012J01274; the Research Grant Council of Huaqiao University under Grant No 09BS515; National Natural Science Fundation of China under Grant No 61202144; Natural Science Foundation of Fujian Province under Grant No 2012J05125.

References

1. Li, H.R.: Reliability and Validity in Qualitative Research. PhD thesis, Harbin Engineering University (2009)
2. Melluish, T., Saunders, C., Nouretdinov, I., Vovk, V.: Comparing the Bayes and Typicalness Frameworks. In: Flach, P.A., De Raedt, L. (eds.) ECML 2001. LNCS (LNAI), vol. 2167, pp. 360–371. Springer, Heidelberg (2001)
3. Elazmeh, W., Japkowicz, N., Matwin, S.: Evaluating misclassifications in imbalanced data. In: Fürnkranz, J., Scheffer, T., Spiliopoulou, M. (eds.) ECML 2006. LNCS (LNAI), vol. 4212, pp. 126–137. Springer, Heidelberg (2006)
4. Li, F., Mi, H., Yang, F.: Exploring the stability of feature selection for imbalanced intrusion detection data. In: 9th IEEE International Conference on Control and Automation, Santiago, pp. 750–754 (2011)
5. Shafer, G., Vovk, V.: A tutorial on conformal prediction. Journal of Machine Learning Research 9, 371–421 (2005)
6. Vovk, V., Gammerman, A., Shafer, A.G.: Algorithmic Learning in a Random World. Springer, New York (2005)
7. Saunders, C., Gammerman, A., Vovk, V.: Transduction with confidence and credibility. In: 16th International Joint Conference on Artificial Intelligence, Stockholm, pp. 722–726 (1999)
8. Gammerman, A., Vovk, V.: Kolmogorov complexity: Sources, theory and applications. The Computer Journal 42(4), 252–255 (1999)
9. Bellotti, T., Luo, Z., Gammerman, A.: Qualified predictions for microarray and proteomics pattern diagnosis with confidence machines. International Journal of Neural Systems 15(4), 247–258 (2005)
10. Vega, J., Murari, A., Pereira, A.: Accurate and reliable image classification by using conformal predictors in the TJ-II Thomson scattering. Review of Scientific Instruments 81, 10–18 (2010)
11. Papadopoulos, H., Vovk, V., Gammerman, A.: Regression Conformal Prediction with Nearest Neighbours. J. Artif. Intell. Res (JAIR) 40, 815–840 (2011)
12. Papadopoulos, H., Haralambous, H.: Reliable Prediction Intervals with Regression Neural Networks. Neural Networks 24(8), 842–851 (2011)
13. Li, F., Kosecka, J., Wechsler, H.: Strangeness based feature selection for part based recognition. In: Computer Vision and Pattern Recognition Workshop, p. 22 (2006)
14. Papadopoulos, H.: Inductive Conformal Prediction: Theory and Application to Neural Networks. In: Tools in Artificial Intelligence, ch.18, pp. 315–330. I-Tech, Vienna (2008)
15. Huazhen, W., Chengde, L., Fan, Y., Jinfa, Z.: An online Algorithm with confidence for Real-Time Fault Detection. Journal of Information and Computational Science 6(1), 305–313 (2009)
16. Fan, Y., Huazhen, W., Hong, M., Weiwen, C.: Using random forest for reliable classification and cost-sensitive learning for medical diagnosis. Bmc Bioinformatics10(1), S22, 14-18 (2009)

17. Devetyarov, D., Nouretdinov, I., Burford, B.: Conformal predictors in early diagnostics of ovarian and breast cancers. In: Progress in Artificial Intelligence, pp. 1–13 (2012)
18. Chawla, N.V., Bowyer, K.W., Hall, L.O.: SMOTE: synthetic minority over-sampling technique. Journal of Artificial Intelligence Research 16(1), 321–357 (2002)
19. Grzymala, J.W., Stefanowski, J.: A comparison of two approaches to data mining from imbalanced data. Journal of Intelligent Manufacturing 16(6), 565–573 (2005)
20. Yen, S.-J., Lee, Y.-S.: Cluster-based under-sampling approaches for imbalanced data distributions. Expert Syst. Appl. 36(3), 5718–5727 (2009)
21. Ji, H., Zhang, H.X.: Classification with Local Clustering in Imbalanced Data Sets. Advanced Materials Research 219, 151–155 (2011)
22. Wu, J., Xiong, H., Chen, J.: COG.: Local decomposition for rare class analysis. Data Mining and Knowledge Discovery 20(2), 191–220 (2010)
23. Prachuabsupakij, W., Soonthornphisaj, N.: Clustering and combined sampling approaches for multi-class imbalanced data classification. In: Zeng, D. (ed.) Advances in Information Technology and Industry Applications. LNEE, vol. 136, pp. 717–724. Springer, Heidelberg (2012)
24. HuaZhen, W., ChengDe, L., Fan, Y., XueQin, H.: Hedged predictions for traditional Chinese chronic gastritis diagnosis with confidence machine. Computers in Biology and Medicine 39(5), 425–432 (2009)
25. Lyman, P., Georgakis, C.: Plant-wide control of the Tennessee Eastman problem. Computers and Chemical Engineering 19(3), 321–331 (1995)
26. Kulkarni, A., Jayaraman, V., Kulkarni, B.: Knowledge incorporated support vector machines to detect faults in tennessee eastman process. Computers and Chemical Engineering 29(10), 2128–2133 (2005)

Osteoporosis Risk Assessment
with Well-Calibrated Probabilistic Outputs

Antonis Lambrou[1,2], Harris Papadopoulos[1,2,3], and Alexander Gammerman[2]

[1] Frederick Research Center, Nicosia, Cyprus
[2] Computer Learning Research Centre, Computer Science Department,
Royal Holloway, University of London, England
{A.Lambrou,A.Gammerman}@cs.rhul.ac.uk
[3] Computer Science and Engineering Department, Frederick University, Cyprus
H.Papadopoulos@frederick.ac.cy

Abstract. Osteoporosis is a disease of bones that results in an increased risk of bone fracture. The diagnosis of Osteoporosis is usually performed by measuring the Bone Mineral Density (BMD) using Dual-Energy X-ray Absorptiometry (DEXA) scanning. In this work, we introduce the use of Venn Prediction in order to assess the risk of Osteoporosis before a DEXA scan, based on known risk factors. Unlike other probabilistic methods, Venn Predictors can provide well-calibrated probabilistic outputs under the assumption that the data used are identically and independently distributed (i.i.d.). Our contribution is two-fold: Firstly, we have collected real-world data from various clinic centres in Cyprus which based on their locality can be used for analysis of Osteoporosis risk factors specifically for Cypriot patients. To the best of our knowledge, local data in Cyprus for Osteoporosis risk assessment have not been collected before. Secondary, our results demonstrate that our method can provide probabilistic outputs that may be practical and trustful to physicians.

Keywords: well calibrated probabilities, osteoporosis, risk assessment, Venn Predictor, Machine Learning.

1 Introduction

Osteoporosis is a disease of bones that results in an increased risk of bone fracture. The diagnosis of Osteoporosis is usually performed by measuring the Bone Mineral Density (BMD) using Dual-Energy X-ray Absorptiometry (DEXA) scanning. A result of BMD that is lower than 2.5 standard deviations from the average of young healthy adults is defined by the World Health Organisation (WHO) as Osteoporosis [11].

We introduce the use of Venn Prediction in order to predict the risk of Osteoporosis before a DEXA scan, based on known risk factors. Unlike other probabilistic methods, Venn Predictors can provide well-calibrated probabilistic outputs under the assumption that the data used are identically and independently distributed (i.i.d.). We have collected real-world data from various

H. Papadopoulos et al. (Eds.): AIAI 2013, IFIP AICT 412, pp. 432–441, 2013.

clinic centres in Cyprus which based on their locality can be used for analysis of Osteoporosis risk factors specifically for Cypriot patients. To the best of our knowledge, local data in Cyprus for Osteoporosis risk assessment have not been collected before. Moreover, our results show that Venn Predictors (VPs) can provide probabilistic outputs that may be practical and trustful to physicians.

The rest of the paper is structured as follows. In Section 2, we outline related work. In Section 3, we give a detailed explanation of the Osteoporosis data that we have collected, and we also explain how Venn Prediction works. In Section 4, we show our experimental results on the Osteoporosis data and discuss. Finally, in Section 5, we conclude and outline our future work.

2   Related Work

The World Health Organisation (WHO) has defined the disease of Osteoporosis as a Bone Mineral Density (BMD) which is lower than 2.5 standard deviations from the average of young healthy adults. Furthermore, BMD that is 1 standard deviation lower is defined as Osteopenia, which is a precursor to Osteoporosis [11]. DEXA stands for Dual Energy X-ray Absorptiometry, and is a standard test for BMD. DEXA scanners throw an X-ray beam at the lumbar vertebrae and measure the shadow cast by the bones. In Fig. 1 we include a sample image of the lumbar spine of a DEXA scan. Software in the machine estimates the amount of calcium in the bone based on the darkness of the shadow. The result is expressed as a number of grams per square centimeter, which is defined as the Bone Mineral Density (BMD). In Table 1, we show how the BMD is mapped to a t-score value compared against the average of young healthy adults.

The Venn Prediction (VP) framework is based on the Conformal Prediction (CP) framework. CP is a novel technique for obtaining reliable confidence measures. The technique is proposed in [7] and later improved in [22] and [24]. CPs are built using classical machine learning algorithms, called underlying algorithms. CPs complement the predictions of the underlying algorithms with measures of confidence. Many CPs have been built to date, based on various algorithms such as Support Vector Machines [22], k-Nearest Neighbours for classification [21] and for regression [18], Random Forests [4], and Genetic Algorithms [8]. The computational efficiency of CPs has also been greatly improved using Inductive Conformal Prediction (ICP) [12], as demonstrated in applications to Ridge Regression [17], and more recently in applications to Neural Networks [19]. The CP framework has been successfully applied to medical problems, such as evaluation of the risk of stroke [9], breast cancer diagnosis [6], classification of leukaemia subtypes [1], and acute abdominal pain diagnosis [15]. Additionally, CPs have been applied to other problems such as Software Effort Estimation in [16].

Venn Prediction has been introduced in [23] where the interested reader can find a detailed description of the framework. Since then, VPs have been developed based on k-Nearest Neighbours [3], Nearest Centroid [2] and Neural Networks [13,14]. Furthermore, VPs based on SVMs have been developed in [10,27],

Table 1. Young Adult (YA) T-score based on the Bone Mineral Density (BMD) according to the Wolrd Health Organisation (WHO)

BMD	1.44	1.32	1.20	1.08	0.96	0.84	0.72	0.60
YA T-Score	2	1	0	-1	-2	-3	-4	-5

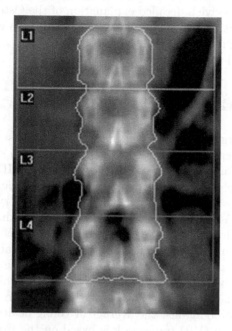

Fig. 1. Image of the Lumbar Spine AP (Anterior Posterior) from a DEXA Scan

and have been compared with Platt's method [20], Binning [5] and Isotonic Regression [26]. As it is shown in [10], such methods do not guarantee that the probabilistic outputs will be well-calibrated.

3   Material and Methods

3.1   Osteoporosis Data

We have collected data from various clinics in Cyprus. In particular, we have collected 389 cases of female patients that have performed a DEXA scan. The data are constructed based on a questionnaire that is given to patients to complete. The patients may have previous history of osteoporosis and may already follow therapy. The questionnaire was constructed by physicians and contains questions that are relevant to Osteoporosis risk factors. Each case is classified as "Normal" or "Risk of Osteoporosis" based on the patient's spine t-score that is given by the DEXA scan. According to the WHO, patients with a t-score above

-1 are diagnosed as healthy, therefore we have classified patients into two classes: "Normal" for patients with t-score above -1, and "Risk of Osteoporosis" otherwise. From the 389 patients, 174 have a t-score above or equal to -1, and 215 have a t-score below -1. In Table 2, we give the list of attributes of our dataset.

3.2   Venn Prediction

In this section, we explain how Venn Prediction (VP) works. Typically, we have a training set[1] of the form $\{z_1, \ldots, z_{n-1}\}$, where each $z_i \in Z$ is a pair (x_i, y_i) consisting of the object x_i and its classification y_i. For a new object x_n, we intend to estimate its probability of belonging to each class $Y_j \in \{Y_1, \ldots, Y_c\}$. The Venn Prediction framework assigns each one of the possible classifications Y_j to x_n and divides all examples $\{(x_1, y_1), \ldots, (x_n, Y_j)\}$ into a number of categories based on a *taxonomy*. A taxonomy is a sequence A_n, $n = 1, \ldots, N$ of finite measurable partitions of the space $Z^{(n)} \times Z$, where $Z^{(n)}$ is the set of all multisets of elements of Z of length n. We will write $A_n(\{z_1, \ldots, z_n\}, z_i)$ for the category of the partition A_n that contains $(\{z_1, \ldots, z_n\}, z_i)$. Every taxonomy $A_1, A_2, \ldots, A_N$ defines a different VP. In this work, we define three VPs based on taxonomies that use the classification output of three underlying algorithms, namely, the J48 decision tree, Random Forests (RF), and Sequential Minimal Optimisation (SMO). Examples are categorized according to the underlying algorithm classifications that are given to them.

After partitioning the examples into categories using a taxonomy, the empirical probability of each classification Y_k in the category τ_{new} that contains (x_n, Y_j) will be

$$p^{Y_j}(Y_k) = \frac{|\{(x^*, y^*) \in \tau_{new} : y^* = Y_k\}|}{|\tau_{new}|}. \tag{1}$$

This is a probability distribution for the class of x_n. So after assigning all possible classifications to x_n we get a set of probability distributions $P_n = \{p^{Y_j} : Y_j \in \{Y_1, \ldots, Y_c\}\}$ that compose the multi-probability prediction of the VP. As proved in [25], these are automatically well calibrated, regardless of the taxonomy used.

The maximum and minimum probabilities obtained for each label Y_k amongst all distributions $\{p^{Y_j} : Y_j \in \{Y_1, \ldots, Y_c\}\}$, define the interval for the probability of the new example belonging to Y_k. We denote these probabilities as $U(Y_k)$ and $L(Y_k)$, respectively. The VP outputs the prediction $\hat{y}_n = Y_{k_{best}}$, where

$$k_{best} = \underset{k=1,\ldots,c}{\arg\max} \; \overline{p(k)}, \tag{2}$$

and $\overline{p(k)}$ is the mean of the probabilities obtained for label Y_k amongst all probability distributions. The probability interval for this prediction is $[L(Y_k), U(Y_k)]$.

---

[1] The training set is in fact a multiset, as it can contain some examples more than once.

Table 2. Table of attributes in the Osteoporosis dataset

#	Attribute name	Type	#	Attribute name	Type
1	Sex	Binary	35	Receive Thyroxine	Binary
2	Age	Numeric	36	Receive Estrogens	Binary
3	Weight	Numeric	37	Neurogenic Anorexia	Binary
4	Height	Numeric	38	Malabsorption syndrome	Binary
5	Start of Menstruation	Numeric	39	Chronic liver diseases	Binary
6	End of Menstruation	Numeric	40	Inflammatory bowel diseases	Binary
7	Pregnacies	Numeric	41	Transplantation	Binary
8	Smoking now	Binary	42	Chronic renal failure	Binary
9	Smoking in the past	Binary	43	Prolonged immobilization	Binary
10	No smoking	Binary	44	Cushing's syndrome	Binary
11	Years of past smoking	Numeric	45	Epilepsy	Binary
12	Years of current smoking	Numeric	46	Insulin Dependent	Binary
13	Cigarettes per day	Numeric	47	Ovariectomy before menopause	Binary
14	Alchohol intake per day	Numeric	48	Chronic gastrointestinal disorders	Binary
15	Caffeine intake per day	Numeric	49	Paget's Disease	Binary
16	History of fracture	Binary	50	Hyperthyroidism	Binary
17	Hip fracture	Binary	51	Parathyroid gland disease	Binary
18	Spine fracture	Binary	52	Receive Steroids	Binary
19	Wrist fracture	Binary	53	Receive Thyroxine	Binary
20	Low energy	Binary	54	Anticonvulsants (for seizures, epilepsy)	Binary
21	High energy	Binary	55	Diuretics	Binary
22	Sports	Binary	56	Heparin	Binary
23	History of osteoporosis	Binary	57	Chemotherapy	Binary
24	Osteoporosis in family	Binary	58	Treatment of osteoporosis	Binary
25	Loss of height	Binary	59	Alendronati	Binary
26	Kyphosis	Binary	60	Risedronati	Binary
27	End of menstrual bleeding	Binary	61	Zoledronati	Binary
28	Arthritis	Binary	62	Raloxifeni	Binary
29	Secondary Osteporosis	Binary	63	Strontio	Binary
30	Breast feeding	Binary	64	Parathormoni	Binary
31	Avoidance of milk	Binary	65	Denosoymapi	Binary
32	Avoidance of sex	Binary	66	Kalsitonini	Binary
33	Diarrhea	Binary	67	Calcium + Bitamin D	Binary
34	Receive Cortisone	Binary	68	Calcium	Binary

Table 3. Results of the six algorithms on the Osteoporosis dataset. We compare the accuracy, the lower and upper probability outputs of the VPs, and the percentages and accuracy rates of examples which have probabilities of at least 75%, 70%, and 60%.

Predictors	J48	RF	SMO	J48-VP	RF-VP	SMO-VP
Accuracy	70.18%	68.89%	67.10%	67.38%	65.17%	65.71%
Lower Probability	-	-	-	64.27%	57.93%	64.21%
Upper Probability	-	-	-	80.62%	78.09%	71.83%
Min probability ≥ 75%	-	-	-	54.73%	34.27%	7.61%
Min probability ≥ 70%	-	-	-	67.10%	60.51%	39.49%
Min probability ≥ 60%	-	-	-	75.53%	64.88%	84.52%
Accuracy at ≥ 75% min. prob.	-	-	-	73.61%	73.43%	74.05%
Accuracy at ≥ 70% min. prob.	-	-	-	72.71%	71.51%	69.01%
Accuracy at ≥ 60% min. prob.	-	-	-	72.06%	71.44%	67.93%

4 Experiments and Results

4.1 Offline Experiments

We perform 10-fold cross validation experiments on our Osteoporosis dataset with the J48 decision tree, Random Forests (RF), Sequential Minimal Optimisation (SMO), J48-VP, RF-VP, and SMO-VP algorithms. Each algorithm performs a Correlation Based Feature Selection (CBFS) during each fold of the experiment. For the RF, we chose the number of trees to 20, and the depth of each tree to 4 nodes. For the SMO, we chose the RBF kernel with a spread parameter of 0.43. These parameters were chosen after several experimental settings. In the results, we show the average accuracy, and for the VPs we also show the average lower probabilities and upper probabilities. Since VPs provide well-calibrated probabilistic outputs, the accuracy of the VPs is expected to fall within the lower and upper probability intervals. We demonstrate that an amount of predictions have significant probabilities of being correct, and such predictions may be practical for physicians. Moreover, we show that the accuracy rates of the predictions do not drop below the minimum probabilities given by the VPs (up to statistical flactuations).

In Table 3, we compare the average accuracy of the four algorithms on the Osteoporosis dataset. The J48 algorithm provides the highest accuracy which is 70.18%. The corresponding J48-VP provides a slightly lower accuracy which is 67.38%. The VPs provide extra information for each prediction, which is the lower and upper probability interval. We show the average lower probabilities and upper probabilities given by the three VPs. and the corresponding accuracy rates of such predictions. The accuracy rates demonstrate the validity of the probabilistic outputs of the VPs. For example, at 75% probability, the accuracy rates should be at least (up to statistical fluctuations) at the 75% which is expected. In our case, the offline results give accuracies of 73.61%, 73.43%, and 74.05%, which are acceptable. In the table, we also show the percentage of examples that have a minimum probability of 75%, 70%, and 60%. The J48-VP

Table 4. Confusion matrices of the three algorithms J48, RF, and SMO on the Osteoporosis dataset

	J48		RF		SMO	
	Normal	Risk of Ost.	Normal	Risk of Ost.	Normal	Risk of Ost.
Normal	113	61	107	67	89	85
Risk of Ost.	55	160	54	161	43	172

gives 54.73% of predictions with a probability of at least 75%. These predictions can be considered significant, since the error rate will be at most 25% in the long run.

In Table 4, we show the confusion matrices of the three algorithms on the Osteoporosis dataset. The confusion matrices demonstrate that the algorithms misclassify examples at a balanced rate between the two classes.

4.2   Online Experiments

In order to show the validity of the probability estimates of the VPs, we conduct experiments in the on-line mode. Initially all examples are test examples and they are added to the training set one by one after a prediction for each one is made. We graph the Cumulative Lower Accuracy Probability (CLAP), the Cumulative Upper Accuracy Probability (CUAP), and the Cumulative Accuracy (CA) curves:

$$CLAP(t) = \frac{1}{t} \sum_{i=1}^{t} U_i(Y_{k_{best}}), \tag{3}$$

$$CUAP(t) = \frac{1}{t} \sum_{i=1}^{t} L_i(Y_{k_{best}}), \tag{4}$$

$$CA(t) = \frac{1}{t} \sum_{i=1}^{t} Acc_i, \tag{5}$$

where t is the number of test examples that have been added to the training set, and $Acc_i = 1$ when the prediction for example x_i is correct and 0 otherwise. Since VPs provide well calibrated probabilistic outputs, it is expected that the CA curve will fall within or near the CLAP and CUAP curves.

In Fig. 2, we show the online results of the three VPs, J48-VP (left), RF-VP (right), and SMO-VP (bottom) on the Osteoporosis dataset. The results demonstrate the validity of the VPs, since the actual accuracy of the predictors always falls within the lower and upper probability outputs of the VPs. Comparing the three VPs, we can see that the RF-VP gives the wider probabilistic outputs, while the SMO-VP gives the narrowest outputs.

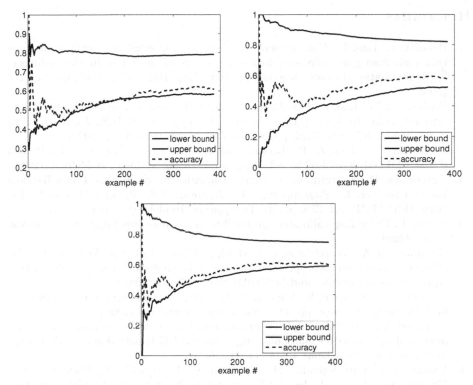

Fig. 2. Online experiments with J48-VP (left), RF-VP (right), and SMO-VP (bottom) on the Osteoporosis dataset

5   Conclusion

In this work, we have applied Venn Prediction to the problem of Osteoporosis Risk Assessment. We have evaluated our method on real-world data that we have collected from various clinics in Cyprus. Our results, demonstrate that our method provides well-calibrated probabilistic outputs in the predictions that can be useful in practice. Precisely, patients may get a prognosis based on Osteoporosis risk factors before the performance of a DEXA scan. In the future, we aim to collect more data and perform supplementary analysis, in order to improve and evaluate further our VPs. Furthermore, we are in the process of building a tool for physicians that will enable them to use our VPs for assessing the risk of Osteoporosis.

Acknowledgments. This work was co-funded by the European Regional Development Fund and the Cyprus Government through the Cyprus Research Promotion Foundation "DESMI 2009-2010", TPE/ORIZO/0609(BIE)/24 ("Development of New Venn Prediction Methods for Osteoporosis Risk Assessment").

References

1. Bellotti, T., Luo, Z., Gammerman, A.: Reliable classification of childhood acute leukaemia from gene expression data using confidence machines. In: Proceedings of IEEE International Conference on Granular Computing (GRC 2006), pp. 148–153 (2006)
2. Dashevskiy, M., Luo, Z.: Reliable probabilistic classification and its application to internet traffic. In: Huang, D.-S., Wunsch II, D.C., Levine, D.S., Jo, K.-H. (eds.) ICIC 2008. LNCS, vol. 5226, pp. 380–388. Springer, Heidelberg (2008)
3. Dashevskiy, M., Luo, Z.: Predictions with confidence in applications. In: Perner, P. (ed.) MLDM 2009. LNCS, vol. 5632, pp. 775–786. Springer, Heidelberg (2009)
4. Devetyarov, D., Nouretdinov, I.: Prediction with Confidence Based on a Random Forest Classifier. In: Papadopoulos, H., Andreou, A.S., Bramer, M. (eds.) AIAI 2010. IFIP AICT, vol. 339, pp. 37–44. Springer, Heidelberg (2010)
5. Drish, J.: Obtaining calibrated probability estimates from Support Vector Machines (1998)
6. Gammerman, A., Nouretdinov, I., Burford, B., Chervonenkis, A., Vovk, V., Luo, Z.: Clinical mass spectrometry proteomic diagnosis by conformal predictors. Statistical Applications in Genetics and Molecular Biology 7(2) (2008)
7. Gammerman, A., Vovk, V., Vapnik, V.: Learning by transduction. In: Uncertainty in Artificial Intelligence, pp. 148–155. Morgan Kaufmann (1998)
8. Lambrou, A., Papadopoulos, H., Gammerman, A.: Reliable confidence measures for medical diagnosis with evolutionary algorithms. IEEE Transactions on Information Technology in Biomedicine 15(1), 93–99 (2011)
9. Lambrou, A., Papadopoulos, H., Kyriacou, E., Pattichis, C.S., Pattichis, M.S., Gammerman, A., Nicolaides, A.: Evaluation of the risk of stroke with confidence predictions based on ultrasound carotid image analysis. International Journal on Artificial Intelligence Tools 21(04), 1240016 (2012)
10. Lambrou, A., Papadopoulos, H., Nouretdinov, I., Gammerman, A.: Reliable probability estimates based on support vector machines for large multiclass datasets. In: Iliadis, L., Maglogiannis, I., Papadopoulos, H., Karatzas, K., Sioutas, S. (eds.) AIAI 2012 Workshops, Part II. IFIP AICT, vol. 382, pp. 182–191. Springer, Heidelberg (2012)
11. World Health Organisation. Prevention and management of Osteoporosis, Geneva (2003)
12. Papadopoulos, H.: Inductive Conformal Prediction: Theory and application to Neural Networks. In: Fritzsche, P. (ed.) Tools in Artificial Intelligence, ch.18, pp. 315–330. InTech, Vienna (2008)
13. Papadopoulos, H.: Reliable probabilistic prediction for medical decision support. In: Iliadis, L., Maglogiannis, I., Papadopoulos, H. (eds.) EANN/AIAI 2011, Part II. IFIP AICT, vol. 364, pp. 265–274. Springer, Heidelberg (2011)
14. Papadopoulos, H.: Reliable probabilistic classification with neural networks. Neurocomputing 107, 59–68 (2013)
15. Papadopoulos, H., Gammerman, A., Vovk, V.: Reliable diagnosis of acute abdominal pain with conformal prediction. International Journal of Engineering Intelligent Systems for Electrical Engineering and Communications 17(2-3), 127–137 (2009)
16. Papadopoulos, H., Papatheocharous, E., Andreou, A.S.: Reliable confidence intervals for software effort estimation. In: Proceedings of the 2nd Workshop on Artificial Intelligence Techniques in Software Engineering (AISEW 2009). CEUR Workshop Proceedings, vol. 475, pp. 211–220. CEUR WS.org (2009)

17. Papadopoulos, H., Proedrou, K., Vovk, V., Gammerman, A.: Inductive confidence machines for regression. In: Elomaa, T., Mannila, H., Toivonen, H. (eds.) ECML 2002. LNCS (LNAI), vol. 2430, pp. 345–356. Springer, Heidelberg (2002)
18. Papadopoulos, H., Vovk, V., Gammerman, A.: Regression conformal prediction with nearest neighbours. Journal of Artificial Intelligence Research 40, 815–840 (2011)
19. Papadopoulos, H., Vovk, V., Gammerman, A.: Conformal prediction with neural networks. In: Proceedings of the 19th IEEE International Conference on Tools with Artificial Intelligence (ICTAI 2007), vol. 2, pp. 388–395. IEEE Computer Society (2007)
20. Platt, J.C.: Probabilistic outputs for Support Vector Machines and comparisons to regularized likelihood methods. In: Advances in Large Margin Classifiers, pp. 61–74. MIT Press (1999)
21. Proedrou, K., Nouretdinov, I., Vovk, V., Gammerman, A.: Transductive confidence machines for pattern recognition. In: Elomaa, T., Mannila, H., Toivonen, H. (eds.) ECML 2002. LNCS (LNAI), vol. 2430, pp. 381–390. Springer, Heidelberg (2002)
22. Saunders, C., Gammerman, A., Vovk, V.: Transduction with confidence and credibility. In: Proceedings of the 16th International Joint Conference on Artificial Intelligence, vol. 2, pp. 722–726. Morgan Kaufmann, Los Altos (1999)
23. Vovk, V., Shafer, G., Nouretdinov, I.: Self-calibrating probability forecasting. In: Thrun, S., Saul, L.K., Schölkopf, B. (eds.) Advances in Neural Information Processing Systems 16, pp. 1133–1140. MIT Press, Cambridge (2004)
24. Vovk, V., Gammerman, A., Saunders, C.: Machine-learning applications of algorithmic randomness. In: Proceedings of the Sixteenth International Conference on Machine Learning, pp. 444–453. Morgan Kaufmann (1999)
25. Vovk, V., Gammerman, A., Shafer, G.: Algorithmic Learning in a Random World. Springer, New York (2005)
26. Zadrozny, B., Elkan, C.: Transforming classifier scores into accurate multiclass probability estimates. In: Proceedings of the 8th ACM International Conference on Knowledge Discovery and Data Mining, pp. 694–699 (2002)
27. Zhou, C., Nouretdinov, I., Luo, Z., Adamskiy, D., Randell, L., Coldham, N., Gammerman, A.: A comparison of venn machine with platt's method in probabilistic outputs. In: Iliadis, L., Maglogiannis, I., Papadopoulos, H. (eds.) EANN/AIAI 2011, Part II. IFIP AICT, vol. 364, pp. 483–490. Springer, Heidelberg (2011)

Tutamen: An Integrated Personal Mobile and Adaptable Video Platform for Health and Protection

David Palma[1], Joao Goncalves[1], Luis Cordeiro[1], Paulo Simoes[2],
Edmundo Monteiro[2], Panagis Magdalinos[3], and Ioannis Chochliouros[4]

[1] OneSource, Consultoria Informatica, Lda.
{palma,john,cordeiro}@onesource.pt
[2] Centre for Informatics and Systems of the University of Coimbra
{psimoes,edmundo}@dei.uc.pt
[3] National and Kapodestrian University of Athens
panagis@di.uoa.gr
[4] Hellenic Telecommunications Organization (OTE) S.A.
ichochliouros@oteresearch.gr

Abstract. A framework for mobile and portable High-Definition Video streaming is proposed, developed and assessed. Suitable for emergency scenarios, involving for instance ambulances and fire-fighters, the presented framework resorts to a state-of-art platform which considers off-the-shelf hardware and available video codecs for High-Definition Video. The obtained results show that the proposed architecture is able to efficiently support rescuing teams in the demanding scenarios where they operate, guaranteeing video quality and ease of use. This solution is particularly useful for situations where experts in the fields can accurately provide their insights and contributions remotely and in a timely fashion.

Keywords: HD video, Streaming, Smart Cities, Mobile First Responders.

1 Introduction

Everyday advances in technology and its availability within an increasing number of communities has motivated new approaches towards technological innovation in many sectors. A valuable contribution in this aspect is the definition of Smart Cities, where technologies can be used to provide dynamic and adaptive services and contexts. Among several topics, the LiveCity project, in which this work has been developed, is concerned with sustainability as well as with a reliable, smart and secure cities. By exploring the advances in wireless communications, in conjunction with the portability of increasingly more powerful computers, this work aims at extending the benefits of integrated personal mobility for demanding scenarios involving aspects such as healthcare and civil protection.

Previous works have already proposed many end-to-end mechanisms connecting hospital experts with ambulances and similar agents in emergency scenarios.

H. Papadopoulos et al. (Eds.): AIAI 2013, IFIP AICT 412, pp. 442–451, 2013.

For instance, not only simulation based but also real-time non-diagnostic systems have been developed, capable of providing trauma and echo-cardiogram videos [1]. Nonetheless, these solutions are strongly dependent on the entire infrastructure provided by ambulances, being limited in terms of their operational area, requiring patients to be inside the vehicle. Moreover, the addition of features such as the monitoring of other vital signals, a clear image of the patients' face or any non-digital medical equipment is non-trivial in these solutions.

Nowadays a big challenge is the design of mobile and ubiquitous telemedicine, versus the known store and forward telemedicine which typically demands tightly-coupled services and rigid architectures [2]. The Tutamen proposal, takes its name from the same latin word which stands for protection, defence and caring. It tackles flaws such as lack of real-time communication between rescuing teams, the portability and flexibility that will allow its usage in a seamless and effective manner, defining a new perspective towards the exploitation of technology for effectively improving the quality of living both in urban and remote areas. Moreover, this proposal is expected not only to allow a physician to be available anywhere at anytime, but also to follow a well defined methodology with field trials and real feedback from the involved professionals, which are typical disregarded by other solutions.

Other recent solutions, such as the eBag [3], require teams operating in the field to carry a suitcase equipped with a full-size laptop and other equipment. This not only creates difficulties in transportation but also in setup. When responding to an emergency scenario, timing is critical and other similar solutions such as the use of robotic agents [4], makes the operation more cumbersome, possibility creating trust issues with patients regarding their efficacy.

The technological challenges raised by the need of providing an ubiquitous and real-time high definition (HD) stream have been addressed in three different approaches, guaranteeing that a modular solution would be achieved. By separating the proposed Tutamen architecture into three blocks - video acquisition, streaming and interfacing - the personal video-to-video (v2v) system was designed taking into account the possibility of future hardware and software updates.

This work presents the Tutamen framework, concerning both wearable and flexible hardware and the software mechanisms used for adaptable real-time video streaming, taking into account the available network conditions.

A general description of the LiveCity project in which the mobile video platform has been developed is presented in Section 2, followed by a detailed presentation of the proposed solution in Section 3, considering a preliminary use case for ambulances, requirements and methodology in Section 4. The assessment of the performance and main characteristics of the Tutamen personal unit are included in Section 5. Finally, Section 6 highlights the main contributions provided by this work.

2   The LiveCity Project

Empowering inhabitants of a city by providing them means to interact with each other in a farther prolific, effectual and socially useful way, by the use of high quality v2v over the Internet, could be presented as the main goal of LiveCity Project. The proposed augmentation could be used to save patients lives, improve city administration, reduce fuel costs, reduce carbon footprint, enhance education and improve city experiences for tourists and cultural consumers.

LiveCity promotes the context of a modern world where live HD interactive v2v can be easily available on a variety of display devices and where a video call with HD quality can be as globally reachable to a city environment as a plain old telephony call; or even a world where any video screen coupled with a video camera in a city could connect an HD full screen video call (where HD video represents moving pictures with at least a resolution similar to 1280 pixels wide and 720 pixels height) at an attractive cost with seamless, utterly simple usability. This is the vision of the essential LiveCity approach. To realize its strategic aims, LiveCity considers standard video encoding already available in off-the-shelf devices.

The LiveCity effort aims to fill the above matters, per separate option, by providing appropriate responses and by suitably promoting experiences and results gained from pilot actions.

3   Camera Data Acquisition Prototype

Initially developed for emergency first responders - namely paramedics units deployed in emergency ambulances - so as to provide a balanced compromise between computational resources (necessary to acquire, encode and transmit live HD video), mobility (wearable device, battery autonomy and hands-free operation) and good performance under extreme conditions, mostly due to the inclusion of rugged devices, video camera able to self-adjust to variable lighting environments and to the possibility of integration with emergency responders equipment, as shown in Figure 1.

Fig. 1. Equipped Member   **Fig. 2.** Micro Computer

Fig. 3. Application layout deployed in the hospital side

The system, with its core component depicted in Figure 2, supports any type of Global System for Mobile Communications (GSM) or Long Term Evolution (LTE) based networks by the means of a Universal Serial Bus (USB) dongle. Composing the system is an x86 compliant dual core board, coupled with an High-Definition Multimedia Interface (HDMI) Mini Card capture device which interfaces with a Mini Peripheral Component Interconnect (PCI) Express expansion card needed by the main board to support this kind of device. External to the main component, we have buttons connected to the machine, allowing the user to control and change states of the current call. Travelling through the back of the user, is a sleeve with cabling to connect the external camera and headset, providing the user with audio feedback and the hospital with a realtime video and audio feed. This setup has been developed within the LiveCity project and its performance results for v2v communication are presented in this paper.

4   Use Case – Ambulances

During the so called "golden hour", patients suffering from an heart attack, on the verge of having a stroke, or suffering from some type of environment induced trauma, could significantly benefit from speed about decision making.

Early recognition of stroke features and opportune referral to a stroke unit by the means of v2v enabled technology should improve overall outcome, saving lives and improving the patient chance of independently functioning, as well as an expedited advice concerning optimal pain management and requirement of evacuation on trauma victims.

4.1   Requirements and Methodology

The most important step in the design and implementation of the prototype was the extraction of a detailed list of requirements from the involved user groups. Due to the focus of the application, this effort was of paramount importance and had to be designed and implemented in every possible detail. Most of these requirements were assessed by working closely with emergency personnel from several different areas, but mostly, first responders.

From a methodological point view we applied an iterative and incremental process (closely resembling the concept of Rational Unified Process [5]) implemented through consecutive cycles of interviews and subsequent software releases. The latter facilitated timely feedback from the involved user groups while in parallel assured that the development process will not result in an application which although functional is useless to the end user.

In the context of this paragraph we will try to briefly present this process focusing primarily in end user requirements, how they were captured and what they implied for the subsequent design and implementation process. The end product is a v2v software application with an end user interface which offers the doctor an opportunity to intervene early in emergency situations and offers the paramedic the opportunity to get medical assistance in a difficult situation.

We look at the three use case scenarios of heart attack, stroke and trauma together for the purposes of the hardware and software development. The storyboard for the emergency use case is the following:

- The paramedic initiates a video-call to the hospital using a portable device.
- In case the doctor is directly available, an alert is sent to his smart phone.
- The doctor proceeds to the emergency room and answers the call.
- The paramedic speaks into the microphone held on his headband and explains the issue.
- The doctor looks at the patient via the monitor and instructs the paramedic to focus on the part of the patient of interest.
- The doctor asks questions to the patient in order to assess his condition.
- The doctor instructs the paramedics on how to proceed.

The latter provided the first basis for design of the application. However, more user specific requirements came after the first release of the application. The involved user groups have different requirements; paramedics need a simple application with a maximum number of two or three buttons that will enable them to communicate with the hospital with minimal overhead; on the other hand the doctors required an application that supports full management of the incoming video and audio streams. Thus, following the well known Model-View-Controller software design concept, two different flavours for the end user application were designed.

From the hospital perspective, the application caters for stream management (video and audio stream management), notifications to the end user (SMS texting in case of absence, highlight of emergency call), capturing of still images from the video source, manual and/or dynamic adaptation of the video/audio quality and encrypted storage of patient information. Finally, the doctor can initiate a call to an ambulance; however the latter is permitted if and only if a call has been received by that ambulance. Figure 3 provides the layout.

From the paramedics' point of view, the application should be simple and easy to use with a minimum number of buttons facilitating communication with the hospital. All notifications and interruptions should take place through a simple beep-code thus enabling the ambulance crew to minimize the amount of time being distracted because of the application.

Not only the application has gone through the scrutiny of the paramedics', but also the hardware configuration of the whole platform. Targeted at personnel with different requirements, one of the most requested features was modularity. Being able to switch the position of the computer box and how the cabling flows through the belt and hoses, became mandatory.

5 Main Results

Aiming to provide a proper evaluation of the platform, measurements regarding the overall performance of the platform during the process of video acquisition, processing and transmission will be performed. Two different types of measurements will be made; firstly the resulting Peak Signal-to-Noise Ratio (PSNR) of the codec while encoding the acquired video into the different bit rates and resolutions shall be measured; secondly a measurement of the transmitted video's PSNR, using different bit rates and resolutions, will be obtained as well. For the sake of simplicity, only the performance of one video codec will be reviewed and the choice will fall upon the open implementation of the H.264/MPEG-4 AVC video codec, x264, since it is the considered to be very robust for video streaming in wireless networks [6].

In order to measure the encoding performance of the codec, its internal measuring tool will be used to retrieve the global PSNR, but with the caveat of not enabling its optimizations for PSNR monitoring. The reason behind this decision has to do with the fact that it is intentional to measure the codec performance with the optimizations that were selected for the Tutamen live streaming, and not the optimizations that achieve a most advantageous PSNR during the encoding process.

An additional quality metric that is typically considered is the Mean Opinion Score (MOS) which indicates the quality of a video and is quantified from 1 (bad quality) up to 5 (excellent quality). However, this metric is a sensorial metric, which means it depends how the quality perceived by humans, therefore not necessarily providing accurate results when calculated [7].

The video quality assessment will be performed taking into account the degradation introduced both by the transcoding and streaming processes. For comparison purposes, the raw FullHD video will be considered as a reference value and compared against the transcoded videos at different bit-rates prior to and after streaming. In fact, variable bit-rates will be used during the streaming of the video, adapting to the available network conditions. However, the reduction of the videos' bit-rate will lower the initial PSNR, therefore the bit-rate reduction must be sensible. In order to measure the PSNR resulting from the transcoding and streaming process, the Evalvid tool [8] and its known methods will be used.

5.1 Usability / Wearable Characteristics

Bearing in mind the use cases proposed for the Tutamen platform, its usability and overall look and feel experience is of paramount importance. In fact, the definition of a comfortable and intuitive device for either physicians, fireman or any other unit working under emergency situations, has influenced not only the design of the wearable communication unit but also of its mechanisms and interaction procedures with the remote units in control centres. Therefore, the results obtained from the process of creating the hardware-based interface and the communication module, provide important insights for future related works.

The developed work was closely followed by doctors and physicians which provided their contribute and knowledge regarding the main requirements and challenges in emergency situations. A key point was the ease in using the entire platform, not only should the device be easily carried, being light in both weight and size, but it should also be intuitively operated, requiring as few buttons as possible. These requirements resulted in a solution slightly exceeding 1kg that can be used by fitting an adjustable belt, capable of being worn with any uniform in any atmospheric conditions. Moreover, taking into account the ease of operation, a simple three button system was developed.

With the available buttons, the units carrying the Tutamen device are able to start and stop the entire system by using the power button as well as to start and stop a direct call to the control centre or hospital. Moreover, since some disaster scenarios lack the desirable lighting, the third button can be used to enable a powerful lighting system, capable of providing good quality video. Finally, due to the harsh conditions in which emergency units usually operate, the buttons module is protected with a cover and each operation (i.e. button press), requires a button press of at least two seconds, thus avoiding accidental operations.

An important requirement identified during the development of this work, in conjunction with emergency teams, was that the equipment should be as hands-free as possible. Therefore, the video acquisition unit, as well as the audio and lighting systems, have considered taking into account this aspect. Regarding the video and lighting systems, they have been designed so that both can be attached to a working helmet or to a headband, allowing the individual wearing the device to transmit and illuminate the desired regions of interest without any adjustments or configurations. Additionally, in order not to distract the person operating Tutamen, each action in the system is transmitted through a headset that is also used to transmit audio to the control centre.

5.2   Technical Performance

In order to correctly address the difficulties found by the operating teams in the field, the performance of the used hardware has been taken into consideration, influencing the choices for each component included in the equipment. The used camera is able to provide FullHD Video at several frame rates, guaranteeing a 170° field of view through a wide angle lens, for up to 2.5 hours of continuous video. The camera has a dedicated power source which can easily be replaced or recharged.

Coupled with the camera acquisition unit, a supportive light system has been developed, providing a 52 Lumen output with a wide viewing angle as well. The lighting system not only has a small power consumption (1.1 Watt), which increases the total working time of the Tutamen solution, but also it is able to operate for 50,000 hours without having to be replaced.

Regarding the overall system's autonomy, the used batteries are able to support the platform up to 3 hours in full load, offering the possibility of being replaced by other charged batteries while the empty batteries can be recharged in a separate module. The running time of the system can be vastly extended if the calls to the coordination centre are only periodic. Moreover, the system can be easily shut-down and booted into working mode in a matter of seconds.

Considering the audio system, which will provide most of the feedback to the physician, a single-sided ear coupling Sennheiser headset is used for system related events (i.e. warnings and notifications such as system ready, new call arrived, among others), while at the same time it is also used for the communication between the physician and the control centre. The headset includes also an adjustable microphone, a button to mute the microphone and a volume regulator for a better experience.

5.3 Calling Performance

The establishment of call between the field teams and the control centres involves the transmission of both audio and FullHD video. Since audio over the internet has already been widely studied, the call performance analysis will focus on the challenges posed by the transmission of high-definition video. Taking into account the variable and limited bandwidth of 3G Networks, different tests were performed with different encoding bit-rates of the video codec, reflecting the available bandwidth at a given time. This feature of the Tutamen architecture aims at guaranteeing the best possible quality at any given time.

Generally, when connecting to a 3G Wireless Network, the device knows the available bandwidth. However, the announced bandwidth does not always correspond to the real available bandwidth, for instance, while the High-Speed Downlink Packet Access (HSDPA) standard may announce 7.2Mb/s bandwidth, this value only represents the maximum bandwidth than can be achieved, providing no actual guarantees. The presented results show the quality of different bit-rate videos, transmitted through the same network conditions of 7.2Mb/s of bandwidth. Since many interferences may occur, the desired transmission bit-rate should be bellow the maximum available bandwidth.

The transcoding process at lower bit-rates may reduce the overall quality of the video. However, the results we present in this paper show that less harmful to previously adapt the video quality to the available network conditions, rather than maintaining a higher quality and incur packet losses that may carry key frames. In fact, the resulting degradation of a higher bit-rate video being transmitted over a lower bandwidth network, may be higher than the the degradation of a video previously transcoded with a lower bit-rate, thus with a forced reduction of the video quality before being streamed, which during the streaming process has few or no losses that compromise the video integrity.

Table 1 presents the percentage of degradation obtained for different bit-rate videos, considering the same network conditions. This illustrates the need to correctly adapt videos' bit-rates during the transcoding process, guaranteeing that the number of losses is reduced and that overall perceived quality is the best

Table 1. Degradation Comparison over a network of 7.2 Mbps

Bit Rates	% of Degradation		
Mb/s	Transcoding	Streaming	Overall
12.0	2.17	33.65	35.09
7.2	4.64	22.47	26.06
5.5	6.02	18.45	23.36
2.0	11.08	16.61	25.85
.384	18.37	0.00	18.37

possible. In particular, by analysing the results in this figure, it becomes clear that bit-rates higher or closer to the network's maximum bandwidth, despite having a lower degradation during the transcoding process, register a higher degradation after being streamed. On the other hand, moderate adjustments of bit-rate are able to provide a better quality after being streamed, presenting a lower degradation.

Nonetheless, as previously mentioned, the sent videos' bit-rate must be enough such that it does not affect the original quality significantly. For example, for a 2Mb/s bit rate, it is clear in Table 1 that even though the lower bit-rate video is within the limits of the network, with few packet losses, it has an initially higher degradation than for 5.5Mb/s, not being worth the using a lower bit rate.

An interesting result occurred for the 384kb/s bit rate as despite having a higher transcoding degradation, there were no registered packet losses. The overall quality is a result of the x264 codec which is able to maintain an acceptable quality regarding the measured PSNR, suggesting that other measurement mechanisms should be used with this codec.

6   Conclusions

The constant development of lighter and more powerful hardware has motivated the creation of new approaches for improving citizens life quality. The work presented in this paper introduces a portable platform for high quality video transmission in emergency scenarios. By resorting to state-of-the-art hardware and video codecs, 3G networks are used to guarantee the expertise of highly qualified professionals in remote locations in timely matter.

Resulting from the feedback obtained from different emergency teams, the designed equipment and architecture stand-out by being extremely portable and simple to use, while allowing High-definition Video to be seamlessly transmitted between teams in the field and experts in the control centre.

In order to assess the quality provided by the Tutamen platform, comprehensive video tests were performed, evaluating the quality of the transmitted videos with different bit-rates in different network conditions. The perceived video quality is of paramount importance as it will be the main input source at the control centre. The obtained results revealed that the platform is capable of successfully encoding and transmitting HD Video with a good MOS (higher than 4), adapting to different network conditions.

An important contribution from the developed work concerns the integration of off-the-shelf hardware with the defined software platform, guaranteeing real-time adaptable video transmission that guarantees the best possible video quality.

Future work involves the extension of this platform to other services such as security and education. Moreover, in order to further improve the adaptable characteristics of the presented solution, an additional feature would be the reduction of the video's resolution when FullHD is not mandatory or whenever network conditions are very poor.

Acknowledgements. This work was performed in project LiveCity which has received research funding from the Community's Seventh Framework programme. This paper reflects only the authors' views and the Community is not liable for any use that may be made of the information contained therein. The contributions of colleagues from LiveCity consortium are hereby acknowledged.

References

1. Panayides, A., Pattichis, M.S., Pattichis, C.S., Schizas, C.N., Spanias, A., Kyriacou, E.: An overview of recent end-to-end wireless medical video telemedicine systems using 3g. In: 2010 Annual International Conference of the IEEE Engineering in Medicine and Biology Society (EMBC), October 31 - September 4, pp. 1045–1048 (2010)
2. El Khaddar, M., Harroud, H., Boulmalf, M., Elkoutbi, M., Habbani, A.: Emerging wireless technologies in e-health trends, challenges, and framework design issues. In: 2012 International Conference on Multimedia Computing and Systems (ICMCS), pp. 440–445 (May 2012)
3. Sugita, N., Yoshizawa, M., Kawata, H., Yambe, T., Konno, S., Saijo, O., Abe, M., Homma, N., Nitta, S.: Telemedicine system necessary in disaster areas. In: 2012 Proceedings of SICE Annual Conference (SICE), pp. 1661–1664 (August 2012)
4. Kumar, S., Krupinski, E.A.: Teleradiology. Springer (2008)
5. Kruchten, P.: The Rational Unified Process: An Introduction, 3rd edn. Addison-Wesley Longman Publishing Co., Inc., Boston (2003)
6. Wang, Y.K., Even, R., Kristensen, T., Jesup, R.: RTP Payload Format for H.264 Video. RFC 6184 (Proposed Standard) (May 2011)
7. Bernardo, V., Sousa, B., Curado, M.: Voip over wimax: Quality of experience evaluation. In: IEEE Symposium on Computers and Communications, ISCC 2009, pp. 42–47 (July 2009)
8. Klaue, J., Rathke, B., Wolisz, A.: EvalVid – A framework for video transmission and quality evaluation. In: Kemper, P., Sanders, W.H. (eds.) TOOLS 2003. LNCS, vol. 2794, pp. 255–272. Springer, Heidelberg (2003)

Defining Key Performance Indicators for Evaluating the Use of High Definition Video-to-Video Services in eHealth

Andreea Molnar and Vishanth Weerakkody

Brunel University Business School
Uxbridge, Middlesex UB8 3PH
{andreea.molnar,vishanth.weerakkody}@brunel.ac.uk

Abstract. This paper examines the process of developing key performance indicators (KPIs) for evaluating the use of high definition video (HD) to video (V2V) communication for tele-monitoring patients in a healthcare setting. The research is performed in the context of the European Commission funded LiveCity project and uses as a case study the monitoring of glaucoma patients using V2V technology. Initially, a set of KPIs are defined for V2V use in eHealth services based on literature and secondary research and these KPIs are then refined based on interviews with medical staff involved in the treatment of glaucoma patients. The obtained KPIs are used to analyse the potential of using V2V services in the context of tele-monitoring glaucoma patients.

Keywords: high definition video, eHealth, tele-monitoring, glaucoma, KPIs.

1   Introduction

In the current economic climate, where governments are under constant pressure to deliver 'better for less', the role and impact that Information and Communication Technology (ICT) has on core public services such as healthcare has increased more than ever before. In particular, most European governments see ICT as a means to facilitate and improve health services and as a mechanism to reduce the cost associated with healthcare service delivery [1, 2, 3, 4]. Among the variety of existing health services, in this paper, we focus on tele-monitoring and we argue that it is possible to improve the quality and acceptance of existing tele-monitoring services by using HD V2V communication. This research is performed in the context of the LiveCity project (EU FP7 CIP grant agreement No. 297291), and focuses on the case of tele monitoring for glaucoma patients.

Although tele-monitoring is not necessarily a new concept, problems are reported due to image noise and jerky sound that may be crucial for the proper diagnosis of patients [5]. In this context, the LiveCity project aims to ensure that HD V2V communication is provided between the hospital and the patient. This is of particular importance for the LiveCity case study, as it involves communication with glaucoma patients, and in this case, for a consultation, high quality images are imperative [6]. However, the technical details on providing high quality video are out of the scope of this paper, as this paper focuses on the evaluation of tele-monitoring services through V2V by defining a set of KPIs.

H. Papadopoulos et al. (Eds.): AIAI 2013, IFIP AICT 412, pp. 452–461, 2013.

In order to do so, the paper is structured as follows: The next section sets the background of this research. It is followed by a brief introduction on how delivery of HD V2V services can be facilitated over public infrastructure. The next section then briefly introduces the evaluation phases of the LiveCity project, followed by an initial set of KPIs for eHealth services. This is followed by the main results of a semi-structured interview that was conducted to collect data from a doctor treating glaucoma patients. The paper then concludes by offering a discussion of the results, conclusions and future work.

2   Background

Using ICT in health services (electronic health – eHealth) has been seen as a solution for reducing the cost associated with healthcare delivery [2], improving decision making to increase the quality of service offered to patients, improving the access to information [7], and contributing to the saving of lives through advanced tools and techniques [8]. However, the adoption of ICT in healthcare has not met initial expectations with reports of poor take up by both practitioners and patients [9, 10]. Prior research has identified several issues with the current use of e-Health such as the communication infrastructure [1, 10], the necessity for the system to be user centric [11], and the need to involve stakeholders from the initial stages of service design [1].

Several services that have been used in e-health, such as Electronic Health Records (EHRs) [12, 13], picture archiving and communication system, and tele-medicine [14] as core services that are enabled by ICT. This research focuses on a particular type of tele-medicine/tele-monitoring, and explores the use of HD V2V between the patient and the hospital. Tele-monitoring has been used for the monitoring of patients suffering from a wide range of diseases such as: diabetes, hypertension, home delivery, home nursing care for elderly and chronically ill, and glaucoma, either through the use of voice or voice and video communication [15, 16]. In this paper we address tele-monitoring for glaucoma patients. Glaucoma is one of the major causes of blindness worldwide and its treatment requires frequent monitoring of patients [17]. When the communication between patient and doctor has been done through V2V, prior research shows that although there is no cost saving between a tele-monitoring consultation and one in the hospital, patients were often satisfied with their personal cost and time savings that came as a consequence of using the system [18]. Among the drawbacks reported for remote consultation, poor image quality appears as one of the biggest issues impeding tele-monitoring.

Problems with the image quality have also been observed in conventional tele-medicine [19]. Such problems have resulted in adequate information not being available for proper diagnosis of patients [5]. In a more recent study, Bradford et al. [20] reported that problems with poor video quality in tele-monitoring contexts have led the nurse "to use a mobile phone for audio and the Internet video call for visual communication only, as the quality of the video could be improved by allowing more bandwidth" [20]. In this respect, LiveCity aims to alleviate the problems that exist with the delivery of video over public internet infrastructure and facilitate the delivery of high definition video for the eHealth context as part of its portfolio.

3   Using High Definition Video in Tele-monitoring: LiveCity

As previous studies performed on video to video communication for tele-monitoring reported various difficulties encountered when delivering video over the Internet, such as poor image and noise quality [5], the LiveCity project aims to alleviate the problems that the transmission of high definition video over public Internet infrastructure faces, through the use of a Virtual Path Slice (VPS) controller (Fig. 1). While it is not our intention to discuss the technical characteristics and functions of the VPS controller in this paper, essentially it is an application that manages bandwidth by avoiding interference from unwanted traffic and therefore ensuring that any loss of delivered data and delays are reduced [21]. It therefore creates a virtual path over the public Internet to allow traffic (data, voice and video) to flow freely without losing quality. This ensures that signals are not interrupted by other external Internet traffic when the connection between two points of contact is established, therefore guaranteeing the transmission of high quality live video images between two points of contact.

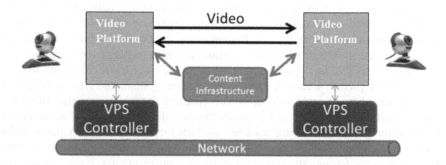

Fig. 1. Video Platform with the Use of VPS Controller [21]

Transmission of HD V2V is particularly important for certain consultations for which the medical staff need to see detailed information. In this respect, the quality of images need to be very good, as is the case of glaucoma patients, where a high quality picture of the eye is necessary for a proper diagnosis [6]. Glaucoma treatment usually consists on applying eye-drops, this being the preferred treatment to an operation. The treatment requires taking the medication in the same dose at the appropriate time and proper application of the eye-drops. However, it is reported that almost half of the patients fail to apply the eye drops correctly [6], resulting in improper treatment of glaucoma leading to complications, side effects, eye-damage and vision loss. In such circumstances the communication between the doctor and the patient is extremely important for the success of the treatment. However, due to various reasons, such as convenience, cost or impossibility to arrive at a hospital premises or private clinic, communication and/or consultation between the patients and the healthcare provider can sometimes be difficult. This is especially significant in cases where patients are unable to physically visit their doctor due to disability or old age, or for those people living in remote locations. In this respect, tele-monitoring could help in consulting glaucoma patients and checking whether they take their treatment properly. In this paper we focus on the first phase of evaluating the tele-monitoring of glaucoma patients using HD V2V in the context of the LiveCity project and defining the KPIs to be used in the second phase for assessing the success of tele-monitoring in this context.

4   Evaluation

The purpose of the evaluation is to determine relevant Key Performance Indicators (KPIs) for successful adoption of video-to-video tele-monitoring for glaucoma patients. The LiveCity evaluation consists of two phases [22]. In the *first phase*, KPIs are determined. During the *second phase*, the obtained KPIs from the first phase are being applied to assess the LiveCity use case (in this case, the treatment of glaucoma patients using V2V tele-monitoring) and revise the KPIs. The revised KPIs will result in final recommendations for designing and implementing V2V tele-monitoring services for glaucoma patients.

The research presented in this paper focuses on the *first phase*. During this phase an initial set of KPIs are identified, followed by a revision of the originally obtained KPIs based on stakeholders consultations (in this case, interviews with doctors). The importance of involving relevant stakeholders in the implementation of ICT in public services in general [23] [12], and understanding doctors' perception of using HD V2V communication in particular is imperative before such services are implemented in healthcare settings to facilitate their adoption.

In this respect, it is important to find out what criteria are necessary to facilitate a successful adoption of HD V2V communication as proposed in the LiveCity project in the context of a healthcare setting. With this aim, we performed a preliminary interview with a doctor treating glaucoma patients at a large University Clinic hospital in Attica, Greece. This is the place where the LiveCity pilot will take place and the interviewed doctor will be involved in tele-monitoring glaucoma patients.

The interview was semi-structured with open-ended questions. Semi-structured interviews are the most common used tool in qualitative research in information systems [24]. The interview was conducted in April 2013 and tape recorded with the doctor's permission. Since the purpose of this study was focused on the initial exploration of the concept of V2V used for the tele-monitoring of glaucoma patients, and not generalisation, a single in-depth interview was deemed adequate to fulfil the overall aim of the study.

4.1   Initial KPIs

In Table 1, we present the initial Use Case related KPIs we perceive to be important for evaluating the LiveCity platform, as drawn from the literature and consultation with LiveCity project partners. The technical KPIs related to the network connectivity and application are not presented in this paper as they are general to every LiveCity services and have been previously covered in [22, 25]. The remaining KPIs have been organised into *user requirements* and *process requirements*.

Table 1. Initial Health KPIs

		Health KPIs
User perspective	Usability/Ease of Use	Ease of use
		Navigation
		Help features
		Background and colour
		User involvement in parameterization
		Menu simplicity
	Satisfaction	Delighted with systems
		Pleased with the system
		Satisfied with the system
		Image quality
	Reliability	Number of interruptions during a session
		Boot time for the application
		Response time
Process Perespective		Anonymity of sensitive data
		Encryption of Sensitive data and communication
		Data storage in a physically secure location
		Data Security
		System development cost
		Time for examination
		Hospital resources committed for the new system
		Learning time for new system use
		Time-to-doctor: meeting the doctor and starting the examination
		Physical presence cost including transfer to the hospital
		Waiting time in the hospital for examination
		Number of patients involved in the pilot
		Number of visits to the hospital (emergency)
		Number of visits to the hospital (outpatient)
		Waiting time in the hospital (emergency visit)
		Waiting time in the hospital (outpatient appointment)
		Number of patients
		Conformance to treatment
		Intraocular pressure control
		Ease of appointment scheduling (time required)
		Ease of appointment scheduling (ease of set up)

User Requirements: these are related to the actual utilisation of the V2V service by the involved users and the respective level of user experience. KPIs are built around:

- *Usability/Ease of Use*: V2V services shall be easy to use by different classes of users,
- *Satisfaction*: captures the different levels of user experience,
- *Reliability*: reflects user point of view regarding reliability aspects of the V2V provision.

Process Requirements: these are related to the end-to-end process tailored to the various V2V provision scenarios. In this case, in Table 1, we outline examples of the specific use case related KPIs for delivering health services using V2V.

4.2   Refining the KPIs: Interview with the Doctor

After defining the initial KPIs from literature and secondary research, we conducted an in-depth interview with the doctor involved in treating glaucoma patients and who will be testing the tele-monitoring HD V2V solution offered by the LiveCity project. The purpose of this interview was to assess the criteria that are most important from his (the doctor's) point of view for the successful adoption of V2V in treating glaucoma patients. The interviewed doctor worked at the University Hospital Clinic responsible for treating (including performing surgical operations) patients with glaucoma in the Attica region as well as from across Greece who commute to the hospital for check-ups and treatment. In this particular case, it is anticipated that tele-monitoring using HD V2V will facilitate better patient conformance to the right treatment and correct application of the eye-drops with the remote assistance from their doctor. Moreover, it is hoped that V2V will help patients access medical information remotely when needed and have their questions or queries answered with regards to their treatment. In the scope of the proposed activities, doctors will be able to interact with their patients and will be able to monitor them allowing frequent patient consultation during the medical treatment process. The high quality of live V2V will eliminate the necessity for patients to travel and visit their hospital and will allow doctors to monitor how patients apply their eye-drops and guide them to take corrective action when necessary. The necessity of refining the KPIs is motivated by the fact that the quality of service is mostly context dependent [26, 27, 28], with prior research showing failed attempts to measure it using general scales in new contexts [29].

The interview with the doctor took place in April 2013. As a result of the interview, additional items were created and qualitative findings were matched with the existing scales to match construct definitions. During the interview, the doctor highlighted the importance of some of the initial KPIs presented in Table 1, such as the importance of patient satisfaction with the system, "....*if they [patients] are happy themselves, the patients who use the technology are satisfied, they will like it, will use it or we may even lose them in the process*". In this process, the Doctor emphasised that the "*video quality is the most important thing*" due to the neccessity to see a detailed picture of the human eye.

Table 2. New KPIs Identified for tele-monioring of Glaucoma Patients using V2V

	Criteria	Interview quote
Process Perespective	Improvement in the patients' health	*"if they do not follow the treatment correctly, the eye will be lost"*
	Reduce the medication cost	*"if they put their drops correctly, maybe they need less bottles, less medicine, less eye drops... this means they do not pay extra money for medicine they do not need to use... often a lot of medicine is wasted because patient do not know how to use it"*
	Improved doctor / patient communication	*"follow our patients more closely"*
	Improved monitoring of patients by doctors	*"We [doctors] are always worried whether they [patients] use the drops correctly"* *"Patients underestimate the difficulty to put a drop into the eye, so we [doctors] can see them [patients] in their home and how they use the eye drops"*
	Improvement in the availability /easiness for patients to reach doctors	*"in case they [patients] want to ask something as an emergency ... they will find it easier this way"* *"patients often complain that when they want to reach us [doctors], it is very difficult to get an appoitment"* *"through the V2V solution, they [patients] find it more easy to reach us"*
	Improve patient perception of doctor availability	*"feeling that we [doctors] are available for them [patients]"*
	Improvement in patient management	*"we [doctors] can decide whether they [patients] need to came or they can stay at home and continue the treatment.... This will help to reduce transportation costs"* *"..So many times they [patients] do not need to come to the hospital...it would be even better for us to decide if they need to come..... I believe that patient management as a result of using the system would be extremely better with V2V... not just better"*
	Improvement in the number of patients treated	*"due to the improved patient management and less appointments and physical consultations, we will be able to treat more patients"*
	Improve the patient skills/knowledge in applying their eye drops	*"Patients underestimate the difficulty of putting a drop into the eye"... "they [patients] need to be trained, educated, and reminded"*
	Patient reliability in getting the treatment correctly	*"Ideally we would realise that patients who have been educated for the first month with more consultations, appoitments, video to video consultations could be more reliable with the treatment afterwards"*

5   Discussions, Conclusions and Future Work

One of the clear benefits of using V2V enabled tele-monitoring is the ability to treat patients without the need for physical consultation and the need to be transported to a hospital. This could be particularly important for patients living in remote locations where access to the hospital is difficult, or for the elderly (who are most prone to be affected by glaucoma and require usually to be accompanied to the hospital) by allowing them to be independent for a longer period of time without the need to commute to the hospital.

The purpose of this paper was to determine a set of KPIs aimed at assessing tele-monitoring services for glaucoma patients. With this aim, we first identified generic KPIs from the literature, secondary sources and through consultation with the involved partners in the LiveCity project. This was followed by an interview with a doctor treating glaucoma patients at a large hospital in Athens, Greece. We presented the KPIs identified initially as well as the new ones obtained as a result of the interview.

From a theoretical perspective this study extends service quality research by developing a set of indicators for tele-monitoring patients who suffer from glaucoma. The implications of this research are highly relevant to the decision makers of eHealth platforms who are offering tele-monitoring services for glaucoma patients. These findings improve the general understanding of how doctors evaluate service quality in the context of glaucoma treatment using tele-monitoring. The proposed indicators should contribute to paving the way for conducting effective design of service delivery systems in the context of eHealth services.

Several limitations in this paper are worth noting. This research was conducted in only one country, and by interviewing only one doctor responsible for treating glaucoma patients. Replication in other contexts would increase the confidence in the proposed indicators; hence, the findings should be treated as indicative rather than confirmatory.

Once the V2V system is implemented for tele-monitoring of glaucoma patients in the context of the LiveCity project, we propose to refine the KPIs during the implementation and testing of the system and consequently propose a framework for evaluating the V2V enabled tele-monitoring of health services. The obtained KPIs and framework will then be used as a source to develop a 'how to' guide for delivering V2V based eHealth services.

Acknowledgments. The authors acknowledge the contributions made to this article by the LiveCity consortium of partners who are funded by the European Commission, especially to NKUA for helping in organising the study presented in the paper.

References

1. Charles, B.L.: Telemedicine can Lower Costs and Improve Access. Healthcare Financial Management 54(4), 66–69 (2000)
2. Dorsey, E.R., Venkataraman, V., Grana, M.J., Bull, M.T., George, B.P., Boyd, C.M., Beck, C.A., Rajan, B., Seidmann, A., Biglan, K.M.: Randomized Controlled Clinical Trial of "Virtual House Calls" for Parkinson Disease "Virtual House Calls" for Parkinson Disease. JAMA Neurology, 1–6 (2013)

3. Porter, M.E., Teisberg, E.O.: Redefining Health Care: Creating Value-based Competition on Results. Harvard Business School Press, USA (2006)
4. Stefanou, C., Revanoglou, A.: ERP Integration in a Healthcare Environment: A Case Study. Journal of Enterprise Information Management 19(4), 115–130 (2006)
5. Shimizu, S., Okamura, K., Nakashima, N., Kitamura, Y., Torata, N., Yamashita, T.: High-Quality Telemedicine using Digital Video Transport System over Global Research and Education Network. In: Georgi Graschew, G. (ed.) Roelofs TAAdvances in Telemedicine: Technologies, Enabling Factors and Scenarios, pp. 87–110. Intech, Croatia (2011)
6. Stamatelatos, M., Katsikas, G., Makris, P., Alonistioti, N., Antonakis, S., Alonistiotis, D., Theodossiadis, P.: Video-to-video for e-health: Use case, concepts and pilot plan. In: Iliadis, L., Maglogiannis, I., Papadopoulos, H., Karatzas, K., Sioutas, S. (eds.) AIAI 2012 Workshops, Part II. IFIP AICT, vol. 382, pp. 311–321. Springer, Heidelberg (2012)
7. Lenz, R., Kuhn, K.A.: Towards a Continuous Evolution and Adaptation of Information Systems in Healthcare. International Journal of Medical Informatics 73(1), 75–90 (2004)
8. Haughom, J.L.: Implementation of an Electronic Health Record. In: BMJ, p. 343 (2011)
9. Atherton, H., Azeem, M.: An Information Revolution: Time for the NHS to Step up to the Challenge. JRSM 104(6), 228–230 (2011)
10. Cresswell, K., Aziz, S.: The NHS Care Record Service (NHS CRS): Recommendations from the Literature on Successful Implementation and Adoption. Informatics in Primary Care 1(3), 153–160 (2009)
11. Westrup, C.: What's in Information Technology. Asian Institute of Technology, Bangkok (1998)
12. Currie, W.L.: Evaluating the Governance Structure for Public Sector IT: The UK National Programme in the Health Service. Evaluating Information Systems, 199–217 (2012)
13. Lafky, D., Tulu, B., Horan, T.: Information Systems and Health Care X: A User Driven Approach to Personal Health Records. Comm. of the Association for IS 17, 1028–1041 (2006)
14. Menachemi, N., Burke, D.E., Ayers, D.J.: Factors Affecting the Adoption of Telemedicine—A Multiple Adopter Perspective. Journal of Medical Systems 28(6), 617–632 (2004)
15. Hjelm, N.M.: Benefits and Drawbacks of Telemedicine. Journal of Telemedicine and Telecare 11(2), 60–70 (2005)
16. Kassam, F., Yogesan, K., Sogbesan, E., Pasquale, L.R., Damji, K.F.: Teleglaucoma: Improving Access and Efficiency for Glaucoma Care. Middle East African Journal of Ophthalmology 20(2), 142 (2013)
17. Poon, C.C.Y., Wang, M.D., Bonato, P., Fenstermacher, D.A.: Editorial: Special Issue on Health Informatics and Personalized Medicine. IEEE Transactions on Biomedical Engineering 60(1), 143–146 (2013)
18. Tuulonen, A., Ohinmaa, A., Alanko, H.I., Hyytinen, P., Juutinen, A., Toppinen, E.: The Application of Teleophthalmology in Examining Patients with Glaucoma: A Pilot Study. Journal of Glaucoma 8(6), 367–373 (1999)
19. Demartines, N., Mutter, D., Vix, M., Leroy, J., Glatz, D., Rösel, F., Harder, F., Marescaux, J.: Assessment of Telemedicine in Surgical Education and Patient Care. Annals of Surgery 231(2), 282
20. Bradford, N., Armfield, N.R., Young, J., Ehmer, M., Smith, A.C.: Safety for Home Care: The Use of Internet Video Calls to Double-Check Interventions. Journal of Telemedicine and Telecare 18(8), 434–437 (2012)

21. Chochliouros, I.P.: Description and Thematic Context of the LiveCity Project. In: Conference of Telecommunication, Media and Internet Techno-Economics. Workshop on Network Operators View on Future Networks Landscape: Will it be Cloudy with a Chance of Data Rain, Internet (2012), http://www.ctte-conference.org/files/2011/CTTE_OperatorsWorkshop_Chochliouros_LiveCity.pdf (cited February 4, 2013)
22. Weerakkody, V., El-Haddadeh, R., Chochliouros, I.P., Morris, D.: Utilizing a high definition live video platform to facilitate public service delivery. In: Iliadis, L., Maglogiannis, I., Papadopoulos, H., Karatzas, K., Sioutas, S. (eds.) AIAI 2012 Workshops, Part II. IFIP AICT, vol. 382, pp. 290–299. Springer, Heidelberg (2012)
23. Goel, S., Dwivedi, R., Sherry, A.: Role of Key Stakeholders in Successful E-Governance Programs: Conceptual Framework. In: Proceedings of the Eighteen American Conference on Information Systems, Seattle, Washington, USA, August 9-11 (2012), http://aisel.aisnet.org/amcis2012/proceedings/EGovernment/19/
24. Myers, M.D., Newman, M.: The Qualitative Interview in IS Research: Examining the Craft. Information and Organization 17(1), 2–26 (2007)
25. Weerakkody, V., Molnar, A., Irani, Z., El-Haddadeh, R.: A Research Proposition for Using High Definition Video in Emergency Medical Services. Health Policy and Technology 2(3), 131–138 (2013), http://dx.doi.org/10.1016/j.hlpt.2013.04.001
26. Babakus, E., Boller, G.W.: An Empirical Assessment of the SERVQUAL Scale. Journal of Business Research 24(3), 253–268 (1992)
27. Carman, J.M.: Consumer Perceptions of Service Quality: An Assessment of the SERVQUAL Dimensions. Journal of Retailing (1990)
28. Dabholkar, P.A., Thorpe, D.I., Rentz, J.O.: A Measure of Service Quality for Retail Stores: Scale Development and Validation. Journal of the Academy of Marketing Science 24(1), 3–16 (1996)
29. Dagger, T.S., Sweeney, J.C., Johnson, L.W.: A Hierarchical Model of Health Service Quality Scale Development and Investigation of an Integrated Model. Journal of Service Research 10(2), 123–142 (2007)

A Novel Rate-Distortion Method in 3D Video Capturing in the Context of High Efficiency Video Coding (HEVC) in Intelligent Communications

Ioannis M. Stephanakis[1], Ioannis P. Chochliouros[2], Anastasios Dagiuklas[3], and George C. Anastassopoulos[4]

[1] Hellenic Telecommunication Organization S.A. (OTE),
99 Kifissias Avenue, GR-151 24, Athens, Greece
stephan@ote.gr
[2] Research Programs Section, Hellenic Telecommunication Organization S.A. (OTE)
99 Kifissias Avenue, GR-151 24, Athens, Greece
ichochliouros@oteresearch.gr
[3] Hellenic Open University, Parodos Aristotelous 18, GR-262 22, Patras, Greece
ntan@teimes.gr
[4] Democritus University of Thrace, Medical Informatics Laboratory, GR-681 00,
Alexandroupolis, Greece
anasta@med.duth.gr

Abstract. *High-Efficiency Video Coding* is currently proposed as the newest and most efficient video coding standard by the ITU-T *Video Coding Experts Group* and the ISO/IEC *Moving Picture Experts Group*. Compression improvement relative to existing standards is estimated in the range of 50%. It is a block-based hybrid video coding algorithm that introduces several novel features compared to MPEG-4 like *Coding Units* associated with *Prediction Units* and *Transform Units*, *Advanced Motion Vector Prediction*, minimization of a rate-distortion Lagrangian cost, directional orientations for intra-picture prediction etc. The core algorithm of *Context-Adaptive Binary Arithmetic Coding* is based upon that of the MPEG-4 standard. View synthesis algorithms for HEVC stereo and 3D encoding are expected to be finalized as a standard extension. A *Multi-view Video Coding* scheme based upon the estimation of correlated parameter sets of elastic models between views is wherein adopted. The order of the tensor equals the number of multiple views. Underlying distributions are updated step-by-step. They are modeled according to context indices.

Keywords: object oriented coding, H264/AVC, High Efficiency Video Coding HEVC/H.265, higher order motion compensation models, CABAC, 3D television, *Rate-Distortion* theory.

1 Introduction

High Efficiency Video Coding (HEVC) [1,2] is a video compression standard that is proposed as a successor to H.264/MPEG-4 AVC (*Advanced Video Coding*) developed

H. Papadopoulos et al. (Eds.): AIAI 2013, IFIP AICT 412, pp. 462–473, 2013.

by the ISO/IEC Moving Picture Experts Group (MPEG) and the ITU-T Video Coding Experts Group (VCEG) under the name ISO/IEC 23008-2 MPEG-H Part 2 and ITU-T H.HEVC [3]. MPEG and VCEG have established a Joint Collaborative Team on Video Coding (JCT-VC) in order to develop the HEVC standard. HEVC standard is said to improve video quality, double the data compression ratio compared to H.264, and support 8K UHD and resolutions up to 8192×4320 pixels [4]. Two profiles are supported, namely Main Profile and High Efficiency 10 (HE10). The HEVC Main Profile (MP) is compared in coding efficiency to H.264/MPEG-4 AVC High Profile (HP), MPEG-4 Advanced Simple Profile (ASP), H.263 High Latency Profile (HLP), and H.262/MPEG-2 Main Profile (MP). As of the end of January 2013, ISO/IEC and ITU-T had approved HEVC as ISO/IEC 23008-2 High Efficiency Video Coding and ITU-T Rec. H.265 respectively, as the final draft international standard.

Simulcast as well as Multiview Video Coding are described in the context of the MPEG-4 AVC standard (Version 8: July 2007 (including SVC extension), Version 9: July 2009 (including MVC extension). [5]. A standard MVC coder consists of N parallelized single view coders. Each of them uses temporal prediction structures and encodes a sequence of successive pictures as intra (I), predictive (P) or bi-predictive (B) frames. Nevertheless MVC is not appropriate for delivering 3-D content for autostereoscopic displays since the bit rate required for coding multiview video with the MVC extension of H.264/AVC increases approximately linearly with the number of coded views. The transmission of 3D video in the Multiview Video plus Depth (MVD) format appears as a promising alternative. ISO/IEC and ITU-T established JCT-3V for the 3D video coding standards as of July 2012. AVC compatible video-plus-depth extension is expected within 2013 [6]. The transmission of 3D video in according to Multiview Video plus Depth (MVD) associates encoded depth data representing the basic geometry of the captured video scene with picture frames. Depth data are estimated based on the acquired pictures. They are obtained by the application of depth estimation algorithms and should not be regarded as ground truth. Based on the transmitted video pictures and depth maps, additional views suitable for displaying 3D video content on autostereoscopic displays can be generated using depth image based rendering (DIBR) techniques at the receiver side. All video pictures and depth maps that represent the video scene at the same time instant build an access unit. The access units of the input MVD signal are coded consecutively similar to MVC. The video picture of the so-called independent view inside an access unit is transmitted first directly followed by the associated depth map. Thereafter, the video pictures and depth maps of other views are transmitted. A video picture is always directly followed by the associated depth map. *Advanced Motion Vector Prediction* (AMVP) is used. Work on HEVC 3D and scalable extensions has been currently under development. It focuses on

- Coding of Independent Views (2D video coding)
- Coding of Dependent Views (DCP, inter-view motion/residual prediction)
- Coding of Depth Maps
- Encoder Control
- View synthesis algorithms

Final draft amendment (FDAM) for HEVC for stereo and 3D is expected in 2015. It improves the compression capabilities for dependent video views and depth data. As of the end of January 2013, ISO/IEC and ITU-T had approved HEVC as ISO/IEC 23008-2 High Efficiency Video Coding and ITU-T Rec. H.265 respectively, as final draft international standard.

2   Higher Order Models for Motion Compensation and Encoder Control

2.1   Novel Video Encoding Features of the HEVC/H. 265 Standard

The video coding layer of HEVC employs the same "hybrid" approach used in all video compression standards since H.261. Each picture is split into block-shaped regions, with the exact block partitioning being conveyed to the decoder. The first picture of a video sequence (and the first picture at each "clean" random access point into a video sequence) is coded using only intra-picture prediction. Transform coefficients from region-to-region within the same picture are spatially predicted but there are no dependencies upon other pictures. The remaining pictures of a sequence or frames between random access points are encoded using temporally-predictive coding modes. The encoding process for inter-picture prediction consists of choosing motion data comprising the selected reference picture and motion vectors (MV) that are applied for predicting the samples of each block. Residual frames are transformed by a linear spatial transform. The transform coefficients are scaled, quantized, entropy coded and transmitted together with the prediction information. Novel features of the HEVC standard are outlined as follows:

- Larger and more flexible coding, prediction, and transform units.
- Improved mechanisms to support parallel encoding & decoding.
- More flexible temporal prediction and scanning structure.
- More accurate intra prediction approach and directions/modes.
- More accurate motion parameters (including merge mode) and sub-pixel prediction.
- Inclusion of non-square transform and allowing asymmetric motion prediction.
- More flexible transform, choice of DST, and no-transform option.
- Rate-distortion optimized quantization (RDOQ).
- Improved in-loop filters, including the new sample adaptive offset

The *Coding Tree Unit* (CTU) in HEVC replaces the *macroblock* structure as known from previous video coding standards. It has a size selectable by the encoder and it can be larger than a traditional *macroblock*. The quadtree syntax of the CTU (see Fig. 1) specifies the size and positions of its luma and chroma *Coding Blocks* (CB). A *Coding Tree Block* (CTB) may contain only one *Coding Unit* (CU) or may be split to form multiple CUs. Thus each CU is characterized by its *Largest* CU (LCU) size and the hierarchical depth in the LCU that the CU belongs to. It has an associated partitioning into *Prediction Units* (PUs) and a tree of *Transform Units* (TUs).

Intra-picture prediction, inter-picture prediction or skip mode are selected at CU level. A PU is basically the elementary unit for prediction and it is defined after the last level of CU splitting. Prediction type and PU splitting type are two concepts that describe the prediction method. Intra prediction allows for symmetric splitting whereas inter prediction allows for both symmetric and asymmetric splitting. At the level of PU, intra-prediction is performed from samples already decoded in adjacent PUs. Such modes as DC (average/flat), angular directions (one from up to 33 as in Fig. 2), *Planar Intra Prediction* (surface fitting), SDIP (*Short Distance Intra Prediction*), MDIS (*Mode Dependent Intra Smoothing*) may be used. Advanced motion prediction (see for example [7]) featuring a "merge" mode or the skip mode may be used for inter prediction. Quarter-pixel precision and 7-tap or 8-tap filters are used for interpolation of fractional-sample positions. Multiple reference pictures are used. A deblocking filter similar to the one used in H.264/MPEG-4 AVC is operated in the inter-picture prediction loop. An adaptive loop filter (ALF) is alternatively employed for higher efficiency. The coefficients of ALF are calculated and transmitted on a frame basis and the MMSE estimator is used. For each degraded frame, ALF can be applied to the entire frame or to local areas. Similar transforms as for H.264 (including the discrete sine transform and the Hadamard transform) are used for encoding the residual data. Three different scanning modes (namely zigzag, horizontal and vertical scan) are used to improve the residual coding.

Such novel features as the option to partition a picture into rectangular regions called *tiles* and *wavefront parallel processing* (WPP) are introduced into the HEVC standard in order to enhance its parallel processing capabilities.

The core algorithm of *Context-Adaptive Binary Arithmetic Coding* (CABAC) is based upon that of the H.264/MPEG-4 standard. The number of contexts used in HEVC is substantially less than the number of contexts in H.264/MPEG-4 AVC whereas the entropy coding design allows for better compression.

2.2 Motion Compensation Using Higher Order Models and Rate-Distortion Control

Higher order motion models and view synthesis techniques are currently investigated in the literature for sub-pixel motion compensation that may be applied 3D and multiview encoding. Higher order motion models such as the *affine model* (AMMCP) [8], the *mesh based* MCP [9], the *elastic* MCP [10] and *View Synthesis by Depth Image Rendering* (DIBR) [11, 12] have been proposed as possible extensions of existing standards. The transformation of one view to another is based on the camera parameters. DIBR uses a view and a depth map for generating arbitrary views. Rough 3D information can be reconstructed based on the depth map so that the transformation due to the disparity effect can be generated. The video compression with DIBR becomes the compression of depth image instead of the correlations between views. The elastic MCP model estimates the spatial transformation parameters between the predicted block with translational MV and the current block. The motion parameters are encoded in the bit stream. The pixel location of the predicted block (x_i, y_i) is transformed to (x'_i, y'_i) by the following equations,

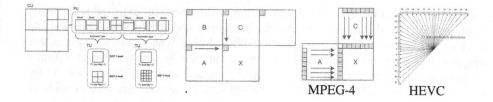

MPEG-4   HEVC

Fig. 1. Structure of Coding Tree Units (CTU). Prediction and Transform Units (PU and TU) are depicted.

Fig. 2. Intra-picture prediction for boundary samples of adjacent blocks (MPEG-4 and HEVC)

$$x'_i = x_i + \sum_{l=1}^{P/2} m_l \varphi_l(x_i, y_i) \quad \text{and} \quad y'_i = y_i + \sum_{l=P/2+1}^{P} m_l \varphi_l(x_i, y_i) \tag{1}$$

where P is the number of elastic parameters used, m_l is the elastic parameter and $\varphi_l(x_i, y_i)$ is the basis function. The elastic MCP model uses discrete cosines as the basis functions and it is defined by

$$\varphi_l(x_i, y_i) = \varphi_{l+P/2}(x_i, y_i) = \cos\left(\frac{(2x_i+1)\pi u}{2\,blocksize\,x}\right)\cos\left(\frac{(2y_i+1)\pi v}{2\,blocksize\,y}\right), \tag{2}$$

where $l = \sqrt{P/2}\; u + v + 1$ and $0 \le (u,v) \le \sqrt{P/2} - 1$. Conventional motion vector is equal to motion parameters $[m_1\; m_{P/2+1}]$ in such a case.

The minimization of a Lagrangian cost function for motion estimation was proposed in [13] (see Fig. 3). Given a reference picture list R and a candidate set M of motion vectors, the motion parameters for a block s_k, which consists of a displacement or motion vector $m=[m_x, m_y]$ and, if applicable, a reference index r, determine the coding mode for coding a block of samples, such as a *macroblock* or a *coding unit*. Additional features of a coding mode are the intra or inter prediction modes or partitions for motion-compensated prediction or transform coding including the quantization step. Given the set of applicable coding modes for a block of samples s_k, the used coding mode is chosen according to

$$c^* = \arg\min_{c \in C_k} (D_k(c) + \lambda \cdot R_k(c)) \tag{3}$$

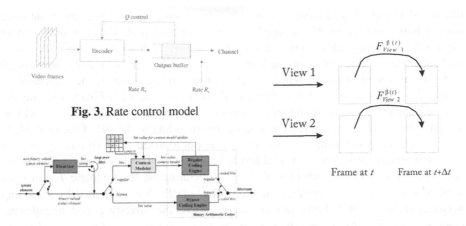

Fig. 3. Rate control model

Fig. 4. Context-based Adaptive Binary Arithmetic Coding

Fig. 5. Updates-update symbols between sequential video frames

where the distortion term $D_k(c)$ represents the SSD between the original block s_k and its reconstruction s'_k, that is obtained by coding the block s_k with the mode c. The term $R_k(c)$ represents the number of bits (or an estimated thereof) that are required for representing the block s_k using the coding c for the given bitstream syntax. It includes the bits required for signaling the coding mode and the associated side information (e.g. motion vectors, reference indices, intra prediction modes and coding modes for sub-blocks of s_k) as well as the bits required for transmitting the transform coefficient levels representing the residual signal. A coding mode is often associated with additional parameters such as coding modes for sub-blocks, motion parameters and transform coefficient levels. While coding modes for sub-blocks are determined in advance, motion parameters and transform coefficient levels are chosen according to Eq. 3. For calculating the distortion and rate terms for the different coding modes, decisions for already coded blocks of samples are taken into account (e.g. by considering the correct predictors or context models).

3   A Scalable Context-Based Adaptive Model for Encoding Motion Compensation Parameters in Multi-view 3D Systems

3.1   The Context-Based Adaptive Binary Arithmetic Coding (CABAC) Model

CABAC (*Context-based Adaptive Binary Arithmetic Coding*) as well as *Rate-Distortion Optimization* are included in the HEVC standard. *Context-based Adaptive Binary Arithmetic Coding* (CABAC) [14] achieves good compression performance through the selection of probability models for syntax elements according to context. It estimates adaptively probabilities based on local statistics and uses arithmetic coding rather than variable-length coding. The following steps [15,16,17,18] are involved in coding a data symbol: 1. *Binarisation*, 2. *Context model selection*, 3. *Arithmetic encoding* and 4. *Probability update*. The above steps are illustrated in Fig. 4. The design of binarization schemes relies on a few basic code trees, whose structure enables a simple on-line computation of all code-words without the need for storing tables. There are four such basic types [16] namely the *unary code*, the *truncated unary code*, the *k*-th *order Exp-Golomb code* and the *fixed-length code*. There are binirization schemes based on a concatenation of these elementary types. The nodes located in the vicinity of the root node of a binary tree are the natural candidates for being modelled individually, whrereas a joint model should be assigned to all nodes on deeper tree levels corrresponding to the tail of the probability density function. According to CABAC, given a predefined set T of past symbols, a so-called *context template*, and a related set $C=\{0,...,C-1\}$ of contexts, contexts are specified by a modeling function $F : T \rightarrow C$ operating on the template T. For each symbol x to be coded, a conditional probability $p(x|F(z))$ is estimated by switching between different probability models according to the already coded neighboring symbols $z \in T$. After encoding x using the estimated conditional probability $p(x|F(z))$, the probability model is updated with the value of the encoded symbol x. Thus $p(x|F(z))$ is estimated on the fly by tracking the actual source statistics. The entity of probability models used in CABAC can be arranged in a linear fashion such that each model can be identified by a unique so-called context index γ.

3.2   A Context-Based Adaptive Model for Correlated Motion Compensation Parameters in Multiple Views

The proposed encoding scheme investigates context-based adaptive models for correlated multilinear object entities between multiple views [19]. It is a model that does not depend upon the parameters of the recording cameras. It is applicable to such object entities as transform elements and/or motion vector parameters and it scales in a straightfoward fashion from stereo to multiple free views. It combines adaptivity regarding the context templates with updates of the correlated orthonormal features of the object entities through all views per GOP. Let us assume an Nth-order tensor A, which resides in the tensor multi-linear space $R^{I_1} \otimes R^{I_2} \otimes \cdots \otimes R^{I_N}$ where $R^{I_1}, R^{I_2} \cdots, R^{I_N}$ are the N vector linear spaces of multiview system featuring N views. The "k-mode vectors" of A are defined as the I_k-dimensional vectors obtained from A by varying its index in k-mode while keeping all other indices fixed [20,21]. Multi-view motion prediction using cross-view prediction vectors may be defined per GOP through a video object that we call *Motion Prediction video-Object* and is denoted as **MPO**. Let us define a group of motion vector parameters pertaining to k-view within some GOP as $\mathbf{M}^{(k)} = \begin{bmatrix} M_1^k & M_2^k & \cdots & M_{I(k)}^k \end{bmatrix}$. Prediction of motion vectors - which are denoted as $\mathbf{v}(m_{block}, n_{block}, t) = [v_x \, v_y]$ for a block indexed by (m_{block}, n_{block}) at t - is carried out once with respect to one of the views or a linear combination of a selected subset. The motion vectors pertaining to k-view according to the elastic model may be decomposed as follows,

$$\mathbf{v}^k(m_{block}, n_{block}, t) = \mathbf{v}(m_{block}, n_{block}, t) + [\sum_{l=1}^{P/2} m_l^k \varphi_l \quad \sum_{l=P/2+1}^{P} m_l^k \varphi_l] = [v_x \, v_y] + \sum_{i=1}^{I(k)} \alpha_i^k (m_{block}, n_{block}, t) M_i^k. \quad (4)$$

Motion Prediction video-Object (**MPO**) holds the orthonormal cross-view prediction vectors. It is defined as,

$$MPO(m_{block}, n_{block}, t) = \sum_{i_1=1}^{I_1} \sum_{i_2=1}^{I_2} \cdots \sum_{i_N=1}^{I_N} s(i_1, i_2, \dots; t) M_{i_1}^1 \circ M_{i_2}^2 \cdots M_{i_N}^N = S(t) \times_1 \mathbf{M}^{(1)} \times_2 \mathbf{M}^{(2)} \times_3 \mathbf{M}^{(3)} \times \cdots \times_N \mathbf{M}^{(N)} \quad (5)$$

Unfolding *MPO* along the k-mode is denoted as

$$\mathbf{MPO}_{(k)} \in R^{I_k \times (I_1 \times \cdots \times I_{k-1} \times I_{k+1} \times \cdots \times I_N)}, \quad (6)$$

where the column vectors of $MPO_{(k)}$ are the k-mode vectors of *MPO*. Unfolding the *Multi-view Video Plane* of an N-view system along the k-mode view results into the following matrix representation

$$\mathbf{MPO}_{(k)}(t) = \mathbf{M}^{(k)} \cdot S_{(k)}(t) \cdot \left(\mathbf{M}^{(k+1)} \otimes \mathbf{M}^{(k+2)} \otimes \cdots \otimes \mathbf{M}^{(N)} \otimes \mathbf{M}^{(1)} \otimes \cdots \otimes \mathbf{M}^{(k-1)} \right)^{\mathsf{T}}, \quad (7)$$

where $\otimes$ denotes the Kronecker product. The core tensor S (in a representation similar to the one described in Eq. 8) is analogous to the diagonal singular value matrix of the traditional SVD and coordinates the interaction of matrices to produce the original tensor. Matrices $\mathbf{M}^{(k)}$ are orthonormal and their columns span the space of the corresponding flattened tensor denoted as $\mathbf{MPO}_{(k)}$. The objective of MPC analysis for predetermined dimensionality reduction is the estimation the N projection matrices $\{ \tilde{\mathbf{M}}^{(k)}(t) \in R^{I_k \times P_k}, k = 1, \dots, N \}$ that maximize the total tensor scatter [22],

$$\left\{ \tilde{\mathbf{M}}^{(k)}(t), k = 1, \dots, N \right\} = \arg \max_{\tilde{\mathbf{M}}^{(1)}, \tilde{\mathbf{M}}^{(2)} \dots \tilde{\mathbf{M}}^{(N)}} \left\| MPO(t) - \Delta(t) \right\|^2, \quad (8)$$

where $\sum_{k=1}^{N} P_k \leq c$. One may estimate adaptively the cross products between motion vector parameters of different video views by maximizing sequentially Eq. 8 for all modes. Its decomposition into non-negative parts and the application of a multiplicative update rule that maintains orthonormality [23,24] is suggested in [19].

The following algorithm outlines in detail the steps of the proposed encoding approach for multi-view systems:

```
Select reference frames for GOP and initialize motion
vector parameters per GOP
  I - Estimate the average motion vectors [vx vy] that
      are common for all views according to Eq. 4.
 II - Subtract average values and estimate the sets of
      motion vector parameters that are strongly
      correlated. Carry out prediction using adjacent
      macroblocks and solve Eq. 8 iteratively (on the
      fly) for optimal MPOs using multiplicative updates
      (or operational updates). Assume (Fig. 5) that
```

$$\tilde{\mathbf{M}}^{(k)}(t + \Delta t) = \Phi_{View\ k}(t + \Delta t, t) \tilde{\mathbf{M}}^{(k)}(t) + \mathbf{b}(t)\mathbf{w}(t) \cong F_{View\ k}^{\beta(t)}(\tilde{\mathbf{M}}^{(k)}; t + \Delta t, t) + \mathbf{b}'(t)\mathbf{w}(t) \quad (9)$$

```
      where t is the frame index, w(t) stands for a
      white process featuring zero mean and (t)denotes
```

some type of operation at *t* like *element-by-element multiplication, component rotation, permutation* etc.

III– Initial sets of motion vector parameters are transmitted as ordered mutivalue sequences indexed by position/order of component. Probability density functions are entropy encoded. A *context index increment* denoted as $\chi_s(order\_of\_component, bin\_index)$ can be assigned. Select the context template for the $\sum_{k=1}^{N} P_k \leq c$ most correlated linear projections of motion parameter sets between views, i.e the joint pdfs yielding the highest values of $S(i_1, i_2, ...; t)$. Address *order_of_component* starting from the maximum *S* in decreasing order. This constitutes an indirect indexing of orthonormal basis vectors. Transmit symbols •(*t*) (multiplicative updates or operations) per view and *order_of_component* in a similar fashion (up to P_k indices for View *k*) as indicated by the structure of **MPOs**.

IV– Encode the symbol levels corresponding to the values of the selected sets of the motion vector parameters, i.e. for the set yielding the smallest distortion error provide *order_of_component* and level. Each level is encoded as a sequence of indexed bins using CABAC models. Find residual distortions.

V– Carry out rate control according to Eq. 3 by selecting the total number of symbols. Select between the *c* most correlated symbols per block/CTU. Continue as long as one gets a valid estimate, i.e. maintain decreasing distortion differences. Otherwise break.

VI– Encode residual frames and **repeat until end of GOP** (go to *Step* II).

4 Numerical Simulations

Numerical simulations for the proposed encoding method have been carried out for the image sequences used for video view interpolation as described in [25]. Each sequence is 100 frames long. The camera resolution is 1024x768 and the capture rate is 15fps. The frames of the ballet sequence in [25] are used to obtain the numerical values presented in this section. The multi-view GOP for the numerical simulations consists of initial four (4) frames. A fixed size macroblock featuring dimensions of 8x8 pixels is used to carry out motion compensation. The average of the first frame of View 1 and of the first frame of View 2 is used as a reference. The elastic MCP model as described in Eq. 2 uses sixteen (16) discrete cosines as the basis functions (parameter P equals 32). The context and the distributions for the first stereo GOP of the sequence are illustrated in Fig. 6 Entropy encoding is used based upon the probabilities depicted in Fig. 6.a. The 32x32 sigma values ordered according to magnitude as obtained from the proposed algorithm are given in Fig. 6.b. The histogram of the highest sigma values over all macroblocks is presented in Fig. 6.c. The magnitude of the residual frames corresponding to *View* 1 and *View* 2 are marginally decreased as linear correlated components between views are taken into account. *Peak-Signal-to-Noise-Ratio* (PSNR) values corresponding to the two views are 31.9 dB and 31.0 dB respectively when the two (c=2) most correlated linear components are used. The additional overhead is estimated to 0.11 bits/pixel. Original and residual frames are given in Fig. 7. PSNR values are slightly improved when the six (c=6) most correlated linear components are taken into account. PSNR value for *View* 1 equals 31.9 dB whereas PSNR value for *View* 2 equals 31.2 dB for an estimated additional overhead of 0.17 bits/pixel. Nevertheless savings on the bits/pixel required for underlying motion compensation could be made should the proposed approach be applied.

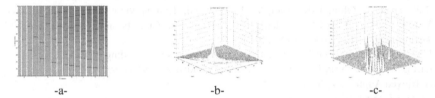

-a-   -b-   -c-

Fig. 6. Contexts and distributions for the $s(i_1, i_2,...;t)$ values for the first four frames of the ballet video sequence (-a- probabilities for one up to sixteen correlation coefficients; -b- the distribution of all 32x32 sigma correlation coefficients and -c- histogram of the highest sigma values over all macroblocks)

Fig. 7. Views 1 and 2 and residual frames for joint motion vector compensation (the two largest $s(i_1, i_2,...;t)$ values have been used)

5   Conclusion

A novel scalable approach to multi-view video encoding based on the so-called MPO structure is proposed. It takes advantage of the correlated linear components between views and it requires no prior knowledge of the capturing cameras in space. Strongly correlated linear components, i.e. lower order projections of tensorial objects, are multiplicative per GOP (or frame or slice). The correlated linear components are assigned initial values per GOP (or frame or slice) whereas subsequent operations (updates and permutations) are defined upon index position. Context templates corresponding to different distributions of the sigma correlation parameters may be selected according to the CABAC model. Updating adaptively indexed orthonormal basis functions in conjunction with CABAC increases encoding efficiency and allows for the incorporation of SVD and MPC transforms into the existing approaches. Initial numerical results indicate that the proposed method yields improvements in encoding multiple views in MPEG standards under development like the HEVC.

Acknowledgments. This research work has been funded by the *LiveCity* European Research Project supported by the Commission of the EC – *Information Society and Media Directorate General* (FP7-ICT-PSP, GA No.297291).

References

1. Ohm, J.-R., Sullivan, G.J., Schwarz, H., Tan, T.-K., Wiegand, T.: Comparison of the Coding Efficiency of Video Coding Standards – Including High Efficiency Video Coding (HEVC). IEEE Trans. on Circuits and Systems for Video Technology 22(12), 1669–1684 (2012)
2. Sullivan, G.J., Ohm, J.-R., Han, W.-J., Wiegand, T.: Overview of the High Efficiency Video Coding (HEVC) Standard. IEEE Trans. on Circuits and Systems for Video Technology 22(12), 1649–1668 (2012)
3. ITU TSB (2010-05-21): Joint Collaborative Team on Video Coding. ITU-T, http://www.itu.int/ITU-T/studygroups/com16/jct-vc/ (retrieved August 24, 2012)
4. http://www.h265.net/
5. ITU-T and ISO/IEC JTC 1: Advanced Video Coding for Generic Audiovisual Services. ITU-T Recommendation H.264 and ISO/IEC 14496-10 (MPEG-4 AVC), Version 1: May 2003, Version 2: May 2004, Version 3: March 2005 (including FRExt extension), Version 4: September 2005, Version 5 and Version 6: June 2006, Version 7: April 2007, Version 8: July 2007 (including SVC extension), Version 9: July 2009 (including MVC extension)
6. Vetro, A., Wiegand, T., Sullivan, G.J.: Overview of the Stereo and Multiview Video Coding Extensions of the H.264/MPEG-4 AVC Standard. Proceedings of the IEEE (Special Issue on 3D Media and Displays) 99(4), 626–642 (2011)
7. Kordasiewicz, R.C., Gallant, M.D., Shirani, S.: Affine Motion Prediction Based on Translational Motion Vectors. IEEE Trans. Circuits Syst. Video Technology 17(10), 1388–1394 (2007)
8. Wiegand, T., Steinbach, E., Girod, B.: Affine Multipicture Motion-Compensated Prediction. IEEE Trans. Circuits Syst. Video Technology 15(2), 197–209 (2005)
9. Nakaya, Y., Harashima, H.: Motion Compensation Based on Spatial Transformations. IEEE Trans. Circuits Syst. Video Technol. 4(3), 339–356, 366–367 (1994)
10. Pickering, M.R., Frater, M.R., Arnold, J.F.: Enhanced Motion Compensation Using Elastic Image Registration. In: Proceedings of IEEE Int. Conference on Image Processing, Atlanta, GA, USA, pp. 1061–1064 (2006)
11. Fehn, C.: A 3D-TV Approach Using Depth-Image-Based Rendering (DIBR). In: Proceedings of Visualization, Imaging And Image Processing (VIIP), pp. 482–487 (2003)

12. Schwarz, H., Bartnik, C., Bosse, S., Brust, H., Hinz, T., Lakshman, H., Marpe, D., Merkle, P., Müller, K., Rhee, H., Tech, G., Winken, M., Wiegand, T.: 3D Video Coding Using Advanced Prediction, Depth Modeling, and Encoder Control Methods. In: IEEE Intl. Conf. on Image Processing (October 2012)
13. Sullivan, G.J., Baker, R.L.: Rate-Distortion Optimized Motion Compensation for Video Compression Using Fixed or Variable Size Blocks. In: Proc. of GLOBECOM 1991, pp. 85–90 (1991)
14. Marpe, D., Blattermann, G., Wiegand, T.: Adaptive Codes for H.26L, ITU-T SG 16/6 Document VCEG-L13, Eibsee, Germany (January 2001)
15. Schwarz, H., Marpe, D., Wiegand, T.: CABAC and slices, JVT document JVT-D020, Klagenfurt, Austria (July 2002)
16. Marpe, D., Schwarz, H., Wiegand, T.: Context-Based Adaptive Binary Arithmetic Coding in the H.264 / AVC Video Compression Standard. IEEE Transactions on Circuits and Systems for Video Technology 13(7), 620–636 (2003)
17. Richardson, I.E.G.: H.264 and MPEG-4 Video Compression – Video Coding for Next Generation Multimedia, pp. 198–207. John Wiley and Sons (2003)
18. Marpe, D., Schwarz, H., Wiegand, T.: Probability Interval Partitioning Entropy Codes. Submitted to IEEE Transactions on Information Theory (June 2010), http://iphome.hhi.de/marpe/download/pipe-subm-ieee10.pdf
19. Stephanakis, I.M., Anastassopoulos, G.C.: A Multiplicative Multi-linear Model for Inter-Camera Predictio. In: Free View 3D Systems. Engineering Intelligent Systems Journal (under print, 2013)
20. de Lathauwer, L., de Moor, B., Vandewalle, J.: A Multilinear Singular Value Decomposition. SIAM Jour. of Matrix Analysis and Appl. 21(4), 1253–1278 (2000)
21. Lu, H.K.N., Plataniotis, K.N., Venetsanopoulos, A.N.: MPCA: Multilinear Principal Component Analysis of Tensor Objects. IEEE Trans. on Neural Networks 19(1), 18–39 (2008)
22. Lu, H.K.N., Plataniotis, K.N., Venetsanopoulos, A.N.: A Survey of Multilinear Subspace Learning for Tensor Data. Pattern Recognition 44(7), 1540–1551 (2011)
23. Yang, Z., Laaksonen, J.: Multiplicative Updates for Non-negative Projections. Neurocomputing 71(1-3), 363–373 (2007)
24. Zhang, Z., Jiang, M., Ye, N.: Effective Multiplicative Updates for Non-negative Discriminative Learning in Multimodal Dimensionality Reduction. Artificial Intelligence Review 34(3), 235–260 (2010)
25. Zitnick, C.L., Kang, S.B., Uyttendaele, M., Winder, S., Szeliski, R.: High-quality video view interpolation using a layered representation. In: ACM SIGGRAPH and ACM Trans. on Graphics, Los Angeles, CA, pp. 600–608 (2004)

(Semi-) Pervasive Gaming Educational and Entertainment Facilities via Interactive Video-to-Video Communication over the Internet, for Museum Exhibits

Ioannis P. Chochliouros[1], Rod McCall[2], Andrei Popleteev[2],
Tigran Avanesov[2], Tomas Kamarauskas[2], Anastasia S. Spiliopoulou[3],
Evangelos Sfakianakis[1], Evangelia Georgiadou[1], Nikoletta Liakostavrou[3],
Ioannis Kampourakis[3], and Ioannis Stephanakis[3]

[1] Research Programs Section,
Hellenic Telecommunications Organization (OTE) S.A.,
99 Kifissias Avenue, GR-151 24, Athens, Greece
{ichochliouros,esfak,egeorgiadou}@oteresearch.gr
[2] SnT, University of Luxembourg,
4, Rue Alphonse Weicker, L-2721 Luxembourg, Luxembourg
{Roderick.McCall,Andrei.Popleteev,Tigran.Avanesov,
Tomas.Kamarauskas}@uni.lu
[3] Hellenic Telecommunications Organization (OTE) S.A.,
99 Kifissias Avenue, GR-151 24, Athens, Greece
{aspiliopoul,nliak,gkabourakis,stephan}@ote.gr

Abstract. Based upon the core concept of the LiveCity Project we focus on the specific *City Cultural Experiences v2v Pilot*, designed to allow for visitors at two defined locations to interact with one another in a joint experience and to get educational/entertainment benefits, originating directly from the museum content delivery. We discuss a set of semi-pervasive games (the so-called *"Twin Cities"* games) which are designed to bring people together at remotely twinned locations through the use of video-to-video communication and multi-touch interaction. We also present an early classification of video-to-video (v2v) interaction games that is designed to inform designers about the potential of such technologies. We classify them as: using video for awareness and communication, interacting with video and video as a game.

Keywords: Content management system (CMS), Internet, gamification, "mixed" reality game, pervasive game, video-to-video (v2v) communication.

1 Introduction

Information and communication technologies (ICTs) can enable learning and educational activities and help people gain new skills in the modern digital-based economies and societies [1]. Among the core priorities for maximizing the social and economic potential of ICT [2] should be the proper development & the dispersion -*as widely as possible*- of a variety of modern infrastructures and/or corresponding facilities, composing the so called *"Future Internet" (FI)* [3]. Thus, the future

H. Papadopoulos et al. (Eds.): AIAI 2013, IFIP AICT 412, pp. 474–485, 2013.

economy will be a network-based knowledge economy, with the Internet being at its center. This option will be critical to support growth and investments as well as to ensure citizens can access the content and services they want, occasionally via new modes and novel educational or communication means.

The aim of the *LiveCity* (*"Live Video-to-Video Supporting Interactive City Infrastructure"*) *Project* is to empower the citizens of a city to interact with each other in a more productive, efficient and socially useful way by using high quality video-to-video (v2v) over the Internet. Video-to-video can be used for a variety of selected activities such as to save patients' lives, improve city administration, reduce fuel costs, reduce carbon footprint, enhance education and improve city experiences for tourists and cultural consumers. To realize its specific targets, LiveCity proposes certain well defined scenarios/use-cases and then it realizes, *during its life-time*, corresponding pilot actions with the involvement of multiple users. According to the original LiveCity context and conformant to the specific description of its WP4 (*"City Cultural & Educational Experiences v2v Pilot"*), two specific scenarios will be developed and the inclusion of live (interactive) v2v communication for use in schools and museums will be examined, *respectively*. In particular, one of the two essential scenarios is the *"Museum Exhibit"* which proposes the establishment a cultural-oriented scenario and it is designed to allow the visitors at two defined locations to interact with one another in a "joint experience"; the related "locations" are the city of Athens (with the participation-involvement of *OTE's Telecommunications Museum*) and the city of Luxembourg (with the participation/involvement of the *Post & Telecoms (P&T) Museum*). LiveCity also aims to "promote" awareness of each city to visitors at both locations. The final design will use live interactive v2v as the "basis" for the interaction along with a multi-touch table and perhaps an external display, operated at both museums. Via this modern kind of application, "games" that are to be deployed will allow people playing together in real time, solving a number of puzzles and clues about related topics as well as promoting a variety of educational aspects for various potential categories of users - museum visitors (i.e.: pupils, students, teachers, educators and the public). Focusing upon an interactive v2v communication directly to the public via the LiveCity Project, can be a decisive option for the promotion of *OTE's Telecommunications Museum* collections to the Hellenic and/or the European public and for the proper and wider dissemination of its programs and/or other related (educational, informative, etc.) initiatives, towards supporting the effort for a more efficient digital cultural and scientific inclusion of local and virtual citizens; this is particularly important for LiveCity as this museum also intends to apply the corresponding facilities to educationally-oriented activities in cooperation with two schools of the Municipality of Vrilissia in Athens, also participating to the LiveCity scope. Moreover, similar beneficiary options stand for the *P&T Museum* in the city of Luxembourg. The proposed activities and the related methodologies will also make all involved people learning more effectively and will support the acquisition of new skills, via the usage of v2v facilities, on a "pure" multimedia-based environment. This will help the effort for ensuring the effective use and exploitation of cultural resources by developing technologies and for making them widely available, usable and re-usable regardless of their form, location, time-sphere, etc. In the scope of LiveCity-based activities, it should be expected that either visitors of the involved museums or other potential users of the proposed innovative solutions may be able to have benefits, directly from the museum content delivery. This will also help to "disperse" cultural and scientific-technological knowledge in a broader multi-media-based environment and will "trigger" or "challenge" new options for enhancing quality of experience (QoE) for the end-users.

2   Background

Modern Internet-based facilities have radically modified the way people can communicate, amuse or even make business (in a variety of sectors [4]) and this has become obvious in a variety of v2v platforms supporting such aspects. Digital distribution of cultural and creative content can enable content providers to reach new and larger audiences while this also permits users to enjoy new experiences. According to the actual EU policy [5], Europe needs to "push ahead" with the creation, production and distribution (on all platforms) of digital content. Yet to date there is little bringing together of museums, cultural venues or twinned cities by using recent technologies. Within the scope of the LiveCity Project we have promoted the realization of a way of interactive v2v communication between citizens being at "twinned museum" sites, intending to support exchange of information and of experiences; museums can reside in different cities as well as in different European countries. This innovative manner of communication also focuses upon specific cultural and educational aspects with the aim of further enhancing interactivity between all potentially involved users. The challenge becomes greater as, apart from museums, other organizations such as schools, educational institutes or municipal authorities can also "join" the effort at later stages. In order to allow for the exhibit to be used after LiveCity has ended at other locations, a content management system (CMS) is being developed which will allow other venues or cities to "add" their own content to the standard deployed games. We have chosen to "base" the game on the underlying definition of "mixed" reality ([6], [7]) which states that a *"mixed reality is one where multiple devices allow views and interactions within a given context"*. This is in contrast to the mixed reality continuum [8] which consists of two opposing "poles", real environments and virtual environments with augmented reality and augmented virtuality exisiting in between. Drawing on the first definition, a "game" could consist of mobile devices, fixed screens and tablet computers. Based on this approach we have developed a platform which embraces displays, multi-touch tables and allows the addition of tablet devices. Given the emphasis on supporting collaboration between two remote locations the game is known as *"Twin Cities"*.

The inspiration for our work comes from two previous EC-funded projects, that is: *IPerG*[1] which developed the "epidemic menace game" and *IPCity*[2] which developed

---

[1]   *IPerG (Integrated Project on Pervasive Gaming)* was an EU-funded project (FP6 - 004457) which started on 01.09.2004 and came to an end on 29.02.2008. Its aim has been the creation of entirely new game experiences, which are tightly interwoven with our everyday lives through the objects, devices and people that surround us and the places we inhabit. The approach has been through the exploration of several showcase games which come under the description of "pervasive games" - *a radically new game form that extends gaming experiences out into the physical world. [http://iperg.sics.se/index.php.].*

the mixed reality tent to allow improved participation in the urban redesign process. In the *IPerG* case, a control room consisting of multiple screens was used where players there could control -*or provide*- advice to players outside using mobile devices. In our system, there are two connected "control" rooms where visitors at each museum can play together a set of common games. From the *IPCity* scope we embraced the idea of using an interactive table-top and video display of a remote location and scouting. In the *IPCity* example, people move objects around on the interactive table and simultaneously these moves are represented on the large screen.

3   Pervasive Gaming Examples in the LiveCity Context

Pervasive games focus on a game play that is embedded in our physical world. Elements of the physical world are inherent parts of the game. Their characteristics and states are sensed and can influence the course of the game. Moreover, pervasive games allow for a game that can be potentially accessed at any time and from any location. Many of these games rely on mobile and pervasive computing technology, such as cellular phones and location sensors and focus on location-based aspects. The *"Twin Cities"* game is semi-pervasive in the sense that it extends two of three facets of pervasive gaming ([9], [10]), namely social and temporal expansion. For example, unlike traditional computer of board games, the *"Twin Cities"* lets anyone take part, thus making it a socially expanded game. Indeed, the only restriction on who can play relates to how many people can physically stand round the table, coupled with any legal considerations (e.g., legal age of consent to use such systems). Additionally, the public location of the game means that others can see and even partially participate in the experience. Due to the "shared nature" of the experience (which is to be outlined in the following parts) this lets people assume many roles [11], including: (i) *Player*: someone who influences the game, e.g. someone who directly takes part by using the table; (ii) *Spectator*: someone who is aware of the game and can influence it, (e.g. standing around but not directly playing the game); (iii) *Bystander*: Someone who is unaware of ongoing game, and has no ability to participate. (These could be other museum visitors who walk around the area but do not pay attention to the game).

In addition to the spatial expansion, *"Twin Cities"* also explores the use of temporal expansion in the context that while individual games can be played, they may take as long as the players wish or indeed be merged into a set of games in which the players collaborate or compete over a period of time - *which again is undefined*. At the time of writing the present paper, we are exploring how to let players save their scores and retain them for future use. This would allow them to return to the venues many times and play against other potential users. We intend to explore the concept of

---

² *IPCity* (FP-2004-IST-4-27571) was an IP EU-funded Sixth Framework program on *Interaction and Presence in Urban Environments*. Its research aim was to investigate analytical and technological approaches to presence in real life settings. Analytically, this includes extending the approaches to presence accounting for the participative and social constitution of presence, the multiplicity and distribution of events in time and space. More information can be found at: *http://www.ipcity.eu/*.

"gamification" [12] where elements such as leader-boards or badges can be used to confer status on people and to encourage them to "remain" within the game. *"Presence"* is a critical component of the experience within the *"Twin Cities"* context, intending to allow people from different cultures to get together via a "game-like" environment. In our scope we can consider two main forms of presence [13]:

- *Social presence:* The feeling of being with another person.
- *Physical presence:* The feeling of being at another place.

The *"Twin Cities"* framework targets the concept of social presence where the video feed is designed to make people feel like they are together [14]. In particular, the emphasis is on improving communication through the use of tangible user interfaces such as the multi-touch table. Prior work in the *IPCity* context has identified that a shared table approach can be used to foster communication between different people and this encourages negotiation & discussion about relevant topics. Our approach extends this perspective to video communication and to remote locations.

3.1   Game Design

This section provides an overview of the selected game designs. Since both involved "parties" are *Telecommunications/Post Museums* (in Athens and Luxembourg) it would be desirable [15] to combine the following key factors: (i) Take advantage of live v2v infrastructure, *where it is possible*, to enhance the experience of the games within the participating museums; (ii) consider and promote awareness of the other museum's exhibits-collections, and; (iii) provide general information about telecommunications and/or related technical facilities. In order to perform an effective design and thus to ensure wider applicability-adoptability of the games, the latter have been designed in a way to be *"as simple as possible"* regarding their perceptive concept without necessitating any specific technical -or other- prior knowledge by the intended users-visitors of the museums. The related concept, per game, has been selected so that to be explicitly relevant to a collection of exhibits and/or to an event-activity promoted by the museum(s). Furthermore, a critical priority of the full game design process was to improve cultural and educational interactivity between users residing in different locations (i.e., cities or countries), in parallel with the inclusion of multiplayer setting so that to have, occasionally, many participating users.

Designing the games ([16], [17]) should also take under consideration the factors discussed as follows: 1) For both involved museums we have considered the next classification regarding visitors: *a)* School groups (i.e.: ages 5-15) cover 90% of visitors (*school groups usually consist of teams of 20-35 students*); *b)* Individual children (i.e.: ages 5-10) cover 5%, and; *c)* Adults (i.e.: ages 35-45) cover the remaining 5% of the visitors' population. 2) Some of the games should mainly address adults in order to "attract" this age group; in this context: *a)* There should be a co-operation mode between the visitors of the two museums; *b)* A game should be playable in both groups and by individuals; *c)* A game should encourage movement and discovery of the related museum's collections. Among the priorities that have influenced the game design process was the inclusion of "multi-player" involvement, although this also depends on other factors like the frequency of visits, any specific events that can attract more people in certain time-slots, etc. In order to overcome this restriction, it would be ideal if the game(s) could be considered as "operational" even in case of the absence of (remote) players ([18], [19]). This simply implicates that it should be possible, for a museum's visitor to "access" a game being at a certain level of progress or to access a new one. The LiveCity platform supports a v2v-based game concept to fulfill this feature, together with the one of the "distant presence". In addition, appropriate equipment (such as cameras) is to be set-up in the participating museums to extend the proximity area of the multi-touch tables that are to be used for the games. The following sections discuss the selected game designs.

3.1.1   Game Conceptualization

As noted earlier, our work is predominantly focused on providing a degree of social presence between game players, our objective being to use multi-touch interaction and a live video feed between the two locations. Our initial game designs focus on supporting social presence from three main game design perspectives where video is used in different ways to encourage play, discussion and cross-cultural awareness. These are outlined in Table 1 below, with the top level "using video for awareness and communication" being the most basic through "Video as the Game", where the objective is to make the video component the goal. A brief description based on each game is provided in the following section; a more thorough description and additional rationale for the concepts used in some of the games can be found in [19].

Table 1. Use of live video-to-video feeds in games

Level	*Example Games*
Using video for awareness and communication	Mosaic, memory and putting things in the right order and quiz and build it together
Interacting with video	Embedded video in mosaics or memory games
Video as the game	TV show with live video mixed with CMS content – create a live broadcast

Using Video for Awareness and Communication

These games use video to encourage discussion *about* the content on each multi-touch table. Rather than the video being the driving factor in the game, video is used passively where players at each location can see and hear the players at the other location while they take part either in local game (one museum only) or multi-player game (two museums playing together). This style of game play was conceived as a way to break down the usual "cultural divide" between twinned locations such as towns or museums where the actual inhabitants or visitors often never meet one another and instead it is on the officials or designated people who get a chance to meet. The approach is also designed to be relatively free form in the sense that while the games should encourage discussion about the content at both museums the visitors are in fact totally free to talk about whatever they wish as there are no content, time or participant restrictions. To a limited extent this allows for a semi-pervasive experience as both players and spectators can take part. Indeed, as the games are ongoing, spectators are free to interact as they wish, perhaps just waving to people or even talking to people at the remote locations. This, in turn, may shape players' behavior.

Fig. 1. Mosaic/Jigsaw game example

For this game type content from each museum is placed in a jigsaw or mosaic which is then put into the wrong order (see Fig.1). Players of the games are then asked to put re-assemble the contents to create the correct image. A more complex version of mosaic based on the "Interacting with Video" level is outlined later.

While mosaic/jigsaw mainly concentrate on visual content play, i.e. they are about reassembling an image the "putting things in the right order" concept operates on both this and the conceptual level. Alternatively, content can be placed in the right order based on conceptual details. For example, at the conceptual level in a telecoms museum, people could be asked to construct a telecoms network across Europe by connecting the correct components such as cables, switches and tele-houses in the right order. Alternative approaches to this include placing things in historical order (see Fig.2).

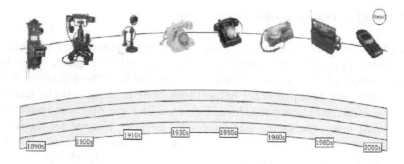

Fig. 2. Put-in-order game with telephone devices

The classic memory game is familiar to many people and is often played with a deck of cards where people turn over two cards and if they match they keep them facing up. The game is played until all cards have been matched in pairs. This approach is used within our system but with tiles, each tile consisting of a piece of relevant content for example telephones or stamps (see Fig.3). The memory game can be played either on a visual basis or again on a conceptual one, where people are asked to memorize related concepts when turning over cards.

Quizzes are another popular format and are again applied with LiveCity, in the example (see Fig.4) (know as "Whose phone is it?") players are asked to match the phone with the right person. In this case, multi-touch is used to drag the images around. While players may be able to guess some of the answers for this, other quizzes' players may also need to spend time finding out about contents in the muscum. Additionally, this approach encourages the exchange of information between the two museums as players could be quizzed on content in the other museum. This would necessitate that players find out and discuss content together.

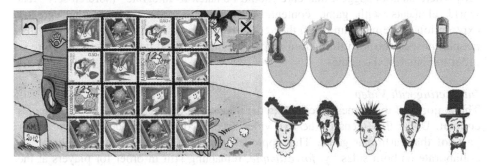

Fig. 3. Memory game example: stamps   **Fig. 4.** The quiz game concept

The building together game starts by showing the players an image of an object (e.g.: a museum exhibit, art piece, building, or something else), which is then split into several parts which, *in turn*, are randomly rotated, resized and mixed. The objective is to "build" the original object from its parts. On the first floor of the *OTE's Museum* in Athens there is a wall-to-wall poster of Alexander Graham Bell, which consists of dozens of the Museum's exhibits and could be easily used for the purposes of this specific game (*as shown in* Fig.5). Multi-touch interaction enables an arbitrary number of players to collaborate in the game. Collaboration with the remote players will be facilitated by the live v2v link between game sites. Multi-touch interaction enables an arbitrary number of players to collaborate in the game. Collaboration with the remote players will be facilitated by the live v2v link between game sites.

Fig. 5. The *Alexander Graham Bell* "build together" game concept

The previous games use the multi-touch table and video feed from a static location in the museum which could result in people being drawn away from the museum content. This problem is not unique to our proposed games and has also been noted in [20] where authors suggest that care should be taken to integrate "more closely" real world elements with game content. Therefore, in order to encourage greater exploration of the museum itself, they could be combined with mobile devices so that people must search for information within the museum and take a video or picture in order to complete the game.

Interacting with Video

The majority of the games described in the previous section can also use live video as content. Under this scenario the live video feed becomes embedded to form all -or part- of that particular game. This approach means that players have to physically collaborate on both sides by, *for example*, remaining still in order for players at the other location so that they can correctly reassemble the mosaic. However, as video is relatively uncontrollable they may adopt other strategies such as using particular movements in order to play a game. These types of games have the potential to increase the level of embodied interaction among players which we hope will, in turn, raise their sense of involvement, engagement and social presence.

In the mosaic game, players at each location could see a live video feed of people from the other location; the video feed is then broken up incorrectly across the mosaic tiles. Under this scenario the players would be confronted with the choice of remaining still at each location -or if they are particularly competitive- perhaps moving around in order to confuse the people at the other location. Such an approach would allow for the exploration of collaborative versus co-operative forms of embodied interaction within such games. Related to this gaming concept is the memory game with a live video feed; under this scenario one of more of the tiles could contain live video from each location which players then have to match. In this setting one tile could represent one feed, or a feed could be spread across many tiles.

Video as the Game

This concept *uses the video feed* as the *objective* of the game. The objective of the TV broadcast game is to create an engaging gamified museum browsing experience which encourages visitors to browse more attentively, discover important, interesting or unusual exhibits selected by the exhibition curators, learn about them and share that knowledge via the live video stream with visitors at the other museum. This approach utilizes the v2v platform, multi-touch table and mobile devices. The game functions in the following way: The visitors of the connected museums can browse through the images of the other museums' exhibits by using multi-touch tables. If the visitors of the Museum A (Team A) want to learn more about a particular exhibit they click a "Show it" button. The multi-touch table at the Museum B displays a message that someone from the Museum A wants to see the particular object. The objective of the Museum B visitors (Team B) is to locate and present the object via live video stream by using provided mobile devices with the game application. Once the challenge is accepted the game starts. First, Team B has to locate the object and take the picture of it by using the mobile application. The picture is instantly displayed in the Museum A and its team confirms that the correct exhibit is photographed. Then Team B has up to 5 minutes to learn as much as possible about the object and prepare to present it. Meanwhile, Team A is shown a presentational video or audio guide based presentation about the same object, they can read additional information and see more images of the particular object. When Team B is ready (or when 5 minute time limit is reached), they start the "broadcast" and have up to 2 minutes to present the object through live video. The provided mobile camera-enabled devices -such as tablets- are used to capture the video, by using the game application. When 2 minutes run out the video is stopped, or can be done manually, if the team mentioned everything they wanted. While the presentation is running, people in Museum A see a live video stream and the list of facts about the object on the multi-touch table. If the broadcasting team mentions a certain fact or feature, the viewers have to tap on the button to confirm that the fact was mentioned correctly. Once the broadcast is over, the teams can swap and the presenter can request to present the object from the other museum too. The points are calculated for the least amount of time taken to locate the exhibit, and prepare for the presentation. Facts about the exhibit can have different point weight depending on their complexity (e.g. a player scores 1 point for mentioning that the object is a "light bulb and 5 points for mentioning "Thomas Edison"). Every participating museum can create or accept the challenges and the roles can be switched instantly once the running activity is finished.

Content Management
In order to let other twinned locations create similar experiences, a content management system has been developed which lets other venues use their content within the game templates.

4   Conclusion

In this paper we have presented a number of games which use video in different ways to promote cross-cultural awareness at twinned locations. Our work was based on the view that video-to-video is largely underutilized within the domain of game playing yet when combined with multi-touch and related forms of intuitive interaction it has a great potential to provide rich experiences. We have expanded upon this by providing three different levels of game which use video-to-video communication in different ways. This classification should provide a useful basis upon which future work can be based. Furthermore, the designs and classifications indicate that at a generic level video to video interaction offers a potential to improve interaction between people and in particular within the proposed cultural contexts.

Acknowledgments. The work presented in this paper is from the LiveCity project which is funded by the European Commission under *DG CONNECT* (FP7-ICT-PSP, Grant Agreement No.297291). Photographs of museum content are from the P&T Post/Musée (Luxembourg) and the OTE's Telecommunications Museum (Athens). We gratefully acknowledge the assistance of the staff from both museums.

References

1. European Commission, Information Society and Media: Education and Information Society: Linking European Policies. Luxembourg (2006)
2. Commission of the European Communities: Communication on Europe 2020: A strategy for smart, sustainable and inclusive growth COM(2010) 2020 final (March 03, 2010)
3. Future Internet Assembly (FIA): Position Paper: Real World Internet (2009)
4. Amit, R., Zott, C.: Value Creation in eBusiness. Strategic Management Journal 22, 493–520 (2001)
5. Commission of the European Communities: Communication on A Digital Agenda for Europe COM(2010) 245 final/2, 26.08.2010. Brussels, Belgium (2010)
6. Koleva, B., Benford, S., Greenhalgh, C.: The Properties of Mixed Reality Boundaries. In: Proceedings of the Sixth European Conference on Computer-Supported Cooperative Work, Copenhagen, Denmark, pp. 119–137 (1999)
7. Benford, S., Giannachi, G.: Performing Mixed Reality. Massachussets Institute of Technology (MIT) Press (2011)
8. Milgram, P., Kishino, F.: A Taxonomy of Mixed Reality Visual Displays. IEICE Transactions on Information Systems E77-D12, 449–455 (1994)
9. Montola, M.: Exploring the Edge of the Magic Circle: Defining Pervasive Games. In: Proceedings of DAC 2005 Conference, Copenhagen, Denmark, December 1-3 (2005)
10. Montola, M., Stenros, J., Waern, A.: Pervasive Games: Theory and Design. Morgan Kaufmann, San Francisco (2009)
11. Montola, M., Waern, A.: Participant Roles in Socially Expanded Games. In: Strang, T., Cahill, V., Quigley, A. (eds.) Pervasive 2006 Workshop Proceedings 165-73, PerGames 2006 Workshop of Pervasive 2006 Conference, May 7-10, pp. 99–106. University College Dublin, Dublin (2006)
12. Deterding, S., Sicart, M., Nacke, L., O'Hara, K., Dixon, D.: Gamification: Using Game Design Elements in Non-Gaming Contexts. In: Proceedings of the 2011 Annual Conference Extended Abstracts on Human Factors in Computing Systems, Vancouver, Canada, pp. 2425–2428. ACM (May 2011)
13. IJsselstein, W., Riva, G.: Being there: The experience of presence in mediated environments. In: Riva, G., Davide, F., IJsselstein, W.A. (eds.) Being there: Concepts, Effects and Measurements of User Presence in Synthetic Environments, pp. 3–16. IOS Press, Amsterdam (2003)

14. Huizinga, J.: Homo Ludens. A Study of Play Element in Culture. Beacon Press (1955)
15. Salen, K., Zimmerman, E.: Rules of Play. Game Design Fundamentals. MIT Press, Massachusetts (2004)
16. Piekarski, W., Thomas, B.: ARQuake: The Outdoor Augmented Reality Gaming System. Communications of the ACM 45(1), 36–38 (2002)
17. Björk, S., Holopainen, J.: Patterns in Game Design. Charles River Media, Massachusetts (2005)
18. Preece, J., Rogers, Y., Sharp, H.: Interaction Design: Beyond Human Computer Interaction. Wiley College (2002)
19. Popleteev, A., McCall, R., Molnar, A., Avanesov, T.: Touch by Touch: Promoting Cultural Awareness with Multitouch Gaming. In: The Proceedings of the LiveCity Workshop on Smart and Pervasive Communications at the 4thInternational Conference on Smart Communications in Network Technologies (SaCoNet), Paris, France, June 17-19 (2013)
20. Herbst, I., Braun, A.-K., McCall, R., Broll, W.: TimeWarp: Interactive Time Travel with a Mobile Mixed Realty Game. In: The Proceedings of the 10th International Conference on Human-Computer Interaction with Mobile Devices and Services, pp. 235–244. ACM (2008)

Modeling and Trading FTSE100 Index Using a Novel Sliding Window Approach Which Combines Adaptive Differential Evolution and Support Vector Regression

Konstantinos Theofilatos[1], Andreas Karathanasopoulos[2], Peter Middleton[3], Efstratios Georgopoulos[4], and Spiros Likothanassis[1]

[1] Department of Computer Engineering and Informatics, University of Patras, Greece
{theofilk,likothan}@ceid.upatras.gr
[2] Business School, University of East London, Unighted Kingdom
a.karathanasopoulos@uel.ac.uk
[3] Management School, University of Liverpool, Unighted Kindgom
peter.william.middleton@gmail.com
[4] Technological Educational Institute of Kalamata, Greece
sfg@teikal.gr

Abstract. The motivation for this paper is to introduce a novel short term trading strategy using a machine learning based methodology to model the FTSE100 index. The proposed trading strategy deploys a sliding window approach to modeling using a combination of Differential Evolution and Support Vector Regressions. These models are tasked with forecasting and trading daily movements of the FTSE100 index. To test the efficiency of our proposed method, it is benchmarked against two simple trading strategies (Buy and Hold and Naïve Strategy) and two modern machine learning methods. The experimental results indicate that the proposed method outperformsall other examined models in terms of statistical accuracy and profitability. As a result, this hybrid approach is established as a credible and worth trading strategy when applied to time series analysis.

Keywords: Differential Evolution, Support Vector Regression, Confirmation Filters, FTSE100, Daily Trading.

1 Introduction

Modeling and trading financial indices remains a very interesting but challenging topic in econometrics Forecasting financial time series is a difficult task because of their complexity and their dynamic and noisy nature. All traditional linear methods and even more sophisticated non-linear machine learning models have failed to capture the complexity and the nonlinearities that exist in financial time series. This is particularly the case during times of uncertainty such as during the credit crisis in 2008.More robust and intuitive models have since been research and applied to financial times series in order to overcome these inefficiencies associated with previous models [1].

H. Papadopoulos et al. (Eds.): AIAI 2013, IFIP AICT 412, pp. 486–496, 2013.

Non-linear machine learning models have three main limitations. The first disadvantage is that most are calibrated to search for a global optimal estimator which in most cases does not exist due to the dynamic nature of financial timeseries. Another drawback is that the algorithms which are used for modeling financial time-series have a lot of parameters which need to be optimized and if this procedure is not performed correctly then the extracted prediction models will produce unsatisfactory results due to the data-snooping effect. Another mistake which is commonly made by practitioners is that, most of the time the training of a prediction model is performed separately from the construction of a trading strategy and thus the overall performance is reduced.

The purpose of this paper is to present a novel methodology which is capable of overcoming the aforementioned limitations. This methodology is based on a sliding window approach for the one day ahead prediction of the FTSE100 returns. To forecast every single return, the proposed trains a machine learning model using a sliding window of contemporary limited historical data. Thus the proposed method searches for the optimal predictor for each day. The machine learning model which was applied is an adaptive hybrid combination of Differential Evolution and the nu-Support Vector Regression (SVR) algorithm [2]. The adaptive Differential Evolution [3] (DE) was used for selecting the optimal feature subset, optimizing the parameters of nu-SVRs and at the same time optimizing the confirmation threshold for the confirmation filters which are used as parameters for our trading strategy. Moreover, a specialized fitness function was designed and used for the evaluation step of the adaptive Differential Evolution Algorithm which takes into account metrics for both statistical accuracy and trading performance.

The performance of the proposed methodology is compared with two traditional linear models and two non-linear machine learning approaches. The empirical analysis reveals that our proposed methodology clearly outperforms the other existing models ranking the highest across all of the examined metrics.

The novelty of the proposed approach is twofold. The first contribution lies in the application of a sliding window training method and the second offers an original adaptive hybrid machine learning methodology. On review of existing literature, the latter is unique and original as this is the first time that an adaptive Differential Evolution algorithm and nu-SVRs are combined into one model used for forecasting financial time series. Moreover, our proposed machine learning method is not only a simple combination of existing methods. The Differential Evolution algorithm optimizes not only the feature subset and the parameters of nu-SVRs but also a parameter of our trading strategy. When this was combined with the usage of a new fitness function specialized in trading tasks, it enabled our proposed method to fill the gap between financial forecasting and trading.

The rest of the paper is organized as follows: Section 2 describes the dataset used for our experiments. In Section 3 the proposed methodology was described in detail and in Section 4 the benchmark models are described and the experimental results are presented. Finally, Section 5 presents concluding remarks and some interesting future directions for research.

2   Related Financial Data

The FTSE 100 index is a weighted according to market capitalization which currently comprises of 101 large cap constituents listed on the London Stock Exchange. For the purpose of our trading simulation the iShares FTSE100 exchange traded fundis traded to capture daily movements of the FTSE100 index. Positions are initiated on the open of a trading day and closed at or around 16:30 GMT.The cash settlement of this index is simply determined by calculating the difference between the traded price on the open and the closing price of the index on the . When the model forecasts a negative return then a short position (sale) is assumed on the open and when it forecasts a positive return a long position (purchase) is taken. Profit / loss is realised on a daily basis and positions are not held overnight.

The FTSE 100 daily time seriesis non-normal (Jarque-Bera statistics confirms this at the 99% confidence interval), containing slight negative skewness and relatively high kurtosis. Arithmetic returns where used to calculate daily returns and they are estimated using equation 1. Given the price level P_1, P_2,…,P_t, the arithmetic return at time t is formed by:

$$R_t = \frac{P_t - P_{t-1}}{P_{t-1}} \tag{1}$$

In table 1 we present the full dataset used.

Table 1. Total dataset

Name of Period	Trading Days	Beginning	End
Total Dataset	1260	1 January 2008	31 December 2012
In sample Dataset	756	1 January 2008	31 December 2010
Out of Sample Set	504	1 January 2011	31 December 2012

As inputs to our algorithms, we selected a combination of autoregressive inputs, moving average time series of the FTSE100 returns, a FTSE100 volume time series, daily highs and lows of FTSE100 index, the VIX realized volatility index, and two metric times series which capture the aggregate advancing and declining volumes as a daily percentage as provided by FACTSET(2013) The inputs used in our modeling process are described in table 2.

In contrast to other non-linear approaches we have incorporated the VIX index to capture volatility. It is believe that this will be of particular benefit during times of higher volatility such as period of crisis. Moreover, some advanced volume metrics where studied as they are considered significant for capturing the markets liquidity. For instance, the advancing volume metric provides a sum of daily volume for those companies with advancing prices during a particular day. The declining volume metric provides a sum of daily volume for those companies with a declining price during a specific day.

Table 2. Explanatory Variables

Number	Variable	Lag
1	Daily High of FTSE100 all share return	1
2	Daily Low of FTSE100 all share return	1
3	FTSE100 all share return	1
4	FTSE100 all share return	2
5	FTSE100 all share return	3
6	FTSE100 all share return	4
7	FTSE100 all share return	5
8	FTSE100 all share return	6
9	FTSE100 all share return	7
10	5 day Moving Average of FTSE100 all share returns	1
11	15 day Moving Average of FTSE100 all share returns	1
12	30 day Moving Average of FTSE100 all share returns	1
13	FTSE100 daily volume	1
14	5 day Moving Average of FTSE100 daily volume	1
15	15 day Moving Average of FTSE100 daily volume	1
16	30 day Moving Average of FTSE100 daily volume	1
17	VIX Realized Volatility Index	1
18	5-day Moving Average of the VIX Index	1
19	15-day Moving Average of the VIX Index	1
20	30-day Moving Average of the VIX Index	1
21	Aggregate Advancing Volume Percentage metric	1
22	5-day Moving Average of Aggregate Advancing Volume Percentage metric	1
23	15-day Moving Average of Aggregate Advancing Volume Percentage metric	1
24	30-day Moving Average of Aggregate Advancing Volume Percentage metric	1
25	Aggregate Declining Volume Percentage metric	1
26	5-day Moving Average of Aggregate Declining Volume Percentage metric	1
27	15-day Moving Average of Aggregate Declining Volume Percentage metric	1
28	30-day Moving Average of Aggregate Declining Volume Percentage metric	1

Advancing volume percentage is defined as:

((Sum of Volume (Day t)) for all companies

where Price (Day t) > Price (Day t-1)/ Sum of Volume for all constituents   (2)

Declining volume percentage is defined as:

((Sum of Volume (Day t)) for all companies

where Price (Day t) < Price (Day t-1)/ Sum of Volume for all constituents   (3)

These calculations use daily prices and volumes for each constituent as provided by the FactSet Pricing Database (2013).

Finally, the set of explanatory variables were normalized in the interval of [-1,1] to avoid overrating inputs which takes higher absolute values.

3   Proposed Method

The proposed methodology is a sliding window approach which is designed to perform daily forecasting and trading. To forecast future values of the FTSE100 returns it trains a machine learning model using a window of a specific amount (name sliding window size) of historical prices for the examined inputs. By this way it outperforms other classical methodologies as it uses most recent data to update prediction models when forecasting next day returns.

The machine learning methodology which was used is based on the nu-Support Vector Regression predictors [4].

Support vector machines (SVM) are a group of supervised learning methods that can be applied to classification or regression. SVMs represent an extension to nonlinear models of the generalized algorithm developed by Vapnik [2]. They have been developed into a very active research area and have already been applied to many scientific problems. For instance, SVM have already been applied in many prediction and classification problems in finance and economics [5, 6] although they are still far from mainstream and the few financial applications so far have only been published in statistical learning and artificial intelligence journals.

SVM models were originally defined for the classification of linearly separable classes of objects. For any particular linear separable set of two-class objects SVM are able to find the optimal hyperplanes that separates them providing the bigger margin area between the two hyperplanes. However, SVM can also be used to separate classes that cannot be separated with a linear classifier. In such cases, the coordinates of the objects are mapped into a feature space using nonlinear functions. Every object is projected in a high-dimensional space feature space in which the two classes can be separated with a linear classifier.

In the task of forecasting financial indexes SVMs can be used for forecasting the directional movement of the examined index. However, these forecasts are not easily transformed to an effective trading strategies as the application of confirmation filters is not straightforward. The introduction of the ε-sensitive loss function by Vapnik(1995) [2] improves the Support Vector Regression approach and provides a robust technique for solving difficult regression problems. An improvement of the classical e-SVM is the more flexible nu-SVR [4] which is adopted for this particular study. When training nu-SVRs, the features which should be used as inputs should be selected carefully to avoid the curse of dimensionality. Moreover, the parameters C (regularization parameter), gamma (Radial Basis Function parameter) and nu should be optimized.

In more recent years a variety of meta-heuristic optimization problems have been proposed such as Genetic Algorithms [7] and Differential Evolution [3]. The most important problem they encounter is the fact that their own parameters should be calibrated in such a manner to enable them to effectively explore the search space while at the same time performing effective local searches.

The DE algorithm is currently one of the most powerful and promising stochastic real parameter optimization algorithms [3]. It belongs to the wider family of evolutionary algorithms as its operation is based on an iterative application of selection, mutation and crossover operators. DE is mainly based on a specific mutation operator. This operator randomly selects individuals from the population based on scaled differences between other randomly selected and distinct population members.

The representation schema which DE uses is the continuous gene representation. Thus, candidate solutions are represented as strings of continuous variables comprising of feature and parameter variables. For every candidate input of our model a feature variable is added to the representation of a member of DE's population. These genes take values from 0 to 2 with values higher than 1 indicating that this feature should be used as input to the SVR model. Four parameter variables are included in the representation of a candidate solution of DEs in the proposed methodology. These are C (values in the interval [0,1024]), gamma (values in the interval [0, 1024]) , nu (values in the interval [0,1]) and the optimal confirmation threshold (values in the interval [0, 0.01]).

The mutation operator which was selected for our proposed methodology selects for every population member X_i three random distinct members of the population ($X_{1,i}$, $X_{2,i}$, $X_{3,i}$) and produces a donor vector using the equation (3):

$$V_i = X_{1,i} + F * (X_{2,i} - X_{3,i})$$ (3)

where F is called mutation scale factor.

The crossover operator applied was the binomial one. This operator combines every member of the population x_i with its corresponding donor vector Vi to produce the trial vector U_i using the equation (4).

$$U_i(j) = \begin{cases} V_i(j) & , \quad \text{if } (rand_{i,j}[0,1] \le Cr) \\ X_i(j) & , \quad otherwise \end{cases}$$ (4)

Where: $rand_{i,j}[0,1]$ is a uniformly distributed random number and Cr is the crossover rate.

Next the selection operator is applied. Every trial vector U_i is evaluated and if it suppresses the corresponding member of the population X_i it takes its position in the population. To evaluate the candidate solutions of the proposed methodology we used the following specialized fitness function:

Fitness Function = Correct_rate - 1000*MSE+Annualized_Return   (5)

Where: Correct_Rate is the percentage of correct predictions, Annualized Return is the annualized return of the extracted trading strategy when taking into account the transaction costs and applying the corresponding conformation filter. MSE is the Mean Square Error (we multiplied it with 1000 to normalize its values in the magnitude of the ones of Correct Rate and Annualized Return). The aforementioned fitness function enables us to achieve high statistical metrics in our forecasting task while at the same time extracting optimized profitable trading strategies.

The termination criterion which was used was a combination of the maximum number of generations to be reached with a convergence criterion. The convergence criterion terminates the algorithm when the performance of the best member of the population is less than 5% away than the mean performance of the population.

The most important control parameters of a DE algorithm are the mutation scale factor F and the crossover rate Cr. Parameter F controls the size of the differentiation quantity which is going to be applied to a candidate solution from the mutation operator. Parameter Cr determines the number of genes which are expected to change in a population member. Many approaches have been developed to control these parameters during the evolutionary process of the DE algorithm [3]. In our adaptive DE version, we deployed one of the most recent promising approaches [8]. This approach randomly selects values during every iteration .For the F parameter selected from a uniform distribution with mean value 0.5 and standard deviation 0.3 and a random value for the parameter Cr from a uniform distribution with mean value Crm and standard deviation 0.1. Crm is initially set to 0.5. The Crm is replaced during the evolutionary process with values that have generated successful trial vectors. As a result, this approach replaces the sensitive user defined parameters F and Cr with less sensitive parameters like their mean values and their standard deviation.

The parameters of our method which needed to be optimized where the population size, the maximum number of generations to reach and the sliding window size. These parameters were optimized with thorough experimentation using only the in-sample dataset and examining the values of the fitness function described in equation 5. The optimal values which were found and used where the following: population size = 30, maximum number of generations to reached = 200, sliding window size = 252.

4   Comparative Experimental Results

4.1   Benchmark Models

The simple benchmark trading strategies which were used for comparison reasons were Naive Strategy and Buy and Hold Strategy. Buy and Hold is a simple strategy, where traders buy the index (asset) at the beginning of the review period and sell it at the end of a predetermined period or once a price target has been reached. The naive strategy simply takes the most recent period change as the best prediction of the future change. Then it goes long if the forecasting is positive and short if it is negative.

From the machine learning benchmark models the ones which were used for comparative reasons, were the hybrid combination of Genetic Algorithms and Artificial Neural Networks proposed in [9] and the hybrid methodology combining Genetic Algorithms and Support Vector Machines [10].

Neural networks exist in several forms in the literature. The most popular architecture is the Multilayer Perceptrons (MLP). Their most important problem is that they require a feature selection step and their parameters are hard to be optimized. For these reasons outline by [9] Genetic Algorithms [7] were used to select suitable inputs. The Levenberg–Marquardt back propagation algorithm [11] is employed during the training procedure which adapts the learning rate parameter during this procedure.

In the second machine learning model which was used for comparative reasons [reference] the authors proposed a hybrid GA and SVM model which is designed to overcome some of the limitations of Artificial Neural Networks and simple SVMs. More specifically in this methodology, a genetic algorithm is used to optimize the SVM parameters and on parallel to find the optimal feature subset. Moreover, this approach used a problem specific fitness function which is believed to produce more profitable prediction models.

4.2   Trading Performance

In this section we present the results of each model from trading the FTSE100 index. The trading performance of all the models considered in the out-of-sample subset is presented in table 3. The trading strategy for the GA-MLP and the GA-SVM is simple and identical for both of them: go long on the open when the forecast return is above zero and go short when the forecast return is below zero. Each position is held for only one trading day. The trading strategy for our model is identical with the previous result except it is more selective when trading as we apply a confirmation filter. The confirmation filter restricts the model for trading when the forecasted value is less than the optimal confirmation threshold for its sliding window period. Because non-linear methodologies are stochastic by nature a single forecast is not sufficient enough to represent a credible result. For this reason, an average of ten estimations where executed to represent each model as presented in table 3.

Table 3. Trading Results

	Naive Strategy	Buy and Hold Strategy	GA-MLP	GA-SVM	Proposed Method
Annualized Return (excluding costs)	6,68%	-1,64%	11,61%	16,27%	**23,56%**
Positions Taken (annualized)	64	0,5	56	60	**47**
Transaction Costs	6,41%	0,05%	5,66%	6,01%	**4,7%**
Annualized Return (including costs)	0,27%	-1,69%	5,94%	10,26%	**18,86%**
Annualized Volatility	17,98%	18,04%	18,04%	18,03%	**13,69%**
Information Ratio (including costs)	0,01	-0,09	0,33	0,57	**1,38**
Maximum Drawdown	-31,29%	-22,42%	-30,15%	-30,15%	**-15,32%**
Correct Directional Change	49,30%	50,10%	53,28	54,08%	**56,75%**

As it was expected the proposed methodology clearly outperformed the existing models with leading results across all the examined metrics.

To further examine the findings of the proposed methodology, table 4 presents the percentage of which each input was selected during the sliding windows training period.

Table 4. Percentage of Selection for each Variable

Number	Variable	Percentage of selection
1	Daily High of FTSE100 all share return	45,02%
2	Daily Low of FTSE100 all share return	49,40%
3	FTSE100 all share return	43,43%
4	FTSE100 all share return	35,06%
5	FTSE100 all share return	58,57%
6	FTSE100 all share return	46,61%
7	FTSE100 all share return	39,44%
8	FTSE100 all share return	45,02%
9	FTSE100 all share return	37,45%
10	5 day Moving Average of FTSE100 all share returns	58,96%
11	15 day Moving Average of FTSE100 all share returns	50,20%
12	30 day Moving Average of FTSE100 all share returns	43,43%
13	FTSE100 daily volume	49,80%
14	5 day Moving Average of FTSE100 daily volume	46,61%
15	15 day Moving Average of FTSE100 daily volume	33,07%
16	30 day Moving Average of FTSE100 daily volume	31,47%
17	VIX Realized Volatility Index	49,40%
18	5-day Moving Average of the VIX Index	51,00%
19	15-day Moving Average of the VIX Index	43,82%
20	30-day Moving Average of the VIX Index	39,04%
21	Aggregate Advancing Volume Percentage metric	52,59%
22	5-day Moving Average of Aggregate Advancing Volume Percentage metric	47,81%
23	15-day Moving Average of Aggregate Advancing Volume Percentage metric	47,81%
24	30-day Moving Average of Aggregate Advancing Volume Percentage metric	45,82%
25	Aggregate Declining Volume Percentage metric	49,40%
26	5-day Moving Average of Aggregate Declining Volume Percentage metric	46,22%
27	15-day Moving Average of Aggregate Declining Volume Percentage metric	34,26%
28	30-day Moving Average of Aggregate Declining Volume Percentage metric	40,24%

From table 5 we see which inputs were more influential in explaining the directional changes of the FTSE100 during the period of January 2008 – December 2012. Although a relatively short time period was examined for the moving average series it can be seen that the shorter term moving average series offer more explanation for daily variations of the FTSE100 index. One possible explanation for this finding is that long term moving averages converge to a constraint value which is slightly varying.

5   Conclusions and Future Work

In the present paper we introduced a novel methodology for acquiring profitable and accurate trading strategies when speculatively trading the FTSE100 index. This methodology is a sliding window combination of an adaptive Differential Evolution with nuSVRs. It not only addresses the limitations of existing non-linear models but it also displays the benefits of using an adaptive hybrid approach to utilizing two algorithms. Furthermore, this investigation also fills a gap in current financial forecasting and trading literature. This was accomplished by using a specialized fitness function and deploying differential evolution to optimize the confirmation threshold of the applied trading strategy on parallel of optimizing the nu-SVR model.

Experimental results proved that the proposed technique clearly outperformed the examined linear and machine learning techniques in terms of an information ratio and net annualized return. This technique is now a proven and profitable technique when applied to forecasting a major equity index. Further applications will be made to test the robustness of our model by trading other equity indices and a wider range of asset classes. In addition, the universe of inputs will be expanded in future research to include returns from specific stocks, fixed income time series, commodities and various other explanatory variables.

References

1. Li, Y., Ma, W.: Applications of Artificial Neural Networks in Financial Economics: A Survey. In: Proceedings of International Symposium on Computational Intelligence and Design (ISCID), vol. 1, pp. 211–214 (2010)
2. Vapnik, V.N.: The Nature of Statistical Learning Theory. Springer, United States of America (2000)
3. Das, S., Suganthan, P.: Differential Evolution: A survey of the State-of-the-Art. IEEE Transactions on Evolutionary Computation 15(1), 4–30 (2011)
4. Chang, C., Lin, C.: Training nu-support vector regression: theory and algorithms. Neural Computation 14(8), 1959–1977 (2002)
5. Ince, H., Trafalis, T.: Short Term Forecasting with Support Vector Machines and Application to Stock Price Prediction. International Journal of General Systems 37(6), 677–687 (2008)
6. Huang, W., Nakamori, Y., Wang, S.: Forecasting Stock Market Movement Direction With Support Vector Machine. Computers & Operations Research 32, 2513–2522 (2005)
7. Holland, J.: Adaptation in Natural and Artificial Systems: An Introductory Analysis with Applications to Biology, Control and Artificial Intelligence. MIT Press, Cambridge (1995)
8. Qin, A., Huang, V., Suganthan, P.: Differential evolution algorithm with strategy adaptation for global numerical optimization. IEEE Transactions of Evolutionary Computation 13(2), 398–417 (2009)

9. Karathanasopoulos, A.S., Theofilatos, K.A., Leloudas, P.M., Likothanassis, S.D.: Modeling the ase 20 greek index using artificial neural nerworks combined with genetic algorithms. In: Diamantaras, K., Duch, W., Iliadis, L.S. (eds.) ICANN 2010, Part I. LNCS, vol. 6352, pp. 428–435. Springer, Heidelberg (2010)

10. Dunis, C., Likothanassis, S., Karathanasopoulos, A., Sermpinis, G., Theofilatos, K.: A hybrid genetic algorithm-support vector machine approach in the task of forecasting and trading. Journal of Asset Asset Management 14, 52–71 (2013)

11. Burney, S., Jilani, T., Ardil, C.: Levenberg–Marquardt algorithm for karachi stock exchange share rates forecasting. World Academy of Science, Engineering and Technology 3, 171–176.5 (2005)

Gene Expression Programming and Trading Strategies

Georgios Sermpinis[1], Anastasia Fountouli[1], Konstantinos Theofilatos[2],
and Andreas Karathanasopoulos[3]

[1] University of Glasgow Business School, University of Glasgow, Adam Smith Building,
Glasgow, G12 8QQ, UK
georgios.sermpinis@glasgow.ac.uk, natfou86@hotmail.com
[2] Pattern Recognition Laboratory, Dept. of Computer Engineering & Informatics,
University of Patras, 26500, Patras, Greece
theofilk@ceid.upatras.gr
[3] Business School, University of East London, Unighted Kingdom
a.karathanasopoulos@uel.ac.uk

Abstract. This paper applies a Gene Expression Programming (GEP) algorithm
to the task of forecasting and trading the SPDR Down Jones Industrial Average
(DIA), the SPDR S&P 500 (SPY) and the Powershares Qqq Trust Series 1
(QQQ) exchange traded funds (ETFs). The performance of the algorithm is
benchmarked with a simple random walk model (RW), a Moving Average
Convergence Divergence (MACD) model, a Genetic Programming (GP) algorithm, a Multi-Layer Perceptron (MLP), a Recurrent Neural Network (RNN)
and a Gaussian Mixture Neural Network (GM). The forecasting performance of
the models is evaluated in terms of statistical and trading efficiency. Three trading strategies are introduced to further improve the trading performance of the
GEP algorithm. This paper finds that the GEP model outperforms all other
models under consideration. The trading performance of GEP is further enhanced when the trading strategies are applied.

Keywords: Genetic Programming, Gene Expression Programming, Daily
Trading, DJIA, S&P500.

1   Introduction

Evolutionary and Genetic Programming are becoming popular forecasting tools with
an increasing number of Finance applications ([1] and [2]). This paper applies a
promising evolutionary algorithm, the GEP, to the task of forecasting and trading the
DIA, SPY and QQQ ETFs. Its performance is benchmarked with a simple random
walk model (RW), a Moving Average Convergence Divergence (MACD) model, a
Genetic Programming (GP) algorithm a Multi-Layer Perceptron (MLP), a Recurrent
Neural Network (RNN) and a Higher Order Neural Network (HONN). All models are
evaluated in terms of statistical accuracy and trading profitability.

The models under study will forecast the one day ahead return of the DIA, SPY
and QQQ ETF. Based on the sign of the forecasted return, a trading signal will be
generated. In order to further improve the trading performance of the GEP algorithm,

H. Papadopoulos et al. (Eds.): AIAI 2013, IFIP AICT 412, pp. 497–505, 2013.

three trading strategies based on one-day ahead volatilities forecasts of the series under study and the size of the forecasts will be introduced.

The five non linear forecasting models will attempt to capture the non linear, non stationary and complex behaviour that dominate financial trading series. The RW and MACD models will act as linear benchmarks. Concerning the trading strategies, the economic rational is twofold. To improve the trading performance of the best performing model (the GEP) and to reduce the risk from the trading signals generated. Complicated trading strategies can be a profitable tool to investors as they consider additional factors that might improve the overall profitability of their portfolio.

The GEP algorithm is a domain-independent technique that runs in various environments. These environments are structured in a manner which approximates problems in order to produce accurate forecasts. GEP is based on the Darwinian principle of reproduction and survival of the fittest. It applies biological genetic operations such as crossover recombination and mutation to identify complex non-linear and non-stationary patterns. [3] and [4] underline that Evolutionary Algorithms address and quantify complex issues as an automated process via programming, which enable computers to process and solve problems.

In financial forecasting although there are several applications of NNs, the empirical evidence of GEP is quite limited with the notable exceptions of [5] and [6]. This can be explained by the complexity of the algorithm compared to NNs (see [7]). Nevertheless, GEP has been applied successfully in other fields of science, such as mining and computing ([8] and [9]). A primitive form of the GEP algorithm, the Genetic Programming (GP) has several financial applications. [10] applied successfully GP in predicting the daily highest and lowest stock prices of six US stocks while [11] applied successfully the same algorithm to the task of forecasting two exchange rates.

NNs are popular forecasting tools in Finance with numerous applications. [12] used NNs to combine volatility forecasts of the US, Canadian, Japanese and UK stock markets and [13] use NNs to forecast and trade successfully the general index of the Madrid Stock Market.

Although there are limited empirical evidence around the forecasting superiority of GEP compared to NNs and GP, their theoretical advantages are great. Genetic Programming (GP) classifies its individuals as non-linear comprising of different shapes and sizes (tree like structures). On the other hand, GEP it also classifies classifies individuals as symbolic strings of fixed size (i.e. chromosomes). GEP clearly distinguishes the differences between the genotype and the phenotype of individuals within a population. In [7] the authors argue that GEP represents not only an individual's genotype, in the form of chromosomes, but also its phenotype as a tree like structure of expressions in order to establish fitness. Compared with NNs, GEP has no risk of getting trapped in local optima and is able to reach the optimal solution faster. The findings of this paper support these arguments as the results show that the GEP algorithm outperforms all other models in terms of statistical accuracy and trading efficiency.

The rest of the paper is organized as follows. Section 2 describes the dataset while section 3 describes the GEP algorithm and the benchmark models while Section 4 displays the empirical results. Section 5 provides some concluding remarks.

2   Dataset

In this study, seven forecasting models are employed to the task of forecasting and trading the one day ahead logarithmic return of the DIA, SPY and QQQ ETFs. ETFs are investment funds that are designed to replicate stock market indices. The DIA, the SPY and the QQQ ETF are designed to mimic the Down Jones Industrial Average, the S&P 500 and NASDAQ 100 stock market indices respectively. ETFs offer investors the opportunity to trade stock market indices with extremely low transaction costs.

The period from 03/09/2002 to 31/12/2008 will act as initial in-sample and the period from 02/01/2009 to 31/12/2012 as out-of-sample period. The parameters of the forecasting models will be optimized during the in-sample period and their performance will be validated to the unknown out-of-sample dataset. This estimation will be rolled forward each year. For example in the start, the models will be trained from 03/09/2002 to 31/12/2008 and validated from 02/01/2009 to 31/12/2009. Then the in-sample period would be rolled forward one year (02/01/2003 to 31/12/2010) and the forecasting models would be validated from 03/01/2011 to 30/12/2011. This rolling forward estimation is conducted three times.

3   Forecasting Models

In this section, the forecasting models under study are discussed.

3.1   The Gene Expression Programming

GEP models are symbolic strings of fixed length that represent the chromosome/genotype (genome) of an organism. They are encoded as non linear entities of different sizes and shapes determining an organism's fitness. GEP chromosomes are consisted by multiple genes with equal lengths across the structure of the chromosome. Each gene is comprised of a head (detailing symbols specific to functions and terminals) and a tail (only includes terminals). These can be represented mathematically by equation [1]:

$$t = (n-1)h + 1 \qquad (1)$$

where:
h = the head length of the gene.
t = the tail length of the gene.
n = total number of arguments within the function (maximum arity)

The set of terminals included within both the heads and tails of the chromosomes is consisted by constants and specific variables. Each gene is equal and fixed in size and they hold the capacity to code for multiple and varied expression trees (ET). The structure of GEP is able to cope in circumstances when the first element of a gene is terminal producing a single node as well as when multiple nodes ('sub-trees' reproduced by functions) are produced in search for eventual terminality. In GEP valid ETs are always generated while in GP this is not guaranteed. Each gene encodes an ET and in situations where multiple generations arise, GEP codes for sub ETs with interlinking functions to enable reproduction of generations. The parameters of our GEP algorithm are based on the guidelines of [5] and [7]. The different steps of the algorithm are explained below:

In the beginning GEP randomly generates an initial population from populations of individuals and all succeeding populations are spawned from this initial population. The size of the initial population in this application is set to 1000. In the spawning of new generations genetic operators evolve each of the individuals by 'mating' them with other individuals in the population. These genetic operators are deciphered by the nature of the problem. In the next step we develop expression trees from our chromosomes. The structure of each ET is in such a way that the root or the first node corresponds with beginning of each gene. The resulting offspring evolved from the first node is dependent on the number of arguments. In this process of evolution the functions may have numerous arguments however the terminals have a parity of zero. Each of the resulting offspring's characteristics is populated in nodes ordered from left to right. This process is concluded once terminal nodes are established. Later the initial population is generated and the resulting ETs are developed. This is explained in detail by [7]. In order to create an accurate model suited to our forecasting requirements it is imperative that a function which minimizes error and improves accuracy is used. In our application, the fitness value is defined as the MSE with the lowest MSE being targeted as the best. The main principal during the process of evolution is the generation of offspring from two superior individuals to achieve 'elitism'. As a consequence the best individuals from the parent generation produce offsprings with the most desirable traits whilst the individuals with less desirable traits are removed. On this basis our model minimizes error and maintains superior forecasting abilities. As explained in greater detail by [7], elitism is the cloning of the best chromosome(s)/individual(s) to the next population (also called generation). The role of 'elitism' (via suited genetic operators) enables the selection of fitter individuals without eliminating the entire population. The selection of individuals based on their 'fitness' is carried out during the 'tournament' selection for reproduction and modification. This process selects the individuals at random with the superior ones being chosen for genetic modification in order to create new generations. The size of each tournament is set to 20. In the reproduction of future generations, we apply the mutation and recombination genetic operators. Then the tournament losers are replaced with the new individuals created by the reproduction in the population. A check is made to determine whether the termination criterion is fulfilled, if it is not we return to the second step. As a termination criterion we used a maximum number of 100,000 generations during which the GEP was left to run. As a result the best individual found during the evolution process is presented.

A RW and a MACD model will act as linear benchmarks to the GEP algorithm while three Neural Network (NN) models and a GP algorithm will act as non linear benchmarks.

3.2   Non Linear Benchmarks

Three Neural Network (NN) models and a GP algorithm will act as non linear benchmarks. All four models are well documented in the literature.

NNs are usually consisted by three or more layers. The first layer is called the input layer (the number of its nodes corresponds to the number of indepedent variables). In this study the inputs were selected among the first 12 autoregressive lags of the forecasting series. The specific choice of each set of inputs was based on a sensitivity analysis in the in-sample period.   The last layer is called the output layer (the number of its nodes corresponds to the number of response variables). An intermediary layer of nodes, the hidden layer, separates the input from the output layer. Its number of nodes defines the amount of complexity the model is capable of fitting. Normally, each node of one layer has connections to all the other nodes of the next layer.   The training of the network (which is the adjustment of its weights in the way that the network maps the input value of the training data to the corresponding output value) starts with randomly chosen weights and proceeds by applying a learning algorithm called backpropagation of errors ([14]).The iteration length is optimised by minimising the MSE in a subset of in-sample dataset (the last year of the in-sample period each time). The most popular architecture NN model is the MLP. RNNs have the ability to embody short-term memory and past errors while HONNs are able to capture higher order correlations (up to the order three or four) within the dataset.

GP are algorithms that evolve algebraic expressions which enable the analysis / optimization of results in a 'tree like structure'. A complete description of GEP is provided by [4].

4   Empirical Results

4.1   Statistical Performance

In the table below, the statistical performance in the out-of-sample period of all models is presented. For the the Root Mean Squared Error (RMSE), Mean Absolute Error (MAE) and Theil-U statistics, the lower the output the better the forecasting accuracy of the model concerned. The Pesaran-Timmermann (PT) test (1992) examines whether the directional movements of the real and forecast values are in step with one another. It checks how well rises and falls in the forecasted value follow the actual rises and falls of the time series. The null hypothesis is that the model under study has no power on forecasting the ETF return series. The Diebold-Mariano (1995) DM statistic for predictive accuracy statistic tests the null hypothesis of equal predictive accuracy. Both the DM and the PT tests follow the standard normal distribution.

Table 1. Out-of-sample statistical performance

		RW	MACD	GP	MLP	RNN	HONN	GEP
DIA	*RMSE*	1.027	0.239	0.073	0.136	0.085	0.069	0.057
	MAE	0.833	0.162	0.059	0.097	0.077	0.054	0.046
	Theil-U	0.989	0.756	0.634	0.673	0.641	0.620	0.601
	PT	0.02	0.35	5.89	5.02	5.68	6.44	6.99
	DM	-14.85	-10.47	-4.62	-5.91	-5.73	-4.45	-
SPY	*RMSE*	1.021	0.287	0.068	0.143	0.072	0.064	0.053
	MAE	0.831	0.195	0.055	0.103	0.061	0.052	0.042
	Theil-U	0.977	0.792	0.621	0.686	0.672	0.613	0.597
	PT	0.05	0.30	6.48	4.85	5.79	6.52	7.11
	DM	-14.56	-11.41	-5.59	-6.65	-5.97	-5.08	-
QQQ	*RMSE*	1.022	0.295	0.069	0.127	0.071	0.067	0.055
	MAE	0.834	0.204	0.058	0.088	0.059	0.055	0.044
	Theil-U	0.987	0.799	0.632	0.658	0.644	0.616	0.599
	PT	0.08	0.25	6.14	5.19	5.73	6.50	7.08
	DM	-15.31	-12.04	-.5.42	-5.81	5.70	-4.77	-

From the table above we note that GEP outperforms all benchmarks for all the statistical measures retained. The HONN model presents the second best performance while the GP algorithm produces the third more statistically accurate forecasts. The PT statistics indicate that all non-linear models under study are able to forecast accurately the directional movements of the three ETF return series while the DM statistics confirm the statistical superiority of the GEP forecasts. On the other hand, the two statistical benchmarks seem unable to provide statistically accurate forecasts for the series and period under study.

4.2   Trading Performance

The trading performance of our models in the out-of-sample period is presented in the table below.

Table 2. Out-of-sample trading performance

		MACD	GP	MLP	RNN	HONN	GEP
DJIA	*Information Ratio*	-0.17	0.76	0.63	0.67	0.79	0.94
	Annualised Return (including costs)	2.16%	15.22%	10.01%	12.94%	15.92%	23.55%
	Maximum Draw-down	-23.24%	-22.04%	-20.91%	-19.95%	-21.42%	-17.87%
S&P 500	*Information Ratio*	-0.24	0.85	0.51	0.74	0.92	1.28
	Annualised Return (including costs)	-2.12%	19.57%	7.65%	16.79%	19.67%	26.68%
	Maximum Draw-down	-36.16%	-30.12%	-35.92%	-34.16%	-32.54%	-26.56%
NASDAQ 100	*Information Ratio*	-0.38	0.78	0.70	0.76	0.86	1.21
	Annualised Return (including costs)	-3.65%	17.57%	14.51%	16.42%	19.54%	24.02%
	Maximum Draw-down	-32.85%	-28.19%	-30.81%	-31.91%	-27.91%	-24.41%

From the table above we note that GEP clearly outperforms its benchmarks are all the trading criteria retained. In the next section, two trading strategies are introduced to further increase the trading performance of GEP.

4.2.1   Time Varying Volatility Leverage

In order to further improve the trading performance of our models we introduce a leverage based on one day ahead volatility forecasts. The intuition of the strategy is to exploit the changes in the volatility. As a first step we forecast the one day ahead exchange rate volatility with a GARCH, GJR, RiskMetrics and EGARCH model in the test and validation sub-periods. Then, we split these two periods into six sub-periods, ranging from periods with extremely low volatility to periods experiencing extremely high volatility. Periods with different volatility levels are classified in the following way: first the average (μ) difference between the actual volatility in day t and the forecast for day t+1 and its 'volatility' (measured in terms of standard deviation (σ) are calculated; those periods where the difference is between μ plus one σ are classified as 'Lower High Vol. Periods'. Similarly, 'Medium High Vol.' (between $\mu + \sigma$ and $\mu + 2\sigma$) and 'Extremely High Vol.' (above $\mu + 2\sigma$ periods can be defined. Periods with low volatility are also defined following the same 1 ζ and 2 σ approaches, but with a minus sign. For each sub-period leverage is assigned starting with 0 for periods of extremely high volatility to a leverage of 2.5 for periods of extremely low volatility.

4.2.2   The Strongest Signal

The second trading strategy is based on the absolute values of our forecasts. The GEP algorithm is forecasting the one-day ahead return of the three indices. In this trading strategy, each day we invest to the index that our GEP algorithm is giving the highest in absolute value return or else the strongest signal. Instead of investing in the three indices based on our GEP forecasts, we invest only to the one that the algorithm indicates that will be the most profitable. These trades are held for one day unless in the following day, our GEP signals indicate the same index with the same sign.

4.3   Empirical Results

In the table below, we present the performance of our trading strategies.

Table 3. Trading Strategies

	GEP-DJIA	GEP-S&P500	GEP-NASDAQ 100	GEP-Level of Confidence
Information Ratio	1.03	1.32	1.29	1.96
Annualised Return (including costs)	26.85%	28.14%	26.98%	34.54%
Maximum Drawdown	-18.95%	-28.95%	-22. 95%	-25.81%

We note that all trading strategies were successful.

5   Conclusions

In this study, a GEP algorithm was applied to the task of forecasting and trading the DJIA, S&P 500 and NASDAQ 100 indices. It was benchmarked against several non linear models. The GEP forecasts outperformed its benchmarks in terms of annualised return and information ratio. This trading performance was further enhanced with the introduction of two trading strategies.

These results should go a step towards convincing a growing number of quantitative fund managers and academics to experiment beyond the bounds of traditional statistical and neural network models. They also validate the importance of trading strategies in improving the profitability of trading signals.

References

1. Chen, S.: Genetic Algorithms and Genetic Programming in Computational Finance. Kluwer Academic Publishers, Amsterdam (2002)
2. Lisboa, P., Edisbury, B., Vellido, A.: 'Business Applications of Neural Networks', Business Applications of Neural Networks: The State-of-the-Art of Real-World Applications. World Scientific, Singapore (2000)
3. Koza, J.R.: Genetic Programming: On the Programming of Computers by Means of Natural Selection. MIT Press, Cambridge (1992)
4. Koza, J.R.: Genetic Programming. In: Williams, J.G., Kent, A. (eds.) Encyclopedia of Computer Science and Technology, vol. 39(suppl. 24), pp. 29–43. Marcel-Dekker, New York (1998)
5. Sermpinis, G., Laws, J., Karathanasopoulos, A., Dunis, C.: Forecasting and Trading the EUR/USD Exchange Rate with Gene Expression and Psi Sigma Neural Networks. Expert Systems with Applications, 8865–8877 (2012)
6. Divsalar, et al.: A Robust Data-Mining Approach to Bankruptcy Prediction. Journal of Forecasting 31, 504–523 (2012)
7. Ferreira, C.: Gene Expression Programming: A New Adaptive Algorithm for Solving Problems. Complex Systems 13, 87–129 (2001)
8. Lopez, H.S., Weinert, W.R.: An Enhanced Gene Expression Programming Approach for Symbolic Regression Problems. International Journal of Applied Mathematics in Computer Science 14, 375–384 (2004)
9. Dehuri, S., Cho, S.B.: Multi-Objective Classification Rule Mining Using Gene Expression. In: Third International Conference on Convergence and Hybrid Information (2008)
10. Kaboudan, M.A.: Genetic Programming Prediction of Stock Prices. Computational Economics 16, 207–236 (2000)
11. Alvarez-Díaz, M., Alvarez, A.: Genetic multi-model composite forecast for non-linear prediction of exchange rates. Empirical Economics 30(3), 643–663 (2005)
12. Donaldson, R.G., Kamstra, M.: Forecast Combining with Neural Networks. Journal of Forecasting 15, 49–61 (1996)
13. Fernandez-Rodríguez, F., González-Martel, C., Sosvilla-Rivero, S.: On the profitability of technical trading rules based on artificial neural networks: Evidence from the Madrid stock market. Economics Letters 69(1), 89–94 (2000)
14. Shapiro, A.F.: A Hitchhiker's Guide to the Techniques of Adaptive Nonlinear Models. Insurance, Mathematics and Economics 26(2), 119–132 (2000)

Kalman Filter and SVR Combinations
in Forecasting US Unemployment

Georgios Sermpinis[1], Charalampos Stasinakis[1], and Andreas Karathanasopoulos[2]

[1] University of Glasgow Business School
georgios.sermpinis@glasgow.ac.uk,
c.stasinakis.1@research.gla.ac.uk
[2] University of East London Business School
a.karatahnasopoulos@uel.ac.uk

Abstract. The motivation for this paper is to investigate the efficiency of a Neural Network (NN) architecture, the Psi Sigma Network (PSN), in forecasting US unemployment and compare the utility of Kalman Filter and Support Vector Regression (SVR) in combining NN forecasts. An Autoregressive Moving Average model (ARMA) and two different NN architectures, a Multi-Layer Perceptron (MLP) and a Recurrent Network (RNN), are used as benchmarks. The statistical performance of our models is estimated throughout the period of 1972-2012, using the last seven years for out-of-sample testing. The results show that the PSN statistically outperforms all models' individual performances. Both forecast combination approaches improve the statistical accuracy, but SVR outperforms substantially the Kalman Filter.

Keywords: Forecast Combinations, Kalman Filter, Support Vector Regression, Unemployment.

1   Introduction

Many applications in the macroeconomic literature aim to derive and compare information from econometric models' forecasts. For that reason, forecasting competitions of linear and non-linear architectures are common and focus on numerous time series, such as unemployment, inflation, industrial production, gross domestic product etc. The artificial NNs are computation models that researchers include in such macroeconomic forecasting schemes, because they embody promising data-adaptive learning and clustering abilities.

The motivation for this paper is to investigate the efficiency of a Neural Network (NN) architecture, the Psi Sigma Network (PSN), in forecasting US unemployment and compare the utility of Kalman Filter and Support Vector Regression (SVR) in combining NN forecasts. An Autoregressive Moving Average model (ARMA) and two different NN architectures, a Multi-Layer Perceptron (MLP) and a Recurrent Network (RNN), are used as benchmarks. The statistical performance of our models is estimated throughout the period of 1972-2012, using the last seven years for out-of-sample testing. The results show that the PSN statistically outperforms all models'

H. Papadopoulos et al. (Eds.): AIAI 2013, IFIP AICT 412, pp. 506–515, 2013.

individual performances. Both forecast combination approaches improve the statistical accuracy, but SVR is substantially better than the Kalman Filter.

Section 2 is a short literature review and Section 3 follows with the description of the dataset used in this application. Sections 4 and 5 give an overview of the forecasting models and the forecast combination methods implemented respectively. The statistical performance of our models is presented in Section 6. Finally, some concluding remarks are summarized in Section 7.

2   Literature Review

Forecasting unemployment rates is a very popular and well documented case study in the literature (see amongst others Rothman [16], Montgomery et al. [14] and Koop and Potter [11]). Swanson and White [20] forecast several macroeconomic time series, including US unemployment, with linear models and NNs. In their approach, NN architectures present promising empirical evidence against the linear VAR models. Johnes [9] reports the results of another forecasting competition between linear autoregressive, GARCH, threshold autoregressive and NNs models, applied to the UK monthly unemployment rate series. In his application, NNs are superior when the forecasting horizon is 18 and 24 months ahead, but fail to outperform the other models in shorter forecasting horizons.

Liang [12] applies Bayesian NNs in forecasting unemployment in West Germany and shows that they present significantly better forecasts than traditional autoregressive models. Teräsvirta et al. [22] examine the forecast accuracy of linear autoregressive, smooth transition autoregressive and NN models for 47 monthly macroeconomic variables, including unemployment rates, of the G7 economies. The empirical results of their study point out the risk for implausible NN forecasts at long forecasting horizons. Nonetheless, their forecasting ability is much improved when they are combined with autoregressive models. This idea of combining forecasts originates from Bates and Granger [1]. Newbold and Granger [15] also suggested combining rules based on variances-covariances of the individual forecasts, while Deutsch et al. [3] achieved substantially smaller squared forecasts errors combining forecasts with changing weights. Harvey [8] and Hamilton [7] both propose using state space modeling, such as Kalman Filter, for representing dynamic systems where unobserved variables (so-called 'state' variables) can be integrated within an 'observable' model. Finally, Terui and Van Dijk [23] also suggest that the combined forecasts perform well, especially with time varying coefficients.

3   US Unemployment Dataset

In this application, we forecast the monthly percentage change of the US unemployment rate (UNEMP), as provided by the online Federal Reserve Economic Data

(FRED) database of the Federal Reserve Bank of St. Louis[1]. The forecasting performance of our models is explored over the period of 1972 to 2012, using the last seven
years for out-of-sample evaluation. The time series is seasonally adjusted and it is
divided into three sub-periods as summarized in Table 1 below:

Table 1. The US Unemployment Dataset - Neural Networks' Training Dataset

PERIODS	MONTHS	START DATE	END DATE
Total Dataset	492	01//01/1972	01/12/2012
Training Dataset *(In-sample)*	324	01//01/1972	01/12/1998
Test Dataset *(In-sample)*	84	01/01/1999	01/12/2005
Validation Dataset *(Out-of-sample)*	84	01/01/2006	01/12/2012

The following graph presents the US unemployment rate for the period under
study:

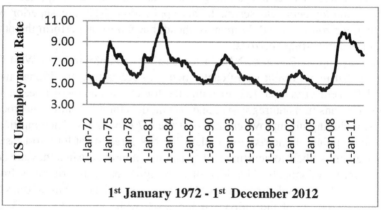

Fig. 1. The US Unemployment Rate

In the absence of any formal theory behind the selection of the inputs of a NN, we
conduct some NN experiments and a sensitivity analysis on a pool of potential inputs
in the in-sample dataset in order to help our decision. Finally, we select as inputs sets
of autoregressive terms of UNEMP that provide the best statistical performance for
each network in the test sub-period. These sets are presented in Table 2 below:

---

[1] Based on the description given by FRED, the US unemployment rate or civilian unemployment rate represents the number of unemployed as a percentage of the labour force. Labour
force data are restricted to people 16 years of age and older, who currently reside in 1 of the
50 states or the District of Columbia, who do not reside in institutions (e.g., penal and mental
facilities, homes for the aged) and who are not on active duty in the Armed Forces.

Table 2. Neural Networks' Inputs

MLP	RNN	PSN
UNEMP (1)*	UNEMP (1)	UNEMP (1)
UNEMP (2)	UNEMP (3)	UNEMP (2)
UNEMP (4)	UNEMP (4)	UNEMP (3)
UNEMP (5)	UNEMP (6)	UNEMP (6)
UNEMP (7)	UNEMP (7)	UNEMP (8)
UNEMP (10)	UNEMP (9)	UNEMP (10)
UNEMP (11)	UNEMP (11)	UNEMP (11)
UNEMP (12)	UNEMP (12)	UNEMP (12)

*UNEMPL UNEMP (1) is the first autoregressive term of the UNEMP series

4   Forecasting Models

4.1   Auto-Regressive Moving Average Model (ARMA)

In this paper an ARMA model is used to benchmark the efficiency of the NNs' statistical performance. Using as a guide the correlogram and the information criteria in the in-sample subset, we have chosen a restricted ARMA (7, 7) model, where all the coefficients are significant at the 95% confidence interval. The selected ARMA model is presented in equation (1) below:

$$\hat{Y}_t = 0.03 + 1.025Y_{t-1} - 0.293Y_{t-2} + 0.511Y_{t-4} - 0.321Y_{t-7} - 1.006\varepsilon_{t-1} + 0.463\varepsilon_{t-2} - 0.545\varepsilon_{t-4} - 0.211\varepsilon_{t-7} \quad (1)$$

where $\hat{Y}_t$ is the forecasted percentage change of the US unemployment rate.

4.2   Neural Networks (NNs)

Neural networks exist in several forms in the literature. The most popular architecture is the Multi-Layer Perceptron (MLP). A standard neural network has at least three layers. The first layer is called the input layer (the number of its nodes corresponds to the number of explanatory variables). The last layer is called the output layer (the number of its nodes corresponds to the number of response variables). An intermediary layer of nodes, the hidden layer, separates the input from the output layer. Its number of nodes defines the amount of complexity the model is capable of fitting. In addition, the input and hidden layer contain an extra node called the bias node. This node has a fixed value of one and has the same function as the intercept in traditional regression models. Normally, each node of one layer has connections to all the other nodes of the next layer.

The network processes information as follows: the input nodes contain the value of the explanatory variables. Since each node connection represents a weight factor, the information reaches a single hidden layer node as the weighted sum of its inputs. Each node of the hidden layer passes the information through a nonlinear activation function and passes it on to the output layer if the calculated value is above a threshold.

The training of the network (which is the adjustment of its weights in the way that the network maps the input value of the training data to the corresponding output value) starts with randomly chosen weights and proceeds by applying a learning algorithm called backpropagation of errors [18].The learning algorithm simply tries to find those weights which minimize an error function (normally the sum of all squared differences between target and actual values). Since networks with sufficient hidden nodes are able to learn the training data (as well as their outliers and their noise) by heart, it is crucial to stop the training procedure at the right time to prevent overfitting (this is called 'early stopping'). This can be achieved by dividing the dataset into three subsets respectively called the training and test sets used for simulating the data currently available to fit and tune the model and the validation set used for simulating future values. The training of a network is stopped when the mean squared forecasted error is at minimum in the test-sub period. The network parameters are then estimated by fitting the training data using the above mentioned iterative procedure (backpropagation of errors). The iteration length is optimised by maximising the forecasting accuracy for the test dataset. Then the predictive value of the model is evaluated applying it to the validation dataset (out-of-sample dataset).

4.2.1   The Multi-Layer Perceptron Model (MLP)
MLPs are feed-forward layered NN, trained with a back-propagation algorithm. According to Kaastra and Boyd [10], they are the most commonly used types of artificial networks in financial time-series forecasting. The training of the MLP network is processed on a three-layered architecture, as described above.

4.2.2   The Recurrent Neural Network (RNN)
The next NN architecture used in this paper is the RNN. For an exact specification of recurrent networks, see Elman [5]. A simple recurrent network has an activation feedback which embodies short-term memory. The advantages of using recurrent networks over feed-forward networks for modeling non-linear time series have been well documented in the past. However, as mentioned by Tenti [21], "the main disadvantage of RNNs is that they require substantially more connections, and more memory in simulation than standard back-propagation networks" (p. 569), thus resulting in a substantial increase in computational time.

4.2.3   The Psi-Sigma Neural Network (PSN)
The PSNs are a class of Higher Order Neural Networks with a fully connected feed-forward structure. Ghosh and Shin [6] were the first to introduce the PSN, trying to reduce the numbers of weights and connections of a Higher Order Neural Network. Their goal was to combine the fast learning property of single-layer networks with the mapping ability of Higher Order Neural Networks and avoid increasing the required number of weights. The price for the flexibility and speed of Psi Sigma networks is that they are not universal approximators. We need to choose a suitable order of

approximation (or else the number of hidden units) by considering the estimated function complexity, amount of data and amount of noise present. To overcome this, our code runs simulations for orders two to six and then it presents the best network. The evaluation of the PSN model selected comes in terms of trading performance.[2]

5   Forecast Combination Techniques

5.1   Kalman Filter

Kalman Filter is an efficient recursive filter that estimates the state of a dynamic system from a series of incomplete and noisy measurements. The time-varying coefficient combination forecast suggested in this paper is shown below:

$$\text{Measurement Equation: } f^t_{c_{NNs}} = \sum_{i=1}^{3} a^t_i f^t_i + \varepsilon_t, \quad \varepsilon_t \sim NID\left(0, \sigma_\varepsilon^2\right) \tag{2}$$

$$\text{State Equation: } a^t_i = a^{t-1}_i + n_t, \quad n_t \sim NID(0, \sigma_n^2) \tag{3}$$

Where:

- $f^t_{c_{NNs}}$ is the dependent variable (combination forecast) at time t
- f^t_i $(i = 1, 2, 3)$ are the independent variables (individual forecasts) at time t
- a^t_i $(i = 1, 2, 3)$ are the time-varying coefficients at time t for each NN
- ε_t, n_t are the uncorrelated error terms (noise)

The alphas are calculated by a simple random walk and we initialized $\varepsilon_1 = 0$. Based on the above, our Kalman Filter model has as a final state the following:

$$f^t_{c_{NNs}} = 13.46 f^t_{MLP} + 16.38 f^t_{RNN} + 41.97 f^t_{PSN} + \varepsilon_t \tag{4}$$

From the above equation we note that the Kalman filtering process also favors the PSN model, which is the model that performs best individually.

5.2   Support Vector Regression (SVR)

Vapnik [24] established Support Vector Regression (SVR) as a robust technique for constructing data-driven and non-linear empirical regression models. They provide global and unique solutions and do not suffer from multiple local minima (Suykens [19]). They also present a remarkable ability of balancing model accuracy and model complexity (and Lu et al.[13]). The SVR function can be specified as:

$$f(x) = w^T \varphi(x) + b \tag{5}$$

---

[2] For a complete description of all the neural network models we used and their complete specifications see Sermpinis et al. [17].

where w and b are the regression parameter vectors of the function and $\varphi(x)$ is the non-linear function that maps the input data vector x into a feature space where the training data exhibit linearity. The ε-sensitive loss $L\varepsilon$ function finds the predicted points that lie within the tube created by two slack variables ξ_i, ξ_i^* :

$$L_\varepsilon(x_i) = \begin{cases} 0 \ if \ |y_i - f(x_i)| \le \varepsilon \\ |y_i - f(x_i)| - \varepsilon \ if \ other \end{cases}, \varepsilon \ge 0 \tag{6}$$

In other words ε is the degree of model noise insensitivity and L_ε finds the predicted values that have at most ε deviations from the actual obtained values y_i. The goal is to solve the following argument[3]:

$$\text{Minimize } C\sum_{i=1}^{n}(\xi_i + \xi_i^*) + \frac{1}{2}\|w\|^2 \text{ subject} \qquad \text{to} \qquad \begin{cases} \xi_i \ge 0 \\ \xi_i^* \ge 0 \\ C > 0 \end{cases} \qquad \text{and}$$

$$\begin{cases} y_i - w^T\varphi(x_i) - b \le +\varepsilon + \xi_i \\ w^T\varphi(x_i) + b - y_i \le +\varepsilon + \xi_i^* \end{cases} \tag{7}$$

The above quadratic optimization problem is transformed in a dual problem and its solution is based on the introduction of two Lagrange multipliers a_i, a_i^* and mapping with a kernel function $K(x_i, x)$:

$$f(x) = \sum_{t=1}^{n}(a_i - a_i^*)K(x_i, x) + b \text{ where } 0 \le a_i, a_i^* \le C \tag{8}$$

Support Vectors (SVs) are called all the xi that contribute to equation (8), thus they lie outside the ε-tube, whereas non-SVs lie within the ε-tube. Increasing ε leads to less SVs' selection, whereas decreasing it results to more 'flat' estimates. The norm term $\|w\|^2$ characterizes the complexity (flatness) of the model and the term

$$\left\{ \sum_{i=1}^{n}(\xi_i + \xi_i^*) \right\}$$

is the training error, as specified by the slack variables. Consequently the introduction of the parameter C satisfies the need to trade model complexity for training error and vice versa (Cherkassky and Ma [2]). In our application, the NN forecasts are used as inputs for a ε-SVR simulation. A RBF kernel[4] is selected and the parameters have been optimized based on cross-validation in the in-sample dataset (ε=0.06, γ= 2.47 and C=0.103).

---

[3] For a detailed mathematical analysis of the SVR solution see Vapnik [24].
[4] The RBF kernel equation is $K(x_i, x) = \exp(-\gamma\|x_i - x\|^2), \gamma > 0$.

6   Statistical Performance

As it is standard in literature, in order to evaluate statistically our forecasts, the RMSE, the MAE, the MAPE and the Theil-U statistics are computed (see Dunis and Williams [4]). For all four of the error statistics retained the lower the output, the better the forecasting accuracy of the model concerned. In Table 3 we present the statistical performance of all our models in the in-sample period.

Table 3. Summary of the In-Sample Statistical Performance

	ARMA	MLP	RNN	PSN	Kalman Filter	SVR
MAE	1.9941	0.0078	0.0077	0.0073	0.0067	0.0062
MAPE	65.25%	52.78%	50.17%	47.73%	45.76%	41.52%
RMSE	2.5903	1.0671	0.9572	0.9045	0.8744	0.8256
Theil-U	0.6717	0.6142	0.5827	0.5325	0.5017	0.4549

We note that from our individual forecasts, the PSN statistically outperforms all other models. Both forecast combination techniques improve the forecasting accuracy, but SVR is the superior model regarding all four statistical criteria. Table 4 below summarizes the statistical performance of our models in the out-of-sample period.

Table 4. Summary of the Out-of-sample Statistical Performance

	ARMA	MLP	RNN	PSN	Kalman Filter	SVR
MAE	0.0332	0.0072	0.0071	0.0061	0.0057	0.0051
MAPE	67.45%	50.17%	48.97%	44.38%	40.21%	34.33%
RMSE	2.4043	0.9557	0.9354	0.8927	0.8549	0.8005
Theil-U	0.5922	0.5654	0.5591	0.4818	0.4657	0.4154

From the results above, it is obvious that the statistical performance of the models in the out-of-sample period is consistent with the in-sample one and their ranking remains the same. All NN models outperform the traditional ARMA model. In addition, the PSN outperforms significantly the MLP and RNN in terms of statistical accuracy. The idea of combining NN unemployment forecasts seems indeed very promising, since both Kalman Filter and SVR present improved statistical accuracy also in the out-of-sample period. Moreover, SVR confirms its forecasting superiority over the individual architectures and the Kalman Filter technique. In other words, SVR's adaptive trade-off between model complexity and training error seems more effective than the recursive ability of Kalman Filter to estimate the state of our process.

7   Concluding Remarks

The motivation for this paper is to investigate the efficiency of a Neural Network (NN) architecture, the Psi Sigma Network (PSN), in forecasting US unemployment and compare the utility of Kalman Filter and Support Vector Regression (SVR) in combining NN forecasts. An Autoregressive Moving Average model (ARMA) and two different NN architectures, a Multi-Layer Perceptron (MLP) and a Recurrent Network (RNN), are used as benchmarks. The statistical performance of our models is estimated throughout the period of 1972-2012, using the last seven years for out-of-sample testing.

As it turns out, the PSN outperforms its benchmark models in terms of statistical accuracy. It is also shown that all the forecast combination approaches outperform in the out-of-sample period all our single models. All NN models beat the traditional ARMA model. In addition, the PSN outperforms significantly the MLP and RNN in terms of statistical accuracy. The idea of combining NN unemployment forecasts seems indeed very promising, since both Kalman Filter and SVR present improved statistical accuracy also in the out-of-sample period. SVR confirms its forecasting superiority over the individual architectures and the Kalman Filter technique. In other, SVR's adaptive trade-off between model complexity and training error seems more effective than the recursive ability of Kalman Filter to estimate the state of our process. The remarkable statistical performance of SVR allows us to conclude that it can be considered as an optimal forecast combination for the models and time series under study. Finally, the results confirm the existing literature, which suggests that nonlinear models, such as NNs, can be used to model macroeconomic series.

References

1. Bates, J.M., Granger, C.W.J.: The Combination of Forecasts. Operational Research Society 20, 451–468 (1969)
2. Cherkassky, V., Ma, Y.: Practical selection of SVM parameters and noise estimation for SVM regression. Neural Networks 17, 113–126 (2004)
3. Deutsch, M., Granger, C.W.J., Teräsvirta, T.: The combination of forecasts using changing weights. International Journal of Forecasting 10, 47–57 (1994)
4. Dunis, C.L., Williams, M.: Modelling and Trading the EUR/USD Exchange Rate: Do Neural Network Models Perform Better? Derivatives Use, Trading and Regulation 8, 211–239 (2002)
5. Elman, J.L.: Finding Structure in Time. Cognitive Science 14, 179–211 (1990)
6. Ghosh, J., Shin, Y.: The Pi-Sigma Network: An efficient Higher-order Neural Networks for Pattern Classification and Function Approximation. In: Proceedings of International Joint Conference of Neural Networks, pp. 13–18 (1991)
7. Hamilton, J.D.: Time series analysis. Princeton University Press, Princeton (1994)
8. Harvey, A.C.: Forecasting, structural time series models and the Kalman filter. Cambridge University Press, Cambridge (1990)
9. Johnes, G.: Forecasting unemployment. Applied Economics Letters 6, 605–607 (1999)
10. Kaastra, I., Boyd, M.: Designing a Neural Network for Forecasting Financial and Economic Time Series. Neurocomputing 10, 215–236 (1996)

11. Koop, G., Potter, S.M.: Dynamic Asymmetries in U.S. Unemployment. Journal of Business & Economic Statistics 17, 298–312 (1999)
12. Liang, F.: Bayesian neural networks for nonlinear time series forecasting. Statistics and Computing 15, 13–29 (2005)
13. Lu, C.J., Lee, T.S., Chiu, C.C.: Financial time series forecasting using independent component analysis and support vector regression. Decision Support Systems 47, 115–125 (2009)
14. Montgomery, A.L., Zarnowitz, V., Tsay, R.S., Tiao, G.C.: Forecasting the U.S. Unemployment Rate. Journal of the American Statistical Association 93, 478–493 (1998)
15. Newbold, P., Granger, C.W.J.: Experience with Forecasting Univariate Time Series and the Combination of Forecasts. Journal of the Royal Statistical Society 137, 131–165 (1974)
16. Rothman, P.: Forecasting Asymmetric Unemployment Rates. The Review of Economics and Statistics 80, 164–168 (1998)
17. Sermpinis, G., Laws, J., Dunis, C.L.: Modelling and trading the realised volatility of the FTSE100 futures with higher order neural networks. European Journal of Finance 1–15 (2012)
18. Shapiro, A.F.: A Hitchhiker's guide to the techniques of adaptive nonlinear models. Insurance: Mathematics and Economics 26, 119–132 (2000)
19. Suykens, J.A.K., Brabanter, J.D., Lukas, L., Vandewalle, L.: Weighted least squares support vector machines: robustness and sparse approximation. Neurocomputing 48, 85–105 (2002)
20. Swanson, N.R., White, H.: A Model Selection Approach to Real-Time Macroeconomic Forecasting Using Linear Models and Artificial Neural Networks. Review of Economics and Statistics 79, 540–550 (1997)
21. Tenti, P.: Forecasting foreign exchange rates using recurrent neural networks. Applied Artificial Intelligence 10, 567–581 (1996)
22. Teräsvirta, T., Van Dijk, D., Medeiros, M.C.: Linear models, smooth transition autoregressions, and neural networks for forecasting macroeconomic time series: A re-examination. International Journal of Forecasting 21, 755–774 (2005)
23. Terui, N., Van Dijk, H.K.: Combined forecasts from linear and nonlinear time series models. International Journal of Forecasting 18, 421–438 (2002)
24. Vapnik, V.N.: The nature of statistical learning theory. Springer (1995)

Particle Swarm Optimization Approach for Fuzzy Cognitive Maps Applied to Autism Classification

Panagiotis Oikonomou[1] and Elpiniki I. Papageorgiou[2,*]

[1] University of Thessaly, Computer and Communication Engineering Dept.,
38221 Volos, Greece
paikonom@uth.gr
[2] Technological Educational Institute of Central Greece,
Informatics & Computer Technology Department, 3rd km Old National Road Lamia-Arthens,
35100 Lamia, Greece
epapageorgiou@teilam.gr
http://users.teilam.gr/~epapageorgiou

Abstract. The task of classification using intelligent methods and learning algorithms is a difficult task leading the research community on finding new classifications techniques to solve it. In this work, a new approach based on particle swarm optimization (PSO) clustering is proposed to perform the fuzzy cognitive map learning for classification performance. Fuzzy cognitive map (FCM) is a simple, but also powerful computational intelligent technique which is used for the adoption of the human knowledge and/or historical data, into a simple mathematical model for system modeling and analysis. The aim of this study is to investigate a new classification algorithm for the autism disorder problem by integrating the Particle Swarm Optimization method (PSO) in FCM learning, thus producing a higher performance classification tool regarding the accuracy of the classification, and overcoming the limitations of FCMs in the pattern analysis area.

1 Introduction

Classification is a data processing technique in which each data set is assigned to a predetermined set of categories. Generally classification goal is the creation of a model which will be used later for the prediction-classification of future unknown data. Classification problems have been aroused the interest of researchers of different domains in the last decade like biology, medical, robotic and so on. Such classification paradigms can be the prediction of cancer cell by characterizing them as benign or malignant, the categorization of bank customers according to their reliability, the determination whether a child suffers from the autism disorder problem and so on [1],[2],[3]. Various learning approaches have been proposed for the classification of input instances and for the comprehension of complex systems function, like Artificial Neural Networks, Clustering methods and Genetic Algorithms.

---

* Corresponding author.

H. Papadopoulos et al. (Eds.): AIAI 2013, IFIP AICT 412, pp. 516–526, 2013.

Fuzzy Cognitive Map [4] is a soft computing technique which is used for modeling and analysis of complex systems. FCM may be considered as a simple mathematical model in which the relations between the elements can be used to compute the "strength of impact" of these elements. FCM can also be considered as an integration of multifold subjects, including neural network, fuzzy logic, semantic network, learning algorithms. It is a dynamic tool involving feedback mechanisms [4], and this dynamicity leads the research community to work on it. Due to its advantageous features, such as simplicity, adaptability to system characteristics, support of inconsistent knowledge, analysis of complex systems, learning from historical data and previous knowledge, FCM has found large applicability in many different scientific fields for modeling, control, management and decision making [5].

The FCM learning, as a main capability of FCM, is a crucial issue in modeling and system analysis. It concerns the adaptation of the connection matrix (known as weight matrix) using diverse adaptive and evolutionary type learning methods, such as unsupervised learning based on the Hebbian method [6,7], supervised ones with the use of evolutionary computation [8-11] and/ or gradient-based methods [12,13].

Up to date to the literature, there is no any previous study on proposing a particle swarm optimization approach for FCM to perform classification. Previous studies related with the FCM application in classification tasks are described. The first work was presented by Papakostas et al. (2008) who implemented FCMs for pattern recognition tasks [14]. In their study, a new hybrid classifier was proposed as an alternative classification structure, which exploited both neural networks and FCMs to ensure improved classification capabilities. A simple GA was used to find a common weight set which, for different initial state of the input concepts, the hybrid classifier equilibrate to different points [14]. Next, Arthi et al. analyzed the performance of FCM using Non-linear hebbian algorithm for the prediction and the classification of autism disorder problem. The classification approach was based on human knowledge and experience, as well as on historical data (patterns). The proposed algorithm presented high classification accuracy of 80% [3]. In order to enhance the learning capabilities of this hebbian-based type of FCM learning, a new learning approach based on the ensemble learning, such as bagging and boosting, was integrated. FCM ensemble learning is an approach where the model is trained using non linear Hebbian learning (NHL) algorithm and further its performance is enhanced using ensemble techniques. This new approach of FCM ensembles, showed results with higher classification accuracy instead of the NHL alone learning technique [15]. Recently, Papakostas et al. (2012) presented some Hebbian-based approaches for pattern recognition, showing the advantages and the limitations of each one [16]. Another study of Zhang et al. [17] proposes a novel FCM, which is automatically generated from data, using Hebbian learning techniques and Least Square methods.

This research work is focused on the application of a new classification technique concerning the autism disorder. The FCM model constructed by physicians to assess three levels of autism (no autism, probable autism and autism) was trained using a new particle swarm optimization (PSO) clustering algorithm for forty real children cases. In other words, the main objective of this study is to present the PSO algorithmm for FCM learning applied to a classification case study.

2   Fuzzy Cognitive Maps

An FCM is a soft computing technique which combines the main aspects of fuzzy logic and neural networks (NN) and avoids the mathematical complexity of system analysis. FCM was originated by Kosko [4] as an extension of cognitive maps in order to create an abstract modeling methodology to model and represent the behavior of a system and the human thinking. Concepts stand for states, variables, inputs, outputs and any other characteristics of the system. Each weight expresses the causal relationship between two interconnected concepts.

Generally there are two main approaches for the creation of a FCM, the expert-based in which the FCM is a manual created and the computational method in which the FCM is made by the processing of historical data. Several scientists have dealt with the computational creation of FCMs in the light of learning algorithms [18].

FCMs have an inference mechanism similar to those of Neural Networks (NN). Combining the Fuzzy Logic and NN, the inference process is accomplished using simple mathematical operations between weight matrices, minimizing in doing so the complexity of a system. The inference process implementation can be described by the following five steps.

Step 1. Read the input vector A.
Step 2. Read the weight matrix W.
Step 3. Calculate the value of each concept by the following equation.

$$A_i(t) = \left(A_i(t-1) + \sum_{j=1, j\neq i}^{n} W_{ji} * A_j(t-1) \right) \tag{1}$$

Step 4. Apply a threshold function, usually sigmoid, to the values which were calculated in Step 3.
Step 5. Until the Concept values reach an equilibrium state (steady state) we continue the process from Step 3.

Concepts and weight matrix values lie between the intervals [0~1] and [-1,+1], respectively. The main difference from NN is the initial determination of the weight matrix and its meaning after estimation. Despite the fact that the main characteristic of both techniques is the weight matrix adaptation, on the NN technique the weight matrix is initialized with random values for all possible connections among nodes and reach to the "global optima", whereas on FCM each weight value has a real meaning for the problem, representing a causal interconnection, so uncertain modification of initial values of weights may converge the system to a "local optima".

3   Learning Algorithms for FCMs

The learning approaches for FCMs are concentrated on learning the connection matrix, based either on expert intervention and/or on the available historical data (like the neural network learning process). In other words we target on finding weights that better represent the relationships between the concepts. Learning approaches for FCMs can be divided into three categories [18]:

1. The hebbian-based algorithms such as NHL, ddNHL, which produce weight matrices based on experts' knowledge that lead the FCM to converge into an acceptable region for the specific target problem.
2. The population-based algorithms such as evolutionary, immune, swarm-based, which compute weight matrices based on historical data that best fit the sequence of input state vectors or patterns.
3. The hybrid algorithms which are focused on computing weight matrices based on experts knowledge and historical data.

Although FCMs have not been widely used on classification tasks, the last decade some researchers have proved that the classification procedure is feasible with FCMs [18]. So far, the usage of FCMs on classification problems has been implemented mainly by hebbian learning approaches [9] and by exploiting both neural networks and FCMs to ensure improved classification capabilities [14]. First, Papageorgiou et al. presented a brain tumour characterization algorithm based on Active Hebbian Learning for FCMs [19]. Next, Papakostas et al. presented a pattern classification algorithm based on FCMs. To map the outputs of the classifier to a problem's classes, three different class mapping techniques were implemented. The first mapping refers to the Class per Output technique where a specific class is assigned to a single output. The second class mapping technique, the Threshold one, works by the extraction of specific output threshold for the output concept values. The last technique consists of the clustering of the values of the output concepts, and for each class the mapping is computed by the calculation of minimum distance of each cluster. Recently, Papakostas et al. used for the classification of the data an idea that stems from NN tactics, which modifies the structure of FCMs by adding Hidden Concept nodes [17]. An extension of FCMs which is also inspired by NN classification theory is also presented at [15] where ensemble learning approaches like bagging or boosting are implemented. One more novel FCM extension for classification of testing instanced has been presented in [17]. The inference process of the LS-FCM model is similar to other FCM approaches but it uses a Least Square methodology to overcome the most weakness of the existing FCM algorithm, namely the heavy calculation burden, convergence and iterative stopping criteria. Song and his coworkers [20] extended the application of the traditional FCMs into classification problems, while keeping the ability for prediction and approximation by translating the reasoning mechanism of traditional FCMs to a set of fuzzy IF–THEN rules. They focused to the contribution of the inputs to the activation of the fuzzy rules and quantified the causalities using mutual subsethood, which works in conjunction with volume defuzzification in a gradient descent-learning framework. In next section, we suggest a population-based algorithm using the Particle Swarm Optimization method in order to achieve higher classification accuracy for the autism disorder problem.

4   Particle Swarm Optimization Algorithm for FCM Classification

Particle Swarm Optimization (PSO) is a computation method based on the social behavior of birds being in a flock. PSO algorithm [21] optimizes a problem by having a population of candidate solutions. The solutions called particles and their existence is at the problem hyperspace. The motion of each particle into the problem hyperspace over time according to a simple mathematical equation defines the Particle position and velocity. Each particle's position is influenced by its local best known position and is also guided toward the best known positions in the search-space, which are updated as better positions found by other particles.

To implement the PSO algorithm for FCM classification two steps are necessary. In the first step, a number of prototypes are positioned, in an unsupervised way, on regions of the input space with some density of the input data. For this, the Particle Swarm Clustering (PSC) [21,22] algorithm is used. In the second step the algorithm must decide about the decision boundaries that partition the underlying output vector from step one into three sets, one for each class. For this purpose, one-dimensional decision boundaries were determined by two methods. The first one is the bayesian statistical decision method [22] and the second one is the minimum Euclidean distance method [23]. The classifier accuracy is estimated by the leave-one-out cross-validation (LOOCV) method [10].

To implement the Particle Swarm Clustering algorithm for FCM we assume that we have a swarm consisting of k particles $\{P_1, P_2, P_3 \ldots P_k\}$. For our approach every particle position is a candidate FCM, meaning a weight matrix. This matrix can be initialized either by random values on the non-zero weights, thus keeping the main problem's signs constrains or by experts' suggestions. There is in general a plethora of weight matrices that lead the concepts to different values according to any input data. Let's consider a data set with T real cases, where each case is represented by a vector. For each estimated vector (which is calculated implementing the eq. (1) for a given weight matrix and an input vector), there is a particle of greater similarity to the input vector, obtained by the Euclidean distance between the particle and the input data. This is the winner particle, and its velocity is updated by eq (2).

$$v_i(t+1) = w * v_i(t) + \phi_1 * \left(p_i^j(t) - x_i(t) \right) + \phi_2 * \left(g^j(t) - x_i(t) \right) \qquad (2)$$

In eq (2), the parameter w, called inertia moment, is responsible for controlling the convergence of the algorithm and it is decreased at each step. The cognitive term $p_i^j(t) - x_i(t)$, associated with the experience of the particle winner, represents the best particle's winner position, in relation to the j_{th} input data so far. The social term $g^j(t) - x_i(t)$ is associated with the particle closest to the input data, that is, the particle that had the smallest distance in relation to the j_{th} input object so far. The parameters w, φ1, φ2, and φ3 (used in eq. (4)) are selected by the practitioner and control the behavior and efficacy of the PSO method. They take values within the range [0,1], to avoid the chaotic behavior of position and velocity vectors, in FCM equilibrium state. The winner particles position is updated by eq (3)

$$x_i(t+1) = x_i(t) + v_i(t+1) \qquad (3)$$

The procedure of the Particle Swarm Clustering Algorithm for FCM is shown in Pseudocode 1. Step 9 of Pseudocode 1 updates all those particles that did not move at iteration t. Thus, after all data sets were presented to the swarm, the algorithm verifies whether some particle did not win in that iteration. These particles are updated using eq (1) in relation to the particle that was elected the winner more often at iteration t. In the last step, the algorithm assigns a label to each estimated data. This task is feasible because we know a priori the correct labels for each data. This knowledge stems from the experts.

$$v_i(t+1) = w * v_i(t) + \phi_3 * (x_{winner} - x_i(t)) \qquad (4)$$

There are two possible termination conditions for the algorithm, which are (a) a maximum number of iterations which is determined empirically and (b) the minimization of a cost function concerning the global optimization methods. In this study the following cost function is found to be appropriate:

$$\frac{1}{N} \sum_i^N (A_i(Out) - Y_i(Out))^2 \qquad (5)$$

Where j is the winner particle, $A_i(Out)$ is the candidate FCM response of the output concept for the i_{th} data set and $Y_i(Out)$ is the given response for the i_{th} data set. N is the number of concepts.

Pseudocode 1. Particle Swarm Clustering Algorithm for FCMs

```
Step 1. At t=0 Initialize the swarm P(0)={P₁,P₂,…,Pₖ} with ran-
dom weight matrices for the Position Vector X and the Velocity
Vector V keeping only the non-zero weights and/or the weights
signs based on the experts knowledge.
D: Data set with T real cases
Y: T-1 training cases of D
C-labels: The correct labels were determined by experts.
While stopping criterion is not met
For each input data row j
For each Particle i
Step 2. Compute the new concept values Aʲₙₑw by eq(1)
Step 3. Compute the distance between Aʲₙₑw and Yʲ
End for
Step 4. Find the Particle with the minimum distance and dec-
lare it as the Winner Particle Pʲₘᵢₙ
Step 5. Compare the distance of Winners Particle position to
its best position thus far.
d1: distance between Pʲₘᵢₙ and Yʲ
d2: distance between the winner's particle best position
pbest_Pʲₘᵢₙ and Yʲ
```

```
        if d1<d2 then
              pbest_P^j_min= P^j_min
```
Step 6. Compare the distance of Winner's Particle position to its global best position thus far.
d3: distance between the winner's particle global best position gbest_P^j_min and Y^j
```
if d1<d3 then
        gbest_P^j_min= P^j_min
```
Step 7. Change the Velocity of winner's particle using eq 2.
Step 8. Change the Position of winner's particle using eq 3.
End for
Step 9. Change the Velocity and the Position for the particles who did not win by eq4 and eq 3.
Step 10. Test the stopping criterion
    End while
Step 11. Assign a label to each data set according to C-labels
Return: The Predicted labels from Step 11 and the new data set which is estimated on step 6.

5   Experimental Analysis and Results

The autism disorder problem was selected as a very complex process and due to its previous use in classification tasks [10,15]. Forty real children cases from an Indian hospital were studied and diagnosed by the experts (doctors). Those forty datasets were collected for classification of three different categories, like twenty three as "Definite Autism" (DA), thirteen as "Probably Autism" (PA) and four as "No-Autism" (NA) children and gathered in [3]. There is previous experience from experts as well as historical data, and the classification objective is to classify these cases into three classes: DA, PA and NA in order to achieve higher classification accuracy. Experts decided about the concepts and their initial interconnections among them and defined that there are twenty three main symptoms for the autism disorder problem, such as climbing on things, bringing objects to parents, etc [3]. The decision concept concerns the autism class.

Table 1. Scenario (I): Classification accuracies of Particle Swarm Classification for FCM. Only the non-zeros weights are initialized by random values.

Boundaries Decision	MED	BSM	MED	BSM	MED	BSM	MED	BSM
Particles	K=20	K=20	K=20	K=20	K=50	K=50	K=50	K=50
Iterations	R=100	R=100	R=500	R=500	R=100	R=100	R=500	R=500
True Positive (All %)	32,12%	33,27%	32%	34.18%	32,12%	34,1%	32,41%	33,18%
Model Accuracy	82,35%	85,3%	82.05%	87,64%	82,35%	87.43%	83.1%	87.64%
Correct Classes (All)	89	89	426	432	89	91	417	432
FCM Classification Accuracy	**89%**	**89%**	**85,2%**	**84,6%**	**89%**	**91%**	**83.4%**	**86,4%**

Table 2. Scenario (II): Classification accuracies of Particle Swarm Classification for FCM. The non-zeros weights are initialized by random values to ± 0.2 of initial values, keeping the problem constrains for weights.

Decision Boundaries	MED	BSM	MED	BSM	MED	BSM	MED	BSM
Particles	K=20	K=20	K=20	K=20	K=50	K=50	K=50	K=50
Iterations	R=100	R=100	R=500	R=500	R=100	R=100	R=500	R=500
True Positive (All %)	32,56%	33,95%	32.37%	33,78%	32,4%	34,2%	32,42%	34,28%
Model Accuracy	83,14%	87,05%	83%	86,43%	83,07%	87.79%	83.14%	87.91%
Correct Classes (All)	81	85	389	401	81	78	399	397
FCM Classification Accuracy	81%	85%	77,8%	80.2%	81%	78%	79,8%	79,4%

The proposed PSO clustering algorithm for FCM was implemented at the 40 records to predict the classification category of each one. Figure 1 illustrates the proposed approach in the case of autism classification problem. Two different scenarios were examined: (I) the first concerns that the initial non-zero weights are initialized by random values and (II) the last concerns that the initial non-zero weights are initialized by random values within a ±0.2 range of their initial values (belong in the interval [Weight-0.2, Weight+0.2]), thus keeping the signs and weight constraints. The classification performance results were gathered in Tables 1 and 2, respectively for each scenario.

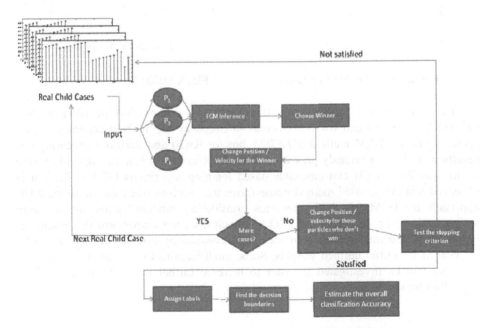

Fig. 1. Particle Swarm Algorithm for FCM Classification

In order to estimate the model accuracy (where all the 40 cases were considered) and the FCM system's accuracy (classification accuracy using the LOOVC method) two different decision boundaries methods were considered: the Minimum Euclidean Distance (MED) and the bayesian statistical decision boundary method (BSM). Additionally different numbers of Particles were considered, 20 and 50 and different numbers of iterations of the algorithm, 100 and 500.

Figures 2 and 3 illustrate the decision boundaries calculated for the decision concepts produced from one algorithm performance for K=20 and R=100.

For the algorithm performance, the row True Positive (TP) represents the average number of the correctly categorized cases according to the decision boundaries chosen. According to LOOCV method 39 random cases were used for the training procedure and the remaining one is used for testing. Thus, for the evaluation of the approach, 39 of the total 40cases were used for training, and only one for testing every time of cross validation. The total model accuracy was calculated by the division of the TP cases with 39. The "Correct Class" represents the total number of cases that have been classified correctly and the "Classification Accuracy" expresses the equivalent proportion.

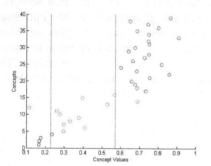 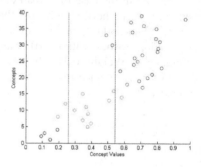

Fig. 2. BSM Classification Lines   **Fig. 3.** MED Classification Lines

The best accuracy in Scenario (I) is derived for the BSM decision method (92.31%), for K=50 and R=100, whereas in Scenario (II) the best accuracy is presented again for BSM method (87.17%), but for K=20 and R=100. Comparing our results with those previously presented using Hebbian-based learning algorithms (the result was 79.9%) [3] and ensemble-based learning algorithms (87.5%) [15], it is observed that the proosed method outperforms the previous one concerning the NHL approach for FCMs, in both scenarios considering random values for non-zero weights. However, the proposed PSO approach does not outperform the ensemble-based FCM learning approach in the cases considering random values in a ±0.2 interval of the initial defined weights. Some modifications to the PSO clustering parameters will be investigated in order to increase further the performance of PSO algorithm for this task.

6 Conclusions

To sum-up, the PSO clustering approach for FCM learning is able to classify autism disorder with reasonably high overall accuracy, sufficient for this application area and therefore, it is established as an efficient learning approach for FCMs. This work presents our first investigation to explore the PSO system characteristics and capabilities in the FCM learning working on classification tasks and the results encourage us to further exploit it. Surely, more research work is needed to be done towards more investigation of the learning methodologies of FCMs and their implementation in pattern recognition.

References

1. Snow, P., Smith, D., Catalona, W.J.: Artificial neural networks in the diagnosis and prognosis of prostate cancer: a pilot study. The Journal of Urology 226, 1923–1926 (1994)
2. Hsieh, N.-C.: An integrated data mining and behavioral scoring model for analyzing bank customers. Expert Systems with Applications 27(4), 623–633 (2004)
3. Arthi, K., Tamilarasi, A., Papageorgiou, E.I.: Analyzing the performance of fuzzy cognitive maps with non-linear hebbian learning algorithm in predicting autistic disorder. Expert Systems with Applications 38(3), 1282–1292 (2011)
4. Kosko, B.: Fuzzy cognitive maps. Int. J. Man-Machine Studies, 65–75 (1986)
5. Papageorgiou, E.I.: Review Study on Fuzzy Cognitive Maps and Their Applications during the Last Decade. In: Glykas, M. (ed.) Business Process Management. SCI, vol. 444, pp. 281–298. Springer, Heidelberg (2013)
6. Papageorgiou, E.I., Stylios, C., Groumpos, P.: Unsupervised learning techniques for fine-tuning Fuzzy Cognitive Map causal links. Intern. Journal of Human-Computer Studies 64(8), 727–743 (2006)
7. Stach, W., Kurgan, L., Pedrycz, W.: Data-Driven Nonlinear Hebbian Learning Method for Fuzzy Congitive Maps. In: IEEE International Conference on Fuzzy Systems IEEE World Congress on Computational Intelligence, FUZZ-IEEE 2008, pp. 1975–1981 (2008)
8. Stach, W., Kurgan, L., Pedrycz, W., Refomat, M.: Genetic learning off fuzzy cognitive maps. Fuzzy Sets and Systems 153(3), 371–401 (2005)
9. Alizadeh, S., Ghazanfari, M., Jafari, M., Hooshmand, S.: Learning FCM by Tabu Search. International Journal of Computer Science (2), 143–149 (2008)
10. Papageorgiou, E.I., Froelich, W.: Multi-step prediction of pulmonary infection with the use of evolutionary fuzzy cognitive maps. Neurocomputing 92, 28–35 (2012)
11. Yesil, E., Urbas, L.: Big Bang – Big Crunch Learning Method for Fuzzy Cognitive Maps. World Academy of Science, Engineering and Technology 47 (2010)
12. Yastrebov, A., Piotrowska, K.: Simulation Analysis of Multistep Algorithms of Relational Cognitive Maps Learning. In: Yastrebov, A., Kuzminska-So?osnia, B., Raczynska, M. (eds.) Computer Technologies in Science, Technology and Education. Institute for Sustainable Technologies - National Research Institute, pp. 126–137 (2012)
13. Madeiro, S.S., Zuben, F.J.V.: Gradient-Based Algorithms for the Automatic Construction of Fuzzy Cognitive Maps. In: 2012 11th International Conference on Machine Learning and Applications (ICMLA), vol. 1, pp. 344–349 (2012)

14. Papakostas, G.A., Boutalis, Y.S., Koulouriotis, D.E., Mertzios, B.G.: Fuzzy cognitive maps for pattern recognition applications. International Journal of Pattern Recognition and Artificial Intelligence 22(8), 1461–1468 (2008)
15. Papageorgiou, E.I., Kannappan, A.: Fuzzy cognitive map ensemble learning paradigm to solve classification problems: Application to autism identification. Applied Soft Computing (2012)
16. Papakostas, G.A., Koulouriotis, D.E., Polydoros, A.S., Tourassis, V.D.: Towards Hebbian learning of Fuzzy Cognitive Maps in pattern classification problems. Expert Systems with Applications 39(12), 10620–10629 (2012)
17. Zhang, Y., Liu, H.: Classification systems based on Fuzzy Cognitive Maps. In: Fourth International Conference on Genetic and Evolutionary Computing (2010)
18. Papageorgiou, E.I.: Learning Algorithms for Fuzzy Cognitive Maps-A review study. IEEE Transactions on Systems Man and Cybernetics (SMC)-Part C 42(2), 150–163 (2012)
19. Papageorgiou, E.I., Spyridonos, P.P., Giotsos, D.T., Stylios, C.D., Ravazoula, P., Niki-foridis, G.N., Groumpos, P.P.: Brain tumor characterization using the soft computing of fuzzy cognitive maps. Applied Soft Computing 8, 820–828 (2008)
20. Song, H.J., Miao, C.Y., Wuyts, R., Shen, Z.Q.: An Extension to Fuzzy Cognitive Maps for Classification and Prediction. IEEE Transactions on Fuzzy Systems 19(1), 116–135 (2010)
21. Kennedy, J., Eberhart, R.: Particle swarm optimization. In: Proceedings of IEEE International Conference on Neural Networks, pp. 1942–1948 (1995)
22. Theodoridis, S., Koutroumpas, K.: Classifiers based on Bayes decision theory. In: Pattern Recognition, 2nd edn., pp. 13–44. Elsevie rScience/Academic Press, USA (2003)
23. Cohen, S.C.M., Castro, L.N.: Data clustering with particle swarms. In: Proceedings of the World Congress on Computational Intelligence, pp. 6256–6262 (2006)

Fuzzy Cognitive Maps with Rough Concepts

Maikel León[*], Benoît Depaire, and Koen Vanhoof

Hasselt University, Diepenbeek, Belgium
maikelleon@gmail.com

Abstract. Artificial Intelligence has always followed the idea of using computers for the task of modelling human behaviour, with the aim of assisting decision making processes. Scientists and researchers have developed knowledge representations to formalize and organize such human behaviour and knowledge management, allowing for easy translation from the real world, so that the computers can work as if they were "humans". Some techniques that are common used for modelling real problems are Rough Sets, Fuzzy Logic and Artificial Neural Networks. In this paper we propose a new approach for knowledge representation founded basically on Rough Artificial Neural Networks and Fuzzy Cognitive Maps, improving flexibility in modelling problems where data is characterized by a high degree of vagueness. A case study about modelling Travel Behaviour is analysed and results are assessed.

Keywords: Rough artificial neural networks, fuzzy cognitive maps, knowledge representation, modelling problems.

1   Introduction

The concept of upper and lower bound has been used in a variety of applications in Artificial Intelligence. In particular, theory of rough sets has demonstrated the usefulness of upper and lower bounds in fields such as rule generation. Additional advances in rough set theory have shown that the concept of upper and lower bounds offer a wider framework that can be suitable for diverse types of applications [1].

On the other hand a Fuzzy Cognitive Map (FCM) is a combination of some aspects from Fuzzy Logic, Neural Networks and other techniques; combining the heuristic and common sense rules of Fuzzy Logic with the learning heuristics of the Neural Networks. They were introduced by Kosko [2], who enhanced cognitive maps with fuzzy reasoning, that had been previously used in the field of socio-economic and political sciences to analyse social decision-making problems.

The use of FCM for many applications in different scientific fields was proposed, they had been apply to analyse extended graph theoretic behaviour, to make decision analysis and cooperate distributed agents, also were used as structures for automating human problem solving skills and as behavioural models of virtual worlds, etc.

This paper proposes rough patterns for simulations using FCM. Each value in a rough pattern is a pair of upper and lower bound. Conventional FCM models

---

[*] Corresponding author.

H. Papadopoulos et al. (Eds.): AIAI 2013, IFIP AICT 412, pp. 527–536, 2013.

generally use a precise input pattern in their estimations. The conventional FCM models need to be modified to accommodate rough patterns. Rough concepts proposed in this paper provide an ability to use rough patterns. Each rough concept stores the upper and lower bounds of the input and output values.

Depending upon the nature of the application, two rough concepts in the net can be connected to each other using either two approaches, using the idea of Rough Artificial Neural Networks (RANN), which have been studied in literature in many aspects. Then, a Fuzzy Cognitive Map with Rough Concepts (RFCM) consists of a combination of rough and conventional concepts connected each other, considering the hybridization as an approach together with other appropriately defined methods, e.g., fuzzy logic and domain-specific analytical techniques.

Hybrid technique such as rough-fuzzy had been attracting great attentions of many researchers since a while, and several examples have shown that the hybrid techniques perform better than the non-hybrid in a huge amount of scientific fields.

2   Rough Artificial Neural Networks

Generalizations of neurons have been followed by generalizations of the entire network structures and corresponding learning mechanisms. New models of neural networks have been studied more and more often as hierarchical structures of complex concepts (granules), which finally resulted in the methodology of rough-neural computing:

- Construction of systems performing complex tasks using simple rough neurons and their straightforward generalizations transforming parameters of concepts.
- Hierarchical structure that represents gradual formation of more complex granules (concepts) modelling complex phenomena or structures, or projection onto simpler granules (concepts) modelling aggregation of information, conflict resolution etc.
- Flexibility and robustness originating in highly adjustable structure of possibly generalized rough neurons, their connections, and intermediate transformations enabling to vary the structures of granules (concepts) throughout the network.
- Ability to learn from examples a desired setting of the network weights, just like in case of standard neural network models, in particular ability to adapt the mechanism of backpropagation for networks involving complex granules and neurons.

Rough set theory introduced by Pawlak in 1982 is a mathematical tool to deal with vagueness and uncertainty of information [3]. The theory of rough sets has demonstrated the usefulness of upper and lower bound, so, the concept of upper and lower bound has been used in a variety of applications in intelligent system. Driven by the idea of decomposing the set of all objects into upper and lower bond, it was introduced the idea of rough neuron to construct RANN.

Each neuron R is a pair, for the upper bound R^* and for the lower bound R_*. Those two neurons can exchange information between each other and between other rough (conventional) neuron. Rough neurons proposed [4] provide an ability to use rough patterns, which are based on the notion of rough values. It is possible to use a rough neuron to successfully characterize a range or values set for variables such as age, weight or temperature.

3   Fuzzy Cognitive Maps

FCM in a graphical illustration seem to be a signed directed graph with feedback, consisting of nodes and weighted arcs (see figure 1).

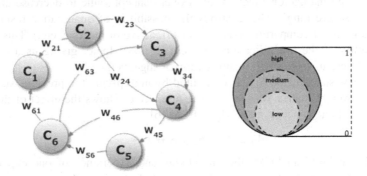

Fig. 1. Simple Fuzzy Cognitive Map. Concept activation level.

Graph nodes place for the concepts that are used to characterize the system behaviour and they are connected by signed and weighted arcs showing the causal relationships that connect the concepts. It must be mentioned that the values in the graph are fuzzy, so concepts take values in the range between [0, 1] and the weights of the arcs are in the interval [-1, 1]. The weights of the arcs between concept C_i and concept C_j could be positive ($W_{ij} > 0$) which means that an augment in the value of concept C_i leads to the increase of the value of concept C_j, and a decrease in the value of concept C_i leads to a reduce of the value of concept C_j. Or there is negative causality ($W_{ij} < 0$) which means that an increase in the value of concept C_i leads to the decrease of the value of concept C_j and vice versa [5].

Observing this graphical representation, it becomes clear which concept influences other concepts showing the interconnections between concepts and it permits updating in the construction of the graph. Each concept represents a characteristic of the system; in general it stands for events, actions, goals, values, trends of the system that is modelled, etc. Each concept is characterized by a number that represents its value and it results from the renovation of the real value of the system's variable.

Beyond the graphical representation of the FCM there is its mathematical model. It consists of a $1 \times n$ state vector A which includes the values of the n concepts and a $n \times n$ weight matrix W which gathers the weights W_{ij} of the interconnections between the n concepts. The value of each concept is influenced by the values of the connected concepts with the appropriate weights and by its previous value. So the value A_i for each concept C_i can be calculated, among other possibilities, by the following rule expressed in (1).

$$A_i = f\left(\sum_{\substack{j=1 \\ j \neq i}}^{n} [A_j \times W_{ji}]\right) \tag{1}$$

Where A_i is the activation level of concept C_i, A_j is the activation level of concept C_j and W_{ij} is the weight of the interconnection between C_j and C_i, it is to say, the value of A_i depends of the weighted sum of its input concepts, and f is a transfer, threshold or normalization function, used over concept value to decrease unbounded inputs to a severe range. This destroys the possibility of quantitative results, but it gives us a basis for comparing nodes (on or off, active or inactive, etc.). This mapping is a variation of the "fuzzification" process in fuzzy logic, giving us a qualitative model and frees us from strict quantification of edge weights.

So the new state vector A_{new} is computed by multiplying the previous state vector A_{old} by the weight matrix W, see (2). The new vector shows the effect of the change in the value of one concept in the whole FCM [6].

$$A_{new} = f(A_{old} \times W) \tag{2}$$

In order to build an FCM, the knowledge and experience of one expert on the system's operation must be used. The expert determines the concepts that best illustrate the system; a concept can be a feature of the system, a state or a variable or an input or an output of the system; identifying which factors are central for the modelling of the system and representing a concept for each one. When the experts have observed which system elements influence others; they must determine the effect among concepts, with a fuzzy value per interconnection, due to it has been reflected that there is a causation fuzzy degree between two connected concepts.

FCM feedback structure also makes a distinguishing from the earlier forward-only acyclic cognitive maps and from modern Artificial Intelligence expert-system search trees. Such tree structures are not dynamical systems because they lack edge cycles or closed inference loops. Nor are trees closed under combination. Combining several trees does not produce a new tree in general because cycles or loops tend to occur as the number of combined trees increases.

4   Introducing Rough Concepts in Fuzzy Cognitive Maps

When a FCM has been constructed, it can be used to model and simulate the behaviour of the system. Firstly, the FCM should be initialized, the activation level of each of the nodes of the map takes a value based on expert's opinion for the current state and then the concepts are free to interact. This interaction between concepts continues until a fixed equilibrium is reached; a limited cycle is reached or a chaotic behaviour is exhibited. So, FCM are a powerful methodology that can be used for modelling systems, avoiding many of the knowledge extraction problems which are usually present in by rule based systems.

It is possible to have better results in the drawing of the FCM, if more than one expert is used. In that case, all experts are polled together and they determine the relevant factors and thus the concepts that should be presented in the map. Then, experts are individually asked to express the relationship among concepts; during the assigning of weights, three parameters must be considered: how strongly concepts influence each other, what is the sign of the weight and whether concepts cause.

But is not always easy the initialization process, and how to fix a value or how to make different experts agree? In a rough pattern, the value of each variable is specified using lower and upper bounds, using the idea from RANN [7]. But also, there is a consequence, related to the links, so figure 2 shows how to solve this problem. If the rough concept A excites the activity of B (i.e. increase in the output of A will result in the increase in the output of B), then A^* will be connected to B^* and A_* will be connected to B_*.

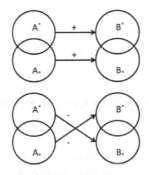

Fig. 2. Connections between rough concepts

On the other hand, if A inhibits the activity of B (i.e. increase in the output of A corresponds to the decrease in the output of B), then A^* will be connected to B_* and A_* will be connected to B^* [8]. So now formula (1) cannot be applied, so a modification for the inference process is needed. Formulas (3), (4) and (5) will describe the necessary readjustment.

$$\text{input}(A_i) = \sum_{\substack{j=1 \\ j \neq i}}^{n} [\text{output}(A_j) \times W_{ji}] \tag{3}$$

$$output(A^*) = \max\left(f\left(input(A^*)\right), f\left(input(A_*)\right)\right) \tag{4}$$

$$output(A_*) = \min\left(f\left(input(A^*)\right), f\left(input(A_*)\right)\right) \tag{5}$$

There are different possibilities to choose a transfer function, one of the most used in literature appears in (6), but many others can be also used.

$$f(x) = \frac{1}{1 + e^{-9(x-0.5)}} \tag{6}$$

To emphasize it is illustrated figure 3, showing the transformation from a classical FCM into a RFCM.

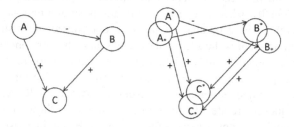

Fig. 3. FCM and the corresponding RFCM

But if the output in general of the rough concept A is desired, it can be computed using formula (7).

$$A = \frac{A^* - A_*}{average(A^*, A_*)} \tag{7}$$

5   Workbench for Modelling Complex Systems Based on FCM and RFCM

The scientific literature shows some software products developed with the intention of drawing FCM by non-expert in computer science, as FCM Modeler [9] and FCM Designer [10]. The first one is a rustic incursion, while the second one is a better implementation, but still hard to interact with and almost without experimental facilities. Figure 4 shows the general architecture of our proposing workbench to model and simulate FCM, having the facility of using also Rough Concepts, in real applications where is extremely difficult to set up an initial point of the system, and a upper and lower bounds are easily defined.

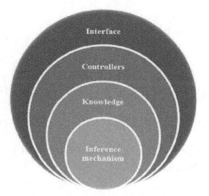

Allows the user-tool interaction through the options to create FCM, definition of parameters and formalization of the information into a knowledge base.

Makes a link between the Interface and the algorithms and data, it is a connectivity layer that guarantees a right manipulation of the information.

Generates the computational representation of the created FCM from an Artificial Intelligence point of view. Processes the input and output data of algorithms in the variables modeling.

Makes the inference process through the mathematical calculus for the prediction of the variable values.

Fig. 4. General architecture of the proposed workbench

In figure 5 it is possible to observe the main window of the workbench, in the interface appears some facilities to create concepts, make relations, and define parameters, also to initialize the execution of the inference process, and visualization options for a better understanding of the simulation process.

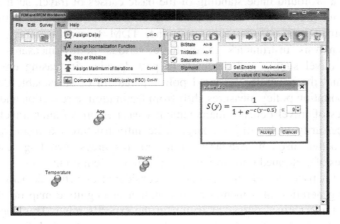

Fig. 5. Main view and Run options of the Workbench

There were defined some facilities and options in the tool, to create, open or save an FCM or RFCM. Through these amenities a non-expert in Computer Science is able to elaborate FCM or RFCM describing systems; we had paid attention to these facilities in order to guaranty that the workbench is usable for simulation and experimentation purposes. It is exposed some important options, where is possible to define the assignment of a delay time in the execution for a better understanding of the running of the FCM or RFCM in the inference process, also it is possible to define the transfer function that the FCM will use in the running.

In simulation experiments the user can compare results using these different functions or just can select the appropriate function depending of the problem to model:

- Binary FCM are suitable for highly qualitative problems where only representation of increase or stability of a concept is required.
- Trivalent FCM are suitable for qualitative problems where representation of increase, decrease or stability of a concept is required.
- Sigmoid FCM are suitable for qualitative and quantitative problems where representation of a degree of increase, a degree of decrease or stability of a concept is required and strategic planning scenarios are going to be introduced.

6   Case Study: Modelling Travel Behaviour

Transport Demand Management (TDM) is of vital importance for decreasing travel-related energy consumption and depressing high weight on urban infrastructure. TDM, or also known as "mobility management", is a term for measures or strategies to make improved use of transportation means by reducing travel demand or distributing it in time and space. Many attempts have been made to enforce TDM that would influence individuals unsustainable travel behaviour towards more sustainable forms, TDM can be effectively and efficiently implemented if they are developed founded on a profound understanding of the basic causes of travel, such as people's reasons and inclinations and comprehensive information of individuals behaviours.

In the process of transportation planning, TDM forecast is one of the most important analysis instruments to evaluate various policy measures aiming at influencing travel supply and demand. In past decades, increasing environmental awareness and the generally accepted policy paradigm of sustainable development made transportation policy measures shift from facilitation to reduction and control.

Objectives of travel demand management measures are to alter travel behaviour without necessarily embarking on large-scale infrastructure expansion projects, to encourage better use of available transport resources avoiding the negative consequences of continued unrestrained growth in private mobility.

Individual activity travel choices can be considered as actual decision problems, causing the generation of a mental representation or cognitive map of the decision situation and alternative courses of action in the expert's mind. This cognitive map concept is often referred to in theoretical frameworks of travel demand models, especially related to the representation of spatial dimensions, but much features can be taken into account (see figure 6).

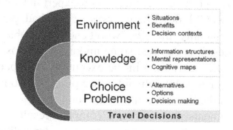

Fig. 6. Abstraction levels of mind related to Travel Behaviour

However, actual model applications are scarce, mainly due to problems in measuring the construct and putting it into the model's operation. The development of the mental map concept can benefit from the knowledge provided by individual tracking technologies. Researches are focusing on that direction, in order to improve developed models and to produce a better quality of systems. At an individual level it is important to realize that the relationship between travel decisions and the spatial characteristics of the environment is established through the individual's perception and cognition of space. As an individual observes space, for instance through travel, the information is added to the individual's mental maps.

Records regarding individual's decision making processes can be used as input to generate mental models. Such models treat each individual as an agent with mental qualities, such as viewpoints, objectives, predilections and inclinations. For the modelling of such models, several artificial intelligence techniques can be used, in this case FCM and RFCM will be study. They try to genuinely simulate individual's decision making processes. Consequently, they can be used not only to understand people's travel behaviours, but also to pretend the changes in their actions due to some factors in their decision atmosphere.

During a decision making process, a decision maker activates a temporary mental representation in his working memory based on his previous experiences or existing knowledge. Therefore, constructing a mental representation requires a decision maker to recall, reorder and summarize relevant information in his long-term memory. It may involve translating and representing this information into other forms, such as a scheme or diagram, supporting coherent reasoning in a connected structure.

In the city of Hasselt, capital of the Flemish province of Limburg, Belgium, a study related to Travel Behaviour was made. The city has a population around 72 000 habitants, with a traffic junction of important traffic arteries from all directions. Hasselt is surrounded by 2 ring roads, the bigger one serves to retain traffic out and the small one helps to keep traffic out of the commercial centre, being almost totally a pedestrian area.

In our experiment more than 220 real habitants were asked to specify how they take into account the transport mode they will use for an imaginary shopping activity: situation, attribute and benefit variables; and starting from that data, a FCM and RFCM structures per person were developed. At the same time, virtual scenarios were presented, and the personal decisions of each individual were stored.

Both models (FCM and RFCM) were created per person, and the scenarios were played. Figure 7 presents the performances of the computational models. Although the FCM models performed 92.68%, and it is considered a good result, RFCM models performed 97.1%, being significantly better.

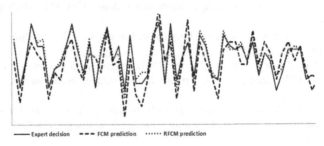

———— Expert decision   ▬ ▬ ▬ FCM prediction   ······ RFCM prediction

Fig. 7. FCM vs. RFCM prediction

In table 1 are indicated other attributes of the comparison, to have more detailed information of the experiment execution. Measures indicate a major discharge of the RFCM models.

Table 1. Attributes of the comparison

	FCM models	RFCM models
Accuracy	92.68%	97.1%
Oversized estimation	11.06%	05.57%
Undersized estimation	13.07%	04.79%
Interpretable	Yes	Yes
Simpler structure	(X)	( )
Easier initialization	( )	(X)

It is considered a substantial result, given by having structures able to simulate how people think when a transport mode decision and area election for specific activity is asked, offering policy makers a tool to play with, in order to test new policies, and to know in advance the possible resounding in the society.

7   Conclusions

In this paper, we presented cooperative rough pattern and FCM techniques. A rough concept can be view as a pair of concepts supporting upper and lower bounds as opposed to precise values, exchanging information with each other during the calculation of their outputs. The development of a workbench based on FCM and RFCM for the modelling of complex systems was presented, showing facilities for the creation of FCM and RFCM, and options to make the inference process comprehensible and used for simulations experiments. A better performance of RFCM over FCM was obtained in a real world modelled problem. In the presented case study a social and politic repercussion is evident, as we offer to policymakers a framework and real data to play with, in order to study and simulate individuals behaviour for city infrastructure development and demographic planning.

References

1. Chandana, S., Mayorga, S.: Rough Approximation based Neuro-Fuzzy Inference System. In: IEEE HIS, pp. 518–521 (2005)
2. Kosko, B.: Fuzzy Cognitive Maps. International Journal of Man-Machine Studies 24, 65–75 (1986)
3. Pawlak, Z.: Rough classification. International Journal of Information and Computer Sciences, 145–172 (1982)
4. Lingras, P.: Rough neural networks. In: IPMU International Conference on Information Processing and Management of Uncertainty in Knowledge-based Systems, pp. 1445–1450 (1996)
5. Wei, Z.: Using fuzzy cognitive time maps for modelling and evaluating trust dynamics in the virtual enterprises. In: Expert Systems with Applications, pp. 1583–1592. Elsevier Ltd. (2008)
6. León, M., Nápoles, G., Rodriguez, C., García, M.M., Bello, R., Vanhoof, K.: A Fuzzy Cognitive Maps Modeling, Learning and Simulation Framework for Studying Complex System. In: Ferrández, J.M., Álvarez Sánchez, J.R., de la Paz, F., Toledo, F.J. (eds.) IWINAC 2011, Part II. LNCS, vol. 6687, pp. 243–256. Springer, Heidelberg (2011)
7. Ming, H., Boqin, F.: Extracting Classification Rules with support rough neural networks. In: Torra, V., Narukawa, Y., Miyamoto, S. (eds.) MDAI 2005. LNCS (LNAI), vol. 3558, pp. 194–202. Springer, Heidelberg (2005)
8. Pal, L., Polkowski, S.: Rough-Neural Computing. Cognitive Technologies Series. Springer (2004)
9. Mohr, S.: Software Design for a Fuzzy Cognitive Map Modelling Tool. Tensselaer Polytechnic Institute (1997)
10. Contreras, J.: Aplicación de MCDD a tareas de supervisión y control. Universidad de los Andes. Mérida, Venezuela (2005)

Intelligent Systems Applied to the Control of an Industrial Mixer

Marcio Mendonça[1], Douglas Matsumoto[1], Lucia V.R. Arruda[2],
and Elpinik I. Papageorgiou[3]

[1] Parana Federal Technological University, Department of Electrical Engineering,
Cornélio Procópio, Brazil
mendonca@utfpr.edu.br, douglas.matsumoto@gmail.com
[2] Parana Federal Technological University, CPGEI, Curitiba, Brazil
lvrarruda@utfpr.edu.br
[3] Technological Education Institute of Lamia, Department of Informatics
and Computer Technology, Lamia, Greece
epapageorgiou@teilam.gr

Abstract. This paper presents the application of intelligent techniques to control an industrial mixer. Control design is based on hebbian evolution of fuzzy cognitive maps. In this context, this paper develops a dynamical fuzzy cognitive map (D-FCM) based on Hebbian Learning algorithms. Two strategies to update FCM weights are derived. Finally, the D-FCM is used to control an industrial mixer. Simulation results of this control are presented. Additionally, results are provided extending some of the algorithms into the Arduino platform in order to acknowledge the performance of the codes reported in this paper.

Keywords: fuzzy cognitive maps, hebbian learning, Arduino platform, process control, fuzzy logic.

1 Introduction

Artificial Intelligence (AI) has applications in various areas of knowledge, such as mathematical biology, neuroscience, computer science and others. The research area of intelligent computational systems aims to develop methods that try to mimic or approach the capabilities of humans to solve problems. These news methods are looking for emulate human's abilities to cope with very complex processes, based on inaccurate and/or approximated information. However, this information can be obtained from the expert's knowledge and/or operational data or behavior of an industrial system [1].

In this context, Fuzzy Cognitive Map (FCM) is a tool for modeling the human knowledge. It can be obtained through linguistic terms, inherent to fuzzy systems, but with a structure similar to the Neural Networks (NN), which facilitates data processing, and has capabilities for training and adaptation. FCM is a technique based on the knowledge that inherits characteristics of Cognitive Maps and Artificial Neural Networks [2], [3], [4], with applications in different areas of knowledge [5], [6], [7],

H. Papadopoulos et al. (Eds.): AIAI 2013, IFIP AICT 412, pp. 537–546, 2013.

[8], [9]. Besides the advantages and characteristics inherited from these primary techniques, FCM was originally proposed as a tool to build models or cognitive maps in various fields of knowledge. It makes the tool easier to abstract the information necessary for modeling complex systems, which are similar in the construction to the human reasoning. Thus, FCM aggregates benefits of the acquisition, processing and adaptability from data and information system to be modeled, with a capacity for intelligent decision making due to its heuristic nature. However, FCM has some limitations, especially in time modeling, restricting for applications where the causes and effects occur simultaneously.

In order to circumvent these drawbacks, dynamical fuzzy cognitive maps (D-FCM), can be developed which have the capability to model and manage behaviors of non-linear time-dependent system and often in real time. Examples of different D-FCMs can be found in the recent literature, as examples, we can cite [10], [11], [12], [13].

Specifically, the work of Mendonça and collaborators [10] presents a type of D-FCM, which aggregates the occurrence of events and other facilities that makes appropriate this type of cognitive map, for the development of intelligent control and automation in an industrial environment.

In this paper, we use the same D-FCM proposes in [10] to control an industrial mixing tank. Different from [10], we use a hebbian algorithm to dynamically adapt the D-FCM weights. In order to validate our D-FCM controller, we compared its performance with a fuzzy logic controller. This comparison is carried out with simulated data. Moreover, to shown the control portability, we embedded the D-FCM controller into a low cost platform based on Arduino.

2   Development

To demonstrate the evolution of the proposal technique (D-FCM) we will use a case study well known in the literature as seen in [3], [14] and [15] to test level controllers. This case was selected to illustrate the need for refinement of a model based on FCM built exclusively with knowledge. The process shown in Figure 1 consists of a tank with two inlet valves for different liquids, a mixer, an outlet valve for removal of liquid produced by mixing and density meter that measures the quality of the produced liquid.

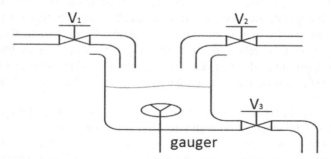

Fig. 1. Mixer Tank (Source: adapted from Stylios, Groumpos, Georgopoulos, 1999)

Valves (V_1) and (V_2) insert two different liquids in the tank. During the reaction of the two liquids, a new liquid characterized by its density value is produced. At this time the valve (V_3) empties the tank in accordance with a campaign output flow, but the liquid mixture should be in the specified levels.

Although relatively simple, this process is a TITO (two inputs two outputs) type with coupled variables. To establish the quality of the control system of the produced fluid, a weighting machine placed in the tank measures the (specific gravity) produced liquid.

When the value of the measured variable G (liquid mass) reaches the range of values between the maximum and minimum [Gmin, Gmax] specified, the desired mixed liquid is ready. The removal of liquid is only possible when the volume (V) is in a specified range between the values [Vmin and Vmax]. The control consists to keep these two variables in their operating ranges, as,

$$V_{min} < V < V_{max} \tag{1}$$

and

$$G_{min} < G < G_{max.} \tag{2}$$

In this study we tried to limit these values from 800 to 850 [mg] for the mass and 850 to 900 [ml] for the volume. According to Papageorgiou et al. [16], through the observation and analysis of operation of the process is possible for experts to define a list of key concepts related to physical quantities involved. The concepts and cognitive model is based on a known FCM model [16], having the following concepts and structure:

• Concept 1 - State of the valve 1 (closed, open or partially open).
• Concept 2 - State of the valve 2 (closed, open or partially open).
• Concept 3 - State of the valve 3 (closed, open or partially open).
• Concept 4 - quantity of fluid (volume) in the tank, which depends on the operational state of the valves V1, V2 and V3.
• Concept 5   - value measured by the G sensor for the density of the liquid.

Considering the initial proposed evolution for FCM we will use a D-FCM to control the mixer which should maintain levels of volume and mass within specified limits.

The process model uses the mass conservation principle to derive a set differential equations representing the process used to test the D-FCM controller. As a result the tank volume is the volume over the initial input flow of the intake valves V1 and V2 minus the outflow valve V3. Similarly, the mass of the tank follows the same principle as shown below. The values used for m_{e1} and m_{e2} were 1.0 and 0.9, respectively.

$$V_{tank} = V_i + V_1 + V_2\text{-}V_3 \tag{3}$$

$$Weight_{tank} = M_i + (V_1 m_{e1}) + (V_2 m_{e2}) - M_{out} \tag{4}$$

The development of the D-FCM is accomplished through two distinct stages. First the D-FCM is developed as a classic FCM where concepts and causal relationships are identified. The concepts can be variables and/or control actions, as already mentioned. However, the heuristic is related to the control condition of volume or weight of the mixture increase, where the inlet valves are closed, thus making it possible to assign inverse causal relationship between the concepts of levels and outlet valves. The output valve defines a positive relationship, when the output flow increases, according to a desired campaign. The flow of the intake valves also increases proportionally. The initial setting is done by using an algorithm based on heuristic optimization method of Simulated Annealing [17], in which an initial solution is cast as an initial guess and then solutions with a certain degree of randomness is tested systematically, until the system get the desired response. The figure 2 shows the schematic graph of a D-FCM controller.

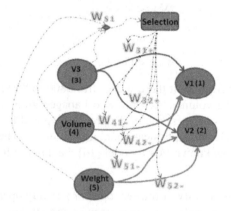

Fig. 2. D-FCM Controller

The second stage of development of the D-FCM is responsible for tuning or refinement of the model for dynamic response of the controller. In this case, when a change of output set-point in the campaign occurs, the weights of the causal relationships are tuned. In this paper two options are showed. First, the switching of the causal relationships in different operating points (set-points). It is shown in Table 1 and require different settings offline. One classic genetic algorithm used for tune off-line with 20 individuals, simple crossing and 1% of mutation, however obtained results similar to the method cited. The second option is online and is in accordance with equations 6-11. These equations are based on Hebb learning that increments or decrements the causal relations in accordance with changes in their associated concepts. To perform this function a new kind of concept and relations were included in the cognitive model. The relation selection assigns rules event-driven, which in this case was the change of set-point, levels of volume and weight of the liquid mixture.

The D-FCM uses the concept of selection for switching the set of causal relations according to the basic rules of relation selection. D-FCM works similarly to a DT-FCM (Decision Tree - FCM) [5], in another way, the D-FCM resembles a hybrid tool between a FCM and a temporal cyclic state machine, for example the work of Acampora and Loia [18], and switched by triggering events. The results of the weight of the causal relations by experts through observation of the dynamic behavior of the system by changing the cause and effect of causal relationships are shown in Table 1.

Table 1. Casual relationship weights

State/Rules (D-FCM)	W13	W14	W23	W24	W53	W54
Initial value	-0.35	-0.40	-0.35	-0,40;	0,00	0,00
Rule 1 (Weight>850 mg)	-0,65	-0,65	-0,65	-0,65	0,10	0,10
Rule 2 (Weight<840 mg)	-0,35	-0,40	-0,35	-0,40	0,20	0,20
Rule 3 (Weight<820 mg)	-0,35	-0,45	-0,35	-0,45	0,68	0,68

In order to establish a correlation and a future comparison between techniques of intelligent systems, a fuzzy controller was also developed. The fuzzy rule base implements weights assignment using the same heuristic control strategy.

As a result, both controllers fuzzy and D-FCM can be run under equal conditions supporting performance comparisons. As an example, we can mention some rules extracted from rule base established by experts:

```
If VOLUME is HIGH, then V1 is LOW, V2 is LOW;
If VOLUME is MEDIAN, then V1 is MEDIAN, V2 is MEDIAN;
If WEIGHT is LOW, then V1 is HIGH, V2 is HIGH;
```

In order to dynamically adapt the D-FCM weights we used the hebbian learning algorithm for FCM that is an adaptation of the classic hebbian method [2]. Different proposals and variations of this method applied in tuning or in learning for FCM are known in the literature [12]. In this paper, the method is used to update the intensity of causal relationships in a deterministic way according to the variation or error in the intensity of the concept or input variable. Specifically, the application of Hebb learning provides control actions as follows: if the weight or volume of the liquid mix increases, the intake valves have a causal relationship negatively intensified and tend to close more quickly. Conversely, if the volume or weight mixture decreases, the intake valves have a causal relationship positively intensified. The mathematical equation is presented in (5).

$$W_i(k) = W_{ij}(k-1) \pm \gamma \Delta A_i \qquad (5)$$

ΔA_i is the concept variation resulting from causal relationship, and it is given by $\Delta A_i = A_i(k)-A_i(k-1)$, γ is the learning rate at iteration k.

Causal relationships that have negative causality have negative sign similarly to positive causal relationships. Equations 6-11 show the implementation of the proposal.

$$W_{51} = W_{51}(k-1) - 0.7\Delta A_i \times k_p \qquad (6)$$

$$W_{52} = W_{52}(k-1) - 0.7\Delta A_i \times k_p \qquad (7)$$

$$W_{41} = W_{41}(k-1) - 0.7\Delta A_i \times k_p \qquad (8)$$

$$W_{42} = W_{42}(k-1) - 0.7\Delta A_i \times k_p \qquad (9)$$

$$W_{31} = W_{31}(k-1) + 0.1\Delta A_i \times k_p \qquad (10)$$

$$W_{32} = W_{32}(k-1) + 0.1\Delta A_i \times k_p \qquad (11)$$

Two variations of the Hebbian learning will be presented. In the first case, a global error is assumed as the sum of the variation only of the volume concept. In the second case, the variation of volume, weight and output valve affect the causal relationships (fig. 2), for example, W_{51} and W_{52} are tuned according to the weight variation $\Delta A_i = V_3(k) - V_3(k-1)$. All values of forgetting factor γ were empirical. Finally, so that variations of the weights had the dynamics needed, any errors or variations of A were multiplied by a factor 20 concerning the gain value.

3   Experimental Results

The results of D-FCM by Hebbian learning with only the volume parameter variation are shown in Fig. 3, which shows the behavior of the controlled variables within the predetermined range for the volume and the weight of the mixture. It is noteworthy that the controller keeps the variables in the control range and pursues a trajectory according to a campaign where output flow is also predetermined. In this initial experiment, a campaign with a sequence of values ranging from 7, 5 and 10 ml/min can be seen as a set-point output flow (outlet valve). Fig. 4 shows the evolution of the weights of the causal relationships during the process.

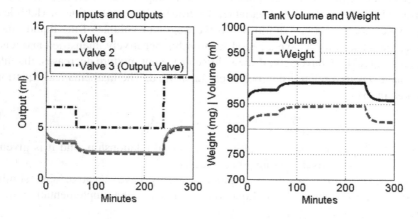

Fig. 3. Inputs and outputs valves, Volume and weight (HL global variation)

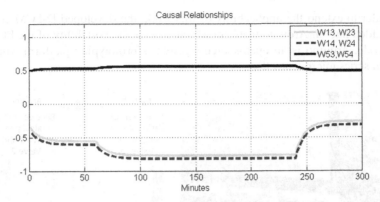

Fig. 4. Causal Relationships in the process (HL global variation)

The fig. 5 shows the results of Hebbian learning algorithm for FCM considering the variations ΔAi of the concepts concerning volume, weight and outlet valve, while in the fig. 6 is displayed the weights of the causal relationship in the process.

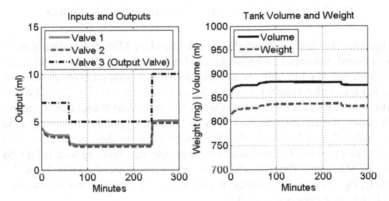

Fig. 5. Inputs and outputs valves, Volume and weight (HL punctual variation)

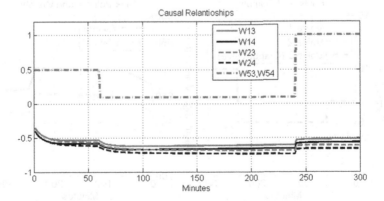

Fig. 6. Causal Relationships in the process (HL punctual variation)

In order to extend the applicability of this work, the developed D-FCM controller is embedded into an Arduino platform which ensures the portability of the FCM generated code. Arduino is an open-source electronics prototyping platform which uses ATMega series microcontrollers.

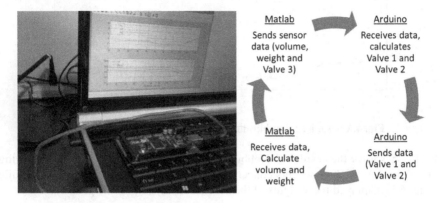

Fig. 7. Cycle and image of the communication Matlab - Arduino

The equations for level and weight are calculated by Matlab simulating the process. Through a Serial communication established with Arduino, Matlab sends the current values of Volume, Weight and output valve to Arduino that receives these data, calculates the values of the concept 1 (valve 1) and concept 2 (valve 2) and then returns these data to Matlab. After this, new values of Volume and Weight are recalculated.

Fig. 8 shows the results obtained with the Arduino platform providing data of the actuators Valve 1 and Valve 2 with Matlab performing data acquisition. The algorithm switches the sets of causal relations that operate similarly to a DT-FCM (decision tree - FCM), where the activation rules and weights are shown in Table 1.

Similarly, fig. 9 shows the results obtained with the Hebbian learning algorithm for FCM with the three parameters of ΔA_i.

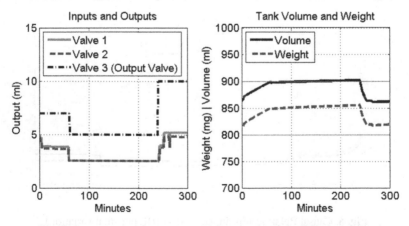

Fig. 8. Inputs and outputs valves, Volume and weight (DT, Arduino)

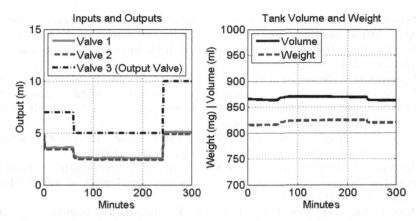

Fig. 9. Inputs and outputs valves, Volume and weight (HL, Arduino)

4 Conclusions

Performing a comparison of the results, we see a decrease in the control range in the cases where there is variation in causal relationships by Hebbian learning algorithms. Figures 7 and 8 show results.

From the data obtained from Arduino by the variations of the D-FCM embedded in the platform, it is observed that the controlled variables are in well behaved ranges, which suggests that the D-FCM codes have low computational complexity due to the simplicity of mathematical processing compared with the classical fuzzy logic, for example. Thus, we can emphasize the portability and the possibility of developing D-FCM controllers on low cost platforms.

Future studies address a comparison with classical PID controllers, weighted Fuzzy controller or other intelligent controller. Finally, our work will be directed to the application of a real mixer controller into a real data environment.

References

1. Passino, M.K., Yourkovich, S.: Fuzzy control. Addison-Wesley, Menlo Park (1997)
2. Kosko, B.: Fuzzy cognitive maps. International Journal Man-Machine Studies 24(1), 65–75 (1986)
3. Glykas, M.: Fuzzy Cognitive Maps: Advances in Theory, Methodologies, Tools and Applications. Springer, Heidelberg (2010)
4. Kosko, B.: Neural networks and fuzzy systems: A dynamical systems approach to machine intelligence. Prentice Hall, New York (1992)
5. Dickerson, J.A., Kosko, B.: Virtual Worlds as Fuzzy cognitive maps. Presence 3(2), 173–189 (1994)
6. Lee, K.C., Lee, S.: A cognitive map simulation approach to adjusting the design factors of the electronic commerce web sites. Expert Systems with Applications 24(1), 1–11 (2003)

7. Papageorgiou, E., Stylios, C., Groumpos, P.: Novel for supporting medical decision making of different data types based on Fuzzy Cognitive Map Framework. In: Proceedings of the 29th Annual International Conference of the IEEE Embs Cité Internationale, Lyon, France, pp. 23–26 (2007)

8. Papageorgiou, E., Stylios, C., Groumpos, P.A.: Combined Fuzzy cognitive map and decision trees model for medical decision making. In: Annual International Conference of the IEEE Engineering in Medicine and Biology Society, vol. 1, pp. 6117–6120. IEEE Engineering in Medicine and Biology Society (2006)

9. Huang, Y.C., Wang, X.Z.: Application of Fuzzy causal networks to waste water treatment plants. Chemical Engineering Science 24(13/14), 2731–2738 (1999)

10. Mendonça, M., Angélico, B., Arruda, L.V.R., Neves, F.: A dynamic fuzzy cognitive map applied to chemical process supervision. Engineering Applications of Artificial Intelligence 26, 1199–1210 (2013)

11. Miao, Y., Liu, Z.Q., Siew, C.K., Miao, C.Y.: Transformation of cognitive maps. IEEE Transactions on Fuzzy Systems 18(1), 114–124 (2010)

12. Papageorgiou, E.: Learning Algorithms for Fuzzy Cognitive Maps. IEEE Transactions on Systems and Cybernetics. Part C: Applications and Reviews 42, 150–163 (2012)

13. Miao, Y., Liu, Z.Q., Siew, C.K., Miao, C.Y.: Dynamical cognitive network - an Extension of fuzzy cognitive. IEEE Trans. on Fuzzy Systems 9(5), 760–770 (2001)

14. Axelrod, R.: Structure of decision: the cognitive maps of political elites. Princeton University Press, New Jersey (1976)

15. Stylios, C.D., Groumpos, P.P., Georgopoulos, V.C.: An Fuzzy Cognitive Maps Approach to Process Control Systems. J. Advanced Computational Intelligence 5, 1–9 (1999)

16. Papageorgiou, E.I., Parsopoulos, K.E., Stylios, C.S., Groumpos, P.P., Vrahatis, M.N.: Fuzzy cognitive maps learning using particle swarm optimization. Journal of Intelligent Information Systems 25, 95–121 (2005)

17. Ghazanfari, M., Alizadeh, S., Fathian, M., Koulouriotis, D.E.: Comparing simulated annealing and genetic algorithm in learning fcm. Applied Mathematics and Computation, 56–68 (2007)

18. Acampora, G., Loia, V.: On the Temporal Granularity in Fuzzy Cognitive Maps. IEEE Transactions on Fuzzy Systems 19(6), 1040–1057 (2011)

Training Fuzzy Cognitive Maps
Using Gradient-Based Supervised Learning

Michal Gregor[1] and Peter P. Groumpos[2]

[1] Department of Control and Information Systems,
Faculty of Electrical Engineering, University of Žilina
`michal.gregor@fel.uniza.sk`
[2] Laboratory for Automation and Robotics,
Department of Electrical and Computer Engineering, University of Patras
`groumpos@ece.upatras.gr`

Abstract. The paper considers a novel approach to learning the weight matrix of a fuzzy cognitive map. An overview of the state-of-the-art learning methods is presented with a specific emphasis on methods initially developed for artificial neural networks, and later adapted for FCMs. These have mostly been based on the concept of Hebbian learning. Inspired by the amount of success these methods have faced in the past, the paper proposes a new approach based on the application of the delta rule and the principle of backpropagation, both of which were originally designed for artificial neural networks as well. It is shown by simulation experiments and comparison with the existing approach based on non-linear Hebbian learning that the proposed approach achieves favourable results, and that these are superior to those of the existing method by several orders of magnitude. Finally, some possible lines of further investigation are suggested.

Keywords: fuzzy cognitive maps, learning, backpropagation.

1 Introduction

In the past several learning methods from the theory of artificial neural networks (ANNs) have been introduced into the theory of fuzzy cognitive maps (FCMs), and have become successful tools for either learning the weight matrix of the FCM from scratch, or tuning an initial weight matrix provided by a group of experts.

Most notable among such approaches were those based on the concept of Hebbian learning. It has been shown before that Hebbian learning can indeed be used to train an FCM from historical data (train it to perform regression). The approach is known as data-driven nonlinear Hebbian learning (DD-NHL). However, Hebbian learning has originally been proposed as an unsupervised learning approach, and it is therefore not necessarily best suited for such application.

In this paper we propose an alternative approach, which too is based on a learning method originally designed for ANNs, that is to say on the delta rule and the backpropagation principle.

H. Papadopoulos et al. (Eds.): AIAI 2013, IFIP AICT 412, pp. 547–556, 2013.

The following sections will provide a brief overview of the DD-NHL approach. It will then proceed to give some essentials concerning the theoretical background concerning the delta rule, and backpropagation. The details concerning the method proposed in this paper will be discussed. Finally, experimental results will be presented and evaluated, and a comparison with DD-NHL will be provided.

2   Fuzzy Cognitive Maps

Fuzzy cognitive maps (FCMs) are a symbolic representation for the description and modelling of complex systems [1]. They can be expressed and visualized using a weighted directed graph. The nodes of such graph represent the concepts associated with the modelled system. Every concept C_i is associated with its activation value A_i.

The edges in the graph are directed and weighted. The weights $w_{ij} \in [-1, 1]$ express causal relationships between the concepts. If $w_{ij} > 0$, we say that concept C_i causes C_j. If $w_{ij} < 0$, concept C_i has negative influence on the activation value of C_j. If $w_{ij} = 0$, there is no link.

At every time step activation values are updated. The update is synchronous. The update rule has several distinct forms. We will make use of the most general one proposed in [2] (the notation has been modified for the sake of consistency):

$$A_i^{(k+1)} = f \left(\sum_{j=1}^{N} A_j^{(k)} w_{ji} \right), \tag{1}$$

where N is the number of concepts, $A_i^{(k)}$ is the activation value of concept C_i at time step k. f is the squashing function, which squashes the dot product $\sum_{j=1}^{N} A_j^{(k)} w_{ji}$ into some convenient interval.

Most often, f is either the sigmoid function, which squashes the dot product into interval $[0, 1]$, or the hyperbolic tangent, which yields the interval $[-1, 1]$. The weight matrix of the FCM is usually constructed by experts. There are several approaches which make the task easier and more reliable – for the discussion of these, the reader may refer to [1, 3] for an instance.

2.1   Fuzzy Cognitive Maps and Learning

There are two main classes of problems in the theory of FCMs to which the existing learning methods apply: (a) *the regression problem*, that is how an FCM can be trained as a regression model for a given dataset; (b) *the attractor problem*, that is to say given an initial FCM, how can we shift its attractor to a desired point, encode a given limit-cycle, etc.

In this paper we shall focus on the regression problem. The methods that can address the regression problem have very useful applications – given data from a real system, they can make the FCM automatically learn to model the system. They can also be used to fine-tune an existing model designed by experts.

Several learning principles originally developed for ANNs have previously been applied to FCMs. These approaches were based on the concept of Hebbian learning. They come in several distinct flavours which will be listed hereinafter.

The first among the approaches inspired by Hebbian learning is the so-called *differential Hebbian learning* (DHL) [2, 4]. It has been shown that the rule is able to encode some sequences into the FCM. However it is not capable of encoding an arbitrary sequence. Furthermore only binary (or bipolar) sequences are considered.

In [5] the authors propose the so-called *active Hebbian learning* (AHL) method, which introduces the idea of the sequence of activation. The expert specifies the sequence in which the concepts are activated. The process starts from a concept which activates concepts linked to it, and thus the activation propagates until all the concepts have been activated at which point the simulation cycle stops and a new one starts. A distinct form of the Hebb rule is used to provide learning.

Finally, there are several papers (e.g. [6–9]) discussing the so-called *nonlinear Hebbian learning* (NHL). The learning rule used in this approach is the Oja rule [6], although in [7] several extensions are added to it. The procedure is as follows: An initial FCM is constructed by the experts. This is run using equation (1). In addition, at every step the rule is applied using the current activation values. Thus the NHL method does not simply learn the proper weight matrix, but rather it also helps to drive the process of convergence in an online manner [8].

A more traditional application of the NHL rule is proposed in [9]. In this case, NHL is used to make an FCM with a randomly initialized weight matrix learn the cause-effect relationships from historical data. This approach is called *data-driven NHL* (DD-NHL). Ideally, historical data from a real system should be used, but [9] suggests that we can create another FCM with a random weight matrix, and use that to generate the historical data instead. Such FCM is run from a randomly selected initial state for a predefined number of steps, and the resulting concept sequence is used as historical data.

3   Delta Rule and Backpropagation

This section will set forth some of the theory concerning the delta rule and the principle of backpropagation – methods originally designed for learning in ANNs, which we now propose to apply to the regression problem of FCM learning.

3.1   The Delta Rule

The delta rule is probably the best known approach to learning weights of an artificial neuron. It has been designed for supervised learning – that is to say learning from a dataset consisting of pairs of the following form: *(input, desired output)*. That is to say, for every sample in the dataset the input as well as the corresponding desired output is specified. Thus it is possible to form an error function [10]:

$$E(W) = \sum_p E^p(W) = \frac{1}{2} \sum_p (D^p - O^p)^2, \qquad (2)$$

where W denotes the weight matrix, and D^p and O^p denote the desired and the real output for input pattern p.

The error function can then be minimized using gradient descent, which leads to the following learning rule [10]:

$$\Delta_p w_j = \gamma \delta^p x_j, \qquad (3)$$

where $\Delta_p w_j$ denotes the prescribed change of weight w_j due to pattern p, γ is the learning rate, and $\delta^p = (D^p - O^p) f'(u^p)$. $f'(p)$ is the first derivative of the squashing function, and $u^p = \sum_{j=1}^{N} w_j x_j^p$ is the inner potential of the neuron (x_j^p is the i-th input of the neuron with pattern p at the input).

3.2   The Backpropagation Principle

The delta rule cannot by itself be used for learning in multi-layer networks, because only the errors of the output neurons can be computed directly – desired outputs of hidden neurons are unspecified.

However, the delta rule can be further generalized to multi-layer networks using the so-called backpropagation principle – in which case the error is propagated back from the output layer to hidden layers. Again, the full derivation of the rule can be found in [10], and we will only state the resulting rule:

$$\delta_h^p = f'(u_h^p) \sum_{o=1}^{N_o} \delta_o^p w_{ho}, \qquad (4)$$

where h refers to a neuron of the hidden layer, and o refers to neurons of the output layer. N_o is the number of neurons in the output layer. If there are several hidden layers, the principle can be applied recursively.

The backpropagation principle has further been extended to perform learning in recurrent neural networks (RNNs) – the approach is known as *backpropagation through time* (BPTT). The idea is that an RNN can be unwrapped in time into a feedforward ANN, and then trained using backpropagation. (For additional details and precise mathematical and algorithmic formulations the reader may refer to [11].)

4   The Proposed Approach

Let us now briefly discuss how we propose to apply the above-mentioned principles to solve the *regression problem* of FCM learning.

4.1   One-Step Delta Rule

First of all, we can directly use the delta rule as given above. The FCM can be considered as a single-layer network, and thus in this version we do not need to employ the backpropagation principle. Also, since we will be learning from historical data where values of concepts are provided for every time step, we should need to do no BPTT either. We will hereinafter refer to this baseline approach as the *one-step delta rule approach* (OSDR). The results, comparison, and evaluation will follow in a separate section.

4.2   Every-Step Delta Rule with Windowed BPTT

A further approach is proposed and studied by the authors: the *every-step delta rule with windowed BPTT* (ESWB) approach. In this case the sequence of historical data is cut into windows of a given size and with a given overlap. For the sake of brevity we will use the notation w[window size, overlap size] to describe windowing. The window size is understood to be the number of samples the window contains, and the overlap size is the number of samples which the window shares with the following one. Thus, windowing of $w[5, 4]$ refers to windowing with window size of 5 samples and overlap of 4 samples.

In the ESWB approach we take the first sample and use that as the input of the FCM. The FCM is then run for *window size* − 1 steps. Concept values from the last step are compared to the last sample in the window. Delta rule is applied to compute the error and to compute weight updates. The updates are stored in a separate vector − they are not applied to the FCM directly.

BPTT is then applied to propagate the error from the last step back in time. In addition to this backpropagated error, we also compute the error using the corresponding samples from the window. These two errors are added together and used to compute weight updates. Weight updates from all steps are accumulated in the same vector and once all steps have been considered, they are applied to the FCM as a batch.

Afterwards the algorithm moves to another window. In this way we make use of both − the BPTT *and* the error computed for that particular sample.

4.3   One-Step Delta Rule with Windowed BPTT

The final approach is proposed mainly for comparison with the one-step delta rule with windowed BPTT (OSWB) approach. This approach is closely related to ESWB except that once the errors are computed using the last sample in the window, *only BPTT* is used to compute weight updates. The other samples from the window are not used to compute error.

5   Simulation Experiments

We conducted several simulation experiments. These were laid out in such manner as to make the results easily comparable to those presented in [9].

Similarly to their work, we first generate an FCM with a random weight matrix. This FCM is run for 20 steps so as to generate the historical data. The data is afterwards used to form *(input, desired output)* pairs for delta rule learning. Learning proceeds on this data for the maximum of 100 epochs (it will also stop if the error goes below 1.10^{-4}). The learning rate is fixed to 0.2.

Testing is first done on the same 20 data steps used in training: these results are used to compute the *in-sample* mean square error (MSE). In addition to that, 10 random initial states are generated and the FCMs are run from each of these for 20 steps. The difference between the outputs of the original and the trained FCM is measured and used to compute the *out-sample* MSE (the procedure taken in [9]).

5.1 Simulation Setup

Unless said otherwise, for every single setting the learning was tested with FCMs of several sizes – with 5 concepts, 10 concepts, and 20 concepts – and with the connection density of 20% and 40% (again the procedure from [9]). The whole process was repeated for 20 independent runs each time. The results were averaged across the runs.

For every configuration the mean square error (MSE) is reported, and also the MSE_{attr}, which specifies how precisely the stable state (the attractor) of the learning FCM corresponds to that of the original FCM – again by giving the mean square error for that. It should be noted, that we compare the values of all concepts – not just of several randomly selected concepts as done in [9]. Also, we did not do restarts in cases when the algorithm did not converge – the algorithm converged to an acceptable solution every time.

5.2 One-Step Delta Rule

The results achieved using one-step delta rule (OSDR) follow in Table 1.

Table 1. Errors using the one-step delta rule

Size	Density	IN-SAMPLE		OUT-SAMPLE	
		MSE	MSE_{attr}	MSE	MSE_{attr}
5	20%	9.96 E-5	3.88 E-6	2.51 E-4	3.88 E-6
	40%	2.14 E-4	8.69 E-6	3.77 E-4	8.69 E-6
10	20%	2.38 E-4	7.23 E-6	1.09 E-3	7.23 E-6
	40%	3.77 E-4	1.89 E-5	1.23 E-3	1.89 E-5
20	20%	4.13 E-4	2.68 E-5	2.89 E-3	2.68 E-5
	40%	6.79 E-4	2.15 E-4	2.22 E-3	2.15 E-4

We also include results achieved DD-NHL [9] for comparison (Table 2). However, it is difficult to give adequate interpretation to some of the results reported there. 100% accuracy is reported, by which it is understood that none of the target concepts differs from its desired value by more than 0.1 once the stable state is reached. However, the policy used by the authors is to restart the algorithm from a new randomly-generated initial weight matrix unless it achieves the accuracy of 100% after a predefined number of iterations. Thus, learning is bound to either achieve 100% accuracy at some point, or else to go on indefinitely. It would be of interest to learn how many restarts were required under any given scenario.

Table 2. Errors using DD-NHL

Size	Density	IN-SAMPLE		OUT-SAMPLE	
		MSE	MSE_{attr}	MSE	MSE_{attr}
5	20%	0.129	?	0.129	?
	40%	0.129	?	0.129	?
10	20%	0.176	?	0.175	?
	40%	0.180	?	0.180	?
20	20%	0.180	?	0.180	?
	40%	0.207	?	0.207	?

In any case, the comparison shows that the results achieved using OSDR are more precise than those achieved using DD-NHL – and that by several orders of magnitude.

We may also note that we have experimented with several levels of connection density for the initial weight matrix of the learning FCM, but this did not appear to make any considerable difference.

5.3 Every-Step Delta Rule with Windowed BPTT

The next experiment was carried out using the ESWB approach. Windowing of $w[5,4]$ was used to cut the signal up. The results are presented in Table 3.

When we compare the results with those achieved using one-step delta rule (OSDR; Table 1), we must conclude that the results achieved using ESWB seem better – except those for the out-sample MSE, which means that generalization has regressed a little.

Table 3. Errors using ESWB with $w[5,4]$

Size	Density	IN-SAMPLE		OUT-SAMPLE	
		MSE	MSE_{attr}	MSE	MSE_{attr}
5	20%	9.08 E-5	1.45 E-7	2.53 E-4	1.45 E-7
	40%	1.88 E-4	2.50 E-7	3.86 E-4	2.50 E-7
10	20%	2.77 E-4	2.42 E-7	1.01 E-3	2.42 E-7
	40%	3.16 E-4	1.01 E-6	1.15 E-3	1.01 E-6
20	20%	3.38 E-4	1.60 E-6	2.86 E-3	1.60 E-6
	40%	5.72 E-4	3.05 E-5	1.98 E-3	3.05 E-5

What we cannot say with certainty yet is whether the difference results from using ESWB, or whether it is simply the effect of doing more training (OSDR is now done several times for most steps due to the overlapping). To ascertain how much effect should be ascribed to that we present Table 4, which provides results for OSDR with the maximum number of epochs set to 500 instead of 100 (because most samples now form part of 5 windows instead of one).

Table 4. Errors using OSDR; and max. of 500 epochs

Size	Density	IN-SAMPLE		OUT-SAMPLE	
		MSE	MSE_{attr}	MSE	MSE_{attr}
5	20%	1.01 E-4	1.31 E-7	2.06 E-4	1.31 E-7
	40%	1.42 E-4	1.99 E-7	3.75 E-4	1.99 E-7
10	20%	2.28 E-4	2.38 E-7	9.99 E-4	2.38 E-7
	40%	3.24 E-4	9.50 E-7	1.17 E-3	9.50 E-7
20	20%	3.82 E-4	1.56 E-6	2.96 E-3	1.55 E-6
	40%	5.34 E-4	2.67 E-5	1.96 E-3	2.67 E-5

The results for in-sample MSE and out-sample are rather mixed in this case – none of the two approaches seems to be decisively the better. Therefore we may conclude that combining the delta rule with BPTT as ESWB approach suggests does not produce any significant improvement.

5.4   One-Step Delta Rule with Windowed BPTT

Finally, let us present the results achieved using the OSWB approach. Although the results of the ESWB approach were not very encouraging, the results of OSWB will be of some theoretical interest even if they prove to be only comparable to those of OSDR with 500 epochs (Table 4) – this will indicate BPTT can efficiently be applied to FCMs, and it even to a certain extent able to supply for computing the actual difference between the desired and the real output at some steps. This property may be useful when data for some of the concepts is not available for all steps, or is not available at all.

The results follow in Table 5. Windowing of $w[5,4]$ was applied.

Table 5. Errors using OSWB with $w[5,4]$.

Size	Density	IN-SAMPLE		OUT-SAMPLE	
		MSE	MSE_{attr}	MSE	MSE_{attr}
5	20%	1.06 E-4	1.07 E-7	2.72 E-4	1.08 E-7
	40%	1.89 E-4	2.23 E-7	4.12 E-4	2.23 E-7
10	20%	2.13 E-4	1.48 E-7	9.41 E-4	1.48 E-7
	40%	3.17 E-4	1.38 E-6	1.05 E-3	1.38 E-6
20	20%	3.77 E-4	1.79 E-6	2.95 E-3	1.79 E-6
	40%	5.35 E-4	2.67 E-5	1.96 E-3	2.67 E-5

We conclude that the results are indeed comparable to those achieved using the 500 epoch OSDR, and using ESWB. In fact, in some cases OSWB even achieves better results than ESWB.

6 Further Work

Several lines of future investigation may be suggested. There is little doubt that learning can be made faster and more precise yet by using some of the more advanced learning methods based on the backpropagation principle, such as Quickprop, Rprop, or by the Levenberg-Marquardt algorithm. To ascertain how much effect such methods will have on learning FCMs may form part of future work.

Also, generalization could be improved by using historical data starting from several initial states instead of just one sequence of data. It is obvious that one sequence may not contain all the data required to learn the corresponding matrix accurately. In cases when more data is available, generalization may be improved considerably.

It should also be noted that the backpropagation algorithm could be used to learn even in cases where the activation values of some of the concepts remain unknown. Simulation results achieved using the OSWB method indicate that backpropagation and BPTT in particular can be used effectively in FCM learning. On the other hand, however, if several concepts are left unspecified, the learning algorithm will not be able to discriminate between them as it has no innate understanding of their meaning whatsoever. Therefore this issue will need some further investigation.

It also remains to be shown how well the learning method will perform when some of the weights are forced to remain fixed to their initial values.

7 Conclusion

The paper has presented an overview of the state-of-the-art methods for learning weights of a fuzzy cognitive map. Special emphasis has been put on methods based on Hebbian learning, which has originally been designed for artificial neural networks.

Inspired by the success of these approaches, we have proposed and presented a new method based on the delta rule, and the backpropagation principle, which has also originally been designed for neural networks. We have given a detailed description of our approach, and of its several varieties. These have been discussed, tested by simulation experiments, and compared with data-driven nonlinear Hebbian learning.

It has been shown that the results achieved using the proposed method surpass the accuracy of nonlinear Hebbian learning by several orders of magnitude.

All the varieties of the approach have been tested in turn. The results seem to indicate that the principle of backpropagation, and especially that of backpropagation through time can be used effectively in FCM learning. This should allow us to train the FCM even in cases, where the activation values of certain concept are not know, or even in cases where values for certain time steps are missing. However, these ideas need further investigation.

In addition to this, several other potential lines of future research and development have been indicated.

References

1. Groumpos, P.P.: Fuzzy Cognitive Maps: Basic Theories and Their Application to Complex Systems. In: Glykas, M. (ed.) Fuzzy Cognitive Maps, pp. 1–22. Springer, Berlin (2010)
2. Dickerson, J.A., Kosko, B.: Virtual Worlds as Fuzzy Cognitive Maps. In: Virtual Reality Annual International Symposium, pp. 471–477 (1993)
3. Stylios, C.D., Christova, N., Groumpos, P.P.: A Hierarchical Modeling Technique of Industrial Plants using Multimodel Approach. In: Proceeding of 10th IEEE Mediterranean Conference on Control and Automation, Lisbon, Portugal (2002)
4. Huerga, A.V.: A Balanced Differential Learning Algorithm in Fuzzy Cognitive Maps. In: Proceedings of the Sixteenth International Workshop on Qualitative Reasoning, pp. 10–12 (2002)
5. Papageorgiou, E.I., Stylios, C.D., Groumpos, P.P.: Active Hebbian Learning Algorithm to Train Fuzzy Cognitive Maps. International Journal of Approximate Reasoning 37(3), 219–249 (2004)
6. Papageorgiou, E.I., Stylios, C.D., Groumpos, P.P.: Fuzzy Cognitive Map Learning based on Nonlinear Hebbian Rule. In: Gedeon, T(T.) D., Fung, L.C.C. (eds.) AI 2003. LNCS (LNAI), vol. 2903, pp. 256–268. Springer, Heidelberg (2003)
7. Papageorgiou, E.I., Groumpos, P.P.: A Weight Adaptation Method for Fuzzy Cognitive Map Learning. Soft Computing 9(11), 846–857 (2005)
8. Papageorgiou, E.I., Stylios, C.D., Groumpos, P.P.: Unsupervised Learning Techniques for Fine-tuning Fuzzy Cognitive Map Causal Links. International Journal of Human-Computer Studies 64(8), 727–743 (2006)
9. Stach, W., Kurgan, L., Pedrycz, W.: Data-driven Nonlinear Hebbian Learning Method for Fuzzy Cognitive Maps. In: IEEE International Conference on Fuzzy Systems, FUZZ-IEEE 2008 (IEEE World Congress on Computational Intelligence), pp. 1975–1981 (2008)
10. Krose, B., Smagt, P.V.D.: An Introduction to Neural Networks. University of Amsterdam (1996), http://www.cs.unibo.it/babaoglu/courses/cas/resources/tutorials/Neural_Nets.pdf (accessed May 10, 2013)
11. Werbos, P.J.: Backpropagation Through Time: What It Does and How to Do It. Proceedings of the IEEE 78(10), 1550–1560 (1990)

An Approach to Hotel Services Dynamic Pricing
Based on the Delphi Method and Fuzzy Cognitive Maps

Dimitris K. Kardaras[1,*], Xenia J. Mamakou[1], Bill Karakostas[2],
and George Gkourakoukis[1]

[1] Business Informatics Laboratory, Dept. of Business Administration,
Athens University of Economics and Business, 76 Patission Street, Athens 10434, Greece
{kardaras,xenia}@aueb.gr, GeorgeGk@outlook.com
[2] Centre for HCI Design, School of Informatics, City University, Northampton Sq.,
London EC1V 0HB, UK
billk@soi.city.ac.uk

Abstract. E-tourism services open up new opportunities for businesses to expand and when possible to gain completive advantage. Dynamic pricing is an area of interest for both researchers and professionals. It's the process of price specification in a way that best suits a tourism organization under certain circumstances that reflect its competitive environment. Many research studies have addressed dynamic pricing from different perspectives. This study suggests that the use of a hybrid approach that combines Delphi method and fuzzy cognitive maps is suitable for it introduces fuzzy logic in order to capture the subjectivity and vagueness involved into evaluating the business settings, but it also provides for the necessary flexibility in analyzing the assumptions and the implications of different pricing scenarios.

Keywords: dynamic pricing, Delphi method, fuzzy cognitive maps, hotel management, e-tourism.

1 Introduction

Tourist arrivals around the world will increase over 200% by 2020 as predicted by the World Tourism Organization [35]. The hotel service has four characteristics [15], [36]: Intangibility: referring to the nature of the service. A service consumer cannot judge the quality of a service until the service is consumed. Inseparability: which implies that both the customer and the service provider should be present so that the service takes place. Variability: implying that the service depends on the provider, the time and the location that is consumed by the customer. Perishability: which refers to the inability of the services to be stored and consumed another time. Heterogeneity: implying that when in contrast to the products, services can be differentiated, especially due to the fact that they are intangible.

Tourism is a highly competitive business but its competitive advantage is no longer natural, but increasingly driven by science, information technology and innovation [5]. The Internet represents already the primary source for tourist to gather

H. Papadopoulos et al. (Eds.): AIAI 2013, IFIP AICT 412, pp. 557–566, 2013.

information for travelers, since 95% of Web users use the Internet to gather travel related information and about 93% indicate that they visited tourism Web sites when planning for vacations let alone the fact that the number of people who search the Internet for tourism related information increases rapidly [5]. Travelers increasingly resort to the Internet to search for tourism offers, to collect destination information and to organize their trips. The available information is there on the web and steadily increasing as well, thus making competition among business more intensive. In such a volatile environment, with well-informed competitors as well as customers, hotel management should adapt their pricing policy in order to meet the requirements of tourists but also to respond to challenges of the competition.

However, one of the most important features in hotel management, also in the tourism industry as a whole, is that many of its products / services are perishable. This makes it more difficult to set the appropriate price for a given business environment at a given point in time [10]. In addition, bearing in mind that tourism is extremely vulnerable to various external pressures and events, such as natural disasters and terrorist attacks, one cannot be sure for its demand. Therefore, dynamic pricing becomes an even more complicated decision problem [34]. As a result, drawing the appropriate pricing policy that can flexibly adjust to current circumstances is of paramount importance for hotel management.

2   Literature Review

One of the many implications that e-tourism has brought to tourism industry is the way that tourism businesses set the price for their services. Dynamic pricing, stems from dynamic packaging, which can be defined as "the combining of different travel components, bundled and priced in real time, in response to the request of the consumer or booking agent" [5]. The problem in the dynamic pricing in the case of the hotel industry is related to the unknown demand distribution of this service [34]. Lewis and Chambers (1989), in Danziger et al., (2004) [10], claim that "pricing in the hotel industry appears to be unscientific, self-defeating, myopic, and not customer-based". Other than the seasonality, hotel service prices are influenced by factors such as unknown demand distribution, income availability, the political stability in a tourism destination, the terrorist attacks, etc. [34].

Dynamic pricing originally introduced in the early 2000s from hotel chains, such as Hilton, InterContinental and Ledra Marriott [23]. Dynamic pricing, which is also known as yield management pricing policy [2], [30] is commonly used in the hotel industry, implying "a method that can help a firm to sell the right inventory unit to the right customers at the right time and at the right price, and thus to help a company optimize its profit". It is also defined as a sophisticated way of managing the offer/demand by manipulating prices and available capacity simultaneously [30]. Dynamic pricing is related to policies such as the Last Room Availability (LRA) and the Best Available Rate (BAR). The LRA policy offers better prices for certain number or types of rooms. A hotel could for example adopt the LRA policy for all room types, 365 days a year, as opposed to a static agreement, where LRA is offered in only 2 room types [29]. On the other hand, BAR ensures customers that the price they pay is the best rate a hotel can offer, given the demand for that particular day [29].

There are two ways for consumers and service providers to reach a dynamic pricing agreement. The first is associated to a client who has a significant volume for a specific hotel. The amount of discount off of the Best Available Rate (BAR) reflects on the one hand the volume that the client brings to this hotel and on the other the travel patterns of the client [29]. The second way to reach a dynamic pricing agreement is associated to the multi-location and the minimal volume of this agreement, in which case, the client offers small volumes for several locations; thus the hotel chain will offer a minimal discount off of the Best Available Rate (BAR) [29]. A blend of these two ways is also possible. Given the fact that the pricing in the case of the hotel industry is based on a constrained supply and a fluctuating demand, the static model of pricing is not realistic. Hence, the dynamic pricing model is regarded as a reasonable solution [29]. Several methods have been applied for hotel services pricing such as the Activity Based Costing –ABC [9], the thumb approach and the Hubbart formula. According to the first, "the room price is equal to 1/1000 of the investment price", whereas according to the Hubbart formula "the room rate equals the satisfied room revenue divided by the anticipated rooms sold, and satisfied room revenue is the cost of the hotel and the owner-desired profit"[6]. Recent studies indicate the value of dynamic pricing in terms the financial but also other tangible or intangible benefits it produces for hotels.

3   Methodology

The aim of this research is to determine the factors that mostly affect the process of dynamic pricing and to develop a model that supports the process of dynamic pricing. This study consists of two phases. The first phase adopts the Delphi method and captures the opinions of a group of 30 experts, with respect to the most influential pricing factors. A two-round Delphi method identified 20 pricing variables which were then included in the dynamic pricing model. In the second phase the same group of experts had to indicate the interrelationships among the factors identified during the first phase. Then, this study utilizes fuzzy cognitive maps in order to model the interrelationships among the factors identified and to provide a model that supports pricing scenarios analysis. The experts were asked to express their beliefs with respect to the strength and polarity of all possible causal relationships among the pricing factors identified from the Delphi method.

3.1   Delphi Method

The Delphi method (DM) was originally developed by Dalkey and Helmer [8]. It can be used to acquire experts' knowledge and beliefs and reach a reliable consensus among the experts [24]. DM rounds (up to four) of experts' questioning provide the experts with important information, like medians, averages and deviation from the previous rounds, so that they can rethink and revise their original beliefs and assumptions. Studies show that the experts' opinions converge towards the average of the group's opinions [4]. DM is applied through a series of recurring questions, usually in the form of questionnaires to a group of experts. After each round of questioning, the questions of each subsequent cycle to each member are accompanied by information on the responses of the other group members, which are presented anonymously. In this way, feedback is given for the experts to revise their opinions. According to Skulmoski et al. [32], the Delphi method is characterised by the Anonymity of the Delphi participants, the Iteration, through which the participants reconsider their opinions, the Controlled feedback, since it provides feedback information to the experts regarding the other members' opinions from previous rounds and Statistical aggregation of the experts' responses, thus producing the consensus of the group. DM is simple and flexible [32], it avoids a direct confrontation among the participants during the application of the method [28] and it also offers the experts feedback in order to review their assumptions and positions [27].

Many methods have been proposed to combine experts' opinions such as mean, median, max, min, mixed operators [20]. This research uses the geometric mean to represent experts' consensus. Thus, the importance of each of the factors identified is calculated by using the geometric mean of all the corresponding answers of the participants. The geometric mean has been used in the literature as one of the best ways to aggregate experts' opinions [16].

According to Mullen [26], there is no consensus regarding the size of the experts panel required by DM. Panel sizes as little as 9 experts [12] have been used in DM, or groups of 10 experts [3], 13 experts[22], or 31 members[16]. DM studies have also engaged groups as large as low hundreds, or even thousands in some studies in Japan [21]. The panel size of 30 experts in the current study is therefore, within the recommend range.

3.2   Fuzzy Cognitive Maps

A Fuzzy Cognitive Map (FCM) is a graph that consists of a number of nodes Ci representing the concepts of the domain in study. These nodes are connected to each other with weighted arcs W(i,j) showing how concept i is causally affected by concept j. The arcs that connect two concepts have weights that correspond to fuzzy qualifiers, such as 'a little', 'moderately', 'a lot'. Furthermore fuzzy numbers can be assigned in order to show the extent to which a concept affects another. FCMs are commonly used to model and study perceptions about a domain, to investigate the interrelationships among its concepts and to draw conclusions based on the implications of specific scenarios. The impact among the concepts of a FCM is estimated using the indirect effect. In other words, the impact caused due to the interrelationships among the concepts along the path from a cause variable (X) to an effect variable (Y) and the total effect, i.e. the sum of all the indirect effects from the cause variable X to the effect variable Y [14].

FCMs are represented by means of an NxN matrix, where N is the number of the concepts in the FCM with i and j representing concepts in the FCM. Every value of this matrix represents the strength and direction of causality between interrelated concepts. The value of causality is assigned values from the interval [-1, +1]. According to [31]:

- > 0 indicates a causal increase or positive causality from node i to j.
- = 0 there is no causality from node i to j.
- < 0 indicates a causal decrease or negative causality from node i to j.

The multiplication between matrices representing FCMs produces the indirect and total effects [37] and allows the study of the impact that a given causal effect D1 is causing. Causal effects can be represented with a 1xN vector [1]. This impact is calculated through repeated multiplications: ExD1 = D2, ExD2 = D3 and so forth, that is, ExDi = Di+1, until equilibrium is reached, which is the final result of the effect D1. Equilibrium is reached when the final result equals to zero, i.e. all cells of the resulting vector are equal to zero (0) and there is no any further causal impact caused by any concept. Different thresholds, depending on the modelling needs, restrict the values that result from each multiplication within the range [-1, +1]. Therefore, if a value is greater than (+1) then it is set to (+1), or it is set to (-1) if the resulting value exceeds the lower limit of (-1). For example, a threshold of (+/-0.5) implies that if the resulting value is greater than (+0.5) or lower than (-0.5) then the value is set to (+1) or (-1) respectively. FCMs have been used in many applications such as in modelling complex dynamic, which are characterized by strong non linearity [33], in personalised recommendations [17], [25], in managing relations in airline services [13], in systems modelling and decision making [14], in EDI design [18] and in EDI performance evaluation [19].

In order to construct the FCM, this study adopts the approach proposed by [3-4], who propose the development of an FCM for ERP tools selection based on experts' consensus, which was reached after a two-round consultation with the use of the Delphi method. The FCM is constructed by considering the median of the experts' responses in order to represent the magnitude of causality among the FCM concepts. As for the sign of each causal relationship, the sign that the majority of the experts propose is selected.

4   Delphi Method Results

The group of experts who agreed to participate in this study had to specify the important factors that influence hotel service prices. The two-round Delphi method resulted in the following list of 20 factors.

The results show that trust is the foremost important factor that influences service price and the decision of a customer to proceed in booking. It is interesting to note that experts find trust even more important than demand. It implies that long term good reputation of the hotel and its highly appreciated services among the customers can provide the foundation for the hotel management to adjust pricing policies even at hard times. Therefore, hotel management should pay special attention to increasing its customers' trust towards their hotel services.

Table 1. List of factors affecting dynamic pricing

Factors affecting dynamic pricing	Geometric Mean
Trust: Hotels ability to reflect all the necessary reassurances to gain customers trust.	4.39
Product's description: All the necessary information that may interest the customer regarding offered services and hotel facilities.	4.16
Awareness and Star Rating: The importance of hotels brand awareness as well as its Star Rating Categorization.	3.60
The distribution channel: The distribution channel that the company uses for its dynamic pricing, and its nature. (Internet, mobile devices, agencies etc).	3.69
Forecast ability: Hotel's ability to forecast future bookings (short and long term).	3.70
Booking incentives: The incentives that hotels offer to its customers in order to increase bookings efficiency. Eg: LRA (Last Room Availability), BRG (Best Rate Guarantee) etc.	3.65
The profile of the customer: The nature of the potential customer. For example, there are high-value customers willing to pay more and low-value customers looking for last minute offers.	3.27
Customer's behavioural trends: The way customers react. For example, buyers tend to request a ceiling or cap rate because they don't like to drive into the unknown.	4.09
Competition: The competition between hotels operating in the same market.	3.72
Market orientation: How clear is the orientation of the market through which the hotel offers its services? There is a variety of markets and most of them, present their prices as the best existing prices. This can confuse customers.	3.70
Heterogeneity among hotels: Usually, many hotels operate in the same area, of the same heterogeneous type of service and ranking.	3.65
Demand and availability: Demand and availability over the region where the hotel operates.	3.43
Economical and political situation: Economical and political situation on the region where the hotel operates.	3.81
Legal constraints: There may exist legal constraints regarding the nature of the offers, such as maximum and minimum possible prices.	3.48
Booking Season: A product may have different price on an ordinary date and different on a holiday season.	3.06
Customer's perceptions: Customer's perceptions of price and satisfaction. The perception of price fairness over offered services etc.	3.70
Customer's preferences: Depending on the product, there might be various preferences that define the final product price (wifi, breakfast/dinner etc).	3.58
Room availability: Room availability in the hotel.	2.46
Historical records: Historical records that allow a company to make price decisions based on earlier records.	3.69
Customer arrival rate: The arrival rate of new customers at the hotel.	3.71

5 Fuzzy Cognitive Mapping

Following the Delphi method, the experts were asked to judge the direction and strength of interrelationships among the pricing factors. The median was calculated in order to specify the strength of factors' interralationships, for it allows for positive or negative signs to be modelled in the FCM. As for the sign of each relationship, following the method by Bueno and Salmeron [3], it is defined according to the majority of the experts' answers. By analyzing experts' responses the following part of the complete FCM was constructed:

	Trust	Demand	Customer Perception	Booking Season	Product Description	Price	Heterogeneity of Hotels	Customer Preferences
Trust	0	0,8	0,6	0	0	0,2	0	0,4
Demand	0,6	0	0,6	0	0	0,2	0,2	0
Customer Perception	0	0	0	0	0	0	0	0
Booking Season	0	0	0	0	0	0	0	0
Product Description	0	0	0,6	0	0	0,4	0	0
Price	0	-0,4	0,2	0,4	0,6	0	-0,2	0,4
Heterogeneity of Hotels	0	0	0	0	0	0,4	0	0
Customer Preferences	0	0,4	0,6	0	0	0,6	0,4	0

Fig. 1. Part of the Dynamic Pricing FCM

By implementing the FCM as a matrix, several pricing scenarios can be investigated. For example, assume that a hotel operates in an area of low heterogeneity, which implies that hotel services are similar to each other, thus intensifying the competition and subsequently increasing the pressure for lower prices. Other assumptions regarding the current situation of the hotel in the scenario are a high trust that customers hold for the hotel, and high demand. The linguistic variables used to describe the scenario are expressed in terms of the following scale [7]:

Table 2. Linguistic variables and corresponding mean of fuzzy numbers

Linguistic Values	The Mean of fuzzy numbers
Very High	1
High	0.75
Medium	0.5
Low	0.25
Very Low	0

Each scenario, which assumes a causal effect, is represented by the Scenario-Vector (SV), which is a vector (1xn), where n is the number of variables that constitute the dynamic pricing FCM. Drawing on the theory of FCM, by multiplying the SV and the FCM, the management can examine the implication on prices and then decide what the most favourable pricing policies can be assumed and followed. More than one multiplication may be needed, until the system produces a final value for the "Price" variable, i.e. the price adjustment (PA). The sign of the value of "Price" indicates that the system suggests a price increase or reduction. The value indicates that magnitude of the price adjustment which in fuzzy terms can be a very high or high, etc. increase.

Assume the following scenario represented by the activation vector shown in Fig. 2:

Trust	Demand	Customer Perception	Booking Season	Product Description	Price	Heterogeneity of Hotels	Customer Preferences
0,7	0,8	0,7	0,4	0,1	0,7	0,2	0,1

Fig. 2. FCM scenario

Specifying the threshold at 0.3 the results of the FCM simulation are the following:

Trust	Demand	Customer Perception	Booking Season	Product Description	Price	Heterogeneity of Hotels	Customer Preferences
0,073728	-0,046592	0,566016	0,188928	0,283392	0,3648	0,057088	0,196608

Fig. 3. FCM result

The results in Fig. 3 indicate that price could be increased by low while at the same time customers' perception of the hotel will increase by medium.

By taking into consideration the current price that hotel management can specify the new-price for example, with the following multiplication:

New-Price = (Old-Price) + ((Old-Price) x (Price-Adjustement)).

For example, if Old-Price=100 euros and Price-Adjustement = +low, then the New-Price=100 + (100*0.3) = 130 euros.

6   Conclusions

By applying a hybrid approach that combines the Delphi method and fuzzy cognitive mapping this research work investigates the potential of developing FCMs in order to support dynamic pricing for hotel management. The combination of the two methods has been used in other research works [3] but not in dynamic pricing of hotel services. The proposed approach to dynamic pricing can provide hotel management with a useful tool in their decision making tasks. As a future work, this study suggests the full development and evaluation of a useable tool based on FCM for dynamic pricing.

References

1. Banini, G.A., Bearman, R.A.: Application of fuzzy cognitive maps to factors affecting slurry rheology. International Journal of Mineral Processing 52, 233–244 (1998)
2. Bayoumi, A.E., Saleh, M., Atiya, A., Aziz, H.A.: Dynamic Pricing for Hotel Revenue Management Using Price Multipliers (2010),
   http://alumnus.caltech.edu/~amir/hotel-dyn-pricing-final.pdf
3. Bueno, S., Salmeron, J.L.: Fuzzy modeling Enterprise Resource Planning tool selection. Computer Standards & Interfaces 30(3), 137–147 (2008)
4. Bueno, S., Salmeron, J.L.: Benchmarking main activation functions in fuzzy cognitive maps. Expert Systems with Applications 36(3), 5221–5229 (2009)
5. Cardoso, J.: E-Tourism: Creating Dynamic Packages using Semantic Web Processes (2005), http://www.w3.org/2005/04/FSWS/
   Submissions/16/paper.html
6. Chen, L.: The Luxury casino hotel dynamic price strategy practices for the FIT customer segment. UNLV Theses/Dissertations/Professional Papers/Capstones. Paper 437 (2009),
   http://digitalscholarship.unlv.edu/thesesdissertations
7. Dagdeviren, M., Yuksel, I.: Developing a fuzzy analytic hierarchy process (AHP) model for behaviour-based safety management. Information Sciences 178, 1717–1733 (2008)
8. Dalkey, N.C., Helmer, O.: An experimental application method to the use of experts. Management Science 9(3), 458–467 (1963)
9. Daly, J.L.: Pricing for profitability: Activity-based pricing for competitive advantage. John Wiley & Sons, New York (2002)
10. Danziger, S., Israeli, A., Bekerman, M.: Investigating Pricing Decisions in the Hospitality Industry Using the Behavioral Process Method. Journal of Hospitality & Leisure Marketing 11(2-3), 5–17 (2004)
11. Farahmand, M., Chatterjee, A.: The case for dynamic pricing. Hospitality Upgrade, 154–155 (Spring 2008)
12. Hsu, T.H., Yang, T.H.: Application of fuzzy analytic hierarchy process in the selection of advertising media. Journal of Management and Systems 7(1), 19–39 (2000)
13. Kang, I., Lee, S., Choi, J.: Using fuzzy cognitive map for the relationship management in airline service. Expert Systems with Applications 26(4), 545–555 (2004)
14. Kosko, B.: Fuzzy cognitive maps. International Journal on Man-Machine Studies 24(1), 65–75 (1986)
15. Kotler, P., Keller, K.L.: Marketing management. Pearson, New York (2006)
16. Kuo, Y.F., Chen, P.C.: Constructing performance appraisal indicators for mobility of the service industries using fuzzy Delphi method. Expert Systems with Applications 35, 1930–1939 (2008)
17. Lee, K., Kwon, S.: A cognitive map-driven avatar design recommendation DSS and its empirical validity. Decision Support Systems 45(3), 461–472 (2008)
18. Lee, S., Han, I.: Fuzzy cognitive map for the design of EDI controls. Information & Management 37(1), 37–50 (2000)
19. Lee, S., Kim, B.G., Lee, K.: Fuzzy cognitive map-based approach to evaluate EDI performance: A test of causal model. Expert Systems with Applications 27(2), 287–299 (2004)
20. Lin, H.Y., Hsu, P.Y., Sheen, G.J.: A fuzzy-based decision-making procedure for data warehouse system selection. Expert Systems with Applications 32, 939–953 (2007)
21. Linstone, H.A., Turoff, M.: Introduction to the Delphi method: techniques and applications. In: Linstone, H.A., Turoff, M. (eds.) The Delphi Method: Techniques and Applications. Addison-Wesley Publishing Company, Reading (1975)

22. Ma, Z., Shao, C., Ma, S., Ye, Z.: Constructing road safety performance indicators using fuzy Delphi metod and Grey Delphi method. Expert Systems with Applications 38, 1509–1514 (2011)
23. Mannix, M.: Dynamic hotel pricing: Signs of a trend? CWT Vision (4), 35–39 (2008)
24. Mereditha, J.R., Amitabh, R., Kwasi, A., Kaplana, B.: Alternative research paradigms in operations. Journal of Operations Management 8(4), 297–326 (1989)
25. Miao, C., Yangb, Q., Fangc, H., Goha, A.: A cognitive approach for agent-based personalized recommendation. Knowledge-Based Systems 20(4), 397–405 (2007)
26. Mullen, P.: Delphi: Myths and Reality. Journal of Health Organisation and Management 17(1), 37–52 (2003)
27. Nasserzadeh, S.M., Reza, M., Hamed, J., Taha, M., Babak, S.: Customer Satisfaction Fuzzy Cognitive Map in Banking Industry. Communications of the IBIMA (2), 151–162 (2008)
28. Okoli, C., Pawlowski, S.: The Delphi method as a research tool: an example, design considerations and applications. Information & Management 42(1), 15–29 (2004)
29. Palamar, L.A., Edwards, V.: Dynamic pricing: Friend or Foe? Buckhiester Management (2007), http://buckhiester.com/wp/assets/Dynamic-Pricing-White-Paper-2007.pdf
30. Rondan-Cataluña, F.J., Ronda-Diaz, I.M.: Segmenting hotel clients by pricing variables and value for money. Current Issues in Tourism, 1–12 (2012)
31. Schneider, M., Shnaider, E., Kandel, A., Chew, G.: Automatic construction of FCMs. Fuzzy Sets and Systems 93(2), 161–172 (1998)
32. Skulmoski, G.J., Hartman, F.T., Krahn, J.: The Delphi Method for Graduate Research. Journal of Information Technology Education 6, 1–21 (2007)
33. Stylios, C.D., Groumpos, P.P.: Fuzzy cognitive maps in modeling supervisory control systems. Journal of Intelligent & Fuzzy Systems 8(2), 83–98 (2000)
34. Wang, X.: Dynamic Pricing with a Poisson Bandit Model. Sequential Analysis: Design Methods and Applications 26(4), 355–365 (2007)
35. WTO. World Tourism Organization (2005)
36. Young, L.: From products to services. Insight and experience from companies which have embraced the service economy. John Wiley & Sons, West Sussex (2008)
37. Yu, R., Tzeng, G.-H.: A soft computing method for multi-criteria decision making with dependence and feedback. Applied Mathematics and Computation 180(1), 63–75 (2006)

Self-tuning PI Controllers via Fuzzy Cognitive Maps

Engin Yesil[1], M. Furkan Dodurka[1,2], Ahmet Sakalli[1],
Cihan Ozturk[1], and Cagri Guzay[1]

[1] Istanbul Technical University, Faculty of Electrical and Electronics Engineering,
Control Engineering Department, Maslak, TR-34469, Istanbul, Turkey
[2] GETRON Bilişim Hizmetleri A. Ş., Yıldız Teknik Üniversitesi Davutpaşa Kampüsü,
Teknopark Binası B1 Blok, Esenler, 34220, Istanbul, Turkey
{yesileng,dodurkam,sakallia,ozturkci,guzay}@itu.edu.tr,
furkan.dodurka@getron.com

Abstract. In this study, a novel self-tuning method based on fuzzy cognitive maps (FCMs) for PI controllers is proposed. The proposed FCM mechanism works in an online manner and is activated when the set-point (reference) value of the closed loop control system changes. Then, FCM tuning mechanism changes the parameters of PI controller according to systems' current and desired new reference value to improve the transient and steady state performance of the systems. The effectiveness of the proposed FCM based self-tuning method is shown via simulations on a nonlinear system. The results show that the proposed self-tuning methods performances are satisfactory.

Keywords: Fuzzy cognitive maps, PI controllers, self-tuning, supervisory control, optimization.

1   Introduction

Although many innovative methodologies have been devised in the past 50 years to handle complex control problems and to achieve better performances, the great majority of industrial processes are still controlled by means of simple proportional-integral-derivative (PID) controllers. PID controllers, despite their simple structure, assure acceptable performances for a wide range of industrial plants and their usage (the tuning of their parameters) is well known among industrial operators. Hence, PID controllers provide, in industrial environments, a cost/benefit performance that is difficult to beat with other kinds of controllers. Astrom states that more than 90% of all control loops utilize PID and most of loops are in fact PI [1].

Cognitive maps were introduced for the first time by Axelrod [2] in 1976 in order to signify the binary cause-effect relationships of the elements of an environment. Fuzzy cognitive maps (FCM) are fuzzy signed directed graphs with feedbacks, and they can model the events, values, goals as a collection of concepts by forging a causal link between these concepts [3]. FCM nodes represent concepts, and edges represent causal links between the concepts. Most widely used aspects of the FCMs are their potential for use in learning from historical data and decision support as a

H. Papadopoulos et al. (Eds.): AIAI 2013, IFIP AICT 412, pp. 567–576, 2013.

prediction tool. Given an initial state of a system, represented by a set of values of its constituent concepts, an FCM can simulate its evolution over time to learn from history and predict its future behavior. For instance, it may stand for that the system would converge to a point where a certain state of balance would exist, and no further changes would occur.

The main advantages of FCMs are their flexibility and adaptability capabilities [4]. As mentioned in [5], [6] and [7], there is a vast interest in FCMs and this interest on the part of researchers and industry is increasing in many areas such as control. In [8], FCM is studied for modeling complex systems and controlling supervisory control systems. In [9], learning approaches based on nonlinear Hebbian rule to train FCMs that model industrial process control problems is performed. A cognitive–fuzzy model, aiming online fuzzy logic controller (FLC) design and self-fine-tuning is implemented [10]. Fuzzy cognitive network (FCN) is used to the adaptive weight estimation based on system operation data, fuzzy rule storage mechanism to control unknown plants [11]. Besides, FCN is used to construct a maximum power point tracker (MPPT) that operates in cooperation with a fuzzy MPPT controller [12]. By combining topological and metrical approaches, an approach to mobile robot map-building that handles qualitatively different types of uncertainty is proposed [13]. A method for neural network FCM implementation of the fuzzy inference engine using the fuzzy columnar neural network architecture (FCNA) is proposed [14].

In this paper, a novel procedure is proposed to design a self-tuning PI controller via FCM particularly for nonlinear systems. Because of nonlinear systems' dissimilar characteristics at different operating points, fixed PI controllers cannot perform successive behaviors. In the proposed self-tuning method, FCM is used to supervise the control system and decide to change the controller parameters when the operating point changes.

2   A Brief Overview of Fuzzy Cognitive Maps

A fuzzy cognitive map F is a 4-tuple (N, W, C, f) [15] where; $N = \{N_1, N_2, ..., N_n\}$ is the set of n concepts forming the nodes of a graph. W: $(N_i, N_j) \rightarrow w_{ij}$ is a function of N×N to K associating w_{ij} to a pair of concepts (N_i, N_j), with w_{ij} denoting a weight of directed edge from N_i to N_j if $i \neq j$ and w_{ij} equal to zero otherwise. Therefore, in brief, $W(N \times N) = (w_{ij}) \in K^{n \times n}$ is a connection matrix. C: $N_i \rightarrow C_i$ is a function that at each concept N_i associates the sequence of its activation degrees such as for $t \in N$, $C_i(t) \in L$ given its activation degree at the moment t. $C(0) \in L^n$ indicates the initial vector and specifies initial values of all concept nodes and $C(t) \in L^n$ is a state vector at certain iteration t. f: $R \rightarrow L$ is a transformation function, which includes recurring relationship on $t \geq 0$ between $C(t + 1)$ and $C(t)$.

The sign of w_{ij} expresses whether the relation between the two concepts is direct or inverse. The direction of causality expresses whether the concept C_i causes the concept C_j or vice versa. Thus, there are three types of weights [16]:

$W_{ij} > 0$, indicates positive causality,

$W_{ij} < 0$, indicates negative causality,

$W_{ij} = 0$, indicates no relation.

Values of concepts change as simulation goes on are calculated by the following formula [17]:

$$C_j(t+1) = f\left(\sum_{\substack{i=1 \\ i \neq j}}^{n} C_i(t)w_{ij}\right) \tag{1}$$

where $C_i(t)$ is the value of ith node at the tth iteration, e_{ij} is the edge weight (relationship strength) from the concept C_i to the concept C_j, t is the corresponding iteration, N is the number of concepts, and f is the transformation (transfer) function.

In general, there are two kinds of transformation functions used in the FCM framework. The first one is the unipolar sigmoid function, where $\lambda > 0$ decides the steepness of the continuous function f and transforms the content of the function in the interval [0,1].

$$f(x) = \frac{1}{1+e^{-\lambda x}} \tag{2}$$

The other transformation function, hyperbolic tangent, that has been used and which transforms the content of the function is in the interval [-1,1],

$$f(x) = tanh(\lambda x) = \frac{e^{\lambda x}-e^{-\lambda x}}{e^{\lambda x}+e^{-\lambda x}} \tag{3}$$

where λ is a parameter used to determine proper shape of the function. Both functions use λ as a constant for function slope.

3   PID Controllers

In industrial environments, PID controllers provide a cost/benefit performance that is difficult to beat with other kinds of controllers. It should be pointed that PID controllers actually possess characteristics of both PI and PD controllers. However, because of their simple structure, PID controllers are particularly suited for pure first or second order processes, while industrial plants often present characteristics such as high order, long time delays, nonlinearities and so on. For these reasons, it is highly desirable to increase the capabilities of PID controllers by adding new features; in this way, they can improve their performances for a wide range of plants while retaining their basic characteristics.

Åström and Hägglund [1] have stated that most of PID controllers that using in industry is PI controller, which means that D term is set to zero. Because of noises and disturbances on the process, derivative actions may affect control performances adversely. In order to avoid this drawback in industrial applications, PID controllers are mostly used as PI controllers without losing steady state performances. However, in some cases, control signal might be generated out of the operating range of actuators due to high-valued integral term. Because of these limitations and saturations, closed loop control structure might be broken for a while. Accordingly, system begins following set point with a short-term steady state error which is called wind-up [18].

PID controller in parallel form includes sum of proportional, integral, and derivative terms of the error signal. It can be written in time-domain as given as follows:

$$u(t) = K_p e(t) + K_i \int e(t)dt + K_d \frac{de(t)}{dt} \tag{4}$$

Laplace form of the (4) is simpler than the time-domain. Transfer function of a parallel connected PID controller is written in (5) as below

$$F(s) = \frac{U(s)}{E(s)} = K_p + \frac{K_i}{s} + K_d s = K_p(1 + \frac{1}{T_i s} + T_d s) \tag{5}$$

where K_p is the proportional, K_i is the integral, K_d is the derivative gain, T_i is the integral T_d is the derivative time constants. The tuning methods of PID controllers for the both linear and nonlinear systems have different approaches.

A system is said to be linear if it obeys the two fundamental principles of homogeneity and additivity. If a given process does not satisfy with the two principles, it can be said that the process is a nonlinear. The controller design for a linear system is more straight-forwardly in comparison with a nonlinear system. It is an obvious fact that the most PID controller design methods are focused on linear systems up to the present. The linear system is not dependent on initial conditions or operating point. [19] states that the essential disadvantage of existing design methods of PI or PID controllers is that desire transient responses cannot be assured for nonlinear systems especially parameter variations and unknown external disturbances. It is a novel idea to overcome this problem that self-tuning methods which determine the controller parameters might be used. As mentioned above, controlling the nonlinear systems with linear PID controllers is not a convenient strategy. Some studies show that combining of the PID controller with gain scheduling gives fine result [20]. The auto-tuning of a PID controller procedure needs gradually less exertion than gain schedule strategy [1]. In these mechanisms, PID parameters are the functions of error (and derivative of error) or/and process states [21]. There are varied prominent studies in self-tuning PID field. Supervisory control or self-tuning can be executed via different methods, for instance optimization based [21-23], fuzzy-logic mechanism [24, 25], neural networks [26, 27]. In this paper, for the first time in the literature and beside of mentioned self-tuning PID strategies, the self-tuning PI controller via FCM method is proposed.

4   Self-tuning Method Based on FCM for PI Controllers

In this section, the proposed self-tuning PI controller via fuzzy cognitive map, which is illustrated in Fig. 1, is presented particularly for the nonlinear systems. When the reference value (r(t)) of the closed-loop control system is changed, FCM tuning mechanism is triggered. Then, FCM self-tuning mechanism changes control parameters, which are the static gain (K_p) and the integral time constant (T_i) of PI controller according to system's current state and destination. Since systems are nonlinear, its dynamics and static characteristics are not same for different working points. So, as the process state changes the optimal parameters of PI control will naturally change nonlinearly. The nonlinear behavior of changing PI control parameters are mimicked with the help of FCM. Therefore, main objective is designing a FCM mechanism that represents nonlinear changing behavior of PI control parameters for different operating conditions. Since K_p and T_i are independent from each other, FCM design can be separated into two sub FCM design. The design methodology is the same for each controller parameter, so only one of the sub-FCM is discussed in details.

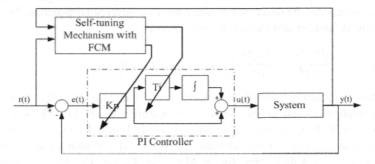

Fig. 1. Self-tuning PI controller structure

In a FCM design, the first step is determining the concepts. In the tuning mechanism, the system's current output and the desired set-point are the input nodes, given as C_1 and C_2. In addition, the output concept is K_p (or T_i), as C_N. To represent nonlinearity of the self-tuning strategy a number of extra inner concepts C_n, which represent the nonlinear behavior of the parameter change, are needed. Here, $n = 3, 4, ..., N\text{-}1$ and N is total of the concepts depending on the system.

The next step is to determine which concepts are connected to each other. In proposed sub FCM, it is assumed that input concepts are affecting to the whole other concepts, moreover output concepts are affected by the whole other concepts. The inner nodes are affected from the previous other nodes, and affects to next nodes with a one iteration delay. Therefore, one of the proposed inner concept,C_n, is affected from the previous concepts $C_{n-1}, C_{n-2}, ..., C_1$ representing nonlinearity. In a similar way, C_n is affecting the further inner concepts $C_{n+1}, C_{n+2}, ..., C_N$ with a one iteration delay.

The third step is to determine the transformation functions f given in (3). In the proposed method all concepts expect inner nodes have their own transformation functions. By putting all together the change of C_n provided that $n > 2$ at $(t+1)^{th}$ iteration will be calculated as follows:

$$C_n(t+1) = f_n\left(C_1(t) + C_2(t) + \sum_{\substack{i=3 \\ i \neq n}}^{N-1} C_i(t - (i-2))\right) \qquad (6)$$

where f_n is corresponding transformation function of C_n with λ_n.

The last step is to determine connection matrix W_{NxN} and λ_n for transformation functions with an appropriate FCM learning methods [28-30] where $n = 3, 4 \dots, N$. Then the built two sub-FCMs are merged into a single FCM to construct the self-tuning mechanism.

5   Simulation Results

A second order nonlinear process with time delay is chosen in order to demonstrate the effectiveness of proposed self-tuning PI controller via FCM. As given in [31], the nonlinear process can be described by the following differential equation. In practical studies, time delay constant (L) is fixed to 5 seconds.

$$\frac{d^2y(t)}{dt^2} + \frac{dy(t)}{dt} + 0.25y^2 = u(t - L) \tag{7}$$

For this simulation example, the number of inner concepts that represent the nonlinear relations in self-tuning is chosen as 6; therefore a FCM with 16 concepts is design. The FCM designed for self-tuning is given in Fig. 2. Concepts C_1 and C_2 are chosen as input concepts systems current state point and destination state point, which is the new reference value, respectively. Moreover, C_9 and C_{16} are chosen as output concepts which represent K_p and T_i.

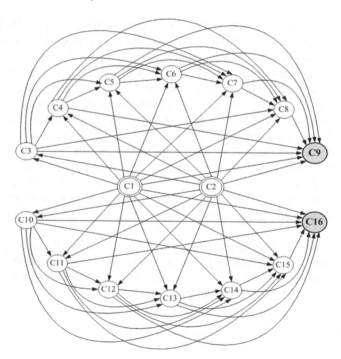

Fig. 2. Illustration of designed FCM for self-tuning mechanism

As a first step, optimal PI controller parameters for different operating points have been gathered using Big Bang – Big Crunch (BB-BC) algorithm [32] via many simulations. Integral square error (ISE) is chosen as the cost function in determination of the optimal PI parameters and historical data for learning of FCM is obtained. Then, all of the concepts values normalized in order to fall within the range [-1, 1] by dividing their maximum values respectively. For determination of the weights between the concepts of the proposed FCM, also, BB-BC learning methodology is used [29] with FCM-GUI [33]. The weight matrix obtained at the end of BB-BC learning is as follows:

$$
W = \begin{bmatrix}
0 & 0 & -0.86 & -1.00 & -1.00 & 0.36 & -0.98 & 0.69 & 0.55 & 0.91 & 0.52 & -0.2 & 0.56 & 0.01 & 0.45 & -0.09 \\
0 & 0 & 0.74 & 0.86 & -0.44 & 0.79 & -0.47 & -0.96 & 0.92 & 0.62 & 0.59 & 1 & 0.12 & -0.97 & -0.19 & 0.5 \\
0 & 0 & 0 & -0.67 & -0.39 & -0.26 & 0.41 & 0.81 & -0.21 & 0 & 0 & 0 & 0 & 0 & 0 & 0 \\
0 & 0 & 0 & 0 & 0.97 & 0.16 & 0.54 & -0.42 & -0.91 & 0 & 0 & 0 & 0 & 0 & 0 & 0 \\
0 & 0 & 0 & 0 & 0 & -0.19 & -0.75 & 0.24 & -0.71 & 0 & 0 & 0 & 0 & 0 & 0 & 0 \\
0 & 0 & 0 & 0 & 0 & 0 & -0.35 & 0.21 & 0.63 & 0 & 0 & 0 & 0 & 0 & 0 & 0 \\
0 & 0 & 0 & 0 & 0 & 0 & 0 & 0.5 & 0.5 & 0 & 0 & 0 & 0 & 0 & 0 & 0 \\
0 & 0 & 0 & 0 & 0 & 0 & 0 & -0.74 & 0 & 0 & 0 & 0 & 0 & 0 & 0 & 0 \\
0 & 0 & 0 & 0 & 0 & 0 & 0 & 0 & 0 & 0 & 0 & 0 & 0 & 0 & 0 & 0 \\
0 & 0 & 0 & 0 & 0 & 0 & 0 & 0 & 0 & -0.33 & 0.2 & -0.70 & 1 & 0.52 & -0.91 \\
0 & 0 & 0 & 0 & 0 & 0 & 0 & 0 & 0 & 0 & -0.01 & -0.48 & 0.22 & 0.03 & 0.92 \\
0 & 0 & 0 & 0 & 0 & 0 & 0 & 0 & 0 & 0 & 0 & -0.34 & 0.07 & -0.07 & -0.95 \\
0 & 0 & 0 & 0 & 0 & 0 & 0 & 0 & 0 & 0 & 0 & 0 & -0.84 & 0.02 & -0.87 \\
0 & 0 & 0 & 0 & 0 & 0 & 0 & 0 & 0 & 0 & 0 & 0 & 0 & 0.37 & 0.37 \\
0 & 0 & 0 & 0 & 0 & 0 & 0 & 0 & 0 & 0 & 0 & 0 & 0 & 0 & 0.41 \\
0 & 0 & 0 & 0 & 0 & 0 & 0 & 0 & 0 & 0 & 0 & 0 & 0 & 0 & 0 \\
\end{bmatrix} \quad (8)
$$

The λ values obtained for the concepts are tabulated in Table 1.

Table 1. Found λ_n values of transformation function of C_n

Consept	C_3	C_4	C_5	C_6	C_7	C_8	C_9
λ Value	0.17	4.41	5.00	3.58	1.41	1.11	0.88
Consept	C_{10}	C_{11}	C_{12}	C_{13}	C_{14}	C_{15}	C_{16}
λ Value	0.88	0.09	4.31	4.84	0.00	1.11	0.97

After obtaining the connection matrix W, four different operating conditions are chosen in order to test the performances of the proposed self-tuning PI controller. While the process is operating at steady state value which is 2, the set point values are changed orderly as 1.8, 2.2, 1.9 and 2.1. Hence, there are four different set points and control regions which means that the process should operate for each operating conditions whether it has different dynamic behaviors. FCM determines the PI controller parameters for each new set point value and triggers controller to changes the controller parameters for optimal behavior at recent operating condition and process dynamics. Table 2 shows the optimal PI parameters and the PI parameters obtained by FCM based self-tuning mechanism. System and controller output due to set point changes are illustrated in Fig. 3. As can be seen from Fig. 3, FCM Self-Tuning PI mechanism determines optimal PI controller parameters when set point changes. The controller parameters produced by the proposed FCM is very close to the optimal parameters. Therefore, the simulations show that the tuned PI controller can perform well in new working condition of system.

Table 2. Comparison of optimal controller parameters and proposed FCM parameters

Change in set-point	2.0-1.8		1.8-2.2		2.2-1.9		1.9-2.1	
	K_P	T_i	K_P	T_i	K_P	T_i	K_P	T_i
Optimal PI	0.91	5.48	3.43	3.82	0.92	5.55	1.07	4.09
FCM PI	0.96	6.125	1.13	1.15	0.95	6.09	1.08	4.10

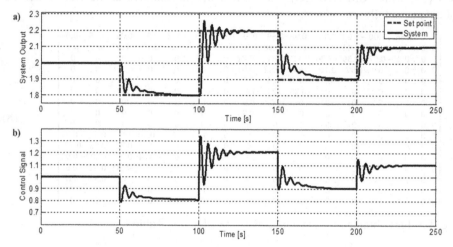

Fig. 3. The illustration of (a) system response, (b) control output

6   Conclusion

In this study, a novel self-tuning PI controller using fuzzy cognitive map is proposed for the first time in the literature. The proposed FCM mechanism tunes the PI controller parameters using system's current and desired operation point (reference value) when a change in set point occurs. For this study, various simulations are studied and one of the simulations based on a nonlinear system is presented. The obtained results show that proposed FCM adopts the controller according to system's new working points. The system output shows that PI controller performance is fairly for various operation conditions since the proposed FCM represent nonlinear self-tuning strategy efficiently. FCM is a resourceful tool for self-tuning mechanism of a PI controller for nonlinear systems.

For the future work, the self-tuning PI with FCM mechanism will be extended for PID controllers. Then, the proposed method will be implemented on a real-time experimental system.

References

1. Åström, K.J., Hägglund, T.: The future of PID control. Control Engineering Practice, 1163–1175 (2000)
2. Axelrod, R.: Structure of Decision: The Cognitive Maps of Political Elites. Princeton University Press, Princeton (1976)
3. Kosko, B.: Fuzzy cognitive maps. International Journal of Man-Machine Studies 24, 65–75 (1986)
4. Aguilar, J.: A survey about fuzzy cognitive maps papers. International Journal of Computational Cognition 3(2), 27–33 (2005)
5. Papageorgiou, E.I.: Review study on fuzzy cognitive maps and their applications during the last decade. In: Proceedings of the 2011 IEEE International Conference on Fuzzy Systems, pp. 828–835. IEEE Computer Society, Taipei (2011)
6. Papageorgiou, E.I.: Learning algorithms for fuzzy cognitive maps: A review study. IEEE Trans. Syst., Man Cybern. C Appl. Rev. 42(2), 150–163 (2011)
7. Papageorgiou, E.I., Salmeron, J.L.: A Review of Fuzzy Cognitive Map research during the last decade. Accepted for publication in IEEE Transactions on Fuzzy Systems 21(1), 66–79 (2013)
8. Stylios, C.D., Groumpos, P.P.: Modeling Complex Systems Using Fuzzy Cognitive Maps. IEEE Transactions on Systems, Man, and Cybernetics, Part A: Systems and Humans 34(1), 155–162 (2004)
9. Papageorgiou, E.I., Stylios, C., Groumpos, P.: Unsupervised learning techniques for fine-tuning fuzzy cognitive map causal links. International Journal of Human-Computer Studies 64, 727–743 (2006)
10. Gonzalez, J.L., Aguilar, L.T., Castillo, O.A.: cognitive map and fuzzy inference engine model for online design and self fine-tuning of fuzzy logic controllers. International Journal of Intelligent Systems 24(11), 1134–1173 (2009)
11. Kottas, T.L., Boutalis, Y.S., Christodoulou, M.A.: Fuzzy cognitive networks: Adaptive network estimation and control paradigms. In: Glykas, M. (ed.) Fuzzy Cognitive Maps. STUDFUZZ, vol. 247, pp. 89–134. Springer, Heidelberg (2010)
12. Kottas, T.L., Karlis, A.D., Boutalis, Y.S.: Fuzzy Cognitive Networks for Maximum Power Point Tracking in Photovoltaic Arrays. In: Glykas, M. (ed.) Fuzzy Cognitive Maps. STUDFUZZ, vol. 247, pp. 231–257. Springer, Heidelberg (2010)
13. Beeson, P., Modayil, J., Kuipers, B.: Factoring the mapping problem: Mobile robot map-building in the hybrid spatial semantic hierarchy. International Journal of Robotics Research 29(4), 428–459 (2010)
14. Ismael, A., Hussien, B., McLaren, R.W.: Fuzzy neural network implementation of self tuning PID control. In: Proc. IEEE Int. Symp. Intelligent Control, pp. 16–21 (1994)
15. Khan, M.S., Chong, A.: Fuzzy cognitive map analysis with genetic algorithm. In: Ind. Int. Conf. Artif. Intell. (2003)
16. Parsopoulos, K.E., Papageorgiou, E.I., Groumpos, P.P., Vrahatis, M.N.: A first study of fuzzy cognitive maps learning using particle swarm optimization. In: Proc. IEEE Congr. Evol. Comput., pp. 1440–1447 (2003)
17. Stach, W., Kurgan, L., Pedrycz, W.: A survey of fuzzy cognitive map learning methods. In: Grzegorzewski, P., Krawczak, M., Zadrozny, S. (eds.) Issues in Soft Computing: Theory and Applications, Exit, pp. 71–84 (2005)
18. Yesil, E., Ozturk, C., Cosardemir, B., Urbas, L.: MATLAB Case-Based Reasoning GUI application for control engineering education. In: IEEE Int. Conf. Information Technology Based Higher Education and Training (2012)

19. Yurkeyich, V.D.: PI/PID control for nonlinear systems via singular perturbation technique. Advances in PID Control (2011)
20. Anh, H. P. H., Nam, N. T.: A new approach of the online tuning gain scheduling nonlinear PID controller using neural network. PID Control, Implementation and Tuning (2011)
21. Mhaskar, P., El-Farr, A.N.H., Christofides, P.D.: A method for PID controller tuning using nonlinear control techniques. In: American Control Conference, pp. 2925–2930 (2004)
22. Liu, G.P., Daley, S.: Optimal-tuning PID control of hydraulic systems. Control Engineering Practice (8), 1045–1053 (2000)
23. He, S.Z., Tan, S., Xu, F.L., Wang, P.Z.: Fuzzy self-tuning of PID controllers. Fuzzy Sets and Systems 56, 37–46 (1993)
24. Yesil, E., Guzelkaya, M., Eksin, I.: Fuzzy logic based tuning of PID controllers for time delay systems. In: Artificial Intelligence and Soft Computing, pp. 236–241 (2006)
25. Soyguder, S., Karakose, M., Ali, H.: Design and simulation of self-tuning PID-type fuzzy adaptive control for an expert HVAC system. Expert Systems with Applications 36(3), 4566–4573 (2009)
26. Fang, M.C., Zhuo, Y.Z., Lee, Z.Y.: The application of the self-tuning neural network PID controller on the ship roll reduction in random waves. Ocean Engineering 37(7), 529–538 (2010)
27. Emilia, G.D., Marra, A., Natale, E.: Use of neural networks for quick accurate auto-tuning of PID controller. Robotics and Computer-Integrated Manufacturing 23(2), 170–179 (2007)
28. Yesil, E., Dodurka, M.F.: Goal-Oriented Decision Support using Big Bang-Big Crunch Learning Based Fuzzy Cognitive Map: An ERP Management Case Study. In: IEEE Int. Conf. Fuzzy Systems (2013)
29. Yesil, E., Urbas, L.: Big Bang - Big Crunch Learning Method for Fuzzy Cognitive Maps. In: International Conference on Control, Automation and Systems Engineering (2010)
30. Yesil, E., Ozturk, C., Dodurka, M.F., Sakalli, A.: Fuzzy Cognitive Maps Learning Using Artificial Bee Colony Optimization. In: IEEE Int. Conf. Fuzzy Systems (2013)
31. Mudi, R.K., Pal, N.P.: A robust self-tuning scheme for PI- and PD-type fuzzy controllers. IEEE Transactions on Fuzzy Systems 7(1), 2–16 (1999)
32. Erol, O.K., Eksin, I.: A new optimization method: Big Bang–Big Crunch. Advances in Engineering Software 37, 106–111 (2006)
33. Yesil, E., Urbas, L., Demirsoy, A.: FCM-GUI: A graphical user interface for Big Bang-Big Crunch Learning of FCM. In: Papageorgiou, E. (ed.) Fuzzy Cognitive Maps for Applied Sciences and Engineering – From Fundamentals to Extensions and Learning Algorithms. Intelligent Systems Reference Library. Springer (2013)

Concept by Concept Learning of Fuzzy Cognitive Maps

M. Furkan Dodurka[1,2], Engin Yesil[1], Cihan Ozturk[1],
Ahmet Sakalli[1], and Cagri Guzay[1]

[1] Istanbul Technical University, Faculty of Electrical and Electronics Engineering,
Control Engineering Department, Maslak, TR-34469, Istanbul, Turkey
[2] GETRON Bilişim Hizmetleri A. Ş., Yıldız Teknik Üniversitesi Davutpaşa Kampüsü,
Teknopark Binası B1 Blok, Esenler, 34220, Istanbul, Turkey
{dodurkam,yesileng,ozturkci,sakallia,guzay}@itu.edu.tr,
furkan.dodurka@getron.com

Abstract. Fuzzy cognitive maps (FCM) are fuzzy signed directed graphs with feedbacks; they are simple and powerful tool for simulation and analysis of complex, nonlinear dynamic systems. However, FCM models are created by human experts mostly, and so built FCM models are subjective and building a FCM model becomes harder as number of variables increases. So in the last decade several methods are proposed providing automated generation of fuzzy cognitive maps from data. The main drawback of the proposed automated methods is their weaknesses on handling with large number of variables. The proposed method brings out a new strategy called concept by concepts approach (CbC) approach for learning of FCM. It enables the generation of large sized FCM models with a high precision and in a rapid way using the historical data.

Keywords: Fuzzy cognitive maps, learning, density, global optimization.

1 Introduction

Cognitive maps were introduced for the first time by Axelrod [1] in 1976 in order to signify the binary cause-effect relationships of the elements of an environment. Fuzzy cognitive maps (FCM) are fuzzy signed directed graphs with feedbacks, and they can model the events, values, goals as a collection of concepts by forging a causal link between these concepts [2]. FCM nodes represent concepts, and edges represent causal links between the concepts. Most widely used aspects of the FCMs are their potential for use in learning from historical data and decision support as a prediction tool.

The main advantages of FCMs are their flexibility and adaptability capabilities [3]. As stated in [4], [5] and [6], there is an enormous interest in FCMs and this interest on the part of researchers and industry is increasing, especially in the areas of control [7], political and social sciences [8], business [9], medicine [10], robotics [11], environmental science [12], agriculture [13] and information technology [14].

Mainly, there are two types of FCMs called manual FCMs and automated FCMs. The unique difference between them is the way used for forming the FCMs. Manual FCMs are produced by experts manually and automated FCMs are produced by other

H. Papadopoulos et al. (Eds.): AIAI 2013, IFIP AICT 412, pp. 577–586, 2013.

information sources numerically [15]. Even sometimes, producing a FCM manually becomes difficult when the experts' interference could not be enough to solve the problem. Because of difficulties in manual FCM generation, the development of computational methods for learning FCM is required for automated FCMs. Lately, a large number of methods for learning FCM model structure have been proposed. These proposed methods can be summed in three groups named Hebbian-type learning methods, population-based (evolutionary) learning methods and hybrid learning algorithms [4].

A simple differential Hebbian learning law (DHL) for FCM is stated in [16]. This has been extended in [17] as a balanced differential learning algorithm for FCM. Further extensions, called nonlinear Hebbian learning (NHL) and Active Hebbian learning algorithm (AHL) are presented in [18] and [19], respectively. An improved version of the NHL method named data driven NHL (DDNHL) is proposed in [20]. Another study to train a FCM is proposed in which a new model for unsupervised learning and reasoning on a special type of cognitive maps that are realized with Petri nets [21]. All Hebbian-type learning methods have the goal to learn the connection matrix with single historical data set.

Rather than Hebbian-type learning methods, population-based learning methods are more in demand. The learning goal of population-based methods can be connection matrix with optimal weights or matching input pattern. Obtaining the connection matrix with optimal weights will lead FCM to its desired activation state values for each concept. Population based learning algorithms with connection matrix goal of learning that are recently studied can be listed as: Particle Swarm Optimization (PSO) [22], Genetic Strategy (GS) [23], Real-coded Genetic Algorithm (RCGA) [24], Simulated Annealing (SA) [15], tabu search [25], immune algorithm [26], Big Bang-Big Crunch (BB-BC) optimization algorithm [27], Extended Great Deluge Algorithm (EDGA) [28], Artificial Bee Colony (ABC) algorithm [29]. In addition, GA [30] and BB-BC [31] learning are used for goal oriented decision support systems on FCM.

The hybrid learning methods are implemented by combining the first two mentioned learning types (Hebbian-based learning (HL) and the population-based learning) for FCMs. There are two hybrid algorithms studied, one has combined NHL and differential evolution (DE) [32] and the other algorithm has combined RCGA and NHL algorithms [33]. The learning goals of these two the hybrid learning methods are the connection matrix and they use single historical data set.

In this study, a novel and comprehensive learning approach called concept by concept (CbC) for the development of fuzzy cognitive maps is proposed. The existing optimization based learning approaches try to find the weights between the concepts at once. The main difference of the proposed approach from the existing learning methods is focusing on only one concept and the links (arcs) to this concept first, and then learning the weights of these connection weights. Then the algorithm searches the next concept and its links and learns the weights. In order to use proposed CbC the historical data of all the concepts must be known. The proposed approach is able to generate a FCM model from input data consisting of a single or multiple sequences of concept state vector values. Proposed CbC is applicable with any of population based optimization algorithms proposed in literature for learning of FCMs. The benefit of this learning approach is presented with two simulation examples.

2   A Brief Overview of Fuzzy Cognitive Maps

A fuzzy cognitive map F is a 4-tuple (N, W, C, f) [30] where;

   $N = \{N_1, N_2, ..., N_n\}$ is the set of n concepts forming the nodes of a graph.

   W: $(N_i, N_j) \rightarrow w_{ij}$ is a function of NxN to K associating w_{ij} to a pair of concepts (N_i, N_j), with w_{ij} denoting a weight of directed edge from N_i to N_j,. Thus $W(N \times N) = (w_{ij}) \in K^{n \times n}$ is a connection matrix.

   C: $N_i \rightarrow C_i$ is a function that at each concept N_i associates the sequence of its activation degrees such as for $t \in N$, $C_i(t) \in L$ given its activation degree at the moment t. $C(0) \in L^n$ indicates the initial vector and specifies initial values of all concept nodes and $C(t) \in L^n$ is a state vector at certain iteration t.

   f: $R \rightarrow L$ is a transformation function, which includes recurring relationship on $t \geq 0$ between $C(t + 1)$ and $C(t)$.

   The sign of w_{ij} expresses whether the relation between the two concepts is direct or inverse. The direction of causality expresses whether the concept C_i causes the concept C_j or vice versa. Thus, there are three types of weights [22]:

   $W_{ij} > 0$, indicates positive causality,

   $W_{ij} < 0$, indicates negative causality,

   $W_{ij} = 0$, indicates no relation.

   Values of concepts change as simulation goes on are calculated by the following formula [34]:

$$C_j(t + 1) = f\left(\sum_{i=1}^{N} e_{ij} C_i(t)\right) \tag{1}$$

where $C_i(t)$ is the value of ith node at the tth iteration, e_{ij} is the edge weight (relationship strength) from the concept C_i to the concept C_j, t is the corresponding iteration, N is the number of concepts, and f is the transformation (transfer) function.

   In general, there are two kinds of transformation functions used in the FCM framework. The first one is the unipolar sigmoid function, where $\lambda > 0$ decides the steepness of the continuous function f and transforms the content of the function in the interval [0, 1].

$$f(x) = \frac{1}{1+e^{-\lambda x}} \tag{2}$$

The second transformation function, hyperbolic tangent, that has been used and which transforms the content of the function is in the interval [-1, 1],

$$f(x) = \tanh(\lambda x) = \frac{e^{\lambda x} - e^{-\lambda x}}{e^{\lambda x} + e^{-\lambda x}} \tag{3}$$

where λ is a parameter used to determine proper shape of the function. Both functions use λ as a constant for function slope.

3   Concept by Concept Learning Methodology

Most of the proposed population based learning approaches become time consuming when the number of concepts of FCM is relatively high. The main reason of this is the quadratic growth of the number of the parameters to be found as concept numbers increasing [35]. For a FCM with N concept, there will be N^2 of weights to be found with learning algorithm if there is no prior knowledge. Therefore, in order to reduce the computational dimension of learning, concept by concept (CbC) method, which is usable with any of the population based algorithms, is proposed.

There are some factors that make the learning of FCM difficult or limit learning method. These factors can be listed as types of input data [5, 29], number of nodes (N), knowledge of concepts' links, density measure, algorithms' search space, input nodes in FCM.

The proposed concept by concept (CbC) learning method develops a candidate FCM with any suitable population based global optimization algorithm from input and output data as given in Fig. 1. The input and output data are given as time series and that consist of a sequence of state vectors which describe a given system at consecutive iteration. The number of these successive iterations of the given historical data is called as the data length, K. Given a FCM with connection matrix (W_{NxN}) and collected data consisting output state vectors matrix of the FCM $(Output_{KxN})$, related to input state vectors matrix $(Input_{KxN})$. According to FCM iterative calculation formula (1), output state vector of t_{th} iteration is $(t+1)_{th}$ iteration of input state vector can be seen by (4)

$$Output(t)_n = Input(t+1)_n \tag{4}$$

Output of a FCM can be found from (1) and this also can be expressed by (5).

$$Output_{KxN} = f(Input_{KxN} \times W_{NxN}) \tag{5}$$

The goal of learning is to determine candidate connection matrix $(\widehat{W}_{NxN})$. As mentioned previously, assuming that all real input and output data is available, so (5) can be stated as follows:

$$Output_{KxN} = f(Input_{KxN} \times \widehat{W}_{NxN}) \tag{6}$$

Therefore, learning problem becomes NxN dimensional optimization problem. Since all real input and output data are known, this problem can be reduced to N total of optimization problem with N dimensional as given in Fig. 1. For example, given $\widetilde{output}_n$ is a n_{th} column of $Output_{KxN}$ is a vector which consists outputs of n_{th} concept C_n. So (6) can be reduced to (7) given as

$$\widetilde{output}_n = f(Input_{KxN} \times \widehat{w}_n) \tag{7}$$

where $\widehat{w}_n$ is a vector of n_{th} column of $\widehat{W}_{NxN}$. Concept C_n is only influenced from the weights in $\widehat{w}_n$, thus, for determining candidate connection matrix $\widehat{W}_{NxN}$ problem can be separated into N sub-learning problems; aiming to determine $\widehat{w}_n$ for $n = 1, 2, ..., N$. Accordingly, $\widehat{W}_{NxN}$ matrix can be found by carrying out N learning problem in (7) for $Output_{(K,n)}$.

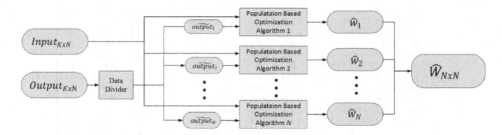

Fig. 1. Concept by Concept (CbC) FCM Learning Scheme

4   Simulation Examples

In this study, two systems with different characteristics are studied. The historical data is generated from the FCMs to use for the proposed CbC learning approach. The first system is a real-world FCM being a relatively easy problem with 13 nodes to show the proposed method's power of convergence accuracy. The second system is a synthetic (randomly generated) FCM with 100 nodes. This FCM is generated with 100 nodes and 100% density in order to show that the proposed method is able to cope with large scaled FCMs. For optimization algorithms in system 1 and 2, different types of cost functions that will be discussed in this section are selected to show the adaptiveness of the proposed method. In the simulation studies, BB-BC optimization algorithm [37] has been used as the global optimization method for CbC learning approach because of its fast convergence.

4.1   System 1: Suspension Viscosity FCM

Suspension viscosity FCM [36] is given in (8), stating what the factors affect suspension viscosity. The 13 concepts (C1 and C2 concepts are the input concepts) of suspension viscosity FCM model are given in Table 1. The transformation function given in (3) is utilized since the values of the nodes may fall within the range [-1, 1] and λ is chosen as 2 for generating the values of all concepts.

Table 1. Concepts of Suspension Viscosity FCM Model

C1:	Gravity	C7:	Liquid viscosity
C2:	Mechanical properties of particles	C8:	Effective particle shape
C3:	Physicochemical interaction	C9:	Effective particle size
C4:	Hydrodynamic interaction	C10:	Temperature
C5:	Effective particle concentration	C11:	Inter-particle attraction
C6:	Particle-particle contact	C12:	Floc structure formation
		C13:	Shear rate

The connection matrix of the FCM built by the experts [36] is as in (8). The density value of this studied FCM is 39%. As mentioned above, it is assumed that the links of concepts nodes are known.

$$W = \begin{bmatrix} 0 & 0 & 0 & 0.03 & 0 & 0 & 0 & 0 & 0 & 0 & -0.13 & 0 & 0 \\ 0 & 0 & 0 & 0.23 & 0 & 0.17 & 0 & 0.33 & 0.37 & 0 & 010 & 0 & 0 \\ 0 & 0 & 0 & 0 & 0.17 & 0.17 & 0 & 0.13 & 0.13 & 0 & 0.23 & 0.27 & 0 \\ 0 & 0 & 0.17 & 0 & 0.17 & 0.03 & 0 & 0 & 0 & 0 & 0 & -0.10 & 0 \\ 0 & 0 & 0 & 0.5 & 0 & 0.60 & 0.23 & 0 & 0 & 0 & 0.33 & 0.13 & -0.13 \\ 0 & 0 & 0.17 & 0.17 & 0.33 & 0 & 0 & 0 & 0.13 & 0 & 0 & 0 & 0 \\ 0 & 0 & 0 & 0.23 & 0 & -0.33 & 0 & 0 & 0 & 0 & -0.13 & -0.23 & -0.10 \\ 0 & 0 & 0 & 0.30 & 0 & 0 & -0.13 & 0 & 0 & 0 & -0.23 & 0.37 & 0 \\ 0 & 0 & 0 & 0.17 & 0 & 0.13 & 0.10 & 0 & 0 & 0 & 0.13 & 0.23 & 0 \\ 0 & 0 & 0.27 & 0.10 & 0.1 & 0.27 & -0.63 & 0.10 & 0 & 0 & 0.10 & 0.10 & 0.10 \\ 0 & 0 & 0.23 & 0 & 0.07 & 0.10 & 0.03 & 0 & 0 & 0 & 0 & 0.53 & 0 \\ 0 & 0 & 0 & 0 & 0.1 & 0 & 0.13 & 0.33 & 0.10 & 0 & 0 & 0 & 0 \\ 0 & 0 & 0 & 0 & 0.03 & 0.13 & -0.33 & -0.13 & 0 & 0.10 & 0 & -0.33 & 0 \end{bmatrix} \quad (8)$$

For input data 5 different historical data sets gathered for different initial 5 state vectors. All weights are searched with in [-1 1] bounds. Chosen cost function for learning is given as

$$f_n = \frac{1}{(K-1)S} \sum_{z=1}^{S} \sum_{t=1}^{K-1} \|C_n(t) - \hat{C}_n\| \qquad (9)$$

where $C_n(t)$ is the given system response, $\hat{C}_n$ is the candidate FCM response of the nth concept for the initial state vector, S is number of gathered historical data sets, which is 5 for this example.

4.2   System 2: Randomly Generated FCM

Rather than, relatively small size FCMs with relatively small value of density parameter, a randomly generated FCM with 100 nodes and 100% density parameter is studied for this example, where no input concept exists. In addition, it is assume that there is no prior knowledge about links between concept nodes. The transformation function given in (3) is utilized since the values of the nodes may fall within the range [-1, 1] and λ is chosen 0.3.

In this simulation example, BB-BC optimization algorithm has been used as in System 1 and single input historical data is gathered. All weights are searched with in [-1 1] bounds. The cost function is selected as follows:

$$f_n = \frac{1}{(K-1)S} \sum_{z=1}^{S} \sum_{t=1}^{K-1} (C_i(t) - \hat{C}_i)^2 \qquad (10)$$

5   Results and Discussions

In order to show the effectiveness of the Concept by Concept (CbC) learning method, two FCMs with different sizes are studied. The study is divided into three phases: Learning phase, generalization capability testing phase and closeness of weights testing phase. All the parameters used in these three phases are summarized in Table 2 where θ is the fitness function coefficient, R is the randomly selected initial state vector and D is the number of parameters to be optimized.

Table 2. Parameters Used in BB-BC Learning and Generalization Capability Test Phases

System No	Population number (N)	Number of iterations	θ	R	D
1	50	4000	10^3	100	66
2	100	10000	10^3	100	10000

For giving the results of the learning phase, the following criteria given in (11) is used.

$$J_1 = \frac{1}{N}\sum_{n=1}^{N} \hat{J}_n \tag{11}$$

where $\hat{J}_n$ is the selected cost function for using in learning phase given in (9) and (10).

In learning phase, 30 simulations are performed via FCM-GUI [38] for the 1st system, and for the 2nd system 10 simulations are performed via FCM-GUI.

In order to normalize, visualize and determine the convergence of the cost function given in (11) and the final value of the cost function, the following fitness function convergence performance which has the value [0, 1] is used [24]:

$$f = \frac{1}{\theta J_1 + 1} \tag{12}$$

where parameter θ is a positive fitness function coefficient. In Fig. 2, the average, worst and best fitness convergence performances of these two different FCM simulations are illustrated.

For the best candidate FCM's due to learning phase results, their generalization capabilities are tested. For this purpose, the real FCM and the candidate FCM are simulated for R randomly chosen initial state vectors [24]. The new criterion is selected as follows:

$$J_2 = \frac{1}{R(K-1)NS}\sum_{r=1}^{R}\sum_{z=1}^{S}\sum_{t=1}^{K}\sum_{n=1}^{N}(C_n^r(t) - \widehat{C_n^r})^2 \tag{13}$$

where $C_n^r(t)$ is the value of nth node at iteration t for the data generated by original FCM model started from the rth initial state vector, similarly, is the value of nth node at iteration t for the data generated by candidate FCM model started from the rth initial state vector.

As the 3th phase, accuracy of the weights is tested. For this criteria, (14) is used where $\widehat{W}_{ij}$ is weights of candidate FCM and W_{ij} is the weights of real FCM.

$$J_3 = \frac{1}{N^2}\sum_{n=1}^{N}(W_{ij} - \widehat{W}_{ij})^2 \tag{14}$$

In Table 3, minimum, maximum, mean and standard deviation values of error functions for learning, generalization capability and accuracy of weights phases are given. It is obvious that the obtained error function values are satisfactory.

Table 3. Learning and Generalization Capability and Accuracy of Weights Phases Results

	System #	Min J_1	Max J_1	Mean and Std. Dev.
Learning Phase	1	0	0	0 ± 0
	2	6.66×10^{-6}	6.98×10^{-6}	$6.87 \times 10^{-5} \pm 1.33 \times 10^{-7}$
		Min J_2	Max J_2	Mean and Std. Dev.
Generalization Phase	1	0	0	0 ± 0
	2	3.95×10^{-3}	1.42×10^{-2}	$8.97 \times 10^{-3} \pm 3.81 \times 10^{-3}$
		Min J_3	Max J_3	Mean and Std. Dev.
Accuracy of Weights Phase	1	0	0	0 ± 0
	2	1.26×10^{-4}	1.39×10^{-4}	$1.34 \times 10^{-6} \pm 1.93 \times 10^{-6}$

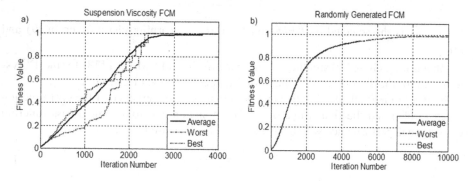

Fig. 2. Fitness functions performance (12) (a) First system; (b) Second system

6   Conclusion

A new approach called concept by concept (CbC) for learning of FCMs is proposed. This approach is applicable when the full historical data of all concepts are available and the approach is independent form the used population based optimization learning method and the chosen cost function. In order to show the benefit of CbC learning two simulation examples are presented. The results show that the proposed learning method is very effective, and able to learn large scaled FCMs with a high accuracy. For the future work, the proposed CbC will be used with popular population based learning methods to compare the performances of these algorithms.

Acknowledgments. This project was supported by TÜBİTAK TEYDEB Industrial Research Funding Program Grant Number 7120842, awarded to GETRON Bilişim Hizmetleri A.Ş.

References

1. Axelrod, R.: Structure of Decision: The Cognitive Maps of Political Elites. Princeton University Press, Princeton (1976)
2. Kosko, B.: Fuzzy cognitive maps. International Journal of Man-Machine Studies 24, 65–75 (1986)

3. Aguilar, J.: A survey about fuzzy cognitive maps papers. International Journal of Computational Cognition 3(2), 27–33 (2005)
4. Papageorgiou, E.I.: Review study on fuzzy cognitive maps and their applications during the last decade. In: 2011 IEEE International Conference on Fuzzy Systems, pp. 828–835. IEEE Computer Society, Taipei (2011)
5. Papageorgiou, E.I.: Learning algorithms for fuzzy cognitive maps: A review study. IEEE Trans. Syst., Man Cybern. C Appl. Rev. 42(2), 150–163 (2011)
6. Papageorgiou, E.I., Salmeron, J.L.: A Review of Fuzzy Cognitive Map research during the last decade. IEEE Transactions on Fuzzy Systems 21(1), 66–79 (2013)
7. Gonzalez, J.L., Aguilar, L.T., Castillo, O.: A cognitive map and fuzzy inference engine model for online design and self fine-tuning of fuzzy logic controllers. International Journal of Intelligent Systems 24(11), 1134–1173 (2009)
8. Andreou, A.S., Mateou, N.H., Zombanakis, G.A.: Soft computing for crisis management and political decision making: the use of genetically evolved fuzzy cognitive maps. Soft Computing Journal 9(3), 194–210 (2006)
9. Glykas, M.: Fuzzy cognitive strategic maps in business process performance measurement. Expert Systems with Applications 40(1), 1–14 (2013)
10. Papageorgiou, E.I., Froelich, W.: Application of Evolutionary Fuzzy Cognitive Maps for Prediction of Pulmonary Infections. IEEE Transactions on Information Technology in Biomedicine 16(1), 143–149 (2012)
11. Motlagh, O., Tang, S.H., Ismail, N., Ramli, A.R.: An expert fuzzy cognitive map for reactive navigation of mobile robots. Fuzzy Sets and Systems 201, 105–121 (2012)
12. Acampora, G., Loia, V.: On the Temporal Granularity in Fuzzy Cognitive Maps. IEEE Transactions Fuzzy Systems 19(6), 1040–1057 (2011)
13. Papageorgiou, E.I., Markinos, A.T., Gemtos, T.A.: Soft Computing Technique of Fuzzy Cognitive Maps to Connect Yield Defining Parameters with Yield in Cotton Crop Production in Central Greece as a Basis for a Decision Support System for Precision Agriculture Application. In: Glykas, M. (ed.) Fuzzy Cognitive Maps. STUDFUZZ, vol. 247, pp. 325–362. Springer, Heidelberg (2010)
14. Lee, K.C., Lee, S.: A causal knowledge-based expert system for planning an Internet-based stock trading system. Expert Systems With Applications 39(10), 8626–8635 (2012)
15. Alizadeh, S., Ghazanfari, M.: Learning FCM by chaotic simulated annealing. Chaos, Solutions & Fractals 41(3), 1182–1190 (2009)
16. Dickerson, J.A., Kosko, B.: Fuzzy virtual worlds. Artif. Intell. Expert 7, 25–31 (1994)
17. Vazquez, A.: A balanced differential learning algorithm in fuzzy cognitive maps. Technical report, Departament de Llenguatges I Sistemes Informatics, Universitat Politecnica de Catalunya, UPC (2002)
18. Papageorgiou, E.I., Stylios, C.D., Groumpos, P.P.: Fuzzy cognitive map learning based on nonlinear Hebbian rule. In: Australian Conference on Artificial Intelligence, pp. 256–268 (2003)
19. Papageorgiou, E.I., Stylios, C.D., Groumpos, P.P.: Active Hebbian learning algorithm to train fuzzy cognitive maps. Int. J. Approx. Reason. 37(3), 219–249 (2004)
20. Stach, W., Kurgan, L., Pedrycz, W.: Data driven nonlinear Hebbian learning method for fuzzy cognitive maps. In: Proc. IEEE World Congr. Comput. Intell., pp. 1975–1981 (2008)
21. Konar, A., Chakraborty, U.K.: Reasoning and unsupervised learning in a fuzzy cognitive map. Inf. Sci. 170, 419–441 (2005)
22. Parsopoulos, K.E., Papageorgiou, E.I., Groumpos, P.P., Vrahatis, M.N.: A first study of fuzzy cognitive maps learning using particle swarm optimization. In: Proc. IEEE Congr. Evol. Comput., pp. 1440–1447 (2003)

23. Koulouriotis, D.E., Diakoulakis, I.E., Emiris, D.M.: Learning fuzzy cognitive maps using evolution strategies: a novel schema for modeling and simulating high-level behavior. In: Proc. IEEE Congr. Evol. Comput., pp. 364–371 (2001)
24. Stach, W., Kurgan, L., Pedrycz, W., Reformat, M.: Genetic learning of fuzzy cognitive maps. Fuzzy Sets Syst. 153(3), 371–401 (2005)
25. Alizadeh, S., Ghazanfari, M., Jafari, M., Hooshmand, S.: Learning FCM by tabu search. Int. J. Comput. Sci. 2(2), 142–149 (2007)
26. Lin, C., Chen, K., He, Y.: Learning fuzzy cognitive map based on immune algorithm. WSEAS Trans. Syst. 6(3), 582–588 (2007)
27. Yesil, E., Urbas, L.: Big Bang - Big Crunch Learning Method for Fuzzy Cognitive Maps. In: International Conference on Control, Automation and Systems Engineering (2010)
28. Baykasoglu, A., Durmusoglu, Z.D.U., Kaplanoglu, V.: Training fuzzy cognitive maps via extended great deluge algorithm with applications. Comput. Ind. 62(2), 187–195 (2011)
29. Yesil, E., Ozturk, C., Dodurka, M.F., Sakalli, A.: Fuzzy Cognitive Maps Learning Using Artificial Bee Colony Optimization. In: IEEE Int. Conf. Fuzzy Systems (2013)
30. Khan, M.S., Chong, A.: Fuzzy cognitive map analysis with genetic algorithm. In: Ind. Int. Conf. Artif. Intell. (2003)
31. Yesil, E., Dodurka, M.F.: Goal-Oriented Decision Support using Big Bang-Big Crunch Learning Based Fuzzy Cognitive Map: An ERP Management Case Study. In: IEEE Int. Conf. Fuzzy Systems (2013)
32. Papageorgiou, E.I., Groumpos, P.P.: A new hybrid learning algorithm for fuzzy cognitive maps learning. Appl. Soft Comput. 5, 409–431 (2005)
33. Zhu, Y., Zhang, W.: An integrated framework for learning fuzzy cognitive map using RCGA and NHL algorithm. In: Int. Conf. Wireless Commun., Netw. Mobile Comput. (2008)
34. Stach, W., Kurgan, L., Pedrycz, W.: A survey of fuzzy cognitive map learning methods. In: Grzegorzewski, P., Krawczak, M., Zadrozny, S. (eds.) Issues in Soft Computing: Theory and Applications, Exit, pp. 71–84 (2005)
35. Stach, W., Kurgan, L., Pedrycz, W.: A divide and conquer method for learning large fuzzy cognitive maps. Fuzzy Sets Syst. 161(19), 2515–2532 (2010)
36. Banini, G.A., Bearman, R.A.: Application of fuzzy cognitive maps to factors affecting slurry rheology. International Journal of Mineral Processing 52, 233–244 (1998)
37. Erol, O.K., Eksin, I.: A new optimization method: Big Bang-Big Crunch. Advances in Engineering Software 37, 106–111 (2006)
38. Yesil, E., Urbas, L., Demirsoy, A.: FCM-GUI: A graphical user interface for Big Bang-Big Crunch Learning of FCM. In: Papageorgiou, E. (ed.) Fuzzy Cognitive Maps for Applied Sciences and Engineering – From Fundamentals to Extensions and Learning Algorithms. Intelligent Systems Reference Library. Springer (2013)

The Relevance of Fuzzy Cognitive Mapping Approaches for Assessing Adaptive Capacity and Resilience in Social–Ecological Systems

Frédéric M. Vanwindekens[1,2], Didier Stilmant[2], and Philippe V. Baret[1]

[1] Université catholique de Louvain – Earth and Life Institute – Agronomy,
Agroecology Croix du Sud 2 bte L7.05.14, B–1348 Louvain-la-Neuve, Belgium
{frederic.vanwindekens,philippe.baret}@uclouvain.be
[2] Walloon Agricultural Research Centre – Agriculture and Natural Environment
Department – Farming Systems, Territory and Information Technologies Unit
Rue du Serpont 100, B–6800 Libramont, Belgium
{f.vanwindekens,stilmant}@cra.wallonie.be

Abstract. Social–Ecological Systems (SES) are complex due to uncertainty related to their nature and their functions. In these systems, decision-making processes and practices of managers are often value-laden and subjective, dominated by their world-views and their own knowledge. People's knowledge are central in building their adaptive capacity but are seldom taken into account by traditional decision-making approaches in modelling SES management. In this paper, we introduce a Fuzzy Cognitive Mapping approach to study the dynamic behaviour of managers' systems of practices. As a case study, we aim to assess farmers' forage management under different climatic scenarios. Results show that summer drought have varying consequences according to farmers' systems of practices. Fuzzy Cognitive Mapping approaches are particularly relevant in studying systems of practices in SES. Their utilisation is promising for the evaluation of adaptive capacity and resilience in SES at local scale (exploitation, community) and regional scale (ecological areas, country).

Keywords: Agriculture, Social–Ecological Systems, Systems of Practices, Fuzzy Cognitive Mapping, Resilience Assessment.

1 Introduction

The management of Social–Ecological Systems (SES, [7]) is complex due to the intricacy of their components, to the uncertainty related to their nature and to the various societal, institutional, physical, ecological, economical processes involved in their functions [1]. Managers' strategies are largely influenced by their perceptions of the ecological, economical and social environments of SES [1]. These influences have been particularly pointed out and studied in the agricultural context [3,4,9]. In order to help farmers in managing their farm, Decision Support Systems (DSS) have been developed by 'management scientists' [13]. But unexpectedly, farmers pay little attention to these DSS [1,4,13,16].

H. Papadopoulos et al. (Eds.): AIAI 2013, IFIP AICT 412, pp. 587–596, 2013.

Recent scientific approaches have been developed to cope with incorporation of human, social and institutional aspects in SES models by explicitly accommodating relations between the natural and human environment [1]. Fuzzy cognitive maps (FCM) are particularly relevant tools in modelling SES based on people explicit knowledge [16] as they can be considered as a model of a belief system [11] constituted by concepts, the key drivers of the system, and edges, causal relationships between concepts. They have been developed by Bart Kosko in 1986 [12] in introducing the notion of 'fuzziness' and 'fuzzy weigth' to relationships of Robert Axelrod's cognitive maps (CM) [2].

In the agricultural context (see [23]), CM and FCM have been successfully applied for (i) analysing people knowledge, beliefs [19] and decision-making on farm [10], (ii) studying adoption of agri-environment measures [15], (iii) modelling farmers perceptions of how their ecosystem works [5] and of the sustainability of farms [6], and (iv) predicting yield production [17,18]. In order to study farmers' systems of practices (SOP) based on their own conceptions, we developed an approach for building cognitive maps by coding people's open-ended interviews. This approach was named CMASOP for 'Cognitive Mapping Approach for Analysing Actors' Systems of Practices' in SES. In a previous paper, we presented the core principles of CMASOP [23]. In this first publication, we applied CMASOP to the general description of farmers' SOP for managing grasslands in two Belgian agroecological areas [23]. In a second step, we developed complementary applications of CMASOP : a comparative one and a typological one. For comparing SOP between groups of managers defined a priori, we coupled CMASOP and descriptive statistical methods. For classifying SOP in a posteriori typological groups, we coupled CMASOP, clustering methods and statistical analysis. Results of the comparative and typological applications of CMASOP are being submitted [22].

In the present paper, we coupled CMASOP and auto-associative neural networks methods [16] for carrying out inferences about farmers' adaptations to climatic uncertainties. The objectives of this development is to model the dynamic behaviour of managers' SOP for assessing their adaptive capacity, and indirectly, the resilience of their SES. The aim of this paper is to present this new development of CMASOP and to demonstrate the relevance of using FCM approaches for assessing adaptive capacity of managers and resilience of their exploitation in social–ecological systems. The management of the second cut in grassland based livestock farming systems of southern Belgium is used as a case-study.

Resilience is defined by Folke et al. as the "capacity of a system to absorb disturbance and reorganize [...]" [7]. Adaptive capacity is defined as the "capacity of actors in a system to influence resilience" [7]. Different studies have used FCM for scenario analysis in SES [16,11,21,26]. They rely on the possibility to compare the steady state calculation under various conditions : (i) current situation, (ii) evolution of some environmental variables (prices, rainfall) or (iii) implementation of different policy options (laws, tax). These concepts are closed to the concept of 'vulnerability' that has been analysed using FCM by Murungweni et al. (2011) in the study of livelihood [14].

2   Materials and Methods

We studied farmers' Systems of practices (SOP) in Ardenne and Famenne, two grassland based agroecological areas in southern Belgium. We collected qualitative and quantitative data during forty-nine open-ended interviews on management of farms systems (structure, technical orientation, world views) and subsystems (forage, herd, grazing). We developed a cognitive mapping based approach for analysing systems of practices (CMASOP[1][23,22]). We applied it for studying grass forage management in our surveyed area.

The core principles of CMASOP consists in coding open-ended interviews of managers in order to create individual cognitive maps (ICM). These ICM can then be used to build a social cognitive map (SCM). As open-ended interviews focus on managers practices in social-ecological systems (SES), the ICM and SCM are considered as inductive models of SOP based on people conceptions [23]. The SCM is *inter alia* constituted by thirteen highly related concepts classed in seven core hubs (First, Second and Third cuts, Silo, Bale wrap, Hay and Cattle movement) and six peripheral hubs (Plot utilization, Plot-Farm distance, Forage quality, Forage quantity, Cutting date and Weather). A quote-retrieving module has been implemented in order to permanently relate each relationships to managers' quotations.

We developed applications for using CMASOP in comparative and typological ways [22]. Differences in SOP between groups of managers can be highlighted in coupling CMASOP and descriptive statistical methods. Typology of systems of practices can be processed by coupling CMASOP, clustering methods and statistical analysis. These developments have been applied to our case study, grass forage management in farming systems. The clustering of SOP in these systems highlighted two contrasted groups of farmers based on the management of their second grass cut(figure 1). The first group of farmers (A, n=24, figure 1(a)) are more prone for silaging (20) than they are for bale wrapping (5) or haying (5). Conversely, the second group of farmers (B, n=25, figure 1(b)) are more prone for bale wrapping (16) and haying (12) than they are for silaging (1). As a result of previous works, the drought has been quoted by farmers as a typical risk in grassland management in Famenne. The potential drought mainly occurs during the summer and have damageable consequences on grass growth and, in parallel, on milk production and animal performances in general. In order to cope with drought, farmers' adaptations are contrasted : grazed area increase or supplementation in grazing plots.

The two SCM (figures 1) have been taken as patterns to build two synthetic FCM (figures 2) for studying systems resilience and farmers' adaptive capacity linked with summer drought. These synthetic FCM show the grassland plots allocated for harvesting (Silo, Bale wrap, Hay) or for grazing (Grazed area). The weights of relationships between Second cut and these four concepts are proportional to their weights in the SCM of the two clusters (0.8, 0.2, 0.2 and

---

[1] CMASOP was developed in R [20]. Figures 1 and 2 were done using Rgraphviz [8]. Figures 3 and 4 were made using ggplot2 [25]

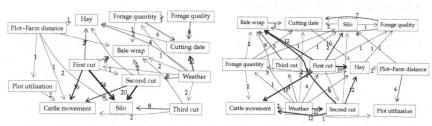

(a) Core of the social map of SOP A cluster, based on silaging

(b) Core of the social map of SOP B cluster, based on bale wrapping and haying

Fig. 1. The 49 farmers' Systems of Practices (SOP) have been classified in two groups using the clustering application of CMASOP [22]. Farmers of the first cluster (A) are specialized and have SOP based on silaging. Farmers of the second cluster (B) are more diversified and have SOP based on bale wrapping and haying. The social maps of two clusters of SOP have been used to build and calibrate two synthetic FCM used in the present study (figure 2)

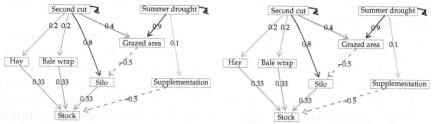

(a) FCM of the SOP A cluster based on silage

(b) FCM of the SOP B cluster based on bale wrap and hay

Fig. 2. FCM of the two different Systems of Practices assessed. Weights of relationships are illustrated by the saturation of the gray : from white (i.e. invisible,0) to black (1). Signs of relationships are illustrated by the type of line : continuous (positive) or dashed (negative). Values of relationships weights are shown besides relationships.

0.4 respectively in A and 0, 0.6, 0.4 and 0.4 in B). We considered that the products harvested on cutting plots constitute the Stock of forage (0.33 for Hay, Bale wrap and Silo in A ; 0.5 for Hay and Bale wrap in B). In case of Summer drought, two adaptations are simulated : the increase of Grazed area (0.9 for A, 0.1 for B) or the Supplementation of forage in grazed plots (0.1 in A, 0.9 in B). The increase of Grazed area involve a decrease of the harvested area (−0.5 for Silo in A and −0.25 for Bale wrap and Hay in B). Two self-reinforcing relationships have been added for the driver concepts Second cut and Summer drought.

Farmers are more prone to distribute forage conditioned in Bale wrap or in Hay, available in individual elements (bales), than to open a whole Silo done with the harvest of the first cut. Therefore, farmers of the cluster B have the possibility to supplement herd in grazing plots because their stock are mostly constituted by Bale wrap and Hay. For the simulations, we supposed that these farmers choose to cope with Summer drought in supplementing. Conversely, farmers of cluster A has only few Bale wrap and Hay in their Stock. For simulations, we supposed then that these farmers choose the other adaptation, in increasing of Grazed area. For the same reason, we supposed that the reduction of Grazed area for these farmers (A) only affect the most important conditioning, Silo.

The simulations have been processed using the auto-associative neural networks method described by Özesmi and Özesmi [16] in order to calculate the activation degrees of each concepts at all time steps till convergence. Activation degrees are semi-quantitative values of concepts that can only be interpreted relative to each other [11]. The two scenarios have been implemented in forcing the activation degree of Summer Drought to 0 ('No Summer Drought') or 1 ('Summer Drought').

3   Results

Figure 3 shows the comparisons of the scenarios 'No Summer Drought' and 'Summer Drought' for the two Systems of Practices (SOP) assessed. The evolution of the two driver concepts ('Second cut' and 'Summer drought') is logically not shown nor analysed.

For SOP A cluster based on silaging, (i) the activation degree of Grazed area strongly increase from 0.380 (No Summer Drought) to 0.862 (Summer Drought), (ii) the activation degree of Supplementation slightly increase from 0.000 to 0.100, (iii) the activation degrees of Silo and of Stock decrease from 0.544 and 0.300 respectively to 0.353 and 0.194 respectively (table 1).

For SOP B cluster based on bale wrapping and haying, (i) the activation degree of Supplementation strongly increase from 0.000 to 0.716 while (ii) the activation degree of Stock strongly decrease from 0.246 to −0.118, (iii) the activation degree of Grazed area increase from 0.380 to 0.462 and (iv) the activation degrees of Bale wrap and Hay slightly decrease from 0.466 and 0.296 respectively to 0.450 and 0.277 respectively (table 1).

The increases of Grazed area for SOP A cluster and of Supplementation for SOP B cluster are adaptations of each groups of farmers in case of Summer drought (figure 4). In the FCM of SOP A cluster, a direct consequence of increasing of Grazed area is the decrease of Silo and a subsequent decrease of Stock that is limited. In the case of SOP B cluster, as a direct consequence of the increase of Supplementation, FCM shows a decrease of the activation degree of Stock more important than for SOP A cluster (figure 4).

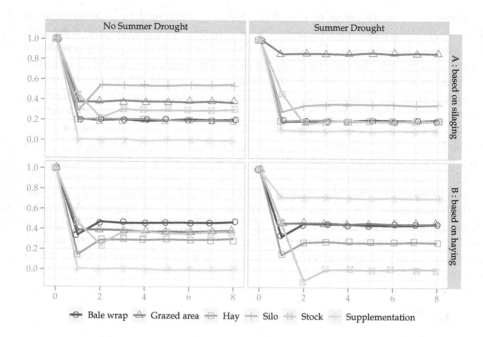

Fig. 3. Evolution of activations degrees of concepts in the four Fuzzy Cognitive Maps. Comparisons of the scenarios 'No Summer Drought' and 'Summer Drought' for the two Systems of Practices (SOP) assessed : A (SOP based on silaging) and B (SOP based on bale wrapping and haying). X-axis: iteration step, Y-axis: activation degree

4   Discussion and Conclusion

This article presents an original method for assessing managers' adaptive capacity under uncertain environmental conditions. This method gain in coherence and relevance through its integration in CMASOP, a complete Cognitive Mapping approach [23,22]. The method is grounded on various kind of qualitative data collected during open-ended interviews. Its descriptive application allows to inductively model SOP of individual and groups of managers based on their own conceptions of their system. Its comparative and typological applications allow to objectively compare and cluster SOP. Finally, CMASOP allows to construct FCM of managers' SOP. In computing the steady states of FCM of various SOP under various environmental conditions, it is relevant in assessing adaptive capacity of managers in complex SES.

This paper present a first application of FCM for studying SOP in SES. The SOP we model are basic in terms of concepts and relations. As a consequence, the dynamic behaviours of the FCMs are elementary. Nevertheless, results confirm influences of farming practices on the whole functioning of the production system. They confirm also that various systems of practices have various effects on the system.

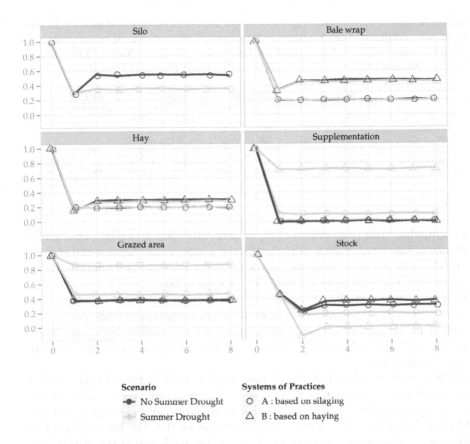

Fig. 4. Evolution of activations degrees of major concepts in the four Fuzzy Cognitive Maps. Comparisons of the adaptations of the two Systems of Practices (SOP) to cope with 'Summer Drought'. X-axis: iteration step, Y-axis: activation degree.

Beyond these results that could appear as relatively evident, results shows the possibility to model a wide variety of SOP in a simple way in order to assess them under various scenarios. The easy way of building model and simulating scenarios is a major advantage of the method presented. We illustrate it in processing further simulations in order to test two other SOP : C, based on a mixed sources of harvested forage (0.33 silo, 0.33 bale wrap and 0.33 hay) and D, whose stock is only constituted by purchased feed. Results are shown in table 1. Technically, our tool is easy-to-use for researchers or even for farmers : the relations and their weights are entered in a spreadsheet that is subsequently processed by an R program [20].

Comparisons of two SOP in case of Summer drought showed differences in terms of Stock between farmers' FCM. These differences could also have significant consequences on concepts not modelled in the present study : feeding of herd during the winter (stocks are smaller), cows selling (stocks are insufficient)

Table 1. Values of activations degrees of major concepts at steady states. Eight simulations are compared. Four Systems of Practices (SOP) : A, based on silageing ; B, based on haying ; C, based on mixed sources and D, feed purchasing. Two climatic scenarios : Normal (No Summer Drought) and Drought (Summer Drought).

	SOP–A		SOP–B		SOP–C		SOP–D	
	Normal	Drought	Normal	Drought	Normal	Drought	Normal	Drought
Silo	0.544	0.353			0.259	0.205		
Bale wrap	0.197	0.197	0.466	0.450	0.259	0.205		
Hay	0.197	0.197	0.296	0.277	0.259	0.205		
Supplementation	0.000	0.100	0.000	0.716	0.000	0.462	0.000	0.462
Grazed area	0.380	0.862	0.380	0.462	0.380	0.716	0.762	0.762
Stock	0.300	0.194	0.364	0.005	0.251	−0.028	0.000	−0.432

or feed purchase (for restoring stocks), treasury (due to purchasing) and, finally, resilience of the whole farms. Although these concepts are beyond the scope of our model, this reasoning illustrates how resilience of farms and adaptive capacity of farmers could be assessed using FCM approaches.

Most of previous studies of managers' practices in SES were conducted in a qualitative way by social scientists or modelled in reductionist DSS. The twofold nature of FCM (qualitative and quantitative) bring another advantage to our method. It allows building SOP model based on people's knowledge and processing simulations.

Further works could include developments of more elaborated FCM including various indicators of economic fields (e.g. production, profit), ecological sciences (e.g. environmental footprint) or social sciences and psychology (e.g. personal fulfilment, happiness). It has not escaped our notice that the broadening of the map could represent an opportunity of measuring resilience of social–ecological systems at local or regional scales.

As asserted by van Vliet et al. (2010, [24]), FCM could be a relevant communication and learning tools between managers and scientists. It would be very interesting to carry out qualitative surveys in order to discuss the results of FCM simulations and adaptive capacity assessment with managers of SES. The results of these surveys could also constitute relevant data for an inductive and qualitative evaluation of resilience and adaptive capacity.

Acknowledgements. The Walloon Agricultural Research Centre provided financial support for this work through the MIMOSA project. We would also like to thank two anonymous reviewers for their helpful comments.

References

1. Ascough II, J.C., Maier, H.R., Ravalico, J.K., Strudley, M.W.: Future research challenges for incorporation of uncertainty in environmental and ecological decision-making. Ecological Modelling 219(3-4, Sp. Iss. SI), 383–399 (2008)

2. Axelrod, R., Ann Arbor, M., Berkeley, C.: Structure of decision: The cognitive maps of political elites. Princeton University Press, Princeton (1976)
3. Darré, J., Mathieu, A., Lasseur, J.: Le sens des pratiques. Conceptions d'agriculteurs et modèles d'agronomes. Science Update, INRA Editions (2004)
4. Edwards-Jones, G.: Modelling farmer decision-making: concepts, progress and challenges. Animal Science 82(pt. 6), 783–790 (2006)
5. Fairweather, J.: Farmer models of socio-ecologic systems: Application of causal mapping across multiple locations. Ecological Modelling 221(3), 555–562 (2010)
6. Fairweather, J., Hunt, L.: Can farmers map their farm system? causal mapping and the sustainability of sheep/beef farms in new zealand. Agriculture and Human Values 28(1), 55–66 (2011)
7. Folke, C., Carpenter, S.R., Walker, B., Scheffer, M., Chapin, T., Rockstrom, J.: Resilience thinking: Integrating resilience, adaptability and transformability. Ecology and Society 15(4) (2010)
8. Gentry, J., Long, L., Gentleman, R., Falcon, S., Hahne, F., Sarkar, D., Hansen, K.: Rgraphviz: Provides plotting capabilities for R graph objects (2010)
9. Ingram, J., Fry, P., Mathieu, A.: Revealing different understandings of soil held by scientists and farmers in the context of soil protection and management. Land Use Policy 27(1), 51–60 (2010)
10. Isaac, M., Dawoe, E., Sieciechowicz, K.: Assessing local knowledge use in agroforestry management with cognitive maps. Environmental Management 43(6), 1321–1329 (2009)
11. Kok, K.: The potential of fuzzy cognitive maps for semi-quantitative scenario development, with an example from brazil. Global Environmental Change 19(1), 122–133 (2009)
12. Kosko, B.: Fuzzy cognitive maps. International Journal of Man-Machine Studies 24(1), 65–75 (1986)
13. McCown, R.L.: Changing systems for supporting farmers' decisions: problems, paradigms, and prospects. Agricultural Systems 74(1), 179–220 (2002)
14. Murungweni, C., van Wijk, M., Andersson, J., Smaling, E., Giller, K.: Application of fuzzy cognitive mapping in livelihood vulnerability analysis. Ecology and Society 16(4) (2011)
15. Ortolani, L., McRoberts, N., Dendoncker, N., Rounsevell, M.: Analysis of farmers' concepts of environmental management measures: An application of cognitive maps and cluster analysis in pursuit of modelling agents' behaviour. In: Glykas, M. (ed.) Fuzzy Cognitive Maps. STUDFUZZ, vol. 247, pp. 363–381. Springer, Heidelberg (2010)
16. Özesmi, U., Özesmi, S.L.: Ecological models based on people's knowledge: a multi-step fuzzy cognitive mapping approach. Ecological Modelling 176(1-2), 43–64 (2004)
17. Papageorgiou, E., Markinos, A., Gemtos, T.: Fuzzy cognitive map based approach for predicting yield in cotton crop production as a basis for decision support system in precision agriculture application. Applied Soft Computing Journal 11(4), 3643–3657 (2011)
18. Papageorgiou, E., Aggelopoulou, K., Gemtos, T., Nanos, G.: Yield prediction in apples using fuzzy cognitive map learning approach. Computers and Electronics in Agriculture 91, 19–29 (2013)
19. Popper, R., Andino, K., Bustamante, M., Hernandez, B., Rodas, L.: Knowledge and beliefs regarding agricultural pesticides in rural guatemala. Environmental Management 20(2), 241–248 (1996)

20. R Development Core Team: R: A Language and Environment for Statistical Computing. R Foundation for Statistical Computing (2009)
21. Soler, L.S., Kok, K., Camara, G., Veldkamp, A.: Using fuzzy cognitive maps to describe current system dynamics and develop land cover scenarios: a case study in the brazilian amazon. Journal of Land Use Science iFirst, 1–27 (2011)
22. Vanwindekens, F.M., Baret, P.V., Stilmant, D.: Using a cognitive mapping approach for studying diversity of farmers' systems of practices in agricultural systems. Working Paper
23. Vanwindekens, F.M., Stilmant, D., Baret, P.V.: Development of a broadened cognitive mapping approach for analysing systems of practices in social–ecological systems. Ecological Modelling 250, 352–362 (2013)
24. van Vliet, M., Kok, K., Veldkamp, T.: Linking stakeholders and modellers in scenario studies: The use of fuzzy cognitive maps as a communication and learning tool. Futures 42(1), 1–14 (2010)
25. Wickham, H.: ggplot2: elegant graphics for data analysis (2009)
26. Wise, L., Murta, A., Carvalho, J., Mesquita, M.: Qualitative modelling of fishermen's behaviour in a pelagic fishery. Ecological Modelling 228, 112–122 (2012)

A Fuzzy Cognitive Map Model for Estimating the Repercussions of Greek PSI on Cypriot Bank Branches in Greece

Maria Papaioannou[1], Costas Neocleous[2], and Christos N. Schizas[1]

[1] University Ave., Department of Computer Science, University of Cyprus, Nicosia, Cyprus
cs03pm2@cs.ucy.ac.cy, schizas@ucy.ac.cy
[2] 31 Archbishop Kyprianos Str, Cyprus University of Technology, Limassol, Cyprus
costas.neocleous@cut.ac.cy

Abstract. Recently, Greece experienced a financial crisis unprecedented in its modern history. In May of 2010 Greece signed a bailout memorandum with Troika (a tripartite committee constituted by the European Central Bank, the European Commission and the International Monetary Fund). In February of 2012, they proceeded to a second bailout package along with a debt restructuring deal that included a private sector involvement (PSI). The overall loss, for the private investors, was equivalent to around 75%. Due to the strong economic ties between Greece and Cyprus, PSI had a substantial impact on the Cypriot economy. A fuzzy cognitive map (FCM) system has been developed and used to study the repercussions of the Greek PSI on the economic dynamics of Cyprus and more specifically on the probability of cutting off the Cypriot Bank branches that operate in Greece. The system allows one to observe how a change on some parameters can affect the stability of the rest of the parameters. Different promising scenarios were implemented, scaling the percentage of PSI from 0% to 80%.

Keywords: Fuzzy Cognitive Maps, Intelligent Systems, Private Sector Involvement.

1   Introduction

In 2008 the subprime mortgage crisis appeared in USA which was a crucial factor that led to the emersion of the recent global financial crisis. Inevitably, this introduced adverse economic effects for most of the European countries. That along with a combination of other factors resulted in the economic victimization of the weakest European economies (e.g. Greece, Portugal, Ireland, and Cyprus). There was a widespread fear that a failure to deal with the national debt crisis could have created a domino effect for the other members of the Eurozone. Greece was the first country which had to deal with the possibility of a national bankruptcy in 2009. As a result of the continuous strong increase in Greek government debt levels, Greece`s sovereign-debt was downgraded by the International Credit Rating Agencies to junk status in April 2010.

H. Papadopoulos et al. (Eds.): AIAI 2013, IFIP AICT 412, pp. 597–604, 2013.

There was a fear that other countries could be affected by the Greek economic crisis. This forced Europe to take important and decisive corrective actions under time pressure. In May of 2010, Troika which is a tripartite committee constituted by the European Central Bank, the European Commission and the International Monetary Fund (IMF), agreed to give Greece a three-year €110 billion loan. As part of the deal with Troika, the Greek government implemented a series of austerity measures. These, led Greece´s recession even deeper. As a result, in February of 2012 Troika decided to provide Greece a second bailout package accompanied with a restructure agreement with private sector involvement (PSI). The debt restructuring deal declared that private holders of Greek government bonds had to accept a 53.5% so-called "haircut" to their nominal values. Eventually, in March 2012, the bond swap was implemented with an approximately 75% nominal write-off.

Since the Greek and the Cypriot economies are strongly related and connected, this led to serious repercussions to the Cyprus economy. More specifically, the Cypriot banking system, which constitutes one of the pillars of the Cyprus economy, suffered from the Greek PSI, since it was highly exposed to the Greek government bonds. Consequently, that had a great impact on Cyprus economy which had already been dealing with economic problems. The future of Cypriot economy and banking system seemed uncertain. The Government of Cyprus and the Cypriot bankers had to deal with an unprecedented situation which was extremely difficult to handle since they could not estimate the overall impact of the Greek PSI on Cyprus economy and Cypriot banking system.

In the current work reported in this paper an attempt is made to model the dynamics of the above problem using the technology of fuzzy cognitive maps (FCM). That is, to study the long term impacts on the Cyprus economy and on the banking system as a result of the Greek PSI.

FCM modelling constitutes an alternative way of building intelligent systems, using uncertain parameters that influence one - another. Such a system is constituted by certain concepts (which are characterized by a state) along with relevant interconnections (which are described by sensitivity values). Essentially, an FCM is developed by integrating the existing experience and knowledge regarding a cause – effect system in a pseudo-dynamic manner. An FCM provides the opportunity to predict the final states of the system caused by a change on the initial concept states. After a change is applied to selected initial states of the concepts of interest, the system is let to evolve for a number of steps until the concept states converge to stable values. Further analysis and work can be done on the converged final states of the system for understanding how indirect cause – effect relations drive the system's behaviour.

There is a wide range of interesting FCM applications [1, 2, 8, 10, and 11] in modelling complex dynamic systems such as medical, environmental, supervisory and political systems.

This study aims to examine the impact of the Greek PSI on the possibility of Cypriot banks cutting off the Cypriot branches which operate in Greece. Such a scenario seemed impossible and far from reality back in April of 2012, at the time when this FCM system was originally developed and tested.

2   The FCM System

A wide range of real life dynamical systems are characterized by causality through their interrelations. Most of the times, experienced and knowledgeable users can identify the important parameters and their interactions in the system. However, the complexity of these systems acts as a prohibitive factor on prognosis of their future states. In many practical situations, it is crucial for the decision makers to have an estimate on the cost of changing a state of a concept and how will affect other concepts of interest. FCMs are a soft computing methodology of modeling dynamic systems, which constitute causal relationships amongst their parameters. They manage to represent human knowledge and experience, in a certain system's domain, into a weighted directed graph model and apply inference presenting potential behaviours of the model under specific circumstances. The parameters are led to interact until the whole system reaches equilibrium. FCMs allow a user to reveal intricate and hidden in the already known causal relationships of the concepts and to take an advantage of the causal dynamics of the system in predicting potential effects under specific initial circumstances.

Two knowledgeable persons were asked to contribute to the development of the current FCM system. In the first place, they had to identify the most important factors that constitute the modeled system. These factors comprise the set of the concepts that were used to simulate the FCM model. Additionally, they had to define the states of the FCM concepts as they were in April of 2012. The states were described by a numeric value from 0 to 1 (0 to 100%), whereas 0 meant that the activation of the particular concept is the minimum possible, and 1 that the concept's activation is maximum possible. Additionally, they had to describe the degree of variation they expect to see, in the final state of every concept after the implementation of the Greek PSI. To be better guided, they first established whether the change will be positive or negative and then defined the intensity of the variation choosing amongst {low, medium, high}.

The next phase of development requested the experts to define the causal relations between the interconnected concepts. Each relation is characterized by a numeric value called sensitivity (or influence coefficient, or weight). In this study a slightly different methodology than that used in the conventional FCMs has been used when defining the sensitivities. The sensitivity of the relation describes the impact of changing the state of C_i on the concept C_j. The equation that was used to calculate the sensitivities, in this study, is given in Equation (1):

$$S_{ij} = \frac{\delta c_j^{t+1,t}}{\delta c_i^{t,t-1}} = \frac{c_j^{t+1} - c_j^t}{c_i^t - c_i^{t-1}} \tag{1}$$

where t is the iteration counter.

Thus, in order to define the sensitivity of every relation, the experts had to make the following assumption: "The states C_i^{t-1} and C_j^t are equal to the initial values of the corresponding concepts as defined in the previous stage of development" and then answer the following question: "If the state of the concept C_i becomes x (in equation (1) is denoted by C_i^t) what will be the new state of concept Cj (denoted by C_j^{t+1})?".

Consider for example the relation between the "Level of Cyprus economy" (C13) and the "Evaluation of the Cyprus Economy by Authoritative Rating Agencies" (C8). The experts defined the initial states of C13 and C8 as 0.5 and 0.2 respectively. Furthermore, they expected a negative relationship of medium intensity of the degree of variation of the level of Cyprus Economy as a result of the Greek PSI. Therefore the question for this sensitivity was finally formed as:

"If the level of Cyprus Economy gets reduced from 0.5 to 0.2, what will be the new state of the concept "Evaluation of the Cyprus Economy by Authoritative Rating Agencies" if now is 0.2?". For this example the experts answered 0.05. However, measuring the change of the two consequent states of a concept, as a percentage of its initial value gives a better feeling about the strength of the variation. So this sensitivity value is calculated as:

$$S_{13,8} = \frac{(0.05-0.2)/0.2}{(0.2-0.5)/0.5} = \frac{-0.15/0.2}{-0.3/0.5} = 1.25 \qquad (2)$$

Finally the sensitivities are divided by the number of iterations the system needs to converge. For this system this number was 10 and yet the final value of $S_{13,8} = 0.125$. That is called the "absolute sensitivity".

Thus, using equations (1) and (2), the equation giving the total accumulated change in the activation of concept C_j due to changes in concepts C_i in discrete time is given by:

$$C_j^{t+1} = C_j^t + \sum_i^n s_{ij} * (C_i^t - C_i^{t-1}) * \frac{C_j^t}{C_i^{t-1}} \qquad (3)$$

where t is the iteration counter, C is the activation strength of the concept of interest and s_{ij} is the sensitivity (weight) which is a measure on how much a change in the current standing of concept C_i affects the changes in the standing of concept C_j.

The methodology used to develop this FCM system differs from other previously proposed methodologies in the following respects. Firstly the weights are defined taking in mind the impact caused on a concept by the change of the other concept when each concept had been rationalized to a scale of 0 to 100%. Secondly there was no use of squashing functions to smooth out the activation levels of the concepts such as the logistic function.

Therefore using only the activation function (3) the system is let to simulate after a change of the initial states for a number of iterations until it converges to stable states. Then the user of the system may observe and make conclusions on the direct or indirect effects of that change reflected on the final states of the system

3   The FCM Concepts

After a series of discussions, the experts concluded that they had to use 15 influencing parameters as shown in Table 1. These are the most significant concepts of the system. The experts had to also decide on the initial value of each concept. Although these concepts are difficult to be objectively quantified, the experts tried to document their decisions, mostly through referring to the press and from relevant statistical information. However, many of the parameters are too complex to be analysed and described based on raw numbers. For example the concept "Level of Cyprus Economy" encompasses characteristics like GDP, unemployment rate, housing, Consumer Price Index, stock market prices, industrial production, etc. In such cases the experts had to define the state of the concept based only on their "feeling" and their understanding of the dynamics of the system.

The initial values describing the states of the concepts as they were in April of 2012 are also given in Table 1. Besides the initial states of the concepts the experts were also asked to define the degree of variation of each concept and the sensitivities of their causal interrelations as described in the previous section. Inevitably, there were some discrepancies between the values given by each expert. Eventually, the average of the sensitivities was used to form the final sensitivity matrix which is presented in Table 2.

4   Results and Conclusions

This study aimed to observe how a change into the percentage of the Greek PSI (concept C3) affects the probability that the Cypriot banks eliminate their branches in Greece (concept C15). Ten scenarios with different percentages of Greek PSI were tested using the FCM system with the initial values (Table 1) and the average sensitivities of Table 2 given by experts. The response of the concept C15 to the variation of C3 is shown in Figure 1.

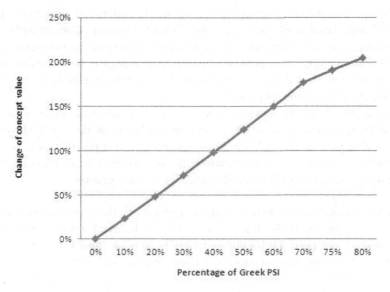

Fig. 1. Effect on the probability of cuttof of the Cypriot Bank branches that operate in Greece based on average sensitivities

In Figure 1 the horizontal axis is showing the locked (not allowed to change) values of the Greek PSI starting from 0% to 80% (where 0% means that there was no implementation of the PSI and 80% means that the PSI was 80%). The vertical axis is showing the degree of the change that happens to the concept C15 in respect to its initial value, due to Greek PSI variations. The changes to the values of C15 are calculated after the system settles and converges to a stable state. Then the percentage change is calculated between the final state value of the concept C15 and its initial value.

When working with an FCM model the actual final values of the concepts are not the objective but rather the trends. This is what a decision maker usually wants. Furthermore, it is important to note that during the developmental phase of the system, the experts expected that the probability of cutting of the Cypriot branches in Greece would have remained low regardless of the degree of the change in Greek PSI. That is why they set a low degree of variation for the corresponding concept. It was surprising though that when the system converged, the system predictions showed an opposite result. That is, the aforementioned probability is significantly increased when the percentage of the Greek PSI is increased. More specifically when the Greek PSI is 75% (which reflects the reality) then the probability of eliminating the Cypriot branches in Greece is increased by about 191% of its initial value (as given in Table 1). The system revealed that there exist strong causal paths connecting the two concepts.

Unfortunately for the Cypriot economy, the results given from the system were fully confirmed in reality few months later. In March of 2013, the Marfin Popular Bank and the Bank of Cyprus, two of the largest banks in Cyprus, sold their branches to Piraeus Bank of Greece.

Nevertheless, it is indicated yet again that the whole system is built based on the two experts' apprehension of the Greek and Cypriot economy. Consequently, this model does not necessarily represent fully the Greek and Cypriot economies or their interrelations. The system could benefit reliability and objectivity by the involvement of more expert opinions and the incorporation of the public opinion (e.g. by giving questionnaires to a big sample of people).

Beyond that, FCMs can be important tools for helping humans to make wiser and more accurate decisions, by presenting them the evolution of the modeled system after a set of changes and the final future consequences of their possible choices. That is why future work must be done to encounter the several open issues [7] concerning FCMs aiming in increasing their credibility and fine-tuned operation.

Acknowledgements. This research was partly supported by the University of Cyprus, the Cyprus University of Technology, and the Cyprus Research Promotion Foundation structural funds (ΤΠΕ/OPIZO/0308(BE)/03). Special thanks to Dr Pambos Papageorgiou, MP for the very valuable discussions we had with him.

References

1. Andreou, A.S., Mateou, N.H., Zombanakis, G.A.: The Cyprus puzzle and the Greek - Turkish arms race: Forecasting developments using genetically evolved fuzzy cognitive maps. Defence and Peace Economics 14(4), 293–310 (2003)
2. Bertolini, M., Bevilacqua, M.: Fuzzy Cognitive Maps for Human Reliability Analysis in Production Systems. In: Kahraman, C., Yavuz, M. (eds.) Production Engineering and Management under Fuzziness. STUDFUZZ, vol. 252, pp. 381–415. Springer, Heidelberg (2010)
3. Carvalho, J.P.: On the semantics and the use of fuzzy cognitive maps and dynamic cognitive maps in social sciences. Soft Computing in the Humanities and Social Sciences 214, 6–19 (2013)
4. Groumpos, P.P.: Fuzzy Cognitive Maps: Basic Theories and Their Application to Complex Systems. In: Glykas, M. (ed.) Fuzzy Cognitive Maps. STUDFUZZ, vol. 247, pp. 1–22. Springer, Heidelberg (2010)
5. Hanafizadeh, P., Aliehyaei, R.: The Application of Fuzzy Cognitive Map in Soft System Methodology. Systemic Practice and Action Research 24(4), 325–354 (2011)
6. Mateou, N.H., Andreou, A.S.: A framework for developing intelligent decision support systems using evolutionary fuzzy cognitive maps. J. Intell. Fuzzy Syst. 19(2), 151–170 (2008)
7. Neocleous, C., Schizas, C., Papaioannou, M.: Important issues to be considered in developing fuzzy cognitive maps. In: 2011 IEEE International Conference on Fuzzy Systems (FUZZ), pp. 662–665 (2011)
8. Neocleous, C., Schizas, C., Papaioannou, M.: Fuzzy cognitive maps in estimating the repercussions of oil/gas exploration on politico-economic issues in Cyprus. In: 2011 IEEE International Conference on Fuzzy Systems (FUZZ), pp. 1119–1126 (2011)
9. Neocleous, C., Schizas, C.N.: Modeling socio-politico-economic systems with time-dependent fuzzy cognitive maps. In: 2012 IEEE International Conference on Fuzzy Systems (FUZZ-IEEE), pp. 1–7 (2012)
10. Papageorgiou, E.I., Aggelopoulou, K.D., Gemtos, T.A., Nanos, G.D.: Yield prediction in apples using Fuzzy Cognitive Map learning approach. Comput. Electron Agric. 91, 19–29 (2013)
11. Sperry, R., Jetter, A.J.: Fuzzy cognitive maps to implement corporate social responsibility in product planning: A novel approach. In: 2012 Proceedings of the Technology Management for Emerging Technologies (PICMET), PICMET 2012, pp. 2536–2541 (2012)

Appendix A

Table 1. The various influencsing parameters that have been studied and their initial values as set by the experts

	CONCEPT NAME	INITIAL VALUE
C1	Cost of Money	50%
C2	Liquidity of Cyprus Banks	60%
C3	Degree of PSI of Greek Government Bonds	0%
C4	Degree of Deposits of Greek citizens and companies in Cyprus Banks	40%
C5	Degree of Deposits of Cypriot citizens and companies in Cyprus Banks	78%
C6	Degree of Success of Bank Recapitalization by Private Equity	20%
C7	Stock Market Value of Banks	40%
C8	Evaluation of the Cyprus Economy by Authoritative Rating Agencies	20%
C9	Confidence of People and companies in Cyprus Banking system	80%
C10	Level of Greek Economic Crisis	80%
C11	Level of Greek workforce that comes to Cyprus for work	60%
C12	Degree of Bank Recapitalization done by the Republic of Cyprus	50%
C13	Level of Cyprus Economy	50%
C14	Probability of the Republic of Cyprus entering EU Support Mechanism	30%
C15	Probability of Cutoff of the Cypriot Bank branches that operate in Greece	20%

Table 2. The sensitivity matrix

	C1	C2	C3	C4	C5	C6	C7	C8	C9	C10	C11	C12	C13	C14	C15
C1	0	0.04	0	0.13	0.01	0	0	0	0	0	0	0	-0.10	0	0
C2	-0.06	0	0	0	0	0	0.08	0.10	0.08	-0.01	0.03	0	0.12	-0.13	-0.40
C3	0	-0.10	0	0.04	0	0	-0.15	-0.11	-0.08	-0.08	0	0	-0.12	0.10	0.45
C4	0	0.13	0	0	0	0	0	-0.40	0.10	0.05	0	0	0.08	0	0
C5	0	0.09	0	0.05	0	0.10	0.03	0	0.03	0	0	0	0.02	0	-0.21
C6	0	0	0	0	0	0	0.05	0.03	0.03	0	0	0	0.01	0	-0.05
C7	0	0	0	0.01	0.01	0.13	0	0	0.03	0	0	0	0	0	-0.03
C8	0	0	0	0.07	0.02	0.07	0.13	0	0.05	0	0.08	0.08	0.03	-0.09	-0.07
C9	0	0.09	0	0.13	0.03	0.07	0	0	0	0	0.07	0	0.05	0	0
C10	0.24	-0.07	0	0.15	-0.04	-0.40	-0.30	-0.20	-0.20	0	0.13	-0.16	-0.24	0.53	0
C11	0	0	0	0.08	0	0	0	0	0	0.04	0	0	0.06	0	0
C12	0.04	0.03	0	0	0	0	0.05	0.04	0.01	0	0	0	-0.02	0	0
C13	-0.07	0	0	0.11	0.05	0.08	0.04	0.13	0.04	-0.02	0.11	0.10	0	-0.28	0
C14	0	0.01	0	-0.01	-0.01	-0.02	-0.02	-0.01	-0.01	0	-0.01	0.02	0.01	0	0
C15	0	0.01	0	0.04	0.00	0.05	0.05	0.03	-0.01	0	0	0	0.01	-0.02	0

A Matlab-CONTAM Toolbox
for Contaminant Event Monitoring
in Intelligent Buildings

Michalis P. Michaelides[1,2,*], Demetrios G. Eliades[2], Marinos Christodoulou[2],
Marios Kyriakou[2], Christos Panayiotou[2], and Marios Polycarpou[2]

[1] Department of Electrical Engineering and Information Technologies
Cyprus University of Technology
30 Archbishop Kyprianos Str., CY-3036 Lemesos, Cyprus
michalis.michaelides@cut.ac.cy
[2] KIOS Research Center for Intelligent Systems and Networks,
and Department of Electrical and Computer Engineering
University of Cyprus
75 Kallipoleos Ave., CY-1678 Nicosia,Cyprus

Abstract. An intelligent building should take all the necessary steps
to provide protection against the dispersion of contaminants from sources
(events) inside the building which can compromise the indoor air quality
and influence the occupants' comfort, health, productivity and safety.
Multi-zone models and software, such as CONTAM, have been widely
used in building environmental studies for predicting airflows and the resulting contaminant transport. This paper describes a developed
Matlab Toolbox that allows the creation of data sets from running
multiple scenarios using CONTAM by varying the different problem
parameters. The Matlab-CONTAM Toolbox is an expandable research
tool which facilitates the implementation of various algorithms related
to contamination event monitoring. In particular, this paper
describes the implementation of state-of-the-art algorithms for detecting
and isolating a contaminant source. The use of the Toolbox is
demonstrated through a building case-study. The Matlab-CONTAM
Toolbox is released under an open-source licence, and is available at
https://github.com/KIOS-Research/matlab-contam-toolbox.

Keywords: Intelligent buildings, multi-zone model, CONTAM, Matlab
Toolbox, contaminant event monitoring, multiple scenarios, detection,
isolation.

1 Introduction

An Intelligent Building is a system that incorporates computer technology to
autonomously govern and adapt the building environment in order to enhance

---

* Corresponding author.

H. Papadopoulos et al. (Eds.): AIAI 2013, IFIP AICT 412, pp. 605–614, 2013.

operational and energy efficiency, cost effectiveness, improve users' comfort, productivity and safety, and increase system robustness and reliability [1, 5]. The dispersion of contaminants from sources (events) inside a building can compromise the indoor air quality and influence the occupants' comfort, health, productivity and safety. These events could be the result of an accident, faulty equipment or a planned attack. Distributed sensor networks have been widely used in buildings to monitor indoor environmental conditions such as air temperature, humidity and contaminant concentrations. Real-time collected data can be used to alert occupants and/or control environmental conditions. Accurate and prompt identification of contaminant sources can help determine appropriate control solutions such as: (i) indicating safe rescue pathways and/or refugee spaces, (ii) isolating contaminated spaces and (iii) cleaning contaminant spaces by removing sources, ventilating and filtering air. Therefore, the accurate and prompt identification of contaminant sources should be an essential part of the Intelligent Building design.

To study the security-related problems in intelligent buildings, large quantities of data are required to be simulated under various conditions, in order to capture the variations in the complex dynamics involved. For the creation of non-stationary datasets related to the presence of contaminants in intelligent buildings, the Matlab-CONTAM Toolbox has been developed and released, which is the main contribution of this work. In the Toolbox, we utilize the computational engine of CONTAM [7], a multi-zone simulation software developed by the US National Institute of Standards and Technology (NIST). With CONTAM, the user can easily create the building outline and specify the zone volumes, the leakage path information and the contaminant sources present. This information can be further utilized for calculating the air-flows and resulting contaminant concentrations in the various building zones. A limitation of the CONTAM v3.1 software is that it can only analyze a single scenario at a time. Furthermore, no algorithms are included within CONTAM for detecting and isolating contaminant sources.

The Matlab-CONTAM Toolbox features a user-friendly Graphical User Interface (GUI) and a modular architecture. It allows the creation of multiple scenarios by varying the different problem parameters (wind direction, wind speed, leakage path openings, source magnitude, evolution rate and onset time) as well as the storage of the computed results in data structures. The data from these scenarios are further analyzed by the developed algorithms for determining solutions for contaminant event monitoring. In this paper, we implement inside the Toolbox, state-of-the-art algorithms for detecting and isolating a contaminant source in the indoor building environment. These are further demonstrated using a 14-zone building case study referred to as the Holmes' house. We should point out that the presented case study is only a small sample of the Toolbox's possibilities. Using the Toolbox, the user can easily create data sets for any building scenario by choosing which parameters to vary. The user can then select an algorithm to analyze these data sets. A key idea behind this work is to provide a software that enables the application of computational intelligence methods in buildings' related research [2–4, 9]. This issue is particularly relevant and challenging at the same time since it is rather difficult to define and propose benchmarks and/or testbeds for learning in non-stationary environments.

The rest of the paper is organized as follows. First, in Section 2, we describe the architecture and the main functionality of the developed Matlab-CONTAM Toolbox. Section 3 describes the model and the implemented algorithms for contaminant event monitoring in intelligent buildings. Then, in Section 4, we demonstrate the Toolbox and how it is applied for the contaminant source detection and isolation problem. Finally, Section 5 provides some concluding remarks and presents our plans for future work.

2   Matlab-CONTAM Toolbox Architecture

The Matlab-CONTAM Toolbox Architecture is depicted in Fig. 1. It features a modular design and a user-friendly Graphical User Interface (GUI). The "Data Module" opens CONTAM Project files (*.prj) and reads the information related to the building parameters. These include the building outline, the zone volumes and the leakage path information. The building zone schematics are extracted from this information and used by the GUI of the Toolbox to plot the building.

The information from the "Data Module" are then used by the "Scenario Construction Module". This module is responsible for specifying the parameters which are to be considered to construct one or multiple simulation scenarios, through the GUI. These include environmental parameters that affect the flow dynamics (i.e. wind speed and direction, temperature, opening status of doors and windows), as well as contaminant source parameters (i.e. magnitude, evolution rate and onset time). The scenarios along with building information are

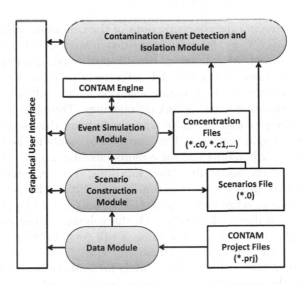

Fig. 1. The software architecture of the Matlab-CONTAM Toolbox

stored in the Scenarios File (*.0). The data stored in the Scenarios File are used by the "Event Simulation Module", which communicates with the CONTAM engine to compute a state matrix for each different flow scenario. When the state matrices have been constructed, the module simulates the different contamination scenarios and computes the contaminant concentration at each zone. All the results are stored in one or more Concentration Files (*.c0, *.c1, ...).

The "Contamination Event Detection and Isolation Module" is comprised of state-of-the-art algorithms for detecting whether a contamination event has occurred in the building, and isolating its zone. The module uses information from the Concentration Files and the Scenarios Files, as computed by the Scenario Construction Module and the Event Simulation Module. The Contamination Event Detection and Isolation Module communicates with the GUI to configure the parameters of the algorithms and to view the results.

3   Contaminant Event Detection and Isolation

In this section, we provide an overview of the theoretical results used in the development of the "Contamination Event Detection and Isolation Module." Next, we describe the state-space model and the implemented algorithms for Contaminant Detection and Isolation (CDI) in intelligent buildings. Let $\mathbb{R}$ represent the set of real numbers and $\mathbb{B} = \{1, 0\}$ the set of binary. Using the multi-zone modeling methodology, the state-space equations for contaminant dispersion in an indoor building environment with n zones can be presented in the following general form,

$$
\begin{aligned}
\dot{x}(t) &= (A + \Delta A)x(t) + Q^{-1}Bu(t) + Q^{-1}Gg(t) \\
y(t) &= Cx(t) + w(t),
\end{aligned}
\tag{1}
$$

where, $x \in \mathbb{R}^n$ represents the concentrations of the contaminant in the building zones, while $A \in \mathbb{R}^{n \times n}$ is the state transition matrix which models changes in contaminant concentration between the different building zones primarily as a result of the air-flows. The term ΔA collectively accounts for the presence of modeling uncertainty in the building envelope as a result of changing wind speed, wind direction and variable leakage openings. Through the Toolbox it is possible to characterize and calculate some bounds on this uncertainty, as we will be demonstrating in Section 4. The controllable inputs in the form of doors, windows, fans and air handling units are represented by $u \in \mathbb{R}^p$ while $B \in \mathbb{B}^{n \times p}$ is a zone index matrix concerning their locations. The final term of the first state-space equation involves the location and evolution characteristics of the contamination sources represented by $G \in \mathbb{B}^{n \times s}$ and $g \in \mathbb{R}^s$ respectively. Note that $Q \in \mathbb{R}^{n \times n}$ is a diagonal matrix with the volumes of the zones, i.e. $Q = \text{diag}(Q_1, Q_2, ..., Q_n)$ where Q_i is the volume of i-th zone. In the second equation, $y \in \mathbb{R}^m$ represents the sensor measurements, $C \in \mathbb{B}^{m \times n}$ is a zone index matrix for the sensor locations and $w \in \mathbb{R}^m$ stands for additive measurement noise. More details on the state-space formulation can be found in [6]. Note that a similar state-space formulation has been commonly used by the fault diagnosis

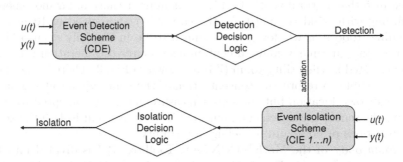

Fig. 2. Architecture of detection and isolation of a single contaminant source

community to represent models of uncertain non-linear systems [10]. In this context, the final term in (1), i.e. $Q^{-1}Gg(t)$, is used to describe process faults that disturb the normal system operation.

The general architecture of the implemented CDI algorithms using state estimation techniques is shown in Fig. 2. For event detection, the Contaminant Detection Estimator (CDE) estimates the contaminant concentration in the different building zones under the "no source" hypothesis. It then compares the actual contaminant concentration as measured by the sensors to the estimated one. Event detection is decided if the difference between actual and estimated concentration (residual) of at least one sensor exceeds the prescribed adaptive threshold. Following detection, a bank of n isolators is activated, one for each building zone. The j-th Contaminant Isolation Estimator (CIE), $j \in \{1, ..., n\}$, estimates the contaminant concentrations in the jth building zone under the "source present" hypothesis. Then, it compares the actual contaminant concentrations as measured by the sensors to the estimated ones. If the residual for the j-th CIE exceeds the threshold, then zone j is *excluded* from being a possible candidate for the source location. Isolation is decided after $n-1$ zones have been excluded, leaving only one possible building zone containing the source. More details on the implemented algorithms including the derivation of the adaptive thresholds can be found in [10].

4   Case Study

In this section, we demonstrate the proposed Matlab-CONTAM Toolbox for detecting and isolating a contaminant source in an indoor building environment through a case study corresponding to the Holmes' house experiment [8]. The interface of the Toolbox is depicted in Fig. 3, loaded with the project file corresponding to the Holmes' house. As shown in the figure, the building is comprised of 14 zones: a garage (Z1), a storage room (Z2), a utility room (Z3), a living room (Z4), a kitchen (Z5), two bathrooms (Z6 and Z13), a corridor (Z8), three bedrooms (Z7, Z9 and Z14) and three closets (Z10, Z11 and Z12). There are a total of 30 leakage path openings corresponding to windows and doors (P1–P30).

It is assumed that natural ventilation is the dominant cause of air movement in the building with wind coming from the east (90°) at a speed of 10 m/s. All the openings (doors or windows) are assumed to be in the fully open position. We assume that at time 3 hours, a contaminant source of generation rate 126.6 g/hr is activated in the utility room (Z3) as shown in Fig. 3. There is one sensor in each zone able to record the concentration of the contaminant at regular intervals at its own location but the sensor measurements are corrupted by noise. Based on the sensor measurements, our goal is to detect and isolate the source under conditions of noise and modeling uncertainty.

The main body of the Matlab-CONTAM Toolbox GUI is divided into four main sections as shown in Fig. 3: Contamination Event, Edit Parameters, Simulation and CDI. In the *Contamination Event* section, the user can specify the contamination source(s) characteristics including the release location(s), the generation rate, the onset time and the duration of the event. From this section, it is also possible to choose a simulation time between 1-24 hr with the appropriate time step. The various problem parameters can be modified in the *Edit Parameters* section, including the Weather Data, the Zone Data, the Path Openings and the Sensor Data. In the *Simulation* section, the user can run the contaminant

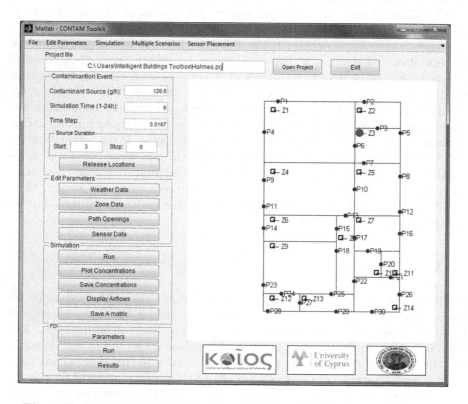

Fig. 3. The Matlab-CONTAM Toolbox displaying the Holmes' House project

transport simulation once all the parameters are appropriately set. This calls CONTAM in the background for calculating the airflows as explained in Section 2, and activates the other options inside the *Simulation* section. From here, the user can view and save the time series of contaminant concentrations in the different building zones (which have sensors installed), display the airflows and save the state transition matrix corresponding to these flows for future reference.

In the *CDI* section, the user can change the parameters, run and view the results for the event detection and isolation algorithms outlined in Section 3. The *Parameters* option displays another interface as shown in Fig. 4, divided into four sub-sections: Uncertainties bound, Detection and Isolation Parameters, Noise Bound and Nominal A matrix. The *Uncertainties bound* sub-section is used to provide a bound on the modeling uncertainty ΔA in model (1). This can be specified in three different ways using the provided dialog box: (i) as a constant known value; (ii) calculated using random sampling within the specified tolerances around the nominal conditions on the wind direction, wind speed, temperature and path openings; and (iii) calculated from an external file (*.mat) of a set of state transition matrices. For example, in the interface depicted in Fig. 4, the bound on the modeling uncertainty is calculated as 0.237 using the second option with 100 random samples within the following intervals from the nominal conditions: wind direction $90 \pm 10°$, wind speed $10 \pm 0,5$ m/s, zone temperature $30 \pm 2°C$ and leakage openings $\pm 10\%$ from the fully open position. In the *Detection and Isolation Parameters* sub-section, the user can modify, if required, the default values used by the detection and isolation algorithms displayed in Fig. 4. These depend on the initial problem assumptions and include the initial maximum state estimation error for detection and isolation, the learning rate and the maximum interval value for estimating the contaminant source, and an initialization value for the estimated source. More information on these parameters can be found in [10]. Next, a bound on the noise needs to be specified in the *Noise Bound* sub-section. Currently, we are assuming uniform, bounded noise, but other types of noise (i.e. Gaussian) can easily be incorporated in the future. Finally, the nominal state matrix A in (1) can be used in its current form or loaded from a file using the *Nominal A matrix* sub-section.

After setting the various parameters, the *Run* option is used in order to run the CDI algorithms and view the results. The sequence of steps performed follows the CDI architecture outlined in Section 3. The user is notified for the progress of the detection and isolation algorithms through user friendly graphical displays and message notification banners. Following the completion of the CDI algorithms, the user can view more detailed results concerning detection and isolation, by using the *Results* option. This opens up a new interface which displays the results of detection and isolation. For the specific test case scenario, the source was detected 3 *min* after the release time and isolated 7 *min* after the release time in the Utility zone. The *Results* interface is fully configurable and allows the user to plot any result concerning the residuals and/or the adaptive thresholds used for detection (CDE) and isolation (CIEs) for the zones of his choice. Some indicative plots of the results are displayed in Fig. 5–6.

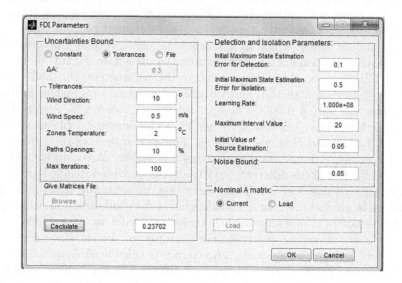

Fig. 4. Interface for setting the CDI parameters and calculating bound on modeling uncertainty

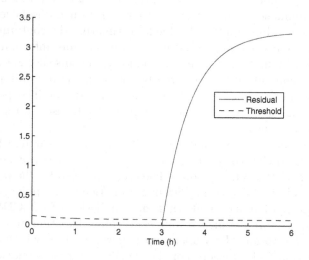

Fig. 5. Contaminant Detection Estimator (CDE) for zone 3 (i.e. sensor in utility room). The output estimation error (residual) is displayed using a solid line while the adaptive threshold is displayed using a dashed line. The contaminant source is detected when the error exceeds the threshold 3 *min* after the release.

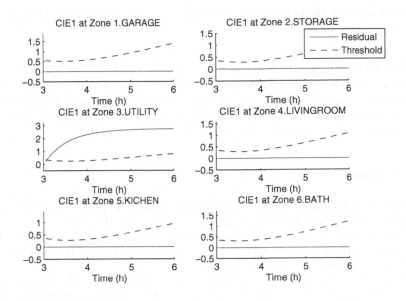

Fig. 6. 1st Contaminant Isolation Estimator (CIE) for zones 1–6. The output estimation error (residual) is displayed using solid lines while the adaptive thresholds are displayed using dashed lines. Zone 1 is excluded as a possible source location when the error exceeds the threshold in zone 3.

5   Conclusions and Future Work

Contaminant event monitoring in intelligent buildings requires large quantities of simulation data under different conditions in order to capture the variations in the complex dynamics involved. To this aim, the Matlab-CONTAM Toolbox has been developed and released, a software that operates within the Matlab environment and provides a programming interface for CONTAM, a multi-zone airflow and contaminant transport analysis tool. To facilitate the interaction of the researchers with the Toolbox, an intuitive graphical user interface has been designed for accessing the different functionalities and a modular architecture has been adopted to assist in the implementation of new methods. In this paper, we describe the implementation of advanced CDI schemes based on the state-space method for detecting and isolating a single contaminant source in an indoor building environment. We demonstrate the use of the Toolbox through a specific 14-zone building case study referred to as the Holmes' house.

In the future, we plan to expand the capabilities of the Toolbox for detecting and isolating multiple contaminant sources. We also plan to incorporate different techniques for event detection. The ultimate goal of the Toolbox is to serve as a common programming framework for research on Intelligent Buildings, by allowing the simulation of multiple contamination scenarios under varying conditions as well as the storage of the results in data structures. These data, can serve as benchmarks in the future for evaluating the performance of the different designed algorithms.

Acknowledgements. This research work has been funded by the European Commission 7th Framework Program, under grant INSFO-ICT-270428 (iSense), and by the Cyprus Research Promotion Foundation's Framework Programme for Research, Technological Development and Innovation, co-funded by the Republic of Cyprus and the European Regional Development Fund.

References

1. Braun, J.: Intelligent building systems-past, present, and future. In: Proc. of American Control Conference (ACC 2007), pp. 4374–4381. IEEE (2007)
2. Dounis, A., Caraiscos, C.: Advanced control systems engineering for energy and comfort management in a building environment—a review. Renewable and Sustainable Energy Reviews 13(6-7), 1246–1261 (2009)
3. Eliades, D., Michaelides, M., Panayiotou, C., Polycarpou, M.: Security-oriented sensor placement in intelligent buildings. Building and Environment 63, 114–121 (2013)
4. Fong, K., Hanby, V., Chow, T.: HVAC system optimization for energy management by evolutionary programming. Energy and Buildings 38(3), 220–231 (2006)
5. Lu, X., Clements-Croome, D., Viljanen, M.: Past, present and future of mathematical models for buildings. Intelligent Buildings International 1(16), 23–38 (2009)
6. Michaelides, M., Reppa, V., Panayiotou, C., Polycarpou, M.M.: Contaminant event monitoring in intelligent buildings using a multi-zone formulation. In: Proc. of Fault Detection, Supervision and Safety of Technical Processes (SAFEPROCESS 2012), vol. 8, pp. 492–497 (2012)
7. Walton, G., Dols, W.: CONTAM 2.4 user guide and program documentation. National Institute of Standards and Technology. NISTIR 7251 (2005)
8. Wang, L., Dols, W., Chen, Q.: Using CFD capabilities of CONTAM 3.0 for simulating airflow and contaminant transport in and around buildings. HVAC&R Research 16(6), 749–763 (2010)
9. Wong, S., Wan, K.K., Lam, T.N.: Artificial neural networks for energy analysis of office buildings with daylighting. Applied Energy 87(2), 551–557 (2010)
10. Zhang, X., Polycarpou, M.M., Parisini, T.: Design and analysis of a fault isolation scheme for a class of uncertain nonlinear systems. Annual Reviews in Control 32(1), 107–121 (2008)

A Hierarchy of Change-Point Methods for Estimating the Time Instant of Leakages in Water Distribution Networks

Giacomo Boracchi[1], Vicenç Puig[2], and Manuel Roveri[1,*]

[1] Dipartimento di Elettronica,
Informazione e Bioingegneria, Politecnico di Milano, Milano, Italy
[2] Department of Automatic Control, Universitat Politècnica de Catalunya, Barcelona, Spain
{giacomo.boracchi,manuel.roveri}@polimi.it,
vicenc.puig@upc.edu

Abstract. Leakages are a relevant issue in water distribution networks with severe effects on costs and water savings. While there are several solutions for detecting leakages by analyzing of the minimum night flow and the pressure inside manageable areas of the network (DMAs), the problem of estimating the time-instant when the leak occurred has been much less considered. However, an estimate of the leakage time-instant is useful for the diagnosis operations, as it may clarify the leak causes. We here address this problem by combining two change-point methods (CPMs) in a hierarchy: at first, a CPM analyses the minimum night flow providing an estimate of the day when the leakage started. Such an estimate is then refined by a second CPM, which analyzes the residuals between the pressure measurements and a network model in a neighborhood of the estimated leakage day. The proposed approach was tested on data from a DMA of a big European city, both on artificially injected and real leakages. Results show the feasibility of the proposed solution, also when leakages are very small.

Keywords: Leakages in Water Distribution Networks, Change-Point Estimation, Nonstationarity Detection.

1 Introduction

Water losses in distribution drinking water networks are an issue of great concern for water utilities, strongly linked with operational costs and water resources savings. Continuous improvements in water losses management are being applied and new technologies are developed to achieve higher levels of efficiency.

Among the wide range of water losses, we consider leakages, a specific type of hydraulic fault, that may be due to pipe breaks, loose joints and fittings, and overflowing water from storage tanks. Some of these problems are caused by the deterioration of the water delivery infrastructure, which is affected by ageing effects and high pressures. Leakages are classified by water utilities as *background* (small undetectable leaks for

---

* This research has been funded by the European Commission's 7th Framework Program, under grant Agreement INSFO-ICT-270428 (iSense).

H. Papadopoulos et al. (Eds.): AIAI 2013, IFIP AICT 412, pp. 615–624, 2013.

which no action to repair is taken), *unreported* (moderate flow-rates which accumulate gradually and require eventual attention), and *reported* (high flow-rates which require immediate attention). In practice, there may be a significant time delay between the time instant when a leakage occurs, when the water utility detects the leakage, and when the leakage is located and repaired [1].

The traditional approach to leakage control is a passive one, whereby the leak is repaired only when it becomes visible. Recently developed acoustic instruments [2] allow to locate also invisible leaks, but unfortunately, their application over a large-scale water network is very expensive and time-consuming. A viable solution is to divide the network into District Metered Area (DMA), where the *flow* and the *pressure* are measured [3, 1], and to maintain a permanent leakage control-system: leakages in fact increase the flow and decrease the pressure measurements at the DMA entrance. Various empirical studies [4, 5] propose mathematical models to describe the leakage flow with respect to the pressure at the leakage location.

Best practice in the analysis of DMA flows consists in estimating the leakage when the flow is minimum. This typically occurs at night, when customers' demand is low and the leakage component is at its largest percentage over the flow [1]. Therefore, practitioners monitor the DMA or groups of DMAs for detecting (and then repairing) leakages by analyzing the minimum night flow, and also employ techniques to estimate the leakage level [1]. However, leakage detection may not be easy, because of unpredictable variations in consumer demands and measurement noise, as well as long-term trends and seasonal effects. Complementary to the minimum flow analysis, pressure loggers at the DMA entrance provide useful information for leak detection and isolation [6]. When a leakage appears in a DMA, the pressure at junctions typically changes, showing the key evidence for the leakage and providing information for its isolation.

In this paper, we address the problem of estimating the time-instant when a leakage has occurred within a DMA. Obtaining accurate estimates of leak time-instant is important, as this information improves the leak-diagnosis operations – including quantifying the leakages effect and understanding the leak causes – as well as the accommodation operations. Peculiarity of the proposed solution is to combine two change-point methods (CPMs) in a hierarchical manner. At first, a CPM analyzes the minimum night flow in the DMA to estimate the day when the leak has occurred. Then, within a range of this specific day, the residuals between the pressure measurements and a network model are analyzed by a second CPM, which estimates the time-instant when the leak has started. Such a coarse-to-fine analysis prevents the use of network models over large time-intervals, where these may be not accurate because of the large dimension of the network. The leak-detection problem is not addressed here, as this can be managed by specific techniques, such as [7]. To illustrate the feasibility of the proposed approach, real data – and a real leakage – coming from a DMA of a big European city are considered.

The structure of the paper is the following: Section 2, states the problem of estimating the leak-time instant in a DMA, while Section 3 presents the hierarchy of CPMs to address this problem. Section 4 presents the results obtained on real data from a DMA. Finally, conclusions are drawn in Section 5.

2 Problem Statement

We focus on a single DMA, connected to the main water supply network through a reduced number of pipes, where the flows and pressure inlets, as well as the pressure of some internal nodes, are recorded. In fault-free conditions, the flow measurements represent the water consumption and follow a *trend* that is stationary, though unknown. After the leakage-time instant T^*, the flow follows a trend affected by a faulty profile. We assume that the leakage induces an abrupt and permanent fault, so that the measured flow at the inflows of the DMA becomes

$$f(t) = \begin{cases} f_0(t), & \text{if } t < T^* \\ f_0(t) + \Delta_f, & \text{if } t \geq T^* \end{cases}, \tag{1}$$

where $\Delta_f > 0$ is the offset corresponding to the leakage magnitude. Similarly, the behavior of the measured pressure can be defined:

$$p(t) = \begin{cases} p_0(t), & \text{if } t < T^* \\ p_0(t) - \Delta_p, & \text{if } t \geq T^* \end{cases}, \tag{2}$$

where $\Delta_p > 0$ is the offset representing the leakage affect the pressure measurements.

Let $\widehat{T}$ be the time instant in which the leakage is detected by a suitable detection method, our goal is to identify T^* by analyzing

$$F = \{f(t), 0 \leq t \leq \widehat{T}\} \text{ and } P = \{p(t), 0 \leq t \leq \widehat{T}\} \tag{3}$$

Thus, the leak is assumed to be constant within the time interval $[T^*, \widehat{T}]$, and we do not consider measurements after $\widehat{T}$ when the accommodation procedures start.

3 A Hierarchy of CPMs to Estimate Leak Time-Instant

Change-Point Methods are hypothesis tests designed to analyze, in an offline manner, whether a given data sequence X contains i.i.d. realizations of a random variable (i.e., null hypothesis) or a change-point that separates X in two subsequences generated from different distributions (alternative hypothesis). Interest reader can refer to [8–10].

In this paper, we illustrate the use of CPMs for estimating T^*, the time instant when a leakage occurs, and we show that, to this purpose, it is convenient to combine two CPMs in a hierarchical scheme yielding coarse-to-fine estimates. At first, a CPM analyses the statistical behavior of the minimum night flows: interestingly, when no leakage affects the network, the minimum night flows are expected to be stationary [7]. Thus, minimum night flow values can be (at least approximatively) modeled as i.i.d. realizations of a random variable. The analysis of the minimum night flow by means of a CPM provides us M_Φ, an estimate of the day when the leakage started, as described in Section 3.1. This is a coarse-grained estimate that can be refined by analyzing the pressure measurements in the few days before and after M_Φ. Unfortunately, CPMs cannot be straightforwardly used on pressure measurements, as these are time-dependent data, while CPMs operate

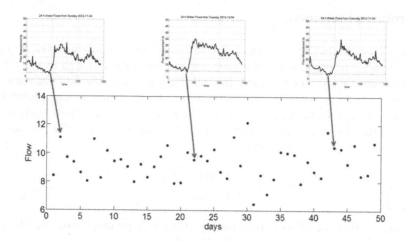

Fig. 1. An illustrative example of the sequence Φ. The values of Φ are displayed in the plot below, and correspond to the minimum night flow of each specific day. The small plots above represent the daily flows: the value of the minimum night flow is plot as an horizontal (red) line.

on i.i.d. realizations of a random variable. To address this problem we run a CPM on residuals of approximating models, as in [11]. Specifically, in Section 3.2 we apply a CPM on pressure residuals measuring the discrepancy between the measurements and estimates provided by a DMA model, see [6]. The use of the ensemble of CPMs [11] instead of conventional CPMs to compensate temporal dependencies in the minimum night flow and in pressure residuals is discussed in Section 3.3.

3.1   CPM on the Minimum Daily Flow

Let us denote by $\phi(\tau)$ the minimum night flow of the day τ, which may be computed as the minimum value or the average flow in a neighborhood of the minimum. Given the flow measurements F in (3) we compute

$$\Phi = \{\phi(\tau), 0 \leq \tau \leq \hat{\tau}\}, \tag{4}$$

the sequence of the minimum night flows, being $\hat{\tau}$ the day containing $\hat{T}$. We say that Φ contains a change-point at τ^* if $\phi(\tau)$ is distributed as

$$\phi(\tau) \sim \begin{cases} \mathcal{P}_0, & \text{if } 0 \leq \tau < \tau^* \\ \mathcal{P}_1, & \text{if } \tau^* \leq \tau \leq \hat{\tau} \end{cases}, \tag{5}$$

where $\mathcal{P}_0$ and $\mathcal{P}_1$ represent the distribution of the minimum night flow without and with a leakage, respectively, and τ^* and $\hat{\tau}$ are the day when the leak occurred and when the leakage has been detected, respectively.

Within the CPM framework, the null hypothesis consists in assuming that all data in Φ are i.i.d., and when the null hypothesis is rejected the CPM provides also an estimate of the change point τ^*, which here corresponds to an estimate of day when the leak has occurred. From the practical point of view, when running a CPM, each time instant $S \in \{1, \ldots, \hat{\tau}\}$ is considered as a candidate change point of Φ, which is accordingly partitioned in two non-overlapping sets

$$\mathcal{A}_S = \{\phi(\tau), \tau = 1, \ldots, S\}, \text{ and } \mathcal{B}_S = \{\phi(\tau), \tau = S + 1, \ldots, \hat{\tau}\},$$

that are then contrasted by means of a suitable test statistic $\mathcal{T}$. The test statistic

$$\mathcal{T}_S = \mathcal{T}(\mathcal{A}_S, \mathcal{B}_S), \tag{6}$$

measures the degree of dissimilarity between $\mathcal{A}_S$ and $\mathcal{B}_S$. Among test statistics commonly used in CPMs, we mention the Mann-Withney [8] (to compare the mean over $\mathcal{A}_S$ and $\mathcal{B}_S$), the Mood [12] (to compare the variance over $\mathcal{A}_S$ and $\mathcal{B}_S$) and the Lepage (to compare both the mean and variance over $\mathcal{A}_S$ and $\mathcal{B}_S$). Other statistics in [13, 14, 10, 15].

The values of $\mathcal{T}_S$ are computed for all the possible partitioning of Φ, yielding $\{\mathcal{T}_S, S = 1, \ldots, \hat{T}\}$; let $\mathcal{T}_{M\Phi}$ denote the maximum value of the test statistic, i.e.,

$$\mathcal{T}_{M\Phi} = \max_{S=1,\ldots,\hat{T}} (\mathcal{T}_S). \tag{7}$$

Then, $\mathcal{T}_{M\Phi}$ is compared with a predefined threshold $h_{l,\alpha}$, which depends on the statistic $\mathcal{T}$, the cardinality l of Φ, and a defined confidence level α that sets the percentage of type I errors (i.e., false positives) of the hypothesis test. When $\mathcal{T}_{M\Phi}$ exceeds $h_{l,\alpha}$, the CPM rejects the null hypothesis, and Φ is claimed to contain a change point at

$$M_\Phi = \underset{S=1,\ldots,\hat{T}}{\text{argmax}} (\mathcal{T}_S). \tag{8}$$

On the contrary, when $\mathcal{T}_{M\Phi} < h_{l,\alpha}$, there is not enough statistical evidence to reject the null hypothesis, and the sequence is considered to be stationary. Summarizing, the outcome of a CPM to estimate the day when the leak occurred is

$$\begin{cases} \text{The leak occurred at day } M_\Phi & \text{if } \mathcal{T}_{M\Phi} \geq h_{l,\alpha} \\ \text{No leak can be found in } \Phi, & \text{if } \mathcal{T}_{M\Phi} < h_{l,\alpha} \end{cases}. \tag{9}$$

3.2 CPM on the Residuals of the Pressure Measurements

A CPM executed on Φ provides an estimate M_Φ of $\hat{\tau}$, the day when the leakage started. To provide an fine-grained estimate of the leakage time-instant, we analyze the pressure measurements in a neighborhood of M_Φ (e.g., one or two days before and after M_Φ). Let $P_{M_\Phi} \subset P$ collects such pressure measurements. It is worth noting that, differently from values in Φ that can be assumed to be i.i.d., the pressure measurements follows their own dynamics, hence CPMs cannot be directly applied to P_{M_Φ}. We address this issue as in [11], and we compute the residuals between the measured pressure and its estimates using the DMA mathematical model f_θ

$$\hat{p}(t) = f_\theta(p(t-1), \ldots, p(t-n_x), u(t), u(t-1), \ldots u(t-n_u)), \tag{10}$$

where θ is the model parameter vector, $p(t) \in \mathbb{R}$ and $u(t) \in \mathbb{R}^m$ represent the output and input of the DMA model, respectively. The parameters $n_x \geq 0$ and $n_u \geq 0$ set the order of the output and input, respectively. The DMA model is based on the hydraulic laws describing the flow balance in the DMA nodes and the pressure drop in the pipes. This model leads to a set of non-linear equations with non-explicit solution that must be solved numerically using a water network simulator (EPANET), as it is done in [6], to obtain $\hat{p}(t)$.

The second CPM of the hierarchy is executed on the residual sequence that refers to pressure measurements in P_{M_Φ}:

$$\mathcal{R}_P = \{p(t) - \hat{p}(t), T_{\text{init}} \leq t \leq T_{\text{end}}\}, \tag{11}$$

where T_{init} and T_{end} are the initial and final time instant of P_{M_Φ}, respectively. Following the CPM formulation in Section 3.1, the CPM on $\mathcal{R}_P$ is immediately obtained by replacing Φ with $\mathcal{R}_P$. We denote by M_P the leakage time-instant estimated form $\mathcal{R}_P$.

3.3   Ensemble of CPMs

Unfortunately, the minimum night flow might suffer from seasonalities or nonstationarities that are difficult to compensate or address, thus, the sequence Φ may not contain truly i.i.d. observations. Moreover, approximating models are never exact, and, because of model bias, sequence $\mathcal{R}_P$ is far from being i.i.d. both before and after T^*, where a large degree of dependency among the residuals is expected. These circumstances violate the hypothesis required by the CPM and explain why the CPMs are not able to properly estimate the change-point on residuals from approximating models, [11].

To reduce the effect of time-dependency in the analyzed sequence, it is possible to use the ensemble of CPMs, which is detailed in [11]. Because of space limitation here, we briefly describe its peculiarities. The ensemble $\mathcal{E}_d$ aggregates d individual estimates provided by CPMs executed on subsequences obtained by randomly sampling the original sequence (either Φ or $\mathcal{R}_P$). Such a random sampling is meant to reduce the temporal dependencies. Experiments in [11] on residuals of ARMA processes show that the ensemble provides better performance in locating the change-point than a single CPM executed on the whole residual sequence.

4   Experimental Results

To illustrate the feasibility of using CPM to estimate the leak time-instant, we consider data from the DMA of a big European city. The DMA is characterized by two inlets where flow an pressure are measured as well as five pressure monitoring sensors right inside. Real records have been collected from 11th November to 22nd December 2012.

4.1   CPM Configuration

Since we expect the leak to induce an abrupt and permanent shift in Φ and $\mathcal{R}_P$, we exploit a nonparametric CPM based on the Mann-Whitney [16] statistics, $\mathcal{U}$, and we

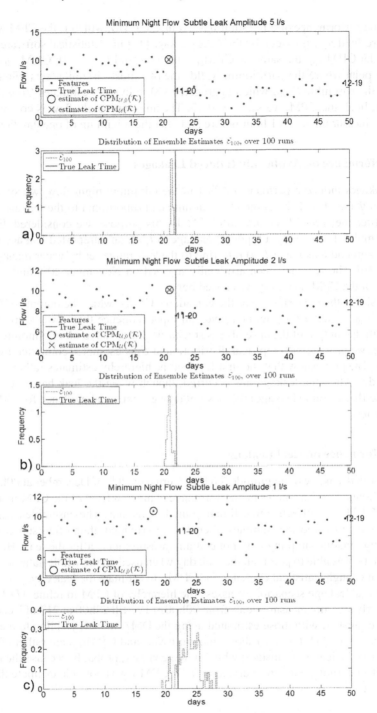

Fig. 2. Analysis of the minimum night flow to estimate the day when a leakage of 5 l/s (a), 2 l/s (b) and 1 l/s (c), appeared. The leak has been artificially injected on November, 20th.

analyze the performance of three solutions. At first, $\text{CPM}_{\mathcal{U}}$, which is the CPM with the proper threshold $h_{\alpha,l}$ provided by the CPM package [17] in R statistical software. Then, we consider $\text{CPM}_{\mathcal{U},0}$, the same as $\text{CPM}_{\mathcal{U}}$ with threshold $h_{n,\alpha} = 0$; $\text{CPM}_{\mathcal{U},0}$ locates a change-point where the partitioning yields the two most dissimilar sets. Finally, we consider the ensemble $\mathcal{E}_{100}$, which aggregates 100 individual estimates from random sampling. In all the CPM, we set $\alpha = 0.05$. The minimum night flow is computed by averaging measurements in 1 hour before and after the minimum of one day flow.

4.2   Performance on Artificially Induced Leakages

Since leakages induce a permanent offset in the minimum night flow, we simulate a leakage in Φ (see Eq. 4) by subtracting an amount proportional to the leakage amplitude (in liters per second) in $\phi(t)$ after T^*. To this purpose, we considered different leakage amplitudes, namely a reported leakage (5 l/s), an unreported leakage (2 l/s) and a background leakage (1 l/s) referring to the leak range set by water management company [6]. The leak has been artificially injected on November 20th, and only the first level of the CPM hierarchy was tested here.

Fig. 2 shows the performance of the considered CPMs: while the outputs of $\text{CPM}_{\mathcal{U}}$ and $\text{CPM}_{\mathcal{U},0}$ are deterministic, i.e., given an input sequence they always provide the same result, the output of the ensemble is stochastic due to the random sampling phase. Therefore, we provide the empirical distribution of $\mathcal{E}_{100}$ estimates computed over 100 iterations. The plots show that the first level of the hierarchy estimates rather successfully the day when the leakage has occurred, though when the leak has a very small magnitude (background leakage) there is not enough statistical evidence for $\text{CPM}_{\mathcal{U}}$ to assess the leak.

4.3   Performance on Real Leakages

A real reported leakage of magnitude 5.6 l/s occurred on 20th of December at 00h30 and lasted 30 hours. Fig. 3 a) shows the performance of the first level of the CPM hierarchy, and only $\text{CPM}_{\mathcal{U},0}$ is effective in estimating the leakage day. The ensemble $\mathcal{E}_{100}$ is not accurate here since too few samples after T^* are provided, thus often subsequences obtained by random sampling may not contain measurements after the leak. However, the $\mathcal{U}$ statistic was able to point out the leak day when used on the whole dataset, though there is not enough statistical evidence for $\text{CPM}_{\mathcal{U}}$ to estimate the leak day.

On the real leakage scenario, we assess the hierarchy of CPM to refine M_Φ by analyzing hourly pressure measurements from 00h00 of 19th till 23h00 of 21st of December, that were compared with those estimated using the DMA model, yielding to the residual sequence $\mathcal{R}_P$ (11). Fig. 3 b) shows that both $\mathcal{E}_{100}$ and $\text{CPM}_{\mathcal{U},0}$ are rather effective in estimating the leak time-instant; while on the contrary, probably because the residuals are not i.i.d. (not even before and after T^*), $\text{CPM}_{\mathcal{U}}$ was not able estimate the leak time-instant.

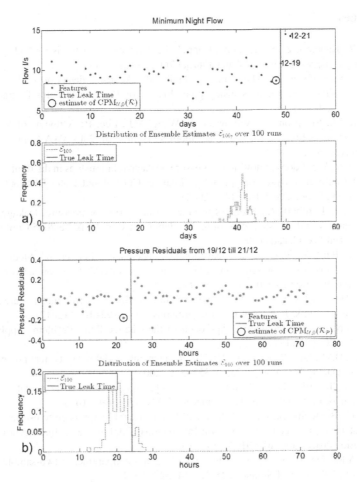

Fig. 3. Analysis of the minimum night flow to estimate the time instant when a leakage appeared in the network. The leakage occurred at 20th of December and lasted for two days.

5 Conclusion

We propose a hierarchy of CPMs to estimate the time instant when a leak has appeared within a DMA. Two CPMs are combined to estimate, first, the day, and then, the exact time when the leakage has occurred. The proposed hierarchy prevents the use of models approximating the pressure measurements over a large time interval, where these could be highly affected by model bias. Application results in a real DMA of a big European city have shown the feasibility of the proposed approach to estimate the leak time-instant also for subtle leaks. Ongoing works concern the use of multivariate CPM to analyze simultaneously the pressure, flow and different indicators that may be affected by the leak (e.g., water billed volumes), as well as providing a complete methodology – including leak detection and isolation – for leak diagnosis based on CPM.

References

1. Puust, R., Kapelan, Z., Savic, D.A., Koppel, T.: A review of methods for leakage management in pipe networks. Urban Water Journal 7(1), 25–45 (2010)
2. Khulief, Y., Khalifa, A., Mansour, R., Habib, M.: Acoustic detection of leaks in water pipelines using measurements inside pipe. Journal of Pipeline Systems Engineering and Practice 3(2), 47–54 (2012)
3. Lambert, M., Simpson, A., Vítkovský, J., Wang, X.J., Lee, P.: A review of leading-edge leak detection techniques for water distribution systems. In: 20th AWA Convention, Perth, Australia (2003)
4. Lambert, A.: What do we know about pressure: leakage relationships in distribution systems? In: IWA Conference System Approach to Leakage Control and Water Distribution System Management, Brno, Czech Rebublic (2001)
5. Thornton, J., Lambert, A.: Progress in practical prediction of pressure: leakage, pressure: burst frequency and pressure: consumption relationships. In: Leakage 2005 Conference Proceedings, Halifax, Canada (2005)
6. Pérez, R., Puig, V., Pascual, J., Quevedo, J., Landeros, E., Peralta, A.: Methodology for leakage isolation using pressure sensitivity analysis in water distribution networks. Control Engineering Practice 19(10), 1157–1167 (2011)
7. Eliades, D.G., Polycarpou, M.M.: Leakage fault detection in district metered areas of water distribution systems. Journal of Hydroinformatics 14(4), 992–1002 (2012)
8. Pettitt, A.N.: A Non-Parametric Approach to the Change-Point Problem. Applied Statistics 28(2), 126–135 (1979)
9. Hawkins, D.M., Qiu, P., Kang, C.W.: The changepoint model for statistical process control. Journal of Quality Technology 35(4), 355–366 (2003)
10. Ross, G.J., Tasoulis, D.K., Adams, N.M.: Nonparametric monitoring of data streams for changes in location and scale. Technometrics 53(4), 379–389 (2011)
11. Alippi, C., Boracchi, G., Puig, V., Roveri, M.: An ensemble approach to estimate the fault-time instant. In: Proceedings of the 4th International Conference on Intelligent Controland Information Processing, ICICIP 2013 (2013)
12. Mood, A.M.: On the asymptotic efficiency of certain nonparametric two-sample tests. The Annals of Mathematical Statistics 25(3), 514–522 (1954)
13. Hawkins, D.M., Zamba, K.D.: A change-point model for a shift in variance. Journal of Quality Technology 37(1), 21–31 (2005)
14. Zamba, K.D., Hawkins, D.M.: A multivariate change-point model for statistical process control. Technometrics 48(4), 539–549 (2006)
15. Ross, G., Adams, N.M.: Two nonparametric control charts for detecting arbitrary distribution changes. Journal of Quality Technology 44(22), 102–116 (2012)
16. Mann, H.B., Whitney, D.R.: On a Test of Whether one of Two Random Variables is Stochastically Larger than the Other. The Annals of Mathematical Statistics 18(1), 50–60 (1947)
17. Ross, G.J.: Parametric and nonparametric sequential change detection in R: The cpm package. Journal of Statistical Software (forthcoming)

EWMA Based Two-Stage Dataset Shift-Detection in Non-stationary Environments

Haider Raza, Girijesh Prasad, and Yuhua Li

Intelligent Systems Research Center, University of Ulster, UK
raza-h@email.ulster.ac.uk, {g.prasad,y.li}@ulster.ac.uk

Abstract. Dataset shift is a major challenge in the non-stationary environments wherein the input data distribution may change over time. In a time-series data, detecting the dataset shift point, where the distribution changes its properties is of utmost interest. Dataset shift exists in a broad range of real-world systems. In such systems, there is a need for continuous monitoring of the process behavior and tracking the state of the shift so as to decide about initiating adaptive corrections in a timely manner. This paper presents a novel method to detect the shift-point based on a two-stage structure involving Exponentially Weighted Moving Average (EWMA) chart and Kolmogorov-Smirnov test, which substantially reduces type-I error rate. The algorithm is suitable to be run in real-time. Its performance is evaluated through experiments using synthetic and real-world datasets. Results show effectiveness of the proposed approach in terms of decreased type-I error and tolerable increase in detection time delay.

Keywords: Non-stationary, Dataset shift, EWMA, Online Shift-detection.

1 Introduction

In the research community of statistics and machine learning, detecting abrupt and gradual changes in time-series data is called shift-point detection [1]. Based on the delay of the detection, shift-point detection methods can be classified into two categories: retrospective detection and real-time detection. The retrospective shift-point detection tends to give more accurate and robust detection; however, it requires longer reaction periods so it may not be suitable for the real-time applications where initiating adaptation closest to the shift-point is of paramount importance. Also, in the real-time systems, each observation coming in the data stream may be processed only once and then discarded; hence retrospective shift-point detection may not be possible to implement. The real-time detection is also called a single pass method and requires an immediate reaction to meet the deadline. One important application of such a method is in pattern classification based on streaming data, performed in several key areas such as electroencephalography (EEG) based brain-computer interface, robot navigation, remote sensing and spam-filtering. Classifying the data stream in non-stationary environments requires the developments of a method which should be computationally efficient and able to detect the shift-point in the underlying distribution of the data stream. The key difference between the conventional classification

H. Papadopoulos et al. (Eds.): AIAI 2013, IFIP AICT 412, pp. 625–635, 2013.

and streaming classification is that, in a conventional classification problem, the input data distribution is assumed not to vary with time. In a streaming case, the input data distribution may change over time during the testing/operating phase [2]. Based on the detected shift, a classifier needs to account for the dataset shift. In non-stationary environments there are several types of dataset shift, a brief review of the dataset shift and the types of dataset shift are presented in the next section.

Some pioneering works [3][4] have demonstrated good shift-point detection performance by monitoring the moving control charts and comparing the probability distributions of the time-series samples over past and present intervals. These methods follow different strategies such as CUSUM (Cumulative SUM) [5], Computational-Intelligence CUSUM (CI-SUSUM) [4], Intersection of Confidence Interval rule (ICI) for change-point detection [6]. However, most of these methods rely on pre-designed parametric models such as underlying probability distribution and auto-regressive models for tracking some specific statistics such as the mean, variance, and spectrum. However, to overcome these weaknesses, recently some researchers have proposed a different strategy, which estimates the ratio of two probability densities directly without density estimation [7]. Thus, the aforementioned methods are not robust against different types of the shift-detection tests because of the delay and the high rate of type-I error (i.e., false-positive). On account of the need for identifying models, this may significantly limit their range of applications in fast data streaming problems.

In this paper we present an approach consisting of a Two-Stage Dataset Shift-detection based on an Exponentially Weighted Moving Average chart (TSSD-EWMA). This TSSD-EWMA in an extension of our work on Dataset Shift-Detection based on EWMA (SD-EWMA) [8], which suffers from the high false-positive rate. In TSSD-EWMA, the stage-I consists of a shift-detection test that activates the stage-II when a shift is alarmed at the stage-I. At stage-II, the suspected shift from stage-I is validated and is either confirmed as a shift or declared as a false alarm. The structure of test is given in Figure 1. It is demonstrated to outperform other approaches in terms of non-stationarity detection with a tolerable time delay and decreased false-positive alarms. So, this scheme can be deployed along with any classifier such as k-nearest neighbor or support vector machine (SVM) in an adaptive online learning framework.

This paper proceeds as follows: Section 2 presents a background of dataset shift and EWMA control chart. Section 3 details the shift-detection algorithm and the structure of two-stage test. Section 4 presents the datasets used in the experiment. Finally, Section 5 presents the results and discussion.

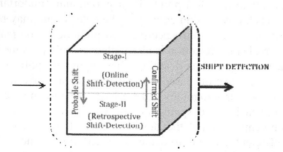

Fig. 1. The structure of the proposed two-stage shift-detection test

2   Background

2.1   Dataset Shift

The term dataset shift [9] was first used in the workshop of neural information processing systems (NIPS, 2006). Assume a classification problem is described by a set of features or inputs x , a target variable y , the joint distribution $P(y, x)$, the prior probability $p(x)$ and conditional probability $p(y|x)$, respectively. The dataset shift is then defined as "cases where the joint distribution of inputs and outputs differs between training and test stages, i.e., when $P_{train}(y, x) \neq P_{test}(y, x)$"[10]. Dataset shift was previously defined by various authors and they gave different names to the same concept such as, concept shift, changes of classification, changing environment, contrast mining, fracture point, and fracture between data. There are three types of dataset shift that usually occur, (i) covariate shift, (ii) prior probability shift and (iii) concept shift; we will briefly describe these shifts in following subsections.

Covariate Shift: It is defined as a case, where $(p_{train}(y|x) = p_{test}(y|x))$ and $(p_{train}(x) \neq p_{test}(x))$ [9] (e.g., the input distribution often changes in the session-to-session transfer of an EEG based brain-computer interface).
Prior Probability Shift: It is defined as a case, where $(p_{train}(x|y) = p_{test}(x|y))$ and $(p_{train}(y) \neq p_{test}(y))$ [9] (e.g., the survivorship in a stock market where there is bias in obtaining the samples).
Concept Shift: It is defined as a case, where $(p_{train}(y|x) \neq p_{test}(y|x))$ [9] (e.g., fraud detection).

The shifts discussed above are the most commonly present in the real-world problems, there are other shifts also that could happen in theory, but we are not discussing those because they appear rarely.

2.2   EWMA Control Chart

An exponentially weighted moving average (EWMA) control chart [11] is from the family of control charts in the statistical process control (SPC) theory. EWMA is an efficient statistical method in detecting the small shifts in the time-series data. The EWMA control chart outperforms the other control charts because it combines current and historical data in such a way that small changes in the time-series can be detected more easily and quickly. The exponentially weighted moving average model is defined as

$$z_{(i)} = \lambda \cdot x_{(i)} + (1 - \lambda) \cdot z_{(i-1)} \tag{1}$$

where λ is the smoothing constant ($0 < \lambda \leq 1$) and z is a EWMA statistics. Moreover, the EWMA charts are used for both uncorrelated and auto-correlated data. We are only considering the auto-correlated data in our study and simulation.

EWMA for Auto-Correlated Data: If data contain a sequence of auto-correlated ob-servations, $x_{(i)}$, then the EWMA statistics in (1) can be used to provide 1-step-ahead prediction model of auto-correlated data. We have assumed that the process observa-tions $x_{(i)}$, can be defined as the equation (2) below, which is a first-order integrated moving average (ARIMA) model. This equation describes non-stationary behavior, wherein the variable $x_{(i)}$ shifts as if there is no fixed value of the process mean.

$$x_{(i)} = x_{(i-1)} + \varepsilon_i - \theta \varepsilon_{i-1} \tag{2}$$

where ε_i is a sequence of independent and identically distributed (i.i.d) random sig-nal with zero mean and constant variance. It can be easily shown that the EWMA with $(\lambda = 1 - \theta)$ is the optimal 1-step-ahead prediction for this process [12]. That is, if $\hat{x}_{(i+1)}(i)$ is the forecast of the observation in the period $(i + 1)$ made at the end of the period i, then, $\hat{x}_{(i+1)}(i) = z_i$, where equation (1) is the EWMA. The 1-step-ahead error $err_{(i)}$ are calculated as [13]:

$$err_{(i)} = x_{(i)} - \hat{x}_{(i)}(i - 1) \tag{3}$$

where $\hat{x}_{(i)}(i - 1)$is the forecast made at the end of period $(i - 1)$. Assume, that the 1-step-ahead prediction errors $err_{(i)}$ are normally distributed. It is given in [12], that it is possible to combine information about the statistical control and process dynam-ics on a single control chart. Then, the control limits of the chart on these errors satis-fy the following probability statements by substituting the right hand side of equation (3) as given below.

$$P\left[-L.\,\sigma_{err} \leq err_{(i)} \leq L.\sigma_{err}\right] = 1 - \alpha$$
$$P\left[\hat{x}_{(i)}(i-1) - L.\sigma_{err} \leq x_{(i)} \leq \hat{x}_{(i)}(i-1) + L.\sigma_{err}\right] = 1 - \alpha$$

where σ_{err} is the standard deviation of the errors, L is the control limit multiplier and $(1-\alpha)$ is a confidence interval. The standard deviation of 1-step-ahead error can be estimated in various ways such as mean absolute deviation or directly calculate the smoothed variance [12]. If the EWMA is a suitable 1-step-ahead predictor, then one could use $z_{(i)}$ as the center line for the period $i + 1$ with UCL (Upper Control Lim-it) and LCL (Lower Control Limit).

$$(UCL_{(i+1)} = z_{(i)} + L.\sigma_{err}) \quad \& \quad (LCL_{(i+1)} = z_{(i)} - L.\sigma_{err}) \tag{4}$$

Whenever, the $x_{(i+1)}$ moves out of $UCL_{(i+1)}$ and $LCL_{(i+1)}$, the process is said to be out of control. This method is also known as *a moving center-line EWMA control chart* [12]. The EWMA control chart is robust to the non-normality assumption if properly designed for the t and gamma distributions [14]. Following the above as-sumption, we have designed a two-stage algorithm for the shift detection in the process observation of auto-correlated data, which is discussed in the next section. So, the two-stage SD-EWMA (TSSD-EWMA) test can be employed when there is con-cern about the normality assumption.

3   Methodology

A control chart is the graphical representation of the sample statistics. Commonly it is represented by three lines plotted along horizontal axis. The center line and two control lines (control limits) are plotted on a control chart, which correspond to target value (μ) and acceptable deviation $(L\sigma)$ from either side of the target value respectively, where L is the control limit multiplier and σ is the standard deviation.

The proposed two-stage shift-detection based on EWMA (TSSD-EWMA) test works at two-stages. In the stage-I, the method employs a control chart to detect the dataset shift in the data stream. The stage-I works in an online mode, which continuously process the upcoming data from the data stream. The Stage-II uses a statistical hypothesis test to validate the shift detected by the stage-I. The stage-II operates in retrospective mode and starts validation once the shift is detected by the stage-I.

3.1   Stage-I

At the stage-I, the test works in two different phases. The first phase is training phase and the second phase is an operation or testing phase. In the first phase, the parameters $(\lambda, z_{(0)}, \sigma_{err(0)}^2)$ are calculated to decide the null hypothesis that there is no shift in the data. To calculate the parameters, first obtain the sequence of observations and calculate the mean. Use the mean as the initial value $z_{(0)}$ and obtain the EWMA statistics by equation (1). The sum of the squared 1-step-ahead prediction error divided by the length of the training dataset is used as an initial value of $\sigma_{err(0)}^2$ for the testing data. The test has been performed on the several values of λ, chosen as suggested in   [12], [14] and the choice of λ is discussed later in the result and discussion section.

In the testing phase, for each observation use equation (1) to obtain the EWMA statistics and follow the steps given in algorithm TSSD-EWMA in Table 1. Next, check if each observation $x_{(i+1)}$ falls within the control limits $[UCL_{(i+1)}, LCL_{(i+1)}]$, otherwise the shift is detected and alarm is raised at the stage-I. Furthermore, the shift detected by the stage-I is passed to the stage-II for validation.

3.2   Stage-II

In stage-II, the shift detected by the stage-I needs to be validated. This phase works in retrospective mode and it executes only when a shift is detected at stage-I. In particular, to validate the shift detected by the stage-I, the available information need to be partitioned into two-disjoint subsequences and then the statistical hypothesis test is applied. In literature, the statistical hypothesis tests are well established. The two-sample Kolmogorov-Smirnov test [15] is used to validate stationarity in the sub-sequences because of its non-parametric nature. The two-sample Kolmogorov-Smirnov test returns a test decision for the null hypothesis that the data in the subsequences are stationary with equal means and equal but unknown variances. The Kolmogorov-Smirnov statistics is briefly described as follows:

$$D_{n,n'} = sup_x \; |F_{1,n}(x) - F_{2,n'}(x)|$$

where sup_x is the supremum and $F_{1,n}(x)$ and $F_{2,n'}(x)$ are the empirical cumulative distribution function on the first and second samples respectively. The n and n' are the lengths of the two samples given as $n = \left((i - (m - 1)) : i\right)$ and $n' = ((i + 1) : (i + m))$ where m is the number of observations in the window of data used for the test. The null hypothesis is rejected at level α and (H=1) is returned if,

$$\sqrt{\frac{n . n'}{n + n'}} D_{n,n'} > K_\alpha$$

where K_α is the critical value and can be found in, e.g., [16].

Table 1. Algorithm-TSSD-EWMA

```
Input: A process x(i), generates independent and identically dis-
tributed observations over time (i).
Output: Shift-detection points
                            Stage-I
Training Phase
    1. Initialize training data x(i) for i=1:n, where n is the num-
       ber of observations in training data
    2. Calculate the mean of x(i) and set as z(0)
    3. Compute the z-statistics for each observation x(i)
                z(i)=lambda*x(i)+(1-lambda)*z(i-1)
    4. Compute the 1-step-ahead prediction error: err(i)=x(i)-z (i-
       1)
    5. Estimate the variance of error for the testing phase
    6. Set lambda by minimizing squared prediction error
Testing Phase
    1. For each data point x(i+1) in the operation/testing phase
    2. Compute z(i)=lambda*x(i)+(1-lambda)*z(i-1)
    3. Compute err(i)= x(i)-z(i-1)
    4. Estimate the variance sigma_hat_err(i)^2=phi*err(i)^2 + (1-
       phi)* sigma_hat_err(i-1)^2
    5. Compute UCL(i+1) and LCL(i+1)
    6. IF(LCL(i+1)< x(i+1)< UCL(i+1))
            THEN (Continue processing)
       ELSE (Go to Stage-II)
                            Stage-II
    1. For each x(i+1)
    2. Wait for m observations after the time i, organise the se-
       quential observations around time i into two partitions, one
       containing x((i-(m-1)):i), another x((i+1):(i+m)).
    3. Execute the hypothesis test on the partitioned data
       IF(H==1)
            THEN (test rejects the null hypothesis): Alarm is raised
            ELSE(The detection received by stage-I is a false-positive)
```

An important point to note here is that we have assumed that non-stationarity occurs due to changes in the input distribution only. So, it is said to be covariate shift-detection in non-stationary time-series by the TSSD-EWMA test. In Table 1, the designed algorithm TSSD-EWMA is presented in the form of pseudo code.

4   Dataset and Feature Analysis

To validate the effectiveness of the suggested TSSD-EWMA test, a series of experimental evaluations have been performed on three synthetic datasets and one real-world dataset. The datasets are described as follows.

4.1   Synthetic Data

Dataset 1-Jumping Mean (D1): The dataset used here is same as the toy dataset given in [1] for detecting change point in a time-series data. The dataset is defined as $y(t)$ in which 5000 samples are generated $(i.e., t = 1, \ldots, 5000)$

$$y(t) = 0.6y(t-1) - 0.5y(t-2) + \varepsilon_t$$

where ε_t is a noise with mean μ and standard deviation 1.5. The initial values are set as $y(1) = y(2) = 0$. A change point is inserted at every 100 time steps by setting the noise mean μ at time t as

$$\mu_N = \begin{cases} 0 & N = 1 \\ \mu_{N-1} + \dfrac{N}{16} & N = 2 \ldots .49 \end{cases}$$

where N is a natural number such that $100(N-1) + 1 \leq t \leq 100N$.

Dataset 2-Scaling Variance (D2): The dataset used here is the same as the toy dataset given in [1], for detecting change-point in time-series data. The dataset is defined as the auto-regressive model, but the change point is inserted at every 100 time steps by setting the noise standard deviation σ at time t as

$$\sigma = \begin{cases} 1 & N = 1,3, \ldots .,49 \\ \ln\left(e + \dfrac{N}{4}\right) & N = 2,4, \ldots 48 \end{cases}$$

where N is a natural number such that $100(N-1) + 1 \leq t \leq 100N$.

Dataset 3-Positive-Auto-correlated (D3): The dataset is consisting of 2000 data-points, the non stationarity occurs in the middle of the data stream, shifting from $N(x; 1,1)$ to $\mathcal{N}(x; 3,1)$, where $\mathcal{N}(x; \mu, \sigma)$ denotes the normal distribution with mean and standard deviation respectively.

4.2   Real-World Dataset

The real-world data used here are from BCI competition-III (IV-b) dataset [17]. This dataset contains 2 classes, 118 EEG channels (0.05-200Hz), 1000Hz sampling rate which is down-sampled to 100Hz, 210 training trials, and 420 test trials. We have selected a single channel (C3) and performed the band-pass filtering on the mu (μ) band (8-12 Hz) to obtain bandpower features. This real-world dataset is a good example to validate if the proposed TSSD-EWMA is able to detect the types of shifts in the data generating process with decreased false-positive. The results and discussion of the experiments are given in the next section.

5   Results

On each dataset, the proposed TSSD-EWMA technique is evaluated by the following metrics to measure the performance of the tests: *False-positive (FP)*: it counts the times a test detects a shift in the sequence when it is not there i.e., (false alarm); *True-positive (TP)*: it counts the times a test detects a shift in the sequence when it is there i.e., (hit alarm); *True*-negative (TN): it counts the times a test does not detect the shift when it is not there; *False-negative (FN)*: it counts the times a test does not detect a shift in the sequence when it is there i.e., (miss); *Not applicable (NA)*: it denotes not applicable situation, where the dataset is not suitable to be executed on the test; *Average delay (AD)*: it measures the average delay in shift-detection, i.e., sum of delay in each shift-detection is divided by the number of shifts detected; *Accuracy (ACC)*: it measures the accuracy of the results i.e.,

$$ACC = \frac{(\#True\ positive + \#Ture\ Negative)}{\left((\#True\ positive + \#False\ negative) + (\#False\ positive + \#True\ Negative)\right)}$$

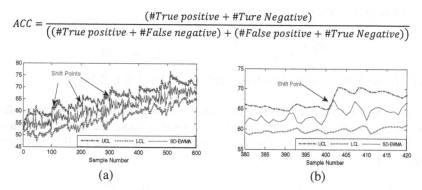

(a)   (b)

Fig. 2. Shift-detection based on TSSD-EWMA: Dataset 1 (jumping mean): (a) the shift point is detected at every 100th point.   (b) Zoomed view of figure a: shift is detected at 401st sample.

It is important to note that the stage-I of the TSSD-EWMA is based on the current observation of the data to detect the shift. So, it is coming without the delay and this is the advantage of this approach that the stage-I of the test always operate in online mode. Once, the shift is detected by the stage-I, the test moves into the retrospective mode to validate the suspected shift in the process. Figure 2 represents the stage-I of the TSSD-EWMA based shift-detection test results from the Dataset 1. The solid line is the observation plotted on the chart and the two dotted lines are the ULC and LCL, whenever the solid line crosses the dotted line (control limits), it is the shift-point detected. The tests have been performed on several values of λ as suggested in [12], [14].

According to Table 2, for the case of D1 with $\lambda = 0.40$, the TSSD-EWMA gives an optimal result with all the shift-detections and no false-positive, however with a delay of 10 samples, whereas, SD-EWMA detects four false-positive with no delay. The delay in TSSD-EWMA is resulting from the stage-II of the test. For the D2, the TSSD-EWMA is not applicable because of the nature of the dataset. For D3 in the TSSD-EWMA, when $\lambda = 0.50$ it detects all the shifts with no false-positive and a low delay of 10 samples, whereas, the SD-EWMA test suffers twelve false-positives and no delay.

Table 2. TSSD-EWMA shift-detection

	Total shifts	Lambda (λ)	SD-EWMA				TSSD-EWMA				
			# TP	# FP	# FN	ACC	# TP	# FP	# FN	AD	ACC
D1	9	0.20	8	1	1	99.8	8	0	1	10	99.9
		0.30	8	3	1	99.6	8	0	1	10	99.9
		0.40	All	4	0	99.6	All	0	0	10	100
D2	5	0.60	All	11	1	99.8	NA	NA	NA	NA	NA
		0.70	All	6	0	99.4	NA	NA	NA	NA	NA
		0.80	4	8	1	99.1	NA	NA	NA	NA	NA
D3	1	0.40	All	13	0	99.1	All	0	0	10	100
		0.50	All	12	0	99.2	All	0	0	10	100
		0.60	All	15	0	99.0	All	0	0	10	100

To assess the performance and compare the results of the TSSD-EWMA, we have chosen other shift-detection methods such as the SD-EWMA [8] and the ICI-CDT [3] because these are state-of-the-art non-parametric sequential shift-point detection tests. Table 3 compares the results of ICI-CDT and SD-EWMA with TSSD-EWMA. The rate of false-positive is much reduced for D1 and D3 datasets. However, in the case of D2 the test is not applicable because of the nature of dataset. The delay in the shift-detection for the TSSD-EWMA is less than that for the ICI-CDT method.

Table 3. Comparison on different shift-detection tests

	ICI-CDT				SD-EWMA				TSSD-EWMA			
	# TP	# FP	# FN	AD	# TP	# FP	# FN	AD	# TP	# FP	# FN	AD
D1	6	0	3	35	9	4	0	0	9	0	0	10
D2	1	0	4	60	5	6	0	0	NA	NA	NA	NA
D3	1	0	0	80	1	12	0	0	1	0	0	10

Table 4. TSSD-EWMA shift-detection in BCI data

Lambda (λ)	Number of Trials	# Shift-points SD-EWMA		# Shift-points TSSD-EWMA	
		Session 2	Session 3	Session 2	Session 3
0.01	1-20	0	0	0	0
	21-45	3	1	2	0
	46-70	4	2	3	1
0.05	1-20	10	4	6	3
	21-45	9	6	6	4
	46-70	7	5	6	4
0.10	1-20	16	9	8	6
	21-45	22	21	15	15
	46-70	25	17	17	13

To perform the shift-detection test, on the real-world dataset we have made use of bandpower features from the EEG data of session one from the aforementioned BCI competition-III (IV-b) and assumed that it is in stationary state. It contains 70 trials and the parameters are calculated in training phase to be used in the testing/operational phase. Next, the test is applied to the sessions two and three and the results are given in Table 4, wherein TSSD-EWMA is compared with the SD-EWMA. For evaluation purposes, we have performed the test on a fixed number of trials from each session and monitored the points where the shifts are detected. For the value of $\lambda=0.01$, and the trials 21-45 and 46-70, one false-positive is reduced by the TSSD-EWMA in each case. With increased value of $\lambda = 0.05$, the number of shifts increased, so the possibility of getting the number of false-positive also increased. Hence, by executing the TSSD-EWMA, the number of false alarms decreased. In case of the EEG signals, no comparison can be done because the results of actual shift points are not provided in [17].

6   Discussion and Conclusion

According to Table 3, the TSSD-EWMA provides better performance in terms of shift-detection with reduced number of false-positives compared to other state-of-the-art methods such as the SD-EWMA[8] and ICI-CDT [3], and it also has much smaller time delay compared to ICI-CDT. The TSSD-EWMA test outperforms over other methods in terms of low false-positive rate for all datasets because of its two-stage structure.

The choice of smoothing constant λ is an important issue in using EWMA control charts. The TSSD-EWMA results show that the detection of shift in the time-series data depends upon the value of the λ. The value of λ can be obtained by minimizing the least mean square prediction error on the datasets. However, selecting the value of λ by minimizing the least mean square prediction error does not always lead to a good choice. Moreover, if the value of the smoothing constant is large the test becomes more sensitive to spurious small shifts and it contains more number of false-positives. Hence, there is a trade-off between the smoothing constant and shift detection.

For the real-world dataset, we have assumed that the data from BCI session 1 is in stationary state and investigated for an optimal value of the smoothing constant λ considering the session 1 data as the training dataset. We have tested several values of λ in the range of (0.01-0.1). As the value of lambda increases, the number of detected shifts increases. Thus the smaller value of λ is a better choice for shift-detection in EEG based BCI, as it avoids shift-detections resulting from noise or spurious changes through much more intense smoothing of the EEG signal. Moreover, for correlated data, the smaller values of λ produce smaller prediction errors thereby resulting in smaller estimated standard error. If λ is too small, the performance of the test results in less false-positive rate but it tends toward getting more false-negatives, because chance of missing the shift is increased by much smoothing of the EEG signal. In summary, the experimental results demonstrate that TSSD-EWMA based test works well for shift-detection in the non-stationary environments.

This paper presented a method of TSSD-EWMA for detecting the shift in data stream based on the two-stage detection architecture. The advantage of using this method over other methods is primarily in terms of having a reduced number of false-positive detections. Our method is focused on auto-correlated data, which contain non-stationarities. Experimental analysis shows that the performance of the approach is good in a range of non-stationary situations. This work is planned to be extended further by employing it into pattern recognition problems involving multivariate data and an appropriate classifier.

References

1. Liu, S., Yamada, M., Collier, N., Sugiyama, M.: Change-Point Detection in Time-Series Data by Relative Density-Ratio Estimation. Neural Networks, 1–25 (2013)
2. Shimodaira, H.: Improving Predictive Inference Under Covariate Shift by Weighting the Log-Likelihood Function. Journal of Statistical Planning and Inference 90(2), 227–244 (2000)
3. Alippi, C., Boracchi, G., Roveri, M.: Change Detection Tests Using the ICI Rule. In: The International Joint Conference on Neural Networks (IJCNN), pp. 1–7 (July 2010)
4. Alippi, C., Roveri, M.: Just-in-Time Adaptive Classifier–Part I: Detecting Nonstationary Changes. IEEE Transactions on Neural Networks 19(7), 1145–1153 (2008)
5. Basseville, M., Nikiforov, I.: Detection of Abrupt Changes: Theory and Application. Prentice-Hall (1993)
6. Alippi, C., Boracchi, G., Roveri, M.: A Just-In-Time Adaptive Classification System Based on the Intersection of Confidence Intervals Rule. Neural Networks: The Official Journal of the International Neural Network Society 24(8), 791–800 (2011)
7. Sugiyama, M., Suzuki, T., Kanamori, T.: Density Ratio Estimation in Machine Learning, p. 344. Cambridge University Press (2012)
8. Raza, H., Prasad, G., Li, Y.: Dataset Shift Detection in Non-Stationary Environments using EWMA Charts. In: IEEE International Conference on Systems, Man, and Cybernetics (accepted, 2013)
9. Moreno-Torres, J.G., Raeder, T., Alaiz-Rodríguez, R., Chawla, N.V., Herrera, F.: A Unifying View on Dataset Shift in Classification. Pattern Recognition 45(1), 521–530 (2012)
10. Sugiyama, M., Schwaighofer, A., Lawrence, N.D.: Dataset Shift in Machine Learning. MIT Press (2009)
11. Roberts, S.W.: Control chart tests Based on Geometric Moving Averages. Technometrics (1959)
12. Dougla, C.M.: Introduction to Statistical Quality Control, 5th edn. John Wiley & Sons (2007)
13. Ye, N., Vilbert, S., Chen, Q.: Computer Intrusion Detection Through EWMA for Autocorrelated and Uncorrelated Data. IEEE Transaction on Reliability 52(1), 75–82 (2003)
14. Connie, M., Douglas, C., George, C.: Robustness of the EWMA Control Chart to Non-Normality. Journal of Quality Technology 31(3), 309 (1999)
15. Snedecor, G.W., Cochran, W.G.: Statistical Methods, Eight. Iowa State University Press (1989)
16. Table of Critical Values for the Two-Sample Test, http://www.soest.hawaii.edu/wessel/courses/gg313/Critical_KS.pdf
17. Klaus-Robert Müller, B.B.: BCI Competition III: Data set IVb (2005), http://www.bbci.de/competition/iii/desc_IVb.html

NEVE: A Neuro-Evolutionary Ensemble
for Adaptive Learning

Tatiana Escovedo, André Vargas Abs da Cruz, Marley Vellasco,
and Adriano Soares Koshiyama

Electrical Engineering Department
Pontifical Catholic University of Rio de Janeiro (PUC-Rio) - Rio de Janeiro – Brazil
{tatiana,andrev,marley,adriano}@ele.puc-rio.br

Abstract. This work describes the use of a quantum-inspired evolutionary algorithm (QIEA-R) to construct a weighted ensemble of neural network classifiers for adaptive learning in concept drift problems. The proposed algorithm, named NEVE (meaning Neuro-EVolutionary Ensemble), uses the QIEA-R to train the neural networks and also to determine the best weights for each classifier belonging to the ensemble when a new block of data arrives. After running eight simulations using two different datasets and performing two different analysis of the results, we show that NEVE is able to learn the data set and to quickly respond to any drifts on the underlying data, indicating that our model can be a good alternative to address concept drift problems. We also compare the results reached by our model with an existing algorithm, Learn++.NSE, in two different nonstationary scenarios.

Keywords: adaptive learning, concept drift, neuro-evolutionary ensemble, quantum-inspired evolution.

1   Introduction

The ability for a classifier to learn from incrementally updated data drawn from a nonstationary environment poses a challenge to the field of computational intelligence. Moreover, the use of neural networks as classifiers makes the problem even harder, as neural networks are usually seen as tools that must be retrained with the whole set of instances learned so far when a new chunk of data becomes available.

In order to cope with that sort of problem, a classifier must, ideally, be able to [1]:

- Track and detect any sort of changes on the underlying data distribution;
- Learn with new data without the need to present the whole data set again for the classifier;
- Adjust its own parameters in order to address the detected changes on data;
- Forget what has been learned when that knowledge is no longer useful for classifying new instances.

A more successful approach consists in using an ensemble of classifiers. This kind of approach uses a group of different classifiers in order to be able to track changes on

H. Papadopoulos et al. (Eds.): AIAI 2013, IFIP AICT 412, pp. 636–645, 2013.

the environment. Several different models of ensembles have been proposed on the literature [2, 3, 4]:

- Ensembles that create new classifiers to each new chunk of data and weight classifiers according to their accuracy on recent data;
- Unweighted ensembles which can cope with new data that belongs to a concept different from the most recent training data;
- Ensembles that are able to discard classifiers as they become inaccurate or when a concept drift is detected.

Most models using weighted ensembles determine the weights for each classifier using some sort of heuristics related to the amount of mistakes the classifier does when working with the most recent data [5]. Although in principle any classifier can be used to build the ensembles, the ones which are most commonly used are decision trees, neural networks and naive Bayes [6].

In this work, we present an approach based on neural networks which are trained by means of a quantum-inspired evolutionary algorithm. Quantum-inspired evolutionary algorithms [7-11] are a class of estimation of distribution algorithms which present, for several benchmarks, a better performance for combinatorial and numerical optimization when compared to their canonical genetic algorithm counterparts. We also use the quantum-inspired evolutionary algorithm for numerical optimization (QIEA-R) to determine the voting weights for each classifier which is part of the ensemble. Every time a new chunk of data arrives, a new classifier is trained on this new data set and all the weights are optimized in order for the ensemble to improve its performance on classifying this new set of data.

Therefore, we present a new approach for adaptive learning, consisting of an ensemble of neural networks, named NEVE (Neuro-Evolutionary Ensemble). To evaluate its performance and accuracy, we used 2 different datasets to execute several simulations, varying the ensemble settings and analysing how do they influence the final result. We also compare the results of NEVE with the results of Learn++.NSE algorithm [2], an existing approach to address adaptive learning problems.

This paper is organized in four additional sections. Section 2 details the Quantum-Inspired Neuro-Evolutionary algorithm and the proposed model, the Neuro-Evolutionary Ensemble Classifier. Section 3 presents and discusses the results of the experiments. Finally, section 4 concludes this paper and present some possible future works.

2   The Proposed Model

2.1   The Quantum-Inspired Neuro-Evolutionary Model

Neuro-evolution is a form of machine learning that uses evolutionary algorithms to train artificial neural networks. This kind of model is particularly interesting for reinforcement learning problems, where the availability of input-output pairs is often difficult or impossible to obtain and the assessment of how good the network performs is made by directly measuring how well it completes a predefined task.

As training the weights in a neural network is a non-linear global optimization problem, it is possible to minimize the error function by means of using an evolutionary algorithm approach.

The quantum-inspired evolutionary algorithm is a class of "estimation of distribution algorithm" (EDA) that has a fast convergence and, usually, provides a better solution, with fewer evaluations than the traditional genetic algorithms [3, 6]. In this model, quantum-inspired genes are represented by probability density functions (PDF) which are used to generate classical individuals through an observation operator. After being observed, the classical individuals are evaluated, as in traditional genetic algorithms, and, by means of using fitness information, a set of quantum-inspired operators are applied to the quantum individuals, in order to update the information they hold in such a way that on the next generations, better individuals will have a better chance to be selected. Further details on how this optimization method works can be found in [7-11].

Based on this algorithm, the proposed quantum-inspired neuro-evolutionary model consists in a neural network (a multilayer perceptron (MLP)) and a population of individuals, each of them encoding a different configuration of weights and biases for the neural network. The training process occurs by building one MLP for each classical individual using the genes from this individual as weights and biases. After that, the full training data set (or the set of tasks to be performed) is presented to the MLP and the average error regarding the data set is calculated for each MLP. This average error is used as the fitness for each individual associated to that MLP, which allows the evolutionary algorithm to adjust itself and move on to the next generation, when the whole process will be repeated until a stop condition is reached.

This subsection presented the quantum-inspired neuro-evolutionary model. This model will be the basis for the algorithm proposed in this paper, to be presented in the next subsection.

2.2   NEVE: The Neuro-Evolutionary Ensemble Classifier

To some applications, such as those that use data streams, the strategy of using simpler models is most appropriate because there may not be time to run and update an ensemble. However, when time is not a major concern, yet the problem requires high accuracy, an ensemble is the natural solution. The greatest potential of this strategy for detecting drifts is the ability of using different forms of detection and different sources of information to deal with the various types of change [4].

One of the biggest problems in using a single classifier (a neural network, for example) to address concept drift problems is that when the classifier learns a dataset and then we need it to learn a new one, the classifier must be retrained with all data, or else it will "forget" everything already learned. Otherwise, using the ensemble, there is no need to retrain it again, because it can "retain" the previous knowledge and still learn new data.

Hence, in order to be able to learn as new chunks of data arrive, we implemented an ensemble with neural networks that are trained by an evolutionary algorithm, presented in section 2.1. This approach makes the ensemble useful for online reinforcement learning, for example. The algorithm works as shown in figure 3 and each step is described in detail on the next paragraphs.

On step 1 we create the empty ensemble with a predefined size equal to s. When the first chunk of data is received, a neural network is trained by means of the QIEA-R until a stop condition is reached (for example, the number of evolutionary generations or an error threshold). If the number of classifiers in the ensemble is smaller than s, then we simply add this new classifier to the ensemble. This gives the ensemble the ability to learn the new chunk of data without having to parse old data. If the ensemble is already full, we evaluate each classifier on the new data set and we remove the one with the highest error rate (including the new one, which means the new classifier will only become part of the ensemble if its error rate is smaller than the error rate of one of the classifiers already in the ensemble). This gives the ensemble, the ability to forget about data which is not needed anymore.

```
1. Create an empty ensemble P
2. Define the ensemble size s
3. For each chunk of data D_i, i = 1, 2, 3, ..., m do
   3.1. Train the classifier using the QIEA-R and a MLP and calculate it's error E' over the data
        chunk
   3.2. If the ensemble is full (number of classifiers = s) then
           i) Calculate the classification error E_j for each classifier c_j in the ensemble
           ii) If (E' > max(E_j))
                A) Replace the classifier with max(E_j) by the new classifier
   3.3. else
           i) Add the new classifier to the ensemble
   3.4. Evolve the voting weights w_j for each classifier in the ensemble using the last chunk of
        data D_i
```

Fig. 1. The neuro-evolutionary ensemble training algorithm

Finally, we use the QIEA-R to evolve a voting weight for each classifier. Optimizing the weights allows the ensemble to adapt quickly to sudden changes on the data, by giving higher weights to classifiers better adapted to the current concepts governing the data. The chromosome that encodes the weights has one gene for each voting weight, and the population is evolved using the classification error as the fitness function. It is important to notice that when the first s-1 data chunks are received, the ensemble size is smaller than its final size and thus, the chromosome size is also smaller. From the s data chunk on, the chromosome size will remain constant and will be equal to s.

In this work, we used only binary classifiers but there is no loss of generality and the algorithm can also be used with any number of classes. For the binary classifier, we discretize the neural network's output as "1" or "-1" and the voting process for each instance of data is made by summing the NN's output multiplied by its voting weight. In other words, the ensemble's output for one instance k from the i-th data chunk is given by:

$$P(D_{ik}) = \sum_{j=0}^{s} w_j c_j(D_{ik}) \tag{1}$$

where $P(D_{ik})$ is the ensemble's output for the data instance D_{ik}, w_j is the weight of the j-th classifier and $c_j(D_{ik})$ is the output of the j-th classifier for that data instance. If $P(D_{ik}) < 0$, we assume the ensemble's output is "-1". If $P(D_{ik}) > 0$, we assume the ensemble's output is "1". If $P(D_{ik}) = 0$, we choose a class randomly.

3   Experimental Results

3.1   Datasets Description

In order to check the ability of our model on learning data sets with concept drifts, we used two different data sets (SEA Concepts and Nebraska also used at [2]) upon which we performed several simulations in different scenarios.

The SEA Concepts was developed by [12]. The dataset consists of 50000 random points in a three-dimensional feature space. The features are in the [0; 10] domain but only two of the three features are relevant to determine the output class. Class labels are assigned based on the sum of the relevant features, and are differentiated by comparing this sum to a threshold.

Nebraska dataset, also available at [13], presents a compilation of daily weather measurements from over 9000 weather stations worldwide by the U.S. National Oceanic and Atmospheric Administration since 1930s, providing a wide scope of weather trends. As a meaningful real world dataset, [2] choosed the Offutt Air Force Base in Bellevue, Nebraska, for this experiment due to its extensive range of 50 years (1949–1999) and diverse weather patterns, making it a longterm precipitation classification/prediction drift problem. Class labels are based on the binary indicator(s) provided for each daily reading of rain: 31% positive (rain) and 69% negative (no rain). Each training batch consisted of 30 samples (days), with corresponding test data selected as the subsequent 30 days. Thus, the learner is asked to predict the next 30 days' forecast, which becomes the training data in the next batch. The dataset included 583 consecutive "30-day" time steps covering 50 years.

3.2   Running Details

On each simulation, we used a fixed topology for the neural networks consisting of 3 inputs for SEA Concepts dataset and 8 inputs for Nebraska dataset, representing the input variables for each dataset. In both datasets, we used 1 output, and we varied the number of the neurons for the hidden layer. Each neuron has a hyperbolic tangent activation function and, as mentioned before, the output is discretized as "-1" or "1" if the output of the neuron is negative or positive, respectively. The evolutionary algorithm trains each neural network for 100 generations. The quantum population has 10 individuals and the classical population 20. The crossover rate is 0:9 (refer to [8, 9] for details on the parameters). The same parameters are used for evolving the weights for the classifiers. The neural network weights and biases and the ensemble weights are allowed to vary between -1 and 1 as those values are the ones who have given the best results on some pre-evaluations we have made.

The first experiment was conducted in order to evaluate the influence of the variation of the parameters values in the results (number of the hidden layer neurons and the size of the ensemble). We used 4 different configurations for each dataset. After running 10 simulations for each configuration, we performed some an Analysis of Variance (ANOVA) [14]. In order to use ANOVA, we tested the Normality assumption for the noise term with Shapiro-Wilk's test [20] and the homogeneity of variances with Bartlett's test [14]. All the statistical procedures were conducted in R package [15], admitting a significance level of 5%.

The second experiment, in turn, aimed to compare the results found by NEVE and Learn++.NSE algorithms and we used, for each dataset, the best configuration found by first experiment (ensemble size and number of neurons at hidden layer values). After running one simulation for each dataset, we made statistical comparisons between the results found by NEVE and Learn++.NSE algorithms. The results of Learn++.NSE can be found at [2]. Then, to evaluate NEVE we made 10 runs for each dataset used, due to the stochastic optimization algorithm used to train NEVE. Based on these runs, we calculate some statistical parameters (mean, standard deviation, etc.) that were used to compute the Welch t-test [14] to evaluate which algorithm had, in average, the best performance in test phase. The normality assumption necessary for Welch t-test was verified using Shapiro-Wilk test [16]. All the statistical analyses were conducted in R statistical package [15].

3.3   First Experiment

Based on the past subsections, we made 40 simulations using SEA dataset and 40 simulations using Nebraska dataset, using 4 different configurations on each dataset. Table 1 displays number of neurons at hidden layer and ensemble size with different levels (5 and 10) and the output (average error in test phase for 10 runs) for each configuration.

In order to evaluate which configuration provided a significant lower error, we have to perform multiple comparisons between the results of each configuration. For each dataset if we decide to use t-test [14] for example, we have to realize 6 comparisons between the configurations, and thus, the probability that all analysis will be simultaneous correct is substantially affected. In this way, to perform a simultaneous comparison between all configurations we fitted a one-way Analysis of Variance (ANOVA) [14] for each dataset, described by:

$$Y_{ij} = \mu + CF_j + \varepsilon_{ij}; \quad \varepsilon_{ij} \sim N(0, \sigma^2) \tag{1}$$

where Y_{ij} is the i-th ouput for the j-th configuration, μ is the global mean, CF_j is the run configuration with j-levels (j =1,2,3,4) for each dataset (A, B, C and D, and E, F, G, H, for SEA and Nebraska respectively) and ε_{ij} is the noise term, Normal distributed with mean zero and constant variance (σ^2). If CF_j is statistically significant, then some configuration demonstrated an average error different from the others. To verify which configuration has the average error less than other configuration we used Tukey's test [14].

Table 1. Results for SEA and Nebraska dataset

SEA dataset				
Neurons Number	Ensemble Size	Config	Error in test phase	
			Mean	Std. dev.
10	10	A	24.99%	0.17%
5	5	B	24.88%	0.19%
10	5	C	25.06%	0.21%
5	10	D	24.75%	0.17%
Nebraska dataset				
Neurons Number	Ensemble Size	Config	Error in test phase	
			Mean	Std. dev.
10	10	E	32.30%	0.48%
5	5	F	32.85%	0.43%
10	5	G	33.04%	0.37%
5	10	H	32.10%	0.46%

Then, we performed the analysis for both dataset and exhibit the main results in Table 2.

Table 2. Results for SEA and Nebraska dataset

ANOVA - SEA dataset		
Method	Test Statistic	p-value
Bartlett's test	0.4443	0.9309
Config	5.4360	0.0035
Shapiro-Wilk's test	0.9260	0.2137
ANOVA - Nebraska dataset		
Method	Test Statistic	p-value
Bartlett's test	0.6537	0.8840
Config	11.5900	< 0.0001
Shapiro-Wilk's test	0.9708	0.3803

Analyzing the results displayed in Table 3, in both datasets the errors variance is homogeneous (Bartlett's test, p-value > 0.05). After verifying these two assumptions (Normal distribution and homogeneity of variances), we fitted the one-way ANOVA. In both datasets, some configurations (A, B, C and D for SEA, and E, F, G and H for Nebraska) demonstrated an average error different from the others (p-value < 0.05). In addition, in both fitted models, the noise term follows a Normal distribution (Shapiro-Wilk's test, p-value > 0.05).

In order to identify which configuration performed, in average, better than other, we made Tukey's test for difference of means. Table 3 present the results of this analysis.

Table 3. Tukey's test for SEA and Nebraska dataset

SEA dataset			Nebraska dataset		
Config	Mean difference	p-value	config	Mean difference	p-value
A-B	0.11%	0.5665	E-F	-0.55%	**0.0145**
A-C	-0.07%	0.8018	E-G	-0.74%	**0.0013**
A-D	0.24%	**0.0317**	E-H	0.20%	0.8849
B-C	-0.18%	0.1399	F-G	-0.19%	0.7434
B-D	0.13%	0.4010	F-H	0.75%	**0.0014**
C-D	0.31%	**0.0029**	G-H	0.94%	**0.0001**

It seems that in SEA dataset the D configuration performed significantly better than A and C, although its results is not statistically different than configuration B. In Nebraska the E and H configurations obtained error measures substantially lower than F and G. In fact, we can choose the configuration D as the best configuration to SEA and H to Nebraska dataset, considering the fixed parameters displayed in table 2. This choice is based on two criterias: lower average error and computational cost to train these models.

3.4   Second Experiment

Aiming to enable a better comparison with the results of the algorithm Learn +. NSE [2] in SEA Concepts dataset, we used 200 blocks of size 25 to evaluate the algorithm in the test phase. The best configuration previously achieved was 5 neurons in hidden layer and the size of the ensemble equal to 5 (see table 4). Also, with the results of Learn +. NSE in Nebraska dataset, we performed similarly to that used in [2]. The best configuration previously achieved was 10 neurons in hidden layer and the size of the ensemble equal to 10. Then, NEVE and Learn++.NSE results were displayed in Table 4.

Table 4. Results of SEA and Nebraska experiments

Dataset	Algorithm	Mean	Standard Deviation
SEA	NEVE	98.21%	0.16%
	Learn++.NSE (SVM)	96.80%	0.20%
Nebraska	NEVE	68.57%	0.46%
	Learn++.NSE (SVM)	78.80%	1.00%

As can be seem, the mean accuracy rate of Learn++.NSE is lower than the best configuration of NEVE, and thus this difference is statistically significant (tcrit = -41.07, p-value < 0.0001), demonstrating that NEVE perrformed better in the test phase on SEA Concepts dataset.

However, in Nebraska the mean accuracy rate of NEVE is lower than the best configuration of Learn++.NSE, and thus this difference is statistically significant (tcrit = 18.26, p-value < 0.0001), demonstrating that NEVE performed better, in average, than Learn++.NSE. Figures 2 and 3 illustrates the hit rate on each test block obtained by NEVE on SEA and Nebraska, respectively.

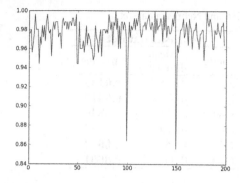
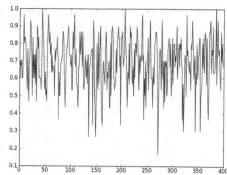

Fig. 2. Evolution of NEVE hit rate in SEA testing set

Fig. 3. Evolution of NEVE hit rate in Nebraska testing set

4   Conclusions and Future Works

This paper presented a model that uses an ensemble of neural networks trained by a quantum-inspired evolutionary algorithm to learn data sets (possibly with concept drifts) incrementally. We analyzed the ability of the model using two different datasets and conducting two different experiments. In the first experiment, we found a good configuration for both datasets and demonstrated how the number of neurons and the ensemble size affect the average error produced by the model. As stated in the results, the ensemble size affected almost two times more the results of NEVE than the number of neurons. In the second experiment, the NEVE algorithm have demonstrated a better performance in SEA dataset compared to Learn++.NSE and yet lower accuracy when comparing with Learn++.NSE in Nebraska dataset.

Although the NEVE algorithm have demonstrated satisfactory performance for the datasets used in the analysis of this study, it is strongly recommended to perform further tests - using different configurations, different datasets and performing different analysis - to confirm the results presented here. We also intend in the future to continue this work, analyzing other existing approaches, such as [17] and [18], and performing new experiments in comparison with these and other algorithms. We still need to investigate other factors related to QIEA-R fine tunning (genetic operators, population size, etc.).

References

1. Schlimmer, J.C., Granger, R.H.: Incremental learning from noisy data. Machine Learning 1(3), 317–354 (1986)
2. Elwell, R., Polikar, R.: Incremental Learning of Concept drift in Nonstationary Environments. IEEE Transactions on Neural Networks 22(10), 1517–1531 (2011)
3. Kuncheva, L.I.: Classifier ensembles for changing environments. In: Roli, F., Kittler, J., Windeatt, T. (eds.) MCS 2004. LNCS, vol. 3077, pp. 1–15. Springer, Heidelberg (2004)

4. Kuncheva, L.I.: Classifier ensemble for detecting concept change in streaming data: Overview and perspectives. In: Proc. Eur. Conf. Artif. Intell., pp. 5–10 (2008)
5. Ahiskali, M.T.M., Muhlbaier, M., Polikar, R.: Learning concept drift in non-stationary environments using an ensemble of classifiers based approach. IJCNN, 3455–3462 (2008)
6. Oza, N.C.: Online Ensemble Learning. Dissertation, University of California, Berkeley (2001)
7. Abs da Cruz, A.V., Vellasco, M.M.B.R., Pacheco, M.A.C.: Quantum-inspired evolutionary algorithms for numerical optimization problems. In: Proceedings of the IEEE World Conference in Computational Intelligence (2006)
8. Abs da Cruz, A.V.: Algoritmos evolutivos com inspiração quântica para otimização de problemas com representação numérica. Ph.D. dissertation, Pontifical Catholic University – Rio de Janeiro (2007)
9. Han, K.-H., Kim, J.-H.: Quantum-inspired evolutionary algorithm for a class of combinatorial optimization. IEEE Trans. Evolutionary Computation 6(6), 580–593 (2002)
10. Han, K.-H., Kim, J.-H.: On setting the parameters of QEA for practical applications: Some guidelines based on empirical evidence. In: Cantú-Paz, E., et al. (eds.) GECCO 2003. LNCS, vol. 2723, pp. 427–428. Springer, Heidelberg (2003)
11. Han, K.-H., Kim, J.-H.: Quantum-inspired evolutionary algorithms with a new termination criterion, He gate, and two-phase scheme. IEEE Trans. Evolutionary Computation 8(2), 156–169 (2004)
12. Street, W.N., Kim, Y.: A streaming ensemble algorithm (SEA) for large-scale classification. In: Proc. 7th ACM SIGKDD Int. Conf. Knowl. Disc. Data Min., pp. 377–382 (2001)
13. Polikar, R., Elwell, R.: Benchmark Datasets for Evaluating Concept drift/NSE Algorithms, http://users.rowan.edu/~polikar/research/NSE (last access at December 2012)
14. Montgomery, D.C.: Design and analysis of experiments. Wiley (2008)
15. R Development Core Team. R: A language and environment for statistical computing. R Foundation for Statistical Computing. Vienna, Austria (2012), Donwload at: http://www.r-project.org
16. Royston, P.: An extension of Shapiro and Wilk's W test for normality to large samples. Applied Statistics 31, 115–124 (1982)
17. Kolter, J., Maloof, M.: Dynamic weighted majority: An ensemble method for drifting concepts. Journal of Machine Learning Research 8, 2755–2790 (2007)
18. Jackowski, K.: Fixed-size ensemble classifier system evolutionarily adapted to a recurring context with an unlimited pool of classifiers. Pattern Analysis and Applications (2013)

Investigation of Expert Addition Criteria for Dynamically Changing Online Ensemble Classifiers with Multiple Adaptive Mechanisms

Rashid Bakirov and Bogdan Gabrys

Smart Technology Research Centre, Bournemouth University, United Kingdom
{rbakirov,bgabrys}@bournemouth.ac.uk

Abstract. We consider online classification problem, where concepts may change over time. A prominent model for creation of dynamically changing online ensemble is used in Dynamic Weighted Majority (DWM) method. We analyse this model, and address its high sensitivity to misclassifications resulting in creation of unnecessary large ensembles, particularly while running on noisy data. We propose and evaluate various criteria for adding new experts to an ensemble. We test our algorithms on a comprehensive selection of synthetic data and establish that they lead to the significant reduction in the number of created experts and show slightly better accuracy rates than original models and non-ensemble adaptive models used for benchmarking.

1 Introduction

Ensemble learning in stationary settings has been extensively researched, and it was shown to often be able to outperform single learners [17], [3], [15]. There has been a number of attempts to apply this ensemble learning paradigm to online learning for non-stationary environments. Many of them try to map experts to the continuous data batches or concepts [10] [9] [5] [13].

We focus on the problem of online classification in incremental fashion, where learner is presented with a single data instance, and after its classification, the true label for this instance is revealed. Our aim in this setting is the creation and maintenance of an experts' ensemble which can adapt to the changes in data.

In "batch learning", the data is presented in chunks providing natural data bases for creation of experts whereas incremental learning makes questions such as when, and on what data basis add an expert, very important. These two learning types share other problems of optimal adjustment of weights and use of suitable criterion for expert removal.

In this work we concentrate on the model of ensemble management introduced in [10]. We analyse the algorithm and focus on its shortcomings, such as undesirable behaviour in noisy environments with regard to addition of new unnecessary experts to ensemble. Larger ensembles require more computational resources and are therefore less desirable. The main purpose of this paper is empirical analysis of how the changes in the following main areas of the model affect the ensemble size and accuracy:

H. Papadopoulos et al. (Eds.): AIAI 2013, IFIP AICT 412, pp. 646–656, 2013.

- *Expert creation criterion* - original model creates a new expert after every misclassification. We explore alternative criteria based on the average accuracy on the defined window of last data instances.
- *Expert quality* - we investigate the possibility of using data windows for the purposes of expert creation.
- *Expert assessment* - even if the expert has a satisfactory accuracy, it might not always be needed to include it in the ensemble. We look into on-the-fly evaluation of the expert, to decide whether its use is beneficial.

Specifically, we use windows of data instances to determine when to add a new expert, or which data basis to use for its training. Our results show that creating expert from larger data bases leads to the highest accuracy rates, while evaluating their performance helps keeping their number lower than the original model and still reach comparable accuracy rates.

2   Related Work

The notion of using more than one expert (ensemble) for making decisions has a long history. Famously, in 1785 Condorcet has established that if the probability p of making a correct decision for each voter is larger than 0.5, then, under certain assumptions, the larger the number of the voters, the higher is the likelihood of reaching the correct decision when choosing among two alternatives [17].

Work of Littlestone and Warmuth [12] is one of the seminal papers on the topic of using expert ensembles for online learning. They consider binary prediction task using multiple experts, with given initial weights, each of whom makes an individual prediction. In the case of wrong prediction, the weights of predictors are multiplied by β such as $0 < \beta < 1$. This work can be considered a special case of [19], which considers continuous prediction and decreasing the weights of experts according to the loss function inversely proportional to their error.

In the last 10 years, there has been an increased interest of data mining community to use ensemble methods while dealing with on-line learning in non-stationary environments. Particularly, the problem of "concept drift" has been often addressed. In this setting, another intuitive reasoning for using ensembles is the intention that each expert should represent a certain concept or a part of it. Among the many classifier algorithms for concept drift scenario we review the most relevant ones for our purposes. A well-known algorithm creating expert ensemble in online mode is DWM [10] , which adapts to drift by creating a new expert each time a datapoint is misclassified by the existing ensemble. New expert gets the weight of one. All experts train online and whenever an expert misclassifies a data instance, its weight is multiplied by $0 < \beta < 1$. After each classification, to reduce the dominance of newly added experts, the weights of existing experts are normalized, so that the highest weighting expert gets a new weight of 1. To reduce the number of experts, they are deleted if their weight is lower than a defined threshold θ. In [11] the same authors present AddExp.D, a variation of this method where the weight assigned to the new experts is the current weight of the ensemble multiplied by a constant γ. Here the authors bound the error of this type of ensemble on the error of the latest created expert, provided that $\beta + 2\gamma < 1$.

CDC algorithm [18] is a similar approach to adaptive ensemble management. CDC has fixed number of experts; it starts with an empty set and adds a new expert every data instance until the set is full. New "immature" experts have the weight of zero and become "mature" after learning on a defined number of instances (usually equal to the size of the ensemble). The experts are weighted based on their performance on a test set, which must be available with every data instance, and removed if the all of the following conditions are satisfied: a) their weight is below the threshold; b) their weight is the smallest among all members of the ensemble; c) they are mature.

An interesting algorithm which aims to minimize the number of experts is discussed in [1]. Here, only two experts are active at a given time; an active predictor, which is trained on the complete set of the instances, and the test predictor which is trained only on the last n instances. If the test predictor starts predicting better than the active one, the active one is deleted, test predictor becomes active, and a new test predictor is started to be trained.

More recently, similar algorithms have been proposed for time series prediction. [7] proposes creating a new expert every instance while deleting some of the older ones and [8] suggests splitting the data stream into epochs and creating an expert which learns on the most recent epoch every τ instances.

3   Elements of Online Expert Ensemble Creation

One of the most researched and well defined reasons for the adaptive models is the problem of concept drift, as introduced in [16]. Concept drift occurs when the statistical distribution of target or input variables (virtual drift), or their relations change over time (real drift) [5]. We are more interested in the real concept drift, more formal definition of which drift can be given as follows. Assume an input vector $\bar{x} = (x_1, x_2, ..., x_n)$ and a function $y = f_t(\bar{x})$ which produces an output y from $\bar{x}$ at time t. If there exists an input vector $\bar{x}'$ and time points t_1, t_2, such that $f_{t_1}(\bar{x}') \neq f_{t_2}(\bar{x}')$, then this is called concept drift.

In this work we concentrate on the model of ensemble management introduced in [10] and [11] (except the pruning part of the latter) which has been empirically shown to be effective and perform well in many cases. Certain bounds on overall number of mistakes are given in [11] as well. However this approach is not entirely problem free, as it becomes clear from following sections.

Reviewed model involves several layers of adaptation - online training, change of exerts' weights, addition and removal of experts. Clearly, the most drastic adaptation method used is adding new experts, which is why we concentrate on this topic. In the following we will analyse the performance of the discussed model and some of its modifications. For this purpose we use the term *reaction time* - the minimum number of observations, after which algorithm will react to observed change by creating an expert and *convergence time* - number of observations, after which algorithm will converge (total weight of the experts which are trained on the new concept is larger than the total weight of the experts) to new concept.

3.1 Condition for Adding of an Expert

Condition for adding an expert largely determines the *reaction time* of an algorithm, and thus plays a significant role in its *convergence time* as well. Reviewed model reacts to misclassifications by creating a new expert. Initially, it is suggested to add an expert every time when the prediction of current ensemble is false. This provides fast reaction time but may, in noisy conditions, result in adding many unnecessary and inaccurate experts. To deal with this problem, in [10] authors suggest that, in noisy domains or for large experiments, only every T-th example could be taken into consideration, which reduces the number of created experts in proportion to T. The drawback of having $T > 1$ is a possibility of slower reaction to the change. This is best manifested during a sudden drift, where, in the worst case, the reaction time is T.

To reduce the effects of noise in a more deterministic way, we propose the averaging window condition for expert creation as an alternative to having $T > 1$ (In [1] a similar condition is used to substitute learners). Here, the strategy is based on the decision of creating an expert from x_n (n-th datapoint) not only as a result of x_n's classification, but on the basis of accuracy in the window of the last l elements, with x_n being the last element of l. We add an expert trained from x_n if the average accuracy of the ensemble in the window is less than fixed threshold value u. If we assume that the change causes algorithm to always misclassify incoming data, then the reaction time to the change in this case can be calculated to be at most $l(1 - u)$ rounded up.

The choice of the threshold may be difficult for unknown data. Also, for the datasets where average accuracy may vary with the time, for example due to changing noise levels, using the above static threshold might result in creation of many unnecessary experts or not creation of experts when needed. We introduce a similar algorithm with dynamic threshold value, which we call "maximum accuracy threshold window" (MTW). The dynamic threshold here is similar to the one used in DDM change detector [6]. While classifying incoming data we record the maximum value of $\mu_{acc} + \sigma_{acc}$ where μ_{acc} is mean accuracy and $\sigma_{acc} = \sqrt{\frac{\mu_{acc}(1-\mu_{acc})}{l}}$ is the standard deviation of the Bernoulli process. We create a new expert when the condition $\mu_{cur} - \sigma_{cur} < \mu_{max} - m * \sigma_{max}$ is met. Here μ_{cur} and σ_{cur} are mean accuracy and standard deviation of current window and μ_{max} and σ_{max} are mean accuracy and standard deviation of the window where the maximum value of $\mu_{acc} + \sigma_{acc}$ was recorded. Parameter m is usually set to 3. After creation of new expert, the maximum values are reset. A possible issue in some cases could be that when the accuracy reaches 1, the new expert will be added when there is a single misclassification. To prevent this it is possible to enforce a certain minimum σ_{max} such as 0.1 or 0.15. Using window based conditioning is illustrated in the Figure 1b.

3.2 Data Basis for New Experts

Assuming uniform class label distribution, training an expert from a single data instance means that this expert will assign the label it has been trained on to all other samples which makes its initial accuracy $1/n$ in the case of n-label classification problem. Low accuracy of experts trained on insufficient amount of data, also discussed in [20], combined with the high weight of the new expert, may result in noticeable negative effect on the accuracy. To counter this we can use a delay in reaction time to train the new expert on more examples before using it for predictions (note: approaches proposed in this section are analysed standalone and not combined with the ones from the section 3.1 at this moment). The simplest option is to train an expert on l datapoints after its creation and only then add it to the ensemble, as in [18]. We call this "mature" experts (MATEX) approach. To prevent multiple reactions to one change, during the time that expert is being "matured", no new experts are introduced. When the expert is added to the ensemble, it is better trained and thus more accurate than the expert which is created from only one datapoint. The reaction time for a change in this case is l. Another advantage of this approach is reducing the effect of noise on the created expert.

One possibility to reduce the number of unnecessary experts created in this way is assessing their performance. A sufficient condition for expert to benefit the ensemble independent of this expert's weight is predicting better than ensemble (another option could be dynamic weighting based on the accuracy assessment, which is not discussed here). So before adding it to the ensemble we can compare it with the performance of the ensemble in the window of size l. Comparison strategy is similarly used in [1]. The comparison can be done in various ways; comparing the prequential accuracies of the expert and ensemble, or constructing certain test and training sets from the datapoints in the window and using cross-validation. If the validation is successful, then the new expert which has been trained on the whole window is added to the ensemble. Here the reaction time is l. To prevent multiple reactions to one change, during the time that expert is being "validated", no new experts are introduced. Here, the effect of noise is further reduced - when the data suddenly becomes noisy, newly created experts will probably not predict better than existing ensemble and thus will be discarded. Validation approach can be combined with MATEX allowing the expert to train on l_{mature} datapoints, before starting the comparison on l_{val} datapoints. This might help prevent the premature removal of experts but will accordingly increase reaction time to $l_{mature} + l_{val}$. It must be noted that this approach requires additional computational effort for the validation. Using window for the data basis of new expert is illustrated in the Figure 1c.

4   Experimental Results

4.1   Methods Description

We have experimented with different variations of the methods described in the Section 3. The implemented window based condition schemes from the section

Fig. 1. Using windows for expert adding condition and data base of a new expert

Table 1. Experiments with window based conditions to add an expert

Type	Threshold u(static) / n(dynamic)	Window length	$min(\sigma_{max})$	Codename
Static	0.5	5	N/A	WIN_5_0.5
Static	0.5	10	N/A	WIN_10_0.5
Static	0.7	10	N/A	WIN_10_0.7
Dynamic	3	5	0	MTW_5
Dynamic	3	10	0	MTW_10
Dynamic	3	10	0.1	MTW_10_0.1
Dynamic	3	10	0.15	MTW_10_0.15

3.1 are presented in Table 1. Implemented methods experimenting with experts' data basis (section 3.2) were MATEX with window sizes of 5 and 10, prequential validation (PVAL) with window sizes of 5 and 10, combination of MATEX and prequential validation each having window of 5 and several variations of cross-validation methods (XVAL) using window size of 10 with different sizes of training and testing sets. We also have experimented with the periodical expert additions with periods T of 5, 7, 10 and 11. In our implementation of original algorithms, WIN and MTW we create a new expert from the single datapoint.

We have used different weighting schemes for all of the experiments, specifically static weighting [10] with new expert weight of 1 with β equal to 0.3, 0.5, 0.7 and dynamic weighting [11] with the same values of β and respective values of γ equal to 0.3, 0.2, 0.1. Unlike dynamic weighting, static weighting makes convergence time n_{conv} dependant on the total weight of ensemble at the moment of expert W creation. This allows implicit control of n_{conv} while limiting its explicit control to some extent. The following results are based on $\beta = 0.5$.

4.2 Results on Synthetic Data

We have synthesised 26 two-dimensional data sets with various properties to examine the behaviour of the algorithms in different situations. We consider rotating hyperplane data and Gaussians with different type of changes - switching between two data sources [14], one Gaussian passing through the other one and returning, Gaussians moving together in one direction and returning (see Figure 2). We have experimented with various magnitudes of changes and levels of artificial noise and decision boundaries overlap (see Table 2).

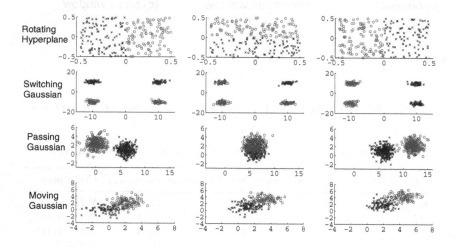

Fig. 2. Changes in experimental datasets. From left to right: data in the start, in the middle, in the end (before possible return to starting position).

We have tested the described methods with Naive Bayes base learner using PRTools 4.2.4 toolbox for MATLAB [4]. MATEX_10 with dynamic weighting showed the best average test and prequential accuracy among all of the methods on 26 datasets, and the best average test accuracy on 6 datasets. Here, the test accuracy is calculated using average predictive accuracy on additional 100 test instances from the same distribution as current training data instance. From the methods with window based expert creation condition (table 1), MTW_10 with dynamic weighting shows the best performance, 0.03% less than the leader. The original DWM showed comparable average test accuracy results (88.4%, 1.3% less than leader) but had 3 times as larger average ensemble size (~15 against ~5). Even in the datasets without noise, average ensemble size of the original DWM is noticeably higher than that of other methods. The same model with dynamic weighting has twice smaller average ensemble size.

In the Table 3 we compare results of different described variations of the original method with the results of original method. Here we use window of 10 for all of the methods. XVAL is leave-one-out cross-validation. Threshold in the WIN is 0.5. Again, MATEX methods with dynamic weighting emerge as a narrow leaders in terms of average test accuracy (see Table 3 for full results). The accuracy rates are quite similar, but we see a noticeable decrease in the number of total created experts and the average ensemble size. Validation methods PVAL and XVAL further reduce the number of total created experts and average ensemble size, while having slightly lower accuracy rates than the leaders and requiring additional computation for validation purposes. To benchmark our results against non-ensemble methods we have run tests with a simple online Naive Bayes classifier without any forgetting, state of the art change detectors DDM [6]

Table 2. Synthetic datasets used in experiments. Column "Drift" specifies number of drifts, the percentage of change in the decision boundary and its type.

Num.	Data type	Instances	Classes	Drift	Noise/overlap
1	Hyperplane	600	2	2x50% rotation	None
2	Hyperplane	600	2	2x50% rotation	10% uniform noise
3	Hyperplane	600	2	9x11.11% rotation	None
4	Hyperplane	600	2	9x11.11% rotation	10% uniform noise
5	Hyperplane	640	2	15x6.67% rotation	None
6	Hyperplane	640	2	15x6.67% rotation	10% uniform noise
7	Hyperplane	1500	4	2x50% rotation	None
8	Hyperplane	1500	4	2x50% rotation	10% uniform noise
9	Gaussian	1155	2	4x50% switching	0-50% overlap
10	Gaussian	1155	2	10x20% switching	0-50% overlap
11	Gaussian	1155	2	20x10% switching	0-50% overlap
12	Gaussian	2805	2	4x49.87% passing	0.21-49.97% overlap
13	Gaussian	2805	2	6x27.34% passing	0.21-49.97% overlap
14	Gaussian	2805	2	32x9.87% passing	0.21-49.97% overlap
15	Gaussian	945	2	4x52.05% move	0.04% overlap
16	Gaussian	945	2	4x52.05% move	10.39% overlap
17	Gaussian	945	2	8x27.63% move	0.04% overlap
18	Gaussian	945	2	8x27.63% move	10.39% overlap
19	Gaussian	945	2	20x11.25% move	0.04% overlap
20	Gaussian	945	2	20x11.25% move	10.39% overlap
21	Gaussian	1890	4	4x52.05% move	0.013% overlap
22	Gaussian	1890	4	4x52.05% move	10.24% overlap
23	Gaussian	1890	4	8x27.63% move	0.013% overlap
24	Gaussian	1890	4	8x27.63% move	10.24% overlap
25	Gaussian	1890	4	20x11.25% move	0.013% overlap
26	Gaussian	1890	4	20x11.25% move	110.24% overlap

and EDDM [2] and Paired Learners method with window size 10 and threshold 0.1 [1]. As expected, online Naive Bayes performs noticeably worse than adaptive methods. Change detectors and paired learners show slightly lower but comparable test accuracy to MATEX methods.

The top performers on some datasets can be different than the average leaders. For instance, validation methods perform better on the dataset with passing Gaussian with 4 drifts. Here, XVAL with dynamic static weighting shows the best accuracy among the methods compared above - 87.2% which is 0.8 % higher than the accuracy of the leader. Intuitively, this can be explained with a large proportion of class intersection area, where the expert creation is not beneficial, and two intersection-free areas where high accuracy experts can be created. In general, expert checking is beneficial for the datasets with variable noise or decision boundary intersection. Figure 3 gives an insight on the performance for selected methods with static weighting from the Table 3 on individual datasets.

Table 3. Results on 26 synthetic datasets, averaged

Method	Average test accuracy	Std. deviation of avg. accuracy	Average total created experts	Average ensemble size
MATEX dynamic weighting	0.898	0.064	54.58	4.96
MATEX static weighting	0.893	0.067	54.62	6.57
MTW dynamic weighting	0.894	0.067	40.38	3.31
MTW static weighting	0.889	0.071	40.54	4.63
DWM periodical dynamic weighting	0.894	0.064	15.58	5.42
DWM periodical static weighting	0.891	0.066	15.81	6.76
XVAL dynamic weighting	0.890	0.068	22.35	2.31
XVAL static weighting	0.893	0.068	21.12	3.31
PVAL dynamic weighting	0.888	0.066	5.23	1.53
PVAL static weighting	0.889	0.066	4.65	1.48
WIN dynamic weighting	0.880	0.073	51.04	3.39
WIN static weighting	0.881	0.072	24.85	4.71
Original dynamic weighting	0.867	0.091	181.15	7.94
Original static weighting	0.884	0.075	156.12	14.97
PAIRED LEARNER	0.891	0.069	4.5	2
DDM	0.88	0.077	2.27	1
EDDM	0.89	0.067	1.92	1
NAIVE_BAYES	0.807	0.137	1	1

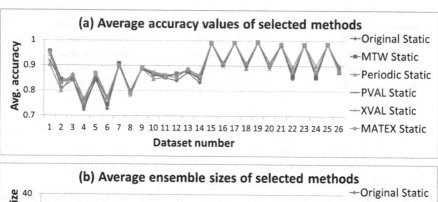

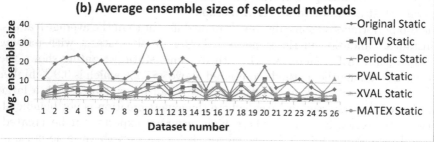

Fig. 3. Average accuracy values and ensemble sizes for selected methods

5   Conclusions

In this work we discuss shortcomings of the investigated dynamic ensemble classification method introduced in [11], and perform analysis of how using data windows for expert creation condition and data basis affect the number of the created experts and predictive accuracy. Our extensive tests with synthetic data show the viability of suggested approaches for different concept drift scenarios. It is not the aim of the paper to present a better novel classification algorithm, however the most of our proposed variations result in slightly better performance and significantly smaller number of experts than the original model. Expert evaluation techniques reduce the average ensemble size to the minimum while retaining comparable performance. We notice that ensemble methods can perform better than non-ensemble methods based on drift detection and there are promising signs that this performance can be improved even further. We conclude that:

- *Window based expert creation criteria* lead to comparable or slightly higher accuracy rates and a reduction of average ensemble size.
- *Window based expert data bases* result in slightly higher accuracy rates and a significant reduction of average ensemble size.
- *Expert validation* leads to comparable accuracy rates and a drastic reduction of ensemble size, but requires more computational effort.

In the future we aim to present the detailed results, discussing why certain methods perform better on certain datasets. Further improvements of the proposed methods, e.g. dynamic starting weights for experts based on their performance during validation phase, dynamic parameter β and explicit handling of recurring concepts can be looked into. Probabilistic analysis of models is currently under way and the analysis of complexity is planned. Another direction of research is intelligent combination of conditioning, data base selection and validation. We intend evaluating the methods on a selection of real datasets.

Acknowledgements. Research leading to these results has received funding from the EC within the Marie Curie Industry and Academia Partnerships and Pathways (IAPP) programme under grant agreement no. 251617. We express our thanks to Indrė Žliobaitė, Damien Fay and anonymous reviewers.

References

1. Bach, S.H., Maloof, M.A.: Paired Learners for Concept Drift. In: 2008 Eighth IEEE International Conference on Data Mining, pp. 23–32 (December 2008)
2. Baena-García, M., del Campo-Ávila, J., Fidalgo, R., Bifet, A., Gavaldà, R., Morales-Bueno, R.: Early drift detection method. In: ECML PKDD 2006 Fourth International Workshop on Knowledge Discovery from Data Streams, Berlin, Germany (2006)
3. Dieterich, T.G.: Ensemble Methods in Machine Learning. In: Kittler, J., Roli, F. (eds.) MCS 2000. LNCS, vol. 1857, pp. 1–15. Springer, Heidelberg (2000)

4. Duin, R.P.W., Juszczak, P., Paclik, P., Pekalska, E., de Ridder, D., Tax, D.M.J., Verzakov, S.: PRTools4.1, A Matlab Toolbox for Pattern Recognition (2007), http://prtools.org
5. Elwell, R., Polikar, R.: Incremental learning of concept drift in nonstationary environments. IEEE Transactions on Neural Networks/A Publication of the IEEE Neural Networks Council 22(10), 1517–1531 (2011)
6. Gama, J., Medas, P., Castillo, G., Rodrigues, P.: Learning with drift detection. In: Bazzan, A.L.C., Labidi, S. (eds.) SBIA 2004. LNCS (LNAI), vol. 3171, pp. 286–295. Springer, Heidelberg (2004)
7. Hazan, E., Seshadhri, C.: Efficient learning algorithms for changing environments. In: ICML 2009 Proceedings of the 26th Annual International Conference on Machine Learning, pp. 393–400 (2009)
8. Jacobs, A., Shalizi, C.R., Clauset, A.: Adapting to Non-stationarity with Growing Expert Ensembles. Tech. rep., Carnegie Mellon University (2010), http://www.santafe.edu/media/cms_page_media/285/AJacobsREUpaper.pdf
9. Kadlec, P., Gabrys, B.: Local learning-based adaptive soft sensor for catalyst activation prediction. AIChE Journal 57(5), 1288–1301 (2011)
10. Kolter, J.Z., Maloof, M.A.: Dynamic weighted majority: An ensemble method for drifting concepts. The Journal of Machine Learning Research 8, 2755–2790 (2007)
11. Kolter, J., Maloof, M.: Using additive expert ensembles to cope with concept drift. In: Proceedings of the 22nd International Conference on Machine Learning, ICML 2005, No. 1990, pp. 449–456 (2005)
12. Littlestone, N., Warmuth, M.: The Weighted Majority Algorithm. Information and Computation 108(2), 212–261 (1994)
13. Minku, L.L., Yao, X.: DDD: A New Ensemble Approach for Dealing with Concept Drift. IEEE Transactions on Knowledge and Data Engineering 24(4), 619–633 (2012)
14. Narasimhamurthy, A., Kuncheva, L.: A framework for generating data to simulate changing environments. In: Proceedings of the 25th IASTED International Multi-Conference: Artificial Intelligence and Applications, pp. 384–389. ACTA Press, Anaheim (2007)
15. Ruta, D., Gabrys, B.: Classifier selection for majority voting. Information Fusion 6(1), 63–81 (2005)
16. Schlimmer, J.C., Granger, R.H.: Incremental Learning from Noisy Data. Machine Learning 1(3), 317–354 (1986)
17. Shapley, L., Grofman, B.: Optimizing group judgmental accuracy in the presence of interdependencies. Public Choice 43(3), 329–343 (1984)
18. Stanley, K.O.: Learning concept drift with a committee of decision trees. Technical report, UT-AI-TR-03-302, Department of Computer Science, University of Texas in Austin (2003)
19. Vovk, V.G.: Aggregating strategies. In: COLT 1990 Proceedings of the Third Annual Workshop on Computational Learning Theory, pp. 371–386. Morgan Kaufmann Publishers Inc., San Francisco (1990)
20. Žliobaitė, I., Kuncheva, L.I.: Theoretical Window Size for Classification in the Presence of Sudden Concept Drift. Tech. rep., CS-TR-001-2010, Bangor University, UK (2010)

Exploring Semantic Mediation Techniques in Feedback Control Architectures

Georgios M. Milis, Christos G. Panayiotou, and Marios M. Polycarpou

KIOS Research Center for Intelligent Systems and Networks,
Department of Electrical and Computer Engineering, University of Cyprus
{milis.georgios,christosp,mpolycar}@ucy.ac.cy

Abstract. Modern control systems implementations, especially in large–scale systems, assume the interoperation of different types of sensors, actuators, controllers and software algorithms, being physical or cyber. In most cases, the scalability and interoperability of the control system are compromised by its design, which is based on a fixed configuration of specific components with certain knowledge of their specific characteristics. This work presents an innovative feedback control architecture framework, in which classical and modern feedback control techniques can be combined with domain knowledge (thematic, location and time) in order to enable the online plugging of components in a feedback control system and the subsequent reconfiguration and adaptation of the system.

Keywords: Control system architecture, interoperability, scalability, semantic knowledge models, plug&play of components.

1 Introduction

Nowadays, systems are designed and built not as monolithic entities but as collection of smaller physical and cyber components that, many times, can be considered as separate systems themselves with their own dynamics and objectives. This *system of systems* paradigm ([15]) necessitates the easy interaction and interoperability of the components that comprise a larger system. Components are expected to take informed decisions and act intelligently towards meeting (or balancing) the system's objectives. This description is valid also for modern control systems, where different types of components, being physical or cyber, interoperate in a larger control system implementation. However, in most cases, the design of feedback control systems is based on a fixed configuration of specific components, with certain knowledge of their specific characteristics. This causes lack of scalability and interoperability for the control system, thus considerably limiting its potential lifetime. There are cases where faulty sensors need to be replaced or additional sensors need to be installed (e.g. due to recent availability of this type of components or due to upgrading to new technology), and this should not require redesign of the overall feedback control system since such action would be impractical and costly.

H. Papadopoulos et al. (Eds.): AIAI 2013, IFIP AICT 412, pp. 657–666, 2013.

The detection and identification of non-modelled events in linear and non-linear systems is currently addressed by the *fault diagnosis* research area. The authors in [8] and [5] provide a thorough overview on the classical algorithms that identify deviations from the normal behaviour of a system, attributed to faults or other external events. The *adaptive* and *fault-tolerant control* research areas address the design of intelligent control type algorithms, that aim to facilitate the flexibility of the control system with respect to on-line adaptation, and accommodation of faults, system uncertainties and/or time variations. Approaches to designing fault tolerant and reconfigurable control systems are presented in [4]. In [10] the author also addresses the issue of fault-tolerant components, while the authors in [7] provide methodologies for designing adaptive approximation-based control systems. Recent efforts in plug&play control ([17] and more recently in [3]), propose methodologies for the online identification of newly introduced dynamics when new components are plugged in a closed-loop system and the subsequent online adaptation of the feedback laws. Also, the authors of the present paper have recently presented initial results of their work [14] on the exploitation of ontology-based semantic mediation techniques in feedback control systems.

The main contribution of our work is the design of an innovative feedback control architecture framework, in which classical and modern feedback control techniques can be combined with domain knowledge (thematic, location and time) in order to enable the online plugging of components in feedback control systems and their subsequent reconfiguration and adaptation. The control system becomes able to make use of and enrich thematic, location and time related structured knowledge about the environment in which it operates.

The rest of the paper is organised as follows: Section 2 formulates the problem, to facilitate the presentation of the solution. Then, section 3 presents the proposed architecture and framework, followed by section 4 where a case-study scenario is given. Finally, section 5 shows a simulation with results and section 6 concludes the report.

2   Problem Formulation

Consider a closed-loop system with sensors measuring plant outputs, actuators acting on controlled inputs following instruction by a control law that considers an error trajectory. The actual plant states are estimated by an observer (e.g. a Luenberger observer [12]), to compensate for the case when some of them are missing, or for redundancy and noise cancellation.

Consider the following cases:

1 A deployed sensor fails and is replaced by a new one having different (and not compliant with the closed-loop system implementation) characteristics.
2 Sensor(s) enter the plant, at different locations and at different times. These sensors measure physical quantities that are already considered as states in the closed-loop system design.

3 Sensor(s) enter the plant as above, but this time some or all of them measure quantities not already taken into consideration for the initial design of the closed-loop system.

When changes happen in the components' synthesis of the closed-loop system, as explained in Section 2, the altered measurement vectors carry new sensing capabilities that can be potentially exploited using different models of the same plant. Therefore, at discrete time steps, the closed-loop system may assume a different model, with different types and/or dimensions of variables and parameters respectively. Without loss of generality, we assume this system is described by the state-space model in (1). The top-pointer $I = 1, 2, ...$, is utilised to distinguish among different models.

$$\dot{x}^{(I)} = A^{(I)}x^{(I)} + B^{(I)}u_a^{(I)} + G^{(I)}d^{(I)}$$
$$y_a^{(I)} = C^{(I)}x^{(I)} + D^{(I)}u_a^{(I)} + H^{(I)}d^{(I)} + \nu^{(I)} \tag{1}$$

where (avoiding the pointer I for simplicity): $x \in \mathcal{R}^n$ is the vector of system states, $u_a \in \mathcal{R}^m$ is the vector of controlled inputs, $d \in \mathcal{R}^q$ is the vector of uncontrolled inputs, $y_a \in \mathcal{R}^p$ is the vector of outputs (measurements), $\nu \in \mathcal{R}^p$ is the vector of measurement noise and A, B, G, C, D, H are the parameter matrices of proper dimensions and content.

The output part in (1), can be written as follows. Note that the signal is split into two parts to facilitate the analysis. Moreover, an extra top-pointer is used to indicate the signals that are changing between cases.

$$y_a^{(0)} = \begin{bmatrix} y_{a1}^{(0)} \\ y_{a2}^{(0)} \end{bmatrix} = \begin{bmatrix} C_1^{(0)} \\ C_2^{(0)} \end{bmatrix} x^{(0)} + \begin{bmatrix} D_1^{(0)} \\ D_2^{(0)} \end{bmatrix} u_a^{(0)} + \begin{bmatrix} H_1^{(0)} \\ H_2^{(0)} \end{bmatrix} g^{(0)} + \begin{bmatrix} \nu_1^{(0)} \\ \nu_2^{(0)} \end{bmatrix} \tag{2}$$

Then, the three cases identified above, lead to the equations:

$$y_a^{(1)} = \begin{bmatrix} y_{a1}^{(0)} \\ y_{a2}^{(1)} \end{bmatrix} = \begin{bmatrix} C_1^{(0)} \\ C_2^{(1)} \end{bmatrix} x^{(0)} + \begin{bmatrix} D_1^{(0)} \\ D_2^{(1)} \end{bmatrix} u_a^{(0)} + \begin{bmatrix} H_1^{(0)} \\ H_2^{(1)} \end{bmatrix} g^{(0)} + \begin{bmatrix} \nu_1^{(0)} \\ \nu_2^{(1)} \end{bmatrix} \tag{3}$$

$$y_a^{(2)} = \begin{bmatrix} y_{a1}^{(0)} \\ y_{a2}^{(0)} \\ y_{a2}^{(0)} \end{bmatrix} = \begin{bmatrix} C_1^{(0)} \\ C_2^{(0)} \\ C_2^{(0)} \end{bmatrix} x^{(0)} + \begin{bmatrix} D_1^{(0)} \\ D_2^{(0)} \\ D_2^{(0)} \end{bmatrix} u_a^{(0)} + \begin{bmatrix} H_1^{(0)} \\ H_2^{(0)} \\ H_2^{(0)} \end{bmatrix} g^{(0)} + \begin{bmatrix} \nu_1^{(0)} \\ \nu_2^{(0)} \\ \nu_2^{(0)} \end{bmatrix} \tag{4}$$

$$y_a^{(3)} = \begin{bmatrix} y_{a1}^{(0)} \\ y_{a2}^{(0)} \\ y_{a2}^{(3)} \end{bmatrix} = \begin{bmatrix} C_1^{(0)}x^{(0)} + D_1^{(0)}u_a^{(0)} + H_1^{(0)}g^{(0)} + \nu_1^{(0)} \\ C_2^{(0)}x^{(0)} + D_2^{(0)}u_a^{(0)} + H_2^{(0)}g^{(0)} + \nu_2^{(0)} \\ C_2^{(3)}x^{(3)} + D_2^{(3)}u_a^{(3)} + H_2^{(3)}g^{(3)} + \nu^{(3)} \end{bmatrix} \tag{5}$$

Equation 3 shows that a part of the sensing signals have been modified, comparing to specifications, resulting in a modified output vector. Equation 4 shows that the output vector has been modified not only in terms of content but also in terms of dimension, while still measuring same quantities. Finally, (5) shows that the output vector has been modified in terms of dimension and the newly introduced part measures different quantities. All described cases need to be properly accommodated in the closed-loop system by utilising available new knowledge and tools.

3   Proposed Architecture and Framework

In an earlier work, [14], the authors presented a basic introduction of the semantic interoperability concepts and the ontologies as a tool to implement knowledge models. Such models have been also used in domestic robotics (DOGont, [6]) to face the interoperation issues by implementing structured representations of domain knowledge. In this work, we adopt knowledge models in combination with control engineering mathematical representations. Efforts to represent the mathematical models in ontological knowledge models can be found in [18] and [11].

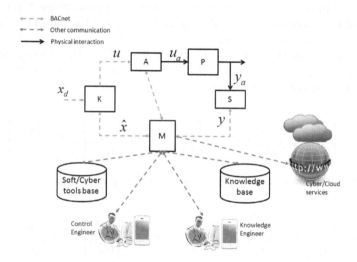

Fig. 1. Block diagram of proposed feedback–control architecture. Details about the content of this figure are given in section 3.

The proposed architecture for the closed-loop system is depicted in fig. 1. As illustrated, the set-up comprises the: i) plant with its parameters and self-dynamics, ii) the physical control system components, like sensors (producing y) and actuators of different types (producing u_a), iii) a tools' base, which stores the implementations of software functions such as observer design implementations producing the state estimation $\hat{x}$, functions performing transformations among measurement units, etc., iv) the humans (e.g. Control Systems Engineer, Knowledge Models Engineer), v) a communication infrastructure (the orange-dashed line shows the

BACnet/IP protocol stack communication [2], whereas the blue-dashed line shows communication through any other protocol), vi) a semantic mediation module, M, which is responsible for the scalability of the control system and vii) a Knowledge Model, implemented as OWL ontology(ies) [1].

A critical component introduced here is M, which has a multi-fold scope as it implements the physical interaction interface among all components. The semantic mediation module strongly relies on the knowledge model to analyse each time's situation and take reasonable and optimal decisions for the operation of the system. It is therefore, of utmost important for the knowledge model to be well designed and defined based on the "closed-world" assumption [13]. We want the knowledge model to support control systems, that might comprise also safety-critical deployments, so the decisions taken should be based on explicit knowledge such as to avoid instability.

3.1   The Knowledge Model

The knowledge model comprises the agreement between all interacting physical and cyber components, about the interpretation of their environment.

This model is implemented as a set of objects' symbols, a set of classes/types for these objects and a set of properties of objects that also implement relations/mappings among them, that is, $\mathcal{A} = \{ \mathcal{T}_H, \mathcal{C}_L, \mathcal{P}_R \}$. For the purpose of this work, we define specific objects, types of objects and properties. In order to keep it simple, we developed our own mini knowledge model. In future practical implementations, this model can be replaced by more complete efforts from the literature, such as combinations of the knowledge models in [9] to describe the environment and interactions of components, and the ones in [18] and [11] to describe the knowledge in mathematical representations.

The set of objects is defined as: $\mathcal{T}_H = \{ o_i \mid i = 1, 2, ..., \}$, where o_i is the reference to an object's literal (e.g. the physical property "temperature") or to the real implementation of the object (e.g. "Sensor1" meaning the device with that identification).

The following classes of objects have been defined:
$\mathcal{C}_L = \{Plant, Model, State, ControlledInput, UncontrolledInput, Output,$
$PlantLocation, PhysicalProperty, MeasurementUnit, Sensor, Actuator,$
$Function\}$ where: $Plant$ is the set of plants served by the knowledge model, $Model$ is the set of system models (e.g. a state-space model of the system), $State$ is the set of system states, $ControlledInput$ is the set of controlled inputs, $UncontrolledInput$ is the set of uncontrolled inputs (disturbances to the plant), $Output$ is the set of measurable outputs of plant, $PlantLocation$ is the set of identified locations in the plant, $PhysicalProperty$ is the set of defined physical properties (e.g. temperature, energy), $MeasurementUnit$ is the set of units for the defined physical properties, $Sensor$ is the set of sensing devices deployed in the plant, $Actuator$ is the set of actuating devices deployed in the plant, $Function$ is the set of functions/mappings defined to represent the mathematical relations among variables.

Then, relations are defined, to represent the properties of objects. A relation is a mapping of the form: $relationName : C_{L(i)} \times C_{L(j)} \mapsto \{\top, \bot\}$. These may define whether an object belongs to a specific class, the relation between a plant and a model, the relation between a model and a state of the plant, the relation between an input/output of the plant and a physical property, the physical property that is measured in a specific unit, the location where a sensing/actuating device is located in, etc.

3.2   The Controller and Observer Implementations

Upon a shift to a different model of the plant, as a result of the inference step, the implementations of the controller and the state-observer change. The new implementations, are either given and retrieved from the knowledge base or they are calculated online. We assume the actuators are driven by a simple proportional controller, while the system states are estimated (mostly for compensation of missing measurements) with a simple Luenberger full-state observer [12], as shown in (6).

$$u = Ke + u_0$$
$$e = x_d - \hat{x} \tag{6}$$
$$\dot{\hat{x}} = A\hat{x} + Bu + G\hat{d} + LW(y - C\hat{x})$$

where $K \in \mathcal{R}^{m \times n}$ is the control gain matrix, $e \in \mathcal{R}^n$ is the error signal (difference between desired and estimated state values), $u_0 \in \mathcal{R}^m$ is the control bias that is used to cancel system disturbances, model uncertainties and retain the system at desired operation, $x_d \in \mathcal{R}^n$ is the desired system states' vector, $\hat{x} \in \mathcal{R}^n$ is the estimated system states' vector, $\hat{d} \in \mathcal{R}^q$ is the estimated uncontrolled inputs' vector, if such option exists, $L \in \mathcal{R}^{n \times n}$ is the observer gain, implemented such as the pair (A, WC) is stable and $W \in \mathcal{R}^{n \times p}$ is a weight matrix that represents the trust on each of the p measurements.

4   Case-Study Scenario

We assume an apartment with three rooms as shown in figure 2a. The apartment is equipped with a central heating installation, however, for budget reasons there is only one heating radiator in the bedroom, accompanied by one temperature sensor in the same room that measures in degrees Celsius. The equipment is used to regulate the temperature of the apartment at desired value. The design also assumes an uncontrolled input to the plant produced by the ambient temperature and modelled by a slightly open window (for simplicity we consider zero transfer of heat through the walls).

The case of replacing a sensor with another one of not compatible specifications, is described by (3) and has been specifically addressed in [14]. Here we consider the owner of the apartment buying a smart phone, which is equipped with temperature sensor. This mobile sensor is entering and leaving the apartment at different occasions during a day, therefore, at discrete sampling times

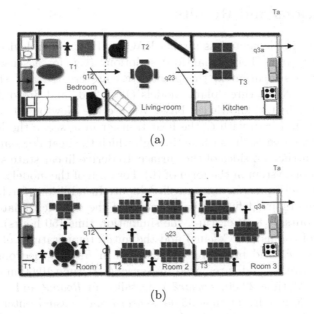

(a)

(b)

Fig. 2. a. The apartment plant. $Q1$ represents the heating input produced by the electric radiator, $T_i, i \in 1, 2, 3, a$ represent the temperature of the three rooms and the ambient respectively, q_{ij}, with $i, j \in 1, 2, 3, a$ representing the flow of heat among the rooms and the ambient, b. The office plant with open doors.

the mediation component is retrieving more than one measurements. This case is described by (4). The knowledge base helps the mediation component to retrieve the measurements and also build the weight matrix W, while the observer continues to producing estimates of the state and the mediation component now feeds the control law with the fused sensors' measurements. This allows benefiting from the availability and accuracy of redundant information.

Later in time, the apartment is bought by an ICT company and is turned into an open-plan, as shown in fig. 2b. Soon after, they notice that people working in Room 3, do not feel comfortable and wear heavy clothes. So, they install temperature sensors in the other two rooms as well. In parallel, a control engineer is asked to design higher-order models of the apartment heating system. For simplicity we consider a manual design of the models, while an alternative would be for an adaptive algorithm like the one in [3] to be used in closed-loop operation. The closed-loop system now fully incorporates the sensing information available (increases the order of the model) and it is now able to maintain better temperature conditions across all rooms (of course with the limited capacity of the single actuator). This case is described by (5).

5   Simulations and Results

The devices are implemented as virtual BACnet/IP-enabled devices, using the BACnet4J API [16]. Their semantic descriptions (e.g. for a sensor, the measurement unit, its location in the plant, etc.) are created and stored in the knowledge base. Next, three plant models (1st-, 2nd- and 3rd-order) are created and stored in the knowledge base. We use the Newton's law of cooling, $Q = cA(x_i - x_j)$, $i \neq j$, with Q the heat transfer in J/sec, c the heat transfer coefficient, A the area of the surface through which the heat flows and x_i, x_j the temperatures in the two sides of the surface, to derive linear state-space models of the closed-loop system in the form of (6). For each of the models, the ambient temperature is acting as an uncontrolled input. In addition, further scenario-related parameters are defined, like the steps of the electric radiator output, a model for the outside temperature, the simulation time (50 hours), the desired temperatures of rooms (25^0Celcius) and the initial temperatures of the rooms.

Initially, *sensor1* and *radiator1* are installed in *Room1* (bedroom). At time 08:00, *sensor1* breaks and is replaced by *sensor4*. At time 10:00 mobile *sensor5* enters *Room1*. At time 22:00, *sensor2* is installed in *Room2* and *sensor3* is installed in *Room3*. Finally, at time 36:00, *sensor6* and *sensor7* enter *Room3* and both leave at time 39:00. The simulation runs in 1-minute steps. During each step, the mediation component reads and stores the sensors' measurements together with their time-stamp. The processing of the measurements and the calculation of the control input is performed at 5-minute intervals. At each such step, the mediation component retrieves information about the current measurements. These are discarded if they were taken more than 2.5 minutes earlier. Moreover, in case a value is in a different unit than the one required by the current control law, the mediation component runs an inference rule [19] and retrieves the literal name of the function to invoke (from those in the Tools base) in order to perform the required transformation. The rule says: *"Find the name of the function that takes as input a real value of the given sensor's measurement unit and produces a real value in the desired measurement unit"*. If no such function is returned, the measurement is discarded. In the implementation of the rule, the given and the desired measurement units are denoted as $o_1, o_2 \in MeasurementUnit$ and any symbols starting with "?" denote a variable that can take as value an object from the knowledge base of the class accepted as argument by the specific relation. The rule is written as:

$z_1 \in \mathcal{Z} = Function(?x) \wedge hasDomain(?x, o_1) \wedge hasRange(?x, o_2)$
$\longmapsto InferredInd(?x)$

At that moment, in case there was any change in the sensors that comprise the closed-loop system, the mediation component retrieves the best available plant model to use for the operation of the controller, given the locations and the measurement properties/units. To this end, several inference rules are executed in the knowledge base. The first one is the:

$z_2 \in \mathcal{Z} = Output(?x) \wedge [associatedWithLocation(?x, o_3)$
$\vee [associatedWithLocation(?x, ?y) \wedge isPartOf(o_3, ?y)]]$
$\wedge isPhysicalProperty(?x, o_4) \longmapsto InferredInd(?x)$

where $o_3 \in Location$ and $o_4 \in PhysicalProperty$ are the given measurement location and the measured physical property of the sensor, respectively. The above rule means: *"Find all available outputs of models, that are either associated directly with the given location or are associated with a different location which is, however, defined as part of the given location, and that are associated with the given physical property (e.g. temperature)"*. At the end, a decision algorithm is invoked which finds the model that is of the highest order, while still controllable and observable under current situation, and returns its constant parameter matrices as defined in (1) and (6). At this moment, the dimensions of all vectors and all parameters of the model to be used, are considered known. Given the new model, the mediation component invokes functions to calculate the observer's and the controller's gain matrices, $L^{(I)}$ and $K^{(I)}$ respectively. This is performed with simple pole placement for observability (pair A, WC) and stability (pair A, B). The new state vector estimation is based each time on the model and the designed observer. The observed state values are used by the controller to compute the next control input value. It is noted that the control input retains the previous value until a new one is produced.

The result is that the closed-loop system is able to transparently integrate any new component and use the new information to operate smoothly despite the events introduced during operation. No downtime or manual re-configuration are required.

6   Conclusions

We have presented a new architecture that can be adopted in the design of feedback control systems, in order to take advantage of the scalability characteristics offered by the combination of the classical control capabilities with a cyber infrastructure and semantic interoperability protocols and interfaces.

The scope of the work was not to advance the control algorithms as such. The current industrial practice suggests using standard controllers (e.g. PID) and applying the interoperability of components at higher application levels. We believe that much more advance intelligent control algorithms, already developed in the literature, can enormously impact the industrial applications if there is a framework for their deployment in large feedback control systems.

There is still lot of work to be done, before we can claim achieving the objectives of this work. Our immediate next steps will be the thorough investigation of the closed-loop system stability within the proposed architecture, as well as, the implementation of a demonstration setup that will pilot test the applicability in real-life scenarios.

Acknowledgments. This work is partially funded by the European Research Council (ERC) under the project "Fault-Adaptive Monitoring and Control of Complex Distributed Dynamical Systems".

References

1. Antoniou, G., Harmelen, F.V.: Web ontology language: Owl. In: Handbook on Ontologies in Information Systems, pp. 67–92. Springer (2003), http://link.springer.com/chapter/10.1007/978-3-540-92673-3_4
2. ASHRAE: BACnet Website, http://www.bacnet.org/
3. Bendtsen, J., Trangbaek, K., Stoustrup, J.: Plug-and-Play Control - Modifying Control Systems Online. IEEE Transactions on Control Systems Technology 21(1), 79–93 (2013)
4. Blanke, M., Kinnaert, M., Lunze, J., Staroswiecki, M.: Diagnosis and fault-tolerant control. Springer (2003)
5. Chen, J., Patton, R.J.: Robust model-based fault diagnosis for dynamic systems. Kluwer Academic Publishers (1999)
6. E-Lite: DogOnt (2012), http://elite.polito.it/dogont
7. Farrell, J., Polycarpou, M.: Adaptive Approximation Based Control: Unifying Neural, Fuzzy and Traditional Adaptive Approximation Approaches. J. Wiley (2006)
8. Gertler, J.: Fault detection and diagnosis in engineering systems. CRC (1998)
9. Holger, N., Compton, M.: The Semantic Sensor Network Ontology: A Generic Language to Describe Sensor Assets. In: 12th AGILE International Conference on Geographic Information Science, Workshop on Challenges in Geospatial Data Harmonisation, Hannover, Germany (2009), http://plone.itc.nl/agile_old/Conference/2009-hannover/shortpaper.htm
10. Isermann, R.: Fault-diagnosis systems: an introduction from fault detection to fault tolerance. Springer (2006)
11. Lange, C.: Ontologies and languages for representing mathematical knowledge on the semantic web. Semantic Web (i) (2013)
12. Luenberger, D.: Observers for multivariable systems. IEEE Transactions on Automatic Control 11(2), 190–197 (1966)
13. Mazzocchi, S.: Closed World vs. Open World: The First Semantic Web Battle (2005), http://www.betaversion.org/~stefano/linotype/news/91/
14. Milis, G.M., Panayiotou, C.G., Polycarpou, M.M.: Towards a Semantically Enhanced Control Architecture. In: IEEE Multi-Conference on Systems and Control, Dubrovnik, Croatia (2012)
15. Samad, T., Parisini, T.: Systems of Systems. In: Samad, T., Annaswamy, A. (eds.) The Impact of Control Technology (2011), http://www.ieeecss.org
16. Serotonin-Software: BACnet I/P for Java (2011), http://sourceforge.net/projects/bacnet4j/
17. Stoustrup, J.: Plug & Play Control: Control Technology Towards New Challenges. European Journal of Control 15(3-4), 311–330 (2009), http://ejc.revuesonline.com/article.jsp?articleId=13584
18. Suresh, P., Joglekar, G., Hsu, S., Akkisetty, P., Hailemariam, L., Jain, A., Reklaitis, G., Venkatasubramanian, V.: OntoMODEL: Ontological Mathematical Modeling Knowledge Management. Computer Aided Chemical Engineering, 985–990 (2008)
19. W3C: SWRL: A Semantic Web Rule Language Combining OWL and RuleML (2004), http://www.w3.org/Submission/SWRL/

Systems Engineering for Assessment
of Virtual Power System Implementations

Slobodan Lukovic and Igor Kaitovic

University of Lugano, Faculty of Informatics, AlaRI
{slobodan.lukovic,igor.kaitovic}@usi.ch

Abstract. In this work we present an adoption of systems engineering methodology for design and assessment of a Virtual Power System (VPS). The VPS has been defined as an aggregation of distributed energy resources, consumers and storages which can operate autonomously, and is presented to the power system as a single unit in technical and commercial terms. The complexity of these critical systems is tackled by means of systems engineering. We have applied our approach in scope of a research project AlpEnergy.

Keywords: systems engineering, modeling, SysML, VPS; assessment.

1   Introduction and Background

In the present work based on achieved results in the course of the international AlpEnergy project we provide description of a methodology used for design, implementation and assessment of Virtual Power System (VPS). VPS integrates, manages and controls distributed energy generators, electrical vehicles (and other controllable loads) and storage capacities and links their technical operation to the demand of consumers and the energy market. Detailed VPS description providing basic definitions and explanation of the concept is presented in [12,14].

One of the main challenges of the project is considered to be tackling of multidisciplinary nature of VPS concept and developing of a model of the supporting ICT structures which are crucial to critical infrastructures. Moreover, an evaluation strategy for pilot implementations was supposed to be developed. To cope with the challenge we have adopted systems engineering methodology already proven as a very efficient instrument to tackle complexity of different heterogeneous systems and different technical backgrounds of stakeholders [8]. The methodology is based on instruments provided by SysML modeling language - widely accepted UML profile for modeling of complex heterogeneous systems. The methodology is presented in details in [1], [13] while similar approach is also applied in [3]. In this work we focus on assessment of different project implementations showing how proposed methodology can be used for that purpose.

VPS concept supposes a flexible structure that embraces widest set of technical and commercial issues. Furthermore, VPS model has to absorb all the changes and adaptations in the system and its environment and for that reason has to be constantly

H. Papadopoulos et al. (Eds.): AIAI 2013, IFIP AICT 412, pp. 667–676, 2013.

updated depending on the newly identified stakeholders or requirements as well as results obtained from results provided by pilot implementations.

VPS design, involves many stakeholders that are not ICT (Information and Communication Technologies) experts. For that reason it was necessary to propose a rather simple, easy to explain - communication and system design strategy. Therefore, we adopted systems engineering methodology, still capable of coping with dynamic model changes [13]. Moreover, the same methodology has to take into account particularities of project and be able to embrace various VPS requirements, results and particularities of implementations from different partners. The main benefit the methodology brings is that it supposes an interactive and not a straight-forward process of system design and that is fully open to changes and upgrades that are inevitable in project development. The ultimate result of the methodology application to the VPS is a reference architecture that is later used for assessment of specific implementations.

2   General Approaches and Methodologies

The first step of the methodology represents context description. It refers to unambiguous definition of system environment in terms of surrounding systems and their relations to the designed system. By correlating information from relevant research projects [2-5] we managed to identify existing and foreseen future stakeholders as well as system context in terms of surrounding systems as well as VPS itself.

A full set of identified stakeholders, includes individuals and organizations: Distribution System Operator (DSO), Electricity retailers, Energy Exchange Stock, Energy market legislation, Local authorities, Equipment producers, VPS shareholders, consumers (i.e. Residential users, Industrial users etc.),as detailed in [13]. Still, as already stated, additional stakeholders could emerge in the future. Moreover, some of the stakeholders identified above are foreseen to just appear or to drastically change in the future (e.g. Energy Exchange Stock). Such changes would consequently trigger additional changes in system requirements and models proposed.

Requirements engineering that includes collecting, tracing, analyzing, qualifying and managing user requirements constitutes the next stage of the system design. The main purpose of this phase is the efficient extraction of system requirements and their refinement trough use cases that are latter traced to specific components.

For each user requirement collected from stakeholders (or in particular case from some of the project partners), a set of system requirements has been derived. The same set of system requirements has been mapped to use cases describing system functionalities and services that are finally traced to one or a group of system components [13]. A conceptual (i.e. general abstract) system model of VPS is supposed to serve as an intermediate step for better orientation in development of reference model (as shown in [11]). The next step represents behavioral modeling of each system functionality presented through use cases and assumed set of components [4]. Interaction based behavioral models are presented by means of sequence diagrams. While progressing with behavioral models of system functionalities, system structure changes in terms of inserting additional components or changing their roles. In some cases, structural change will induce additional behavioral changes. The outcome of this step is set of consistent structural and behavioral models. It is important to notice that these of models are developed in parallel and they impact and complement each other. Based on these models methodology finally yields reference model.

Being flexible and iterative, methodology described can be used in different stages of project development. That makes it suitable for the application to the research projects such as VPS design is. Since not all stakeholders are known at the moment (e.g. energy market), some requirements are tailored to project goals and realistic assumptions.

3   VPS Requirements and Use Cases Definition

Having stakeholders identified, requirements engineering phase starts by collecting user requirements from them. Once initial set of user requirements has been obtained, initial abstract model of the system has to be defined. This refers to identification of components, so that system requirements related with components and their interaction can be identified. For each user requirement separate sequence diagram has been developed representing the way that requirement can be satisfied through interaction of initially assumed VPS system components and surrounding systems. While focusing on functional ones, we have carefully collected and organized user requirements in three packages representing communication, commercial and technical requirements. For each interaction that appear among VPS and surrounding systems as well as so far identified internal components, system requirement is defined. Each user requirement has been mapped to one or more use cases describing VPS functionalities (as detailed in [13]). As the system modeling progress, set of system components could change, and those sequence diagrams could be re-engineered after validation.

Applying the same principle to all other user requirements, the final set of system requirements has been derived. In fact, user requirements are mapped into system ones. The definition of use cases initiates behavioral modeling of the system. Use cases development is based on system requirements, through grouping and refinement process. A few basic use cases, describing the main functionalities of the VPS are defined at first and related with the actors and internal components involved in operations they describe. While defining those main use cases relating them with actors, major user requirements were taken into account (as shown in [11]). This marks the starting point for understanding the system functionalities. Each stakeholder can identify use cases that he is interested in, and latter follow decomposition only of these use cases, unburden from the functionalities of the rest of the system. So, managing system complexity is done through its functional and hierarchical decomposition.

3.1   Use-Cases Decomposition

Each of the major use cases is further decomposed through more fain-grained diagrams with atomic use cases directly traced (i.e. related) to system requirements. Still, further extension of all diagrams is possible and even expected with project evolution as new requirements could evolve. The model represents a result of an iterative process taking place during entire course of the project. As an demonstrator of the further steps of the methodology we focus on energy trading scenario (i.e. use case) which implies external and internal contracting. In later steps these use cases will be elaborated in details. Use cases are related by proper requirements as shown in the figure.

Applying a comprehensive approach based on initial findings on system functionalities and composition (coming from related projects and knowledge of the project partners) a general abstract model of the VPS has been developed as an intermediate modeling result (see [11]). It defines initial set of components and their interactions and serves for orientation in further phases of the modeling methodology.

3.2   Basic System Components

Brief description of the main components in the model is given in the following:

- Control Center are manages the system assuring the power quality.
- Advanced Measuring Infrastructure (AMI) goes way beyond consumption metering and includes measuring of additional parameters and their collecting
- Actuating Embedded Systems (AES) are extending the SCADA functionalities and directly control field devices (smart load, DER or storage).
- *Trading Agent* is envisioned as component that handles all the commercial issues of the VPS including biding, energy selling etc.
- *VPS operation manager* is the brain of the system that manages energy balance, verifies the feasibility of predicted production, etc.

Different kinds of scenarios (i.e. use cases) have been developed. In order to show effectiveness of the approach, we use an example of energy trading. Energy trading for AlpEnergy VPS considers 'day-ahead' market contracting and it may concern amount and prices of bulk energy, ancillary services and so forth. This major use case has been decomposed in two   phases (i.e. sub-cases): External contracting- negotiation between SmartGrid; Internal contracting - negotiation - between VPS and its clients. We will show main actors and describe their interaction in the following.

3.3   External and Internal Trade Contracting

Basically the process starts by a request from the VPS manager (which keeps portfolio with preferences and statistics of all VPS consumers and producers) to the Smart Grid for obtaining the energy price. The trading agent of VPS communicates with the AMI and the predictions manager for current meter data and energy balance predictions, respectively, in order to verify affordability of the contract (e.g. offered price and amount of energy) obtained from the Smart Grid. Eventually, negotiation takes place between the VPS manager and Smart Grid and the process iterates until an optimal price is reached. The sequence diagram describing this process is given in Fig.1.

Ones the contract is set at the level of entire VPS it is also necessary to negotiate in the context of local contracts (for each VPS member in particular). In this case VPS manager requests from the tariff manager to calculate the dynamic prices for each individual consumer (or producer). AMI and predictions manager are contacted again to check current and expected energy balance in the system and per each VPS member. Demand Side Management (DSM) is involved to customize offers for each VPS member individually based on their statistics and portfolios. The price is then negotiated with the local energy manage (LEM - which is a component assigned to each member of VPS), since the price in this case is a local (i.e. internal VPS matter), rather than external. The sequence diagram in Fig. 1 (on the right) describes the process.

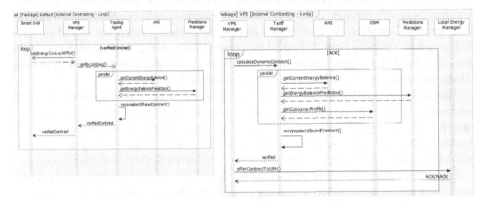

Fig. 1. Internal and external contracting

4   Structural VPS Model

Applying the methodology, we continue with defining system behavior and structure needed for achieving each of the use cases previously defined. For each VPS functionalities/use case a specific sequence diagram has been derived describing how particular use case can be satisfied through interaction of various components. Initially each component has some predefined set of functionalities that is being extended and additionally defined through each sequence diagram. Basics of behavioral model of the system are set in the previous Section where sequence diagrams were given (Fig. 4 and 5) denoting information flow.

Starting from the general abstract model defined as a tentative system structure at early phase of the project and iteratively improving it through new findings from developed use cases we build a set of main VPS components together with identified functionalities.

Eventually we present simplified VPS internal block diagram providing details about information flow between components which facilitates defining the communication standard inside VPS. With a clear identification of components and it's functionalities, same methodology described here can be used for deriving further fine-grain model of the components. It should be noted that model presented is derived from the requirements of AlpEnergy project while considering contemporary results of other relevant projects as well [2-4]. Thus, with additional upgrades, it can be used as a reference point for other similar projects dealing with Smart Grid aggregations.

4.1   Building Reference Model

General, abstract model of proposed ICT supporting structure for VPS (given in [11]) has been developed according to existing state-of-the-art solutions and inputs coming from different pilot implementations. The model is a result of discussions among all the partners and wide compromise about all its elements has been achieved. The later findings obtained through described systems engineering process have resulted in VPS structural models. As a very next step in the project a reference model of VPS ICT structure has been built. It is used as a central point of the AlpEnergy project and it is considered to be one of its main achievements as it elaborates the result of mutual efforts and gives an overview on the system infrastructure.

The reference model shown in Fig.1 is intended to be a starting point for ICT assessment process. It represents a suggested solution that would also serve as a reference for evaluation of all pilot implementations. Apart of that, as a visual description, it can assist project dissemination activities as it has been made to be general, comprehensive and understandable using the widely accepted SysML modeling language.

The reference model has been adopted to identified specific requirements of AlpEnergy and regularly upgraded according to proven solutions coming from exact pilot implementations. As lighthouse guidelines for abstract model we have considered solutions provided by related relevant international projects like [2-4], respectable international institutions and initiatives [5] and also cutting-edge research achievements [6,7]. The basic abstract model developed from such a theoretical approach is developed in SysML modeling language, it represents an intermediate step.

Still such a model served just for orientation and provided good guidelines for other phases of the project development. The input from existing pilot implementations and experiences gained in field testing are another precious source of information needed for development of the precise reference model. We interviewed AlpEnergy implementing partners and studied their technical documentation. Nevertheless, without an unified describing method and in absence of defined mutual understanding platform for VPS design we faced many problems in extracting technical details from available informal descriptions.

4.2   System Composition

It is possible to notice, from these figures, basic similarities and differences in VPS design from one to another case. Nevertheless, the lack of unified and standardized description method aggravates communication, knowledge and best practice transfer, and at the same time assessment process. Therefore, the need for establishing common mutual describing instrument is quite obvious.

As already stated we have approached the problem using System Engineering methodologies based on SysML modeling language that has been developed during the project and described in previous annual reports and publications [10,11].

Finally bringing all together, abstract model and selected best-practice implemented solutions, the reference model has been built. The working version of the model has been reconsidered by the partners several times (according to Systems Engineering practice [8]) and eventually the VPS reference model in Fig. 2 has been obtained as the recommended optimal solution accepted by entire AlpEnergy consortium.

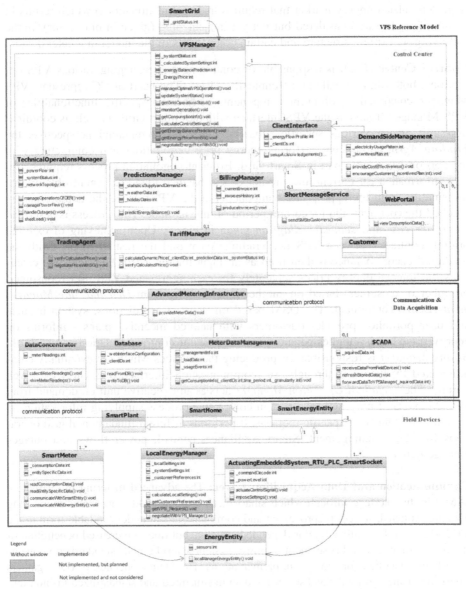

Fig. 2. VPS ICT infrastructure – general reference model

The entire ICT structure of VPS is logically encapsulated in three groups represented in reference model in form of three layers: Power Systems Management Applications (Control Center), Communication and Data Acquisition Field Devices.

Each of these layers represents a functional block in VPS design (note that functionalities that are grayed out are deemed optional in VPS operation). Considering the model in Fig. 2, it should be noted that it considers even foreseen future scenarios (e.g. deregulated energy market that requires trading functionalities) which results in components that are considered but not yet implemented (given in blue color). In the following we explain structure and certain components of the model in greater detail:

Control Center (CC): is responsible of controlling and managing various VPS operations, both commercially and technically related. The heart of CC represents VPS Manager composed of different components performing specific functionalities of VPS Manager. It acts as the VPS interface with the Smart Grid, as well as coordination unit among various specific subunits. From the commercial perspective, the *Trading Agent* component performs energy price negotiations and contracting with the appropriate counterpart in Smart Grid. Efficient negotiation process requires precise information on current and predicted grid balance (Trading Agent obtains this information from the appropriate components of the VPS). Once the compromise is reached the contract valid for certain period of time is made. This process is described using *use case* and *sequence diagram* given in Fig. 4. Upon achieving of external pricing contract (between VPS i.e. Trading Agent and the rest of Smart Grid) the predicted energy balance is then used as an input for the *Tariff Manager* that calculates the internal energy prices separately for each aggregated VPS user (consumer, producer or prosumer) based on another input coming from *Demand Side Management* (DSM) component. The DSM based on statistics, predictions, system balance and user portfolios provides customers with tailored incentive plans - informs on energy price trends; applies penalty plans in cases of contract violation and so forth. *Billing Manager* is responsible for producing the invoices which are eventually communicated to customers through user communication interface (e.g. SMS or a web portal). The *Technical Operation Manager* acts as an executive unit on demand from VPS Manager, it may also take part in contracting process supplying information on technical feasibility of specific requests. It performs different fundamental grid operations like DER management, Smart Appliances control, power flows and outages management etc.

Communication and Data Acquisition Layer: The advanced metering infrastructure (AMI) is the interface through which smart meter readings and other status monitoring data are gathered, stored, preprocessed and communicated in understandable form to the VPS Manager. It is rather vendor depended structure but due to increased penetration of standardization also in this segment of power system several crucial standard components have been identified for most of implementations. The Data Concentrator (DC) gathers smart meter measurements and stores the data in structured and standardized fashion into the Database which maintains the web portal configuration as well as administrative data arranged according to customers IDs. In some implementations, the database block can be part of the CC, rather than data acquisition one. *Meter Data Management* (**MDM**) further process the data on different granularity levels (i.e. from appliances to DER level), filters them by time period and presents them in a required structured form to CC components. VPS deals with integration of many different technologies so that it should support system like Supervisory Control and Data Acquisition (*SCADA*) module to work as a data bridge/interface between field devices and the VPS manager.

Field Devices – Backend Layer: This layer considers backend embedded systems which measure, control or actuate electro-mechanical devices directly attached to the energy entities (energy entities may be - generators, home appliances, storages etc.). These embedded systems include smart meters, PLC (Programmable Logic Controllers), RTU (Remote Terminal Units), 'smart sockets' and so forth. The devices from this layer translate the control signals received directly from the VPS manager (or Local Energy Management in case of distributed control) into an action (e.g. reduce power level, disconnect/reconnect customer etc.). The Local Energy Management unit provides local control decision based on user preferences and constraints but also requests from VPS Manager.

5   Assessment of ICT Solutions

The assessment process has been agreed to be done in three phases:

- Development of a reference model
- Development of specific implementations' models according to same principles and standards
- Comparison of these two kinds of models by mapping a model of specific case into reference model according to defined standards
- Bearing in mind the reference model represented in Fig. 7. and details provided by each partner, appropriate models of pilot implementations have been developed. It can be noted that all the considered case have some differences in the implementation but still the basic structure was preserved. The exact models are result of technical documentation studding and direct interviews with implanting partners.

We present here assessment of one implementation that we consider as the most demonstrative. The assessment models of all other implementations are given in AlpEnergy Report [11].The implementation to be assessed has been the one from the Allgäu region [11]. It has been considered as the most mature pilot and like this also served for the reference model development. The implementation reflects very well conceptual solutions even though some considered solutions like Trading Agent and Local Energy Manager are not fully implemented. The implemented VPS comprises different kind of generators (like PV cells, windmills etc.), Smart Homes, dynamic tariffs management, some elementary DSM and so forth. The detailed assessment report for all the pilot implementations in SysML fashion, is given in [11].

6   Conclusion

In this paper we have presented how system engineering methodology can be efficiently adapted and applied for the design and assessment of a critical structure that is a virtual power system. Even though, due to the size of the system we focus only on functional requirements and structural model of the system in this work, similar methodology can be applied to ensure that critical non-functional properties of the system have been achieved. Given reference model is highly scalable and easily can be extended to support more both functional and non-functional requirements. As a future work, we plan to asses availability and safety of implemented solutions.

References

1. Lukovic, S., Kaitovic, I., Bondi, U.: Adopting system engineering methodology to Virtual Power Systems design flow. In: Proc. of the 1st Workshop on Green and Smart Embedded System Technology: Infrastructures, Methods and Tools (CPSWEEK/GREEMBED 2010) (2010)
2. E-ENergy project, http://www.e-energy.de
3. Fenix project, http://www.fenix-project.org/
4. ADDRESS project, http://www.addressfp7.org/
5. SmartGrids Technology Platform, http://www.smartgrids.eu/
6. Li, F., Qiao, W., Sun, H., Wan, H., Wang, J., Xia, Y., Xu, Z., Zhang, P.: Smart Transmission Grid: Vision and Framework. IEEE Transactions on Smart Grid 1(2), 168–177 (2010)
7. Sauter, T., Lobashov, M.: End-to-End Communication Architecture for Smart Grids. IEEE Transactions on Industrial Electronics 58(4), 1218–1228 (2010)
8. Weilkiens, T.: Systems Engineering with SysML/UML: Modeling, Analysis, Design. Kaufmann OMG Press (2007)
9. Vukmirovic, S., Lukovic, S., Erdeljan, A., Kulic, F.: A Smart Metering Architecture as a step towards Smart Grid realization. In: 2010 IEEE International Energy Conference and Exhibition (EnergyCon), pp. 357–362 (2010)
10. Vukmirovic, S., Lukovic, S., Erdeljan, A., Kulic, F.: Software architecture for Smart Metering systems with Virtual Power Plant. In: MELECON 2010 - 15th IEEE Mediterranean Electrotechnical Conference, pp. 448–451 (2010)
11. Lukovic, S., Kaitovic, I., Sami, M.G.: AlpEnergy: Virtual Power System (VPS) as an Instrument to Promote Transnational Cooperation and Sustainable Energy Supply in the Alpine Space - Rapport final. Bundesamt für Energie, Pub. no. 290677 (2012)
12. AlpEnergy White Book, http://www.alpine-space.eu/uploads/tx_txrunningprojects/AlpEnergy_White_Book_Virtual_Power_Plants.pdf
13. Kaitovic, I., Lukovic, S.: Adoption of Model-Driven methodology to aggregations design in Smart Grid. In: 2011 9th IEEE International Conference on Industrial Informatics (INDIN), pp. 533–538 (2011)
14. Lukovic, S., Kaitovic, I., Mura, M.: Virtual Power Plant as a bridge between Distributed Energy Resources and Smart Grid. In: 2010 43rd Hawaii International Conference on System Sciences (HICSS), pp. 1–8 (2010)

Design of Attack-Aware WDM Networks
Using a Meta-heuristic Algorithm

Konstantinos Manousakis and Georgios Ellinas

KIOS Research Center for Intelligent Systems and Networks,
Department of Electrical and Computer Engineering, University of Cyprus, Cyprus
{manousakis.konstantinos,gellinas}@ucy.ac.cy

Abstract. Transparent optical Wavelength Division Multiplexing (WDM) networks are vulnerable to physical layer attacks. This work proposes a meta-heuristic based algorithm for the planning phase of optical WDM networks while considering the impact of high-power in-band jamming attacks. The proposed heuristic algorithm serves sequentially the connections in a particular order and the meta-heuristic algorithm, called Harmony Search, is used to find better orderings in order to establish the requested connections. The objective of the proposed algorithm is to establish the requested connections in a specific way that minimizes the impact of high-power jamming signals through in-band channel crosstalk.

Keywords: physical layer attacks, routing and wavelength assignment, harmony search, optical networks.

1   Introduction

Optical networks nowadays rely on Wavelength Division Multiplexing (WDM) in order to increase their capacity. WDM enables different connections to be established concurrently through a common fiber, subject to the distinct wavelength assignment constraint; that is, the connections sharing a fiber must occupy separate wavelengths. All-optical WDM channels that may span multiple consecutive fibers are called lightpaths. In the absence of wavelength conversion, a lightpath must be assigned a common wavelength on each link it traverses; this restriction is referred to as the wavelength continuity constraint. Since lightpaths are the basic switched entities of a WDM optical network architecture, their effective establishment and usage is crucial. It is thus important to propose efficient algorithms to select the routes for the connection requests and to assign wavelengths on each of the links along these routes, among the possible choices, so as to optimize a certain performance metric. This is known as the routing and wavelength assignment (RWA) problem [1], that is usually considered under two alternative traffic models. Offline (or static) lightpath establishment addresses the case where the set of connections is known in advance, usually given in the form of a traffic matrix that describes the number of lightpaths that have to be established between each pair of nodes. Dynamic (or online) lightpath establishment considers the case where connection requests arrive at random time instants, over a

H. Papadopoulos et al. (Eds.): AIAI 2013, IFIP AICT 412, pp. 677–686, 2013.

prolonged period of time, and are served upon their arrival, on a one-by-one basis. Offline RWA is usually used during the network design and planning phase, while online RWA is used during the network operation phase.

Offline RWA is known to be an NP-hard optimization problem and several heuristics and meta-heuristics have been proposed to solve the problem. A meta-heuristic is a procedure designed to find a good solution to a difficult optimization problem. Meta-heuristic algorithms can be classified as swarm intelligent techniques and evolutionary algorithms. Swarm intelligence algorithms are heuristic search methods that mimic the metaphor of natural biological evolution and/or the social behavior of species. Evolutionary algorithms use iterative progress, such as growth or development in a population and use mechanisms inspired by biological evolution, such as reproduction, mutation, recombination, and selection. Ant Colony Optimization (ACO) and Particle Swarm Optimization (PSO) are well-known and successful swarm intelligence optimization algorithms, while Genetic Algorithms (GAs) and Harmony Search (HS) are evolutionary optimization techniques. Meta-heuristic techniques that have been proposed to solve the offline RWA problem use techniques from ACO [2], GA [3], and PSO [4].

Optical telecommunication networks, providing services to users such as companies, governmental institutions, and private citizens, are considered one of the critical infrastructures of a country. In all-optical transparent networks, where a data signal remains in the optical domain for the entire path, there exist several vulnerabilities in the network that enable malicious signals to propagate through several parts of the network. Optical networks need to be able to detect and locate failures (faults or attacks) and degradations as fast and as accurately as possible, in order to restore lost traffic and repair the failure. An attack can be defined as an intentional action against the ideal and secure functioning of the network.

In this work, a meta-heuristic algorithm is used to solve the static attack-aware RWA problem. In the static case, the set of connections is known in advance and path selection and wavelength assignment are performed offline, aiming at the joint optimization of the lightpaths used by all the connection requests. The objective of the proposed offline attack-aware RWA algorithm is to design an optical network that minimizes the impact of a network attack. The algorithm uses a nature inspired meta-heuristic, mimicking the improvisation process of music players [5], known as a harmony search (HS) algorithm. The proposed attack-aware heuristic algorithm serves sequentially the connections in a particular order, and the HS algorithm is used to find better orderings.

The rest of the paper is organized as follows. Section 2 describes the physical layer attacks. In Section 3, the Harmony Search meta-heuristic and Harmony Search based attack-aware RWA algorithm are presented, followed by simulation results in Section 4. Finally, Section 5 presents some concluding remarks and avenues for future research.

2   Network Attacks

There are several physical layer attacks that can occur in transparent optical networks as presented in [6-7]. One of the most important attacks is *in-band jamming* that is the result of intra-channel crosstalk between the same wavelengths in optical switches.

Another attack is the *out-of-band jamming*, where high-power signals can introduce nonlinearities, causing crosstalk effects between channels on different wavelengths in the same fiber (inter-channel crosstalk). Moreover, gain competition in optical amplifiers is another possible form of attack in optical networks, where a high-power jamming signal can increase its own power, thus resulting in reduction in the gain of the rest of the channels.

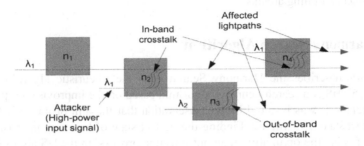

Fig. 1. High-power jamming attack propagation

In Fig. 1, the effects of in-band and out-of-band crosstalk propagation are depicted, where an attacker uses a high-power jamming signal on wavelength λ_1 to perform his attack from node n_2. Due to in-band crosstalk, this signal will affect any lightpath that crosses node n_2 and uses wavelength λ_1. Then, the affected lightpaths will perform further attacks through the nodes they cross. As can be seen from Fig. 1, the affected lightpath in node n_2 will spread the attack in node n_4 and will affect other lightpaths that use the same wavelength (λ_1). A lightpath that uses a different wavelength than the lightpath used by the attacker will not be affected, regardless of the fact that the lightpaths cross the same node. However, due to out-of-band crosstalk, a lightpath at one wavelength will affect any other lightpath using an adjacent wavelength of the same fiber. For example, in Fig. 1, the lightpath crossing node n_3 and wavelength λ_2, will be affected by the attacker, assuming that the two lightpaths use the same fiber after node n_3 (λ_1 and λ_2 are adjacent wavelengths).

When planning an optical network, the basic problem that has to be addressed is the RWA problem. Since in optical networks security of the established connections is critical, it is of paramount importance to protect the network from possible attacks. The concept of preventive, attack-aware RWA problem was proposed in [8]. In that work, the authors formulated the routing sub-problem of RWA as an integer linear program with the objective of decreasing the potential damage of jamming attacks causing out-of-band crosstalk in fibers and gain competition in optical amplifiers. A tabu search heuristic was subsequently proposed to cover larger network instances. Furthermore, in [9], authors proposed a wavelength assignment approach to limit the potential propagation of in-band crosstalk attacks through the network. Authors in [10] extended upon their work in [9] by considering a more realistic case where crosstalk attacks can maximally spread in one or two steps. This means that secondary attacked signals are not strong enough for the attack to propagate further. In [10], authors proposed the use of wavelength-selective optical attenuators as power equalizers inside the network nodes to limit the propagation of high-power jamming attacks. They presented heuristic algorithms with the objective of minimizing the number of power equalizers needed to reduce, to a desired level, the propagation of high-power jamming attacks.

This work considers that an attack is performed by high-power jamming signals through in-band crosstalk and the objective of the proposed offline attack-aware RWA algorithm is to design an optical network that minimizes the impact of a high-power in-band jamming attacks.

3   Harmony Search Algorithm

This section describes the Harmony Search (HS) meta-heuristic algorithm as presented in [5]. HS is a search heuristic algorithm based on the improvisation process of jazz musicians. It was inspired by the observation that the aim of music is to search for a perfect state of harmony. Finding the perfect state of harmony in music is analogous to finding the optimality in an optimization process. In the HS algorithm, each musician (= decision variable) plays (= generates) a note (= a value) for finding the best harmony (= global optimum determined by objective function evaluation).   The core part of the HS algorithm is the Harmony Memory (HM), where HM is a collection of random generated harmonies (=vectors). When a musician is improvising, there are three possible choices: (1) play any famous piece of music exactly from harmony memory; (2) play something similar to a known piece; or (3) compose new or random notes. Authors in [5] formalized these three options into a quantitative optimization process, and the three corresponding components become: usage of harmony memory, pitch adjusting, and randomization. Thus, harmony search tries to find a vector X which optimizes a certain objective function $f(X)$. The steps of the harmony search algorithm are as follows:

1. Initialize the harmony memory (HM) matrix:

$$\text{HM}=\begin{bmatrix} x_1^1 & \cdots & x_S^1 \\ \vdots & \ddots & \vdots \\ x_1^{HMS} & \cdots & x_S^{HMS} \end{bmatrix} \Rightarrow \begin{bmatrix} f(X^1) \\ \vdots \\ f(X^{HMS}) \end{bmatrix}$$

Generate *HMS* (*Harmony Memory Size*) random vectors $X^i = \left(x_1^i, x_2^i, \ldots, x_S^i\right)$ of size S and store them in the harmony memory (HM) matrix, where S is the number of musical instruments (decision variables, x_j^i) and *HMS* is the number of random harmonies (vectors). $f(X)$ is the objective function that needs to be optimized.

2. Improvise a new harmony (Generate a new vector X^{new})
   (a) For every variable x_j^{new} of the new vector X^{new} choose a value from the HM matrix $x_j^{new}, 1 \leq new \leq HMS$ with probability *HMCR* (*Harmony Memory Considering Rate*) and with probability 1- *HMCR*, choose a random value.
   (b) Change the value of the variable x_j^{new} with probability *PAR* (*Pitch Adjusting Rate*), by a small amount Δ, $x_j^{new} = x_j^{new} \pm \Delta$.

3. Update the HM:
   Evaluate the objective function $f(X^{new})$ of the vector X^{new}. If the value of $f(X^{new})$ is better than the worst value $f(X^{worst})$ in HM, then replace the vector X^{worst} with X^{new} in HM.
4. Check the stopping criterion:
   Repeat steps 2 and 3 until the maximum iteration criterion is met.

The exact values of the parameters are discussed in Section 5.

4   Attack-Aware RWA Using a Meta-Heuristic Algorithm

The proposed algorithm solves the static RWA problem with the objective to minimize the impact of high-power jamming signals that spread in the network through in-band channel crosstalk. The *attack-aware heuristic* algorithm establishes the requested connections one-by-one in the form of lightpaths following a specific order. For every demand, the algorithm computes a set of candidate lightpaths. From this set, the lightpath with the minimum in-band crosstalk interaction is chosen. By minimizing the in-band channel crosstalk interactions, the spread of high-power jamming signals through in-band crosstalk is also minimized. The harmony search meta-heuristic is used to define different orderings for the requested connections. The evaluation of the objective function of the harmony search algorithm $f(X)$ is performed through simulation of the *attack-aware heuristic* algorithm. The ordering with the best value of the objective function is chosen to establish the demands. In the following subsections the main steps of the algorithm are described.

4.1   Initialization of Harmony Memory (HM)

For solving the attack-aware RWA problem using HS (Harmony Search), *HMS* (*Harmony Memory Size*) random music harmonies (random harmony vectors -X^i, $1 \le i \le HMS$) are initially constructed and stored in the Harmony Memory (HM) matrix. Each harmony vector $X^i = (x_1^i, x_2^i, ..., x_S^i)$ in the harmony memory represents the order that the demands are considered. Several values for *HMS* are discussed in the simulation section. The size S of each vector X^i is equal to the number of (s,d) pairs ($S=N^2-N$), where N is the number of network nodes, and each element x_j^i of the vector X^i has an integer value between 1 and (N^2-N). The sequence of the demands represents a music harmony in HM terminology. In order to have valid sequences for each harmony, each harmony should contain all the integers from 1 to (N^2-N) and each integer should appear only once. Based on these constraints, all the requested connections are considered. A sequence is invalid if there are two or more integers with the same value.

4.2   Improvisation of a New Harmony

A new harmony is improvised (a New Harmony vector X^{new} is generated) based on the *HMCR* (*Harmony Memory Considering Rate*) and *PAR* (*Pitch Adjusting Rate*) described in Section 3. Each New Harmony vector $X^{new} = \left(x_1^{new}, x_2^{new}, ..., x_{N^2-N}^{new}\right)$ must comply with the constraints provided in Section 4.1. For this reason, when considering a new value for each variable x_j^{new} from the set of integer values from 1 to $(N^2\text{-}N)$, the values that have already assigned to variables $x_1^{new}, ..., x_{j-1}^{new}$ are removed from the available set. The *HMCR*, which ranges from 0 to 1, defines the probability to choose a variable value from HM and the *PAR* defines the probability of shifting to neighboring values within a range of possible values. If the New Harmony vector is better, in terms of the objective function cost $f(X^{new})$ as presented in Section 4.3 that follows, than the worst harmony vector in HM, then the New Harmony vector is included in HM while the worst is excluded.

4.3   Objective Function – Attack-Aware Heuristic Algorithm

This section describes the evaluation of the objective function $f(X)$ used by the harmony search algorithm. Each link l of the network is characterized by a Boolean wavelength availability vector $\overline{w_l} = [w_{li}] = (w_{l1}, w_{l2}, ..., w_{lW})$ whose i^{th} element w_{li} is equal to 0 if the ith wavelength of link l is utilized by a connection, and equal to 1, otherwise.

In this phase, k candidate paths P_{sd}, that have been pre-computed in the first step of the algorithm by employing a k-shortest path algorithm, are given as input in the *attack-aware heuristic* algorithm for serving each requested connection (s,d). The wavelength availability vector of a path p consisting of links $l \in p$ is defined as follows: $\overline{W_p} = [W_{pi}] = \&_{l\in p}\overline{W_l} = [\&_{l\in p}w_{li}]$, where "&" denotes the Boolean AND operation. Thus, the element W_{pi} is equal to 1 if wavelength i is available for transmission over path p. The above equation enforces the wavelength continuity constraint among the links comprising a path.

The connections are sequentially established one-by-one in the form of lightpaths. The demands are served according to the order defined by the Harmony Vector. For each demand, the candidate lightpath with the smallest number of in-band channel interactions is chosen. The objective of the attack-aware RWA heuristic algorithm is to minimize the number of lightpaths that interact with other lightpaths through in-band channel crosstalk and thus to minimize the propagation of high-power jamming signal attacks.

4.4   Flowchart of the Algorithm

The flowchart of the algorithm described in the previous subsections is given in Fig.2. The generation of a new vector X^{new} is achieved following the second step of the algorithm described in section 3, taking into account the constraints of section 4.2. Moreover, the evaluation of the objective function $f(X)$ is performed following the description of section 4.3.

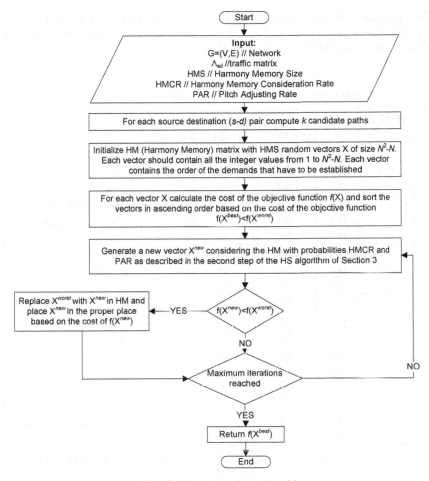

Fig. 2. Flowchart of the algorithm

5   Simulation Results

To evaluate the performance of the proposed algorithm, a number of simulation experiments were performed using the generic Deutsche Telekom backbone network (DTnet) as shown in Fig. 3. The network consists of 14 nodes and 23 links with an average nodal degree of 2.85. The traffic load of the network is defined as the ratio of the number of requested connections to the total number of possible connections. For example, traffic load equal to 0.5 corresponds to the case were half of the entries of the traffic matrix are equal to 1 and half equal to 0.

The aim of this section is to study the impact of high-power jamming signals through in-band crosstalk and the evolution of the algorithm solution under different settings of the three important parameters: the pitch adjusting rate (PAR), the harmony memory size (HMS), and the harmony memory considering rate (HMCR). The number of maximum iterations was considered equal to 100, that corresponds to the stopping criterion of the algorithm. First, a sensitivity analysis of the harmony search model parameters was performed. Table 1 shows the analysis results for several values of the parameters (HMS = {1, 5, 10}, HMCR = {0.5, 0.8, 0.9}, and PAR = {0.1, 0.4, 0.7}).

Table 1. Results of sensitivity analysis with HS parameters

HMS	HMCR	PAR	In-band W=30	In-band W=40
1	0.5	0.1	52	14
		0.4	54	16
		0.7	50	14
	0.8	0.1	50	18
		0.4	50	14
		0.7	54	**12**
	0.9	0.1	50	14
		0.4	56	18
		0.7	50	14
5	0.5	0.1	52	14
		0.4	54	18
		0.7	56	16
	0.8	**0.1**	**48**	**14**
		0.4	52	16
		0.7	56	16
	0.9	0.1	52	16
		0.4	52	14
		0.7	52	16
10	0.5	0.1	50	18
		0.4	56	16
		0.7	54	16
	0.8	0.1	54	16
		0.4	54	18
		0.7	52	20
	0.9	0.1	56	14
		0.4	50	18
		0.7	56	16

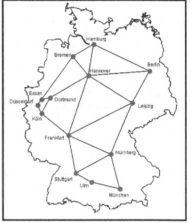

Fig. 3. DT network topology

The term "In-band" in Table 1, represents the total number of lightpaths' interactions through in-band crosstalk. The network load was equal to 0.6 and the analysis was performed for the cases of 30 and 40 available wavelengths per fiber. It is evident, that the larger the HMCR value, the less exploration is achieved; and the algorithm further relies on stored values in HM and this potentially leads to the algorithm reaching and remaining at a local optimum. On the other hand, choosing too small a value of HMCR will decrease the algorithm efficiency and the HS will behave like a pure random search, with less assistance from the historical memory.

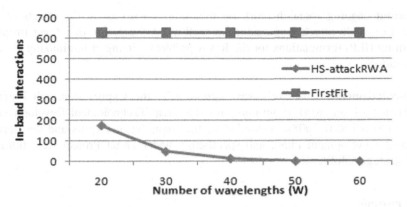

Fig. 4. In-band interactions vs number of available wavelengths

In Fig. 4, the HS attack-aware RWA algorithm was also compared to the *first-fit* algorithm, an algorithm that assigns the first available wavelength to the connections. The performance metric in this case was the number of in-band interactions versus the available number of wavelengths per fiber. For these simulations the network load was again equal to 0.6 and the three parameters used for the HS meta-heuristic, i.e., the pitch adjusting rate (PAR), the harmony memory size (HMS), and the harmony memory considering rate (HMCR) were set to the values 0.1, 5, and 0.8 respectively. These values were obtained from the sensitivity analysis as shown in Table 1 above. As can be seen from the figure, the first-fit algorithm has the same performance irrespective of the number of available wavelengths and the number of interactions is quite high (more than 600 interactions) compared to the case were the proposed HS attack-aware RWA algorithm was used. For the latter case it is shown that the proposed technique clearly outperforms the traditional RWA approach, minimizing the in-band channel interactions based on the available wavelengths. Thus, with the proposed approach the attack is contained, and it is not allowed to propagate extensively in the network, resulting in degradation of the established network connections.

6   Conclusions

This work proposed an algorithm for solving the RWA problem during the design phase of a transparent WDM optical network with the objective of minimizing the high-power in-band crosstalk propagation which is caused when a high-power jamming signal is maliciously introduced in the network at a specific network node. A meta-heuristic approach based on the harmony search technique is utilized to obtain an ordering of the connection requests and this ordering is used during the RWA algorithm to establish all requested connections in the network. Performance results indicate that the proposed solution outperforms the traditional RWA technique that does not account for the propagation of the jamming signal due to intra-channel crosstalk, significantly minimizing this propagation in the network and thus drastically containing the effect of the attack on the network infrastructure.

Current ongoing research work on this subject focuses on the effects of inter-channel crosstalk attack propagation, as well as on the design of integer linear programming (ILP) formulations for the RWA problem, aiming at minimizing the attack propagation.

Acknowledgment. This work was supported by the Cyprus Research Promotion Foundation's Framework Programme for Research, Technological Development and Innovation (DESMI 2008), co-funded by the Republic of Cyprus and the European Regional Development Fund, and specifically under Grant Project New Infrastructure/ Strategic/0308/26).

References

1. Ramaswami, R., Sivarajan, K.: Routing and wavelength assignment in all-optical networks. IEEE/ACM Transactions on Networking 3(5), 489–500 (1995)
2. Haijin, Z., Naifu, Z.: Ant colony optimization for dynamic RWA in WDM networks with partial wavelength conversion. J. Photonic Network Communications 11(2), 229–236 (2006)
3. Monoyios, D., Vlachos, K.: Multiobjective genetic algorithms for solving the impairment-aware routing and wavelength assignment problem. IEEE/OSA Journal of Optical Communications and Networking 3(1), 40–47 (2010)
4. Martins-Filho, J., Chaves, D., Bastos-Filho, C., Aguiar, D.: Intelligent and fast IRWA algorithm based on power series and Particle Swarm Optimization. In: Proc. of International Conference on Transparent Optical Networks, Athens, Greece (2008)
5. Geem, Z.W., Kim, J.H., Loganathan, G.V.: A new heuristic optimization algorithm: Harmony search. Simulation 76(2), 60–68 (2001)
6. Médard, M., Marquis, D., Barry, R., Finn, S.: Security issues in all-optical networks. IEEE Network 11(3), 42–48 (1997)
7. Mas, C., Tomkos, I., Tonguz, O.: Optical networks security: A failure management framework. In: Proc. of SPIE ITCom, Orlando, FL, USA, pp. 230–241 (2003)
8. Skorin-Kapov, N., Chen, J., Wosinska, L.: A new approach to optical networks security: Attack-aware routing and wavelength assignment. IEEE/ACM Transactions on Networking 18(3), 750–760 (2010)
9. Furdek, M., Skorin-Kapov, N., Grbac, M.: Attack-aware wavelength assignment for localization of in-band crosstalk attack propagation. IEEE/OSA Journal of Optical Communications and Networking 2(11), 1000–1009 (2010)
10. Skorin-Kapov, N., Furdek, M., Pardo, R., Pavón Mariño, P.: Wavelength assignment for reducing in-band crosstalk attack propagation in optical networks: ILP formulations and heuristic algorithms. European Journal of Operational Research 222(3), 418–429 (2012)

Exploring Artificial Intelligence Utilizing BioArt

Panagiota Simou[1], Konstantinos Tiligadis[1], and Athanasios Alexiou[2]

[1] Department of Audio and Visual Arts,
Ionian University, Plateia Tsirigoti 7, 49100 Corfu, Greece
{simou,gustil}@ionio.gr
[2] Department of Informatics
Ionian University, Plateia Tsirigoti 7, 49100 Corfu, Greece
alexiou@ionio.gr

Abstract. While artificial intelligence combined with Bioinformatics and Nano-technology offers a variety of improvements and a technological and healthcare revolution, Bioartists attempt to replace the traditional artistic medium with biological materials, bio-imaging techniques, bioreactors and several times to treat their own body as an alive canvas. BioArt seems to play the role of a new scientific curator in order to manipulate laboratory aesthetics and bridge trans-humanism creations with culture and tradition.

Keywords: Artificial Intelligence, BioArt, Bioethics, Post humanism.

1 Introduction

"…υπό δ΄ αμφίπολοι ρώοντο άνακτι χρύσειαι, ζωῆσι νεήνισσιν εϊκυίαι της εν μεν νοός εστί μετά φρεσίν, εν δε και αυδή και σθένος…".

As indicated in the Greek mythology, Talos is said to be the first Artificial Intelligence Robot that Hephaestus created in his ancient lab, as well as golden maids who could move, speak and think like human beings and tripod tables for the automatic transfer of food and drink to gods. It seems that the necessity of creating metallic assistants-robots constitutes a spontaneous attribute of human culture and temperament with roots reaching back to the dawn of our civilization.

According to Humanism, human beings have the right and responsibility to give meaning and shape to their own lives, building a more humane society through an ethic based on human and other natural values in the spirit of reason and free inquiry through human capabilities. The moral person guided from his evolving social behavior, can easily comprehend and be committed to laws and principles that a scientific field, such as Artificial Intelligence (AI) or Nanoscience set as a precondition, in order to improve the structural elements of human biological existence [1-2].

The fact of the upcoming era of Post humanism, hypothetical raises various questions and bioethical issues, such as the degree of influence of human consciousness, dignity, rights and fundamental freedoms by merging human beings and machines. Additionally, the Universal Declaration of Human Rights on genetic data defines personal identity as a combination of distinctive genetic makeup with educational, environmental, personal, emotional, social, spiritual and cultural factors.

H. Papadopoulos et al. (Eds.): AIAI 2013, IFIP AICT 412, pp. 687–692, 2013.

Modern sciences that complement Medicine and Biology, such as Nanotechnology, Bioinformatics, stem cell technology, transgenic organisms, neural prosthetics and AI applications, have enhanced human intelligence, offering new possibilities for the human body, documenting the potential immortality of our biological existence. Thus a series of philosophical questions and ethical considerations rises through these innovative applications of human intellect. Bioethics, as a field of philosophical and critical approach, have an essential duty to contribute to the regulatory investigation of ethical issues rising from biomedical applications, but also to clarify the relationship between the life sciences and art, as is now expressed through biotechnology.

Artists involved in biological sciences, have now adopted a very unique way of expression, while laboratory components have already become an integral part of the artistic process or art masterpieces, a few times. It is a fact, the presence of artists in modern biological laboratories, in order to overcome the well-known effects and decorative of Fine Arts and manipulate life. BioArt is not a simple theme or artistic movement, but a complex tool for creating projects such as: use and transplant of mechanical devices in humans, creation of robot-clones etc.

2   Super Intelligence in the Post Mondo

By definition super intelligence is any intellectual component that generates the best human brains, including scientific creativity, general wisdom, and social skills [3]. Obviously super intelligence can produce new knowledge and solutions to hard i.e. a smart Nano machine can recognize cancer cells using the tunneling phenomenon, or a machine learning algorithm could give right decisions in the forecast of neurogenerative diseases like epilepsy [1].

On the other hand the parallel innovating structure of the so called 'convergent technologies', referring to the NBIC tools and including Nano science and nanotechnology, biotechnology, biomedicine and genetic engineering, information technology and cognitive science, seems to remove any barrier in scientific and technological achievement [4]. The Nano devices which can repair cells, promise great improvements in longevity and quality of life, involving radical modifications of the human genome and leading to the old but diachronic issue of human immortality [5].

Unlike mankind, AI applications can use and manipulate the storage knowledge through scientific or social networks in a more efficient way. Therefore one of the best ways to ensure that these super intelligent creations will have a beneficial impact on the world is to endow it with philanthropic values and friendliness [6]. Additionally, super intelligence could give us indefinite lifespan, either by stopping and reversing the aging process through the use of Nano medicine [7].

Often Art is prior to Science and has the ability to envision and create projects which is not yet feasible to be implemented, or be applied in human societies. For all innovative artists, art often works in a Post Mondo ("Perfect Future"), which constitutes the past, from a given starting point onwards.

In other cases Science comes to inspire Art, give stimuli for new ideas and techniques and give new directions in the use of innovative tools and materials. While

Science and Art are the most common human types of knowledge and emotional expressions, they have to be perpetually in a dynamic interdependence, interactive communication and exchange principles in order to be more understandable and representative of human existence and dimension. It is imperative to answer a series of questions before establishing the hypothetical era of Post Mondo and Super Intelligence:

- How the world would like if the scientific achievements were accessible to all people (cloning, production of human organs, personalized gene therapy etc.)?
- If for any illness or any form of disability there was a proper treatment, what would be the structure of any future Perfect World?
- Human societies will tolerate smart, intelligent and strong members and on what life expectancy?
- In human societies of perfect and potentially immortal beings is it possible to adjust the principles of economics and politics and the traditional values of relationships?
- While the achievement of human perfection, the progress of genetics, the use of mechanical additives and the creation of transgenic animals improve cell physiology, could this lead to people who will not obey to laws as we know them now?

In order to develop the Perfect-Human, all the human structural elements have to be immortal, overcoming the Central Dogma of Genetics and upsetting the fundamental principles of human evolution. If we also take into consideration the studies of modern genetics on the non-mental and spiritual identification of clones with their original organism (i.e. the case of twins) another important question raises:

- Are there any psycho-spiritual disruptions in cloned organisms or in organisms with prosthetic biological implants?
- Have we already excluded the possibility of cultured tissues reaction (i.e. BioArt Semi-Living) with the environment and human actions [8]?

The study on perfection's definition and principles consist of a major philosophical enquiry for thousands of years. Is there perfection in Nature and Evolutionary Laws? Perfect creations follow the natural law of evolution or they are kept unaltered in order to remain perfect? Perfection's evolution does ultimately equate with the supernatural?

3   Seeking for a New Role - Ethical Challenges

The development of nanotechnology is moving rapidly, and without any clear public guidance or leadership as to the moral tenor of its purposes, directions and outcomes; where nanotechnology is leading and what impact it might have on humanity is anyone's guess [9]. What appears to be missing at the present time is a clearly articulated prognosis of the potential global social benefits and harms that may develop from further scientific and technological advances in all of these areas [10].

Super Intelligence should be comprehensible to the public, including activities that benefit society and environment, guided by the principles of free participation to all decision-making processes; AI techniques in Biomedicine, should respect the right of access to information, having the best scientific standards and encouraging creativity, flexibility and innovation with accountability to all the possible social, environmental and human health impacts.

BioArt even today, represents all those forms of art, associated in any way with the life sciences, and uses the life and living things both as a medium and as an objects' expression. In the transhuman's era, BioArt requires close cooperation between artists and scientists, and also a thorough understanding of the research process, from the artists mainly. Obviously, laboratory techniques associated with biotechnology, such as stem cell research and its applications have already been implemented by specialized scientists around the world. Bio Artists have to communicate and share this scientific knowledge with the society, in the way that they perceive and accept it as part of human evolution. Sometimes hybrid art offers to the audient macabre and repulsive artistic creations, using human body as the basic material like Stelarc, Orlan and Franko-B [11-13], transgenic art creations like Kac [14], Cybernetics technology like Haraway and Warwick [15-16], hyper-real sculptures like Piccinini [17].

BioArt searches for the meaning of human existence in relation to the natural environment and enhances the value of parts of the human body as autonomous and interdependent subsets. Judging from the already registered events and the historical development of Arts today, there is a strong possibility that BioArt outline a total new future serving as a communication channel between science and society, while humans are still trying to conquer the torrent of scientific knowledge.

According to philosopher Nietzsche, the distance that separates Perfect-Man from man is identical to that which separates man from monkey. The transition in a supernatural state involves an evolutionary leap and a profound change in human nature. BioArt is willing to lead this evolutionary perspective, through the expressions of Nanotechnology, Neuroengineering, Bioinformatics, Molecular Biology and Cellular Therapy, but also to establish new rules through the methodological approaches of bioethics.

BioArt must highlight and promote the uniqueness of the individual, to strengthen the role of man in the AI and probably to propose solutions for the improvement and protection of life according to human ideals. In order to achieve this, a dialogue about what is ethical and legal to be expressed through BioArt must start immediately, such as the use of human organs, tissues, blood, bacteria and viruses or even more the intentional infliction of pain.

- Can we assume that BioArt main objectives are the selection of a new way of achieving Post humanity and biological immortality, by not following the common paths of aesthetic expression and traditional -for human senses- beauty?
- Harmony and aesthetic which are features of our nature, is it possible to remain stationary into the acceptable social contexts or will adapt new data and scientific challenges without the discrete and finite limits of human consciousness?
- Could harmony and aesthetics reflect human vision and its finite capabilities (human perception of three-dimensions) or are there independent natural rules which are not influenced by our evolution?

It is clearly that BioArt can apply innovative techniques for example on exploiting or creating new sensations in cases of disabled people, in the technological application of the phenomenon of quantum entanglement, in the education process and in support the AI products integration into society (e.g. Robot-clones). BioArt seems to play the role of scientific curator presenting the products of AI in an acceptable way for the human brain, providing also a set of principles and rules such as: aesthetics, behavioural properties, boundaries and functions, adaptation and harmonization in customs. Additionally, BioArt should examine the developmental stages of human cognitive and determine ethical and adaptation rules for artificial intelligence's products in human evolution.

Appears however that this new Art, exists only in technologically advanced societies where science works as a tool, raising ethical questions about equality and accessibility on super intelligence among people and nations.

4   Conclusion

For over 100 years, the scientific activity experiencing such growth, that seems to replace the entire culture. Initially this is an illusion caused by the speed of this development and this qualification triumph that characterizes Science, gave the right to dominate the entire culture. Few researchers also, fearing the domination of society by the science, predict the destruction of culture [18].

Science, however changes the 'DNA of our thinking', expression, perception and our aesthetics. Additionally through Art's manipulation, we can compete and imitate nature using AI applications, only if we manage to discover and model mechanisms and structural elements.

According to Zarr and Catts [8] it is important to mention that BioArt is a pluralist practice with its artists occupying different ethical positions such as the creation of public acceptance for biotech developments or the generation of heated public debate about their uses.

Therefore it is obviously that many ethical and legal challenges seek for answers concerning BioArt and hybrid art in general and their different approaches from artists, art theorists, curators, ethicists and philosophers, scientists and engineers.

References

1. Alexiou, A., Psixa, M., Vlamos, P.: Ethical Issues of Artificial Biomedical Applications. In: Iliadis, L., Maglogiannis, I., Papadopoulos, H. (eds.) EANN/AIAI 2011, Part II. IFIP AICT, vol. 364, pp. 297–302. Springer, Heidelberg (2011)
2. Alexiou, A., Vlamos, P.: Ethics at the Crossroads of Bioinformatics and Nanotechnology. In: 7th International Conference of Computer Ethics and Philosophical Enquiry (2009)
3. Bostrom, N.: How Long Before Super intelligence? International Journal of Futures Studies 2 (1998)
4. Roco, M.C., Bainbridge, W.S.: Converging technologies for improving human performance. Journal of Nanoparticle Research 4, 281–295 (2002)

5. Drexler, E.: Engines of Creation. Bantam, New York (1986)
6. Kurzweil, R.: The Age of Spiritual Machines: When Computers Exceed Human Intelligence. Viking, New York (1999)
7. Moravec, H.: Robot: Mere Machine to Transcendent Mind. Oxford University Press, New York (1999)
8. Zurr, I., Catts, O.: The ethical claims of Bioart: Killing the Other or Self Cannibalism. Art and Ethics 4(2) (2003)
9. Berne, R.W.: Towards the conscientious development of ethical nanotechnology. Science and Engineering Ethics 10, 627–638 (2004)
10. Sweeney, E.A.: Social and Ethical Dimensions of Nanoscale Research. Science and Engineering Ethics 12(3), 435–464 (2006)
11. Stelarc website, http://stelarc.org
12. Orlan website, http://www.orlan.eu
13. Franko-B website, http://www.franko-b.com
14. Kac, E.: Transgenic Art. Leonardo Electronic Almanac 6(11) (1998)
15. Haraway, D.: A cyborg manifesto. Science, Technology and socialist-feminism in the late twentieth century. In: Simians, Cyborgs and Women: The reinvention of nature, pp. 149–181. Routledge, New York (1991)
16. Warwick, K.: Future issues with robots and cyborgs. Studies in Ethics, Law, and Technology 4(3), 1–18 (2010), doi:10.2202/1941-6008.1127
17. Piccinini website, http://www.patriciapiccinini.net
18. Prigogine, I., Stengers, I.: Order out of Chaos. Bantam Books, University of Michigan (1984)

Can Machines Make Ethical Decisions?

Iordanis Kavathatzopoulos and Ryoko Asai

Department of IT – VII, Uppsala University, Sweden
{iordanis,ryoko.asai}@it.uu.se

Abstract. Independent systems and robots can be of great help to achieve goals and obtain optimal solutions to problems caused by the quantity, variation and complexity of information. However, we always face ethical issues related to the design as well as to the running of such systems. There are many problems, theoretical and practical, in integrating ethical decision making to robots. It is impossible to design or run such systems independently of human wish or will. Even if we create totally independent decision making systems, we would not want to lose control. Can we create really independent ethical decision systems? Recent research showed that emotions are necessary in the process of decision making. It seems that it is necessary for an independent decision system to have "emotions." In other words, a kind of ultimate purpose is needed that can lead the decision process. This could make a system really independent and by that ethical.

Keywords: robots, systems, autonomous, independent, decision making, ethics, moral.

1   Introduction

The development of Information Technology, systems, robots, etc., that are capable of processing information and acting independently of their human operators, has been accelerated as well as the hopes, and the fears, of the impact of those artifacts on environment, market, society, on human life generally. Many ethical issues are raised because of these systems being today, or in the future, capable of independent decision making and acting. Will these IT systems or robots decide and act in the right way or will they cause harm?

In situations where humans have difficulties perceiving and processing information, or making decisions and implementing actions, because of the quantity, variation and complexity of information, IT systems can be of great help to achieve goals and obtain optimal solutions to problems. One example of this is financial transactions where the speed and volume of information makes it impossible for human decision makers to take the right measures, for example in the case of a crisis. Another example is dangerous and risky situations, like natural disasters or battles in war, where the use of drones and military robots may help to avoid soldier injuries and deaths. A third example comes from human social and emotional needs, for example in elderly care where robots may play an important role providing necessary care as well as to be a companion to lonely elderly people.

H. Papadopoulos et al. (Eds.): AIAI 2013, IFIP AICT 412, pp. 693–699, 2013.

It is clear that such IT systems have to make decisions and act to achieve the goals for which they had been built in the first place. Will they make the right decisions and act in a proper way? Can we guarantee this by designing them in a suitable way? But if it is possible, do we really want such machines given the fact that their main advantage is their increasing independence and autonomy, and hence we do not want to constrain them too much?

There are many questions around this, most of which converge on the issue of moral or ethical decision making. The definition of what we mean by ethical or moral decision making or ethical/moral agency is a very much significant precondition for the design of proper IT decision systems. Given that we have a clear definition we will be able to judge whether an IT system is, 1) capable of making ethical decisions, and 2) able to make these decisions independently and autonomously.

2   Focus on the Process of Ethical Decision Making

Ethics and morals have originally the same meaning in Greek and Latin. However, today, in philosophy as well as in psychology we usually give them different meanings. "Ethics" is often used in connection to meta-philosophy or to psychological processes of ethical decision making, whereas the term "moral" is adopted when we talk about normative aspects or about the content of a decision. The distinction between content and process is important in the effort to define ethical or moral decision making.

In common sense, ethics and morals are dependent on the concrete decision or the action itself. Understanding a decision or an action being ethical/moral or unethical/immoral is based mainly on a judgment of its normative qualities. The focus on values and their normative aspects is the basis of the common sense definition of ethics.

Despite its dominance, this way of thinking causes some difficulties. We may note that bad or good things follow not only from the decisions of people but also from natural phenomena. Usually sunny weather is considered a good thing, while rainy weather is not. Of course this is not perceived as something related to morality. But why not? What is the difference between humans and nature acting in certain ways? The answer is obvious: Option, choice.

Although common sense does realize that, people's attachment to the normative aspects is so strong that it is not possible for them to accept that ethics is an issue of choice and option. If there is no choice, or ability of making a choice, then there is no issue of ethics. However this does not solve our problem of the definition of Autonomous Ethical Agents, since IT systems are actually making choices.

Now if ethics are connected to choice then the interesting aspect is how the choice is made, or not made; whether it is made in a bad or in a good way. The focus here is on how, not on what; on the process not on the content or the answer. Indeed, regarding the effort to make the right decision, philosophy and psychology point to the significance of focusing on the process of ethical decision making rather on the normative content of the decision.

Starting from one of the most important contributions, the Socratic dialog, we see that *aporia* is the goal rather than the achievement of a solution to the problem investigated. Reaching a state of no knowledge, that is, throwing aside false ideas, opens up for the right solution. The issue here for the philosopher is not to provide a ready answer but to help the other person in the dialog to think in the right way [1, 2]. Ability to think in the right way is not easy and apparently has been supposed to be the privilege of the few able ones [3]. For that, certain skills are necessary, such as Aristoteles's *phronesis* [4]. When humans are free from false illusions and have the necessary skills they can use the right method to find the right solution to their moral problems [5].

3   Skills for Ethical Decision Making

This philosophical position has been applied in psychological research on ethical decision making. Focusing on the process of ethical decision making psychological research has shown that people use different ways to handle moral problems. According to Piaget [6] and Kohlberg [7], when people are confronted with moral problems they think in a way which can be described as a position on the heteronomy-autonomy dimension. Heteronomous thinking is automatic, purely emotional and uncontrolled thinking or simple reflexes that are fixed dogmatically on general moral principles. Thoughts and beliefs coming to mind are never doubted. There is no effort to create a holistic picture of all relevant and conflicting values in the moral problem they are confronted with. Awareness of own personal responsibility for the way one is thinking or for the consequences of the decision are missing.

Autonomous thinking, on the other hand, focuses on the actual moral problem situation, and its main effort is to search for all relevant aspects of the problem. When one is thinking autonomously the focus is on the consideration and investigation of all stakeholders' moral feelings, duties and interests, as well as all possible alternative ways of action. In that sense autonomy is a systematic, holistic and self-critical way of handling a moral problem.

Handling moral problems autonomously means that a decision maker is unconstrained by fixations, authorities, uncontrolled or automatic thoughts and reactions. It is the ability to start the thought process of considering and analyzing critically and systematically all relevant values in a moral problem situation. This may sound trivial, since everybody would agree that it is exactly what one is expected to do in confronting a moral problem. But it is not so easy to use the autonomous skill in real situations. Psychological research has shown that plenty of time and certain conditions are demanded before people can acquire and use the ethical ability of autonomy [8].

Nevertheless, there are people who have learnt to use autonomy more often, usually people at higher organizational levels or people with higher responsibility. Training and special tools do also support the acquisition of the skill of autonomy. Research has shown that it is possible to promote autonomy. It is possible through training to acquire and use the skill of ethical autonomy, longitudinally and in real life [9].

4   Tools for Ethical Decision Making

IT systems have many advantages that can be used to stimulate autonomy during a process of ethical decision making. For example EthXpert and ColLab [10, 11, 12] are intended to support the process of structuring and assembling information about situations with possible moral implications. Analogous with the deliberation of philosophers throughout history as well as with the findings of psychological research on ethical decision making, we follow the hypothesis that moral problems are best understood through the identification of authentic interests, needs and values of the stakeholders in the situation at hand.

Since the definition of what constitutes an ethical decision cannot be assumed to be at a fix point, we further conclude that this kind of system must be designed so that it does not make any assertions of the normative correctness in any decisions or statements. Consequently, the system does not make decisions and its sole purpose is to support the decision maker (a person, a group or the whole organization) when analyzing, structuring and reviewing choice situations.

In the system, interests of each imaginable stakeholder are identified in a systematic procedure over six steps. 1) Define stakeholders: The system's focus on interests leads to an associative process of identifying related stakeholders. For each stakeholder that is directly involved in the situation there may be third party stakeholders that could influence it. The simple question of who is affected by a specific interest of a stakeholder will help the user to become aware of these. In EthXpert and ColLab the addition of stakeholders is very straightforward and therefore does not provide any obstacle to widening the scope of the problem. 2) Define for each stakeholder its interests: The user determines a set of relevant interests, specifically for each stakeholder. All interests that might relate and affect other stakeholders are important to consider and in the process of scrutinizing interests additional stakeholders will naturally become involved in the analysis. 3) Define how interests relate to other stakeholders: Determining how the interests and values of the stakeholders relate to other stakeholders draws a picture of the dynamics and dependencies in the situation. The considerations that are brought up when an interest is facing another stakeholder may therefore reveal important conflicts. Further, as described above, this approach may help to track down previously unidentified stakeholders, since the topics that are brought up in one relation are not necessarily unique to that and therefore will raise the inclusion of other stakeholders. 4) Define main options: The most apparent alternatives for handling the moral problem can be immediately stated. Usually main alternatives are to their character mutually excluding in some aspect, similar to answering a question with "Yes" or "No". There is no obligation to apply such a polarization, but to make full use of the later stage of formulating compromise options it can be useful to consider whether such patterns exist. 5) Translate considerations: For each optional strategy the user is urged to state how the interests of the stakeholders are affected by the option if that option would be the final decision. The considerations from the interest-stakeholder matrix will not be automatically copied to the decision matrix. Instead the interest-stakeholder relationships will serve as background and incentive for considering how the different decision alternatives affect the stakeholders. 6) Define compromise options: To counter problems in the main options, i.e. unacceptable negative effects, compromise solution candidates can be forked from main alternatives. A compromise option will inherit considerations from its parent, but the user should revise these and determine the difference in effect between them. The feature is useful for considering many options that only differ partly. The intention is to allow any user to easily get an overview of the strengths and weaknesses of similar alternatives.

5   Non-independent Ethical Agents

Ethical decision support programs, like EthXpert and ColLab, can be integrated into robots and other decision making systems to secure that decisions are made according to the basic theories of philosophy and to the findings of psychological research. This would be the ideal. But before we are there we can see that ethical decision making support systems based on this approach can be used in two different ways.

During the development of a non-independent decision making system, support tools can be used to identify the criteria for making decisions and for choosing a certain direction of action. This means that the support tool is used by the developers, they who make the real decisions, and who make them according to the previous named philosophical/psychological approach [13].

Another possibility is to integrate a support tool, like EthXpert and ColLab, into the non-independent decision system. Of course, designers can give to the system criteria and directions, but they can also add the support tool itself, to be used in the case of unanticipated future situations. The tool can then gather information, treat it, structure it and present it to the operators of the decision system in a way which follows the requirements of the above mentioned theories of autonomy. If it works like that, operators of non-independent systems make the real decisions and they are the users of the ethical support tool.

A non-independent system that can make decisions and act in accordance to the hypothesis of ethical autonomy is a system which 1) has the criteria already programmed in it identified through an autonomous way in an earlier phase by the designers, or 2) prepares the information of a problem situation according to the theory of ethical autonomy, presents it and stimulates the operators to make the decision in a way compatible with the theory of ethical autonomy.

All this can work and it is possible technically. But how could we design and run a really independent ethical decision making system? However, before we can speculate on that it is important to address some issues shortly, regarding the criteria for independence.

6   Independent Ethical Agents

One is the issue of normative quality of the decisions made. Can we use this criterion for the definition of an independent ethical decision system? As we have already discussed this is not possible although it is inherently and strongly connected to common sense, and sometimes into research [14]. Normative aspects can be found in the consequences of obviously non-independent natural phenomena. Besides, there are always good arguments supporting opposite normative positions. So this cannot be a working criterion [15].

The alternative would be the capability of choice. Connected to this is the issue of free will. We could say that really independent systems are those that are free to decide whatever they want. However, this has many difficulties. There is theoretical obscurity around the definition of free will as well as practical problems concerning its description in real life situations. Furthermore, it is obvious that many systems can make "choices." Everything from simple relays to complex IT systems is able to choose among different alternatives, often in arcane and obscure ways, reminiscent of the way humans make choices. Then the problem would be where to put the threshold for real choice making.

If the ability to make choices cannot be the criterion to determine the independence of a decision system, then the possibility to control the system by an operator becomes interesting. Wish or effort to control, external to the system, may be something that has to be involved and considered. The reason of the creation of IT systems is the designers' and the operators' wish to give them a role to play. These systems come to existence and are run as an act of will to control things, to satisfy needs. It is an execution of power by the designers and the operators. We can imagine a decision system as totally independent, but even this cannot be thought without a human wish or will behind it. It is always a will for some purpose. It can be a simple purpose, for example to rescue trapped people in collapsed buildings, or an extremely complex purpose, like to create systems able of making independent decisions! In any case the human designer or operator wants to secure the fulfillment of the main purpose and does not want to lose control.

So the issue could be about possession of an original purpose, a basic feeling, an emotion. Indeed recent research in neurobiology and neuropsychology shows that emotions are necessary in the decision making process [16]. It seems that a rational decision process requires uninterrupted connection to emotions. Without this bond the decision process becomes meaningless. Another effect of the "primacy" of emotions and purposes is that very often heteronomous or non-rational ways to make ethical decisions are adopted, despite the human decision maker being able to think autonomously and rationally.

Thus the criterion for a really independent decision system could be the existence of an emotional base that guides the decision process. Human emotions and goals have been evolved by nature seemingly without any purpose. That may happen in decision systems and robots if they are left alone, but designers, operators, and humans would probably not want to lose control. So what is left? Can we create really independent ethical decision systems?

7   Non-independent Ethical Agents

The criterion of such a system cannot be based on normative aspects, or on the ability to make choices, or on having own control, or on ability of rational processing. It seems that it is necessary for an independent decision system to have "emotions" too. That is, a kind of ultimate purposes that can lead the decision process, and depending on the circumstances, even make the system react automatically, or alternatively, in a rational way.

Well, this is not easy to achieve. It may be impossible. However, if we accept this way of thinking we may be able to recognize a really independent or autonomous ethical agent, if we see one, although we may be not able to create one. This could work like a Turing test for robot ethics because we would know what to look for: A decision system capable of autonomous ethical thinking but leaning most of the time toward more or less heteronomous ways of thinking; like humans who have emotions leading them to make decisions in that way.

References

1. Πλάτων [Platon]: Θεαίτητος [Theaitetos]. Ι.Ζαχαρόπουλος [I. Zacharopoulos], Αθήνα [Athens] (1981)
2. Πλάτων [Platon]: Απολογία Σωκράτους [Apology of Socrates]. Κάκτος [Kaktos], Αθήνα [Athens] (1992)
3. Πλάτων [Platon]: Πολιτεία [The Republic]. Κάκτος [Kaktos], Αθήνα [Athens] (1992)
4. Αριστοτέλης [Aristoteles]: Ηθικά Νικομάχεια [Nicomachean Ethics]. (Πάπυρος [Papyros], Αθήνα [Athens]) (1975)
5. Kant, I.: Grundläggning av Sedernas Metafysik [Groundwork of the Metaphysic of Morals]. Daidalos, Stockholm (1785/2006)
6. Piaget, J.: The Moral Judgement of the Child. Routledge and Kegan Paul, London (1932)
7. Kohlberg, L.: The Just Community: Approach to Moral Education in Theory and Practice. In: Berkowitz, M., Oser, F. (eds.) Moral Education: Theory and Application, pp. 27–87. Lawrence Erlbaum Associates, Hillsdale (1985)
8. Sunstein, C.R.: Moral Heuristics. Behavioral and Brain Sciences 28, 531–573 (2005)
9. Kavathatzopoulos, I.: Assessing and Acquiring Ethical Leadership Competence. In: Prastacos, G.P., et al. (eds.) Leadership through the Classics, pp. 389–400. Springer, Heidelberg (2012)
10. Kavathatzopoulos, I., Laaksoharju, M.: Computer Aided Ethical Systems Design. In: Arias-Oliva, M., et al. (eds.) The "Backwards, Forwards, and Sideways" Changes of ICT, pp. 332–340. Universitat Rovira i Virgili, Tarragona (2010)
11. Laaksoharju, M.: Let us be Philosophers! Computerized Support for Ethical Decision Making. Uppsala University, Department of Information Technology, Uppsala (2010)
12. Laaksoharju, M., Kavathatzopoulos, I.: Computerized Support for Ethical Analysis. In: Botti, M., et al. (eds.) Proceedings of CEPE 2009 – Eighth International Computer Ethics and Philosophical Enquiry Conference. Ionian University, Kerkyra (2009)
13. Kavathatzopoulos, I.: Philosophizing as a usability method. CEPE 2013, Ambiguous Technologies: Philosophical Issues, Practical Solutions, Human Nature. Universidade Autónoma de Lisboa, Lisbon (in press, 2013)
14. Kohlberg, L.: The Philosophy of Moral Development: Moral Stages and the Idea of Justice. Harper and Row, San Francisco (1984)
15. Wallace, W., Allen, C.: Moral Machines: Teaching Robots Right from Wrong. Oxford University Press, New York (2009)
16. Koenigs, M., Tranel, D.: Irrational Economic Decision-Making after Ventromedial Prefrontal Damage: Evidence from the Ultimatum Game. The Journal of Neuroscience 27, 951–956 (2007)

Ethical Issues in Neuroinformatics

Athanasios Alexiou, Georgia Theocharopoulou, and Panayiotis Vlamos

Department of Informatics, Ionian University, Plateia Tsirigoti 7, 49100 Corfu, Greece
{alexiou,zeta.theo,vlamos}@ionio.gr

Abstract. Scientific progress in Artificial Intelligence contributed to generate concepts of information processing and representation that provide a general sense of a unified theory of how the brain works. Comprehending the neurobiological basis of consciousness and memory mechanisms in the brain using information technology techniques involves mainly brain data analysis. Nowadays, there is an increasingly progress in the wide area of information technology applications to our understanding of neurological and psychological disorders. Neuroinformatics can enhance the development of systems that mimic brain storage and retrieval of information. In this paper we take into consideration the ethical implications that arise in the application of information-based methods in neuroscience research processes.

Keywords: Neuroinformatics, Artificial Intelligence, Bioethics.

1   Introduction

The exploration of brain functions from a variety of fields has led to scientific developments in Artificial Intelligence (AI), like computational methods and systems for learning and knowledge discovery. Neuroinformatics is a research field concerned with the study of neuroscience with the use of information technology. Neuroinformatics organizes and integrates neuroscience data and applies computational tools and mathematical models in order to understand human brain physiology and functionality (Fig 1). According to the International Neuroinformatics Coordinating Facility (INCF) there are three main fields where Neuroinformatics have been applied [1]:

- Development of tools and databases for management and sharing of neuroscience data at all levels of analysis
- Development of tools for analyzing and modelling neuroscience data
- Development of computational models of the nervous system and neural processes.

These areas of research are important for the integration and analysis of increasingly large-volume, high-dimensional, and fine-grain experimental data. Neuroinformaticians provide and create interoperable databases for clinical and research scientists. Therefore, it is necessary to be able to share data and findings. The field of Neuroinformatics came into existence when the Human Brain Project began in the early 1990's. The primary goal of the Human Brain Project and Neuroinformatics is the

H. Papadopoulos et al. (Eds.): AIAI 2013, IFIP AICT 412, pp. 700–705, 2013.
© IFIP International Federation for Information Processing 2013

development of new technologies for creating databases and database search tools, for information exchange of neuroscience data, in order scientists and researchers to build data models of the brain and also create simulations of brain functions [2].

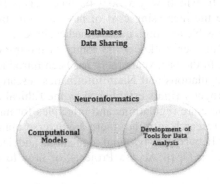

Fig. 1. Neuroinformatics Research Fields

Neuroinformatics has a variety of applications and uses in both the field of neuroscience and also information technology. Like other new technologies, there are legal and ethical issues that need to be accounted for in order to perform Neuroinformatics' research. There are several opinions on data sharing, and there are legal and ethical issues that need to be accounted for regarding four major key issues:

- Gathering data in order to create Neuroscience Databases
- Data sharing used in developing tools for analyses, simulation and modelling neuronal functions
- Data access in order to protect the subjects included in the data
- Legal and ethical issues that need to be accounted for performing research and clinical trials on patients.

Obviously, we have to distinguish the ethics of the productive technologies as a human's risk, from the ethics concerning the scientific and the AI's progress. The use of AI techniques in Neuroinformatics is totally separated from religious sentiments, human dignity or essential emotional states [3]. In this article we outline future directions for research regarding ethic challenges in the above mentioned issues.

2   Legal and Ethical Issues in Neuroinformatics

Data sharing projects include databases designed to archive data sets (e.g. neuroimaging results), and tools for neuroscience data. One major effort supporting integrative neuroscience research is the Neuroscience Information Framework (NIF). Few important aspects to the legal issues associated with Neuroinformatics regarding data and tool sharing are: protection for the creator of the data or tool, protection for the user and protection for the subjects included in the data [4].

Considering the availability and accessibility of data, it is very important for researchers and users to protect themselves and give credits to the appropriate researchers or institutions responsible for the research. But is this enough in order to achieve balance between the objective of scientific progress and the usage from sensitive groups of population? In which way will it become beneficial for all humans and more feasible to merge the increasing cost of new methods in health diagnosis and treatment, with the accessible and high quality medical care [5]?

It is also important for users to know the source of the information in order to be amenable to validation. In the past years there has been introduced national and international data-privacy regulations and Neuroinformatics researchers follow the copyright and intellectual property rights established by the Ethical Review Boards across the world [4]. While there are regulations and principles formulated with biomedical and social-behavioural research in mind, there are diverse opinions expressed by investigators about the fit between the national regulations and data sharing like: Should public money be spent on Open Access Projects, or how to share legally massive amounts of data?

In some cases further consideration is required; therefore several factors should be considered amongst others:

- proprietary information that must be protected
- legal and institutional requirements or regulations that need to be established
- collection of data, analysis of data, integration of the results into a conceptual model
- actual modelling of the experimental result, and development of a scientific manuscript

The above mentioned concepts should be considered in creating principles adequate for enabling collaborative and integrative information sharing in the field of neuroscience studies. The legal issues regarding sharing data have led to more strict and precise rules in order to ensure human subject anonymity and avoid inappropriate use of information. On the other hand international collaborations generated new concerns regarding fulfilment of International Ethical Review Boards requirements.

It is necessary that these cooperative efforts are carried out with similar and clearly defined fields, terminology description of data, models, experimental procedures etc., as well as developing appropriate common guidelines regarding legal issues of sharing and analyzing neuroscience data. Furthermore, credits should be appropriately addressed for both data owner, as well as data user. Some of the challenges regard also issues like the rights of each researcher the purpose of the data sharing, access to data, allocation of rights/ownership etc. Policy makers and developers should consider ethics and policy challenges regarding the following:

- Regulating the content of databases
- Access to databases
- Use of databases

The researchers should aim to ensure that data remains confidential and simultaneously develop models that fulfil their technical requirements in order to satisfy their expectations.

3   Regulating the Content of Databases

Image-based data are primary drivers for Neuroinformatics efforts [6]. For example the functional Magnetic Resonance Imaging Data Center (fMRIDC) was established with the objective of creating a mechanism by which members of the Neuroscientific community might easily share functional neuroimaging data. Additionally, the Journal of Cognitive Neuroscience between 2000 and 2006 required that all authors who published in the journal submit their supplementary data to the fMRIDC [7]. Authors whose papers are based on results from datasets obtained from the Data Center are expected to provide descriptive meta-information for data use, credit original study authors and acknowledge the fMRIDC and accession number of the data set [8].

It is obvious that researchers using data from digital databases may encounter the risk of discovering violation of the providers' privacy, or disclosure of sensitive information. Since Neuroinformatics worldwide repositories will be populated by data, standardization of protocols for managing incidental findings, and limitation to the risk of intrinsic and consequential harm of donor are necessary. Ethical principles should enhance regulations for protecting human research participants, so as data sharing will always be safe and efficient.

Another issue that should be considered in neuroimaging sharing is the quality control of brain imaging data. Appropriate cautions should be applied in the use of these data, when predicting for example, diseases of the CNS. Concerns about the use of functional Magnetic Resonance Imaging (fMRI) scans of adolescents to predict later onset psychiatric disorders are referred in [9] regarding the reliability of neuroimaging due to the complexity and plasticity of the brain.

It is obvious that the hybrid ethical codes concern early diagnosis of CNS diseases differ from Neuroinformatics to traditional medicine approach due to the influence of other ethical frameworks and perspectives on their basic research and development. For some, this information could be helpful, empowering or enlightening and may enhance human health. For others, it is likely that such information could result in fear, anxiety and other mental health issues [3].

4   Access and Use of Databases

Artificial Intelligence research is undergoing a revolution in the area of understanding the mechanisms underlying intelligent behavior. Internet-accessibility of databases has enabled the usage of data from providers, patients, and research participants everywhere. Research studies involve a degree of trust between subjects and researchers, the former giving their consent and the latter protecting their privacy [10]. Different countries have various regulations and procedures regarding ethical issues of human data sharing:

- The federal human subject protection law in the United States mandates that all identifying information be removed from data prior to submission for sharing [11].
- Additionally, Institutional Review Boards (IRB) are administrative bodies in the U.S., established to protect the rights and welfare of human research subjects and are designated to review clinical investigations regulated by U.S. Food and Drug Administration (FDA).

A major continuing goal of these boards is to ensure that national guidelines for protection of privacy and informed consent are best applied in the development of neuroscience data repositories. However, technical issues might arise, like the case where researchers wish to share retrospective data and process the way to obtain a new written informed consent. For example, the steering committee of the 1000 Functional Connectomes Project (FCP), a data sharing initiative project, realizing the necessity for privacy, agreed to a full anonymization of all datasets in accordance with the U.S. Health Insurance Portability and Accountability Act (HIPAA) [12].

Another aspect that raises further ethical issues is commercialization of databases. What details are necessary to share in data exchange and scientific results? The challenge is how to prevent exploitation of vulnerable populations, and avoid conflicts of interest. Data storing and data sharing can be both a challenge and a desire in for scientific researchers. The research of hybrid ethics in Neuroinformatics tools has to examine the malpractices during the past years in order to explore efficient and effective ways to accomplish data organization with ethical use of data. There is a worldwide debate about principles and regulations for the scientific processes. Guidance is needed to protect against discriminatory practices in case of commercialization of data sharing, as well as for issues regarding ownership and intellectual property of results and follow on innovation. A 'clear' definition on using and sharing data and meta-data seems to be necessary.

5   Conclusion

The next generation of neuroscientists contributes to a worldwide research network of excellence sharing data, tools, and information. Success on data sharing depends on the participation of the different types of neuroscience researchers. There should be considered a more critical thinking on searching for neuroscience relevant resources, including the ethical aspect. There is a major challenge in developing large-scale international collaborations and there is significant effort in developing integrated models of neuroscience processes. The more informed our brains are by science at all levels of analysis, the better will be our brains theoretical evolution [12]. Neuroinformatics can play a pivotal role in human brain research leading to innovations in neuroscience, informatics and treatment of brain disorders. Guidance to investigators, IRB representatives, public health officials and others regarding bioethics are key issues in neuroscience data sharing.

References

1. International Neuroinformatics Coordinating Facility,
   http://www.incf.org/documents/incf-core-documents/
   INCFStrategyOverview
2. Shepherd, G.M., Mirsky, J.S., Healy, M.D., Singer, M.S., Skoufos, E., Hines, M.S., Nadarni, P.M., Miller, P.L.: The Human Brain Project: Neuroinformatics tools for integrating, searching and modeling multidisciplinary neuroscience data. J. Neurophysiology 74, 1810–1815 (1998)

3. Alexiou, A., Psixa, M., Vlamos, P.: Ethical Issues of Artificial Biomedical Applications. In: Iliadis, L., Maglogiannis, I., Papadopoulos, H. (eds.) EANN/AIAI 2011, Part II. IFIP AICT, vol. 364, pp. 297–302. Springer, Heidelberg (2011)
4. Amari, S., Beltrame, F., Bjaalie, J.G., Dalkara, T., De Schutter, E., Egan, G.F., Goddard, N.H., Gonzalez, C., Grillner, S., Herz, A., Hoffmann, K.P., Jaaskelainen, I., Koslow, S.H., Lee, S.Y., Matthiessen, L., Miller, P.L., Da Silva, F.M., Novak, M., Ravindranath, V., Ritz, R., Ruotsalainen, U., Sebestra, V., Subramaniam, S., Tang, Y., Toga, A.W., Usui, S., Van Pelt, J., Verschure, P., Willshaw, D., Wrobel, A.: Neuroinformatics: the integration of shared databases and tools towards integrative neuroscience. Journal of Integrative Neuroscience 1(2), 117–128 (2002)
5. Alexiou, A., Vlamos, P.: Ethics at the Crossroads of Bioinformatics and Nanotechnology. In: 7th International Conference of Computer Ethics and Philosophical Enquiry (2009)
6. Martone, M.E., Amarnath, G., Ellisman, M.: E-neuroscience: challenges and triumphs in integrating distributed data from molecules to brains. Nature Neuroscience 7(5), 467–472 (2004)
7. Journal of Cognitive Neuroscience Instructions for Authors, J. Cognitive Neuroscience, http://jocn.mitpress.org/misc/ifora.shtml
8. Illes, J., Lombera, S.: Identifiable neuro ethics challenges to the banking of neuro data. Minn. JL Sci. & Tech. 10, 71 (2008)
9. Fuchs, T.: Ethical issues in neuroscience. Current Opinion in Psychiatry 19(6), 600–607 (2006)
10. International Neuroinformatics Coordinating Facility, http://www.incf.org/documents/incf-core-documents/INCFStrategyOverview
11. Wolf, L.E., Lo, B.: Untapped Potential: IRB Guidance for the Ethical Research Use of Stored Biological Materials. IRB: Ethics & Human Res. 26(1), 1–8 (2004)
12. Maarten, M., Bharat, B.B., Castellanos, F.X., Milham, M.P.: Making data sharing work: The FCP/INDI experience. NeuroImage (2012), doi:10.1016/j.neuroimage.2012.10.064

Author Index

Printed in the United States
By Bookmasters